This is a volume in the series
CHEMISTRY OF

The organic compounds of germanium

This is a volume in the series
THE CHEMISTRY OF ORGANOMETALLIC COMPOUNDS

THE CHEMISTRY OF ORGANOMETALLIC COMPOUNDS

A Series of Monographs

Dietmar Seyferth, *editor*

Department of Chemistry,
Massachusetts Institute of Technology
Cambridge, Massachusetts

PUBLISHED:

Chemistry of the iron group metallocenes

Part One: ferrocene, ruthenocene, osmocene

By Myron Rosenblum

The organic compounds of lead

By Hymin Shapiro and F. W. Frey

The organic chemistry of tin

By Wilhelm P. Neumann

The organic compounds of germanium

MICHEL LESBRE
PIERRE MAZEROLLES
JACQUES SATGÉ
Paul Sabatier University
TOULOUSE

Interscience Publishers
a division of
JOHN WILEY & SONS
London · New York · Sydney · Toronto

Library of Congress catalog card no. 73–116163

ISBN 0 471 52780 7

Printed by J. W. Arrowsmith Ltd., Bristol, England

PREFACE

This volume continues the series of monographs *The Chemistry of Organometallic Compounds* edited by Professor Dietmar Seyferth.

We have undertaken to describe from a critical point of view the chemistry of organic germanium derivatives whose development in the course of the last dozen years has been considerable. With this in mind, the present volume has been divided into chapters, each of which corresponds to the study of the derivatives in which germanium is linked to another element: germanium–carbon bond, germanium–hydrogen bond, germanium–halide bond, germanium–oxygen bond, germanium–sulphur, selenium and tellurium bonds, germanium–nitrogen and phosphorus bonds. The last two chapters have been devoted to the di- and polygermanes and the derivatives containing a germanium-metal bond, which represent exciting areas of great current activity.

We have made comparisons with the chemistry of the analogous silicium and tin compounds wherever these are appropriate and these give evidence that the often "unique" character of germanium chemistry is a consequence of the "high" electronegativity of this element.

The literature references should be complete through 1968 and many references from 1969 are also included.

We wish to express our gratitude to Professor Dietmar Seyferth, who was kind enough to read and correct our manuscript, for suggestions that proved to be very useful.

MICHEL LESBRE
PIERRE MAZEROLLES
JACQUES SATGÉ

CONTENTS

1 General aspects of organogermanium chemistry

1-1 HISTORICAL DEVELOPMENT

Tetraethylgermane, synthesized in 1887 by Winkler (71) one year after the discovery of the "ekasilicium", remained for a long time the only known organogermanium compound. After 1925, organogermanium chemistry began its development with the important contributions of Dennis (14–16), Kraus (39–45), Johnson (36, 37), Rochow (63–65, 67), Anderson (3–5) and others. The masterly publication of Krause and von Grosse, *Die Chemie der metall-organischen Verbindungen* which appeared in 1937, contained only 25 references dealing with organogermanes (46). From that time, interest in organogermanium chemistry increased rapidly, parallel to the fast-growing importance of organosilicon and organotin chemistry.

Original publications are now appearing at the rate of about 150 per year.

A number of reviews has been published which provide a good background to organogermanium chemistry. In 1951, Johnson presented the first important review; approximately 230 compounds were tabulated (35). A listing of the physical constants of approximately 260 organogermanium compounds was given in Kaufman's *Handbook of Organometallic Compounds* in 1961 (38). A monograph by Rijkens (61), sponsored by the Germanium Research Committee, was published in 1960 (survey of the literature from 1950 to 1960) while in 1963 Quane and Bottei described over 700 compounds (59).

Good compilations of organogermanes were published later by Dub (18) in 1966, Glockling (*Quarterly Reviews*, 1966) (29), and Mironov and Gar (52) in 1967.

1

More recently, the book by Glockling *The Chemistry of Germanium* in 1969 covers many aspects of the inorganic and organic chemistry of germanium (30).

For the sake of completeness, two short general surveys included in the monographs on organometallic chemistry by Rochow, Hurd and Lewis (1957) (64) and later by Coates (1960) (12) should be mentioned.

1-2 CHARACTERISTIC FEATURES OF ORGANOGERMANIUM CHEMISTRY

Germanium lies between silicon and tin in Group IV B of the Periodic Table, and its organic chemistry shows many similarities to that of silicon. Carbon–germanium bonds in alkylgermanes, like the analogous carbon–silicon bonds, are much more stable thermally and less reactive chemically than carbon–tin bonds. However, the analogy between silanes and germanes has been overemphasized; as a matter of fact, very many chemical differences are known. Among these are the following:

(a) West reported the reduction of the halides, R_3GeX, to the hydrides by amalgamated zinc in hydrochloric acid (70). This reaction does not occur with the corresponding silicon or tin compounds.

(b) Halides of type R_3GeX or R_2GeX_2 are hydrolyzed to the hydroxides or oxides with much more difficulty than the analogous silicon compounds. Chipperfield and Prince found considerable differences in the reactions of Ge—X and Si—X bonds to nucleophilic attack (10). The Si—O bond is clearly stronger than the Ge—O bond and is more easily formed. Kinetic observations show that triphenylchlorosilane reacts more rapidly than triphenylchlorogermane (20), perhaps because π-interactions between the phenyl groups and silicon are more important than are π-interactions between phenyl groups and germanium.

(c) Trialkylsilanes are much more sensitive to acid or base solvolysis than are trialkylgermanes (22, 40). Particularly potassium hydroxide is virtually without action on the trialkylgermanes, whereas the trialkyl-silanes react readily with this reagent (66). The benzyl–germanium bond is not cleaved by aqueous methanolic alkali, in contrast to the $PhCH_2$—Si bond. The ease of alkaline solvolysis in aqueous ethanol of hydrides R_3MH increases in the order Ge < Si < Sn (6).

(d) Organogermanium hydrides behave quite differently from silicon hydrides when treated with organolithium reagents: triphenylsilane or triethylsilane reacts to form tetrasubstituted silanes in an alkylation reaction (25, 26):

$$R_3SiH + R'Li \rightarrow R_3SiR' + LiH$$

Triphenylgermane, on the other hand, like triphenylmethane (72), reacts in the sense of a metalation reaction:

$$Ph_3GeH + R'Li \rightarrow Ph_3GeLi + R'H$$

$$Ph_3CH + R'Li \rightarrow Ph_3CLi + R'H$$

(e) Organogermanium hydrides are much more reactive than the corresponding silanes in free radical additions to various olefins (48, 51).

(f) Many silicon—carbon bonds are readily cleaved by nucleophilic reagents such as lithium aluminium hydride in refluxing THF. The analogous germanium compounds are not cleaved under similar conditions (24, 56). Kinetic studies confirmed this difference in behaviour of the C—Si and C—Ge bonds with respect to other nucleophilic reagents (6). The same difference is found between the Si—Si and Ge—Ge bonds (60).

(g) Although the energies of the Si—Si and Ge—Ge bonds are quite similar in magnitude, germanium seems to be less suitable than silicon for forming high polymers (57). Also, it has not been possible hitherto to synthesize $(-Ge-O-)^x$ chains analogous to silicones which contain the backbone $(-Si-O-).^x$

(h) The action of certain sterically hindered Grignard reagents differs fundamentally when they react with germanium tetrachloride or, on the other hand, with silicon tetrachloride (67).

(i) α-Hydroxysilanes undergo a classical rearrangement to silyl-ethers, in alkali or pyridine medium (7):

$$Ph_3Si\overset{|}{\underset{OH}{-}}CPh_2 \xrightarrow{\text{base}} Ph_3SiOCHPh_2$$

No reaction was observed with analogous α-hydroxygermanes under similar conditions (54).

(j) The base strength of the ketones of type R_3MCOPh (M = C, Si, Ge, R = Me or Ph) follows the sequence Si > Ge > C (69).

Different authors have pointed out that in some chemical reactions, germanium appears to be more electronegative than silicon. A study of the infrared spectra of structurally similar compounds of Group IV elements also indicated the higher electronegativity of germanium as compared with silicon (21). The large difference observed between Si—X and Ge—X bond energies implies that germanium behaves as if it were more electronegative than silicon when bound to an electronegative atom.

As a matter of fact, the problem of the electronegativity to germanium compared with that of other elements of Group IV B of the Periodic System has been the subject of many discussions. A study of the recent literature shows remarkable divergences between the different values proposed for the electronegativity of element 32 (1, 2, 17, 23, 31–34, 47, 55, 58, 68):

1.70   1.71   1.79   1.80   1.90   2.00   2.01   2.02   2.03   2.06   2.10

2.20   2.31

In Pauling's original scale (55) there is a progressive decrease of electronegativities with increasing atomic numbers of the elements:

C   2.5   Si   1.8   Ge   1.7

But the agreement between the experimental values of germanium–oxygen or germanium–halogen bonds and those calculated from Pauling's electronegativity is not satisfactory; for this reason, Fineman and Daignault (23) modified Pauling's classical equation for the calculation of the electronegativities and in accordance with the thermochemical data proposed the following scale in which germanium and silicon are reversed:

C   2.58   Ge   2.02   Si   1.90

In fact, thermochemical data do not give self-consistent results for germanium (Pauling 1.7 (55), Drago 1.8 (17), Huggins 1.9 (33), Fineman and Daignault 2.02 (23), Johnson 2.20 (34).

This is why several authors related electronegativity to measured properties other than thermochemical ones, such as force constants (Gordy) (31), bond moments (McKean) (49), ESR spectra (Curtis and Allred) (13), etc. Allred and Rochow have used the relative PMR chemical shifts in the tetramethyl derivatives $M(CH_3)_4$ (M = Group IV B elen ent) as the basis of a novel scale of electronegativity and proposed the values (1, 2):

C   2.60   Ge   2.0   Si   1.9

From the coupling constants, Cawley and Danyluck draw essentially the same conclusions (9).

These figures which have been confirmed later by Spiesecke and Schneider (68) are in good agreement with the values deduced respectively by Gordy (32) from the nuclear quadrupole coupling constants (RQN) (Ge 2.03) and by Brook (8) from the specific rotations of the (+) enantiomers of the type:

$$C_6H_5 \overset{\displaystyle CH_3}{\underset{\displaystyle C_{10}H_7}{-MH}} \qquad (Ge \quad 2.06)$$

On the basis of the most recent work (47, 50), it seems to have been established that the electronegativity of germanium is slightly higher than that of silicon and the following approximate values can be considered:

$$\text{C} \quad 2.4 \text{ to } 2.5 \qquad \text{Ge} \quad 2.0 \text{ to } 2.1 \qquad \text{Si} \quad 1.8 \text{ to } 1.9$$

If it is admitted—in agreement with most authors—that the electronegativity of hydrogen is approximately 2.1, it becomes obvious that of all elements of Group IV B, germanium is the one having the electronegativity nearest in magnitude to that of hydrogen. Consequently, the germanium–hydrogen bond is evidently less polar in the sense $\text{Ge}^{+\delta}-\text{H}^{-\delta}$ than the silicon–hydrogen bond. In this way, a certain number of singularities found in the chemistry of the organogermanium compounds can be explained, such as the action of organolithium reagents on triphenylgermane (27, 28) or the reversal of polarity of the germanium–hydrogen bond which have been found experimentally (50) in the sequence of derivatives

$$\text{R}_3\text{GeH} \qquad \text{R}_2(\text{Cl})\text{GeH} \quad \text{R}(\text{Cl}_2)\text{GeH} \qquad \text{Cl}_3\text{GeH}$$

On the other hand, the availability of the germanium 4 d-orbitals for the formation of additional σ- or π-bonds must be taken into account.

1-3 NOMENCLATURE

While inorganic nomenclature is now commonly employed in organotin and organolead chemistry, rules of organic nomenclature are still used in organosilicon and also organogermanium chemistry.

In 1952, IUPAC'S commission for the reform of organic nomenclature adopted a series of 19 rules to provide one systematic and unique name for each of the organosilicon compounds. The roots "silane" and "germane" are widely used; monovalent radicals are derived (silyl, germyl). Compound radical names may be found in the usual manner, for example $\text{Ph}_3\text{GeCH}_2\text{COOH}$ is the triphenylgermylacetic acid (cf. IUPAC'S report 1966: *Tentative Rules for Nomenclature of Organometallic Compounds*).

REFERENCES

1. Allred, A. L., and E. G. Rochow, *J. Am. Chem. Soc.*, **79**, 5361 (1957).
2. Allred, A. L., and E. G. Rochow, *J. Inorg. Nucl. Chem.*, **5**, 269 (1958).
3. Anderson, H. H., *J. Org. Chem.*, **20**, 900 (1955).
4. Anderson, H. H., *J. Am. Chem. Soc.*, **78**, 1692 (1956).
5. Anderson, H. H., *J. Org. Chem.*, **21**, 869 (1956).
6. Bott, R. W., C. Eaborn, and T. W. Swaddle, *J. Chem. Soc.*, 2342 (1963).
7. Brook, A. G., M. Warner, and E. M. Griskin, *J. Am. Chem. Soc.*, **81**, 781 (1959).

8. Brook, A. G., *J. Am. Chem. Soc.*, **85**, 3052 (1963).
9. Cawley, S., and S. S. Danyluck, *Can. J. Chem.*, **41**, 1850 (1963).
10. Chipperfield, J. R., and R. H. Prince, *J. Chem. Soc.*, 3567 (1963).
11. *Chem. Eng. News*, **30**, 4517 (1952).
12. Coates, G. E., *Organometallic Compounds*, Wiley, New York, 1957.
13. Curtis, M. D., and A. L. Allred, *J. Am. Chem. Soc.*, **87**, 2554 (1965).
14. Dennis, L. M., and F. E. Hance, *J. Am. Chem. Soc.*, **47**, 37 (1925).
15. Dennis, L. M., *Z. Anorg. Chem.*, **174**, 97 (1928).
16. Dennis, L. M., and W. J. Patnode, *J. Am. Chem. Soc.*, **52**, 2779 (1930).
17. Drago, R. S., *J. Inorg. Nucl. Chem.*, **15**, 237 (1960).
18. Dub, M., *Organometallic Compounds*, Vol. 2, Springer Ed., Berlin, 1966.
19. Dzurinskaya, N. G., V. F. Mironov, and A. D. Petrov, *Dokl. Akad. Nauk SSSR*, **138**, 1107 (1961).
20. Eaborn, C., and D. R. M. Walton, *J. Organometal. Chem.*, **4**, 217 (1965).
21. Egorov, Y. P., G. G. Karei, L. A. Leites, V. F. Mironov, and A. D. Petrov, *Izv. Akad. Nauk SSSR*, 1880 (1962).
22. Emeléus, H. J., and S. F. A. Kettle, *J. Chem. Soc.*, 2444 (1958).
23. Fineman, M. A., and R. Daignault, *J. Inorg. Nucl. Chem.*, **10**, 205 (1959).
24. Gilman, H., and W. H. Atwell, *J. Am. Chem. Soc.*, **86**, 2087 (1964).
25. Gilman, H., and C. W. Gerow, *J. Am. Chem. Soc.*, **78**, 5435 (1956).
26. Gilman, H., and H. W. Melvin, *J. Am. Chem. Soc.*, **71**, 4050 (1949).
27. Gilman, H., and C. Wu, *J. Am. Chem. Soc.*, **75**, 2935 (1953).
28. Gilman, H., and A. Zuech, *J. Am. Chem. Soc.*, **71**, 4050 (1949).
29. Glockling, F., *Quart. Rev. (London)*, **20**, 45 (1966).
30. Glockling, F., *The Chemistry of Germanium*, Academic Press, London, 1969.
31. Gordy, W., *J. Chem. Phys.*, **14**, 305 (1946).
32. Gordy, W., *Discussions Faraday Soc.*, **19**, 14 (1953).
33. Huggins, M. C., *J. Am. Chem. Soc.*, **75**, 4123 (1953).
34. Johnson, D. A., *Some Thermodynamic Aspects of Inorganic Chemistry*, Cambridge University Press, 1969.
35. Johnson, O. H., *Chem. Rev.*, **48**, 259 (1951).
36. Johnson, O. H., and W. H. Nebergall, *J. Am. Chem. Soc.*, **71**, 1720 (1949).
37. Johnson, O. H., and E. A. Schmall, *J. Am. Chem. Soc.*, **80**, 2931 (1958).
38. Kaufman, H. C., *Handbook of Organometallic Compounds*, Van Nostrand, Princeton, 1961.
39. Kraus, C. A., and C. L. Brown, *J. Am. Chem. Soc.*, **52**, 3690 (1930).
40. Kraus, C. A., and S. Carney, *J. Am. Chem. Soc.*, **56**, 765 (1934).
41. Kraus, C. A., and C. A. Flood, *J. Am. Chem. Soc.*, **54**, 1635 (1932).
42. Kraus, C. A., and L. S. Foster, *J. Am. Chem. Soc.*, **49**, 457 (1927).
43. Kraus, C. A., and H. S. Nutting, *J. Am. Chem. Soc.*, **54**, 1622 (1932).
44. Kraus, C. A., and C. S. Sherman, *J. Am. Chem. Soc.*, **55**, 4694 (1933).
45. Kraus, C. A., and C. B. Wooster, *J. Am. Chem. Soc.*, **52**, 372 (1930).
46. Krause, E., and A. von Grosse, *Die Chemie der metall-organischen Verbindungen*, Borntraeger, Berlin, 1937.
47. Labarre, J. F., M. Massol, and J. Satgé, *Bull. Soc. Chim. France*, 736 (1967).
48. Lesbre, M., and J. Satgé, *Compt. Rend. Paris*, **247**, 471 (1958).
49. McKean, D. C., *Chem. Commun.*, 146 (1966).
50. Massol, M., J. Satgé, and M. Lesbre, *Compt. Rend. Paris*, 1806 (1966).
51. Meen, R. H., and H. Gilman, *J. Org. Chem.*, **22**, 684 (1957).
52. Mironov, V. F., and T. K. Gar, *Organogermanium Compounds*, Nauka, Moscow, 1967.

53. Mironov, V. F., and A. L. Kravchenko, *Izv. Akad. Nauk SSSR Otd. Khim. Nauk*, 1563 (1963).
54. Nicholson, D. A., and A. L. Allred, *Inorg. Chem.*, **4**, 1752 (1965).
55. Pauling, L., *Nature of the Chemical Bond*, 2nd ed., Cornell University Press, New York, 1948.
56. Peddle, G. J. D., and D. N. Roark, *Can. J. Chem.*, **46**, 2507? (1968).
57. Petrov, A. D., *Compt. Rend. 17° Congrés International de Chimie Pure*, Munich, 1959.
58. Pritchard, H. O., and H. A. Skinner, *Chem. Rev.*, **55**, 745 (1955).
59. Quane, D., and R. S. Bottei, *Chem. Rev.*, **63**, 403 (1963).
60. Rahinivich, I. B., V. I. Tel'nei, N. V. Hayahin, and G. A. Razuvaev, *Dokl. Akad. Nauk SSSR*, **149**, 216 (1963).
61. Rijkens, F., *Organogermanium Compounds*, Germanium Research Committee, Utrecht, 1960.
62. Rochow, E. G., and A. L. Allred, *J. Inorg. Nucl. Chem.*, **5**, 269 (1958).
63. Rochow, E. G., R. Didchenko, and R. C. West, *J. Am. Chem. Soc.*, **73**, 5486 (1951).
64. Rochow, E. G., D. T. Hurd, and R. N. Lewis, *The Chemistry of Organometallic Compounds*, Wiley, New York, 1957.
65. Rochow, E. G., D. Seyferth, and A. C. Smith, *J. Am. Chem. Soc.*, **77**, 907 (1955).
66. Schott, G., and C. Harzdorf, *Z. Anorg. Allgem. Chem.*, **367**, 105 (1961).
67. Seyferth, D., and E. G. Rochow, *J. Org. Chem.*, **20**, 250 (1955).
68. Spiesecke, H., and W. C. Schneider, *J. Chem. Phys.*, **35**, 722 (1961).
69. Yates, K., and F. Angolini, *Can. J. Chem.*, **44**, 2229 (1966).
70. West, R., *J. Am. Chem. Soc.*, **75**, 6080 (1953).
71. Winkler, C., *J. Prakt. Chem.*, **36**, 177 (1887).
72. Ziegler, K., *Z. Anorg. Chem.*, **49**, 459 (1936).

2 Germanium–carbon bond aliphatic and aromatic

2-1 TETRAALKYLGERMANES, TETRAARYLGERMANES AND MIXED DERIVATIVES

2-1-1 Preparations

2-1-1-1 From Germanium Halides. Several methods are available for the conversion of germanium–halogen bonds to germanium–carbon bonds:

(a) Grignard reaction

This reaction is the most widely used for the synthesis of tetra-alkyl derivatives. It is particularly suitable for the preparation of symmetrical compounds from germanium tetrahalides:

$$GeX_4 + 4\,RMgX \longrightarrow R_4Ge + 4\,MgX_2$$

The reaction is exothermic and is usually carried out in ether medium for the saturated derivatives; to alkylate all four Ge—X bonds, it is advisable to use a large excess of Grignard reagent (100%) and to heat the reaction mixture to about 100°C after having displaced the diethyl ether with a less volatile hydrocarbon solvent, (benzene, toluene or xylene). Under these conditions, the reaction is normally effected in yields of $\sim 80\%$. The tetraalkylgermanes with long alkyl chains, however, appear to form in markedly lower yields (105). For tetramethylgermane (b.p. 43.7°C) which forms an azeotrope with diethyl ether, the use of di-*n*-butyl ether as solvent has been recommended (119, 208, 270, 424, 433, 439).

More recently, Zablotna (440) resumed the systematic study of the preparation of tetramethylgermane in different solvents. It was observed that, in the action of germanium tetrachloride on methylmagnesium iodide, the yield was lower the higher the boiling point of the solvent, and that all the organic fractions always contained germanium tetraiodide, the yield of which increased with increasing boiling point of the solvents used. This is due to the fact that, instead of the reaction intended, exchange of halogen between germanium tetrachloride and methylmagnesium iodide occurs:

$$GeCl_4 + 4\,CH_3MgI \longrightarrow GeI_4 + 4\,CH_3MgCl$$

Thus the poor yields of tetramethylgermane are not due to the difficulties of its separation but rather to an inappropriate trend of the reaction. As a matter of fact, when germanium tetraiodide is made to react with methylmagnesium iodide, an 80 % yield of tetramethylgermane is obtained:

$$GeI_4 + 4\,CH_3MgI \longrightarrow (CH_3)_4Ge + 4\,MgI_2$$

The Grignard reaction can also be applied for the aromatic series; thus, tetraphenylgermane is obtained in good yield (120, 122, 135, 183):

$$GeCl_4 + 4\,C_6H_5MgBr \longrightarrow (C_6H_5)_4Ge + 2\,MgCl_2 + 2\,MgBr_2$$

in total absence of free magnesium. If free magnesium is present, hexa-phenyldigermane is formed in considerable amount (69 %) (122). When the ratio of $GeCl_4$ to phenylmagnesium bromide is 1:4, the major product is triphenylchlorogermane (40 %) but tetraphenylgermane is obtained with good yields (70–75 %) when this ratio is 1:5 (214).

Hindered derivatives. The symmetrical R_4Ge tetraalkylgermanes and tetraarylgermanes the R groups of which occupy a large volume are difficult or impossible to obtain: in the action of the germanium tetra-chloride on the 1-naphthylmagnesium bromide, the reaction stops short at the monobromide derivative (432):

$$GeBr_4 \xrightarrow{\text{1-}C_{10}H_7MgBr} (1\text{-}C_{10}H_7)_3GeBr$$

The action of the germanium tetrachloride on the 1-naphthylmagnesium chloride in excess also leads to the tri-substituted derivative (46).

In the aliphatic series, the preparation of the tetraisopropyl derivative is particularly difficult. This very complex reaction has been studied recently in a comprehensive way by Carrick and Glockling (37). By the action of an excess of isopropylmagnesium chloride on germanium chloride, these authors obtained, after hydrolysis, besides the compounds $i\text{-}Pr_3GeOH$, $(i\text{-}Pr_2GeO)_3$, $i\text{-}Pr_3GeH$ which had already been observed (9, 221), considerable amounts (5–20 %) of tetraisopropylgermane, $(i\text{-}C_3H_7)_4Ge$, and triisopropyl n-propylgermane $(i\text{-}C_3H_7)_3Ge(n\text{-}C_3H_7)$ (4–12 %); the latter compound seems to be due to the isomerization of the Grignard reagent with catalytic impurities contained in the magnesium.

Secondary reactions. In these Grignard reactions, besides the expected tetraalkylgermane, a small amount of higher-molecular-weight poly-germanes is formed; especially in the tetraethylgermane preparation, Gilman (116) has been able to isolate 8 % of hexaethyldigermane $(C_2H_5)_3Ge-Ge(C_2H_5)_3$. Köster and coworkers (163) also obtained in the synthesis of tetraethyl-, tetrapropyl- and tetrabutylgermane, by the mag-nesium method, the following polygermanes: Pr_6Ge_2 ($b_{0.04} = 123 - 27°$), Bu_6Ge_2 ($b_2 = 182–86°$), Et_8Ge_3 ($b_{0.05} = 133–37°$), Pr_8Ge_3 ($b_{0.01} = 138–42°$) and Bu_8Ge_3 ($b_{0.04} = 194–99°$) and recorded even the formation of tetragermanes. These reactions, which are of the same type as those recorded by Seyferth (349) in the $GeCl_4$–vinylmagnesium bromide reaction, imply the transitory existence of an intermediate containing a Ge—Mg bond capable of reacting with the germanium halides:

$$R_3GeX + RMgX \rightarrow R_3GeMgX$$

$$R_3GeMgX + R_3GeX \rightarrow R_3Ge-GeR_3 + MgX_2$$

$$2\,R_3GeMgX + R_2GeX_2 \rightarrow R_3Ge-GeR_2-GeR_3$$

The reaction of an ethereal $GeCl_4$ solution with sterically hindered Grignard reagents such as cyclohexylmagnesium chloride stops at the trisubstituted germanium derivative. After hydrolysis of the mixture an appreciable amount of tricyclohexylgermane was found. This compound, the percentage of which increases with the excess of Grignard reagent, results from the hydrolysis of $(C_6H_{11})_3GeMgCl$ formed (237).

An extensive study of this kind of reaction (236) showed that the last compound results from the reduction of germanium tetrachloride to germanium dichloride followed by the insertion of $GeCl_2$ into the germanium–magnesium bond.

The mixed derivatives can be obtained from the symmetrical deriva-tives by cleavage followed by alkylation or arylation (79, 197, 327, 328). The great difference in reactivity of the (aliphatic) Ge—C linkages and the (aromatic) Ge—C linkages with respect to bromine can be used several times in this type of synthesis (81, 83):

$$(C_6H_5)_4Ge \xrightarrow{Br_2} (C_6H_5)_3GeBr \xrightarrow{C_2H_5MgBr} (C_6H_5)_3GeC_2H_5 \xrightarrow{2\,Br_2}$$

$$(C_6H_5)Ge(C_2H_5)Br_2 \xrightarrow{1\text{-}C_{10}H_7MgBr} (C_6H_5)(C_2H_5)Ge(1\text{-}C_{10}H_7)_2 \ldots etc.$$

Owing to steric hindrance, failures have been recorded in the synthesis of mixed tricyclohexylgermanes $(C_6H_{11})_3GeR$ with R = i-C_3H_7, C_6H_5 (154).

Mendelsohn and coworkers (238) observed that the action of an excess of some Grignard reagents on triorganogermanium chlorides at high

temperatures leads to a mixture of mixed tetraorganogermane, triorganogermane, saturated and olefinic hydrocarbons:

$$R_3GeCl + R'MgX \xrightarrow[\text{(2) H}_2\text{O}]{\text{(1) 160°}} R_3GeR' + R_3GeH + R'(+H) + R'(-H)$$

Triorganogermane is not formed in this case by hydrolysis of a Ge—Mg
compound but results from the reduction of a Ge—Cl bond by magnesium
dihydride:

$$2\,R_3GeCl + MgH_2 \rightarrow 2\,R_3GeH + MgCl_2$$

(b) Organolithiums

Organolithiums are less frequently used than the organomagnesiums
for the preparation of tetraalkylgermanes from tetrahalides as the yields
usually are lower (104, 116). Thus, the action of germanium tetrabromide
on ethyllithium:

$$GeBr_4 + 4\,C_2H_5Li \rightarrow 4\,LiBr + (C_2H_5)_4Ge$$

only yields 12% of tetraethylgermane (116); in this reaction, 8.6% of
hexaethyldigermane is also formed, in addition to a large amount of
polymeric material. However, the action of 1-naphthyllithium on tri-
(1-naphthyl)bromogermane has allowed the preparation of the sterically
highly hindered tetra(1-naphthyl)germane, which is inaccessible by other
methods (432):

$$1\text{-}C_{10}H_7Li + (1\text{-}C_{10}H_7)_3GeBr \rightarrow LiBr + (1\text{-}C_{10}H_7)_4Ge \quad (38\%)$$

Mixed derivatives can also be prepared by this method (79, 296):

$$2\,(p\text{-}C_6H_5{-}C_6H_4)Li + (CH_3)_2GeCl_2 \rightarrow$$
$$2\,LiCl + (CH_3)_2Ge(C_6H_4{-}C_6H_5)_2$$

However, failures have been recorded:

$$2\,(p\text{-}C_6H_5CH_2C_6H_4)Li + (CH_3)_2GeCl_2 \rightarrow \text{No reaction}$$

(c) Organosodium derivatives and Würtz reaction

Brook and Gilman (28) prepared triphenylmethyltriphenylgermane in
50% yield by the reaction of triphenylmethylsodium with triphenylbromogermane in ethereal solution:

$$(C_6H_5)_3CNa + BrGe(C_6H_5)_3 \rightarrow (C_6H_5)_3C{-}Ge(C_6H_5)_3 + NaBr$$

but, as with the silicon analogue, its preparation was accompanied by the
formation in 25% yield of the isomeric p-triphenylgermylphenyldiphenyl-
methane $p\text{-}(C_6H_5)_3Ge{-}C_6H_4{-}CH(C_6H_5)_2$. Coupling of organic halides

with germanium halides in the presence of alkali metal is frequently used for the preparation of tetraaryl- and aralkylgermanium compounds (79):

$$4\,ArX + GeCl_4 + 8\,Na \rightarrow 4\,NaCl + 4\,NaX + Ar_4Ge$$

$$(C_2H_5)_3GeBr + XC_6H_4Cl + 2\,Na \rightarrow$$

$$NaCl + NaBr + (C_2H_5)_3GeC_6H_4X \qquad X = o\text{-Me},\ m\text{-Me}$$

$$p\text{-Et},\ o\text{-Ph}$$

and toluene is commonly used as a solvent in these reactions. Eaborn (74) had pointed out that the initially formed organosodium compounds RNa might undergo metal–hydrogen exchange with the solvent and that benzylic compounds might be formed; as a matter of fact, he has been able to demonstrate in the action of a trialkylsilane on chlorides of the type XC_6H_4Cl (X = o-Me, H, p-Me, p-OMe), with sodium in the toluene, the formation of trialkylbenzylsilanes (8 %) (85). Identical reactions are likely to occur with the organic germanium derivatives and for this reason, a more inert solvent, such as a saturated paraffin, is preferable to toluene (or xylene) for these reactions.

(d) Organozincs, organoaluminiums

Tetraethylgermane, the first known organic derivative of germanium, has been synthesized by Winkler (435) by the action of germanium tetra-chloride on diethylzinc:

$$GeCl_4 + 2\,Zn(C_2H_5)_2 \rightarrow 2\,ZnCl_2 + Ge(C_2H_5)_4$$

This method has been adopted again recently by Lengel and Dibeler (195) to prepare a sample of pure tetramethylgermane. The reactions between germanium tetrachloride and aluminium alkyls, sometimes with added sodium chloride, have been reported to give the tetraalkylgermanes $(C_2H_5)_4Ge$ and $(i\text{-}C_4H_9)_4Ge$ in about 70 % yield (444, 446, 448).

The alkylation of $GeCl_4$ by trimethylaluminium, methylaluminium sesquichloride, triethylaluminium and tri-isobutylaluminium has been examined by Glockling and Light (123): tetraalkylgermanes are formed, together with di- and polygermanes Ge_nR_{2n+2}. The authors found that in these reactions the rate of monoalkylation of $GeCl_4$ is slow compared with subsequent alkylation stages, so that the intermediate alkylchloro-germanes are not normally isolable, and that addition of sodium chloride increases the rate but not the yield of $(CH_3)_4Ge$ in the $(CH_3)_3Al—GeCl_4$ system. In these reactions the use of donor solvents (ethers) leads to undesirable complications.

2-1-1-2   From Germanium Hydrides

(a) Hydrogermylation of olefins

Organogermanium hydrides, R_3GeH and R_2GeH_2, add to olefins with peroxide catalyst or, better, with chloroplatinic acid catalyst:

$$(C_6H_5)_3GeH + RCH=CH_2 \xrightarrow{Bz_2O_2} (C_6H_5)_3GeCH_2CH_2R$$

$$(C_2H_5)_3GeH + CH_2=CH(CH_2)_5CH_3 \xrightarrow{PtCl_6H_2} (C_2H_5)_3Ge(CH_2)_7CH_3$$

The hydrogermylation yields are much better with chlorohydrogeno-germanes, the addition of which to the olefinic bonds proceeds quantitatively without a catalyst; afterwards the halogen can be eliminated by alkylation, by means of a Grignard reagent (204):

$$R_2(Cl)GeH + R'CH=CH_2 \rightarrow R_2(Cl)GeCH_2CH_2R' \xrightarrow{RMgX}$$

$$R_3GeCH_2CH_2R'$$

(b) Action of diazo derivatives

Aryl- and alkylgermanes react with diazomethane in the presence of Cu powder (180, 335a):

$$(C_2H_5)_3GeH + CH_2N_2 \rightarrow (C_2H_5)_3GeCH_3 + N_2$$

(c) Action of organolithium reagents

Tetrabenzylgermane is obtained in 60% yield by the reaction of tribenzylgermane on benzyllithium (50):

$$(C_6H_5CH_2)_3GeH + C_6H_5CH_2Li \rightarrow (C_6H_5CH_2)_4Ge + LiH$$

2-1-1-3   Other Preparations

(a) From germanium–metal derivatives

The action of halogenated derivatives on the germyllithiums (116) allows the preparation of R_4Ge compounds in good yield:

$$(C_6H_5)_3GeLi + C_{18}H_{37}Br \rightarrow$$

$$LiBr + (C_6H_5)_3GeC_{18}H_{37} \qquad (70\%)$$

$$(C_6H_5)_3GeLi + BrCH_2CH_2C_6H_5 \rightarrow$$

$$LiBr + (C_6H_5)_3GeCH_2CH_2C_6H_5 \quad (73\%)$$

To relate the configurations of ethyl(1-naphthyl)phenylgermane and methyl(1-naphthyl)phenylgermane, whose optical resolutions have been reported respectively by Eaborn (21, 81) and Brook (29), these authors (76) have prepared optically-active ethylmethyl(1-naphthyl)phenylgermane from these derivatives through stereospecific metalation with butyl-lithium followed by the action of the appropriate alkyl iodide.

Their results are summarized in the following stereochemical representations with the assumptions that metalation involves retention and the coupling with the alkyl iodides involves inversion of configuration (75):

$$
\begin{array}{ccc}
& C_6H_5 & \\
& | & \\
CH_3-Ge-1\text{-}C_{10}H_7 & & 1\text{-}C_{10}H_7-Ge-C_2H_5 \\
& | & \\
& H & H \\
\end{array}
$$

$$
\begin{array}{c}
C_6H_5 \\
| \\
1\text{-}C_{10}H_7-Ge-C_2H_5 \\
| \\
H
\end{array}
$$

S (−) $\xrightarrow[(2)\ C_2H_5I]{(1)\ C_4H_9Li}$ $\xleftarrow[(2)\ CH_3I]{(1)\ C_4H_9Li}$ R (+)

$$
\begin{array}{c}
C_2H_5 \\
| \\
CH_3-Ge-1\text{-}C_{10}H_7 \\
| \\
C_6H_5
\end{array}
$$

(+)

At reflux temperature and in the presence of an excess of lithium, tribenzylgermyllithium reacts with ethylene glycol dimethyl ether, giving a progressive series of reactions involving demethylation of the ether and further cleavage of benzyl groups, leading finally to trimethylbenzyl-germane (50):

$(C_6H_5CH_2)_3GeLi \xrightarrow{C_4H_{10}O_2} (C_6H_5CH_2)_3GeCH_3 \xrightarrow{Li}$

$(C_6H_5CH_2)_2Ge(CH_3)Li \xrightarrow{C_4H_{10}O_2} (C_6H_5CH_2)_2Ge(CH_3)_2 \xrightarrow{Li}$

$C_6H_5CH_2Ge(CH_3)_2Li \xrightarrow{C_4H_{10}O_2} C_6H_5CH_2Ge(CH_3)_3$

The composition of the mixture depends on the reflux time: tribenzyl-methylgermane was isolated after refluxing for 4 hours, while dibenzyl-dimethyl- and benzyltrimethylgermanes were present when the reaction was continued for 19 hours.

Triethylgermyllithium reacts at 20° with olefinic compounds in benzene medium (426, 430). The action on ethylene is a novel synthesis of tetra-ethylgermane:

$$(C_2H_5)_3GeLi + CH_2=CH_2 \xrightarrow[OH_2]{C_6H_6} (C_2H_5)_4Ge \quad (67.8\%)$$

With styrene the reaction is exothermic:

$$(C_2H_5)_3GeLi + CH_2=CH-C_6H_5 \xrightarrow{H_2O} (C_2H_5)_3GeCH_2CH_2C_6H_5$$

(43%)

1-Hexene reacts very slowly (21 hours at 90°) and the product yields are very poor (8.6%).

Triethylgermylpotassium, prepared by reaction of potassium metal on hexaethyldigermane in hexamethylphosphotriamide, gives with methyl chloride 94% yield of triethylmethylgermane (33):

$$(C_2H_5)_3GeK + CH_3Cl \xrightarrow{THF} (C_2H_5)_3GeCH_3 + KCl$$

Bis(triethylgermyl)cadmium, which is formed by reaction of triethylgermane on diethylcadmium, is cleaved by benzyl chloride to give triethylbenzylgermane (428):

$$[(C_2H_5)_3Ge]_2Cd + 2 C_6H_5CH_2Cl \rightarrow 2 (C_2H_5)_3GeCH_2C_6H_5 + CdCl_2$$

Tetraethylgermane is obtained in 45% yield by extended heating of a mixture of aluminium chloride and hexaethyldigermane (425):

$$(C_2H_5)_3Ge{-}Ge(C_2H_5)_3 + AlCl_3 \xrightarrow{200°;\ 15\,h} (C_2H_5)_4Ge$$

(b) From Ge–O, Ge–S and Ge–P derivatives

Triethylgermanium oxide is converted into tetraethylgermane by the action of ethylmagnesium bromide (225):

$$[(C_2H_5)_3Ge]_2O + 2 C_2H_5MgBr \rightarrow MgO + MgBr_2 + 2 (C_2H_5)_4Ge$$

Methyllithium cleaves bis-(trimethylgermanium) oxide, giving a mixture of lithium trimethylgermanolate and tetramethylgermane (350):

$$(CH_3)_3GeOGe(CH_3)_3 + CH_3Li \rightarrow (CH_3)_3GeOLi + (CH_3)_4Ge$$

while the attack of organolithium reagents on mixed germasilaoxide occurs at the more electrophilic Group IV atom (350):

$$(CH_3)_3SiOGe(CH_3)_3 + C_6H_5Li \rightarrow (CH_3)_3GeC_6H_5 + (CH_3)_3SiOLi$$

The germanium–sulphur bond is cleaved by Grignard (332) and organolithium (147) reagents:

$$(C_2H_5)_3GeSCH_3 + C_2H_5MgBr \rightarrow (C_2H_5)_4Ge + CH_3SMgBr$$

$$(C_6H_5)_3GeSCH_3 + C_4H_9Li \rightarrow (C_6H_5)_3GeC_4H_9 + CH_3SLi$$

In the same way, germanium–phosphorus compounds react with organolithium reagents (31):

$$(C_2H_5)_3Ge{-}P(C_6H_5)_2 + n\text{-}C_4H_9Li \rightarrow$$

$$(C_2H_5)_3Ge(n\text{-}C_4H_9) + (C_6H_5)_2PLi$$

(c) By transformation of substituted compounds

The reduction of 4-triphenylgermyl 1-butanol tosylate by lithium aluminium hydride leads to triphenylbutylgermane (118):

$$(C_6H_5)_3Ge(CH_2)_4OSO_2-p-C_6H_4CH_3 \xrightarrow{\text{AlLiH}_4} (C_6H_5)_3GeC_4H_9$$

2-1-2   Physical properties and other characteristics

Saturated tetraalkylgermanes are very stable liquids which can be heated without any appreciable decomposition to about 300°C, at atmospheric pressure. The dissymmetry of mixed tetraalkylgermanes does not affect this thermal stability. The decomposition of aikylgermanes occurs above 400°C only; Geddes and Mack (108) observed in the pyrolysis of tetraethylgermane at 420–450°C the formation of metallic germanium and various hydrocarbons which are also obtained by the pyrolysis of butane. Tetraarylgermanes are usually crystalline solids which melt without decomposition.

The heat of formation of tetraethylgermane and $\bar{D}$ (Ge—C) in this compound were determined by combustion calorimetry by different authors. Values reported were:

$\bar{D}$ (Ge—C)   58.9 $\pm$ 4.5 kcal   Rabinovich and coworkers (310)
                            58 $\pm$ 2 kcal   Bills and Cotton (16)
                        56.6 $\pm$ 2 kcal   Pope and Skinner (308)

The latter authors found 56.7 kcal/mol for tetra n-propylgermane.

2-1-2-1   Magneto-optical Study. Magneto-optical study of the first eight symmetrical tetraalkylgermanes (201) showed that the increase in rotation of the CH_2 group presents in these compounds an alternated variation (Figure 2-1) which attenuates quickly and leads, after the n-butyl derivative, to a value equal to the CH_2 modulus in the alkanes.

As the additivity law applies normally from tetrabutylgermane, it has been possible to determine the (Ge—C) linkage constant as well as the germanium modulus (linked with 4 carbon atoms):

$$[\rho]_{(GeC_4)} = \frac{160}{4} \cdot 10^{-2} \text{ min} \quad \text{and} \quad [\rho]_{Ge} = 122 \cdot 10^{-2} \text{ min}$$

For the analogous stannanes, the difference $[\rho]_{M(n+1)} - [\rho]_{M(n)}$ from one term to the other is also variable; for the first stannanes, however, the damping is less fast than for the alkylgermanes (422).

2-1-2-2   Magnetic Study. Measurement of the magnetic susceptibility of the tetra-n-alkylgermanes and the atomic germanium (234) leads to the

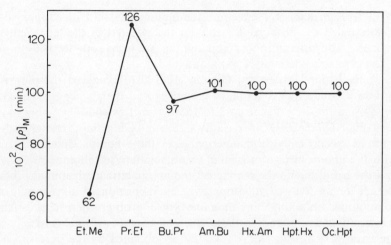

FIG. 2-1   Increase in rotation of CH_2 group in tetraalkylgermanes

same observation which had already been made for the analogous lead, mercury, and tin compounds: the share of the central element decreases as the molecular weight increases to reach asymptotically the value of the atomic susceptibility of the metal itself (Figure 2-2).

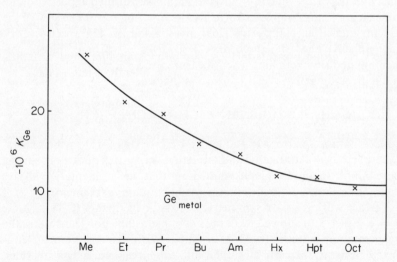

FIG. 2-2   Variation of K_{Ge} in tetraalkylgermanes $(C_nH_{2n+1})_4Ge$

2-1-2-3   Raman and UV Spectra. First Young (439) and later Lippincott (208) studied the absorption bands in the infrared of the tetramethyl-germane. The latter authors compared the spectrum obtained with those of the analogous derivatives of tin and lead.

Fuchs, Moore, Miles and Gilman (105) also compared the infrared spectra of some higher tetraalkylgermanes and their analogous silicon compounds in the range from 2 to 16 micrometres and they found only very small differences. More recently, Cross and Glockling (51) studied the infrared spectra of 80 organogermanes in the range of 2500–200 cm^{-1} with the main object of assigning characteristic group frequencies that give the bands due to the symmetric and asymmetric germanium–carbon stretch for methyl-, ethyl-, isopropyl-, butyl- and benzylgermanes. For compounds containing more than one type of alkyl group specific assignments are uncertain and, in these cases, all bands attributable to $v_{(Ge-C)}$ are given.

The Raman spectrum of tetramethylgermane made up by Siebert (389) has been examined in detail by Lippincott and Tobin (208) and compared with the IR and Raman spectra of the tetramethylated compounds of tin and lead.

The ultraviolet spectra of the compounds 4-$(CH_3)_3MC_6H_4C_6H_5$ and 4,4'-$(CH_3)_3MC_6H_4-C_6H_4M(CH_3)_3$ (M = C, Si, Ge, Sn) have been studied by Curtis and coworkers (57). The spectra are characterized by a broad band at 250 nm (*p* band) and a strong band at 200 nm (*β* band). The heteroatom parameters previously used in the interpretation of the ESR spectra of the silicon- and germanium-substituted biphenyl anion radicals are found to predict correctly the ordering of the energies of the 250 nm transition. Nagy *et al.* (269) investigated the UV spectra of tri-methylphenyl and trimethylbenzyl derivatives of carbon, silicon, germanium and tin in ethanol. Their measurements show that the inductive and hyperconjugative effects increase in the order:

$$t\text{-}C_4H_9CH_2 < (CH_3)_3SiCH_2 < (CH_3)_3GeCH_2 < (CH_3)_3SnCH_2$$

2-1-2-4   Mass Spectra. De Ridder and coworkers (313) in their mass-spectral study of tetramethyl and tetraethyl derivatives of carbon, silicon, germanium, tin and lead have pointed out that the fragmentation always leaves the electron deficiency on the metal-containing fragment.

Further work (314) on symmetrical tetraalkylgermanes $(C_nH_{2n+1})_4Ge$ (n = 1, 2, 3, 4, 5, 6) allowed, with the aid of metastable ion transitions, the fragmentation patterns to be determined. With the exception of tetra-methylgermane, these are all similar and can be explained by a few rules, the first being the tervalence of the germanium after primary ionization. In all spectra the base peak corresponds to the ions $HGe(alkyl)_2$ while for tetramethylgermane the base peak is the ion $Ge(CH_3)_3{}^+$.

Carrick and Glockling (37) studied the mass spectra of various isopropylgermanium derivatives and gave the detailed fragmentation pattern of tetraisopropylgermane with masses, formulae and relative intensities of the different metastable ions.

2-1-2-5 Labelled Tetraalkylgermanes. Germanoorganic compounds labelled with ^{75}Ge and ^{77}Ge were prepared by Nowak (288) by activation in the thermal column of the nuclear reactor E.W.A. The compounds of radioactive germanium were obtained in a pure state, after separation of arsenic compounds formed, by extraction with aqueous solution of inactive tracers or by high-voltage paper electrophoresis.

Akerman and coworkers (1, 2) used the favorable radiochemical properties of the ^{77}Ge isotope ($t_{1/2}$: 11.3 h; γ : 1.750 MeV) to study the flux circulation in the different parts of an oil refinery.

2-1-2-6 Toxicity of Alkylgermanes. The germanium derivatives are far less toxic than the analogous derivatives of tin (40).

Saturated aliphatic germanes were examined for toxicity after enteral or intraperitoneal administration to rats and mice by Caujolle and coworkers (41). Molecular asymmetry and branching increased the toxicity; the intraperitoneal LD_{50} values of Pr_4Ge, i-Pr_4Ge, Et_4Ge, Et_3GePr, Et_3GeBu, Et_3GeAm and $Et_3Ge—CH_2CH{=}CH_2$ were respectively 12; 0.43; 0.59; 1.43; 2.11; 4.69 and 0.022 g/kg in rats.

Alkylhalogermanes are markedly more toxic.

2-1-3 Chemical properties

Tetraalkyl- and tetraarylgermanes are very stable compounds which are insensitive to the action of water and the oxygen of air. Their chemical inertness resembles that of alkanes, as Winkler ascertained in 1880 for tetraethylgermane.

Only the most reactive electrophilic agents are able to cleave the Ge—C bond and, in the case of the alkylated derivatives, this cleavage often takes place only in the presence of a catalyst.

Catalytic hydrogenation of tetraphenylgermane virtually gives no result, whereas tetraphenylsilane is readily converted into tetracyclohexylsilane (90% yield). However, the silicon derivative is the more sterically crowded. Spialter and coworkers (400) suggest that the central atom of germanium causes poisoning of the catalyst.

2-1-3-1   Cleavage Reactions

(a) Halogens, acids

Halogens react with tetraalkylgermanes with difficulty in the absence of a catalyst (182) and a substitution reaction on the alkyl chains often superimposes itself on the cleavage reaction. However, trimethylbromogermane is obtained in excellent yield when tetramethylgermane and bromine are boiled together for 20 hours in the presence of propyl bromide (258):

$$(CH_3)_4Ge + Br_2 \rightarrow (CH_3)_3GeBr \quad (98\%)$$

or when the mixture tetramethylgermane–bromine is kept in a sealed tube for a week at room temperature (32).

In presence of an acid Lewis catalyst, the cleavage reaction occurs readily and allows access to the Ge-halogenated aliphatic derivatives of germanium: in the presence of aluminium iodide, iodine quickly cleaves the Ge—C bonds:

$$R_4Ge + I_2 \rightarrow R_3GeI + RI$$

With mixed alkylgermanes there is a preferential cleavage of the lightest radicals.

The Ge—C (aromatic) bonds are more sensitive than the Ge—C (aliphatic) bonds to the action of the halogens and especially of bromine (94, 156, 183, 290) and this difference in reactivity is often used to effect selective cleavages in the preparation of mixed derivatives:

$$Ar_4Ge \rightarrow Ar_3GeBr \rightarrow Ar_3GeR \rightarrow Ar_2Ge(Br)R \rightarrow Ar_2GeRR'$$

Iodine cleaves tetraphenylgermane only at high temperature (214):

$$(C_6H_5)_4Ge + I_2 \xrightarrow{\text{decalin}} C_6H_5I + (C_6H_5)_3GeI \quad (30\% \text{ yield})$$

Hydracids in aqueous solution, which cleave the tetraalkylstannanes easily, seem to have no action on the saturated tetraalkylgermanes: it is possible, without any trouble, to carry out in the course of their preparation the hydrolysis of the Grignard reagent excess by means of a dilute solution of hydrochloric acid.

On the other hand, the Ge—C(aryl) bonds are sensitive to the action of hydracids and strong organic acids; the phenylated derivatives are cleaved by 48% hydrobromic acid at boiling temperature:

$$(C_2H_5)_3GeC_6H_5 + HBr \rightarrow (C_2H_5)_3GeBr + C_6H_6$$

and by trifluoroacetic acid (328a):

$$(C_6H_5)_4Ge + CF_3COOH \rightarrow CF_3COOGe(C_6H_5)_3 + C_6H_6$$

Simons (393) studied the action of hydrogen bromide in chloroform at room temperature on various tetraarylgermanes. He found that the cleavage rate increases in the order:

$$C_6H_5CH_2- < C_6H_5- < m\text{-}CH_3C_6H_4- < p\text{-}CH_3C_6H_4-$$

$$(C_6H_5)_3Ge(p\text{-}CH_3C_6H_4) \xrightarrow{\text{HBr}} (C_6H_5)_3GeBr + C_6H_5-CH_3$$

$$(C_6H_5)_3Ge(m\text{-}CH_3C_6H_4) \xrightarrow{\text{HBr}} (C_6H_5)_3GeBr + C_6H_5-CH_3$$

$$(m\text{-}CH_3C_6H_4)_3Ge(p\text{-}CH_3C_6H_4) \xrightarrow{\text{HBr}}$$

$$(m\text{-}CH_3C_6H_4)_3GeBr + C_6H_5-CH_3$$

In same experimental conditions $(o\text{-}CH_3C_6H_4)_4Ge$ gives $(o\text{-}CH_3C_6H_4)_3GeBr$ while $(m\text{-}CH_3C_6H_4)_4Ge$ leads to the dibromide $(m\text{-}CH_3C_6H_4)_2GeBr_2$. Hydrogen halides can cleave the Ge—C bond of the saturated tetraalkylgermanes: anhydrous hydrofluoric acid attacks tetramethylgermane at $-10°C$ and the trimethylfluorogermane transformation is almost achieved at $20–25°C$ (119). Triethylfluorogermane is also obtained from tetraethylgermane with a 71% yield. Gaseous hydrobromic acid reacts with tetraalkylgermanes in the presence of a catalyst: Dennis and Patnode (67) used this method for the preparation of trimethylbromogermane:

$$(CH_3)_4Ge + HBr \xrightarrow{Al_2Br_6} (CH_3)_3GeBr + CH_4$$

Hydrogen chloride reacts in the same manner (126).

Dilute sulphuric acid is without action and can be used, in the same way as hydrochloric acid, in the hydrolysis of Grignard reagent excess. Concentrated sulphuric acid is without action in the cold on the symmetrical or mixed saturated tetraalkylgermanes and it is used in the purification of these derivatives. On warming, it reacts with decomposition.

Fuming nitric acid reacts strongly, and the sulphonitric mixture which transforms the organic derivatives of germanium into GeO_2 is used for the gravimetric determination of germanium.

(b) Mineral halides

Mercuric chloride and stannic chloride cleave without catalyst the Ge—C bond in the tetraalkylgermanes (209, 221):

$$Et_4Ge + HgCl_2 \xrightarrow{150°C} Et_3GeCl + EtHgCl$$

$$Bu_4Ge + SnCl_4 \xrightarrow{200°C} Bu_3GeCl + BuSnCl_3$$

The germanium tetrachloride reacts on the tetraphenylgermane (345) under drastic conditions (heating at 350°C for 36 hours being required):

$$Ph_4Ge + 3\,GeCl_4 \;\longrightarrow\; 4\,PhGeCl_3$$

The polybrominated derivatives are very easily obtained by the action of $GeBr_4$ on the tetraalkylgermanes in the presence of aluminium bromide (223a).

Aluminium chloride cleaves one Ge—C bond of tetrabutylgermane above 150°C (316):

$$Bu_4Ge + AlCl_3 \;\longrightarrow\; Bu_3GeCl + BuAlCl_2$$

The cleavage of tetramethylgermane by gallium chloride (338) has been used for the preparation of dichloromethylgallium (yield: 92%):

$$Me_4Ge + GaCl_3 \;\longrightarrow\; Me_3GeCl + MeGaCl_2$$

(c) Alkyl halides

In the presence of 2 weight % of $AlCl_3$ tetraethylgermane is cleaved by isopropyl halides (312):

$$(C_2H_5)_4Ge + i\text{-}C_3H_7X \xrightarrow{\;AlCl_3\;} (C_2H_5)_3GeX \qquad \begin{array}{ll} 60\%\ \text{yield} & X = Cl \\[4pt] 86.6\% & X = Br \end{array}$$

The gaseous products formed consisted of a mixture containing 7–10% ethane, 35–45% ethylene, 42–54% propane and 1–3% of a propylene–butane fraction. Mironov (258) cleaved tetramethylgermane in the same manner and obtained excellent yields of monohalogenated products:

$$(CH_3)_4Ge + i\text{-}C_3H_7X \xrightarrow{\;AlX_3\;} (CH_3)_3GeX \qquad (92\text{–}95\%) \qquad X = Cl, Br$$

With trimethylphenylgermane the phenyl group is split off:

$$(CH_3)_3GeC_6H_5 + i\text{-}C_4H_9Cl \xrightarrow{\;AlCl_3\;} (CH_3)_3GeCl \quad (85\%)$$

A similar reaction was observed with tetrabutylgermane and isoamyl halide (59):

$$(C_4H_9)_4Ge + i\text{-}C_5H_{11}X \xrightarrow{\;AlX_3\;} (C_4H_9)_3GeX \quad (57\text{–}67\%) \qquad X = Cl, Br$$

The carboxylic acid chlorides also produce cleavage of the Ge—C (alkyl) bond either at high temperatures (221) or at room temperature in the presence of aluminium chloride (324):

$$(CH_3)_4Ge + CH_3COCl \xrightarrow{\;AlCl_3\;} (CH_3)_3GeCl \quad (74\%)$$

$$(CH_3)_4Ge + 2CH_3COCl \xrightarrow{\;AlCl_3\;} (CH_3)_2GeCl_2 \quad (70\%)$$

(d) Oxygen, sulphur

Tetraalkyl and tetraarylgermanes are very stable in air and can be heated at high temperatures (250–280°) without appreciable change.

Strong oxidizing agents (as the sulphonitric mixture) destroy organo-germanes, giving germanium dioxide.

Sulphur reacts with tetrabutylgermane at 230° giving a six-membered ring sulphide (336):

$$3\,(C_4H_9)_4Ge + 6\,S \rightarrow [(C_4H_9)_2GeS]_3$$

Schmidt explains the formation of this compound by an insertion reaction of sulphur in the Ge—C bond, followed by elimination of dibutyl sulphide:

$$(C_4H_9)_4Ge \xrightarrow{S_8} (C_4H_9)_2Ge(SC_4H_9)_2 \rightarrow [(C_4H_9)_2GeS]_3 + (C_4H_9)_2S$$

Tetraphenylgermane is degraded to the metal above 270° (336):

$$(C_6H_5)_4Ge + 2\,S \rightarrow Ge + 2\,(C_6H_5)_2S$$

(e) Cleavage by alkali metals

Kraus and Foster (183) observed in 1925 that metallic sodium in liquid ammonia reacts slowly with tetraphenylgermane:

$$(C_6H_5)_4Ge + 2\,Na \xrightarrow{NH_3} (C_6H_5)_3GeNa + NaNH_2 + C_6H_6$$

When excess sodium is used, the orange solution turns red; the colour change is said to correspond to a second cleavage.

Brook and Gilman (28) had cleaved triphenylmethyltriphenylgermane with sodium–potassium alloy (1:5) in ether. After carbonation and treatment by diazomethane they obtained the corresponding esters:

$$(C_6H_5)_3C—Ge(C_6H_5)_3 + Na/K \xrightarrow[(2)\ CH_2N_2]{(1)\ CO_2} (C_6H_5)_3GeCOOCH_3$$
$$+ (C_6H_5)_3CCOOCH_3$$

Gilman (116) effected many cleavage experiments (tetraethyl-, tetrabutyl-, tetraoctyl-, tetrabenzyl-, tetra-2-phenylethyl- and tri-n-hexylphenyl-germane) by the alkali metals in different solvents; with GeEt$_4$, when Li was used, there was little evidence of cleavage but with sodium–potassium alloy, triethylgermylpotassium was formed and immediately reacted with the solvent. Cleavage of Ge—C bond in tetraphenylgermane is possible by use of Li wire in ethereal solvents such as THF or ethylene glycol dimethyl ether (110):

$$(C_6H_5)_4Ge + 2\,Li \xrightarrow{CH_3OCH_2CH_2OCH_3} (C_6H_5)_3GeLi + C_6H_5Li$$

(phenyllithium decomposes by reaction with the solvent). Similar results were obtained with n-octadecyltriphenylgermane:

$$(n\text{-}C_{18}H_{37})Ge(C_6H_5)_3 \xrightarrow[\text{GDME}]{\text{Li}} (C_6H_5)_2(n\text{-}C_{18}H_{37})GeLi$$

Ge—benzyl bonds are actually much easier than Ge—phenyl bonds to cleave. Thus, lithium shot in GDME medium readily cleaves tetrabenzylgermane, giving tribenzylgermyllithium and a small amount of the dilithium dibenzyl derivative (50):

$$(C_6H_5CH_2)_4Ge \xrightarrow[0°]{\text{Li, }C_4H_{10}O_2} (C_6H_5CH_2)_3GeLi + (C_6H_5CH_2)_2GeLi_2$$
$$+ C_6H_5CH_2Li$$

(f) Nucleophilic reagents

Scission of the tetraalkylgermanes by nucleophilic reagents is extremely rare. Carbon–tin bonds are cleaved more readily than carbon–silicon bonds while carbon–germanium bonds are not or only partially cleaved (20). Rühlmann and Heine (323) have reported the partial cleavage of C—Si bonds in benzyl triphenylsilane by sodium hydride in cyclohexane but no reaction was observed with the analogous compound of germanium.

It seems clear that nucleophilic attack is easier on silicon than on germanium, probably because of increased nuclear shielding (149).

It was recently shown that many Si—C bonds are readily broken by lithium aluminium hydride either in ether (109) or in refluxing THF; for example the benzyl group of benzyltriphenylsilane is cleaved to give triphenylsilane:

$$(C_6H_5)_3SiCH_2C_6H_5 \xrightarrow[\text{(b) }H_3O^+]{\text{(a) LiAlH}_4} (C_6H_5)_3SiH$$

In similar conditions benzyltriphenylgermane is not cleaved while the benzyl group is rapidly removed from benzyltriphenyltin at room temperature (294).

However, the benzylgermanium bond in $C_6H_5CH_2GeR_3$ and $ClC_6H_4CH_2GeR_3$ compounds is partially cleaved by alkali in 80% methanol. The approximate relative reactivities for chlorobenzyl-compounds $ClC_6H_4CH_2MR_3$ are:

$$
\begin{array}{ll}
M = Ge & 10^{-3} \\
Si & 1 \\
Sn & 17
\end{array}
$$

These results accord with those of alkaline solvolysis in aqueous ethanol of R_3MH compounds and of hydrolysis of $(C_6H_5)_3MCl$ compounds (46, 342).

2-1-3-2 Redistribution Reactions. Redistribution reactions in organogermanium chemistry involving random intermolecular exchange of alkyl groups have not been studied in detail until very recently. They occur much less easily than in organotin chemistry and require a catalyst. Pollard and coworkers (303) investigated redistribution reactions between pairs of symmetrical tetraalkyls, $GeR_4 + GeR'_4$, obtaining mixed tetraalkylgermanes R_3GeR', $R_2GeR'_2$, $RGeR'_3$. Reactions were carried out by refluxing the mixtures of germanes using 1/800 molar quantities of reagents and approximately 2 mol % of aluminium chloride as catalyst. Aluminium bromide was found to be less efficient as a catalyst than aluminium chloride in the alkylgermane redistribution reactions.

Quantitative data were obtained by CPV and NMR measurements (265). The reported data show that the redistribution of the two kinds of alkyl groups in the pairs $GeR_4 + GeR'_4$ is a purely statistical redistribution, the "random equilibrium mixture"; with the pair $GeEt_4 + GePr_4$ in equimolar proportions the relative amounts of alkylgermanes are shown next:

	Et_4Ge	Et_3GePr	Et_2GePr_2	$EtGePr_3$	Pr_4Ge
mol % theor.	6.2	25	37.5	25	6.2
found	8	27	35	23	9

As in the silicon analogues, the interchange of methyl and ethyl groups occurs more rapidly and under milder conditions than any other alkyl interchange. A characteristic feature of these exchange reactions as compared to silicon compounds is the much faster rate of equilibration found for germanium compounds: at 130°C reaction took one hour to go to completion, but at 170°C, a degree of redistribution of 100% was reached in 2 or 3 min only for the pairs $GeEt_4 + GePr_4$, $GePr_4 + GeBu_4$, $GeEt_4 + GeAm_4$, etc.

When the redistribution reactions were carried out at 200°C with greater proportion of aluminium chloride, cleavage of simple alkyl groups by the catalyst takes place, giving a mixture of different trialkylchlorogermanes:

$$(CH_3)_4Ge + (C_3H_7)_4Ge \xrightarrow{5\% \text{ AlCl}_3} (CH_3)_xGe(C_3H_7)_{3-x}Cl$$

In the reaction involving tetra *n*-butylgermane, exchange of *n*-butyl groups and chlorine led to a mixture of trialkylmonochloro- and dialkyldichlorogermane, for example:

$$(CH_3)_4Ge + (C_4H_9)_4Ge \xrightarrow{5\% \text{ AlCl}_3} (CH_3)_xGe(C_4H_9)_{3-x}Cl$$
$$+ (CH_3)_xGe(C_4H_9)_{2-x}Cl_2$$

Simple tetraalkylgermanes and tetraalkylsilanes redistribute readily on refluxing with catalytic amounts of aluminium chloride, giving a mixture of ten products: with the pair $Si(C_2H_5)_4 + Ge(C_3H_7)_4$ equilibrium was reached in less than 90 min and gas chromatography revealed the existence of the ten compounds: $SiEt_4$, $SiPr_4$, $GeEt_4$, $GePr_4$, $SiEt_3Pr$, $GeEt_3Pr$, $SiEt_2Pr_2$, $GeEt_2Pr_2$, $SiEtPr_3$, $GeEtPr_3$. Further, two tetraalkylgermanes redistribute with germanium tetrabromide in the presence of aluminium bromide to give mixed mono-, di- and tribromogermanes:

$$Pr_4Ge + Bu_4Ge + Br_4Ge \xrightarrow{\text{AlBr}_3} GePr_xBu_yBr_{4-x-y}$$

but the mixture contains also ethyl derivatives like $PrBu_2GeEt$, $Pr_2BuGeEt$, $PrBuGeEt_2$, Pr_3GeEt, etc.

A broad study of the catalysed alkylhalogens redistribution reactions on germanium has been reported from the TNO Institut at Utrecht (160, 316). Tributylchlorogermane and dibutyldichlorogermane were obtained in good yields from germanium tetrachloride and tetrabutylgermane. Butyltrichlorogermane seems to be more reactive than germanium tetrachloride and is easily alkylated in the presence of aluminium chloride:

$$Bu_4Ge + BuGeCl_3 \xrightarrow{\text{AlCl}_3} Bu_3GeCl + Bu_2GeCl_2$$

The slowest of the equilibrium reactions was the exchange of methyl groups for chlorine atoms on germanium. At 300°C in the presence of 0.1 % of aluminium chloride, Burch and van Wazer (33a) observed the following reactions:

$$GeMe_4 + GeCl_4 \rightleftarrows Me_3GeCl + MeGeCl_3$$

$$Me_3GeCl + MeGeCl_3 \leftrightarrows 2\,Me_2GeCl_2 \quad \text{(about 100-fold slower)}$$

$$GeMe_4 + MeGeCl_3 \rightarrow Me_3GeCl + Me_2GeCl_2$$

2-1-3-3   Insertion Reactions. Sulphur trioxide, which is a strongly electrophilic reagent, reacts with tetraalkyl and tetraarylgermanes (18, 341) giving organogermanium derivatives of sulphonic acids by an insertion reaction:

$$p\text{-}(C_2H_5)_3GeC_6H_4Ge(C_2H_5)_3 + SO_3 \rightarrow$$

$$p\text{-}(C_2H_5)_3GeC_6H_4SO_2OGe(C_2H_5)_3$$

$$(CH_3)_4Ge + SO_3 \rightarrow (CH_3)_3Ge\text{-}O\text{-}SO_2CH_3$$

Under suitable conditions organogermasulphate and sulphonic acid anhydride can be detected as products of a side reaction:

$$2\,(CH_3)_4Ge + 3\,SO_3 \rightarrow [(CH_3)_3GeO]_2SO_2 + CH_3SO_2OSO_2CH_3$$

2-1-3-4   Homolytic Scissions. Vyazankin and coworkers (427) investigated some homolytic reactions of tetraethylgermane; in the presence of organic peroxides, tetraethylgermane acts as a hydrogen donor (homolysis of carbon-hydrogen bonds):

$$R^{\cdot} + (C_2H_5)_4Ge \longrightarrow RH + (C_2H_5)_3Ge\overset{\cdot}{C}_2H_4$$

$$(A)$$

The radical A reacts by recombination, giving a digermane $C_{16}H_{38}Ge_2$, and by decomposition, leading to ethylene:

$$(C_2H_5)_3Ge\overset{\cdot}{C}_2H_4 \longrightarrow C_2H_4 + (C_2H_5)_3Ge^{\cdot}$$

Homolytic chlorination of tetraethylgermane with sulphuryl chloride was carried out in the presence of benzoyl peroxide; the yield of α-chloroethyl-triethylgermane was 26 % (in the same conditions, tetraethyl silane was converted to α-chloroethyltriethylsilane in 50 % yield (306). Triethyl-chlorogermane resulting from decomposition of β-chloroethyltriethyl-germane was also formed (16.8 % yield).

2-1-4   Saturated GeR₄ derivatives

Table 2-1 showing some physical properties of saturated GeR_4 derivatives appears on the following pages.

Table 2-1
Saturated GeR₄ Derivatives

COMPOUND	EMPIRICAL FORMULA	M.P. (°C)	B.P. (°C/mm)	n_D^{20}	d_4^{20}	REFERENCES
$(CH_3)_4Ge$	$C_4H_{12}Ge$	—	—	—	0.9781 (18°)	47, 66, 119, 123, 195, 201, 208, 221, 258, 270, 314, 389, 424, 433, 440, 441, 449
		—	—	1.3896	0.9758	
		−88	43.2/736	1.3863 (25°)	0.9661 (25°)	
$(CH_3)_3GeC_2H_5$	$C_5H_{14}Ge$	—	80	1.4090	0.9843	193, 197, 222, 257, 442
			77	1.4085	0.9815	
$(CH_3)_3GeCH\big<\!\!{}^{CH_2}_{CH_2}$	$C_6H_{14}Ge$	—	—	1.4321	—	352
$(CH_3)_2Ge(C_2H_5)_2$	$C_6H_{16}Ge$	—	109	1.4221	0.9885	193, 197, 221
$(CH_3)_3Ge(n\text{-}C_3H_7)$	$C_6H_{16}Ge$	—	87.5	—	—	193, 441, 442
			99.5			
$(CH_3)_3GeCH_2\!-\!CH\!-\!CH_2\big|_{CH_2}$	$C_7H_{16}Ge$	—	102/753	1.4158	0.9662	70, 71
			126/745	1.4355	1.0290	
$(CH_3)_3GeCH_2CH(CH_3)_2$	$C_7H_{18}Ge$	—	121/740	—	—	340
$(CH_3)_3C\!-\!Ge(CH_3)_3$	$C_7H_{18}Ge$	95–95.5	111/758	—	—	250
$(CH_3)_3Ge(n\text{-}C_4H_9)$	$C_7H_{18}Ge$	—	118	1.4225	0.9597	200, 442
			128/755			

Compound	Formula	M.p.	B.p./mm	n	d	References
$CH_3Ge(C_2H_5)_3$	$C_7H_{18}Ge$	—	135/757	1.4332	0.9906	193, 197, 222
$(CH_3)_3GeCH_2Ge(CH_3)_3$	$C_7H_{20}Ge_2$	—	156–57/750	1.4502	1.1544	248, 261
$(CH_3)_3GeCH$ (cyclopentyl: $\begin{array}{c}CH_2-CH_2\\ \diagdown\\ CH_2-CH_2\end{array}$)	$C_8H_{18}Ge$	—	34.5/8	1.4552	1.0422	253
$(C_2H_5)_4Ge$	$C_8H_{20}Ge$	−92.7	163.5/760; 72.7/34	1.4428	0.9931; 0.9882 (25°)	6, 51, 64, 116, 182, 201, 221, 308, 314, 412, 426, 427, 435, 344
$(CH_3)_2Ge(n\text{-}C_3H_7)_2$	$C_8H_{20}Ge$	—	141	1.4322	0.9660	193, 441, 445
$(CH_3)_3GeCH_2CH_2Ge(CH_3)_3$	$C_8H_{22}Ge_2$	—	149.5/746	1.4532	1.1375	256, 272
$(CH_3)_3\,GeC_6H_5$	$C_9H_{14}Ge$	—	65–66/18; 182–83	1.5075	1.1167	12, 94, 106, 269, 329, 350
$C_3H_7Ge(C_2H_5)_3$	$C_9H_{22}Ge$	—	77.5–78/22.5	1.4460	0.9810	193, 197, 221
$(CH_3)_3Ge(CH_3)_3Ge(CH_3)_3$	$C_9H_{24}Ge_2$	—	73–74/20	1.4500	1.1075	241
$(CH_3)_3GeCH_2C_6H_5$	$C_{10}H_{16}Ge$	—	90/28	1.5140	1.1011	51, 241, 269
$C_4H_9Ge(C_2H_5)_3$	$C_{10}H_{24}Ge$	—	94–95/28; 91–92/20	1.4483	0.9711	31, 51, 116, 197, 221, 222, 335
$CH_3Ge(C_3H_7)_3$	$C_{10}H_{24}Ge$	—	181–81.5; 189–91	1.4429	0.9582	193, 221
$(C_2H_5)_3Ge(C_5H_9)$	$C_{11}H_{24}Ge$	—	84–85/20	1.4756	1.0343	199
$C_5H_{11}Ge(C_2H_5)_3$	$C_{11}H_{26}Ge$	—	133/40; 104–5/20	1.4493	0.9625	197, 221
$(C_2H_5)_3Ge(C_6H_5)$	$C_{12}H_{20}Ge$	—	116–17/13; 117–18/13.5	1.5147	—	12, 31, 51, 79, 429
$(CH_3)_3GeC_6H_4Ge(CH_3)_3$	$C_{12}H_{22}Ge_2$	105–7	—	—	—	206

Table 2-1—continued

COMPOUND	EMPIRICAL FORMULA	M.P. (°C)	B.P. (°C/mm)	n_D^{20}	d_4^{20}	REFERENCES
$(C_3H_7)_4Ge$	$C_{12}H_{28}Ge$	-73	81/11	—	0.9539	37, 193, 201, 221, 302, 308, 314, 412, 420
$i\text{-}(C_3H_7)_4Ge$	$C_{12}H_{28}Ge$	—	225	1.4537	0.9690	37, 444
$(C_2H_5)_2Ge(C_4H_9)_2$	$C_{12}H_{28}Ge$	—	86-87/5	1.4515	0.9513	197, 221
$(n\text{-}C_3H_7)_2Ge(i\text{-}C_3H_7)_2$	$C_{12}H_{28}Ge$	—	159-60/50	1.4447 (25°)	0.9491 (25°)	37
$i\text{-}C_3H_7Ge(C_3H_7)_3$	$C_{12}H_{28}Ge$	—	109-10/14	1.4580	0.9364	37
$n\text{-}C_3H_7Ge(i\text{-}C_3H_7)_3$	$C_{12}H_{28}Ge$	—	227-28/758	1.4516	0.9547	37
$n\text{-}C_6H_{13}Ge(C_2H_5)_3$	$C_{12}H_{28}Ge$	—	117/20	1.4580	—	426, 430
$(CH_3)_3GeCH_2CH_2CH[Ge(CH_3)_3]_2$	$C_{12}H_{32}Ge_3$	—	104-5/7	1.4940	1.2368	250
$(C_2H_5)_3Ge(o\text{-}C_6H_4CH_3)$	$C_{13}H_{22}Ge$	—	126/11	1.5200	—	79
$(C_2H_5)_3Ge(m\text{-}C_6H_4CH_3)$	$C_{13}H_{22}Ge$	—	128/12	1.5140	—	79
$(C_2H_5)_3Ge(p\text{-}C_6H_4CH_3)$	$C_{13}H_{22}Ge$	—	136/16	1.5134	—	12, 79
$(C_2H_5)_3GeCH_2C_6H_5$	$C_{13}H_{22}Ge$	—	78-81/1	1.5178	—	20, 51, 116, 428
$CH_3Ge(C_4H_9)_3$	$C_{13}H_{30}Ge$	—	118-19/4	1.4493	0.9386	221
$(n\text{-}C_3H_7)_3Ge(n\text{-}C_4H_9)$	$C_{13}H_{30}Ge$	—	113/10	1.4595	1.0183	442
$(C_2H_5)_3GeCH_2Ge(C_2H_5)_3$	$C_{13}H_{32}Ge_2$	—	54.5/0.35	1.4822	—	33
$(C_6H_5)_2Ge(CH_3)_2$	$C_{14}H_{16}Ge$	—	138-39/14	1.5730	1.180	193
$(C_2H_5)_3GeCH_2CH_2C_6H_5$	$C_{14}H_{24}Ge$	—	145/10	1.5078	—	426, 430
$p\text{-}CH_3C_6H_4CH_2Ge(C_2H_5)_3$	$C_{14}H_{24}Ge$	—	107/1.5	1.5167	—	20
$p\text{-}C_2H_5C_6H_4Ge(C_2H_5)_3$	$C_{14}H_{24}Ge$	—	131-33/11-12	1.5136	—	79
$C_8H_{17}Ge(C_2H_5)_3$	$C_{14}H_{32}Ge$	—	143/17	1.4540	0.938	196, 329

Compound	Molecular formula	m.p.	b.p./mm	n_D	d	References
$(C_2H_5)_3Ge(CH_2)_2Ge(C_2H_5)_3$	$C_{14}H_{34}Ge_2$	—	126/1.5	1.4775	1.0935	196, 228, 329
$(p\text{-}C_6H_5C_6H_4)Ge(CH_3)_3$	$C_{15}H_{15}Ge$	—	—	—	—	57
$(C_2H_5)_3Ge(p\text{-}C_6H_4i\text{-}C_3H_7)$	$C_{15}H_{26}Ge$	—	136/7	1.5095	—	79
$(C_2H_5)_3Ge[2,4,6\text{-}C_6H_2(CH_3)_3]$	$C_{15}H_{26}Ge$	—	152/9	1.5333	—	79
$(C_3H_7)_3GeC_6H_5$	$C_{15}H_{26}Ge$	—	101–2/20	1.4680	0.9784	328
$n\text{-}C_3H_7Ge(C_4H_9)_3$	$C_{15}H_{34}Ge$	—	82/0.45	1.4685	0.9876	442
$(C_2H_5)_3Ge(CH_2)_3Ge(C_2H_5)_3$	$C_{15}H_{36}Ge_2$	—	128–29/1.4	1.4759	1.0807	196, 228, 329
$(C_6H_5)_2Ge(C_2H_5)_2$	$C_{16}H_{20}Ge$	—	316/760	—	—	51, 96, 418
$(C_6H_5CH_2)_2Ge(CH_3)_2$	$C_{16}H_{20}Ge$	53–55	82–100/0.001	—	—	50, 51
$(CH_3)_3GeCH(C_6H_5)_2$	$C_{16}H_{20}Ge$	73.5	—	—	—	20
$(C_2H_5)_3Ge(1\text{-}C_{10}H_7)$	$C_{16}H_{22}Ge$	—	153/4	1.5120	—	79
$(C_2H_5)_3Ge(2\text{-}C_{10}H_7)$	$C_{16}H_{22}Ge$	—	153/3–4	1.5785	—	79
$(C_2H_5)_3Ge(m\text{-}C_6H_4t\text{-}C_4H_9)$	$C_{16}H_{28}Ge$	—	116–17/4–5	1.5050	—	79
$(C_2H_5)_3Ge(p\text{-}C_6H_4t\text{-}C_4H_9)$	$C_{16}H_{28}Ge$	—	125/2	1.5105	—	79
[fused-ring structure bearing two $Ge(CH_3)_3$ groups]	$C_{16}H_{28}Ge_2$	—	110–13/0.6	1.5379	1.2096	171, 172, 173
$(C_4H_9)_4Ge$	$C_{16}H_{36}Ge$	—	127–28/4; 274–78/760	1.4563	0.9327	7, 51, 160, 169, 201, 221, 290, 314
$(i\text{-}C_4H_9)_4Ge$	$C_{16}H_{36}Ge$	—	134–35/17	1.4594	0.9374	123, 197, 221
$(C_2H_5)_3Ge(CH_2)_4Ge(C_2H_5)_3$	$C_{16}H_{38}Ge_2$	—	136/0.8	1.4765	1.0715	224, 427
$(C_6H_5)_2Ge(C_2H_5)(i\text{-}C_3H_7)$	$C_{17}H_{22}Ge$	—	175–90; 104–5/0.15	1.5641 (25°)	—	81, 83, 343
[fused-ring structure bearing two $Ge(CH_3)_3$ groups and a CH_3 group]	$C_{17}H_{30}Ge_2$	—	101–4/0.2	—	—	171, 172

Table 2-1—continued

COMPOUND	EMPIRICAL FORMULA	M.P. (°C)	B.P. (°C/mm)	n_D^{20}	d_4^{20}	REFERENCES
$(C_6H_{11})_3GeCH_3$	$C_{17}H_{36}Ge$	48–48.5	—	—	—	154
$C_2H_5Ge(C_5H_{11})_3$	$C_{17}H_{38}Ge$		148–49/7	1.4565	0.9274	197, 221
$(C_2H_5)_3Ge(o\text{-}C_6H_4C_6H_5)$	$C_{18}H_{24}Ge$		159/3	1.5701		79
$(C_2H_5)_3Ge(p\text{-}C_6H_4C_6H_5)$	$C_{18}H_{24}Ge$		158–59/0.7	1.5080		79
$(C_2H_5)_3Ge(2\text{-}C_{12}H_9)$	$C_{18}H_{24}Ge$		150–52/3.3	1.5697		116
$(C_6H_5)_2Ge(C_3H_7)_2$	$C_{18}H_{24}Ge$		158–60/6	1.5583	1.1138	327
$(C_6H_5)_2Ge(C_2H_5)(n\text{-}C_4H_9)$	$C_{18}H_{24}Ge$					51
$(CH_3)_3GeC_6H_4{-}C_6H_4Ge(CH_3)_3$	$C_{18}H_{26}Ge_2$					57
$(C_4H_9)_3GeC_6H_5$	$C_{18}H_{32}Ge$		136–37/5	1.4661	0.9415	328
$(i\text{-}C_4H_9)_3GeC_6H_5$	$C_{18}H_{32}Ge$		125–26/8	1.4783	0.9703	328
$m\text{-}(C_2H_5)_3GeC_6H_4Ge(C_2H_5)_3$	$C_{18}H_{34}Ge_2$		149/1.8	1.5177		77
$p\text{-}(C_2H_5)_3GeC_6H_4Ge(C_2H_5)_3$	$C_{18}H_{34}Ge_2$		157–58/2–3	1.5218		77
$(C_6H_5)_3GeCH_3$	$C_{19}H_{18}Ge$	70.5–71	—			29, 51, 52, 112, 184
$(CH_3)(C_2H_5)(C_6H_5)(1\text{-}C_{10}H_7)Ge$	$C_{19}H_{20}Ge$			1.6228 (21°)		76
$(C_2H_5)_3GeCH(C_6H_5)_2$	$C_{19}H_{26}Ge$		148/0.9	1.5685		20
$(n\text{-}C_3H_7)_3Ge(1\text{-}C_{10}H_7)$	$C_{19}H_{28}Ge$		101–2/10	1.4734	0.9622	311
$(C_6H_5CH_2)Ge(C_4H_9)_3$	$C_{19}H_{34}Ge$		100–10/0.001			50, 51
$(C_6H_{11})_3GeCH_3$	$C_{19}H_{36}Ge$	48–48.5				154
$(C_6H_5)_3GeC_2H_5$	$C_{20}H_{20}Ge$	75–76				51, 81, 184, 290
		77–78.5				
$(C_6H_5)_2Ge(C_4H_9)_2$	$C_{20}H_{28}Ge$		173–75/4	1.5508	1.0922	51, 285, 327
			135–40/0.02	1.5494		
$(C_6H_5)_2Ge(i\text{-}C_4H_9)_2$	$C_{20}H_{28}Ge$		160–62/4	1.5498	1.0821	327
$(CH_3)_2Ge[C_6H_4Ge(CH_3)_3]_2$	$C_{20}H_{32}Ge_3$	94–96	—			206

Compound	Formula	m.p. (°C)	b.p. (°C/mm)	n_D	d	Ref.
$C_2H_5Ge(C_6H_{11})_3$	$C_{20}H_{38}Ge$	38.5–39	173/4	1.4582	0.9152	154
$(C_5H_{11})_4Ge$	$C_{20}H_{44}Ge$	—	163–64/10	1.457 (17.5°)	0.9147	201, 221, 314
$(i-C_5H_{11})_4Ge$	$C_{20}H_{44}Ge$	—	165/11	1.4560	0.9131	412
$(C_6H_5)_3Ge-i-C_3H_7$	$C_{21}H_{22}Ge$	126–27	—	—	—	214
$(C_6H_5)_3GeC_3H_7$	$C_{21}H_{22}Ge$	86–86.5	—	—	—	185
$(C_5H_{11})_3GeC_6H_5$	$C_{21}H_{38}Ge$	—	154–55/5	1.4675	0.9273	328
$(i-C_5H_{11})_3GeC_6H_5$	$C_{21}H_{38}Ge$	—	141–42/5	1.4697	0.9369	328
$(C_6H_{11})_3GeC_3H_7$	$C_{21}H_{40}Ge$	124–25	—	—	—	154
$(C_6H_5)_3GeC_4H_9$	$C_{22}H_{24}Ge$	84.5–85.5	—	—	—	51, 118, 185
$(C_6H_5)_3Ge(t-C_4H_9)$	$C_{22}H_{24}Ge$	80–81	—	—	—	317
$(C_6H_5)_3Ge(sec-C_4H_9)$	$C_{22}H_{24}Ge$	160–62	160/0.01	—	—	214, 317
$(p-CH_3C_6H_4)_2(C_6H_5)GeC_2H_5$	$C_{22}H_{24}Ge$	70–71	164–65/3	—	—	343
$(C_6H_5CH_2)_3GeCH_3$	$C_{22}H_{24}Ge$	82–85	220/13	—	—	50, 51
$(CH_3)_3GeC(C_6H_5)_3$	$C_{22}H_{24}Ge$	180.5–81	—	—	—	20
$(C_6H_5)_2Ge(i-C_5H_{11})_2$	$C_{22}H_{32}Ge$	—	187–89/4	1.5392	1.0592	327
$(C_6H_5CH_2)_2Ge(C_4H_9)_2$	$C_{22}H_{32}Ge$	—	130–33/0.001	—	—	50, 51
$[(CH_3)_2CH-CH_2]_3Ge(1-C_{10}H_7)$	$C_{22}H_{34}Ge$	124–25	123–24/7	1.4674	0.9460	311
$(n-C_4H_9)_3Ge(1-C_{10}H_7)$	$C_{22}H_{34}Ge$	—	139/4	1.4700	0.9591	311
$(C_2H_5)_2(C_6H_5)Ge(CH_2)_2Ge(C_6H_5)(C_2H_5)_2$	$C_{22}H_{34}Ge_2$	—	195/0.8	1.5521	1.1789	225
$(C_6H_{11})_3GeC_4H_9$	$C_{22}H_{42}Ge$	55	152–53	1.5691 (70°)	—	154
$(C_6H_5)_3Ge-i-C_5H_{11}$	$C_{23}H_{26}Ge$	65–66	—	—	—	214
$(C_6H_5)_3GeC_5H_{11}$	$C_{23}H_{26}Ge$	42–43	170–83/0.001	—	—	185
$(C_6H_5CH_2)_3GeC_2H_5$	$C_{23}H_{26}Ge$	34–35	183–84/3	—	—	13, 50, 51
$(C_2H_5)_2(C_6H_5)Ge(CH_2)_3Ge(C_6H_5)(C_2H_5)_2$	$C_{23}H_{36}Ge_2$	56–57	170/0.2	1.5458	1.1629	229a
$(n-C_5H_{11})Ge(C_6H_{11})_3$	$C_{23}H_{44}Ge$	78–79	—	—	—	154

Table 2-1—continued

COMPOUND	EMPIRICAL FORMULA	M.P. (°C)	B.P. (°C/mm)	n_D^{20}	d_4^{20}	REFERENCES
$(C_6H_5)_4Ge$	$C_{24}H_{20}Ge$	225–26 230–31 232 225–28 230	— —	— —	— —	112, 122, 135, 169, 183, 412
$(C_6H_5)_3Ge(C_6H_{11})$	$C_{24}H_{26}Ge$	143–46 95–96	220–21/8	—	—	103, 143, 214
$(C_6H_5)_3GeC_6H_{13}$	$C_{24}H_{28}Ge$	65–66 76	200–2/5 —	1.5560 (71°)	—	214, 318
$(C_6H_5)_2Ge(C_6H_{13})_2$	$C_{24}H_{36}Ge$	—	203–5/4	1.5331	1.0326	327
$(C_6H_{11})_3GeC_6H_5$	$C_{24}H_{38}Ge$	210–11	—	—	—	79, 155
$(C_2H_5)_2(C_6H_5)Ge(CH_2)_4Ge(C_6H_5)(C_2H_5)_2$	$C_{24}H_{38}Ge_2$	—	200/1	1.5441	1.1516	229a
$(C_6H_{13})_3GeC_6H_5$	$C_{24}H_{44}Ge$	—	160–5/0.5	1.4984 (22°)	0.972 (25°)	104
$(n\text{-}C_6H_{13})_4Ge$	$C_{24}H_{52}Ge$	—	192/3 158–61/0.5	1.4608 1.4567 (27°)	0.9077 0.908 (27°)	104, 201, 221, 314
$(C_6H_5)_3GeCH_2C_6H_5$	$C_{25}H_{22}Ge$	85–86.5	—	—	—	30, 51, 185, 413
$(C_6H_5)_3Ge(o\text{-}C_6H_4CH_3)$	$C_{25}H_{22}Ge$	110–11	—	—	—	214
$(C_6H_5)_3Ge(m\text{-}C_6H_4CH_3)$	$C_{25}H_{22}Ge$	136–38	—	—	—	15, 393, 399
$(C_6H_5)_3Ge(p\text{-}C_6H_4CH_3)$	$C_{25}H_{22}Ge$	123–24	—	—	—	290, 393
$(C_6H_5)_3GeC_7H_{15}$	$C_{25}H_{30}Ge$	51–52	204–5/5	1.5586 (65°)	—	214
$(C_6H_5CH_2)_3Ge(C_4H_9)$	$C_{25}H_{30}Ge$	—	150–63/0.001	—	—	50, 51
$(C_2H_5)_3GeC(C_6H_5)_3$	$C_{25}H_{30}Ge$	—	206–8/0.23	1.5915	—	20
$(n\text{-}C_5H_{11})_3Ge(1\text{-}C_{10}H_7)$	$C_{25}H_{40}Ge$	—	169–70/5	1.4706	0.9355	311
$(i\text{-}C_5H_{11})_3Ge(1\text{-}C_{10}H_7)$	$C_{25}H_{40}Ge$	—	154–55/5	1.4671	0.9301	311

Compound	Formula	mp	bp	n_D	d	References
$(C_6H_{11})_3GeCH_2C_6H_5$	$C_{25}H_{40}Ge$	54.5	199–200/0.3	1.5685	1.1255	154, 236
m-$CH_3C_6H_4Ge(C_6H_{11})_3$	$C_{25}H_{40}Ge$	124	—	—	—	79
p-$CH_3C_6H_4Ge(C_6H_{11})_3$	$C_{25}H_{40}Ge$	135	—	—	—	79
$(C_6H_5)_3GeCH_2CH_2C_6H_5$	$C_{26}H_{24}Ge$	147–49	—	—	—	116, 143
$(C_6H_5)_2Ge(o$-$C_6H_4CH_3)_2$	$C_{26}H_{24}Ge$	165–66	—	—	—	326
$(p$-C_6H_5—$C_6H_4)_2Ge(CH_3)_2$	$C_{26}H_{24}Ge$	176	—	—	—	296
$(C_6H_5)_3GeC_8H_{17}$	$C_{26}H_{32}Ge$	72–73	210–11/3	1.5495 (75°)	—	103, 214
$(C_6H_5)_2Ge(C_7H_{15})_2$	$C_{26}H_{40}Ge$	63–64	215–17/3	1.5252	1.0148	327
p-$C_2H_5C_6H_4Ge(C_6H_{11})_3$	$C_{26}H_{42}Ge$	138	—	—	—	79
$(p$-$CH_3C_6H_4)_3GeC_6H_5$	$C_{27}H_{26}Ge$	191	—	—	—	343
$(n$-$C_9H_{19})Ge(C_6H_5)_3$	$C_{27}H_{34}Ge$	48–49	216–17/5	1.5558 (65°)	—	214
$p(i$-$C_3H_7)C_6H_4Ge(C_6H_{11})_3$	$C_{27}H_{44}Ge$	129.5	—	—	—	79
$(C_7H_{15})_3GeC_6H_5$	$C_{27}H_{50}Ge$	—	242–43/10	1.4683	0.9143	328
$(1$-$C_{10}H_7)_2(C_6H_5)GeC_2H_5$	$C_{28}H_{24}Ge$	166.5–67	—	—	—	81
$(o$-CH_3—$C_6H_4)_4Ge$	$C_{28}H_{28}Ge$	175–76	—	—	—	393, 394
$(m$-CH_3—$C_6H_4)_4Ge$	$C_{28}H_{28}Ge$	146–50.1	—	—	—	51, 120, 393, 394
$(p$-CH_3—$C_6H_4)_4Ge$	$C_{28}H_{28}Ge$	148–49	—	—	—	51, 120, 343, 394, 412
$(C_6H_5$—$CH_2)_4Ge$	$C_{28}H_{28}Ge$	224	—	—	—	13, 51, 120, 290, 393
$(C_6H_5)_2Ge(p$-$C_6H_4C_2H_5)_2$	$C_{28}H_{28}Ge$	227–29	—	—	—	326
$(o$-$CH_3C_6H_4)Ge(p$-$CH_3C_6H_4)_3$	$C_{28}H_{28}Ge$	107–8	—	—	—	393
$(p$-$CH_3C_6H_4)Ge(m$-$CH_3C_6H_4)_3$	$C_{28}H_{28}Ge$	110	—	—	—	393
$(n$-$C_{10}H_{21})Ge(C_6H_5)_3$	$C_{28}H_{36}Ge$	111–12	220–22/3	1.5205 (65°)	0.9980	214
$(CH_3)_3Ge[C_6H_4Ge(CH_3)_2]_3CH_3$	$C_{28}H_{42}Ge_4$	164–66	—	—	—	206
$(C_6H_5)_2Ge(C_8H_{17})_2$	$C_{28}H_{44}Ge$	98.5–100.5	225–26/3	1.5274	—	327
p-$(t$-$C_4H_9)C_6H_4Ge(C_6H_{11})_3$	$C_{28}H_{46}Ge$	51–52	—	—	—	79
$(C_6H_{13})_3Ge(1$-$C_{10}H_7)$	$C_{28}H_{46}Ge$	137–39	204–5/6	1.4740	0.9352	311
$(C_6H_{11}CH_2)_4Ge$	$C_{28}H_{52}Ge$	66	—	—	—	236

Table 2-1—continued

COMPOUND	EMPIRICAL FORMULA	M.P. (°C)	B.P. (°C/mm)	n_D^{20}	d_4^{20}	REFERENCES
$(n-C_7H_{15})_4Ge$	$C_{28}H_{60}Ge$	—	217/3	1.4625	0.8985	201, 221
$(p-CH_3C_6H_4)_2Ge(C_2H_5)(i-C_3H_7)$	$C_{29}H_{30}Ge$	—				343
$(C_6H_5)_2Ge(p-C_6H_4-i-C_3H_7)_2$	$C_{30}H_{32}Ge$	127–28				326
$(C_6H_5)_2Ge(C_9H_{19})_2$	$C_{30}H_{48}Ge$		229–30/3	1.5171	0.9913	327
$(C_8H_{17})_3GeC_6H_5$	$C_{30}H_{56}Ge$		270–72/8	1.4685	0.9053	328
$(C_6H_5)_3GeCH(C_6H_5)_2$	$C_{31}H_{26}Ge$	156–59				30
$(C_7H_{15})_3Ge(1-C_{10}H_7)$	$C_{31}H_{52}Ge$		230–31/4	1.4705	0.9233	311
$(C_6H_5)_3GeCH_2CH(C_6H_5)_2$	$C_{32}H_{28}Ge$	95–97				114
$(C_6H_5—CH_2CH_2)_4Ge$	$C_{32}H_{36}Ge$	53–54 57				36, 105
$(C_6H_5)_2Ge(C_{10}H_{21})_2$	$C_{32}H_{52}Ge$	—	234–35/6	1.5051	0.9761	327
$(C_6H_{11}CH_2CH_2)_4Ge$	$C_{32}H_{60}Ge$	135.5–38.5				105
$(n-C_8H_{17})_4Ge$	$C_{32}H_{68}Ge$	—	236/2 210–13/0.1	1.4640 (28°) 1.4630 (24°)	0.8938 0.891 (24°)	105, 201, 221
$[CH_3(CH_2)_3CH—CH_2]_4Ge$ $\quad\quad\quad\quad\quad\mid$ $\quad\quad\quad\quad\quad C_2H_5$	$C_{32}H_{68}Ge$	—	195/0.15	1.4688 (24°)	0.909 (26°)	105
$(C_9H_{19})_3GeC_6H_5$	$C_{33}H_{62}Ge$	—	290–92/8	1.4808	0.9197	328
$(C_8H_{17})_3Ge(1-C_{10}H_7)$	$C_{34}H_{58}Ge$	—	251–52/8	1.4760	0.9279	311
$(C_6H_5CH_2CH_2CH_2)_4Ge$	$C_{36}H_{44}Ge$	—	240–50/0.05	1.5704	1.106 (25°)	105
$(C_6H_5)_3GeC_{18}H_{37}$	$C_{36}H_{52}Ge$	76–77 76.5–77.5				114, 116
$(C_{10}H_{21})_3GeC_6H_5$	$C_{36}H_{68}Ge$	—	310–12/8	1.4740	0.9052	328
$(C_6H_5)_3GeC(C_6H_5)_3$	$C_{37}H_{30}Ge$	342–44				28

Compound	Formula	mp	bp	n	d	Ref.
$p\text{-}(C_6H_5)_3GeC_6H_4\text{—}CH(C_6H_5)_3$	$C_{37}H_{30}Ge$	208–10	—	—	—	28
$[(C_6H_5)_3Ge]_2CH_2$	$C_{37}H_{32}Ge_2$	134	—	—	—	184
$(C_9H_{19})_3Ge(1\text{-}C_{10}H_7)$	$C_{37}H_{64}Ge$	—	285–86/6	1.4730	0.9105	311
$(p\text{-}C_6H_5\text{—}C_6H_4)_3GeC_2H_5$	$C_{38}H_{32}Ge$	154–56	—	—	—	343
$(C_6H_5)_2(C_{18}H_{37})GeCH_2CH_2C_6H_5$	$C_{38}H_{56}Ge$	34.5–36	—	—	—	116
$(C_6H_5)_3Ge(CH_2)_3Ge(C_6H_5)_3$	$C_{39}H_{36}Ge_2$	135–35.5	—	—	—	115, 396
$(1\text{-}C_{10}H_7)_4Ge$	$C_{40}H_{28}Ge$	270	—	—	—	432
$(n\text{-}C_{10}H_{21})_4Ge$	$C_{40}H_{84}Ge$	−15	230–40/0.015 245/0.25	1.4643 (25°) 1.4660	0.879 (31°) 0.8871	105, 221
$p\text{-}(C_6H_5)_3GeC_6H_4Ge(C_6H_5)_3$	$C_{42}H_{34}Ge_2$	349–50	—	—	—	206
$(p\text{-}C_6H_5\text{—}C_6H_4)_4Ge$	$C_{48}H_{36}Ge$	270–72	—	—	—	343
$(n\text{-}C_{12}H_{25})_4Ge$	$C_{48}H_{100}Ge$	—	275/0.4	1.4675	0.8837	105, 221
$(C_{14}H_{29})_4Ge$	$C_{56}H_{116}Ge$	—	330–40/0.0005	1.4654 (31°)	0.879 (31°)	105
$(C_{16}H_{33})_4Ge$	$C_{64}H_{132}Ge$	37–38	360–70/0.0005	1.4640 (31°)	0.880 (31°)	105
$(C_{18}H_{37})_4Ge$	$C_{72}H_{148}Ge$	43–45	—	—	—	105

2-2 OLEFINIC DERIVATIVES OF GERMANIUM

2-2-1 Preparations

Olefinic derivatives of germanium may be prepared by the following procedures:
   linking an olefinic chain to germanium;
   creating an olefinic bond within a chain already linked to a germanium atom;
   linking a chain and simultaneously creating the olefinic bond;
   transforming an olefinic derivative.

2-2-1-1 Linking an Olefinic Chain. An olefinic chain can be linked to a germanium atom:

(a) By direct synthesis

Allyl chloride or methallyl chloride will react with a germanium and copper powder mixture (ratio 25:10), respectively yielding allyltrichlorogermane or methallyltrichlorogermane (300).

(b) By condensation of a germanous halide with an allyl halide

Germanous iodide will easily react with allyl iodide, producing triiodoallylgermane (233); this reaction has a 100% yield:

$$I_2Ge + ICH_2-CH=CH_2 \rightarrow I_3GeCH_2-CH=CH_2$$

Germanous bromide reacts with allyl bromide in the same manner (107).

(c) From organomagnesium compounds

This is the usual process for preparing vinyl and allyl derivatives. Vinylmagnesium bromide in THF solution (280) reacts easily with alkyl- or arylgermanium halides (228, 270, 349):

$$(C_2H_5)_3GeBr + CH_2=CHMgBr \xrightarrow{THF} MgBr_2$$
$$+ (C_2H_5)_3GeCH=CH_2$$
$$(C_6H_5)_2GeBr_2 + 2 CH_2=CHMgBr \xrightarrow{THF} 2 MgBr_2$$
$$+ (C_6H_5)_2Ge(CH=CH_2)_2$$

The reaction of germanium tetrachloride on vinylmagnesium bromide produces tetravinylgermane, and some hexavinyldigermane resulting from side reactions:

$$GeCl_4 + 4(CH_2=CH)MgBr \xrightarrow{THF}$$
$$2 MgBr_2 + 2 MgCl_2 + (CH_2=CH)_4Ge$$

Seyferth (349) suggested the following reactions for digermane formation:

$$GeCl_4 + 2\,CH_2{=}CHMgBr \rightarrow$$

$$GeCl_2 + CH_2{=}CH{-}CH{=}CH_2 + MgCl_2 + MgBr_2$$

$$GeCl_2 + 2\,CH_2{=}CHMgBr \rightarrow$$

$$(CH_2{=}CH)_2Ge + MgCl_2 + MgBr_2$$

$$(CH_2{=}CH)_2Ge + (CH_2{=}CH)MgBr \rightarrow$$

$$(CH_2{=}CH)_3GeMgBr$$

$$(CH_2{=}CH)_3GeMgBr + (CH_2{=}CH)_3GeCl \rightarrow$$

$$(CH_2{=}CH)_3Ge{-}Ge(CH{=}CH_2)_3 + \tfrac{1}{2}MgCl_2 + \tfrac{1}{2}MgBr_2$$

We can note that this side reaction does not occur when vinylating tin and silicon tetrahalides: with these elements only the expected tetravinyl derivative is produced.

The allyl derivatives are prepared in diethyl ether solution by action of germanium halides upon allylmagnesium bromide (114, 144, 214, 228, 230):

$$R_3GeBr + CH_2{=}CH{-}CH_2MgBr \rightarrow MgBr_2 + CH_2{=}CH{-}CH_2GeR_3$$

The preparation of tetraallylgermanium from germanium tetrachloride requires the repeated action of allylmagnesium bromide in large excess (228).

The olefinic derivatives of higher homologous compounds also are prepared in good yield from corresponding organomagnesium compounds.

Long chain ω olefinic derivatives of germanium can also be prepared by the action of allyl bromide on a germanium-containing organomagnesium compound (224, 230). Traces of cuprous chloride catalyze the reaction:

$$R_3Ge(CH_2)_nMgBr + CH_2{=}CH{-}CH_2Br \rightarrow$$

$$MgBr_2 + R_3Ge(CH_2)_nCH_2CH{=}CH_2$$

This reaction is generalized for $n > 3$.

Nametkin (271) obtained trimethylpropenyl- and trimethyl-β-styryl germanes by the action of trimethyliodogermane with the corresponding Normant's reagents.

Perfluorovinyl derivatives of germanium had been obtained by adding a mixture of organogermanium halide and bromotrifluoroethylene to magnesium in tetrahydrofuran (363):

$$R_3GeX + BrCF{=}CF_2 \xrightarrow{Mg} R_3GeCF{=}CF_2 \qquad \begin{cases} R = C_2H_5, X = I \\ R = C_6H_5, X = Br \end{cases}$$

Phenylstyryl derivatives of germanium, tin and lead have been synthesized by Noltes and coworkers (281):

$$(C_6H_5)_{4-n}MX_n \xrightarrow[\text{THF}]{p\text{-ClMgC}_6\text{H}_4\text{CH}=\text{CH}_2}$$

$$(C_6H_5)_{4-n}M(C_6H_4CH{=}CH_2)_n \qquad \begin{cases} M = Ge, Sn, Pb \\ n = 1, 2 \end{cases}$$

under comparable conditions, triphenylchlorosilane is unreactive towards the p-styrenyl Grignard reagent. Trimethylstyrenyl- and trimethyl-α-methylstyrenyl derivatives are prepared in the same way (282):

$$(CH_3)_3MX + p\text{-ClMgC}_6H_4{-}\overset{\overset{\textstyle R}{\displaystyle |}}{C}{=}CH_2 \xrightarrow{\text{THF}}$$

$$p\text{-}(CH_3)_3MC_6H_4{-}\overset{\overset{\textstyle R}{\displaystyle |}}{C}{=}CH_2 \qquad \begin{cases} M = Si, Ge, Sn, Pb \\ R = H, CH_3 \end{cases}$$

Sarankina and Manulkin (325) have also prepared a great variety of phenylpropenyl and phenylsubstituted styrylgermanes by using Normant's reagent.

(d) From organolithium compounds

These are less used than organomagnesium compounds in the different syntheses of germanium olefinic derivatives. However Seyferth obtained a 75 % yield of trimethylpropenylgermane (70 % cis and 30 % $trans$) from propenyllithium (364):

$$(CH_3)_3GeCl + CH_3CH{=}CHLi \rightarrow LiCl + (CH_3)_3GeCH{=}CH{-}CH_3$$

and prepared triphenylvinylgermane from vinyllithium (365):

$$(C_6H_5)_3GeBr + CH_2{=}CHLi \rightarrow LiBr + (C_6H_5)_3GeCH{=}CH_2 \quad (76\%)$$

Nametkin obtained trimethylvinylgermane with 50 % yield (270) and triphenylallylgermane was produced in the same way in 54 % yield (366).

Nesmeyanov (277) prepared tetra-cis- and tetra-$trans$-propenylgermanes by the reaction of germanium tetrachloride with cis- and $trans$-propenyllithium respectively.

A similar condensation, when heating a mixture of ethereal solutions of trimethylgermanium iodide and isopropenyllithium, yields trimethyl-*i*-propenylgermane (271):

$$CH_3-CLi{=}CH_2 + (CH_3)_3GeI \rightarrow LiI + (CH_3)_3Ge{-}\underset{\underset{CH_3}{|}}{C}{=}CH_2$$

(e) From organomercuric compounds

Vinyltrichlorogermane has been prepared (24) by the action of divinyl-mercury upon germanium tetrachloride:

$$GeCl_4 + (CH_2{=}CH)_2Hg \rightarrow CH_2{=}CHGeCl_3 + CH_2{=}CHHgCl$$

(f) By way of Würtz reactions

Nametkin (271) prepared α-styryltrimethylgermane by the action of sodium upon a mixture of trimethyliodogermane and α-styryl bromide:

$$C_6H_5CBr{=}CH_2 + (CH_3)_3GeI \xrightarrow{2Na} NaBr + NaI + (CH_3)_3Ge\underset{\underset{C_6H_5}{|}}{C}{=}CH_2$$

and pure trimethylvinylgermane by Würtz reaction between trimethyl-iodogermane and vinyl bromide in absolute ether (yield 30%) (270). Birr and Kräft (17) prepared in the same way tetra-β-styrylgermane from germanium tetrachloride.

(g) By adding a germanium hydride to a diene or an allene (94, 246, 334)

$$R_2\underset{\underset{X}{|}}{Ge}{-}H + H_2C{=}\underset{\underset{R'}{|}}{C}{-}\underset{\underset{R'}{|}}{C}{=}CH_2 \rightarrow R_2\underset{\underset{X}{|}}{Ge}{-}CH_2{-}\underset{\underset{R'}{|}}{C}{=}\underset{\underset{R'}{|}}{C}{-}CH_3$$

2-2-1-2 Creating an Olefinic Bond within a Chain already linked to a Germanium Atom

(a) By dehydrohalogenation

Petrov and coworkers (176, 245) obtained vinyltrichlorogermane by dehydrohalogenation of $ClCH_2CH_2GeCl_3$; this compound was prepared by benzoyl peroxide catalyzed side-chain chlorination of ethyltrichloro-germane with sulphuryl chloride. The dehydrohalogenation was effected by heating in quinoline at 200°C or by distillation from $AlCl_3$.

Mironov and coworkers (242) dehydrohalogenated $ClCH_2CHClGeCl_3$; the nature of the end product depended on the agent used: quinoline and aluminium chloride lead to two monochlorinated position isomers:

$$ClCH_2CHClGeCl_3 \underset{\xrightarrow{AlCl_3}}{\xrightarrow{quinoline}} \begin{array}{l} CH_2{=}CClGeCl_3 \\ CHCl{=}CHGeCl_3 \end{array}$$

Elimination of one HX molecule from a halogenated chain linked to a germanium atom depends on the nature of the nucleophilic agent used, on the nature of the halogen, and on the position of the halogen with reference to the germanium atom (198). The reaction of sodium ethoxide on triethyl-ω-bromo-alkylgermanes yields nearly exclusively the substituted derivative:

$$(C_2H_5)_3Ge(CH_2)_nBr + C_2H_5ONa \longrightarrow$$

$$(C_2H_5)_3Ge(CH_2)_nOC_2H_5 + NaBr \qquad (n = 4, 5, 6)$$

Alcoholic KOH, sodium phenoxide, triethylamine and sodium amide also lead to substitution reactions.

The action of potassium tertiobutylate on these same derivatives, $(C_2H_5)_3Ge(CH_2)_nBr$ ($n = 3, 4, 5$), leads to a mixture of the olefin and the ether:

$$(C_2H_5)_3Ge(CH_2)_nBr + t\text{-}C_4H_9OK \longrightarrow$$

$$(C_2H_5)_3Ge(CH_2)_{n-2}CH{=}CH_2 + t\text{-}C_4H_9OH + KBr$$

$$(C_2H_5)_3Ge(CH_2)_nBr + t\text{-}C_4H_9OK \longrightarrow$$

$$(C_2H_5)_3Ge(CH_2)_nOC_4H_9\text{-}t + KBr$$

The yield of olefin rises when changing the halogen atom in the order $Cl < Br < I$

	OLEFIN PRODUCED (%)		
X	$n = 3$	$n = 4$	$n = 5$
Cl	5	19	23
Br	25	36	46
I	30	62	74

As n gets smaller, the elimination reaction is gradually replaced by the Sn^2 reaction.

(b) By dehydration

Dehydration of germanium alcohols may be carried out by letting phosphorus oxychloride react in the presence of pyridine (202, 223):

$$3\,(C_2H_5)_3GeC{\equiv}C{-}COH(CH_3)_2 \xrightarrow{\text{POCl}_3}$$

$$3\,(C_2H_5)_3GeC{\equiv}C{-}\underset{\underset{CH_3}{|}}{C}{=}CH_2$$

$$(C_4H_9)_3GeCH=CH-COH(CH_3)_2 \xrightarrow{\text{POCl}_3}$$

$$(C_4H_9)_3GeCH=CH-C(CH_3)=CH_2$$

or by letting phosphorus tribromide react in boiling benzene (280). Some α-hydroxygermanes give the corresponding vinylgermanes: 2-triphenyl-germyl-2-propanol gives 2-triphenylgermylpropene in 63% yield:

$$(C_6H_5)_3Ge-COH(CH_3)_2 \xrightarrow{\text{PBr}_3} (C_6H_5)_3Ge-C(CH_3)=CH_2$$

while 2-triphenylgermyl-2-butanol leads to a mixture of olefin isomers:

$$(C_6H_5)_3GeC(OH)(CH_3)C_2H_5 \xrightarrow{\text{PBr}_3}$$

$$(C_6H_5)_3GeC(CH_3)=CHCH_3 + (C_6H_5)_3GeC(C_2H_5)=CH_2$$

On the other hand, under the same experimental conditions, the reaction of 1-triphenylgermyl-1-phenyl-2-methylpropanol yields the corresponding bromide:

$$(C_6H_5)_3GeC(OH)(C_6H_5)CH(CH_3)_2 \xrightarrow{\text{PBr}_3}$$

$$(C_6H_5)_3GeC(Br)(C_6H_5)CH(CH_3)_2$$

and with 3-triphenylgermyl-3-pentanol, the only product isolated was hexaphenyldigermoxane:

$$(C_6H_5)_3GeC(OH)(C_2H_5)_2 \xrightarrow{\text{PBr}_3; \text{OH}_2} [(C_6H_5)_3Ge]_2O$$

The dehydration of these α-hydroxygermanes is a suitable method for the synthesis of monosubstituted vinylgermanes in which the substituent is bonded to the same carbon atom as germanium. Indeed, Grignard syntheses of such derivatives have not been reported, and addition of tri-phenylgermane to monosubstituted acetylenes gives the position isomer:

$$(C_6H_5)_3GeC(OH)(CH_3)(C_6H_5) \xrightarrow{\text{PBr}_3} (C_6H_5)_3GeC(C_6H_5)=CH_2$$

$$(C_6H_5)_3GeH + C_6H_5C\equiv CH \longrightarrow (C_6H_5)_3GeCH=CH-C_6H_5$$

Several olefinic germanium compounds have been prepared in good yields (74–82%) by Gverdtsiteli and coworkers (132) by dehydration of dienic germa-carbinols with potassium hydrogen sulphate in the presence of a small quantity of dithizone (as antioxidant):

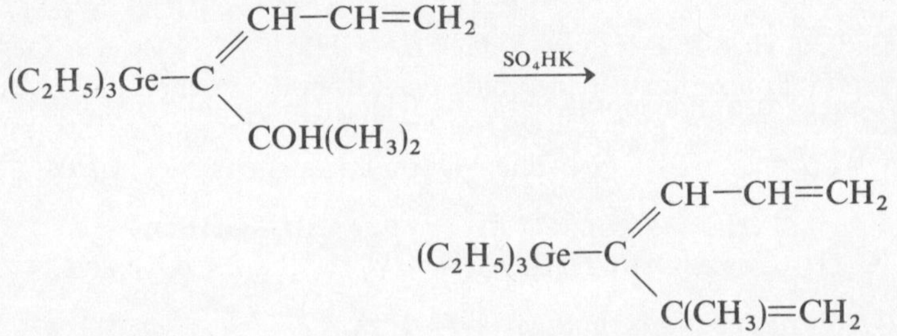

$$(C_2H_5)_3Ge-C \overset{CH-CH=CH_2}{\underset{COH(CH_3)_2}{\Big\langle}} \xrightarrow{SO_4HK}$$

$$(C_2H_5)_3Ge-C \overset{CH-CH=CH_2}{\underset{C(CH_3)=CH_2}{\Big\langle}}$$

However, compounds containing cyclopentyl and cyclohexyl radicals partially decomposed during distillation and the yields of their olefinic dehydration product is lower (33–50%).

Good results have also been achieved by heating the alcohol in the presence of *para*-toluenesulphonic acid (230); the alkenyne

$$(C_2H_5)_3Ge(CH_2)_5C\equiv C-C(CH_3)=CH_2$$

is thus obtained from the corresponding alcohol in 60% yield.

(c) By partial saturation of an acetylenic derivative of germanium

When using a selective catalyst (Lindlar's catalyst) it is possible to interrupt the hydrogenation of the acetylenic chain linked to the germanium atom at the ethylenic stage (229):

$$(C_4H_9)_3GeC\equiv CH + H_2 \rightarrow (C_4H_9)_3GeCH=CH_2$$

$$(C_2H_5)_3GeCH_2C\equiv C-CH_2CH=CH_2 + H_2 \rightarrow$$

$$(C_2H_5)_3GeCH_2CH=CH-CH_2CH=CH_2$$

Eisch and Foxton (92) have recently reported that trialkylgermyl terminal alkynes can be reduced easily and stereospecifically in a *cis*- or *trans*-fashion by diisobutylaluminium hydride:

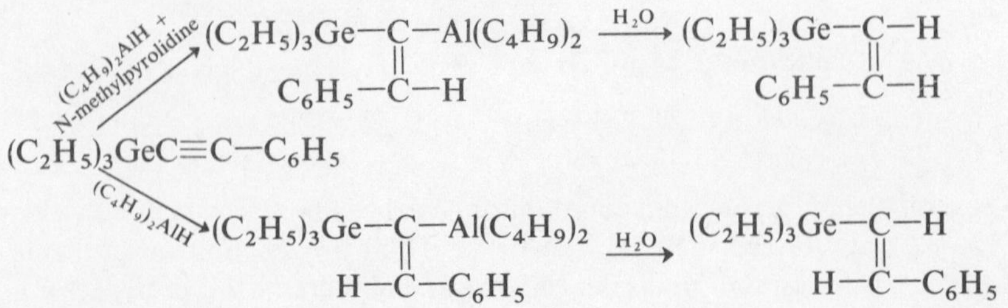

$$(C_2H_5)_3GeC\equiv C-C_6H_5$$

$$\xrightarrow{(C_4H_9)_2AlH^+ \; N\text{-methylpyrrolidine}} (C_2H_5)_3Ge-\underset{\overset{\|}{C_6H_5-C-H}}{C}-Al(C_4H_9)_2 \xrightarrow{H_2O} (C_2H_5)_3Ge-\underset{\overset{\|}{C_6H_5-C-H}}{C}-H$$

$$\xrightarrow{(C_4H_9)_2AlH} (C_2H_5)_3Ge-\underset{\overset{\|}{H-C-C_6H_5}}{C}-Al(C_4H_9)_2 \xrightarrow{H_2O} (C_2H_5)_3Ge-\underset{\overset{\|}{H-C-C_6H_5}}{C}-H$$

Hydrogenosilanes, germanes and stannanes add easily to acetylenic germanium derivatives:

$$R_3GeC\equiv CH + R'_3MH \rightarrow R_3GeCH=CHMR'_3 \qquad M = Si, Ge, Sn$$

Action of bromine upon tetraethynylgermane leads to the expected dibromovinyl derivative (61, 230):

$$(HC\equiv C)_4Ge + 4\,Br_2 \rightarrow Ge(CBr=CHBr)_4$$

(d) By Wittig reactions on α-germyl ketones (27)

Benzoyltriphenylgermane, when treated with methylene- or ethylidene-triphenylphosphorane gives α(triphenylgermyl)styrene and α(triphenylgermyl)-*trans*-β-methylstyrene

$$(C_6H_5)_3GeCOC_6H_5 + (C_6H_5)_3P^+{-}CH^-R \rightleftarrows$$

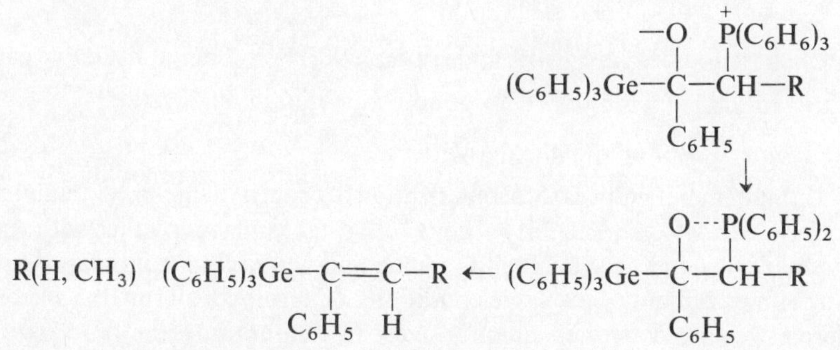

$$(C_6H_5)_3Ge{-}\underset{\underset{C_6H_5}{|}}{\overset{\overset{-O}{|}}{C}}{-}\underset{}{\overset{\overset{+}{P(C_6H_6)_3}}{\underset{}{CH}}}{-}R$$

$$\downarrow$$

$$R(H, CH_3)\quad (C_6H_5)_3Ge{-}\underset{\underset{C_6H_5}{|}}{\overset{}{C}}=\underset{\underset{H}{|}}{\overset{}{C}}{-}R \leftarrow (C_6H_5)_3Ge{-}\underset{\underset{C_6H_5}{|}}{\overset{\overset{O\cdots P(C_6H_5)_2}{|}}{C}}{-}\underset{}{\overset{}{CH}}{-}R$$

2-2-1-3   Linking a Chain to the Germanium Atom and simultaneously creating the Ethylenic bond. Anti-Markownikow addition of a germanium hydride to acetylenic derivatives is a general procedure for preparing organogermanium compounds with a substituted vinyl group (4, 54, 127, 132, 159, 202, 204, 211, 374):

$$(C_4H_9)_3GeH + HC\equiv C{-}C_6H_5 \rightarrow (C_4H_9)_3GeCH=CHC_6H_5$$

$$(C_4H_9)_3GeH + HC\equiv C{-}Ge(C_4H_9)_3 \rightarrow$$

$$(C_4H_9)_3GeCH=CHGe(C_4H_9)_3$$

$$Cl_3GeH + HC\equiv CH \xrightarrow{ether} Cl_3GeCH=CHGeCl_3$$

$$(C_4H_9)_3GeH + CH_3CH_2CH_2{-}CHOH{-}C\equiv CH \xrightarrow{H_2PtCl_6}$$

$$(C_4H_9)_3Ge{-}CH=CHCH(OH){-}C_3H_7$$

By a carbon–carbon insertion reaction into the germanium–nitrogen bond an olefinic derivative is produced (43):

$$(CH_3)_3GeN(CH_3)_2 + C_2H_5OCOC\equiv C-COOC_2H_5 \rightarrow$$

$$\underset{\displaystyle COOC_2H_5}{\overset{\displaystyle COOC_2H_5}{(CH_3)_3Ge-C=C-N(CH_3)_2}}$$

2-2-1-4 Transforming an Olefinic Derivative. Trichloro derivatives obtained by direct synthesis may be transformed into trialkyl (or triaryl) derivatives by reaction with an organomagnesium compound (69, 70):

$$Cl_3GeCH_2CH=CH_2 + 3\,RMgX \rightarrow$$

$$R_3GeCH_2-CH=CH_2 + 3\,MgXCl$$

Olefinic triiodides and tribromides prepared from GeI_2 and $GeBr_2$ can be alkylated in similar fashion.

(a) Isomerization of olefinic derivatives

Lithium metal induces the isomerization of propenyl-silicon, germanium and tin derivatives (360, 361). The reaction takes place in ether with the stannanes whereas the analogous silicon and germanium compounds are not isomerized under the same conditions. In tetrahydrofuran, the silicon derivatives are converted quantitatively in 6 hours into the pure *trans*-isomer. For germanium derivatives, whether the initial compound is the *cis*-isomer or the *trans*-isomer, an equilibrium mixture containing 92% of the *trans*- and 8% of the *cis*-isomer is obtained in the same solvent after 20 hours.

A mechanism involving radical-anion intermediates about whose C—C bonds rotation is possible and which are stabilized by $C(2p) \rightarrow$ metal (nd) π-bonding is suggested for the observed isomerization:

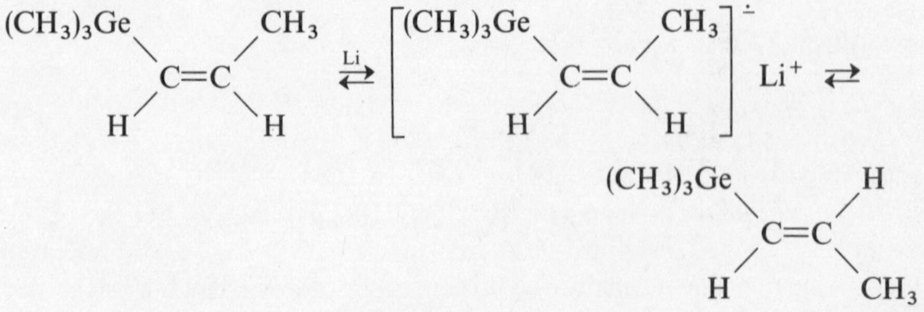

Seyferth suggests a mechanism for the alkali metal displacement reaction involving electron transfer from the alkali metal to the metal alkyl followed by migration of an organic group to the reagent metal:

$$M + RM' \rightarrow M^+[R-M']^- \rightarrow R-M + M'$$

2-2-2   Physical properties and characteristics

2-2-2-1   IR Spectra. The valence vibration frequency of the $C{=}C$ bond in olefinic alkylgermanes depends on the position of the double bond in reference to the position of the germanium atom (88, 144, 217). There is about a 40 cm^{-1} shift for vinyl derivatives and a 40 to 50 cm^{-1} shift for substituted vinyl derivatives, with respect to $v_{C=C}$ of simple terminal olefins.

When, with reference to the germanium atom, the double bond is located in the β position (allylic derivatives) the shift is of about 15 cm^{-1}; when the double bond is farther removed, the $v_{C=C}$ frequency is normal.

	VINYL DERIVATIVES $(C_2H_5)_3GeCH{=}CH_2$ $(C_2H_5)_2Ge(CH{=}CH_2)_2$	ALLYL DERIVATIVES $(C_4H_9)_3GeCH_2CH{=}CH_2$ $(CH_2{=}CH-CH_2)_4Ge$	OLEFINS R$-CH{=}CH_2$
$v_{C=C}$	1600	1630	1643
v_{C-H}	$\begin{cases} 3046 \\ \\ 3000 \end{cases}$	$\begin{cases} 3075 \\ 3056 \\ 3030 \end{cases}$	$\begin{cases} 3075\text{–}3095 \\ \\ 3010\text{–}3040 \end{cases}$
$\delta{=}C{\raise2pt\hbox{$\scriptstyle H$}}{\lower2pt\hbox{$\scriptstyle H$}}$ in plane	1400	1430	1410–1420
$\delta{=}C{\raise2pt\hbox{$\scriptstyle H$}}{\lower2pt\hbox{$\scriptstyle H$}}$ out of plane	$\begin{cases} 1000 \\ \\ 940 \end{cases}$	$\begin{cases} 995 \\ \\ 930 \end{cases}$	$\begin{cases} 995\text{–}985 \\ \\ 915\text{–}905 \end{cases}$

The intensity of the $v_{C=C}$ absorption band is greater in allylic derivatives than in vinylic derivatives. In the case of the symmetrical derivative, bis(tributylgermyl)ethylene, $(C_4H_9)_3GeCH{=}CHGe(C_4H_9)_3$, the $v_{C=C}$ absorption band disappears. In an investigation on organometallic compounds with a propenyl substituent (347, 364), Seyferth pointed out the characteristic absorption bands of the two trimethylpropenylgermane isomers: 1610 and 980 cm^{-1} for the *trans* isomer, and 1610 cm^{-1} for the *cis* isomer.

Henry and Noltes (144) have studied the infrared spectra of compounds of the type $(C_6H_5)_4M$, $(C_6H_5)_3M-CH=CH_2$, $(C_6H_5)_2M(CH=CH_2)_2$, $(C_6H_5)_3M-CH_2CH=CH_2$ and $(C_6H_5)_2M(CH_2CH=CH_2)_2$ in the region of 3500–680 cm^{-1} (M = Si, Ge, Sn and Pb) and have made empirical assignments for specific vibrations. Comparison of the spectra reveals a shift of some characteristic band frequencies to longer wavelength in the series M = Si, Ge, Sn and Pb. Nametkin (271) gives the characteristic absorption bands of propenyl and styryl germanes.

Infrared and Raman spectra of 3-alkyn-1-yl organogermanium compounds are reported in detail by Stadnichuk (404); the frequency shifts observed in these spectra relative to those of hydrocarbon analogues are interpreted in terms of interaction p of the π-electrons of multiple bonds with the d-orbitals of the Ge atoms.

The infrared spectra of triphenyl-*p*-styrenyl- and diphenyl-di-*p*-styrenyl-germanes, stannanes and plumbanes have been recorded and discussed by Noltes and coworkers (281). Nesmeyanov (277) gives the IR spectra of tetra-*cis*- and *trans*-propenylgermanes.

2-2-2-2 NMR Spectra. The magnetic resonance investigation of propenyltrimethylgermanes enabled Seyferth (347, 364) to identify the two following isomers:

$$trans \begin{cases} \text{multiplet at 5.9 ppm } (J \text{ est. } 19.2 \text{ Hz}) \text{ H}_\alpha + \text{H}_\beta \\ \text{doublet at 1.8 ppm } (J \text{ 4.8 Hz}) \text{ CH}_3 \end{cases}$$

$$cis \begin{cases} \text{sextet at 6.40 } (J \text{ 6.6 Hz}) \text{ H}_\beta \\ \text{doublet at 5.75 } (J \text{ 14.2 Hz}) \text{ H}_\alpha \\ \text{doublet at 1.75 } (J \text{ 6.6 Hz}) \text{ CH}_3 \end{cases}$$

The NMR investigation carried out at 100 MHz (421) revealed a positive difference between *trans*-propenyltrimethylgermane ($J_{AB} = 18.0$ Hz) and the *cis* isomer ($J_{AB} = 13.2$ Hz).

2-2-2-3 Raman Spectra. In a study of the physical properties of alkenyl-silanes, Egorov (88a, 298) proved that the intensity of the Raman line corresponding to the $C=C$ vibration in compounds of the type $R_3Si(CH_2)_nCH=CH_2$ is at a maximum for the allylsilanes: by combining the spectral and chemical data, the conclusion was drawn that σ-π conjugation between the Si—C bond and the double bond was present in the allylsilanes (298). Later (88, 91) a study of allylic compounds

Table 2-2

Frequency and Intensity of the Line corresponding to C=C
Vibration in the Raman Spectra of Alkenylgermanes

ALKENYL COMPOUND	ν (cm^{-1})	I_∞
$RCH=CH_2$	1640	400
$(CH_3)_3GeCH_2CH=CH_2$	1630	1170
$(CH_3)_3GeCH_2CH=CH-CH_2Ge(CH_3)_3$	1650	4520
$\begin{array}{c}CH\!\!=\!\!CH\\[-2pt]\big\vert\qquad\diagdown\\[-2pt]\qquad\qquad CHGeCl_3\\[-2pt]\big\vert\qquad\diagup\\[-2pt]CH_2\!-\!CH_2\end{array}$	1606	1790
$\begin{array}{c}CH\!\!=\!\!CH\\[-2pt]\big\vert\qquad\diagdown\\[-2pt]\qquad\qquad CH\!-\!CH_3\\[-2pt]\big\vert\qquad\diagup\\[-2pt]CH_2\!-\!CH_2\end{array}$	1614	370
$\begin{array}{c}CH\!-\!CH_2\\[-2pt]\big\Vert\qquad\diagdown\\[-2pt]\qquad\qquad GeCl_2\\[-2pt]\big\Vert\qquad\diagup\\[-2pt]CH\!-\!CH_2\end{array}$	1613	290
$\begin{array}{c}CH\!-\!CH_2\\[-2pt]\big\Vert\qquad\diagdown\\[-2pt]\qquad\qquad Ge(CH_3)_2\\[-2pt]\big\Vert\qquad\diagup\\[-2pt]CH\!-\!CH_2\end{array}$	1605	270

$(CH_3)_nM(CH_2-CH=CH_2)_{4-n}$ (M = C, Si, Ge, Sn) with a double bond in the β-position with respect to the heteroatom showed anomalous chemical reactivity, high intensity $I_{C=C}$ of the Raman vibration, bathochromic shift in the UV spectrum and considerable exaltation of molar refraction. These remarkable properties, characteristic of the "β-effect", were attributed to σ-π conjugation. In the analogous series an increase of the $I_{C=C}$ Raman line intensity was observed in the ratio 1:2:4:8 from carbon to tin.

A comparative study of Raman spectra of allyl compounds $X_3MCH_2CH=CH_2$ (M = Si or Ge; X = CH_3, F, Cl or Br) (90) showed that the frequency increases with increasing electronegativity of the substituent in the normal series CH_3 < Br < Cl < F. A completely different relation was shown by the intensity of the Raman lines which increases to a greater or lesser extent in the series F < Cl < CH_3 < Br.

The changes in the frequencies and intensities observed in the spectra of the $X_3MCH_2CH=CH_2$ compounds when the substituent on the metal is changed are due to the changes within the redistribution of the electron density at the M atom under the influence of the polar substituents.

Leites (194) investigated the Raman spectra of the different types of β-alkenylgermanes: according to her, the condition for the "β effect" is that the σ-π bonds be conjugated (this effect results in an amplified intensity of the characteristic $v_{C=C}$ bands). When the orbitals are orthogonal this "β effect" disappears.

Table 2-2 gives an account of the differences observed.

2-2-2-4 Magneto-optical Measurements. Magneto-optical measurements (189) of olefinic germanium derivatives are in good agreement with the calculated values.

2-2-3 Chemical properties

Depending on the experimental conditions, olefinic derivatives of germanium can take part in addition or cleavage reactions.

2-2-3-1 Addition Reactions

(a) Hydrogen

Hydrogenation of olefinic alkylgermanes is possible in the presence of a catalyst: with Raney nickel, trialkylvinyl- and trialkylallylgermanes are quantitatively transformed respectively into trialkylethyl- and trialkyl-propylgermanes, the rate being higher for the allylic derivative (229).

In the same way, hydrogenation of triethylcyclopentadienylgermane leads to triethylcyclopentylgermane (199). Hydrogenation of $Me_3ZCH=CH_2$ and $Me_3ZCH_2CH=CH_2$, where Z is C, Si, Ge or Sn, over Raney nickel at room temperature occurs at rates that are comparable to the general reactivities of the compounds. The rate declines in the series of Si > Ge > C > Sn for the central atom (100).

(b) Organometallic hydrides

In the presence of chloroplatinic acid, derivatives of the R_3MH type (M = Si, Ge, Sn) are easily added to linear ethylenic derivatives of germanium (202, 228, 284):

$$(C_2H_5)_3GeH + (C_2H_5)_3GeCH_2CH=CH_2 \rightarrow$$

$$(C_2H_5)_3Ge(CH_2)_3Ge(C_2H_5)_3$$

This is an anti-Markownikow type addition. The chloroplatinic acid-catalyzed addition of the Si—H link to the olefinic double bond of vinyl-germanium compounds is used in the preparation of polymers (176, 304):

$$H(CH_3)_2SiOSi(CH_3)_2H + (C_2H_5)_2Ge(CH{=}CH_2)_2 \xrightarrow[H_2PtCl_6]{120^\circ}$$

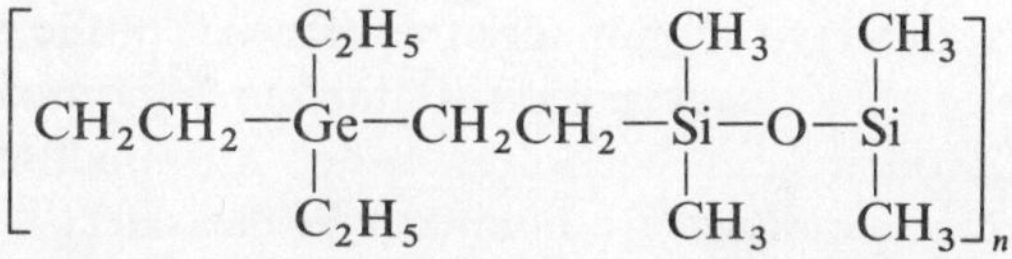

$$\left[CH_2CH_2-\overset{\overset{\displaystyle C_2H_5}{|}}{\underset{\underset{\displaystyle C_2H_5}{|}}{Ge}}-CH_2CH_2-\overset{\overset{\displaystyle CH_3}{|}}{\underset{\underset{\displaystyle CH_3}{|}}{Si}}-O-\overset{\overset{\displaystyle CH_3}{|}}{\underset{\underset{\displaystyle CH_3}{|}}{Si}} \right]_n$$

Hydrogenostannanes add on styryl derivatives of germanium (284). With the distyryl compound, triphenylstannane gives a product containing three metal atoms:

$$(C_6H_5)_2Ge(C_6H_4{-}CH{=}CH_2)_2 + 2(C_6H_5)_3SnH \rightarrow$$

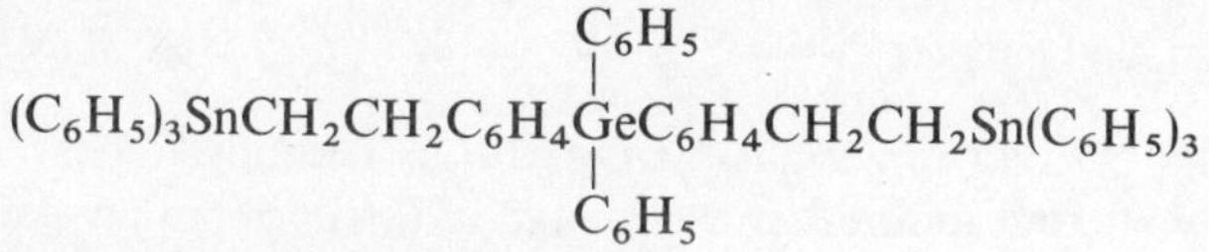

$$(C_6H_5)_3SnCH_2CH_2C_6H_4\overset{\overset{\displaystyle C_6H_5}{|}}{\underset{\underset{\displaystyle C_6H_5}{|}}{Ge}}C_6H_4CH_2CH_2Sn(C_6H_5)_3$$

while diphenylstannane leads to polymers:

$$(C_6H_5)_2SnH_2 + (C_6H_5)_2Ge(C_6H_4CH{=}CH_2)_2 \rightarrow$$

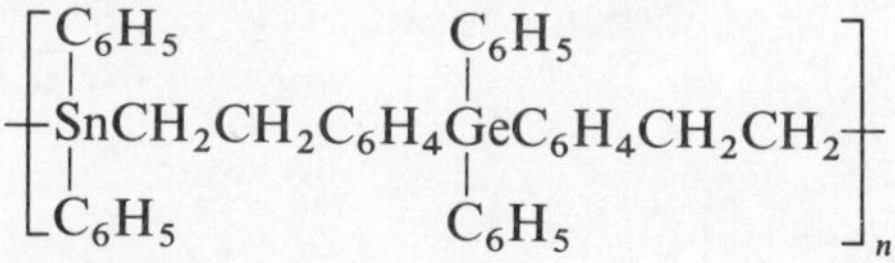

$$\left[-\overset{\overset{\displaystyle C_6H_5}{|}}{\underset{\underset{\displaystyle C_6H_5}{|}}{Sn}}CH_2CH_2C_6H_4\overset{\overset{\displaystyle C_6H_5}{|}}{\underset{\underset{\displaystyle C_6H_5}{|}}{Ge}}C_6H_4CH_2CH_2- \right]_n$$

Henry and Noltes (146) obtained a six-membered ring by heating diphenyl-stannane with diphenyldivinylgermane:

$$(C_6H_5)_2SnH_2 + (C_6H_5)_2Ge(CH{=}CH_2)_2 \rightarrow$$

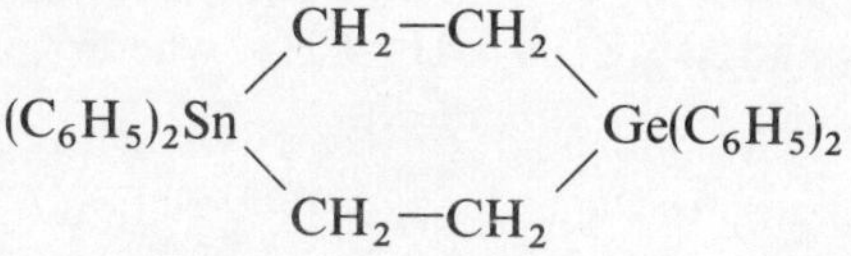

$$(C_6H_5)_2Sn \underset{CH_2-CH_2}{\overset{CH_2-CH_2}{\diagup \diagdown}} Ge(C_6H_5)_2$$

and a 1,4-digermacyclohexane is formed by action of dibutylgermane with diethyldivinylgermane.

(c) Halogens, pseudo halogens, halides

At low temperature, olefinic derivatives of the

$$(C_2H_5)_3Ge(CH_2)_nCH{=}CH_2 \qquad\qquad n \neq 1 \text{ or } 2$$

type will add a bromine molecule in 100% yield in the reaction:

$$(C_2H_5)_3Ge(CH_2)_nCH{=}CH_2 + Br_2 \rightarrow (C_2H_5)_3Ge(CH_2)_nCHBr{-}CH_2Br$$

Under the same experimental conditions, for derivatives with an allyl, methallyl or butenyl chain ($n = 1$ or $n = 2$) cleavage of the olefinic chain occurs (230). The substituted styrenes $p\text{-}R_3MC_6H_4CH{=}CH_2$ when treated with thiocyanogen were shown to undergo the reaction with decreasing ease in the series of $Sn > C > Ge > Si$ substituents (45). The degree of thiocyanation attained in 24 hours is 92% for Me_3Ge and 95.5% for Et_3Ge.

Cyclopentadienyl groups are easily cleaved (199):

Styrylgermanes which are vinylic compounds surprisingly show similar behaviour (202, 329):

$$(C_4H_9)_3GeCH{=}CHC_6H_5 + Br_2 \rightarrow (C_4H_9)_3GeBr + C_6H_5CH{=}CHBr$$

With hexavinyldigermane an addition reaction is not observed with bromine and iodine but a cleavage of the germanium–germanium bond occurs (349):

$$(CH_2{=}CH)_3Ge{-}Ge(CH{=}CH_2)_3 + Br_2 \rightarrow 2\,(CH_2{=}CH)_3GeBr$$

Tetrachlorodiborane adds to the double bond of vinyl derivatives of boron, silicon, germanium and tin to form 1,2-bis(dihaloboryl)ethyl derivatives (49):

$$Cl_3GeCH{=}CH_2 + B_2Cl_4 \rightarrow Cl_3Ge{-}\underset{\underset{BCl_2}{|}}{CH}{-}CH_2{-}BCl_2$$

For the trichlorovinyl compounds the rate of the addition is in the order $Sn > Si > Ge$.

(d) Mercaptans

Thioglycolic acid easily reacts with trialkylvinyl and trialkylallylgermanes. In the latter case the reaction is exothermic (228):

$$(C_2H_5)_3GeCH_2CH{=}CH_2 + SHCH_2COOH \rightarrow$$
$$(C_2H_5)_3Ge(CH_2)_3SCH_2COOH$$

A similar reaction has been observed with analogous silanes (34), whereas the corresponding stannanes are cleaved.

Mercaptoethanol also leads to an addition reaction.

Free radical addition of C_2H_5SH to enynic germahydrocarbons occurs mainly at the double bond and in the 1,4 position, the yields of the latter products being somewhat greater than those obtained from analogously constituted silanes (411). Heating $(C_2H_5)_3GeC{\equiv}C{-}CH{=}CH_2$ with C_2H_5SH in the presence of $t\text{-}Bu_2O_2$ gives 75% adduct containing 86%

acetylenic, 11% allenic and 3% dienic products. Both inductive and steric effects affect the reaction: with $(C_2H_5)_3GeC\equiv C-CH=CH-CH_3$ the adduct contains 97% of acetylenic and 3% allenic material, while $(C_2H_5)_3GeC\equiv C-C(CH_3)=CH_2$ gives only 100% of acetylenic compound.

(e) Diazoacetic ester

Carboethoxycarbene which has electrophilic properties gives addition reactions with trimethylallylgermane (69); ethylester of 2-(trimethyl-germylmethyl) cyclopropane carboxylic acid is formed:

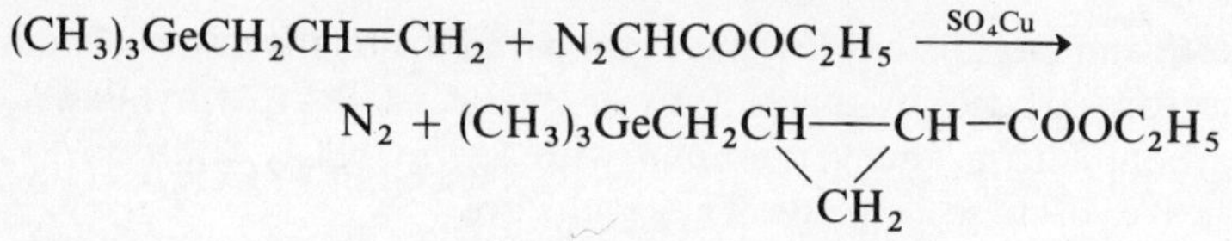

$$(CH_3)_3GeCH_2CH=CH_2 + N_2CHCOOC_2H_5 \xrightarrow{SO_4Cu}$$

$$N_2 + (CH_3)_3GeCH_2CH\underset{\underset{CH_2}{\diagdown\diagup}}{\underline{\hspace{1cm}}}CH-COOC_2H_5$$

The yield (71.5%) is higher than for the silicon (66.5%) and carbon (59%) isologues.

(f) Simmons Smith's reaction

Seyferth (352) obtained the first cyclopropyl derivatives of silicon, germanium and tin by the reaction of the trimethylvinyl compounds with methylene iodide in the presence of a zinc–copper couple prepared by the Shank and Shechter method (369):

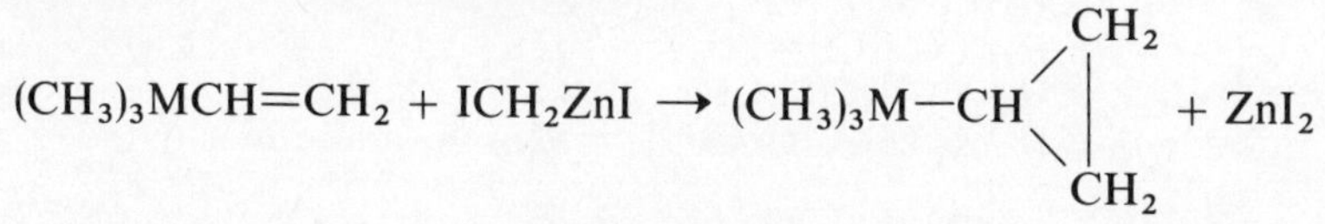

$$(CH_3)_3MCH=CH_2 + ICH_2ZnI \rightarrow (CH_3)_3M-CH\underset{\underset{CH_2}{\diagdown}}{\overset{\overset{CH_2}{\diagup}}{|}} + ZnI_2$$

$$(M = Si, Ge, Sn)$$

The yield of the cyclopropyl derivative is higher with trimethylvinyl-silane (50%) than with the germanium (29%) and tin (27%) analogues and in the latter case extensive redistribution occurs giving a mixture of products.

Dolgii (70) prepared in the same manner silicon and germanium hydrocarbons of the cyclopropane series and noted that the position of the heteroatom with respect to the double bond had not an important effect on the yield of the reaction; (trimethylgermylmethyl) cyclopropane was obtained in 32.3% yield from trimethylallylgermane:

$$(CH_3)_3GeCH_2CH=CH_2 + CH_2I_2 \xrightarrow{Zn/Cu} (CH_3)_3GeCH_2-CH\underset{\underset{CH_2}{\diagdown}}{\overset{\overset{CH_2}{\diagup}}{|}}$$

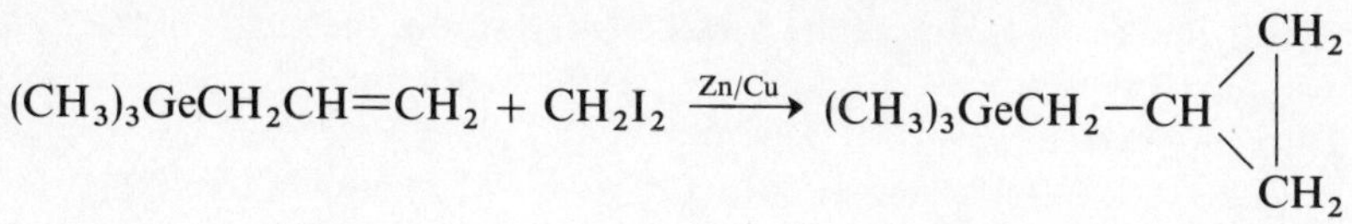

(g) Dihalocarbenes

The addition of dichlorocarbene to vinylic and allylic derivatives of Group IV elements was studied by Seyferth and coworkers (357) using his general procedure. The corresponding cyclopropyl compounds are formed with an excellent yield:

$$(CH_3)_3GeCH_2CH=CH_2 + C_6H_5HgCCl_2Br \rightarrow$$

$$(CH_3)_3GeCH_2-CH\underset{\underset{CCl_2}{\diagdown \diagup}}{\qquad}CH_2 + C_6H_5HgBr \quad (94\%)$$

Seyferth and Dertouzos (354) have compared the relative rates of addition of dichlorocarbene (by thermolysis of $PhHgCCl_2Br$) to vinyl derivatives of carbon, silicon and germanium with k_{rel} for CCl_2 addition to allyl derivatives of these elements. The results are:

OLEFIN	k_{rel}	TEMP. (°C)
1-Heptene	1.00	80
$CCl_2=CHCl$	0.14	80
$C_2H_5(CH_3)_2CCH=CH_2$	0.043	80
$C_2H_5(CH_3)_2SiCH=CH_2$	0.069	80
$(C_2H_5)_3SiCH=CH_2$	0.048	80
$(C_2H_5)_3GeCH=CH_2$	0.064	80
$(CH_3)_3CCH_2CH=CH_2$	0.78	71
$(CH_3)_3SiCH_2CH=CH_2$	4.2	71
$(CH_3)_3GeCH_2CH=CH_2$	5.7	71

In the allyl derivatives where the steric factors are not very important the order of reactivity $Me_3C < Me_3Si < Me_3Ge$ is the order of increasing (+I) inductive effect for Me_3M^{IV} groups (19, 23, 431). The low reactivity of silicon and germanium vinyl compounds compared with the allyl derivatives can be explained by steric hindrance and $C-M^{IV}$ π-bonding between the filled π orbital and vacant M d-orbitals (lower for Ge than for Si).

The action of sodium iodide on trimethyl(trifluoromethyl) tin in 1,2-dimethoxyethane in the presence of triethylvinylgermane gives gem-difluorocyclopropyltriethylgermane according to the equation:

$$(CH_3)_3SnCF_3 + NaI + (C_2H_5)_3GeCH=CH_2 \rightarrow$$

$$(CH_3)_3SnI + NaF + (C_2H_5)_3Ge-CH_2\underset{\underset{CH_2}{\diagdown \diagup}}{\qquad}CF_2$$

Seyferth (355) suggests that the mechanism of this reaction involves nucleophilic attack by iodide ion on tin to displace the trifluoromethyl anion, which then loses fluoride ion to form difluorocarbene which subsequently adds to triethylvinylgermane.

(h) Radical additions

In the presence of benzoyl peroxide, extended heating of alkylvinyl-germanes (in the same way as vinylsilanes (358) and vinylstannanes) leads to anti-Markownikow addition reactions with carbon tetrachloride, chloroform and ethyl bromoacetate (228):

$$(C_4H_9)_3Ge(CH{=}CH_2) + CHCl_3 \rightarrow (C_4H_9)_3GeCH_2{-}CH_2CCl_3$$

$$(C_2H_5)_3Ge(CH{=}CH_2) + CH_2BrCOOC_2H_5 \rightarrow$$

$$(C_2H_5)_3GeCHBr(CH_2)_2COOC_2H_5$$

(i) Organolithium compounds

Cason and Brooks (38) reported the addition of phenyllithium to the double bond of triphenylvinylsilane. Seyferth and coworkers (362, 367, 368) in a study of the reaction of phenyllithium with the analogous germanium derivative noted a similar process:

$$(C_6H_5)_3GeCH{=}CH_2 + C_6H_5Li \rightarrow (C_6H_5)_3GeCHLi{-}CH_2{-}C_6H_5$$

The effect of this reagent upon vinylic tin and lead derivatives is different; the tetraphenyl derivative and vinyllithium are formed:

$$4\,C_6H_5Li + (CH_2{=}CH)_4Sn \rightarrow (C_6H_5)_4Sn + 4\,CH_2{=}CHLi$$

$$(C_6H_5)_3PbCH{=}CH_2 + C_6H_5Li \rightarrow (C_6H_5)_4Pb + CH_2{=}CHLi$$

Stadnichuk investigated the action of butyllithium upon different alkyl-germanes with an enyne chain (401).

(j) Oxidizing agents

Dilute (10%) potassium permanganate solution at 0°C reacts with triethylallylgermane forming an α-glycol (228):

$$2\,KMnO_4 + 3\,(C_2H_5)_3GeCH_2{-}CH{=}CH_2 + 4H_2O \rightarrow$$

$$2\,MnO_2 + 3\,(C_2H_5)_3GeCH_2CH(OH)CH_2OH + 2\,KOH$$

Triethylallylgermane absorbs ozone at $-10°C$ forming an ozonide which may be decomposed by water.

(k) Diels–Alder reactions

Germanium derivatives with a conjugated dienic chain take part in the usual diene syntheses (202):

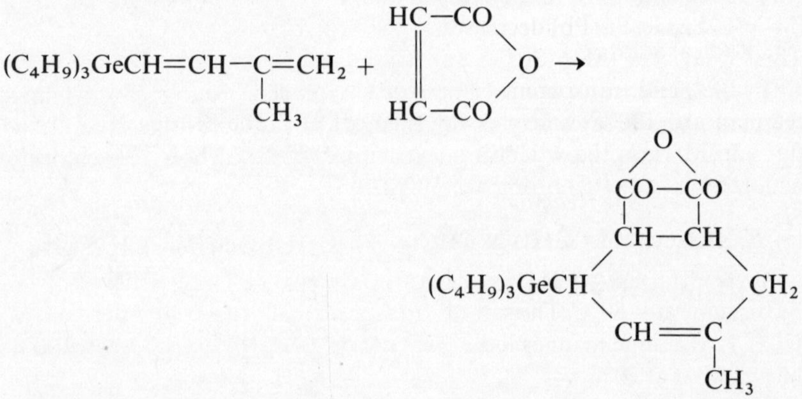

$$(C_4H_9)_3GeCH=CH-\underset{\underset{CH_3}{|}}{C}=CH_2 + \underset{HC-CO}{\overset{HC-CO}{\|}}\!\!\!>\!\!O \rightarrow$$

$$(C_4H_9)_3GeCH \underset{CH=C}{\overset{CH-CH}{\diagup}}\!\!\!\!CH_2$$

vinyltrichlorogermane which has marked dienophilic properties gives cyclic adducts with many conjugated dienes; cyclopentadiene gives an exothermic reaction (225):

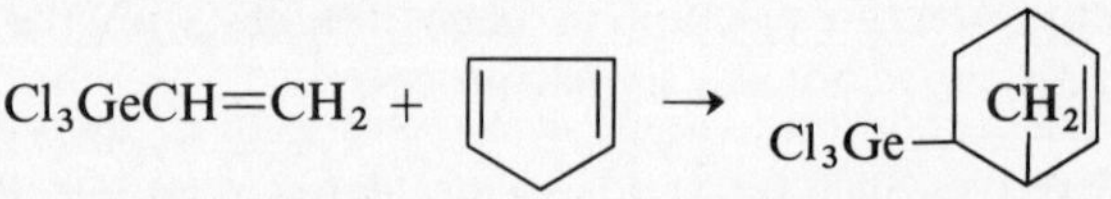

$$Cl_3GeCH=CH_2 + \text{[diene]} \rightarrow Cl_3Ge-\text{[adduct with CH}_2\text{]}$$

while 1.3 butadiene, isoprene and 2.3 dimethylbutadiene require heating.

(l) Polymerization

Olefinic derivatives of germanium hardly tend to polymerize. Korshak and coworkers (177) who studied methylallylgermanes and triethylallyl-germane, observed polymerization of these compounds under 6000 atm pressure at 120°C in the presence of a catalyst. The polymers produced are either solid or liquid. On the other hand, under the same conditions the branched $(CH_3)_3GeCH_2-\underset{\underset{CH_3}{|}}{C}=CH_2$ derivative does not polymerize.

Tetraallylgermane, which is more reactive, will polymerize under 1 atm at 200°C in the presence of benzoyl peroxide, yielding a solid and elastic light yellow polymer that will not dissolve in benzene (228). The mercury-sensitized polymerization of vinylgermane gives a white, insoluble solid which was assumed to be $(-GeH_2CH_2CH_2-)_n$ formed by a mechanism similar to that postulated for vinylsilane polymerization (24).

Noltes, Budding and van der Kerk (282) studied the thermal or radical-initiated polymerization of germylstyrenes and obtained transparent, solid germyl-substituted polystyrenes. They observed that the rates of radical-initiated polymerization for the monomers

$$p\text{-}(CH_3)_3M{-}C_6H_4{-}CH{=}CH_2$$

(M = C, Si, Ge, Sn, Pb) decreases in the order Pb > Si > C > Ge > Sn. These polymers (M = Si, Ge, Sn) have a softening point in the range 170–180°C and transparent colourless films can be obtained at elevated temperature (190°C) and pressure (200 kg/cm^2). The brittleness of these films decreases in the order Sn > Ge > Si.

2-2-3-2   Cleavage Reactions

(a) Cleavage of the Ge—C bond

$R_3Ge(CH_2)_nCH{=}CH_2$ type derivatives (with $n = 1$ or 2) will be cleaved by bromine at $-80°C$. The same is true for methallyl and styryl derivatives (202). In the same manner iodine will cleave triethylallylgermane without any catalyst at 20°C.

$$(C_2H_5)_3GeCH_2CH{=}CH_2 + I_2 \rightarrow CH_2{=}CHCH_2I + (C_2H_5)_3GeI$$

At $-80°C$ hydrogen halides easily cleave the Ge—(allyl) bond.

$$(C_4H_9)_3GeCH_2{-}CH{=}CH_2 + HBr \rightarrow (C_4H_9)_3GeBr + CH_3CH{=}CH_2$$

Under the same conditions, vinylic derivatives seem to yield an unstable addition derivative which it has not been possible to isolate.

Strong organic acids also cleave Ge-(vinyl) and Ge-(allyl) bonds. Figures 2-3 and 2-4 show respectively the comparative reactivities of tributylvinylgermane and tributylallylgermane with different organic acids. The greater reactivity of the allylic derivatives is emphasized on the one hand by a higher reaction speed and on the other hand by greater extent of cleavage. (Formic acid will only cleave the allylic chain derivative.)

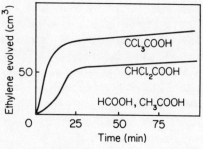

FIG. 2-3 Reaction rates of tributylvinyl-germane with organic acids

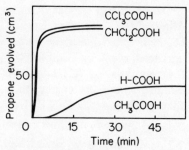

FIG. 2-4 Reaction rates of tributylallyl-germane with organic acids

Inorganic acid chlorides easily react with allylic derivatives:

$$(C_2H_5)_3GeCH_2CH{=}CH_2 + SO_2Cl_2 \rightarrow$$

$$SO_2 + ClCH_2CH{=}CH_2 + (C_2H_5)_3GeCl$$

Organic acid chlorides show similar reactivity. Mercuric chloride in alcoholic solution at 20°C will cleave tributylallylgermane (228):

$$(C_4H_9)_3GeCH_2CH{=}CH_2 + HgCl_2 \rightarrow$$

$$CH_2{=}CH{-}CH_2HgCl + (C_4H_9)_3GeCl$$

The mechanism of alkali cleavage of 3-phenylallyl derivatives of silicon, germanium and tin by mercury salts in aqueous alcoholic media has been studied spectrophotometrically by Roberts and El'Kaissi (320). The product of reaction is β-methylstyrene. This reaction appears to be subject to large steric effects. In aqueous DMSO the order of reactivity is $Et_3Sn \gg Et_3Si > Et_3Ge$. The greater reactivity of 3 phenylallyl compounds compared with their benzyl analogues is interpreted in terms of homo-allylic interactions. The kinetics and mechanism of this reaction have been studied by Roberts (319) in various solvents. Cleavage in acetonitrile is a simple bimolecular, electrophilic substitution of mercury for germanium. The reaction is more than one hundred times as fast as that of the corresponding silane; traces of water in the solvent have a profound effect on reaction rates.

In ethanol the comparative reactivity of silicon, germanium and tin derivatives is $Et_3Sn \gg Et_3Ge > Et_3Si$, corresponding to the relative electron-releasing properties of the groups.

(b) Cleavage of the C=C bond

A 5% potassium permanganate water-acetone solution reacts with triethylallylgermane to cause rupture of the chain by oxidizing the double bond.

Ozonolysis of triphenylvinylgermanes was studied by Nicholson and Allred (280). 1-Triphenylgermyl-1-phenylethylene gave a low yield of benzoyltriphenylgermane but with the other vinylgermanes examined unreacted olefin or hexaphenyldigermoxane was recovered.

Ozonolysis of $(C_2H_5)_3GeCH_2C(CH_3){=}C(CH_3)_2$ gives a complex mixture where the corresponding ketone $(C_2H_5)_3GeCH_2COCH_3$ was characterized as the D.N.P.H. (M.P. 105°C) (334).

2-2-4   Olefinic GeR_4 derivatives

Table 2-3 showing some physical properties of olefinic GeR_4 derivatives appears on the following pages.

Table 2-3
Olefinic GeR₄ Derivatives

COMPOUND	EMPIRICAL FORMULA	M.P. (°C)	B.P. (°C/mm)	n_D^{20}	d_4^{20}	REFERENCES
$(CH_3)_3GeCF=CF_2$	$C_5H_9GeF_3$	—	—	1.5330 (25°)	—	3
$(CH_3)_3GeCCl=CCl_2$	$C_5H_9GeCl_3$	—	—	—	—	353
$(CH_3)_3GeCCl=CH_2$	$C_5H_{11}GeCl$	—	120.5/755	1.4517	1.1749	260
$(CH_3)_3GeCH=CHCl$	$C_5H_{11}GeCl$	—	134/756	1.4600	1.1844	240
$Cl_3GeCH=CHGe(CH_3)_3$	$C_5H_{11}Ge_2Cl_3$	—	100-3/12	—	—	260
$(CH_3)_3Ge(CH=CH_2)$	$C_5H_{12}Ge$	—	70.6/735 69-71	1.4153 1.4168	0.9970	100, 176, 245
$(CH_3)_2Ge(CF=CF_2)_2$	$C_6H_6GeF_6$	—	—	—	—	3
$(CH_3)_3GeCH_2CH=CH_2$	$C_6H_{14}Ge$	—	101/764 100/757 97-98/760	1.4333 1.4327	0.9952 0.9930	69, 70, 88, 94, 100, 177, 271, 300, 301
$(CH_3)_3GeCH=CH-CH_3$ cis	$C_6H_{14}Ge$	—	100-3	1.4278 (25°)	—	271, 347, 361, 364
$(CH_3)_3GeCH=CH-CH_3$ trans	$C_6H_{14}Ge$	—	100-3	1.4319 (25°)	—	271, 347, 361, 364, 421
$(CH_3)_3GeC(CH_3)=CH_2$	$C_6H_{14}Ge$	—	97-98/760 97.5/747	1.4260 (25°) 1.4294	1.0006	94, 271
$CH_2=CHCOOCH_2Ge(CH_3)_3$	$C_7H_{14}GeO_2$	—	50.5-52/14	1.4410	1.1294	261
$(CH_3)_3GeCH_2CH_2CH=CH_2$	$C_7H_{16}Ge$	—	124/755.5	1.4363	0.9907	261
$(CH_3)_3GeCH_2CH=CH-CH_3$	$C_7H_{16}Ge$	—	126-28/760 68/94	1.4428 1.4435	0.9907 0.9912	246, 250
$(CH_3)_3GeCH_2CH=CH-CH_3$ cis	$C_7H_{16}Ge$	—	125-27/760	1.4430 (25°)	—	94

Table 2-3—continued

COMPOUND	EMPIRICAL FORMULA	M.P. (°C)	B.P. (°C/mm)	n_D^{20}	d_4^{20}	REFERENCES
$(CH_3)_3GeCH_2CH=CH—CH_3$ trans	$C_7H_{16}Ge$	—	125–27/760	1.4375 (25°)	—	94
$(CH_3)_3GeCH_2C(CH_3)=CH_2$	$C_7H_{16}Ge$	—	121/733	1.4416	0.9908	177, 300
$(CH_3)_3GeCH_2CHOHCH=CH_2$	$C_7H_{16}GeO$	—	68–69/14	1.4595	1.1123	261
$Ge(CHBr=CBr)_4$	$C_8H_4GeBr_8$	131, 123	—	—	—	61, 230
$Ge(CH=CH_2)_4$	$C_8H_{12}Ge$	—	52–54/27	1.4676 (25°)	1.040 (25°)	349
$(CF_2=CF)_4Ge$	$C_8F_{12}Ge$	—	123–24	1.3662 (18°)	1.7719 (18°)	409
$C_2H_5Ge(CH=CH_2)_3$	$C_8H_{14}Ge$	—	55–57/28	1.4605 (25°)	—	349
$(C_2H_5)_3GeCF=CF_2$	$C_8H_{15}GeF_3$	—	45/10	1.4138 (25°)	—	363
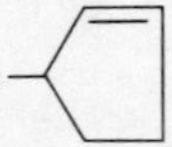 (cyclopentenyl)$Ge(CH_3)_3$	$C_8H_{16}Ge$	—	32/9	1.4697	1.0642	94, 194, 253
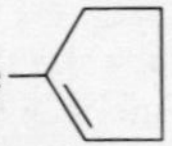 (cyclopentenyl)$Ge(CH_3)_3$	$C_8H_{16}Ge$	—	52–54/15	1.4657 (25°)	—	94
$(C_2H_5)_2Ge(CH=CH_2)_2$	$C_8H_{16}Ge$	—	59/28; 149.8/745	1.4540 (25°); 1.4575	1.0193	217, 266, 299, 349
$(CH_3)_2Ge(CH_2CH=CH_2)_2$	$C_8H_{16}Ge$	—	146/737	1.4645	1.0337	166, 177, 301
$(CH_3)_3GeCH_2OCOC(CH_3)=CH_2$	$C_8H_{16}GeO_2$	—	53.5–54/8	1.4490	1.1104	261
$(CH_3)_3GeCH=CH—CH_2CH_3$	$C_8H_{18}Ge$	—	79/90	1.4500	0.9884	246
$(CH_3)_3GeCH(CH_3)CH=CH—CH_3$	$C_8H_{18}Ge$	—	146.5/760	1.4491	0.9737	231
$(CH_3)_3GeCH_2—CH=C(CH_3)_2$	$C_8H_{18}Ge$	—	78–81/9	1.4643	—	246
$(CH_3)_3GeCH_2—C(CH_3)=CH—CH_3$; $(C_2H_5)_3Ge(CH=CH_2)$	$C_8H_{18}Ge$	—	61/28	1.4501	1.0048	164, 189, 217, 220, 228, 229

Compound	Formula		bp/mm	n_D	d	References
$(CH_3)_3SiCH=CHGe(CH_3)_3$	$C_8H_{20}GeSi$	—	159.5/855	1.4498	0.9642	240, 258, 259
$(CH_3)_3Si$ $\diagdown$ $C=CH_2$ $(CH_3)_3Ge$ $\diagup$	$C_8H_{20}GeSi$	—	164.5/757	1.4523	0.9717	258, 259
$(CH_3)_3GeCH=CH-Ge(CH_3)_3$	$C_8H_{20}Ge_2$	—	79–79.5/35	1.4628	1.1480	247, 259, 260
$(CH_3)_3Ge$ $\diagdown$ $C=CH_2$	$C_8H_{20}Ge_2$	—	72–73/28	1.4655	1.1646	259, 260
$(CH_3)_2(C_2H_5)Ge-$ 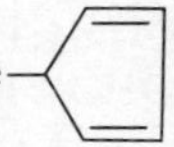 (cyclopentadienyl)	$C_9H_{16}Ge$	—	31.5–32.5/4	1.4930	1.0931	253
$(CH_3)_3Ge-$ 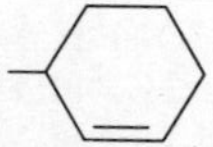 (cyclohexenyl)	$C_9H_{18}Ge$	—	38–39/1.5	1.4795 (25°)	—	94
$(C_2H_5)_3GeCH=CH-CH_3$ cis	$C_9H_{20}Ge$	—	58–60/11	1.4625	1.0342	277
$(C_2H_5)_3GeCH=CH-CH_3$ trans	$C_9H_{20}Ge$	—	60–62/10	1.4590	1.0108	277
$(C_2H_5)_3Ge-C=CH_2$ $\quad CH_3$	$C_9H_{20}Ge$	—	55–56/11	1.4581	—	277
$(C_2H_5)_3GeCH_2CH=CH_2$	$C_9H_{20}Ge$	—	180/732 50/6	1.4594	1.0004	177, 241, 300, 319
$(CH_3)_3GeC(CH_3)_2C(CH_3)=CH_2$	$C_9H_{20}Ge$	—	79.5–80/27	1.4580	0.9891	252
$(C_2H_5)_3GeCH=CH-CH_2OH$	$C_9H_{20}GeO$	—	112–13/12	1.4808	1.0980	73
$(CH_3)_3GeCH=CH-CH_2Ge(CH_3)_3$	$C_9H_{22}Ge_2$	—	78–79/18	1.4715	1.1473	250
$(C_2H_5)_3Ge-C=C(CF_3)CF_2$	$C_{10}H_{15}GeF_5$	—	160	—	—	53
$(C_2H_5)_3GeC(CF_3)=CHCF_3$	$C_{10}H_{16}GeF_6$	—	82/26	—	—	54
$CH_3Ge(CH_2CH=CH_2)_3$	$C_{10}H_{18}Ge$	—	190/747	1.4867	1.0222	177, 301

Table 2-3—continued

COMPOUND	EMPIRICAL FORMULA	M.P. (°C)	B.P. (°C/mm)	n_D^{20}	d_4^{20}	REFERENCES
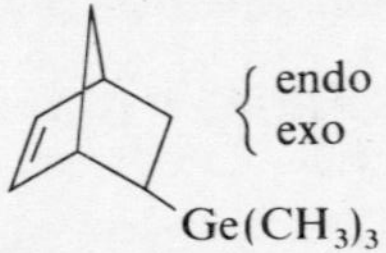 (bicyclo[2.2.1]heptenyl)–Ge(CH₃)₃ endo	$C_{10}H_{18}Ge$	—	83–84/1.7	1.4834 (25°)	—	94
(bicyclo[2.2.1]heptenyl)–Ge(CH₃)₃ exo	$C_{10}H_{20}Ge$	—	—		—	217
$(C_2H_5)_2Ge(CH_2CH=CH_2)_2$	$C_{10}H_{22}Ge$	—	88.5/18	1.4655	0.9971	334, 335
$(C_2H_5)_3GeCH_2CH=CHCH_3$	$C_{10}H_{22}Ge$	—	88/18	1.4667	1.0002	335
$(C_2H_5)_3GeCH_2CH=CHCH_3$ cis	$C_{10}H_{22}Ge$	—	—	1.4608	0.9872	335
$(C_2H_5)_3GeCH_2CH=CHCH_3$ trans	$C_{10}H_{22}Ge$	—	77/11	1.4648	0.9973	230
$(C_2H_5)_3GeCH_2C(CH_3)=CH_2$	$C_{10}H_{22}Ge$	—	86/17	1.4587	0.9888	230, 233
$(C_2H_5)_3Ge(CH_2)_2CH=CH_2$	$C_{10}H_{22}Ge$	—	85/11.5	1.4713	1.1190	194
$(CH_3)_3GeCH_2CH=CHCH_2Ge(CH_3)_3$	$C_{10}H_{24}Ge_2$	—	39–41/0.05	1.5400	1.1060	44, 45, 282
$p\text{-}(CH_3)_3GeC_6H_4CH=CH_2$	$C_{11}H_{16}Ge$	—	57/0.33	1.5403	1.1030	
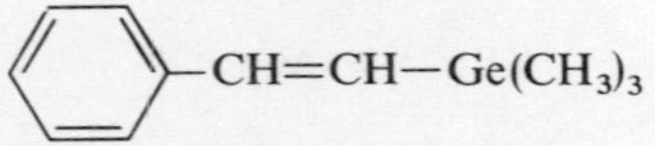 (phenyl)–CH=CH–Ge(CH₃)₃	$C_{11}H_{16}Ge$	—	83/4 ; 93/7	1.5425	1.1441	271
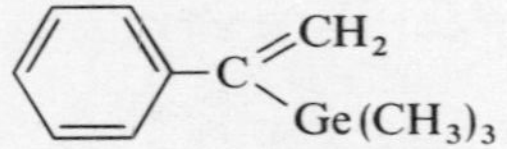 (phenyl)–C(=CH₂)–Ge(CH₃)₃	$C_{11}H_{16}Ge$	—	57/2.5	1.5365	1.1379	271
(cyclopentadienyl)–Ge(C₂H₅)₃	$C_{11}H_{20}Ge$	—	59–60/4 ; 105/16	1.5005 ; 1.5029	1.0773 ; 1.0740	199, 253
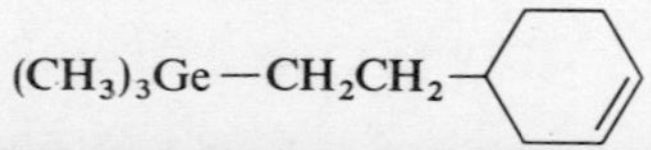 $(CH_3)_3Ge{-}CH_2CH_2{-}$(cyclohexenyl)	$C_{11}H_{22}Ge$	—	93–94/0.3	1.4756 (25°)	—	94

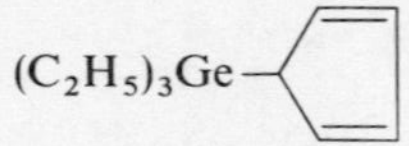

Compound	Formula		bp/mm	n_D	d	Ref.
$(CH_3)_3Ge$ (on cyclooctene ring)	$C_{11}H_{22}Ge$	—	78–78.5/0.75	1.4869 (25°)	—	94
$CH_3COOCH_2CH=CHGe(C_2H_5)_3$	$C_{11}H_{22}GeO_2$	—	87–89/4	1.4655	1.0841	254
$(C_2H_5)_3GeCH_2CH=C(CH_3)_2$	$C_{11}H_{24}Ge$	—	116.5/38	1.4678	0.9824	231
$(C_2H_5)_3Ge(CH_2)_3CH=CH_2$	$C_{11}H_{24}Ge$	—	93/12	1.4597	0.9785	230
$(CH_3)_2Ge[\text{ring}]_2$	$C_{12}H_{16}Ge$	—	71–73/4	1.5490	1.1629	102, 253
$(C_2H_5)_2(C_6H_5)GeCH=CH_2$	$C_{12}H_{18}Ge$	—	111/11	1.5224	1.0996	225
$(CH_3)_3GeC_6H_4C(CH_3)=CH_2\text{-}p$	$C_{12}H_{18}Ge$	—	55–57/0.3	1.5344	1.0936	282
$(CH_2=CH)_3Ge—Ge(CH=CH_2)_3$	$C_{12}H_{18}Ge_2$	—	55/0.3	1.5217 (25°)	1.171 (25°)	349
$Ge(CH_2CH=CH_2)_4$	$C_{12}H_{20}Ge$	—	106/14	1.5030	1.0094	189, 217, 228
$(cis\ CH_3CH=CH)_4Ge$	$C_{12}H_{20}Ge$	—	77.5–78.5/4	1.5040	1.0377	277
$(trans\ CH_3CH=CH)_4Ge$	$C_{12}H_{20}Ge$	—	64–66/1	1.4930	1.0074	277
$[CH_2=C(CH_3)]_4Ge$	$C_{12}H_{20}Ge$	—	60–62/2	1.4935	1.0245	277
		—	90–91/11	1.5110 (18°)	1.061 (18°)	419
		—	105/10	1.4771	0.9901	229
$(C_2H_5)_3Ge-C\big(CH_2CH=CHCH_2CH=CH_2\big)\big(CH{-}CH=CH_2\big)$	$C_{12}H_{24}Ge$	—	99/2	1.5012	1.0725	134
$(C_2H_5)_3Ge-C\big(CHOH{-}CH_3\big)\big(CH{-}CH=CH_2\big)$	$C_{12}H_{24}GeO$	—				
$(C_2H_5)_3Ge-C\big(CH_2CH_2OH\big)\big(CH{-}CH=CH_2\big)$	$C_{12}H_{24}GeO$	—	129–31/4	1.5030	1.0471	134
$(C_2H_5)_3GeCH=CHCH_2OCH_2OCH_2CH_2Cl$	$C_{12}H_{25}GeO_2Cl$	—	127–29/2	1.4745	1.1324	4
$(C_2H_5)_3GeCH=CH(CH_2)_3CH_3$	$C_{12}H_{26}Ge$	—	102/10	1.4623	0.9655	196, 202, 329
$(C_2H_5)_3GeCH_2—C(CH_3)=C(CH_3)_2$	$C_{12}H_{26}Ge$	—	103/10	1.4748	0.9859	334, 335

Table 2-3—continued

COMPOUND	EMPIRICAL FORMULA	M.P. (°C)	B.P. (°C/mm)	n_D^{20}	d_4^{20}	REFERENCES
$(C_3H_7)_3GeCH=CH-CH_3$	$C_{12}H_{26}Ge$	—	80-1/3	1.4562 (18°)	0.9697 (18°)	419
$(C_3H_7)_3GeC(CH_3)=CH_2$	$C_{12}H_{26}Ge$	—	95-96/5	1.4536 (18°)	0.9667 (18°)	419
$(n\text{-}C_3H_7)_3GeCH_2CH=CH_2$	$C_{12}H_{26}Ge$	—	132-33/23	1.4527	0.9426	417
$(i\text{-}C_3H_7)_3GeCH_2CH=CH_2$	$C_{12}H_{26}Ge$	—	90-91/23	1.4885	1.1988	236, 417
			112-14/25	1.4847	0.9952	
$(C_2H_5)_3GeCH=CHCHOHC_3H_7$	$C_{12}H_{26}GeO$	—	88-90/0.7	1.4740	1.0300	380
$(C_2H_5)_3GeCH=CHCHOHCH(CH_3)_2$	$C_{12}H_{26}GeO$	—	73-74/0.2	1.4670	1.0350	380
$(C_2H_5)_2(C_6H_5)GeCH_2CH=CH_2$ $C(CH_3)=CH_2$	$C_{13}H_{20}Ge$	—	92/1.4	1.5263	1.0875	229a
$(C_2H_5)_3GeC$≡ $CH-CH=CH_2$	$C_{13}H_{24}Ge$	—	71/2	1.5016	1.0216	132
$(CH_3)_3Ge-C$≡$C-N(CH_3)_2$ $COOEtCOOEt$ $COH(CH_3)_2$	$C_{13}H_{25}GeO_4N$	—	80/0.02	—	—	43
$(C_2H_5)_3GeC$≡ $CH-CH=CH_2$ $CH=CH$	$C_{13}H_{26}GeO$	—	112/6	1.4954	1.0390	132
$(CH_3)_2C$ $CH-CH_3$ O	$C_{13}H_{26}GeO$	—	83-85/6	1.4703	1.0121	129
$(C_3H_7)_3GeCH=C(CH_3)_2$	$C_{13}H_{27}Ge$	—	93-94/4	1.4558 (18°)	0.9566 (18°)	419
$(C_2H_5)_3Ge(CH_2)_5CH=CH_2$	$C_{13}H_{28}Ge$	—	122/13	1.4620	0.9770	224

Compound	Formula		bp/mm	n_D	d	Ref.
$(C_3H_7)_3GeCH=CHC_2H_5$ $\quad\quad COH(CH_3)_2$	$C_{13}H_{28}Ge$	—	87–88/3	1.4608 (18°)	0.9663 (18°)	419
$(C_2H_5)_3GeC\!\equiv\!C$—CHC_2H_5	$C_{13}H_{28}GeO$	—	86/1	1.4795	1.0415	133
$(C_3H_7)_3GeCH=CH-OC_2H_5$	$C_{13}H_{28}GeO$	—	76.5–77/1	1.4600	0.9968	159
$(C_2H_5)_3GeCH=CHCOH$—C_3H_7	$C_{13}H_{28}GeO$	—	81–83/0.3	1.4646	1.0064	378
$(C_2H_5)_3GeC\!\equiv\!C$—CH—$CHOHCH_3$ $\quad COH(CH_3)_2$	$C_{13}H_{28}GeO_2$	—	155/5	1.4870	1.030	129
$H[CH_2CH=CHCH_2Ge(CH_3)_2]_2]_2$—$CH_3$	$C_{13}H_{28}Ge_2$	—	65–67/0.2	1.4875	1.1212	274
$(C_2H_5)_3GeC_6H_4CH=CH_2$-$p$	$C_{14}H_{22}Ge$	—	85/0.3	1.5472	1.0742	44, 45, 164, 297
$(C_2H_5)_3GeC\!\equiv\!C$—$C(CH_3)=CHCH_3$	$C_{14}H_{26}Ge$	—	107–8/3	1.4965	1.0367	132
$(C_2H_5)_3GeC\!\equiv\!C$ $\quad CH$—$CH=CH_2$ $\quad CH$—$CH=CH_2$	$C_{14}H_{26}GeO_2$	—	106–8/1	1.4836	1.0763	133
$(C_2H_5)_3GeC\!\equiv\!C$ $\quad CH_2CH_2OCOCH_3$ $\quad CH$—$CH=CH_2$	$C_{14}H_{26}GeO_2$	—	111–12/2	1.4800	1.0699	133
$(C_2H_5)_3GeCH_2CH_2$ 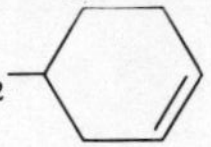 $\quad CH(CH_3)OCOCH_3$ (cyclohexenyl)	$C_{14}H_{28}Ge$	—	154/18	1.4845	1.0234	335

Table 2-3—continued

COMPOUND	EMPIRICAL FORMULA	M.P. (°C)	B.P. (°C/mm)	n_D^{20}	d_4^{20}	REFERENCES
$(C_2H_5)_3GeC$ with CHOH–C$_3$H$_7$ / CH–CH=CH$_2$	$C_{14}H_{28}GeO$	—	123–24/4	1.4970	1.0457	134
$(C_2H_5)_3GeC$ with COH(CH$_3$)C$_2$H$_5$ / CH–CH=CH$_2$	$C_{14}H_{28}GeO$	—	98–100/3	1.4972	1.0430	132
$(C_4H_9)_3GeCBr=CH_2$	$C_{14}H_{29}GeBr$	—	—	—	—	222
$(C_4H_9)_3Ge(CH=CH_2)$	$C_{14}H_{30}Ge$	—	109/2	1.4598	0.9479	189, 217, 228, 229
$(C_2H_5)_3GeC$ with COH(CH$_3$)C$_2$H$_5$ / CH–C$_2$H$_5$	$C_{14}H_{30}GeO$	—	99–100/1	1.4835	1.0358	133
$(C_2H_5)_3GeCH=CHCOH$ with CH$_3$ and t-C$_4$H$_9$	$C_{14}H_{30}GeO$	—	84–86/0.2	1.4662	1.0079	378
$(C_2H_5)_3GeCH=CHCOH$ with CH$_3$ and n-C$_4$H$_9$	$C_{14}H_{30}GeO$	—	88–90/0.2	1.4690	1.0127	378

Structure	Formula		B.p.			References
$(C_2H_5)_3Ge-C=CH-COH(CH_3)_2$, $COH(CH_3)_2$	$C_{14}H_{30}GeO_2$	70	133/0.6	—	—	196, 202, 329
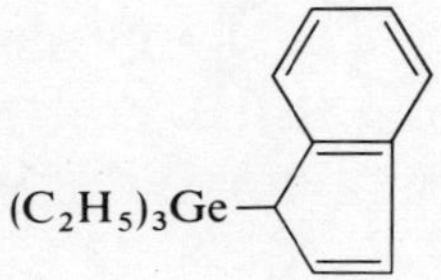 indenyl $(C_2H_5)_3Ge$	$C_{15}H_{22}Ge$	—	161/4	1.5589	1.1230	199
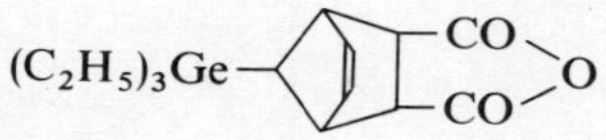 $(C_2H_5)_3Ge$ bicyclic $CO-O-CO$	$C_{15}H_{22}GeO_3$	56	—	—	—	199, 253
$(C_2H_5)_3GeC_6H_4-C=CH_2$, CH_3	$C_{15}H_{24}Ge$	—	129–30/3	1.5368	1.0685	305
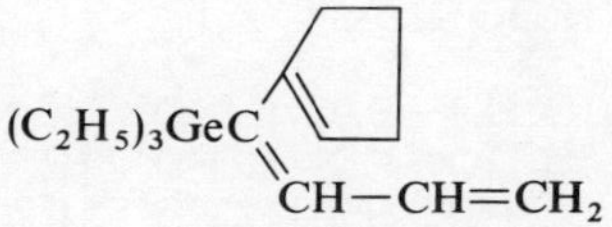 $(C_2H_5)_3GeC$ cyclopentenyl $CH-CH=CH_2$	$C_{15}H_{26}Ge$	—	120–22/3	1.5160	1.0530	132
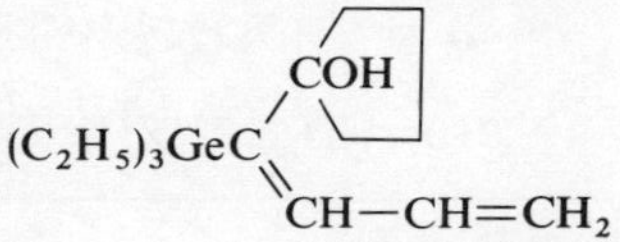 $(C_2H_5)_3GeC$ COH $CH-CH=CH_2$	$C_{15}H_{28}GeO$	—	116–18/1	1.5120	1.0788	132
$(C_2H_5)_3GeC$ $C(CH_3)_2OCOCH_3$ $CH-CH=CH_2$	$C_{15}H_{28}GeO_2$	—	95–96/1.5	1.4820	1.0405	133, 167

Table 2-3—continued

COMPOUND	EMPIRICAL FORMULA	M.P. (°C)	B.P. (°C/mm)	n_D^{20}	d_4^{20}	REFERENCES
$(C_4H_9)_3GeCH_2CH=CH_2$	$C_{15}H_{32}Ge$	—	117/2	1.4664	0.9629	189, 217, 228, 229
$(C_4H_9)_3GeC(CH_3)=CH_2$	$C_{15}H_{32}Ge$	—	138–39/11	1.4588 (18°)	0.9469 (18°)	419
$(C_4H_9)_3GeCH=CH—CH_3$	$C_{15}H_{32}Ge$	—	127–28/3	1.4598 (18°)	0.9443 (18°)	419
$(C_3H_7)_3GeCH=CH—OC_4H_9$	$C_{15}H_{32}GeO$	—	93.5–94/1	1.4610	0.9512	159
$(C_4H_9)_3GeCH=CH—CH_2OH$	$C_{15}H_{32}GeO$	—	105/0.2	1.4780	1.0080	196, 202, 329
$(C_2H_5)_3GeC$ ⟨CHCH$_2$OC$_2$H$_5$ / CH(CH$_3$)OC$_2$H$_5$⟩	$C_{15}H_{32}GeO_2$	—	136–37/1	1.4782	1.1232	131
Ge⟨furan⟩$_4$ 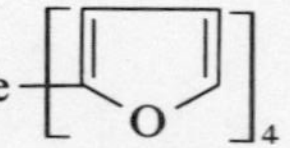	$C_{16}H_{12}GeO_4$	99–100	163/1	—	—	117, 191
Ge⟨thiophene⟩$_4$ 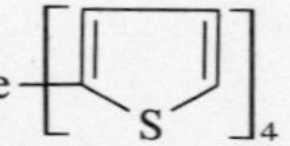	$C_{16}H_{12}GeS_4$	149–50	—	—	—	186
$(C_6H_5)_2Ge(CH=CH_2)_2$	$C_{16}H_{16}Ge$	—	111–13/0.003 145–47/4	1.5867	1.1757	144, 327
(fluorenyl)$(CH_3)_3Ge$ 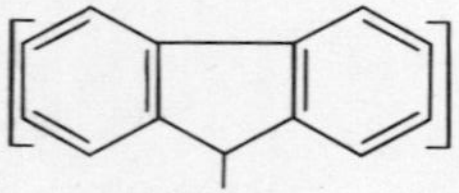	$C_{16}H_{18}Ge$	192	—	—	—	18
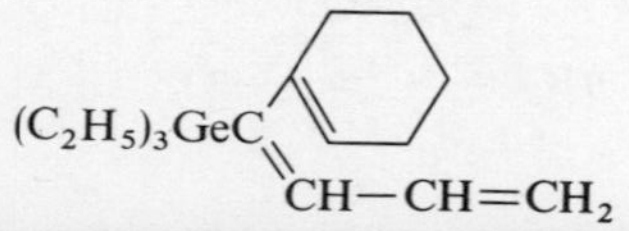 $(C_2H_5)_3GeC$⟨cyclohexenyl⟩CH—CH=CH$_2$	$C_{16}H_{28}Ge$	—	139–40/4	1.5180	1.0332	132

Structure	Formula		B.p. (°C/mm)	n_D	d	Ref.
$Ge[CH=C(CH_3)_2]_{2.4}$	$C_{16}H_{28}Ge$	—	139–40/11	1.5180 (18°)	1.018 (18°)	419
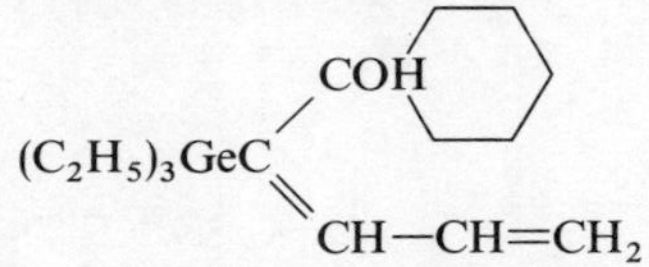 $(C_2H_5)_3GeC(\text{cyclohexyl-COH})=CH-CH=CH_2$	$C_{16}H_{30}GeO$	—	114–16/1	1.5140	1.0777	132
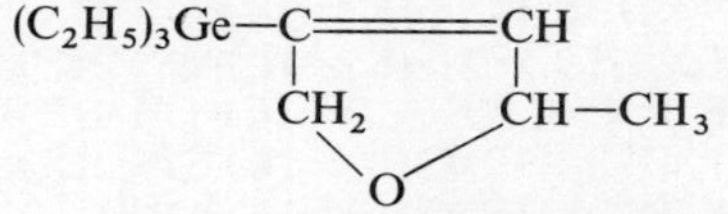 $(C_2H_5)_3Ge-C(=CH,\ CH-CH_3,\ CH_2,\ O)$	$C_{16}H_{30}GeO$	—	128–29/8	1.5008	1.0380	129
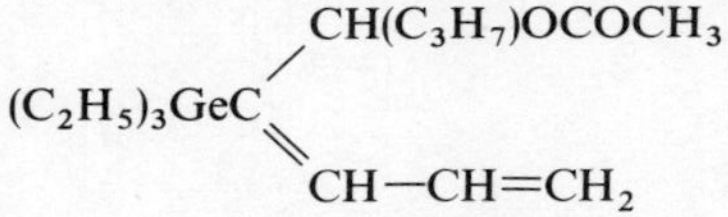 $CH(C_3H_7)OCOCH_3$ … $(C_2H_5)_3GeC=CH-CH=CH_2$	$C_{16}H_{30}GeO_2$	—	126–27/3	1.4812	1.0390	133
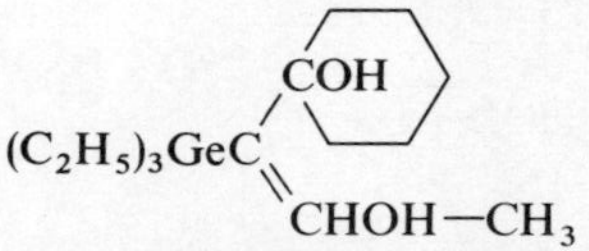 $(C_2H_5)_3GeC(\text{cyclohexyl-COH})CHOH-CH_3$	$C_{16}H_{32}GeO_2$	—	153–54/8	1.5040	1.1225	129
$(C_4H_9)_3GeCH=CH-C_2H_5$	$C_{16}H_{34}Ge$	—	123–24/3	1.4605 (18°)	0.9423 (18°)	419
$(C_4H_9)_3GeCH=C(CH_3)_2$	$C_{16}H_{34}Ge$	—	127–28/4	1.4602 (18°)	0.9431 (18°)	419
$HCGe(C_2H_5)_3=CGe(C_2H_5)_3COOCH_3$	$C_{16}H_{34}GeO_2$	—	120/0.7	—	—	188
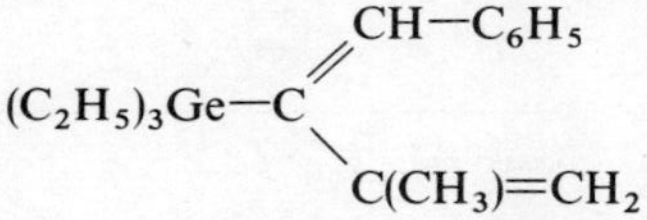 $(C_2H_5)_3Ge-C(CH-C_6H_5)=C(CH_3)=CH_2$	$C_{17}H_{26}Ge$	—	138–40/2	1.5360	1.0638	127
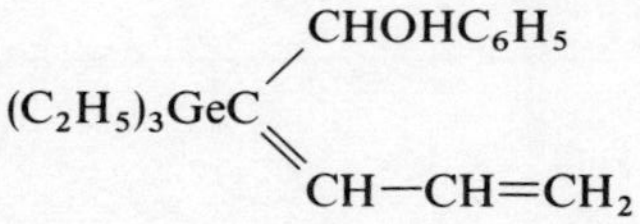 $CHOHC_6H_5$ … $(C_2H_5)_3GeC=CH-CH=CH_2$	$C_{17}H_{26}GeO$	—	163–64/3	1.5432	1.1075	134

Table 2-3—continued

COMPOUND	EMPIRICAL FORMULA	M.P. (°C)	B.P. (°C/mm)	n_D^{20}	d_4^{20}	REFERENCES
(C₂H₅)₂GeC with CH—C₆H₅ / COH(CH₃)₂	C₁₇H₂₈GeO	—	144-45/2	1.5240	1.0816	127
(C₂H₅)₃GeC=... C—O—COCH₃ (cyclopentane)	C₁₇H₃₀GeO₂	—	138/3	1.5108	1.0933	133
(C₄H₉)₃Ge (cyclopentadiene)	C₁₇H₃₂Ge	—	109/0.4	1.4942	1.0115	199
(C₂H₅)₃GeCH=CHC(CH₃)(C₄H₉)O(CH₂)₂CN	C₁₇H₃₃GeON	—	101-2/0.2	1.4700	1.0270	378
(C₂H₅)₃GeCH=CHC(CH₃)(C₄H₉)O(CH₂)₂CN	C₁₇H₃₃GeON	—	111-12/0.2	1.4740	1.0329	378
(C₄H₉)₃GeCH=CH—C(CH₃)=CH₂	C₁₇H₃₄Ge	—	98/0.3	1.4825	0.9585	202, 329
(C₄H₉)₃GeCH=CH—CH₂CH=CH₂	C₁₇H₃₄Ge	—	98/0.3	1.4689	0.9462	58, 217
(C₄H₉)₃GeCH=CH—COH(CH₃)₂	C₁₇H₃₆GeO	—	120/1	1.4731	0.9811	196, 202, 329
(C₆H₅)₂Ge[—C—CH / C=N / CH / NH]₂	C₁₈H₁₆GeN₄	215	—	—	—	61
(C₆H₅)₂Ge(CH₂CH=CH₂)₂	C₁₈H₂₀Ge	—	158-60/3 / 117-20/0.007	1.5859	1.1462	144, 327

$(C_6H_5)_2Ge(CH=CH-CH_3)_2$	$C_{18}H_{20}Ge$	—	179–80/9	1.5789	1.1363	325
$(C_2H_5)_3GeC$ $\begin{smallmatrix}CH-C_6H_5\\C(CH_3)=CHCH_3\end{smallmatrix}$	$C_{18}H_{28}Ge$	—	135–37/2	1.5280	1.0751	127
$(C_2H_5)_3GeC$ $\begin{smallmatrix}CH-C_6H_5\\COH(CH_3)C_2H_5\end{smallmatrix}$	$C_{18}H_{30}GeO$	—	152–53/2	1.5230	1.0759	127
$cis\text{-}trans\text{-}(C_2H_5)_3GeC(C_6H_5)=CHP(C_2H_5)_2$	$C_{18}H_{31}GeP$	—	—	—	—	331
$cis\text{-}trans\text{-}(C_2H_5)_3GeCH=C(C_6H_5)P(C_2H_5)_2$	$C_{18}H_{31}GeP$	—	—	—	—	331
$(C_4H_9)_3Ge-CH=CH-C_4H_9$	$C_{18}H_{38}Ge$	—	—	—	—	217
$(C_4H_9)_3GeCH_2-C(CH_3)=C(CH_3)_2$	$C_{18}H_{38}Ge$	—	164/13	1.4728	0.9498	334, 335
$(n\text{-}C_5H_{11})_3GeCH_2CH=CH_2$	$C_{18}H_{38}Ge$	—	162–63/23	1.4587	0.9198	417
$(C_4H_9)_3GeCH=CHCHOH-C_3H_7$	$C_{18}H_{38}GeO$	—	146–48/1	1.4757	0.9864	374
$(C_4H_9)_3GeCH=CHCOH(CH_3)C_2H_5$	$C_{18}H_{38}GeO$	—	130–31/2	1.4696	0.9760	374
$(C_2H_5)_3GeCH_2CH=CHCH(CH_2CH_2OH)Ge(C_2H_5)_3$	$C_{18}H_{40}Ge_2O$	—	153–54/1	1.5180	1.1423	134
$(C_2H_5)_3GeCH_2CH=CHCH(CHOHCH_3)Ge(C_2H_5)_3$	$C_{18}H_{40}GeO_2$	—	147–50/2	1.4980	1.1269	134
$(C_2H_5)_3GeCH=CH(CH_2)_4Ge(C_2H_5)_3$	$C_{18}H_{40}Ge_2$	—	148/0.8	1.4818	1.0614	230
(fluorenyl) $(C_2H_5)_3Ge-$ 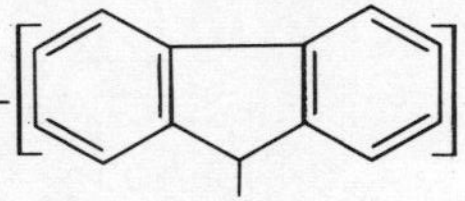 …$CH-C_6H_5$	$C_{19}H_{24}Ge$	—	165/0.5	1.6071	1.1761	20, 199
$(C_2H_5)_3GeC$ $\begin{smallmatrix}CH-C_6H_5\\C(CH_3)_2OCOCH_3\end{smallmatrix}$	$C_{19}H_{30}GeO_2$	—	135–36/2	1.5160	1.0906	127

Table 2-3—continued

COMPOUND	EMPIRICAL FORMULA	M.P. (°C)	B.P. (°C/mm)	n_D^{20}	d_4^{20}	REFERENCES
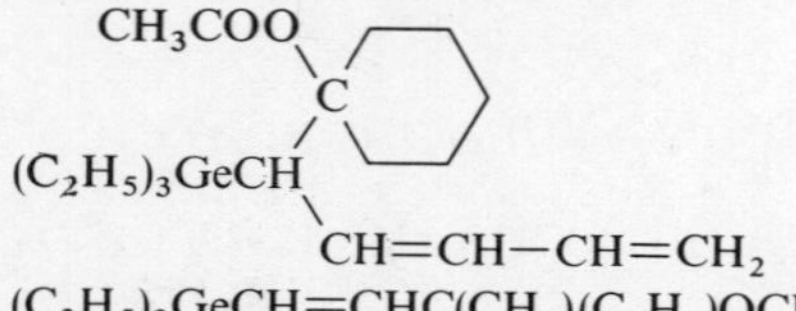 CH$_3$COO—C(ring)(C$_2$H$_5$)$_3$GeCH—CH=CH—CH=CH$_2$	$C_{19}H_{34}GeO_2$	—	146–48/1	1.5062	1.0872	133
$(C_2H_5)_3GeCH=CHC(CH_3)(C_3H_7)OCH(CH_3)OC_4H_9$	$C_{19}H_{40}GeO_2$	—	114–16/0.2	1.4600	0.9782	378
$(C_6H_5)_3GeCF=CF_2$	$C_{20}H_{15}GeF_3$	84	—	—	—	351, 363
$(C_6H_5)_3Ge(CH=CH_2)$	$C_{20}H_{18}Ge$	62–64	—	—	—	144, 365
$(C_6H_5)_2Ge[CH=C(CH_3)_2]_2$	$C_{20}H_{24}Ge$	—	180–81/7	1.5717	1.1126	325
$(C_2H_5)_3GeC\equiv C$—CH—C$_6$H$_5$ C(CH$_3$)(C$_2$H$_5$)OCOCH$_3$	$C_{20}H_{32}GeO_2$	—	142–45/2	1.5195	1.0952	127
$(C_4H_9)_3GeCH=CH—C_6H_5$ CH—C$_6$H$_5$	$C_{20}H_{34}Ge$	—	127/0.2	1.5140	1.0221	196, 202, 329
$(C_2H_5)_3GeC$—C(CH$_3$)$_2$OSi(CH$_3$)$_3$	$C_{20}H_{36}GeOSi$	—	152–55/5	1.5020	1.0227	127
$(C_4H_9)_3GeCH=CHCH_2OCH_2OC_4H_9$	$C_{20}H_{42}GeO_2$	—	149–50/0.5	1.4711	0.9925	374
$(C_2H_5)_3GeCH_2CH=CH$—Ge(C$_2$H$_5$)$_3$ CHOHC$_3$H$_7$	$C_{20}H_{44}Ge_2O$	—	188–89/4	1.4990	1.0934	134

Compound	Formula	mp (°C)	bp (°C/mm)	n_D	d	References
$(C_6H_5)_3GeCH=CH-CH_3$	$C_{21}H_{20}Ge$	67–68	208–10/6	—	—	325
$(C_6H_5)_3Ge-C(CH_3)=CH_2$	$C_{21}H_{20}Ge$	96–97	—	—	—	280
$(C_6H_5)_3GeCH_2CH=CH_2$	$C_{21}H_{20}Ge$	59–60, 90–91.5, 87–89	173–75/3	1.5989 (75°)	—	114, 115, 214, 317, 366
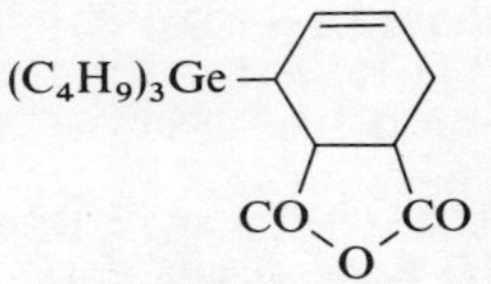 $(C_4H_9)_3Ge$ [cyclohexene peroxide diester]	$C_{21}H_{36}GeO_3$	—	193/2	1.500	1.125	329
$(C_2H_5)_3GeC(=CH-C_6H_5)\,C(CH_3)(C_2H_5)OSi(CH_3)_3$	$C_{21}H_{38}GeOSi$	—	156–58/2	1.5180	1.0468	127
$(n\text{-}C_6H_{13})_3GeCH_2CH=CH_2$	$C_{21}H_{44}Ge$	—	195–96/10	1.4627	0.9137	417
$(C_6H_5)_3GeC(CH_3)=CHCH_3$	$C_{22}H_{22}Ge$	116–20	—	—	—	280
$(C_6H_5)_3GeC(C_2H_5)=CH_2$	$C_{22}H_{22}Ge$	—	—	—	—	280
$(C_6H_5)_3GeCH=C(CH_3)_2$	$C_{22}H_{22}Ge$	119–20	—	—	—	280, 325
$(C_4H_9)_3GeCH=CHC(CH_3)_2OCH_2OC_4H_9$	$C_{22}H_{46}GeO_2$	—	158–60/1	1.4661	0.9740	374
$\left[(C_2H_5)_3GeC=\!\!\begin{array}{l}CH-\\ \text{[cyclohexyl]}\,COH(CH_3)_2\end{array}\right]_2$	$C_{22}H_{46}Ge_2O_2$	—	163–65/2	1.4893	1.0921	128
$(C_2H_5)_3GeC=\!\!\begin{array}{l}CH-\,\text{[cyclohexanol]}\\ CH-CH(CH_3)OGe(C_2H_5)_3\end{array}$	$C_{22}H_{46}Ge_2O_2$	—	198/8	1.4990	1.1256	129

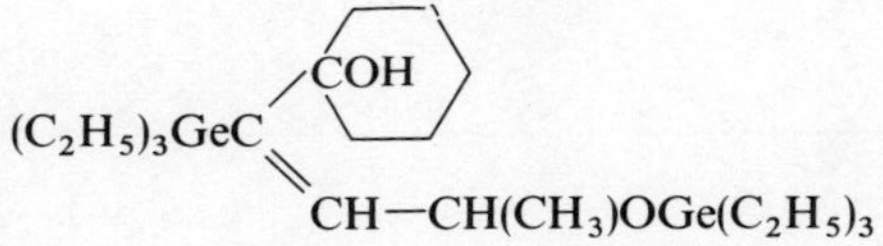

Table 2-3—continued

COMPOUND	EMPIRICAL FORMULA	M.P. (°C)	B.P. (°C/mm)	n_D^{20}	d_4^{20}	REFERENCES
$(C_6H_5)_3Ge$-⬠ 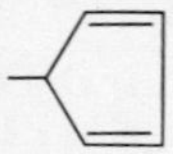	$C_{23}H_{20}Ge$	176–77	—	—	—	199
$(C_6H_5)_3GeCH{=}CH{-}COH(CH_3)_2$ $\;CHOHC_6H_5$	$C_{23}H_{24}GeO$	90.5–91.5	—	—	—	143
$(C_2H_5)_3GeCH$ $CH{=}CH{-}CH_2Ge(C_2H_5)_3$	$C_{23}H_{42}Ge_2O$	—	198–200/2	1.5320	1.1329	134
$(C_4H_9)_3GeCH{=}CHCH(C_3H_7)OCH_2OC_4H_9$	$C_{23}H_{48}GeO_2$	—	177–80/0.5	1.4685	0.9702	374
$(C_4H_9)_3GeCH{=}CHC(CH_3)(C_2H_5)OCH_2OCH_2OC_4H_9$	$C_{23}H_{48}GeO_2$	—	169–71/1	1.4610	0.9612	374
$(n{-}C_7H_{15})_3GeCH_2CH{=}CH_2$	$C_{24}H_{50}Ge$	—	225–27/10	1.4656	0.9109	417
$[(C_2H_5)_3GeCH{=}CHCH_2OSi(C_2H_5)(CH_3)]_2O$	$C_{24}H_{54}GeO_3$	—	188–89/0.2	1.4641	1.0527	375
$(C_6H_5)_3GeC(C_6H_5){=}CH_2$	$C_{26}H_{22}Ge$	119–20 120.5–22.5	—	—	—	280
$(C_6H_5)_3GeC_6H_4CH{=}CH_2{-}p$	$C_{26}H_{22}Ge$	116–17	—	—	—	281
$(C_6H_5)_3GeCH{=}CH{-}C_6H_5$	$C_{26}H_{22}Ge$	147	—	—	—	143
$(C_6H_5)_3GeCF{=}CFSi(C_2H_5)_3$	$C_{26}H_{30}GeSiF_2$	64–66	—	—	—	362
$(C_2H_5OCO)_2C[Ge(C_2H_5)_3]_3C{=}C[OGe(C_2H_5)_3]OC_2H_5$ $\;COOC_2H_5$	$C_{26}H_{50}GeO_8$	—	—	—	—	188
$(C_4H_9)_3GeCH{=}CHGe(C_4H_9)_3$	$C_{26}H_{56}Ge_2$	—	150/0.2	1.4798	0.9958	196, 202, 217, 222, 347
$[(C_2H_5)_3GeCH{=}CHCH_2Si(CH_3)(C_3H_7)]_2O$	$C_{26}H_{58}GeO$	—	199/0.26	1.4662	1.0449	375

Structure	Formula	mp / bp	n	d	Ref.
(C6H5)3Ge[indene]	C27H22Ge	126	—	—	199
(C6H5)3GeC6H4—CH2CH=CH2	C27H24Ge	91–92	—	—	325
(C6H5)3GeC6H4C(CH3)=CH2	C27H24Ge	95–96	—	—	325
(C6H5)3GeC(C6H5)=CH—CH3	C27H24Ge	89–91.5	—	—	27
(n-C8H17)3GeCH2CH=CH2	C27H56Ge	251–52/10	1.4684	0.9097	417
(C6H5)2Ge(C6H4CH=CH2-p)2	C28H24Ge	139–40	—	—	281
(C6H5)3GeC6H4—CH=CHC2H5	C28H26Ge	99–100	—	—	325
(C6H5)3GeC6H4C(CH3)=CH—CH3	C28H26Ge	102–3	—	—	325
C7H15Ge(CH=CH—C5H11)3	C28H54Ge	174/0.4	1.4798	0.9194	196, 202, 329
(C6H5)2Ge[indene]2	C30H22Ge	144–45	—	—	199
(C6H5)2Ge(C6H4CH2CH=CH2)2	C30H28Ge	Failed to melt up to 200	—	—	325
(C6H5)2Ge[C6H4—C(CH3)=CH2]2	C30H28Ge	117–18	—	—	325
(C6H5)3Ge[fluorene]	C31H24Ge	214	—	—	199
(C6H5—CH=CH)4Ge	C32H28Ge	198–200	—	—	17
(C6H5)2Ge[C6H4C(CH3)=CH—CH3]2	C32H32Ge	130–31	—	—	325
(C9H7)4Ge	C36H28Ge	196–98	—	—	199
[(C2H5)3GeCH=CH—C(CH3)Si(CH3)(C3H7)]2O, C4H9	C36H78GeO	205–7/0.2	1.4538	0.9816	375
[(C6H5)3Ge]2C=CH2	C38H32Ge2	145–46	—	—	280
[(C6H5CH=CH)3Ge]2	C48H42Ge2	230–32	—	—	17

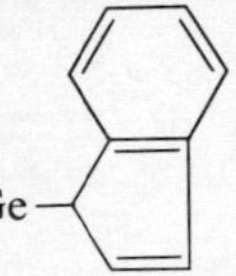

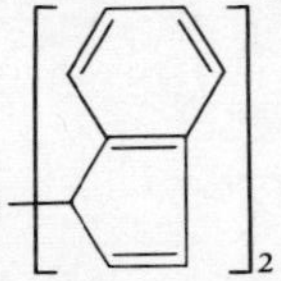

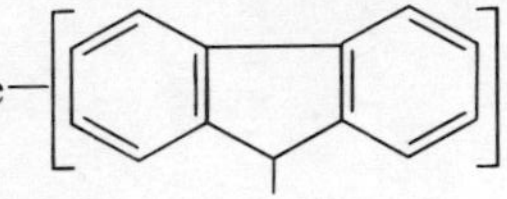

2-3   ACETYLENIC AND ENYNIC DERIVATIVES OF GERMANIUM

2-3-1   Preparations

2-3-1-1   Acetylenic Derivatives. Four types of true acetylenic derivatives are to be considered:

(a)   Derivatives of type $\diagdown\diagup GeC\equiv CH$

Tributylethynylgermane was isolated first through dehydrobromination of tributyl-1,2-dibromoethylgermane (222):

$$(n\text{-}C_4H_9)_3GeCHBrCH_2Br \xrightarrow{-2\,HBr} (n\text{-}C_4H_9)_3GeC\equiv CH$$

The elimination of the first HBr was accomplished by means of anhydrous diethylamine, and of the second by means of sodium amide.

Ethynyl derivatives are now more easily obtained by reacting trialkyl-halogermanes with the acetylenic Grignard reagent in tetrahydrofuran (157):

$$R_3GeBr + HC\equiv CMgBr \xrightarrow{THF} MgBr_2 + R_3GeC\equiv CH$$

This method allows several ethynyl groups to be attached to a single germanium atom (230, 395)

$$Cl_4Ge + 4\,HC\equiv CMgBr \rightarrow 2\,MgBr_2 + 2\,MgCl_2 + (HC\equiv C)_4Ge$$

An alternative is available: it consists in reacting an alkylgermanium halide with sodium acetylide in a suitable solvent (61). Using this procedure, however, a drop in efficiency occurs owing to side-reactions:

$$(n\text{-}C_4H_9)_2GeCl_2 \xrightarrow{NaC\equiv CH,\ THF}$$

$$(n\text{-}C_4H_9)_2Ge(C\equiv CH)_2 + HC\equiv CGe(n\text{-}C_4H_9)_2C\equiv CGe(Bu_2)C\equiv CH$$

$$+ \text{ polymers}$$

$$Cl_4Ge \xrightarrow{NaC\equiv CH,\ THF} (HC\equiv C)_4Ge + (HC\equiv C)_3GeC\equiv CGe(C\equiv CH)_3$$

Similar reactions had been observed to take place by Hartmann (137) when triphenyltin chloride is made to react with sodium acetylide:

$$(C_6H_5)_3SnCl + NaC\equiv CH \rightarrow (C_6H_5)_3SnC\equiv CH + NaCl$$

$$(C_6H_5)_3SnC\equiv CH + NaC\equiv CH \rightarrow (C_6H_5)_3SnC\equiv CNa + HC\equiv CH$$

$$(C_6H_5)_3SnC\equiv CNa + (C_6H_5)_3SnCl \rightarrow$$

$$ClNa + (C_6H_5)_3SnC\equiv CSn(C_6H_5)_3$$

$$2\,(C_6H_5)_3SnC\equiv CH \;\rightarrow$$

$$HC\equiv CH + (C_6H_5)_3SnC\equiv CSn(C_6H_5)_3$$

According to the work of Davidsohn and Henry (61), however, on account of the greater thermal stability of germanium ethynyl derivatives, the mechanism suggested by Hartmann seems to be hardly applicable, and it is more likely that the resulting dimers in the aforesaid reactions arise from reacting halogermane with the disodium acetylide which has been formed by disproportionation of sodium acetylide:

$$2\,HC\equiv CNa \;\rightarrow\; HC\equiv CH + NaC\equiv CNa$$

$$2\,(C_6H_5)_3GeCl + NaC\equiv CNa \;\rightarrow\; 2\,NaCl + (C_6H_5)_3GeC\equiv CGe(C_6H_5)_3$$

Steingross and Zeil (408) obtained trimethylethynylgermane by hydrolyzing the corresponding germanium lithium derivative:

$$((CH_3)_3GeC\equiv CCl + LiC_6H_5 \;\rightarrow\; (CH_3)_3GeC\equiv CLi + C_6H_5Cl$$

$$(CH_3)_3GeC\equiv CLi + H_2O \;\rightarrow\; (CH_3)_3GeC\equiv CH + LiOH$$

Nesmeyanov (278) observed that tributylethynylgermane is also formed by exchanging the hydrogen bonded to the germanium atom of tributylgermane for the ethynyl group of dibutylethynylstibine:

$$(C_4H_9)_3GeH + (C_4H_9)_2Sb-C\equiv CH \;\rightarrow$$

$$(C_4H_9)_3Ge-C\equiv CH + (C_4H_9)_2SbH$$

(b) Derivatives of type $\diagdown\!\!\diagup Ge CH_2C\equiv CH$

Propargylic derivatives are obtained by reaction of germanium halide with 2-propynyl Grignard reagent in ether solution (215, 215a, 217, 222, 229, 395). Owing to the propargylic transposition, (309, 436) the allenic isomer is formed simultaneously in the process:

$$R_3GeX + BrMgCH_2C\equiv CH \;\rightarrow\; MgBrX + \begin{cases} R_3GeCH_2C\equiv CH \\ R_3GeCH=C=CH_2 \end{cases}$$

In order to isolate the acetylenic derivative, the mixture is treated with ammoniacal cuprous chloride, and the alkyne is regenerated through decomposition of the cuprous alkynide by a concentrated potassium cyanide solution:

$$R_3GeCH_2C\equiv CH + Cu(NH_3)_nCl + (n\text{-}1)H_2O \;\rightarrow$$

$$R_3GeCH_2C\equiv CCu + NH_4Cl + (n\text{-}1)NH_4OH$$

$$R_3GeCH_2C{\equiv}CCu + 3\,KCN + H_2O \rightarrow$$

$$R_3GeCH_2C{\equiv}CH + K_2[Cu(CN)_3] + KOH$$

(c) Derivatives of type $\diagdown\!\!\!\diagup\!\!Ge(CH_2)_nC{\equiv}CH$

Long-chain acetylenic derivatives of germanium may be obtained by three different methods.

Dehydrobromination of the dibromo derivative of an ω-ethylenic alkyl germane. Unlike allyl and butenyl derivatives whose chains are liable to cleavage by bromine:

$$R_3Ge(CH_2)_nCH{=}CH_2 + Br_2 \rightarrow R_3GeBr + Br(CH_2)_nCH{=}CH_2$$

$$(n = 1 \text{ or } 2)$$

ω-ethylenic alkyl germanes add bromine quantitatively at low temperature:

$$R_3Ge(CH_2)_nCH{=}CH_2 + Br_2 \rightarrow R_3Ge(CH_2)_nCHBrCH_2Br$$

Dehydrobromination of these dibromides carried out by means of alcoholic potassium hydroxide, sodamide in liquid ammonia or in xylene leads to the expected alkynes:

$$R_3Ge(CH_2)_nCHBrCH_2Br + 2\,KOH \rightarrow$$

$$2\,KBr + R_3Ge(CH_2)_nC{\equiv}CH + H_2O$$

$$R_3Ge(CH_2)_nCHBrCH_2Br + 2\,NH_2Na \rightarrow$$

$$2\,NaBr + 2\,NH_3 + R_3Ge(CH_2)_nC{\equiv}CH$$

The sodamide process in xylene is the only one directly capable of yielding a chromatographically pure product. The other processes result in the formation of a small percentage of isomers which require further purification of the alkyne as previously described in the processing of propargylic derivatives.

Reaction of an aliphatic ω-bromo germanium derivative with sodium acetylide. It has been shown by Normant (287) that alkyl halides readily react with $HC{\equiv}CNa$ in hexamethylphosphotriamide to give the corresponding alkynes. This reaction has been extended to the ω-bromoalkylgermanium derivatives (230):

$$(C_2H_5)_3Ge(CH_2)_4Br + HC{\equiv}CNa \xrightarrow{\text{HMPT}}$$

$$(C_2H_5)_3Ge(CH_2)_4C{\equiv}CH + NaBr$$

A 60% product yield is achieved, but owing to the presence of a small amount of allenic isomer, the product requires further chemical purification.

Reaction of propargyl bromide with aliphatic germaorganomagnesium compounds. This reaction leads to the formation of a mixture of the acetylenic derivative and its allenic isomer:

$$R_3Ge(CH_2)_nMgBr + HC\equiv CCH_2Br \rightarrow$$

$$MgBr_2 + R_3Ge(CH_2)_nCH=C=CH_2$$

Chromatographically pure allenic derivatives can be obtained by extracting the acetylenic derivative with an alcoholic silver nitrate solution.

(d) Germoxy acetylenic derivatives

Two derivatives of the $R_3GeO—CH_2—C\equiv CH$ type have been obtained by means of a transalkoxylation reaction between methoxytrialkylgermane and propargylic alcohol (330). This reaction is quantitative:

$$R_3GeO—CH_3 + HC\equiv C—CH_2OH \rightarrow$$

$$CH_3OH + R_3GeO—CH_2—C\equiv CH \qquad (R = C_2H_5, C_4H_9)$$

2-3-1-2 Disubstituted Acetylenic Derivatives

(a) Derivatives of type $\diagdown\!\!\!\!-GeC\equiv C—R$ /

These derivatives are generally obtained by reaction of a germanium halide with an acetylenic Grignard reagent (84, 151, 215a, 222, 223, 408) or organolithium (17, 84, 86, 140, 292, 352) compound:

$$(n\text{-}C_4H_9)_3GeBr + BrMgC\equiv CC_4H_9 \rightarrow$$

$$MgBr_2 + (n\text{-}C_4H_9)_3GeC\equiv CC_4H_9$$

$$(C_2H_5)_3GeBr + BrMgC\equiv C—C\equiv CC_2H_5 \rightarrow$$

$$MgBr_2 + (C_2H_5)_3GeC\equiv C—C\equiv C—C_2H_5$$

$$GeCl_4 + 4C_6H_5C\equiv CLi \rightarrow 4LiCl + (C_6H_5C\equiv C)_4Ge$$

$$GeCl_4 + 4(CH_3)_3CC\equiv C—Li \rightarrow 4LiCl + [(CH_3)_3C—C\equiv C]_4Ge$$

Alkynylation of germanium halides can also be accomplished by complex acetylides of aluminium (125)

$$NaAl(C\equiv C—C_4H_9)_4 + 4(C_4H_9)_3GeCl \rightarrow$$

$$4(C_4H_9)_3GeC\equiv C—C_4H_9 + AlCl_3 + NaCl$$

Several types of alkylgermanes with an acetylenic chain bearing other functions in it have been prepared either from the Grignard reagent (84, 223, 373, 381, 383, 386, 387):

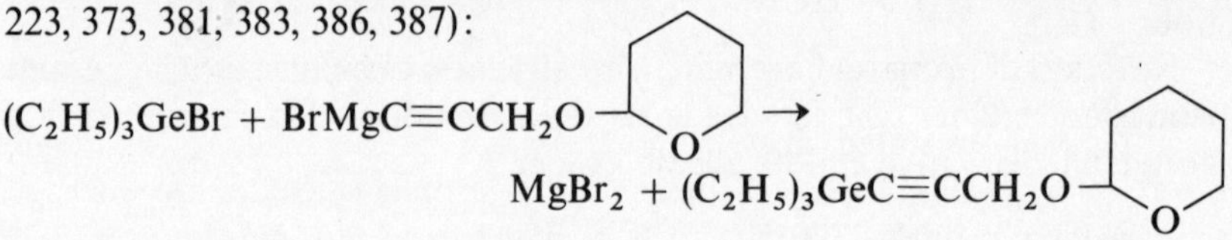

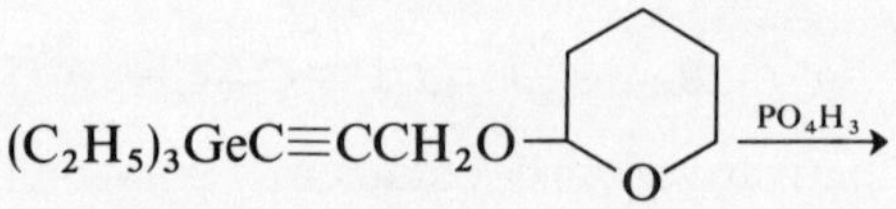

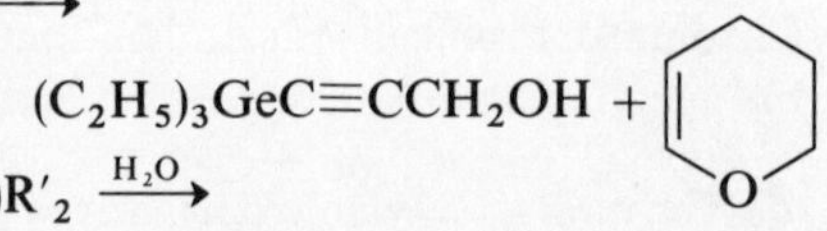

$$(C_2H_5)_3GeBr + BrMgC\equiv CCH_2O{-}\underset{O}{\bigcirc} \rightarrow$$

$$MgBr_2 + (C_2H_5)_3GeC\equiv CCH_2O{-}\underset{O}{\bigcirc}$$

$$(C_2H_5)_3GeC\equiv CCH_2O{-}\underset{O}{\bigcirc} \xrightarrow{PO_4H_3}$$

$$(C_2H_5)_3GeC\equiv CCH_2OH + \underset{O}{\bigcirc}$$

$$R_3GeBr + BrMgC\equiv C{-}C(OMgBr)R'_2 \xrightarrow{H_2O}$$

$$MgBr_2 + R_3GeC\equiv C(OH)R'_2$$

$$(C_2H_5)_3GeBr + BrMgC\equiv CCH_2OCH(CH_3)OC_4H_9 \rightarrow$$

$$MgBr_2 + (C_2H_5)_3GeC\equiv CCH_2OCH(CH_3)OC_4H_9$$

$$(C_2H_5)_3GeBr + BrMgC\equiv C{-}C_6H_4{-}X \rightarrow$$

$$MgBr_2 + (C_2H_5)_3GeC\equiv C{-}C_6H_4{-}X$$

organolithium derivatives (390):

$$(C_6H_5)_2P{-}C\equiv CLi + (CH_3)_3GeBr \rightarrow$$

$$LiBr + (C_6H_5)_2P{-}C\equiv C{-}Ge(CH_3)_3$$

or from germacetylenic Grignard reagents (223):

$$(C_2H_5)_3GeC\equiv CMgBr + C_3H_7CHO \xrightarrow{H_2O}$$

$$\tfrac{1}{2}MgBr_2 + \tfrac{1}{2}Mg(OH)_2 + (C_2H_5)_3GeC\equiv CCHOH{-}C_3H_7$$

$$(C_2H_5)_3GeC\equiv CMgBr + CH_2{=}CHCH_2Br \rightarrow$$

$$MgBr_2 + (C_2H_5)_3GeC\equiv CCH_2CH{=}CH_2$$

Shikhiev (377) obtained organogermanium diacetylenic alcohols containing isolated triple bonds by the reaction of the corresponding Iotsich reagents with 3-chloro-3-methyl-1-pentynyltriethylgermane:

$$BrMgOC(R_2)C\equiv CMgBr + (C_2H_5)_3GeC\equiv C{-}C(CH_3)(C_2H_5)Cl \xrightarrow{H_2O}$$

$$(C_2H_5)_3GeC\equiv C{-}C(CH_3)(C_2H_5)C\equiv C{-}CR_2OH$$

Disubstituted alkynylgermanium compounds can be obtained by another route, using the reaction of trialkylgermylamine with an acetylenic derivative (333):

$$(C_4H_9)_3GeN(CH_3)_2 + HC{\equiv}C{-}C_6H_5 \rightarrow$$

$$(C_4H_9)_3GeC{\equiv}C_6H_5 + (CH_3)_2NH$$

Lastly, a disubstituted acetylenic derivative of germanium has been obtained by cleaving a tin derivative with germanium tetrachloride at a temperature of 200°C (370):

$$(C_2H_5)_3SnC{\equiv}C{-}CH_3 + GeCl_4 \rightarrow (C_2H_5)_3SnCl + Cl_3GeC{\equiv}C{-}CH_3$$

(b) Derivatives of type $\diagdown\!\!\!-Ge(CH_2)_nC{\equiv}CR$

Acetylenic germano Grignard reagent compounds resulting from the reaction of acetylenic derivatives of germanium with ethylmagnesium bromide react readily with aldehydes, ketones, and reactive halogen derivatives (230):

$$(C_2H_5)_3Ge(CH_2)_4C{\equiv}CMgBr + C_3H_7CHO \xrightarrow{OH_2}$$

$$\tfrac{1}{2}MgBr_2 + \tfrac{1}{2}Mg(OH)_2 + (C_2H_5)_3Ge(CH_2)_4C{\equiv}CCHOHC_3H_7$$

$$(C_2H_5)_3Ge(CH_2)_5C{\equiv}CMgBr + CH_3COCH_3 \xrightarrow{OH_2}$$

$$\tfrac{1}{2}MgBr_2 + \tfrac{1}{2}Mg(OH)_2 + (C_2H_5)_3Ge(CH_2)_5C{\equiv}CCOH(CH_3)_2$$

Sodium derivatives of germanoalkynes also lend themselves to the synthesis of disubstituted derivatives:

$$(C_2H_5)_3Ge(CH_2)_5CHBrCH_2Br + 3\,NaNH_2 \rightarrow$$

$$(C_2H_5)_3Ge(CH_2)_5C{\equiv}CNa + 2\,NaBr + 3\,NH_3$$

$$(C_2H_5)_3Ge(CH_2)_5C{\equiv}CNa + CH_3I \rightarrow$$

$$NaI + (C_2H_5)_3Ge(CH_2)_5C{\equiv}CCH_3$$

(c) Germoxy disubstituted acetylenic derivatives (330)

Acetylenic alcohols are dehydrocondensed with trialkylgermanes in the presence of reduced copper in rather poor yields. Moreover, partial catalytic hydrogenation of the triple bond of the resulting alkynoxyalkyl-germane can hardly be avoided. It is much more advantageous to prepare these compounds by way of alcoholysis of alkoxytributylgermane:

$$(C_4H_9)_3GeOCH_3 + CH_3(CH_2)_3C{\equiv}C{-}CH_2CH_2OH \rightarrow$$

$$(C_4H_9)_3GeOCH_2CH_2C{\equiv}C(CH_2)_3CH_3 + CH_3OH$$

(d) Halogen-substituted acetylenic germanium compounds

Steingross and Zeil (407) have synthesized chloroethynyltrimethyl-germane according to the reactions:

$$trans\text{-}ClCH{=}CHCl + 2\,CH_3Li \rightarrow LiC{\equiv}CCl + LiCl + 2\,CH_4$$

$$(CH_3)_3GeBr + LiC{\equiv}CCl \rightarrow LiBr + (CH_3)_3GeC{\equiv}CCl$$

A bromoacetylenic compound is also obtained by adding bromine to germa-alkynyl Grignard reagent at low temperature (225):

$$(C_2H_5)_3Ge(CH_2)_3C{\equiv}CMgBr + Br_2 \xrightarrow{-80°C}$$

$$(C_2H_5)_3Ge(CH_2)_3C{\equiv}CBr + MgBr_2$$

2-3-2 Physical properties and characteristics

2-3-2-1 Terminal Acetylenic Derivatives. With the exception of tetra-ethynylgermane which is a crystallized colourless solid, melting at about 93° and possessing highly explosive properties on impact or friction, terminal acetylenic aliphatic derivatives containing one germanium atom are stable colourless liquids which can be distilled without decomposition.

Their infrared spectra are characterized by absorption bands in the ranges of $2100\,\text{cm}^{-1}$ and $3300\,\text{cm}^{-1}$ corresponding to the valency vibrations $v_{C{\equiv}C}$ and $v_{{\equiv}C-H}$ respectively.

The location of the germanium atom relative to the triple carbon–carbon bond affects the frequency of valency vibrations $v_{C{\equiv}C}$ and $v_{{\equiv}C-H}$. The displacement is at a maximum when germanium is in the α position (61, 217, 408).

	FREQUENCY $v_{{\equiv}CH}$	FREQUENCY $v_{C{\equiv}C}$
$(HC{\equiv}C)_4Ge$	3303*	2059
α derivatives	3290*	2027
$\quad\left\{\begin{array}{l}(CH_3)_3GeC{\equiv}CH\\(C_2H_5)_3GeC{\equiv}CH\\(C_4H_9)_3GeC{\equiv}CH\end{array}\right.$	3276	
β derivatives	3310*	2110
$\quad(C_2H_5)_3GeCH_2C{\equiv}CH$	3304	
Terminal alkynes	3320*	2119–2120
$\quad C_4H_9C{\equiv}CH$	3315	
$\quad C_5H_{11}C{\equiv}CH$		

* In CCl_4 solution.

During an NMR investigation of $R_3MC \equiv CH$ type organometallic derivatives (391) Simonnin noted that the chemical shift of the acetylenic proton depends not only on the nature of the heteroatom, but also on the nature of the radicals linked to this heteroatom. For instance, when $M = Ge$, δ varies from 2.10 ppm when $R = C_2H_5$ to 2.51 when $R = C_6H_5$. In the aliphatic series, $\delta(\equiv CH)$ for trimethylethynylsilane and triethylethynylgermane occurs at lower fields than for *tert*-butylacetylene: this unexpected fact might be explained by the p_π–d_π interaction between the vacant silicon or germanium d orbital and the π electrons of the triple bond.

2-3-2-2 Disubstituted Acetylenic Derivatives. Disubstituted aliphatic derivatives are usually either liquids or low melting point substances: aromatic derivatives are often obtained in a crystalline state. In their infrared spectra, a drop in the frequency of valency vibration $v_{C \equiv C}$ is also observed.

		FREQUENCY $v_{C \equiv C}$
$\alpha\alpha$ derivatives $-GeC \equiv CGe-$	(Raman)	2095
α derivatives $R_3GeC \equiv C-R$	one band	2165–2170
β derivatives $R_3GeCH_2C \equiv CR$	two bands	2235 2225
disubstituted alkynes $RC \equiv CR'$	two bands in the range from	2230 2300

The infrared spectrum of the $(C_2H_5)_3Ge-C \equiv C-C \equiv C-C_2H_5$ derivative shows the existence of three bands in the range of 2100 cm^{-1}. From the foregoing the 2237–2223 cm^{-1} doublet may be traced on the triple bond in the γ position relative to the germanium atom, whereas the 2098 cm^{-1} band may depend on the triple bond in the α position: there is a general decrease in the frequencies owing to conjugation.

From investigations carried out with about ten acetylenic aliphatic derivatives of germanium (218), it has been found that the integrated intensity of the absorption band $v_{C \equiv C}$ is considerably higher than that of acetylenic derivatives of carbon and seems to depend on the electrical disymmetry existing between the groups attached to the $-C \equiv C-$ system. With the $R_3Ge-C \equiv C-R'$ type of compounds, it is a decreasing linear function of the Taft coefficients σ_R^*, applied to R' substituents.

With disubstituted derivatives of the $R_3Ge-C\equiv C-GeR'_3$ type, intensity of the band $\nu_{C\equiv C}$ is practically zero owing to the molecule's symmetrical arrangement. UV and IR spectra of several acetylenic compounds of Group IV B elements were investigated by Masson and coworkers (215a); a bathochromic effect and increasing intensity were observed from carbon to lead. Spectra of $(CH_3)_3MC\equiv C-Cl$ compounds (M = Si, Ge, Sn and Pb) were investigated by Steingross and Zeil (407) and the IR and Raman frequencies interpreted in terms of the normal vibrations. Carbon-13 proton couplings of two acetylenic germanium derivatives have been measured by Simonnin (392)

$$(C_2H_5)_3Ge-C\equiv C-CH_3 \qquad J_{13C-H} = 131\,Hz \pm 1$$

$$(C_2H_5)_3Ge-C\equiv C-H \qquad J_{\equiv 13C-H} = 236\,Hz \pm 2;$$

$$J_{13C\equiv C-H} = 42\,Hz \pm 0.5$$

2-3-3   Chemical properties

Acetylenic germanium derivatives can undergo addition reactions to the carbon–carbon triple bond, cleavage reactions of the carbon–germanium bond, and substitution reactions of the acetylenic hydrogen (in the case of terminal alkynes).

2-3-3-1   Addition Reactions.

Hydrogen: Acetylenic germanium derivatives can undergo hydrogenation in the presence of a catalyst. The hydrogenation rate and the nature of the resulting products depend on the catalyst used and the position of the germanium atom relative to the triple bond (229). As a rule, the presence of a germanium atom in proximity to a triple bond has a hindering effect on catalytic hydrogenation: reluctance to react is greater when the germanium atom is situated in the α position relative to the $-C\equiv C-$ bond than when it is placed in the β position.

For any given position of the heteroatom relative to the triple bond, hydrogenation occurs more readily with a terminal acetylenic triple bond than with a disubstituted one.

Terminal acetylenic derivatives $(C_2H_5)_3GeC\equiv CH$ and

$$(C_2H_5)_3GeCH_2C\equiv CH$$

can be completely hydrogenated at a temperature of 20°C and under ordinary pressure in the presence of Raney nickel.

Hydrogenation of disubstituted derivatives of the $R_3GeC\equiv C-R'$ type encounters greater difficulty; the nature of the R' grouping has an influence

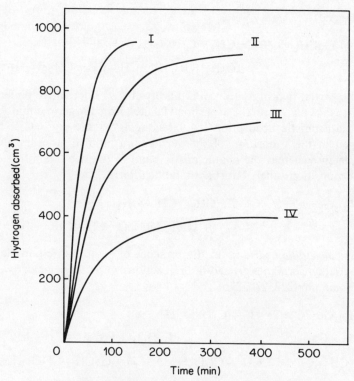

FIG. 2-5 Hydrogenation of acetylenic germanium compounds—
I: $(C_2H_5)_3GeC\equiv CH$; II: $(C_2H_5)_3GeC\equiv C-CH_2OH$;
III: $(C_4H_9)_3GeC\equiv C-C_3H_7$; IV: $(C_4H_9)_3GeC\equiv CGe(C_2H_5)_3$

on the hydrogenation rate (Figure 2-5). Among these derivatives, 1-tri-butylgermyl-2-triethylgermylacetylene, in which the disubstituted triple bond is in the α position relative to any two germanium atoms, possesses the lowest hydrogenation rate.

It does not seem possible to stop hydrogenation at the ethylenic stage, in the presence of Raney nickel; by using an inactivated palladium catalyst (Lindlar catalyst (207)) triethylethynylgermane and triethylpropargyl-germane have been successfully converted to triethylvinyl- and triethyl-allylgermane respectively:

$$(C_2H_5)_3Ge(CH_2)_n-C\equiv CH + H_2 \xrightarrow{\text{Pd/Pb}}$$

$$(C_2H_5)_3Ge(CH_2)_nCH=CH_2 \qquad (n = 0, 1)$$

The partial hydrogenation of 6-triethylgermyl-1-hexene-4-yne leads to the formation of 6-triethylgermyl-1,4-hexadiene, the double bond being unchanged:

$$(C_2H_5)_3GeCH_2-C{\equiv}C-CH_2-CH{=}CH_2 + H_2 \xrightarrow{Pd/Pb}$$

$$(C_2H_5)_3GeCH_2-CH{=}CH-CH_2-CH{=}CH_2$$

Lastly, partial hydrogenation of a diacetylenic α, γ derivative preferably tends to take place with the triple bond farthest from the germanium atom; the characteristic absorption band belonging to the $-C{\equiv}C-$ bond situated in the γ position of the germanium atom disappears almost completely, whereas the characteristic band of the triple bond situated in the α position suffers but slight modification:

$$(C_2H_5)_3Ge-C{\equiv}C-C{\equiv}C-CH_2-CH_3 + H_2 \xrightarrow{Pd/Pb}$$

$$(C_2H_5)_3GeC{\equiv}C-CH{=}CH-CH_2-CH_3$$

Organometallic hydrides: In the presence of a catalyst (chloroplatinic acid) trialkyl germanes give addition reactions with acetylenic derivatives of germanium (222, 230, 276):

$$(C_4H_9)_3Ge-C{\equiv}C-H + (C_4H_9)_3GeH \rightarrow$$

$$(C_4H_9)_3Ge-CH{=}CH-Ge(C_4H_9)_3$$

$$(CH_3)_3Ge-C{\equiv}CH + Cl_3GeH \rightarrow (CH_3)_3GeCH{=}CHGeCl_3$$

$$(CH_3)_3Ge-C{\equiv}CH + Cl_3SiH \rightarrow (CH_3)_3GeCH{=}CHSiCl_3$$

$$(C_4H_9)_3GeC{\equiv}CH + (C_4H_9)_3SnH \rightarrow (C_4H_9)_3GeCH{=}CHSn(C_4H_9)_3$$

Diisobutylaluminium hydride adds to trialkylgermanium alkynes giving, after hydrolysis, the corresponding olefins. In hydrocarbon media the *trans*-isomer is obtained while the *cis*-isomer is formed in the presence of one equivalent of strong donor, e.g. N-methylpyrolidine (92).

Bromine: Acetylenic derivatives of germanium react instantly with bromine, at low temperatures; according to the operating conditions and the position of the triple bond with respect to the germanium atom, one can observe either an addition reaction to the triple bond or a cleavage reaction of the Ge$-$C bond (acetylenic chain). Tetraethynylgermanium adds four bromine molecules and leads to a crystallized dibromovinylic derivative (NMR: singlet at 7.52 ppm) (61, 230):

$$(H-C{\equiv}C)_4Ge + 4 Br_2 \rightarrow (H-CBr{=}CBr)_4Ge$$

Under the same experimental conditions, bromine is found to cleave propargylic derivatives.

Water: In the presence of mercuric sulphate, it is possible to convert an acetylenic derivative into the corresponding ketone (230):

$$(C_2H_5)_3Ge(CH_2)_3-C\equiv CH + H_2O \rightarrow (C_2H_5)_3Ge(CH_2)_3-CO-CH_3$$

Diazo derivatives: Reacting diazomethane with diphenyldiethynyl-germane, at room temperature in ether, gives rise to a dipyrazolylgermane (61):

$$(C_6H_5)_2Ge(C\equiv CH)_2 + 2\ CH_2N_2 \rightarrow (C_6H_5)_2Ge\left[\begin{array}{c} -C\text{------}CH \\ \parallel \qquad\quad \parallel \\ HC \qquad\quad N \\ \diagdown \qquad \diagup \\ NH \end{array}\right]_2$$

Dihalocarbenes: 1-Triethylgermyl-2-trifluoromethyl acetylene reacts at 140° with difluorocarbene (generated *in situ* from trimethyl trifluoromethyl-stannane) (53):

$$(C_2H_5)_3Ge-C\equiv C-CF_3 + (CH_3)_3SnCF_3 \rightarrow$$

$$(C_2H_5)_3Ge-C\!\!=\!\!=\!\!=\!\!C-CF_3 + (CH_3)_3SnF$$
$$\diagdown \quad \diagup$$
$$CF_2$$

2-3-3-2   Cleavage Reactions of the Germanium–Carbon Bond (Unsaturated Chain). When bromine is allowed to react at 20° with tri-*n*-butylpentynyl-germane, a cleavage of the Ge—C (acetylenic chain) bond occurs and tri-*n*-butylbromogermane is formed.

Aliphatic derivatives of germanium containing either ethynyl or propargyl chains are easily cleaved in the cold by trichloracetic acid:

$$CCl_3COOH + R'_3Ge-C\equiv CH \rightarrow CCl_3COOGeR'_3 + HC\equiv CH$$

$$CCl_3COOH + R'_3GeCH_2-C\equiv CH \rightarrow CCl_3COOGeR'_3 + CH_3C\equiv CH$$

With regard to kinetics (Figure 2-6), one will notice that this acid accomplishes the cleavage of the propargyl chain much more rapidly than that of the ethynyl chain derivative; the same amount of difference in reactivity can be observed when this acid is reacted with vinyl and allyl chain derivatives. The foregoing results emphasize the lability of the chains linked with a germanium atom that have an unsaturated carbon–carbon bond in the β position relative to the heteroatom (222). Diethyldihex-1-ynylgermane is unaffected by hot and cold water and even by hot concentrated hydrochloric acid (151).

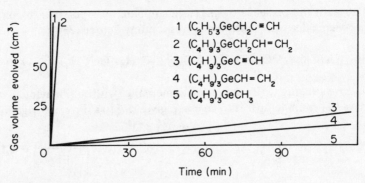

FIG. 2-6   Action of trichloracetic acid on some unsaturated mixed alkyl-germanes

Bott, Eaborn and Walton (22) measured the rates of cleavage of a series of $X-C_6H_4C\equiv C-Ge(C_2H_5)_3$ compounds by a mixture of methanol and aqueous perchloric acid at 29°C:

$$XC_6H_4C\equiv C-Ge(C_2H_5)_3 + YOH \xrightarrow{\text{H}^+}$$

$$XC_6H_4C\equiv CH + (C_2H_5)_3GeOY \qquad Y = H \text{ or } CH_3$$

They found that the silicon analogue required markedly more concentrated acid than the germyl compound, while the stannyl compound was cleaved even by neutral aqueous methanol.

The order of ease of cleavage is thus Si < Ge ≪ Sn, as in acid cleavage of $ArMR_3$ (78) (M = Si, Ge, Sn). In contrast to benzyl–germanium bonds, phenylethynyl–germanium bonds are readily cleaved in aqueous methanolic alkali (86):

$$YC_6H_4C\equiv C-GeR_3 + HOZ \xrightarrow{\text{OZ}^-} YC_6H_4C\equiv CH + R_3GeOZ$$

Phenylethynyltrimethylgermane and phenylethynyltriethylgermane are less reactive than the corresponding silicon compounds, but corresponding stannanes react more easily, even with neutral aqueous methanol at room temperature; the order of ease of cleavage Ge < Si ≪ Sn is different from that of acid cleavage of the same compounds.

Eaborn's study shows that in alkali cleavage of the

$$(C_2H_5)_3GeC\equiv C-C_6H_4Y$$

compounds, the reaction is facilitated by electron withdrawal and retarded by electron release from the substituents Y, and that the effects of the substituents Y can be correlated approximately with their Hammett σ constants.

2-3-3-3 Substitution Reactions. Terminal acetylenic derivatives of germanium give rise to substitution reactions of the acetylenic hydrogen:

Silver salts yield a white alkynide precipitate. This reaction, which is quantitative, allows any trace of terminal acetylenic derivative to be removed from a mixture of isomers.

The reaction involving terminal acetylenic alkylgermanes with ammoniacal cuprous chloride is slower and does not seem to be quantitative. However, owing to the ready regeneration of the alkyne from the cuprous derivative (through the action of a concentrated solution of potassium cyanide) this reaction is used to achieve the isolation of a terminal germanoalkyne from a mixture of isomers.

Grignard reagents react as expected with elimination of alkane:

$$R_3Ge(CH_2)_n\!-\!C\!\equiv\!CH + R'MgX \;\rightarrow\; R'H + R_3Ge(CH_2)_n\!-\!C\!\equiv\!C\!-\!MgX$$

The resulting reactive acetylenic germano Grignard reagents may serve as starting materials for the synthesis of numerous products.

Eaborn (82) has measured the rates of detritiation at 25°C in a mixture of methanol (1 vol.) and an aqueous buffer solution (4 vol., pH 8.05) of some $XC\!\equiv\!C^3H$ compounds in which $X = R_3M$ or R_3MCH_2, with $M = C$, Si and Ge. The results of these experiments show that the electropositivity of silicon is slightly less than that of germanium, and markedly greater than that of carbon.

In the compounds of R_3M type, silicon and germanium derivatives are all several times as reactive as the analogous carbon compounds, whereas if only inductive effects operated the Me_3C compound would be the most reactive. The high reactivity of the silicon and germanium compounds is to be attributed to $p_\pi\!-\!d_\pi$ interaction between the alkynyl group and the metalloid atom; an electron-withdrawing effect which stabilizes the negatively-charged transition state more than the initial state.

2-3-4 Acetylenic GeR₄ derivatives
Enynic GeR₄ derivatives

Tables 2-4 and 2-5 showing some physical properties of acetylenic GeR_4 derivatives and enynic GeR_4 derivatives respectively appear on the following pages.

Table 2-4
Acetylenic GeR₄ Derivatives

COMPOUND	EMPIRICAL FORMULA	M.P. (°C)	B.P. (°C/mm)	n_D^{20}	d_4^{20}	REFERENCES
(CH₃)₃GeC≡C—Cl	C₅H₉GeCl	—	51/65	1.4525	—	407
(CH₃)₃GeC≡C—D	C₅H₉GeD	—	—	—	—	408
(CH₃)₃GeC≡C≡CH	C₅H₁₀Ge	—	72.5	1.4180	1.0391	258, 408
(CH₃)₃GeC≡CCF₃	C₆H₉GeF₃	—	—	—	—	53, 55
(CH₃)₃GeC≡C—CH₃	C₆H₁₂Ge	—	—	—	—	408
Ge(C≡CH)₄	C₈H₄Ge	91–92 94	—	—	—	61, 230
(CH₃)₂Ge(C≡CCF₃)₂	C₈H₆GeF₆	—	—	—	—	55
(C₂H₅)₃GeC≡CH	C₈H₁₆Ge	—	70–71/65	1.4485 1.4450 (25°)	1.0241	82, 189, 217, 222, 229, 287, 391
(CH₃)₃GeC≡C—COH(CH₃)₂	C₈H₁₆GeO	30	72–73/7	—	—	386, 387
(CH₃)₃GeC≡CSi(CH₃)₃	C₈H₁₈GeSi	25	150–51	1.4405 (28°)	0.9642 (28°)	93, 258
(CH₃)₃GeC≡C—Sn(CH₃)₃	C₈H₁₈GeSn	41–44	47/1	—	—	93
(CH₃)₃GeC≡CGe(CH₃)₃	C₈H₁₈Ge₂	35	163–64	—	—	258
(C₂H₅)₃GeC≡C—CF₃	C₉H₁₅GeF₃	—	98–100/105	—	—	53
(C₂H₅)₃GeCH₂C≡CH	C₉H₁₈Ge	—	76–77/18 100/60 89/30	1.4669 1.4664	1.0326	82, 215, 217, 218, 222, 229
(C₂H₅)₃GeC≡C—CH₃	C₉H₁₈Ge	—	72/15	1.4661	—	215a, 391
(C₂H₅)₃GeC≡CCH₂OH	C₉H₁₈GeO	—	81/11 94/3 112/9	1.4825 1.4822	1.1172 1.1150	73, 189, 217, 223, 229, 381, 437

(CH$_3$)$_3$GeC≡C—COH(C$_2$H$_5$)CH$_3$	C$_9$H$_{18}$GeO	—	67–69/4	1.4355	1.0150	383
(C$_2$H$_5$)$_2$Ge(CH$_2$C≡CH)$_2$	C$_{10}$H$_{16}$Ge	—	71–73/4	—	—	395
(C$_2$H$_5$)$_3$GeC≡C—OC$_2$H$_5$	C$_{10}$H$_{20}$GeO	—	60/1	1.4632	1.0765	225
(CH$_3$)$_3$GeC≡C—C$_6$H$_5$	C$_{11}$H$_{14}$Ge	—	70/1.5	1.5429 (25°)	—	86, 264
		—	64/1	1.5429	1.1474	
(C$_2$H$_5$)$_3$GeC≡C—CH$_2$OCOCH$_3$	C$_{11}$H$_{20}$GeO$_2$	—	87–88/2	1.4692	1.1052	254
(C$_2$H$_5$)$_3$GeC≡C—C(CH$_3$)$_2$Cl	C$_{11}$H$_{21}$GeCl	—	65–67/2	1.4533	1.0484	385
(C$_2$H$_5$)$_3$Ge(CH$_2$)$_3$—C≡C—Br	C$_{11}$H$_{21}$GeBr	—	89.5/0.9	1.4956	1.2810	225
(C$_2$H$_5$)$_3$Ge(CH$_2$)$_3$C≡CH	C$_{11}$H$_{22}$Ge	—	107/20	1.4643	1.0060	230
(C$_2$H$_5$)$_3$GeC≡C—COH(CH$_3$)$_2$	C$_{11}$H$_{22}$GeO	—	90–92/5	1.4670	1.0670	189, 217, 223, 383
(C$_2$H$_5$)$_3$GeC≡C—CSi(CH$_3$)$_3$	C$_{11}$H$_{24}$Ge	—	80/1	1.4686	1.0183	264
(C$_6$H$_5$)(C$_2$H$_5$)Ge(C≡CH)$_2$	C$_{12}$H$_{12}$Ge	—	210–11	1.4551	0.9709	179, 395
Ge(CH$_2$C≡CH)$_4$	C$_{12}$H$_{12}$Ge	—	133–36/28	—	—	215
Ge(C≡C—CH$_3$)$_4$	C$_{12}$H$_{12}$Ge	168	115/2	—	—	215a
(C$_4$H$_9$)$_2$Ge(C≡CH)$_2$	C$_{12}$H$_{20}$Ge	—	46/0.5	—	—	61
(C$_2$H$_5$)$_3$GeC≡C—C≡C—C$_2$H$_5$	C$_{12}$H$_{20}$Ge	—	118/5	1.5116	1.0320	189, 217, 218, 223, 229
(CH$_3$)$_2$Ge[C≡CCOH(CH$_3$)$_2$]$_2$	C$_{12}$H$_{20}$GeO$_2$	97	—	—	—	387
(C$_2$H$_5$)$_3$GeC≡CC(CH$_3$)(C$_2$H$_5$)Cl	C$_{12}$H$_{23}$GeCl	—	70–71/2	1.4560	1.0600	377
(C$_2$H$_5$)$_3$Ge(CH$_2$)$_4$C≡CH	C$_{12}$H$_{24}$Ge	—	113/15	1.4644	0.9943	230
(C$_2$H$_5$)$_3$GeC≡C—CHOH—C$_3$H$_7$	C$_{12}$H$_{24}$GeO	—	106/1	1.4722	1.0355	189, 217, 223, 218, 381
(C$_2$H$_5$)$_3$GeC≡C—COH(C$_2$H$_5$)(CH$_3$)	C$_{12}$H$_{24}$GeO	—	95–97/3	1.4690	1.0350	383
(C$_2$H$_5$)$_3$GeC≡CCH$_2$OCOCH$_2$COCH$_3$	C$_{13}$H$_{22}$GeO$_3$	—	136–37/4	1.4805	1.1397	255
(C$_2$H$_5$)$_3$Ge(CH$_2$)$_5$C≡CH	C$_{13}$H$_{26}$Ge	—	98/1.9	1.4647	0.9876	230
(C$_2$H$_5$)$_3$GeC≡C—COH(C$_2$H$_5$)$_2$	C$_{13}$H$_{26}$GeO	—	105–7/2	1.4713	1.0380	383
(HC≡C)$_3$GeC≡CGe(C≡CH)$_3$	C$_{14}$H$_6$Ge$_2$	143	—	—	—	61
(C$_2$H$_5$)$_3$GeC≡CC$_6$H$_4$Br-o	C$_{14}$H$_{19}$GeBr	—	138/1.5	1.5570 (25°)	—	22, 84, 86
(C$_2$H$_5$)$_3$GeC≡CC$_6$H$_4$Br-m	C$_{14}$H$_{19}$GeBr	—	116/0.3	1.5580 (25°)	—	22, 84, 86
(C$_2$H$_5$)$_3$GeC≡CC$_6$H$_4$Br-p	C$_{14}$H$_{19}$GeBr	—	139/1.5	1.5612 (25°)	—	22, 84, 86

Table 2-4—continued

COMPOUND	EMPIRICAL FORMULA	M.P. (°C)	B.P. (°C/mm)	n_D^{20}	d_4^{20}	REFERENCES
$(C_2H_5)_3GeC{\equiv}CC_6H_4Cl\text{-}o$	$C_{14}H_{19}GeCl$	—	138/2.8	1.5451 (25°)	—	22, 84, 86
$(C_2H_5)_3GeC{\equiv}CC_6H_4Cl\text{-}m$	$C_{14}H_{19}GeCl$	—	128/1.2	1.5430 (25°)	—	22, 84, 86
$(C_2H_5)_3GeC{\equiv}CC_6H_4Cl\text{-}p$	$C_{14}H_{19}GeCl$	—	118/0.9	1.5466 (25°)	—	22, 84, 86
$(C_2H_5)_3GeC{\equiv}CC_6H_4F\text{-}p$	$C_{14}H_{19}GeF$	—	106/1.6	1.5188 (25°)	—	22, 84, 86
$(C_2H_5)_3GeC{\equiv}CC_6H_4I\text{-}p$	$C_{14}H_{19}GeI$	—	160/1.8	1.5882 (25°)	—	22, 84, 86
$(C_2H_5)_3GeC{\equiv}CC_6H_5$	$C_{14}H_{20}Ge$	—	139–40/2; 115/1.5; 132/7	1.5321 (19°); 1.5360 (25°); 1.5280	1.0981	22, 84, 86, 142, 264
$(C_2H_5)_2Ge[C{\equiv}C{-}COH(CH_3)_2]_2$	$C_{14}H_{24}GeO_2$	56–57	—	—	—	376, 383
$(CH_3)_2Ge[C{\equiv}C{-}COH(CH_3)C_2H_5]_2$	$C_{14}H_{24}GeO_2$	82–83	—	—	—	383
$(C_4H_9)_3GeC{\equiv}CH$	$C_{14}H_{28}Ge$	—	100/1.7; 105–6/2.5; 112/8	1.4581	0.9602	189, 217, 218, 222, 276, 278
$(C_2H_5)_3GeC{\equiv}CCH_2CH_2(C_4H_9)$	$C_{14}H_{28}Ge$	—	90–91/1	1.4570	0.9374	401
$(C_2H_5)_3Ge(CH_2)_5C{\equiv}C{-}CH_3$	$C_{14}H_{28}Ge$	—	116/1.6	1.4721	0.9688	230
$(C_2H_5)_3GeC{\equiv}C{-}COH(CH_3){-}C(CH_3)_3$	$C_{14}H_{28}GeO$	—	90–91/2	1.4697	0.9787	379
$(C_2H_5)_3GeC{\equiv}CGe(C_2H_5)_3$	$C_{14}H_{30}Ge_2$	—	50/14	1.4630	1.1100	136
$(C_2H_5)_3GeC{\equiv}CC_6H_4CF_3\text{-}m$	$C_{15}H_{19}GeF_3$	—	112/2.6	1.4898 (25°)	—	22, 84, 86
$(C_2H_5)_3GeC{\equiv}CC_6H_4CH_3\text{-}o$	$C_{15}H_{22}Ge$	—	119/1.9	1.5330 (25°)	—	22, 84, 86
$(C_2H_5)_3GeC{\equiv}CC_6H_4CH_3\text{-}m$	$C_{15}H_{22}Ge$	—	115/1.2	1.5336 (25°)	—	22, 84, 86
$(C_2H_5)_3GeC{\equiv}CC_6H_4CH_3\text{-}p$	$C_{15}H_{22}Ge$	—	147/5.8	1.5375 (25°)	—	22, 84, 86
$(C_2H_5)_3GeC{\equiv}CC_6H_4OCH_3\text{-}o$	$C_{15}H_{22}GeO$	—	138/2.0	1.5300 (25°)	—	22, 84
$(C_2H_5)_3GeC{\equiv}CC_6H_4OCH_3\text{-}m$	$C_{15}H_{22}GeO$	—	140/2.0	1.5390 (25°)	—	22, 84, 86
$(C_2H_5)_3GeC{\equiv}CC_6H_4OCH_3\text{-}p$	$C_{15}H_{22}GeO$	—	135/1.2	1.5450 (25°)	—	22, 84, 86
$(C_2H_5)_3GeC{\equiv}CCH_2CH(CH_3)C_4H_9$	$C_{15}H_{30}Ge$	—	95–97/0.5	1.4738	0.9675	401

Compound	Empirical formula	M.p.	B.p./mm	n_D	d	References
$(C_2H_5)_3GeC\equiv C-CH_2-O-CH(CH_3)-O-C_4H_9$	—	—	146–48/4	1.4519	1.1130	381, 382, 438
$(CH_3)_3GeC\equiv CC(C_2H_5)(CH_3)OCH(CH_3)OC_4H_9$	$C_{15}H_{30}GeO_2$	—	123–24/12	1.431	0.9640	383
$(C_2H_5)_3GeOCH_2C\equiv CGe(C_2H_5)_3$	$C_{15}H_{32}Ge_2O$	49	173/15	1.4789	1.1351	330
$(C_6H_5)_2Ge(C\equiv CH)_2$	$C_{16}H_{12}Ge$	43–45	—	—	—	61, 395
$(C_2H_5)_3GeC\equiv CC_6H_5,2,3-(CH_3)_2$	$C_{16}H_{24}Ge$	—	160/5	1.5291 (25°)	0.981 (21°)	22, 84, 86
$(CH_3)Ge[C\equiv C-COH(CH_3)_2]_3$	$C_{16}H_{24}GeO_3$	208	—	—	—	386, 387
$(CH_3)_2Ge[C\equiv CC(CH_3)_2OCOCH_3]_2$	$C_{16}H_{24}GeO_4$	54	132/1.8	—	—	387
$(C_2H_5)_2Ge(C\equiv C-C_4H_9)_2$	$C_{16}H_{28}Ge$	—	142–43/7	1.4729 (21°)	1.0005	151
$(C_2H_5)_3GeC\equiv CC(CH_3)_2C\equiv CC(CH_3)_2OH$	$C_{16}H_{28}GeO$	45–47	98/0.15	—	—	385
$(C_2H_5)_2Ge[C\equiv C-COH(CH_3)C_2H_5]_2$	$C_{16}H_{28}GeO_2$	—	99–100/3	1.4700	—	383
$(C_2H_5)_3Ge(CH_2)_4C\equiv C-CHOHC_3H_7$	$C_{16}H_{32}GeO$	—	113/0.1	1.4794	1.0148	230
$(C_2H_5)_3Ge(CH_2)_5C\equiv C-COH(CH_3)_2$	$C_{16}H_{32}GeO$	—	113/0.4	1.4768	1.0094	230
$(C_6H_5)_2PC\equiv CGe(CH_3)_3$	$C_{17}H_{19}GeP$	35	126–27/0.05	—	—	390
$(C_6H_5)_2P(S)C\equiv CGe(CH_3)_3$	$C_{17}H_{19}GePS$	173	—	—	—	390
$(C_3H_7)_3GeC\equiv CC_6H_5$	$C_{17}H_{26}Ge$	—	147–49/4	1.5274 (19°)	—	142
$(C_2H_5)_3GeC\equiv CC_6H_2,2,4,6(CH_3)_3$	$C_{17}H_{26}Ge$	—	128/0.8	1.5330 (25°)	—	22, 84, 86
$(C_2H_5)Ge[C\equiv C-COH(CH_3)_2]_3$	$C_{17}H_{26}GeO_3$	171	—	—	—	376
$(C_2H_5)_3GeC\equiv CC(CH_3)(C_2H_5)C\equiv CC(CH_3)_2OH$	$C_{17}H_{30}GeO$	—	105–7/3	1.4740	1.0027	377
$(C_4H_9)_3GeC\equiv CC(CH_3)_2Cl$	$C_{17}H_{33}GeCl$	—	95–96/0.3	1.4810	1.0295	373
$(C_4H_9)_3GeC\equiv CC_3H_7$	$C_{17}H_{34}Ge$	—	117/0.5	1.4645	0.9425	189, 217, 218, 222, 229
$(C_4H_9)_3GeC\equiv C-C(CH_3)_2OH$	$C_{17}H_{34}GeO$	—	107–8/0.3	1.4705	0.9960	373
$(C_2H_5)_3GeC\equiv C-C(CH_3)_2OCH(CH_3)OC_4H_9$	$C_{17}H_{34}GeO_2$	—	138–39/13	1.4598	1.0100	383
$(C_2H_5)_3GeC\equiv C(CH_2)_3Ge(C_2H_5)_3$	$C_{17}H_{36}Ge_2$	—	140/0.8	1.4815	1.0780	230
$(C_6H_5)_2Ge(CH_2C\equiv CH)_2$	$C_{18}H_{16}Ge$	—	—	—	—	395
$(C_2H_5)_3GeC\equiv CC_6H_4(t-C_4H_9)-p$	$C_{18}H_{28}Ge$	—	120/0.01	1.5290 (25°)	—	22, 84, 86
$(C_2H_5)_2Ge[C\equiv C-C(CH_3)_2-OCOCH_3]_2$	$C_{18}H_{28}GeO_4$	31–32	147/1.9	—	—	383

Table 2-4—continued

COMPOUND	EMPIRICAL FORMULA	M.P. (°C)	B.P. (°C/mm)	n_D^{20}	d_4^{20}	REFERENCES
$(CH_3)_2Ge[C{\equiv}C{-}C(C_2H_5)(CH_3){-}OCOCH_3]_2$	$C_{18}H_{28}GeO_4$	43–44	—	—	—	383
$(C_2H_5)_3GeC{\equiv}CC(CH_3)(C_2H_5)C{\equiv}CC(CH_3)(C_2H_5)OH$	$C_{18}H_{32}GeO$	—	111/3	1.4770	1.0040	377
$(CH_3)_2Ge[C{\equiv}C{-}COH(CH_3)C(CH_3)]_2$	$C_{18}H_{32}GeO_2$	—	140–41/4	1.4735	1.0011	379
$(C_4H_9)_3GeC{\equiv}C{-}C(CH_3)(C_2H_5)Cl$	$C_{18}H_{35}GeCl$	—	105–6/13	1.4840	0.9995	373
$(C_4H_9)_3GeC{\equiv}C{-}C_4H_9$	$C_{18}H_{36}Ge$	—	132/0.7	1.4648	0.9381	125, 189, 217,
			130/1	1.4670	—	218, 222
$(C_4H_9)_3GeC{\equiv}C{-}C(CH_3)(C_2H_5)OH$	$C_{18}H_{36}GeO$	—	115–16/3	1.4785	0.9905	373
$(C_2H_5)_3GeC{\equiv}CCH(C_3H_7)OCH(CH_3){-}OC_4H_9$	$C_{18}H_{36}GeO_2$	—	152–53/2	1.4600	1.0230	381, 382
$(C_2H_5)_3GeC{\equiv}C{-}C(C_2H_5)(CH_3)OCH(CH_3)OC_4H_9$	$C_{18}H_{36}GeO_2$	—	141–43/12	1.4610	0.9960	383
$(C_2H_5)_3GeC{\equiv}C{-}C(C_2H_5)_2OCH(CH_3)OC_4H_9$	$C_{19}H_{38}GeO_2$	—	151–53/14	1.4658	1.0200	383
$CH_3Ge[C{\equiv}C{-}COH(CH_3)C_2H_5]_3$	$C_{19}H_{30}GeO_3$	165–67	—	—	—	376
$(C_6H_5)_3GeC{\equiv}CH$	$C_{20}H_{16}Ge$	—	150/0.5	—	—	61, 215a, 319
$(C_2H_5)_2Ge(C{\equiv}C{-}C_6H_5)_2$	$C_{20}H_{20}Ge$	—	139–40/0.01	1.5982 (21°)	1.129 (21°)	142, 151
			191–93/2	1.5935 (19°)	—	
$(C_4H_9)_3GeC{\equiv}C{-}C_6H_5$	$C_{20}H_{32}Ge$	—	153/2	1.5110 (19°)	—	142, 333
			150/1.4	1.5195	1.0078	
$C_2H_5Ge[C{\equiv}C{-}COH(CH_3)C_2H_5]_3$	$C_{20}H_{32}GeO_3$	152–53	—	—	—	376
$(C_2H_5)_2Ge[C{\equiv}C{-}COH(CH_3)C(CH_3)_3]_2$	$C_{20}H_{36}GeO_2$	—	151–52/2	1.4800	1.0100	379
$[(CH_3)_3SiC{\equiv}C]_4Ge$	$C_{20}H_{36}GeSi_4$	160	—	—	—	174

Compound	Formula	M.p. (°C)	B.p. (°C/mm)	n_D	d	Ref.
$(C_4H_9)_3GeC\equiv C-C(CH_3)_2OCH_2CH_2CN$	$C_{20}H_{37}GeON$	—	132–33/0.1	1.4795	1.0304	373
$(C_2H_5)_3GeC\equiv CGe(C_4H_9)_3$	$C_{20}H_{42}Ge_2$	—	119/0.1	1.4749	1.0314	217, 229
$(C_6H_5)_3GeCH_2C\equiv CH$	$C_{21}H_{18}Ge$	88	—	—	—	215
$(C_6H_5)_3GeC\equiv C-CH_3$	$C_{21}H_{18}Ge$	106	—	—	—	215a, 319
$(C_6H_5)_3GeC\equiv C-C\equiv CH$	$C_{22}H_{16}Ge$	—	—	—	—	215a, 319
$(C_6H_5)_3GeCH(CH_3)C\equiv CH$	$C_{22}H_{20}Ge$	104	—	—	—	215
$(C_6H_5)_3GeCH_2C\equiv C-CH_3$	$C_{22}H_{20}Ge$	82	—	—	—	215
$(C_6H_5)_2Ge(C\equiv C-C_3H_7)_2$	$C_{22}H_{24}Ge$	—	136/0.01	1.5702 (21°)	1.188 (21°)	151
$(i-C_3H_7)_2Ge(C\equiv CC_6H_5)_2$	$C_{22}H_{24}Ge$	146–47	205/2	1.5840 (19°)	—	142
$CH_3Ge[C\equiv C-COH(C_2H_5)_2]_3$	$C_{22}H_{36}GeO_3$	—	—	—	—	376
$HC\equiv CGe(C_4H_9)_2C\equiv CGe(C_4H_9)_2C\equiv CH$	$C_{22}H_{38}Ge_2$	—	130/0.50	—	—	61
$(C_2H_5)_3GeC\equiv CC(CH_3)_2C\equiv CC(CH_3)OCH(CH_3)OC_4H_9$	$C_{22}H_{40}GeO_2$	—	152–54/3	1.4680	1.0140	385
$(C_4H_9)_3GeC\equiv CC(CH_3)(C_2H_5)C\equiv CC(CH_3)_2OH$	$C_{23}H_{42}GeO$	—	125–26/5	1.4917	1.0038	384
$(C_2H_5)_3GeC\equiv C-C(C_2H_5)C\equiv CC(CH_3)_2OCH-OC_4H_9$ (CH_3)	$C_{23}H_{42}GeO_2$	—	158–60/3	1.4710	1.0020	377
$(C_4H_9)_3GeC\equiv C-C(CH_3)_2OCH(CH_3)-OC_4H_9$	$C_{23}H_{46}GeO_2$	—	126–27/1	1.4625	0.9505	373
$[(CH_3)_3C-C\equiv C]_4Ge$	$C_{24}H_{36}Ge$	190–91	—	—	—	292
$(C_4H_9)_3GeC\equiv CC(CH_3)(C_2H_5)-O-CH(CH_3)-OC_4H_9$	$C_{24}H_{48}GeO_2$	—	144–45/0.2	1.4505	0.9415	373
$(C_4H_9)_3GeC\equiv CC(CH_3)(C_2H_5)OSi(C_2H_5)_3$	$C_{24}H_{50}GeOSi$	89	150–51/2	1.4590	0.9083	373
$(C_6H_5)_3GeC\equiv C-C_6H_5$	$C_{26}H_{20}Ge$	92.5–93	—	—	—	86, 142; 215a
$(C_2H_5)Ge(C\equiv C-C_6H_5)_3$	$C_{26}H_{20}Ge$	72–73	—	—	—	178, 395
$(C_6H_{11})_3GeC\equiv C-C_6H_5$	$C_{26}H_{38}Ge$	102	—	—	—	142

Table 2-4—continued

COMPOUND	EMPIRICAL FORMULA	M.P. (°C)	B.P. (°C/mm)	n_D^{20}	d_4^{20}	REFERENCES
(C₄H₉)₃GeC≡C—C(C₂H₅)C≡CC(CH₃)₂O(CH₂)₂CN with CH₃	C₂₆H₄₅GeON	—	103–4/0.3	1.4962	1.0267	384
(C₆H₅)₂Ge(C≡C—C₆H₅)₂	C₂₈H₂₀Ge	82.5–83.5; 82	—	—	—	142, 151
C₆H₅Ge(C≡CC₆H₅)₃	C₃₀H₂₀Ge	108–9	—	—	—	142
(C₆H₅)₃GeC≡C—C≡C—C₆H₅	C₂₈H₂₀Ge	114	—	—	—	215a
C₆H₅Ge(C≡CC₆H₅)₃	C₃₀H₂₀Ge	108–9	—	—	—	142
Ge(C≡C—p-C₆H₄Br)₄	C₃₂H₁₆GeBr₄	266	—	—	—	140
Ge(C≡C—C₆H₅)₄	C₃₂H₂₀Ge	187–88	—	—	—	17
(C₆H₅)₂PC≡CGe(C₆H₅)₃	C₃₂H₂₅GeP	102–3	—	—	—	390
(C₆H₅)₂P(O)C≡CGe(C₆H₅)₃	C₃₂H₂₅GePO	136–37	—	—	—	390
(C₆H₅)₂P(S)C≡CGe(C₆H₅)₃	C₃₂H₂₅GePS	155	—	—	—	390
Ge(C≡C—C₆H₁₁)₄	C₃₂H₄₄Ge	146	—	—	—	141
(C₆H₅)₃GeC≡CPb(C₆H₅)₃	C₃₈H₃₀GePb	134–35	—	—	—	93
(C₆H₅)₃GeC≡CSi(C₆H₅)₃	C₃₈H₃₀GeSi	151	—	—	—	93
(C₆H₅)₃GeC≡CSn(C₆H₅)₃	C₃₈H₃₀GeSn	149	—	—	—	93
(C₆H₅)₃GeC≡CGe(C₆H₅)₃	C₃₈H₃₀Ge₂	148–49; 127	—	—	—	61,93,136,138
(C₆H₁₁)₃GeC≡CGe(C₆H₁₁)₃	C₃₈H₆₆Ge₂	158	—	—	—	136
(C₆H₅)₃GeC≡CC≡CGe(C₆H₅)₃	C₄₀H₃₀Ge₂	294	—	—	—	139

Table 2-5
Enynic GeR₄ Derivatives

COMPOUND	EMPIRICAL FORMULA	M.P. (°C)	B.P. (°C/mm)	n_D^{20}	d_4^{20}	REFERENCES
$(CH_3)_3GeC\equiv C-CH=CH_2$	$C_7H_{12}Ge$	—	133–34	1.4655	1.0711	258
$(C_2H_5)_3GeC\equiv C-CH=CH_2$	$C_{10}H_{18}Ge$	—	62–63/5	1.4850	1.0333	401, 402, 404
$(C_2H_5)_3GeC\equiv C-CH=CH-CH_3$	$C_{11}H_{20}Ge$	—	78–79/5	1.4950	1.0173	401, 402, 404
$(C_2H_5)_3GeC\equiv C-C(CH_3)=CH_2$	$C_{11}H_{20}Ge$	—	97/23	1.4792	1.0068	217, 223, 404
			70–71/5	1.4808	1.0048	
$(C_2H_5)_3GeCH_2-C\equiv C-CH_2CH=CH_2$	$C_{12}H_{22}Ge$	—	111/11	1.4815	1.0295	189, 217, 223, 229
$(C_2H_5)_3GeC\equiv C-CH=C(CH_3)_2$	$C_{12}H_{22}Ge$	—	80–82/3	1.4682	0.9750	379
$(C_2H_5)_2Ge[C\equiv C-C(CH_3)=CH_2]_2$	$C_{14}H_{20}Ge$	—	108–10/4	1.4874	1.0100	376
$(C_2H_5)_3GeC\equiv C-C(t\text{-}C_4H_9)=CH_2$	$C_{14}H_{26}Ge$	—	80–82/3	1.4682	0.9750	379
$(C_2H_5)_3Ge(CH_2)_3C\equiv C-CH_2-CH=CH_2$	$C_{14}H_{26}Ge$	—	112/1.2	1.4787	0.9969	230
$(C_2H_5)_3Ge-C\big(C(CH_3)=CH_2\big)\!=\!CH-C\equiv C-C(CH_3)=CH_2$	$C_{16}H_{26}Ge$	—	121/2	1.5100	1.0560	128
$(C_2H_5)_3Ge(CH_2)_5C\equiv C-C(CH_3)=CH_2$	$C_{16}H_{30}Ge$	—	126/1.2	1.4848	0.9780	230
$(C_2H_5)_3Ge-C\big(COH(CH_3)_2\big)\!=\!CH-C\equiv C-COH(CH_3)_2$	$C_{16}H_{30}GeO_2$	—	152/2	1.4990	1.0630	128
$(C_4H_9)_3GeC\equiv C-CH_2-CH=CH_2$	$C_{17}H_{32}Ge$	—	95/0.2	1.4723	0.9582	189, 217, 218, 223

Table 2-5—continued

COMPOUND	EMPIRICAL FORMULA	M.P. (°C)	B.P. (°C/mm)	n_D^{20}	d_4^{20}	REFERENCES
$(CH_3)_2Ge\left[C \equiv C - C = CH_2\\ \quad\quad\quad C(CH_3)_3\right]_2$ $C(CH_3) = CH - CH_3$	$C_{18}H_{28}Ge$	—	118–19/4	1.4810	0.9814	379
$(C_2H_5)_3GeC\begin{array}{l}CH - C \equiv C - C(CH_3) = CH - CH_3\\ C(CH_3) = CH - CH_3\end{array}$	$C_{18}H_{30}Ge$	—	125/2	1.5260	1.0309	128
$(C_2H_5)_3GeC\begin{array}{l}CH - C \equiv C - COH(CH_3)C_2H_5\\ COH(CH_3)C_2H_5\end{array}$	$C_{18}H_{32}GeO$	—	146–48/2	1.5200	1.0428	128
$(C_2H_5)_3GeC\begin{array}{l}CH - C \equiv C - COH(CH_3)C_2H_5\end{array}$	$C_{18}H_{34}GeO_2$	—	164/2	1.4964	1.0534	128
$(C_2H_5)_2Ge\left[C \equiv C - C = CH_2\\ \quad\quad\quad C(CH_3)_3\right]_2$	$C_{20}H_{32}Ge$	—	131–32/4	1.4870	0.9880	379
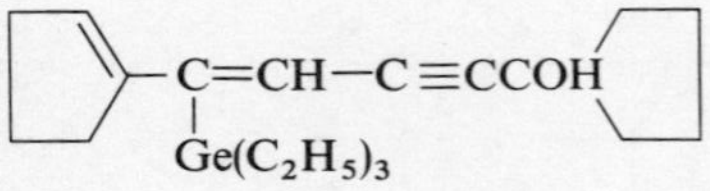 $C = CH - C \equiv CCOH$ $Ge(C_2H_5)_3$ (cyclopentenyl)	$C_{20}H_{32}GeO$	—	184/3	1.5412	1.1050	130

Structure	Formula		B.p./mm	n	d	
$(C_2H_5)_3GeC$ with $C(CH_3)=C(CH_3)_2$	$C_{20}H_{34}Ge$	—	156/10	1.5250	0.9874	130
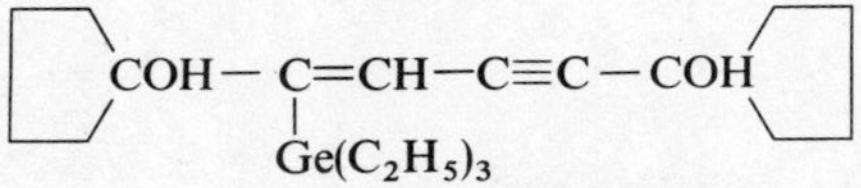 $COH-C=CH-C\equiv C-COH$, $Ge(C_2H_5)_3$ (cyclopentane ring)	$C_{20}H_{34}GeO_2$	—	192/3	1.5332	1.1232	130
$(C_4H_9)_2Ge$ with $CH=CH(CH_2)_3CH_3$ and $C\equiv C(CH_2)_3CH_3$	$C_{20}H_{38}Ge$	—	125/0.2	1.4723	0.9201	205
$(C_2H_5)_3GeC$ with $C\equiv C-COH(CH_3)CH(CH_3)_2$ and $CH-COH(CH_3)-CH(CH_3)_2$	$C_{20}H_{38}GeO_2$	—	151–53/1	1.5025	1.0350	130
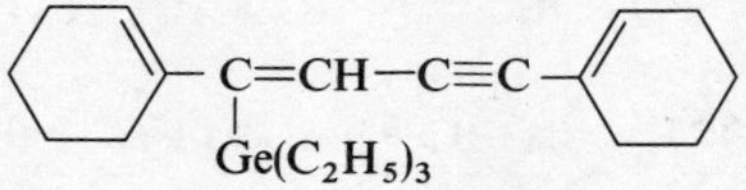 $C=CH-C\equiv C$, $Ge(C_2H_5)_3$ (cyclohexene)	$C_{22}H_{34}Ge$	—	161/7	1.5480	1.0860	130
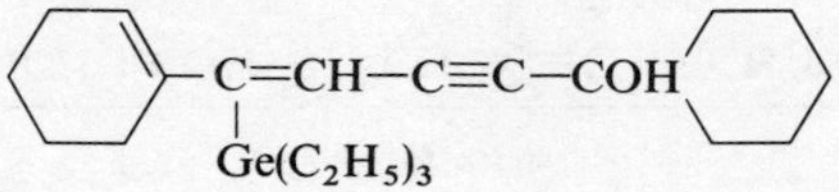 $C=CH-C\equiv C-COH$, $Ge(C_2H_5)_3$ (cyclohexane)	$C_{22}H_{36}GeO$	—	170–71/2	1.5420	1.0917	130
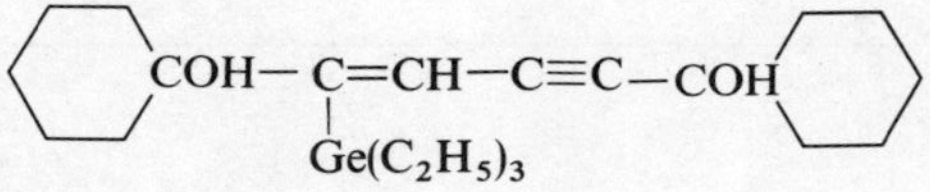 $COH-C=CH-C\equiv C-COH$, $Ge(C_2H_5)_3$ (cyclohexane)	$C_{22}H_{38}GeO_2$	—	201–2/3	1.5440	1.0995	130

Table 2-5—continued

COMPOUND	EMPIRICAL FORMULA	M.P. (°C)	B.P. (°C/mm)	n_D^{20}	d_4^{20}	REFERENCES
(CH₃)₂COH—C≡CH—C≡C—C(CH₃)₂OSi(C₂H₅)₃ │ Ge(C₂H₅)₃	$C_{22}H_{44}GeO_2Si$	—	172/3	1.4940	1.0027	128
(n-C₄H₉)₃GeC≡C—C(CH₃)(C₂H₅)C≡CC(CH₃)=CH₂	$C_{23}H_{40}Ge$	—	115–16/1	1.4800	0.9805	384
C₂H₅(CH₃)COH—C≡CHC≡C—C(C₂H₅)(CH₃)OSi(C₂H₅)₃ │ Ge(C₂H₅)₃	$C_{24}H_{48}GeO_2Si$	—	167/1	1.5100	1.0305	128
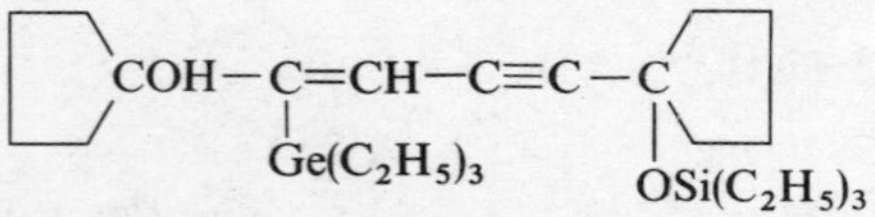 ⬠—COH—C≡CH—C≡C—C—OSi(C₂H₅)₃ │ Ge(C₂H₅)₃	$C_{26}H_{48}GeO_2Si$	—	183/1	1.4690	1.0363	130
(CH₃)₂CHCOH(CH₃)—C≡CH—C≡C—C(CH₃)—CH(CH₃)₂ OSi(C₂H₅)₃ │ Ge(C₂H₅)₃	$C_{26}H_{52}GeO_2Si$	—	147–49/2.	1.4886	1.0036	130
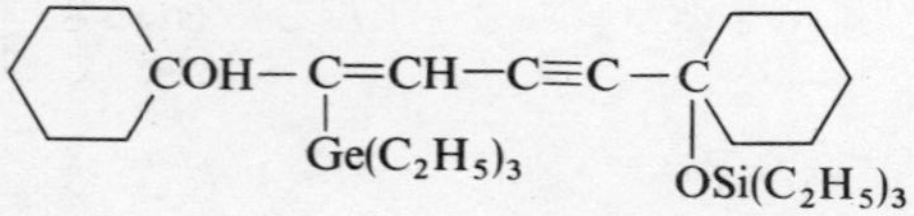 ⬡—COH—C≡CH—C≡C—C—OSi(C₂H₅)₃ │ Ge(C₂H₅)₃	$C_{28}H_{52}GeO_2Si$	—	216–18/2	1.5315	1.0643	130

2-4   ALLENIC DERIVATIVES OF GERMANIUM

These compounds are formed together with the acetylenic isomers by the reactions of germanium halides with propargylic Grignard reagents (215, 217, 222):

$$R_3GeX + BrMgCH_2-C{\equiv}CH \rightarrow \begin{cases} R_3GeCH_2-C{\equiv}CH \\ R_3GeCH=C=CH_2 \end{cases}$$

The yield of the allenic derivative is generally about 20%.

Masson and coworkers (215) in their study of propargylic and allenic derivatives of Group IV B elements observed that the allenic carbon chain on the derivative produced always displays a structure which has been reversed from that of the propargylic bromide used:

$$R_3MX \xrightarrow[\text{Mg}]{BrCHR'-C{\equiv}C-R''} \begin{cases} R_3M-CHR'-C{\equiv}C-R'' \\ R_3M-CR''=C=CHR' \\ R_3M-CR'=C=CHR'' \end{cases}$$

$$M = C, Si, Ge, Sn, Pb$$

$$R = CH_3, C_2H_5, C_6H_5$$

$$R' = H, CH_3$$

$$R'' = H, CH_3, C_6H_5$$

In the Group IV B series, the percentage yield of allenic compound increases with the metal's atomic weight:

M	Si	Ge	Sn	Pb
Allene %	10	20	70	95

Allenic derivatives can be prepared in high yield by the action of propargyl bromide on germano Grignard reagents (230):

$$R_3Ge(CH_2)_nMgBr + HC{\equiv}C-CH_2Br \rightarrow \begin{cases} R_3Ge(CH_2)_nCH=C=CH_2 \\ R_3Ge(CH_2)_nCH_2-C{\equiv}CH \end{cases}$$

Finally, it has been found (401) that, as in the case of the corresponding silicon enynes, but at higher temperature, triethylgermavinyl-, -propenyl- and isopropenylacetylenes add butyllithium giving, after hydrolysis, a mixture of acetylenic and allenic derivatives:

$$(C_2H_5)_3Ge-C\equiv C-CH=CH_2 \xrightarrow[\text{(2) H}_2\text{O}]{\text{(1) C}_4\text{H}_9\text{Li}}$$

$$\begin{cases} (C_2H_5)_3Ge-C\equiv C-CH_2CH_2C_4H_9 \\ (C_2H_5)_3Ge-CH=C=CH-CH_2C_4H_9 \end{cases}$$

With 1-triethylgermylpent-3-en-1-yne an intramolecular rearrangement leads to a vinylallenic compound (401):

$$(C_2H_5)_3Ge-C\equiv C-CH=CH-CH_3 \xrightarrow[\text{(2) H}_2\text{O}]{\text{(1) C}_4\text{H}_9\text{Li}}$$

$$(C_2H_5)_3Ge-CH=C=CH-CH=CH_2$$

Allenic derivatives can be obtained in pure state from a terminal acetylenic-allenic mixture by treatment with an excess of alcoholic silver nitrate solution (217, 230) to remove the acetylene. Their IR spectra are characterized by a strong absorption band in the range 1930–1950 cm^{-1} (215, 217, 401).

2-4-1   Allenic GeR$_4$ derivatives

Table 2-6 showing some physical properties of allenic GeR$_4$ derivatives appears on the following page.

Table 2-6
Allenic GeR$_4$ Derivatives

COMPOUND	EMPIRICAL FORMULA	M.P. (°C)	B.P. (°C/mm)	n_D^{20}	d_4^{20}	REFERENCES
$(C_2H_5)_3GeCH=C=CH_2$	$C_9H_{18}Ge$	—	89/30	—	—	215, 217, 229
$(CH_2=C=CH)_4Ge$	$C_{12}H_{12}Ge$	—	115/2	—	—	215
$(C_2H_5)_3Ge(CH_2)_4CH=C=CH_2$	$C_{13}H_{26}Ge$	—	122/8	1.4785	0.9947	230
$(C_2H_5)_3GeCH=C=CH-CH_2C_4H_9$ (CH$_3$)	$C_{14}H_{28}Ge$	—	90–91/1	1.4721	0.9688	401
$(C_2H_5)_3GeCH=C=CH-CH$ (C$_4$H$_9$)	$C_{15}H_{30}Ge$	—	95–97/0.5	1.4738	0.9675	401
$(C_2H_5)_3GeCH=C=C-CH_2C_4H_9$ (CH$_3$)	$C_{15}H_{30}Ge$	—	95–97/1	1.4740	0.9619	401
$(C_6H_5)_3GeCH=C=CH_2$	$C_{21}H_{18}Ge$	88	—	—	—	215
$(C_6H_5)_3GeCH=C=CH-CH_3$	$C_{22}H_{20}Ge$	104	—	—	—	215
$(C_6H_5)_3Ge-C(CH_3)=C=CH_2$	$C_{22}H_{20}Ge$	82	—	—	—	215

REFERENCES

1. Akerman, K., and A. Szuchnik, *Intern. J. Appl. Radiation Isotopes*, **15**, 319 (1964).
2. Akerman, K., W. Zablotny, R. Zablotna, and A. Szuchnik, *Isotopenpraxis*, **3**(5), 192 (1967).
3. Akhtar, M., and H. C. Clark, *Can. J. Chem.*, **46**, (12), 2165 (1968).
4. Ali-Zade, I. G., M. I. Shikhieva, M. A. Salimov, N. D. Abdullaev, and I. A. Shikhiev, *Dokl. Akad. Nauk SSSR*, **173** (1), 89 (1967).
5. Allen, P. W., and L. E. Sutton, *Acta Cryst.*, **3**, 46 (1950).
6. Anderson, H. H., *J. Am. Chem. Soc.*, **71**, 1799 (1949).
7. Anderson, H. H., *J. Am. Chem. Soc.*, **73**, 5800 (1951).
8. Anderson, H. H., *J. Am. Chem. Soc.*, **74**, 2370 (1952).
9. Anderson, H. H., *J. Am. Chem. Soc.*, **75**, 814 (1953).
10. Anderson, H. H., *J. Org. Chem.*, **21**, 869 (1956).
11. Bauer, H., and K. Burschkies, *Chem. Ber.*, **65**, 956 (1932).
12. Bauer, H., and K. Burschkies, *Chem. Ber.*, **66**, 1156 (1933).
13. Bauer, H., and K. Burschkies, *Chem. Ber.*, **67**, 1041 (1934).
14. Baukov, Yu. I., and I. F. Lutsenko, *Zh. Obshch. Khim.*, **32**, 3838 (1962).
15. Benkeser, R. A., C. E. Boer, R. E. Robinson, and D. M. Sauve, *J. Am. Chem. Soc.*, **78**, 682 (1956).
16. Bills, J. L., and F. A. Cotton, *J. Phys. Chem.*, **68**, 806 (1964).
17. Birr, K. H., and D. Kräft, *Z. Anorg. Allgem. Chem.*, **311**, 235 (1961).
18. Bott, R. W., C. Eaborn, and T. J. Hashimoto, *J. Chem. Soc.*, 3906 (1963).
19. Bott, R. W., C. Eaborn, K. C. Pande, and T. W. Swaddle, *J. Chem. Soc.*, 1217 (1962).
20. Bott, R. W., C. Eaborn, and T. W. Swaddle, *J. Chem. Soc.*, 2342 (1963).
21. Bott, R. W., C. Eaborn, and I. D. Varma, *Chem. Ind. (London)*, 614 (1963).
22. Bott, R. W., C. Eaborn, and D. R. M. Walton, *J. Organometal. Chem.*, **1**, 420–26 (1964).
23. Bott, R. W., C. Eaborn, and D. R. M. Walton, *J. Organometal. Chem.*, **2**, 154 (1964).
24. Brinckman, F. E., and F. G. A. Stone, *J. Inorg. Nucl. Chem.*, **11**, 24 (1959).
25. Brockway, L. O., *Rev. Mod. Phys.*, **8**, 231 (1936).
26. Brockway, L. O., and H. O. Jenkins, *J. Am. Chem. Soc.*, **58**, 2036 (1936).
27. Brook, A. G., and S. A. Fieldhouse, *J. Organometal. Chem.*, **10**, 235 (1967).
28. Brook, A. G., and H. Gilman, *J. Am. Chem. Soc.*, **76**, 77 (1954).
29. Brook, A. G., and G. J. D. Peddle, *J. Am. Chem. Soc.*, **85**, 1869 (1963).
30. Brook, A. G., M. A. Quigley, G. J. D. Peddle, N. V. Schwartz, and M. Warner, *J. Am. Chem. Soc.*, **82**, 5102 (1960).
31. Brooks, E. H., F. Glockling, and K. A. Hooton, *J. Chem. Soc.*, 4283 (1965).
32. Brown, M. P., and G. W. A. Fowles, *J. Chem. Soc.*, 2811 (1958).
33. Bulten, E. J., and J. G. Noltes, *Tetrahedron Letters*, (16), 1443 (1967).
33a. Burch, G. M., and J. R. van Wazer, *J. Chem. Soc.*, A, 586 (1966).
34. Burkhard, C. A., *J. Am. Chem. Soc.*, **72**, 1078 (1950).
35. Burlachenko, G. S., *Dissertation* (MGU) (1966).
36. Burschkies, K., *Chem. Ber.*, **69**, 1143 (1936).
37. Carrick, A., and F. Glockling, *J. Chem. Soc.*, A, 623 (1966).
38. Cason, L. F., and H. G. Brooks, *J. Am. Chem. Soc.*, **74**, 4582 (1952).
39. Cason, L. F., and H. G. Brooks, *J. Org. Chem.*, **19**, 1278 (1954).
40. Caujolle, F., D. Caujolle, and H. Bouissou, *Compt. Rend.*, **257**, 551 (1963).
41. Caujolle, F., D. Caujolle, Dao-Huy Giao, J. L. Foulquier, and E. Maurel, *Compt. Rend.*, **262** (11), 1302 (1966).
42. Cawley, S., and S. S. Danyluk, *J. Chem. Phys.*, **38**, 285 (1963).

43. Chandra, G., T. A. George, and M. F. Lappert, *Chem. Commun.*, (31), 116 (1967).
44. Chernyshev, E. A., A. D. Petrov, and T. L. Krasnova, *Synthesis and Properties of Monomers*, Ed. "Nauk" (1964) p. 103.
45. Chernyshev, E. A., A. A. Zelenetskaya, and T. L. Krasnova, *Izv. Akad. Nauk SSSR, Ser. Khim. Nauk*, (6), 1118–20 (1966).
46. Chipperfield, J. R., and R. H. Prince, *Proc. Chem. Soc.*, 385 (1960).
47. Clark, E. A., and A. Weber, *J. Chem. Phys.*, **45**, 1759 (1966).
48. Cohen, S. C., M. L. N. Reddy, and A. G. Massey, *Chem. Commun.*, (9), 451 (1967).
49. Coyle, T. D., and J. J. Ritter, *J. Organometal. Chem.*, **12**, 269 (1968).
50. Cross, R. J., and F. Glockling, *J. Chem. Soc.*, 4125 (1964).
51. Cross, R. J., and F. Glockling, *J. Organometal. Chem.*, **3**, 146 (1965).
52. Cross, R. J., and F. Glockling, *J. Chem. Soc.*, 5422 (1965).
53. Cullen, W. R., and W. R. Leeder, *Inorg. Chem.*, **5** (6), 1004 (1966).
54. Cullen, W. R., and G. E. Styan, *J. Organometal. Chem.*, **6**, 117 (1966).
55. Cullen, W. R., and M. C. Waldman, *Inorg. Nucl. Chem. Lett.*, **4** (4), 205–7 (1968).
56. Curtis, M. D., and A. L. Allred, *J. Am. Chem. Soc.*, **87**, 2554 (1965).
57. Curtis, D., R. K. Lee, and A. L. Allred, *J. Am. Chem. Soc.*, **89**, 5150 (1967).
58. Dao Huy-Giaó, Thèse de Spécialité Toulouse, 75 (1961).
59. Dao Huy-Giao, *Compt. Rend.*, **260**, 6937 (1965).
60. Davidsohn, W., and M. C. Henry, *2nd Internat. Symp. Organometallic Chem.*, Madison, Wis., 1965.
61. Davidsohn, W., and M. C. Henry, *J. Organometal. Chem.*, **5**, 29–34 (1966).
62. Davidsohn, W. E., and M. C. Henry, *Chem. Rev.*, **67** (1), 73 (1967).
63. Davidsohn, W. E., K. Hills, and M. C. Henry, *J. Organometal. Chem.*, **3**, 285 (1965).
64. Dennis, L. M., *Z. Anorg. Chem.*, **174**, 97 (1928).
65. Dennis, L. M., and F. E. Hance, *J. Am. Chem. Soc.*, **47**, 370 (1925).
66. Dennis, L. M., and F. E. Hance, *J. Phys. Chem.*, **30**, 1055 (1926).
67. Dennis, L. M., and W. I. Patnode, *J. Am. Chem. Soc.*, **52**, 2779 (1930).
68. Dibeler, V. H., *J. Res. Nat. Bur. Std.*, **49**, 235 (1952).
69. Dolgii, I. E., A. P. Mesheryakov, and G. K. Gaivoronskaya, *Izv. Akad. Nauk SSSR, Otd. Khim. Nauk*, (3), 572 (1963).
70. Dolgii, I. E., A. P. Mesheryakov, and G. K. Gaivoronskaya, *Izv. Akad. Nauk SSSR, Otd. Khim. Nauk*, (6), 1111 (1963).
71. Dolgii, I. E., A. P. Mesheryakov, and I. B. Shvedova, *Izv. Akad. Nauk SSSR*, 192 (1965).
72. Dubac, J., Dissertation, Toulouse, 1969.
73. Dzhurinskaya, N. G., V. F. Mironov, and A. D. Petrov, *Dokl. Akad. Nauk SSSR*, **138**, 1107 (1961).
74. Eaborn, C., *Organosilicon Compounds*, Butterworths, London, 1960, p. 27.
75. Eaborn, C., R. E. E. Hill, and P. Simpson, *J. Organometal. Chem.*, **15**, p. 1 (1968).
76. Eaborn, C., R. E. E. Hill, P. Simpson, A. G. Brook, and D. MacRae, *J. Organometal. Chem.*, **15**, 241 (1968).
77. Eaborn, C., A. Leyshon, and K. C. Pande, *J. Chem. Soc.*, 3423 (1960).
78. Eaborn, C., and K. C. Pande, *J. Chem. Soc.*, 1566 (1960).
79. Eaborn, C., and K. C. Pande, *J. Chem. Soc.*, 3200 (1960).
80. Eaborn, C., and K. C. Pande, *J. Chem. Soc.*, 297 (1961).
81. Eaborn, C., P. Simpson, and I. D. Varma, *J. Chem. Soc.*, (8), 1133 (1966).
82. Eaborn, C., G. A. Skinner, and D. R. M. Walton, *J. Organometal. Chem.*, **6** (4), 438 (1966).
83. Eaborn, C., and J. D. Varma, *J. Organometal. Chem.*, **9** (2), 377 (1967).

84. Eaborn, C., and D. R. M. Walton, *J. Organometal. Chem.*, **2**, 95 (1964).
85. Eaborn, C., and D. R. M. Walton, *J. Organometal. Chem.*, **3**, 168 (1965).
86. Eaborn, C., and D. R. M. Walton, *J. Organometal. Chem.*, **4** (3), 217 (1965).
87. Egorochkin, A. N., M. L. Khidekel, G. A. Razuvaev, V. F. Mironov, and A. L. Kravchenko, *Izv. Akad. Nauk SSSR Ser. Khim.*, 1312 (1964).
88. Egorov, Yu. P., *Izv. Akad. Nauk SSSR Otd. Khim. Nauk* 124 (1957).
88a. Egorov, Yu. P., and P. A. Bazhulin, *Dokl. Akad. Nauk SSSR*, **88** (4), 647 (1953).
89. Egorov, Yu. P., G. G. Kirei, L. A. Leites, V. F. Mironov, and A. D. Petrov, *Izv. Akad. Nauk SSSR Otd. Khim. Nauk*, (10), 1880 (1962).
90. Egorov, Yu. P., L. A. Leites, I. D. Kravtsova, and V. F. Mironov, *Izv. Akad. Nauk SSSR Otd. Khim. Nauk*, (6), 1114 (1963).
91. Egorov, Yu. P., L. A. Leites, and V. F. Mironov, *Zh. Strukt. Khim.*, **2** (5), 562 (1961).
92. Eisch, J. J., and M. W. Foxton, *J. Organometal. Chem.*, **11**, 24 (1968).
93. Findeiss, W., W. Davidsohn, and M. C. Henry, *J. Organometal. Chem.*, **9** (3), 435 (1967).
94. Fish, R. H., and H. G. Kuivila, *J. Org. Chem.*, **31** (8), 2445 (1966).
95. Flitcroft, N., P. J. Harbourne, I. Paul, P. M. Tucker, and F. G. A. Stone, *J. Chem. Soc.*, A, 1130 (1966).
96. Flood, E. A., *J. Am. Chem. Soc.*, **54**, 1663 (1932).
97. Flood, E. A., and L. Horvitz, *J. Am. Chem. Soc.*, **55**, 2534 (1933).
98. Florin, R. E., and T. W. Mears, *U.S. At. Energy Comm. B.N.L.*, **89**, 2446 (1955).
99. Florinskii, F. S., *Zh. Obshch. Khim.*, **32**, 1443 (1962).
100. Frejdlin, L. K. H., I. E. Zhukova, and V. F. Mironov, *Izv. Akad. Nauk SSSR*, 2258 (1960).
101. French, F. A., and R. S. Rasmussen, *J. Chem. Phys.*, **14**, 389 (1946).
102. Fritz, H. P., and C. G. Kreiter, *J. Organometal. Chem.*, **4**, 313 (1965).
103. Fuchs, R., and H. Gilman, *J. Org. Chem.*, **22**, 1009 (1957).
104. Fuchs, R., and H. Gilman, *J. Org. Chem.*, **23**, 911 (1958).
105. Fuchs, R., L. O. Moore, D. Miles, and H. Gilman, *J. Org. Chem.*, **21**, 1113 (1956).
106. Gar, T. K., Dissertation, M, (1965).
106a. Gar, T. K., L. A. Leites and V. F. Mironov, *Izvest. Akad. Nauk SSSR, Ser. Khim.*, No. 6, 1336 (1968).
107. Gar, T. K., and V. F. Mironov, *Izv. Akad. Nauk SSSR*, (5), 855, (1965).
108. Geddes, R. L., and E. Mack, *J. Am. Chem. Soc.*, **52**, 4372 (1930).
109. Gilman, H., and W. H. Atwell, *J. Am. Chem. Soc.*, **86**, 2687 (1964).
110. Gilman, H., and C. W. Gerow, *J. Am. Chem. Soc.*, **77**, 4675 (1955).
111. Gilman, H., and C. W. Gerow, *J. Am. Chem. Soc.*, **77**, 5509 (1955).
112. Gilman, H., and C. W. Gerow, *J. Am. Chem. Soc.*, **78**, 5435 (1956).
113. Gilman, H., and C. W. Gerow, *J. Am. Chem. Soc.*, **78**, 5823 (1956).
114. Gilman, H., and C. W. Gerow, *J. Am. Chem. Soc.*, **79**, 342 (1957).
115. Gilman, H., and C. W. Gerow, *J. Org. Chem.*, **22**, 334 (1957).
116. Gilman, H., M. B. Hughes, and C. W. Gerow, *J. Org. Chem.*, **24**, 352 (1959).
117. Gilman, H., and R. W. Leeper, *J. Org. Chem.*, **16**, 466 (1951).
118. Gilman, H., and E. A. Zuech, *J. Org. Chem.*, **26**, 3035 (1961).
119. Gladshtein, B. M., V. V. Rode, and L. Z. Soborovski, *Zh. Obshch. Khim.*, **29**, 2155 (1959).
120. Glockling, F., and K. A. Hooton, *J. Chem. Soc.*, 3509 (1962).
121. Glockling, F., and K. A. Hooton, *J. Chem. Soc.*, 1849 (1963).
122. Glockling, F., and K. A. Hooton, *Inorg. Syn.*, **8**, 31 (1966).
123. Glockling, F., and J. R. C. Light, *J. Chem. Soc.*, A, 623, (1967).
124. Glockling, F., and J. R. C. Light, *J. Chem. Soc.*, A, 717 (1968).

125. Gravilenko, V. V., L. L. Ivanov, and L. I. Zakharkin, *Zh. Obshch. Khim.*, **37** (3), 550, (1967).
126. Griffiths, J. E., and M. Onyszchuk, *Can. J. Chem.*, **39**, 339 (1961).
127. Gverdtsiteli, I. M., L. V. Baramidze, and M. V. Chelidze, *Zh. Obshch. Khim.*, **37** (12), 2654 (1967).
128. Gverdtsiteli, I. M., and M. A. Buachidze, *Dokl. Akad. Nauk SSSR*, **158**, 147 (1964).
129. Gverdtsiteli, I. M., and M. A. Buachidze, *Soobshch. Akad. Nauk Gruz.*, **37**, 59 (1965).
130. Gverdtsiteli, I. M., and M. A. Buachidze, *Soobshch. Akad. Nauk Gruz.*, **37**, 323 (1965).
131. Gverdtsiteli, I. M., and M. A. Buachidze, *Soobshch. Akad. Nauk Gruz.*, **48** (3), 571–4 (1967).
132. Gverdtsiteli, I. M., T. L. Guntsadze, and A. D. Petrov, *Dokl. Akad. Nauk SSSR*, **153**, 107 (1963).
133. Gverdtsiteli, I. M., T. L. Guntsadze, and A. D. Petrov, *Soobshch. Akad. Nauk Gruz.*, **36**, 579 (1964).
134. Gverdtsiteli, I. M., T. L. Guntsadze, and A. D. Petrov, *Dokl. Akad. Nauk SSSR*, **157**, 607 (1964).
135. Harris, D. M., W. A. Nebergall, and O. H. Johnson, *Inorg. Syn.*, **5**, 70 (1957).
136. Hartmann, H., and J. U. Ahrens, *Angew. Chem.*, **70**, 75 (1958).
137. Hartmann, H., and C. Beermann, *Z. Anorg. Allgem. Chem.*, **276**, 20 (1954).
138. Hartmann, H., and C. Beermann, B.R.D. Pat. 1,062,244 (1960).
139. Hartmann, H., E. Dietz, K. Komorniczyk, and W. Reiss, *Naturwiss.*, **48**, 570 (1961).
140. Hartmann, H., and M. K. El A'ssar, *Naturwiss.*, **52** (11), 304 (1965).
141. Hartmann, H., and K. Meyer, *Naturwiss.*, **52** (11), 303 (1965).
142. Hartmann, H., H. Wagner, B. Karbstein, M. K. El A'ssar, and W. Reiss, *Naturwiss.*, **51**, 215 (1964).
143. Henry, M. C., and M. F. Downey, *J. Chem. Soc.*, **26**, 2299 (1961).
144. Henry, M. C., and J. G. Noltes, *J. Am. Chem. Soc.*, **82**, 555 (1960).
145. Henry, M. C., and J. G. Noltes, *J. Am. Chem. Soc.*, **82**, 558 (1960).
146. Henry, M. C., and J. G. Noltes, *J. Am. Chem. Soc.*, **82**, 561 (1960).
147. Hooton, K. A., and A. L. Allred, *Inorg. Chem.*, **4**, 671 (1965).
148. Huggins, M. L., *J. Am. Chem. Soc.*, **75**, 4123 (1953).
149. Hughes, E. D., *Quart. Rev. (London)*, **5**, 245 (1951).
150. Ibekwe, S. D., and M. J. Newlands, *Chem. Commun.*, 114 (1965).
151. Ibekwe, S. D., and M. J. Newlands, *J. Chem. Soc.*, 4608 (1965).
152. Johnson, O. H., and D. M. Harris, *J. Am. Chem. Soc.*, **72**, 5565 (1950).
153. Johnson, O. H., and D. M. Harris, *Inorg. Syn.*, **5**, 74 (1957).
154. Johnson, O. H., and W. H. Nebergall, *J. Am. Chem. Soc.*, **70**, 1706 (1948).
155. Johnson, O. H., and W. H. Nebergall, *J. Am. Chem. Soc.*, **71**, 1720 (1949).
156. Johnson, O. H., W. H. Nebergall, and D. M. Harris, *Inorg. Syn.*, **5**, 76 (1957).
157. Jones, E. R. H., L. Skatteböl, and M. C. Whithing, *J. Org. Chem.*, **21**, 4765 (1956).
158. Jones, K., and M. F. Lappert, *J. Organometal. Chem.*, **3**, 295 (1965).
159. Kazankova, M. A., N. P. Protsenko, and I. F. Lutsenko, *Zh. Obshch. Khim.*, **38** (1), 106 (1968).
160. van der Kerk, G. J. M., F. Rijkens, and M. J. Janssen, *Rec. Trav. Chim.*, **81**, 764 (1962).
161. Kettering, C. F., and W. W. Sleator, *Physics*, **4**, 39 (1933).
162. Khmel'nitskii, R. A., A. A. Polyakova, A. A. Petrov, F. A. Medvedev, and M. D. Stadnichuk, *Zh. Obshch. Khim.*, **35** (5), 773 (1965).
163. Köster-Pflugmacher, A., and A. Hirsch, *Naturwiss.*, **54** (24), 645 (1967).
164. Kolesnikov, G. S., and S. L. Davydova, *Zh. Obshch. Khim.*, **29**, 2042 (1959).

165. Kolesnikov, G. S., S. L. Davydova, and T. I. Ermolaeva, *Vysokomolecul. Soed.*, **1**, 591 (1959).

166. Kolesnikov, G. S., S. L. Davydova, and T. I. Ermolaeva, *Vysokomolekul. Soed.*, **1**, 1493 (1959).

167. Kolesnikov, G. S., S. L. Davydova, and N. V. Klimentova, *Vysokomolekul. Soed.*, **2**, 563 (1960).

168. Kolesnikov, G. S., S. L. Davydova, and N. V. Klimentova, *J. Polymer Sci.*, **52**, 55 (1961).

169. Kolesnikov, G. S., S. L. Davydova, and N. V. Klimentova, *Vysokomolekul. Soed.*, **4**, 1098 (1962).

170. Kolesnikov, G. S., S. L. Davydova, and N. V. Klimentova, *Intern. Congress of Macromolecular Chemistry*, Ed. "Akad. Nauk SSSR" (1960), p. 156; R. Zh. Kh., (1961) 5 R 69.

171. Kolesnikov, S. P., O. M. Nefedov, *Angew. Chem.*, **77**, 345 (1965).

172. Kolesnikov, S. P., O. M. Nefedov, *Zh. V.K.O. im. D.I. Mendel*, **10**, 478 (1965).

173. Kolesnikov, S. P., O. M. Nefedov, and V. I. Sheiichenko, *Izv. Akad. Nauk SSSR*, 443 (1966).

174. Komarov, N. V., and O. G. Yarosh, *Zh. Obshch. Khim.*, **37** (1), 264 (1967).

175. Korshak, V. V., A. M. Polyakova, V. F. Mironov, and A. D. Petrov, *Izv. Akad. Nauk SSSR Otd. Khim. Nauk*, 178 (1959).

176. Korshak, V. V., A. M. Polyakova, V. F. Mironov, and A. D. Petrov, *Dokl. Akad. Nauk SSSR*, **128**, 960 (1959).

177. Korshak, V. V., A. M. Polyakova, A. D. Petrov, and V. F. Mironov, *Dokl. Akad. Nauk SSSR*, **112**, 436 (1957).

178. Korshak, V. V., A. M. Sladkov, and L. K. Luneva, *Izv. Akad. Nauk SSSR, Otd. Khim. Nauk*, 728 (1962).

179. Korshak, V. V., A. M. Sladkov, and L. K. Luneva, *Izv. Akad. Nauk SSSR*, 2251 (1962).

180. Kramer, K., and N. A. Wrigh, *J. Chem. Soc.*, 3604 (1963).

181. Kraus, C. A., *J. Chem. Educ.*, **29**, 488 (1952).

182. Kraus, C. A., and E. A. Flood, *J. Am. Chem. Soc.*, **54**, 1635 (1932).

183. Kraus, C. A., and L. S. Foster, *J. Am. Chem. Soc.*, **49**, 457 (1927).

184. Kraus, C. A., and H. S. Nutting, *J. Am. Chem. Soc.*, **54**, 1622 (1932).

185. Kraus, C. A., and C. S. Sherman, *J. Am. Chem. Soc.*, **55**, 4694 (1933).

186. Krause, E., and G. Renwanz, *Chem. Ber.*, **65**, 777 (1932).

187. Kross, R. D., and V. A. Fassel, *J. Am. Chem. Soc.*, **77**, 5858 (1955).

188. Kuehlein, K., W. P. Neumann, and H. P. Becker, *Angew. Chem. Intern. Ed.*, **6** (10), 876 (1967).

189. Labarre, J. F., and P. Mazerolles, *Compt. Rend.*, **254**, 3998 (1962).

190. Laubengayer, A. W., and R. B. Corey, *J. Phys. Chem.*, **30**, 1043 (1926).

191. Leeper, R. W., *Iowa State Coll. J. Sci.*, **18**, 57 (1943).

192. Leites, L. A., Yu. P. Egorov, G. S. Kolesnikov, and S. L. Davydova, *Izv. Akad. Nauk SSSR Otd. Khim. Nauk*, 1976 (1961); Engl. Transl., 1844.

193. Leites, L. A., Yu. P. Egorov, G. Ya. Zueva, and V. A. Ponomarenko, *Izv. Akad. Nauk SSSR*, 2132 (1961).

194. Leites, L. A., T. K. Gar, and V. F. Mironov, *Dokl. Akad. Nauk SSSR*, **158** (2), 400 (1964).

195. Lengel, J. H., and V. H. Dibeler, *J. Am. Chem. Soc.*, **74**, 2683 (1952).

196. Lesbre, M., *Chem. Weekblad*, **58**, 351 (1962).

197. Lesbre, M., and P. Mazerolles, *Compt. Rend.*, **246**, 1708 (1958).

198. Lesbre, M., P. Mazerolles, and M. T. Barthès, unpublished work.

199. Lesbre, M., P. Mazerolles, and G. Manuel, *Compt. Rend.*, **255**, 544 (1962).
200. Lesbre, M., P. Mazerolles, and G. Manuel, *Compt. Rend.*, **257**, 2303 (1963).
201. Lesbre, M., P. Mazerolles, and D. Voigt, *Compt. Rend.*, **240**, 622 (1955).
202. Lesbre, M., and J. Satgé, *Compt. Rend.*, **250**, 2220 (1960).
203. Lesbre, M., and J. Satgé, *Compt. Rend.*, **254**, 4051 (1962).
204. Lesbre, M., J. Satgé, and M. Massol, *Compt. Rend.*, **256**, 1548 (1963).
205. Lesbre, M., J. Satgé, and M. Massol, *Compt. Rend.*, **257**, 2665 (1963).
206. Leusink, A. J., J. G. Noltes, H. A. Budding, and G. J. M. van der Kerk, *Rec. Trav. Chim.*, **83**, 844 (1964).
207. Lindlar, H., *Helv. Chim. Acta*, **35**, 446 (1952).
208. Lippincot, E. R., and M. C. Tobin, *J. Am. Chem. Soc.*, **75**, 4141 (1953).
209. Luijten, J. G. A., and F. Rijkens, *Rec. Trav. Chim.*, **83**, 857 (1964).
210. Luneva, L. K., *Usp. Khim.*, **36** (7), 1140 (1967).
211. Luneva, L. K., A. M. Sladkov, and V. V. Korshak, *Vysokomolekul. Soed.*, **7**, 427 (1965).
212. Lutsenko, I. F., Yu. I. Baukov, and G. S. Burlachenko, *J. Organometal. Chem.*, **6**, 496 (1966).
213. MacKay, K. M., and R. Watt, *J. Organometal. Chem.*, **6**, 336 (1966).
214. Manulkin, Z. M., A. B. Kuchkarov, and S. A. Sarankina, *Dokl. Akad. Nauk SSSR*, **149** (2), 318 (1963).
215. Masson, J. C., Minh le Quan, and P. Cadiot, *Bull. Soc. Chim. France*, (3), 777 (1967).
215a. Masson, J. C., Minh le Quan, and P. Cadiot, *Bull. Soc. Chim. France*, (3), 1085 (1968).
216. Mathis-Noël, R., M. Constant, J. Satgé, and F. Mathis, *Spectrochim. Acta*, **20**, 515 (1964).
217. Mathis-Noël, R., P. Mazerolles, and F. Mathis, *Bull. Soc. Chim. France*, 1955 (1961).
218. Mathis-Noël, R., M. C. Sergent, P. Mazerolles, and F. Mathis, *Spectrochim. Acta*, **20** (9), 1407 (1964).
219. Maxwell, L. R., *J. Opt. Soc. Am.*, **30**, 374 (1940).
220. Mays, J. M., and B. P. Dailey, *J. Chem. Phys.*, **20**, 1695 (1952).
221. Mazerolles, P., Dissertation, Toulouse, 1959.
222. Mazerolles, P., *Bull. Soc. Chim. France*, 856 (1960).
223. Mazerolles, P., *Compt. Rend.*, **251**, 2041 (1960).
223a. Mazerolles, P., *Bull. Soc. Chim. France*, 1911 (1961).
224. Mazerolles, P., *Compt. Rend.*, **257**, 1481 (1963).
225. Mazerolles, P., *et al.* Unpublished work.
226. Mazerolles, P., and J. Dubac, *Compt. Rend.*, **257**, 1103 (1963).
227. Mazerolles, P., J. Dubac, and M. Lesbre, *J. Organometal. Chem.*, **5**, 35 (1966).
228. Mazerolles, P., and M. Lesbre, *Compt. Rend.*, **248**, 2018 (1959).
229. Mazerolles, P., M. Lesbre, and Dao Huy-Giao, *Compt. Rend.*, **253**, 673 (1961).
229a. Mazerolles, P., M. Lesbre, and M. Joanny, *J. Organometal. Chem.*, **16**, 227 (1969).
230. Mazerolles, P., M. Lesbre, and S. Marre, *Compt. Rend.*, **261**, 4134 (1965).
231. Mazerolles, P., and G. Manuel, *Bull. Soc. Chim. France*, (1), 327 (1966).
232. Mazerolles, P., and G. Manuel, *Bull. Soc. Chim. France*, (7), 2511 (1967).
233. Mazerolles, P., and G. Manuel, *Compt. Rend.*, **267**, 1158 (1968).
234. Mazerolles, P., and D. Voigt, *Compt. Rend.*, **240**, 2144 (1955).
235. McRae, D. M., and S. A. Fieldhouse, *J. Am. Chem. Soc.*, **89** (3), 704 (1967).
236. Mendelsohn, J. C., F. Metras, J. C. Lahournere, and J. Valade, *J. Organometal. Chem.*, **12**, 327 (1968).
237. Mendelsohn, J. C., F. Metras, and J. Valade, *Compt. Rend.*, **261**, 756 (1965).
238. Mendelsohn, J. C., F. Metras, and J. Valade, *Bull. Soc. Chim. France*, 2133 (1966).
239. Metlesics, W., and H. Zeiss, *J. Am. Chem. Soc.*, **82**, 3321 (1960).

240. Mironov, V. F., and N. G. Dzhurinskaya, *Izv. Akad. Nauk SSSR*, 75 (1963).
241. Mironov, V. F., N. G. Dzhurinskaya, T. K. Gar, and A. D. Petrov, *Izv. Akad. Nauk SSSR Otd. Khim. Nauk* (3), 460 (1962).
242. Mironov, V. F., N. G. Dzhurinskaya, and A. D. Petrov, *Dokl. Akad. Nauk SSSR*, **131**, 98 (1960).
243. Mironov, V. F., N. G. Dzhurinskaya, and A. D. Petrov, *Izv. Akad. Nauk SSSR Otd. Khim. Nauk* (11), 2066 (1960).
244. Mironov, V. F., N. G. Dzhurinskaya, and A. D. Petrov, *Izv. Akad. Nauk SSSR*, 2095 (1961).
245. Mironov, V. F., Yu. P. Egorov, and A. D. Petrov, *Izv. Akad. Nauk SSSR Otd. Khim. Nauk*, 1400 (1959).
246. Mironov, V. F., and T. K. Gar, *Dokl. Akad. Nauk SSSR*, **152** (5), 1111 (1963).
247. Mironov, V. F., and T. K. Gar, *Izv. Akad. Nauk SSSR Ser. Khim.*, 1515 (1964).
248. Mironov, V. F., and T. K. Gar, *Izv. Akad. Nauk SSSR*, 1887 (1964).
249. Mironov, V. F., and T. K. Gar, *Izv. Akad. Nauk SSSR*, 755 (1965).
250. Mironov, V. F., and T. K. Gar, *Izv. Akad. Nauk SSSR Ser. Khim.*, (2), 291 (1965).
251. Mironov, V. F., and T. K. Gar, *Izv. Akad. Nauk SSSR Ser. Khim.*, (3), 482 (1966).
252. Mironov, V. F., T. K. Gar, V. Z. Anisimova, and E. M. Berliner, *Zh. Obshch. Khim.*, **37** (10), 2323 (1967).
253. Mironov, V. F., T. K. Gar, and L. A. Leites, *Izv. Akad. Nauk SSSR Otd. Khim. Nauk*, 1387 (1962).
254. Mironov, V. F., V. P. Kozyukov, and V. D. Sheludyakov, *Zh. Obshch. Khim.*, **37**, 1669 (1967).
255. Mironov, V. F., V. P. Kozyukov, and V. D. Sheludyakov, *Zh. Obshch. Khim.*, **37**, 1915 (1967).
256. Mironov, V. F., and A. L. Kravchenko, *Izv. Akad. Nauk SSSR*, 1563 (1963).
257. Mironov, V. F., and A. L. Kravechenko, *Zh. Obshch. Khim.*, **34**, 1356 (1964).
258. Mironov, V. F., and A. L. Kravchenko, *Izv. Akad. Nauk SSSR Otd. Khim. Nauk*, (6), 1026 (1965).
259. Mironov, V. F., A. L. Kravchenko, and A. L. Leites, *Izv. Akad. Nauk SSSR*, 1177 (1966).
260. Mironov, V. F., A. L. Kravchenko, and A. D. Petrov, *Dokl. Akad. Nauk SSSR*, **155**, 843 (1964).
261. Mironov, V. F., A. L. Kravchenko, and A. D. Petrov, *Izv. Akad. Nauk SSSR*, 1209 (1964).
262. Mironov, V. F., A. D. Petrov, and N. L. Maximova, *Izv. Akad. Nauk SSSR*, 1954 (1959).
263. Mironov, V. F., V. A. Ponomarenko, G. Vzenkova, I. E. Dolgii, and A. D. Petrov, *Chemistry and Practical Use of Organosilicon Compounds*, 1 L., Es. Ts. B.T.I. (1958), 189.
264. Mironov, V. F., E. S. Sobolev, and L. M. Antipin, *Zh. Obshch. Khim.*, **37** (8), 1707 (1967).
265. Moedrizer, K., *Organometal. Chem. Rev.*, **1**, 179 (1966).
266. Morgan, G. T., and H. D. K. Drew, *J. Chem. Soc.*, **125**, 1261 (1924).
267. Morgan, G. T., and H. D. K. Drew, *J. Chem. Soc.*, 1760 (1925).
268. Müller, R., and L. Heinrich, *Chem. Ber.*, **95**, 2276 (1962).
269. Nagy, J., J. Reffy, A. Borbely-Kuszmann, and K. Becker-pálossy, *Intern. Symp. Organosilicon Chem. Sci. Commun. Prague*, 241 (1965).
270. Nametkin, N. S., S. G. Durgar'yan, and L. I. Tikhonova, *Dokl. Akad. Nauk SSSR*, **172** (3), 615 (1967).

271. Nametkin, N. S., S. G. Durgar'yan, and L. I. Tikhonova, *Dokl. Akad. Nauk SSSR*, **172** (4), 867 (1967).
272. Nefedov, O. M., and S. P. Kolesnikov, *Izv. Akad. Nauk SSSR*, 2068 (1963).
273. Nefedov, O. M., and S. P. Kolesnikov, *Vysokomolekul. Soed.*, **7**, 1857 (1965).
274. Nefedov, O. M., S. P. Kolesnikov, A. S. Khachaturov, and A. D. Petrov, *Dokl. Akad. Nauk SSSR*, **154**, 1389 (1964).
275. Nesmeyanov, A. N., K. N. Anisimov, N. E. Kolobova, and F. S. Denisov, *Izv. Akad. Nauk SSSR*, 2246 (1966).
276. Nesmeyanov, A. N., and A. E. Borisov, *Dokl. Akad. Nauk SSSR*, **174** (1), 96 (1967).
277. Nesmeyanov, A. N., A. E. Borisov, and N. V. Novikova, *Dokl. Akad. Nauk SSSR*, **165** (2), 333 (1965).
278. Nesmeyanov, A. N., A. E. Borisov, and N. V. Novikova, *Dokl. Akad. Nauk SSSR*, **172** (6), 1329 (1967).
279. Neumann, W. P., and K. König, *Angew. Chem. Intern. Ed.*, **1**, 212 (1962).
280. Nicholson, D. A., and A. L. Allred, *Inorg. Chem.*, **4** (12), 1751 (1965).
281. Noltes, J. G., H. A. Budding, and G. J. M. van der Kerk, *Rec. Trav. Chim.*, **79**, 408 (1960).
282. Noltes, J. G., H. A. Budding, and G. J. M. van der Kerk, *Rec. Trav. Chim.*, **79** (9–10), 1076 (1960).
283. Noltes, J. G., M. C. Henry, and M. J. Janssen, *Chem. Ind. (London)*, 298 (1959).
284. Noltes, J. G., and G. J. M. van der Kerk, *Rec. Trav. Chim.*, **80** (7), 623 (1961).
285. Noltes, J. G., and G. J. M. van der Kerk, *Rec. Trav. Chim.*, **81**, 41 (1962).
286. Normant, H., *Compt. Rend.*, **239**, 1510 (1954).
287. Normant, J. F., *Bull. Soc. Chim. France*, 859 (1965).
288. Nowak, M., *Intern. J. Appl. Radiation Isotopes*, **16**, 649 (1965).
289. Obreimov, I. V., and N. A. Chumaevskii, *Zh. Strukt. Khim.*, **5**, 59 (1964).
290. Orndorff, W. K., D. L. Tabern, and L. M. Dennis, *J. Am. Chem. Soc.*, **49**, 2512 (1927).
291. Osipov, O. A., B. L. Shelepina, and O. E. Shelepin, *Zh. Obshch. Khim.*, **36**, 264 (1966).
292. Pant, B. C., and H. F. Reiff, *J. Organometal. Chem.*, **15**, 65 (1968).
293. Patil, H. R. H., and W. A. G. Graham, *J. Organometal. Chem.*, **5** (8), 1401–5 (1966).
294. Peddle, G. J. D., and D. N. Roark, *Can. J. Chem.*, **46**, 2507 (1968).
295. Petrov, A. D., Lecture A. 110, 17° *Int. Kongress Reine u. Angew. Chemie, München*, 1959.
296. Petrov, A. D., M. A. Chel'tsova, and S. D. Komarov, *Izv. Akad. Nauk SSSR*, (3), 550–52 (1965).
297. Petrov, A. D., E. A. Chernyshev, and T. L. Krasnova, *Dokl. Akad. Nauk SSSR*, **140**, 837 (1961).
298. Petrov, A. D., Yu. P. Egorov, V. F. Mironov, G. I. Nikishin, and A. A. Bugorkova, *Izv. Akad. Nauk SSSR Otd. Khim. Nauk*, 50 (1956).
299. Petrov, A. D., and V. F. Mironov, *Izv. Akad. Nauk SSSR Otd. Khim. Nauk*, 1941 (1957).
300. Petrov, A. D., V. F. Mironov, and I. E. Dolgii, *Izv. Akad. Nauk SSSR Otd. Khim. Nauk*, 1146 (1956).
301. Petrov, A. D., V. F. Mironov, and N. G. Dzhurinskaya, *Dokl. Akad. Nauk SSSR*, **128**, 302 (1959).
302. Pike, R. M., and A. A. Lavigne, *Rec. Trav. Chim.*, **83**, 883 (1964).
303. Pollard, F. H., G. Nickless, and P. C. Uden, *J. Chromatog.*, **19**, 28 (1965).
304. Polyakova, A. M., and N. A. Chumaevskii, *Dokl. Akad. Nauk SSSR*, **130**, 1037 (1960).

305. Polyakova, A. M., V. V. Korshak, and E. S. Tambovtseva, *Vysokomolekul. Soed.*, **3**, 662 (1961).
306. Ponomarenko, V. A., and V. F. Mironov, *Dokl. Akad. Nauk SSSR*, **94**, 485 (1954).
307. Ponomarenko, V. A., G. Ya. Zueva, and N. S. Andreev, *Izv. Akad. Nauk SSSR*, 1758 (1961).
308. Pope, A. E., and H. A. Skinner, *Trans. Faraday Soc.*, **60**, 1404 (1964).
309. Prevost, Ch., M. Gaudemar, and J. Honiberg, *Compt. Rend.*, **230**, 1186 (1950).
310. Rabinovich, I. B., V. I. Tel'noi, N. V. Karyakin, and G. A. Razuvaev, *Dokl. Akad. Nauk SSSR*, **149** (2), 324 (1963).
311. Rashkes, A. M., Z. M. Manulkin, and A. B. Kuchkarev, *Zh. Obshch. Khim.*, **37** (5), 1046 (1967).
312. Razuvaev, G. A., N. S. Vyazankin, O. S. D'yachkovskaya, I. G. Kiseleva, and Yu. I. Dergunov, *Zh. Obshch. Khim.*, **31**, 4056 (1961).
313. Ridder, J. J. de, and G. Dijkstra, *Rec. Trav. Chim.*, **86**, 737 (1967).
314. Ridder, J. J. de, G. van Koten, and G. Dijkstra, *Rec. Trav. Chim.*, **86**, 1325 (1967).
315. Rijkens, F., M. J. Janssen, and G. J. M. van der Kerk, *Rec. Trav. Chim.*, **84**, 1597 (1965).
316. Rijkens, F., and G. J. M. van der Kerk, *Rec. Trav. Chim.*, **83**, 723 (1964).
317. Rijkens, F., and G. J. M. van der Kerk, *Investigation in the Field of Organogermanium Chemistry*, Germanium Research Committee, Utrecht, 1964.
318. Rivière, P., and J. Satgé, *Bull. Soc. Chim. France.*, (11), 4039 (1967).
319. Roberts, R. M. G., *J. Organometal. Chem.*, **12**, 97 (1968).
320. Roberts, R. M. G., and F. El'Kaissi, *J. Organometal. Chem.*, **12**, 79 (1968).
321. Rochow, E. G., *J. Am. Chem. Soc.*, **70**, 1801 (1948).
322. Rothermundt, M., and K. Z. Burschkies, *Imminitätsforsch*, **87**, 445 (1936).
323. Rühlmann, K., and H. Heine, *Z. Chem.*, **6**, 421 (1966).
324. Sakurai, H., K. Tominaga, T. Watanabe, and M. Kumada, *Tetrahedron Letters*, **43**, 5493 (1966).
325. Sarankina, S. A., and Z. M. Manulkin, *Uzbeksk. Khim. Zh.*, **9** (3), 30 (1965).
326. Sarankina, S. A., and Z. M. Manulkin, *Zh. Obshch. Khim.*, **35** (5), 845 (1965).
327. Sarankina, S. A., Z. M. Manulkin, and A. B. Kuchkarov, *Dokl. Akad. Nauk Uz SSR*, **22** (7), 35–37 (1965).
328. Sarankina, S. A., Z. M. Manulkin, and A. B. Kuchkarov, *Zh. Obshch. Khim.*, **37** (1), 217 (1967).
328a. Sartori, P., and M. Weidenbruch, *Chem. Ber.*, **100** (6), 2049 (1967).
329. Satgé, J., *Ann. Chim. (Paris)*, **6**, 519 (1961).
330. Satgé, J., *Bull. Soc. Chim. France*, 630 (1964).
331. Satgé, J., and C. Couret, *Compt. Rend.*, **264** (26), 2169 (1967).
332. Satgé, J., and M. Lesbre, *Bull. Soc. Chim. France*, 2578 (1965).
333. Satgé, J., M. Lesbre, and M. Baudet, *Compt. Rend.*, **259**, 4733 (1964).
334. Satgé, J., and M. Massol, *Compt. Rend.*, **261**, 170 (1965).
335. Satgé, J., M. Massol, and M. Lesbre, *J. Organometal. Chem.*, **5**, 241 (1966).
335a. Satgé, J., and P. Rivière, *Bull. Soc. Chim. France*, 1773 (1966).
336. Schmidt, M., and H. Schumann, *Z. Anorg. Allgem. Chem.*, **325**, 130 (1963).
337. Schmidt, M., and I. Ruidisch, *Z. Anorg. Allgem. Chem.*, **311**, 342 (1961).
338. Schmidbaur, H., and W. Findeiss, *Chem. Ber.*, **99** (7), 2187–96 (1966).
339. Schmidbaur, H., and W. Tronich, *Chem. Ber.*, **100** (3), 1032 (1967).
340. Schmidbaur, H., and S. Waldmann, *Chem. Ber.*, **97**, 3381 (1964).
341. Schmidbaur, H., L. Sechser, and M. Schmidt, *J. Organometal. Chem.*, **15**, 77 (1968).
342. Schott, G., and C. Harzdorf, *Z. Anorg. Allgem. Chem.*, **307**, 105 (1960).
343. Schwarz, R., and M. Lewinsohn, *Chem. Ber.*, **64**, 2352 (1931).

344. Schwarz, R., and W. Reinhardt, *Chem. Ber.*, **65**, 1743 (1932).
345. Schwarz, R., and M. Schmeisser, *Chem. Ber.*, **69**, 579 (1936).
346. Semlyen, J. A., C. R. Walker, and C. S. C. Phillips, *J. Chem. Soc.*, 1197 (1965).
347. Seyferth, D., *Record Chem. Progr.*, **26**, 87 (1965).
348. Seyferth, D., *J. Org. Chem.*, **22**, 478 (1957).
349. Seyferth, D., *J. Am. Chem. Soc.*, **79**, 2738 (1957).
350. Seyferth, D., and D. L. Alleston, *Inorg. Chem.*, **2**, 418 (1963).
351. Seyferth, D., K. Brändle, and G. Raab, *Angew. Chem.*, **72**, 77 (1960).
352. Seyferth, D., and H. M. Cohen, *Inorg. Chem.*, **1**, 913 (1962).
353. Seyferth, D., R. J. Cross, and B. Prokai, *J. Organometal. Chem.*, **7** (2), 20–21 (1967).
354. Seyferth, D., and H. Dertouzos, *J. Organometal. Chem.*, **11**, 263 (1968).
355. Seyferth, D., H. Dertouzos, R. Suzuki, and J. Yick-Pui Mui, *J. Org. Chem.*, **32**, 2980 (1967).
356. Seyferth, D., H. P. Hoffman, R. Burton, and J. F. Helling, *Inorg. Chem.*, **1**, 227 (1962).
357. Seyferth, D., T. F. Jula, H. Dertouzos, and M. Pereyre, *J. Organometal. Chem.*, **11**, 63 (1968).
358. Seyferth, D., and E. G. Rochow, *J. Org. Chem.*, **20**, 250 (1955).
359. Seyferth, D., and E. G. Rochow, *J. Am. Chem. Soc.*, **77**, 907 (1955).
360. Seyferth, D., G. Singh, and R. Suzuki, *Pure Appl. Chem.*, **13** (1, 2), 159 (1966).
361. Seyferth, D., R. Suzuki, and L. G. Vaughan, *J. Am. Chem. Soc.*, **88**, 286 (1966).
362. Seyferth, D., and Tadashi Wada, *Inorg. Chem.*, **1**, 78 (1962).
363. Seyferth, S., Tadashi Wada, and G. E. Maciel, *Inorg. Chem.*, **1**, 232 (1962).
364. Seyferth, D., and L. G. Vaughan, *J. Organometal. Chem.*, **1**, 138 (1963).
365. Seyferth, D., and M. A. Weiner, *J. Am. Chem. Soc.*, **83**, 3583 (1961).
366. Seyferth, D., and M. A. Weiner, *J. Org. Chem.*, **26**, 4797 (1961).
367. Seyferth, D., and M. A. Weiner, *J. Am. Chem. Soc.*, **84**, 361 (1962).
368. Seyferth, D., M. A. Weiner, L. G. Vaughan, G. Raab, D. E. Welch, H. M. Cohen, and D. L. Alleston, *Bull. Soc. Chim. France*, 1364 (1963).
369. Shank, R. S., and H. Shechter, *J. Org. Chem.*, **24**, 1825 (1959).
370. Sharanina, L. G., V. S. Zavgorodnii, and A. A. Petrov, *Zh. Obshch. Khim.*, **36**, 1154 (1966).
371. Shchembelov, G. A., and Yu. A. Ustynyuk, *Dokl. Akad. Nauk SSSR*, **173** (4), 847 (1967).
372. Sheline, R. K., *J. Chem. Phys.*, **18**, 602 (1950).
373. Shikhiev, I. A., and N. D. Abdullaev, *Zh. Obshch. Khim.*, **35** (8), 1348–50 (1965).
374. Shikhiev, I. A., N. D. Abdullaev, and M. I. Aliev, *Zh. Obshch. Khim.*, **36**, 942 (1966).
375. Shikhiev, I. A., G. F. Askerov, and Sh. V. Garaeva, *Zh. Obshch. Khim.*, **38**, 639 (1968).
376. Shikhiev, I. A., and I. A. Aslanov, *Zh. Obshch. Khim.*, **34** (2), 397–8 (1964).
377. Shikhiev, I. A., I. A. Aslanov, and N. T. Mekhmandarova, *Zh. Obshch. Khim.*, **35**, 459 (1965).
378. Shikiev, I. A., I. A. Aslanov, and N. T. Mekhmandarova, *Zh. Obshch. Khim.*, **36**, 1295 (1966).
379. Shikhiev, I. A., I. A. Aslanov, and N. T. Mekhmandarova, *Zh. Obshch. Khim.*, **36** (7), 1297 (1966).
380. Shikhiev, I. A., I. A. Aslanov, N. T. Mekhmandarova, and S. Sh. Verdieva, *Azerb. Khim. Zh.*, (4), 42 (1965).
381. Shikhiev, I. A., I. A. Aslanov, and B. G. Yusufov, *Zh. Obshch. Khim.*, **31**, 3647–48 (1961).
382. Shikhiev, I. A., I. A. Aslanov, and B. G. Yusufov, *Avt. svid.* 140.056 (1961) *Byull. Izobr.*, (15), 19 (1961).

383. Shikhiev, I. A., I. A. Aslanov, and B. G. Yusufov, *Zh. Obshch. Khim.*, **32** (10), 3148 (1962).
384. Shikhiev, I. A., B. M. Guseinzade, and N. D. Abdullaev, *Dokl. Akad. Nauk Azerb. SSSR*, **20**, 13 (1964).
385. Shikhiev, I. A., B. M. Guseinzade, T. N. Mekhmandarova, and I. A. Aslanov, *Zh. Obshch. Khim.*, **34**, 394 (1964).
386. Shikhiev, I. A., M. F. Shostakovskii, N. V. Komarov, and I. A. Aslanov, Avt. svid. 117.493 (1959) *Byull. izobr.*, No. 2, 28 (1959).
387. Shikhiev, I. A., M. F. Shostakovskii, N. V. Komarov, and I. A. Aslanov, *Zh. Obshch. Khim.*, **29**, 1549 (1959).
388. Shomaker, V., and D. P. Stevenson, *J. Am. Chem. Soc.*, **63**, 37 (1941).
389. Siebert, H., *Z. Anorg. Chem.*, **263**, 82 (1950).
390. Siebert, W., W. E. Davidsohn, and M. C. Henry, *J. Organometal. Chem.*, **15**, 69 (1968).
391. Simonnin, M. P., *J. Organometal. Chem.*, **5**, 155 (1966).
392. Simonnin, M. P., *Bull. Soc. Chim. France*, 1774 (1966).
393. Simons, J. K., *J. Am. Chem. Soc.*, **57**, 1299 (1935).
394. Simons, J. K., E. C. Wagner, and J. H. Muller, *J. Am. Chem. Soc.*, **55**, 3705 (1933).
395. Sladkov, A. M., and L. K. Luneva, *Zh. Obshch. Khim.*, **36** (3), 553 (1966).
396. Smith, F. B., and C. A. Kraus, *J. Am. Chem. Soc.*, **74**, 1418 (1952).
397. Smith, J. A. S., and E. J. Wilkins, *J. Chem. Soc.*, A, 1749 (1966).
398. Sommer, R., W. P. Neumann, and B. Schneider, *Tetrahedron Letters*, **51**, 3875 (1964).
399. Spanier, E. J., *Dissertation Abstr.*, **25**, 6950 (1965).
400. Spialter, L., G. R. Buell, and Ch. W. Harris, *J. Org. Chem.*, **30**, 375 (1965).
401. Stadnichuk, M. D., *Zh. Obshch. Khim.*, **36** (5), 937 (1966).
402. Stadnichuk, M. D., and A. A. Petrov, *Zh. Obshch. Khim.*, **35**, 451 (1965).
403. Stadnichuk, M. D., and A. A. Petrov, *Zh. Obshch. Khim.*, **35**, 700 (1965).
404. Stadnichuk, M. D., T. V. Yakorleva, and A. A. Petrov, *Zh. Obshch. Khim.*, **37** (1), 222 (1967).
405. Staveley, L. A. K., H. P. Paget, B. B. Goalby, and J. B. Warren, *Nature*, **164**, 787 (1949).
406. Staveley, L. A. K., H. P. Paget, B. B. Goalby, and J. B. Warren, *J. Chem. Soc.*, 2290 (1950).
407. Steingross, W., and W. Zeill, *J. Organometal. Chem.* **6**, 109 (1966).
408. Steingross, W., and W. Zeil, *J. Organometal. Chem.*, **6** (5), 464 (1966).
409. Sterlin, R. N., and S. S. Dubov, *Zh. Vses. Khim. obshch. im. D.I. Mendeleeva*, **7**, 117 (1962) *C.A.*, **57** (1962) 294e, *Zh. Khim.*, (1962) 22 B 80.
410. Strohmeier, W., and K. Miltenberger, *Chem. Ber.*, **91**, 1357 (1958).
411. Sulimov, I. G., and M. D. Stadnichuk, *Zh. Obshch. Khim.*, **37** (10), 2329 (1967).
412. Tabern, D. L., W. K. Orndorff, and L. M. Dennis, *J. Am. Chem. Soc.*, **47**, 2039 (1925).
413. Tamborski, C., F. E. Ford, W. L. Lehu, G. J. Moore, and E. J. Soloski, *J. Org. Chem.*, **27**, 619 (1962).
414. Tamborski, C., E. J. Soloski, and S. M. Dec, *J. Organometal. Chem.*, **4** (6), 446–54 (1965).
415. Teal, G. K., and C. A. Kraus, *J. Am. Chem. Soc.*, **72**, 4706 (1950).
416. Thompson, H. W., and J. J. Frewing, *Nature*, **135**, 507 (1935).
417. Tillyaev, K. S., and Z. M. Manulkin, *Tr. Tashkent Farmasevt. Inst.*, **4**, 349 (1966).
418. Trautman, C. E., and H. A. Ambrose, U.S. Pat. 2,416,360, Feb. 25 (1947).
419. Tursunbaev, T. L., and Z. M. Manulkin, *Zh. Obshch. Khim.*, **37** (1), 219 (1967).

420. Ulbricht, K., M. Kakoubkova, and V. Chvalovsky, *Collection Czech. Chem. Commun.*, **33** (6), 1693 (1968).
421. Vaughan, L. G., and D. Seyferth, *J. Organometal. Chem.*, **5**, 295 (1966).
422. Voigt, D., M. Lesbre, and F. Gallais, *Compt. Rend.*, **239**, 1485 (1954).
423. Vol'pin, M. E., Yu. T. Struchkov, L. B. Vilkov, V. S. Mastryukov, V. G. Dulova, and D. N. Kursanov, *Izv. Akad. Nauk SSSR*, 2063 (1963).
424. Vondel, D. F. van de, *J. Organometal. Chem.*, **3**, 400 (1965).
425. Vyazankin, N. S., E. N. Gladyshev, S. P. Korneva, and G. A. Razuvaev, *Zh. Obshch. Khim.*, **34**, 1645 (1964).
426. Vyazankin, N. S., E. N. Gladyshev, S. P. Korneva, and G. A. Razuvaev, *Zh. Obshch. Khim.*, **36** (11), 2025 (1966).
427. Vyazankin, N. S., E. N. Gladyshev, and G. A. Razuvaev, *Dokl. Akad. Nauk SSSR*, **153** (1), 104, (1963).
428. Vyazankin, N. S., G. A. Razuvaev, and V. T. Bychakov, *Dokl. Akad. Nauk SSSR*, **158**, 382 (1964).
429. Vyazankin, N. S., G. A. Razuvaev, V. T. Bychakov, and V. L. Zvezdin, *Izv. Akad. Nauk SSSR*, 562 (1966).
430. Vyazankin, N. S., G. A. Razuvaev, E. N. Gladyshev, and S. P. Korneva, *J. Organometal. Chem.*, **7** (2), 353 (1967).
431. Walton, D. R. M., *J. Organometal. Chem.*, **3**, 438 (1965).
432. West, R., *J. Am. Chem. Soc.*, **74**, 4363 (1952).
433. West, R., H. R. Hunt, and R. O. Whipple, *J. Am. Chem. Soc.*, **76**, 310 (1954).
434. Wieber, M., and C. D. Frohning, *Angew. Chem. Intern. Ed.*, **4** (12), 1096 (1965).
435. Winkler, C., *J. Prakt. Chem.*, **36** (2), 177 (1887).
436. Wotiz, J. H., *J. Am. Chem. Soc.*, **72**, 1639 (1950).
437. Yusufov, B. G., I. A. Shikhiev, and I. A. Aslanov, Avt. svid. 140.426 (1961).
438. Yusufov, B. G., I. A. Shikhiev, and I. A. Aslanov, Avt. Svid. 140.056 (1961) Byull. izobr. (15), 19 (1964).
439. Young, C. W., J. S. Koehler, and D. S. McKinney, *J. Am. Chem. Soc.*, **69**, 1410 (1947).
440. Zablotna, R., *Bull. Acad. Polon. Sci. Ser. Chim.*, **12** (7), 475 (1964).
441. Zablotna, R., K. Akerman, and A. Szuchnik, *Bull. Acad. Polon. Sci. Ser. Chim.*, **12** (10), 695 (1964).
442. Zablotna, R., K. Akerman, and A. Szuchnik, *Bull. Acad. Polon. Sci. Ser. Chim.*, **14**, 731 (1966).
443. Zhakharkin, L. I., V. I. Bregadze, and O. Yu. Okhlobystin, *J. Organometal. Chem.*, **4**, 211 (1965).
444. Zhakharkin, L. I., and O. Yu. Okhlobystin, *Zh. Obshch. Khim.*, **31**, 3662 (1961).
445. Zharkharkin, L. I., O. Yu. Okhlobystin, and B. N. Strunin, *Izv. Akad. Nauk SSSR*, 2254 (1961).
446. Zhakharkin, L. I., O. Yu. Okhlobystin, and B. N. Strunin, *Zh. Prikl. Khim.*, **36**, 2034 (1963).
447. Zueva, G. Ya., I. F. Manucharova, I. P. Yakoulev, and V. A. Ponomarenko, *Izv. Akad. Nauk SSSR, Neorg. Mater.*, **2**, 229 (1966).
448. Ger. Pat. 1,028,576 (1958). Brit. Pat. 820,146 (1959).
449. Ger. Pat. 1,239,687, May 3 (1967), Appl. date March 12 (1965).

3 Germanium–carbon bond cyclic

3-1 "SMALL" RING GERMANIUM DERIVATIVES

3-1-1 Three-membered rings

Germacyclopropane or germacyclopropene derivatives are not yet known. Vol'pin and coworkers made an interesting attempt at producing the latter. Relying on the structural similarity between dihalocarbenes and germanous iodide GeI_2, they spent considerable time investigating the action of this compound on acetylenic derivatives (53, 58, 141, 143–145, 148). With acetylene in particular, they produced a stable crystalline derivative they thought to be a cyclopropene:

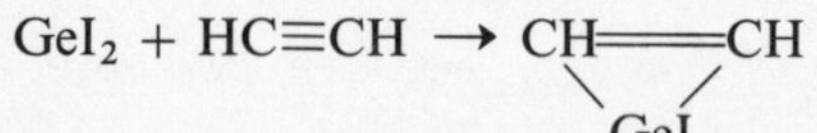

$$GeI_2 + HC\equiv CH \rightarrow CH\!\!=\!\!\!=\!\!CH$$
$$\underset{GeI_2}{\diagdown\ \diagup}$$

The structure of this derivative seemed to be confirmed first by the fact that the two iodine atoms are hydrolysable and secondly by the IR, NMR and Raman spectra: molecular symmetry on both sides of the double bond agrees with the absence of any absorption band around $1600\,cm^{-1}$ in the IR spectrum; the NMR spectrum reveals but one type of proton; lowering to $1557\,cm^{-1}$ of the Raman frequency for the $C\!\!=\!\!C$ bond with the dimethyl derivative could be due to the presence of the germanium atom in the α position with reference to the double bond; it was possible to explain the fact that the structure remained unchanged when this derivative was treated with base or with a Grignard reagent by an "aromatic" stability of this germacyclopropenic ring.

116

But in their more recent and more thorough investigation of these derivatives, Johnson and coworkers (49, 50), and then Vol'pin (149) firmly established that this kind of reaction produces a cyclohexadienic derivative with two germanium atoms:

$$2\,GeI_2 + 2\,R-C\equiv C-R \;\rightarrow\; \underset{\displaystyle I_2Ge}{}\;\;
\begin{array}{c}
R \quad\; R \\
| \qquad | \\
C=C \\
\diagup \qquad \diagdown \\
\qquad\qquad\qquad GeI_2 \\
\diagdown \qquad \diagup \\
C=C \\
| \qquad | \\
R \quad\; R
\end{array}$$

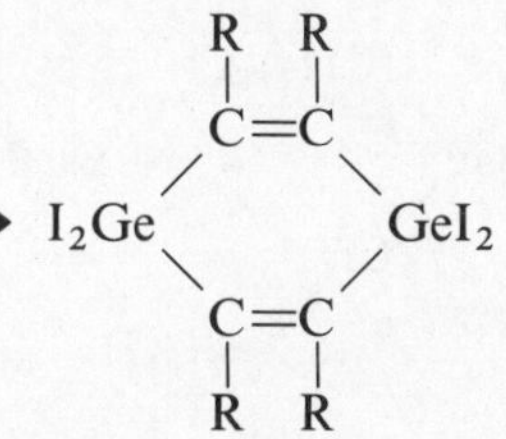

with $R = H, C_6H_5$.

3-1-2 Four-membered rings

3-1-2-1 Preparations.

Dialkylgermacyclobutanes are produced by the following reactions:

Sodium condensation of an equimolecular mixture of 1,3-dibromopropane and dialkyldibromogermane (75, 82):

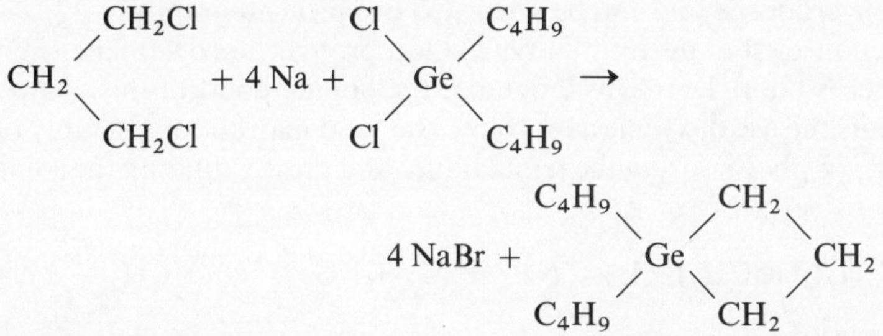

$$\underset{CH_2\diagdown CH_2Cl}{\overset{\diagup CH_2Cl}{}} \;+\; 4\,Na \;+\; \underset{Cl\diagdown}{\overset{Cl\diagup}{}}Ge\underset{\diagdown C_4H_9}{\overset{\diagup C_4H_9}{}} \;\rightarrow$$

$$4\,NaBr \;+\; \underset{C_4H_9\diagdown}{\overset{C_4H_9\diagup}{}}Ge\underset{\diagup CH_2}{\overset{\diagdown CH_2}{}}\underset{\diagdown CH_2}{\overset{\diagup CH_2}{}}$$

Even when the reaction is carried out in dilute medium, polymers remain the principal product. However, it is possible to isolate dialkylgermacyclobutanes by means of vapour phase chromatography. The yields are around 10%.

Ring closure of a dihalogenated derivative belonging to the $R_2Ge(X)CH_2CH_2CH_2X$ series.

The ring closure by magnesium that Sommer (135) proposed for preparing silacyclobutane derivatives:

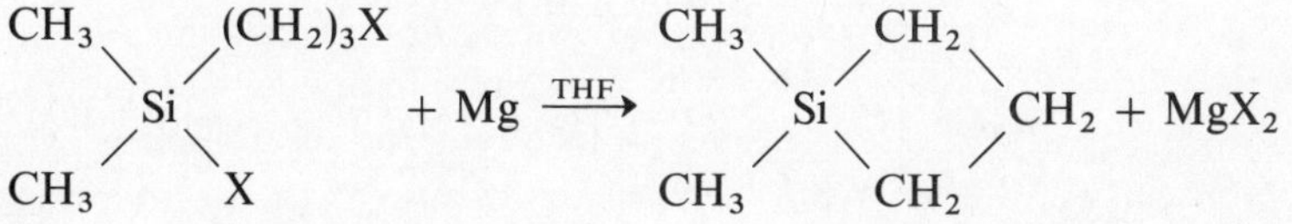

$$\underset{CH_3\diagdown}{\overset{CH_3\diagup}{}}Si\underset{\diagdown X}{\overset{\diagup (CH_2)_3X}{}} \;+\; Mg \;\xrightarrow{\;THF\;}\; \underset{CH_3\diagdown}{\overset{CH_3\diagup}{}}Si\underset{\diagup CH_2}{\overset{\diagdown CH_2}{}}\underset{\diagdown CH_2}{\overset{\diagup CH_2}{}} \;+\; MgX_2$$

cannot be repeated with the analogous germanium derivative either in THF or in ether solution.

Ring closure by alkali metals (26, 75, 82):

Diethylchloro(3-chloropropyl)germane reacts with a sodium–potassium alloy (23% Na, 77% K) in boiling toluene, producing diethylgermacyclobutane in 35% yield:

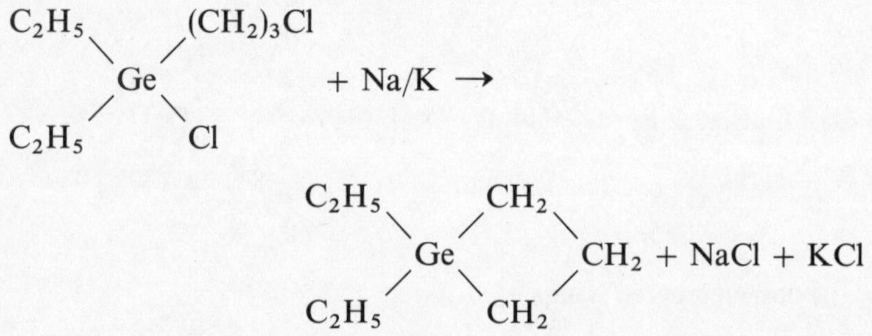

$$
\begin{array}{c}
C_2H_5 \diagdown \qquad \diagup (CH_2)_3Cl \\
\qquad Ge \\
C_2H_5 \diagup \qquad \diagdown Cl
\end{array}
\quad + \; Na/K \; \rightarrow
$$

$$
\begin{array}{c}
C_2H_5 \diagdown \qquad \diagup CH_2 \diagdown \\
\qquad Ge \qquad\qquad CH_2 \; + \; NaCl \; + \; KCl \\
C_2H_5 \diagup \qquad \diagdown CH_2 \diagup
\end{array}
$$

Besides the polymers formed in the same reaction, a little germane

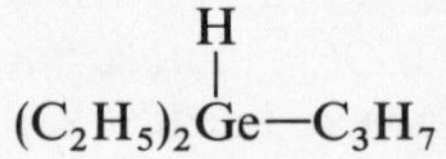

$$
\begin{array}{c}
\overset{\displaystyle H}{\underset{\displaystyle |}{}} \\
(C_2H_5)_2Ge{-}C_3H_7
\end{array}
$$

also is produced and can be separated by chromatography.

Action of sodium on dibutyl(3-chloropropyl)chlorogermane similarly produces dibutylgermacyclobutane: the boiling point of this compound permits the use of xylene as solvent. The yield may be considerably raised to 75% by both raising the temperature and greatly diluting the solution:

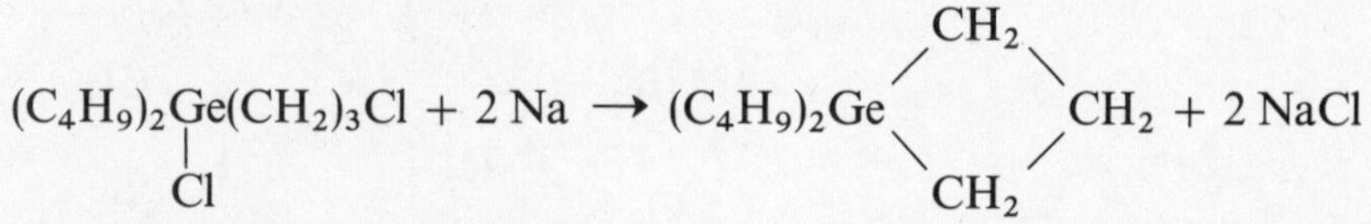

$$
(C_4H_9)_2Ge(CH_2)_3Cl \; + \; 2\,Na \; \rightarrow \; (C_4H_9)_2Ge
\begin{array}{c}
\diagup CH_2 \diagdown \\
\qquad\qquad CH_2 \; + \; 2\,NaCl \\
\diagdown CH_2 \diagup
\end{array}
$$
$$
\underset{\displaystyle Cl}{|}
$$

Some traces of dibutylpropylgermane are also produced and may be eliminated by chromatography.

This synthesis procedure has been extended to the ring closure of branched chains:

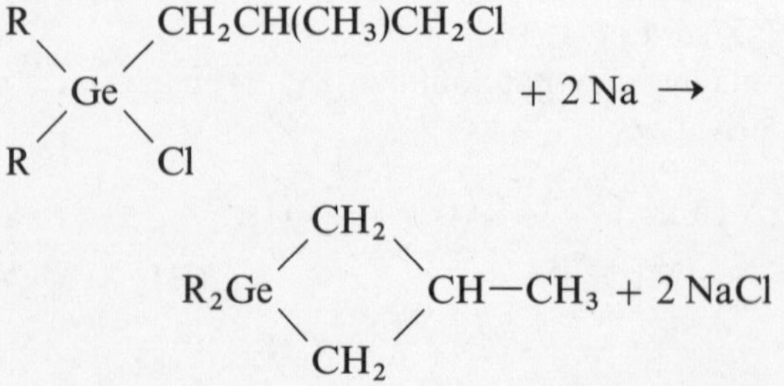

$$
\begin{array}{c}
R \diagdown \qquad \diagup CH_2CH(CH_3)CH_2Cl \\
\qquad Ge \\
R \diagup \qquad \diagdown Cl
\end{array}
\quad + \; 2\,Na \; \rightarrow
$$

$$
R_2Ge
\begin{array}{c}
\diagup CH_2 \diagdown \\
\qquad\qquad CH{-}CH_3 \; + \; 2\,NaCl \qquad \text{(yield: 70\%)} \\
\diagdown CH_2 \diagup
\end{array}
$$

$$
R = C_2H_5 \text{ (solvent toluene)}
$$
$$
R = C_4H_9 \text{ (solvent xylene)}
$$

and allows the synthesis of germaspiranes containing a four-membered ring (75):

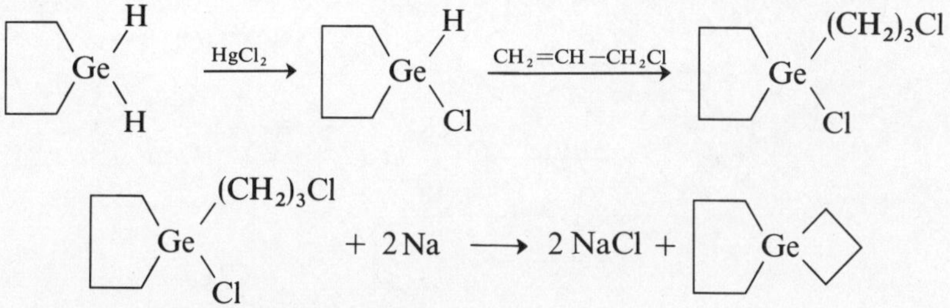

Desulphuration of a germathiacyclopentane

Action of sodium in boiling octane on α-thiagermacyclopentanes, by ring contraction, leads to the corresponding germacyclobutanes in good yields (70):

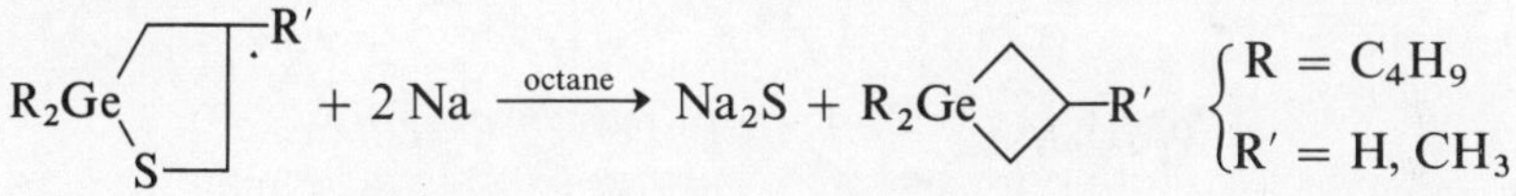

Using toluene as solvent, the yield decreases because of a side reaction:

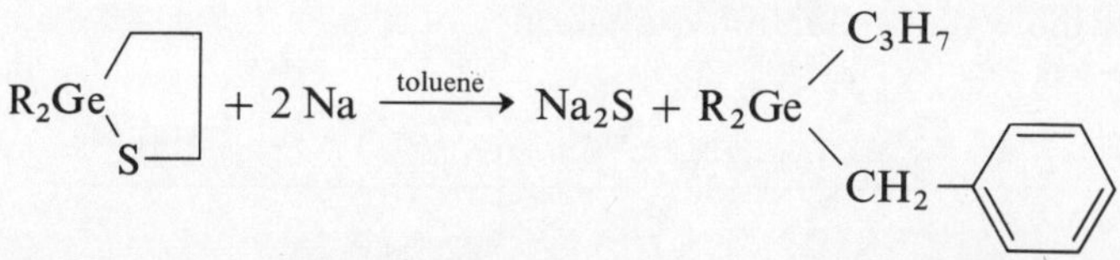

Hydrogenogermacyclobutanes are produced by using sodium to build a ring out of hydrogeno-1,4-dichlorogermane in toluene solution (78).

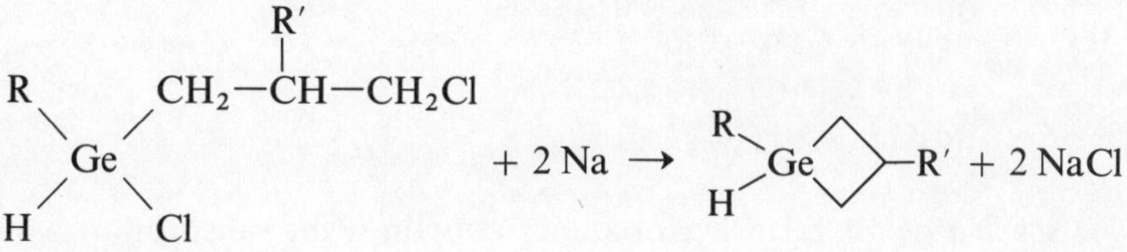

$$\begin{cases} R = C_4H_9 \\ R' = H \end{cases} \text{(yield: 45\%)} \quad \begin{cases} R = C_4H_9 \\ R' = CH_3 \end{cases} \text{(yield: 20\%)}$$

The yields are lower than with dialkyl series. This seems due to the thermal instability of these derivatives which even without any catalyst are easily condensed into polymers.

Halogenogermacyclobutanes (78) are produced by action of the derivatives mentioned above on different halogenated compounds:

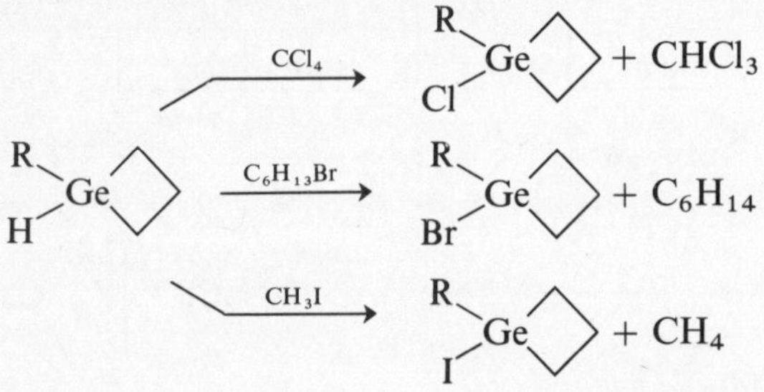

The yields are over 90%.

3-1-2-2   Physical Properties and Characteristics. The germacyclobutane derivatives so far known are colourless liquids, the boiling points of which are higher than those of the linear derivatives with the same number of carbons:

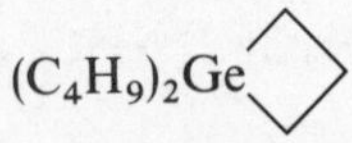

$(C_4H_9)_2Ge$ ◁▷   B.P. 111–12°C/18 mm

$(C_2H_5)_3Ge(C_5H_{11})$ B.P. 104–5°C/20 mm

Like cyclopropanes and cyclobutanes (20), germacyclobutanes display a notable molecular refraction exaltation:

	ΔMR_D	R_D (Ge—C cyclobutane)
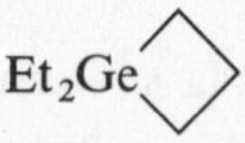 Et_2Ge ◁▷	+0.53	3.24
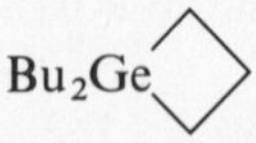 Bu_2Ge ◁▷	+0.53	3.24
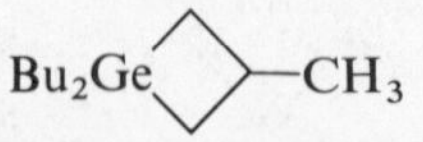 Bu_2Ge ◁▷—CH_3	+0.60	3.27

IR spectra of all germacyclobutanes contain a fine absorption band around $1120 \, cm^{-1}$ which disappears in the spectra of the compounds resulting from ring opening. Silacyclobutanes are characterized by an absorption band in the same region (136).

NMR spectra: Protons located on the germacyclobutane frame undergo chemical shift which depends on their position in the ring in reference to the heteroatom. The chemical shift is towards weaker fields which permits the differentiation of protons belonging to cyclobutane rings from those belonging to an aliphatic chain or a five- or six-membered cyclane:

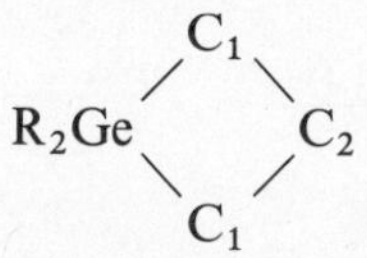

$$R_2Ge \underset{C_1}{\overset{C_1}{<}} C_2$$ Chemical shift (Ref. TMS (downfield from tetramethyl-silane))

$\delta H(C_1)$ 1.40 ppm $\delta H(CH_2$ in α of Ge)
  0.8–1.2 ppm

$\delta H(C_2)$ 2.20 ppm $\delta H(CH_2$ in β of Ge)
  1.2–1.3 ppm

The magnitude of spin–spin coupling between the protons on the 1 and 2 carbon atoms ($J = 7$ to 8 cps) is about the same as that reported for analogous cyclobutane systems.

3-1-2-3   Chemical Properties. The very high chemical reactivity of germacyclobutane derivatives, greater than that of cyclobutanes, seems due to the angular tension (ring strain) of the molecule and to the polarizability of the intracyclic Ge—C bonds.

Ring-opening reactions. Electrophilic reagents usually produce rapid quantitative ring-opening reactions:

$$\underset{R}{\overset{R}{>}}\underset{\delta^+}{Ge}\overset{}{<}\!\!\!\rangle_{\delta^-} \xrightarrow{+A^-B^+} \underset{R \quad A}{\overset{R}{>}}Ge-CH_2CH_2CH_2-B$$

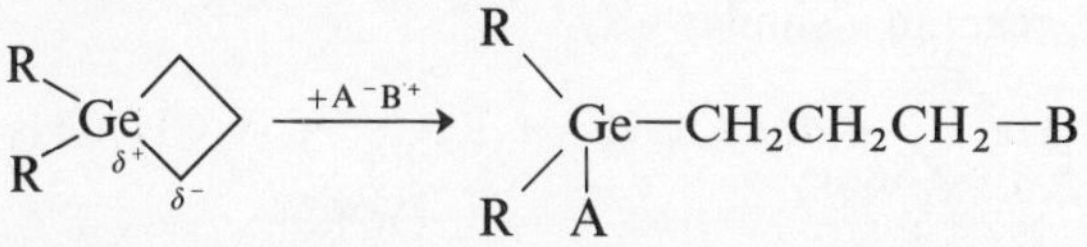

Action of halogens. Like the analogous silacyclobutanes (34, 51, 101), germacyclobutanes immediately react with bromine, one intracyclic germanium–carbon bond being cleaved. This quantitative reaction occurs even at $-60°C$ without any catalyst:

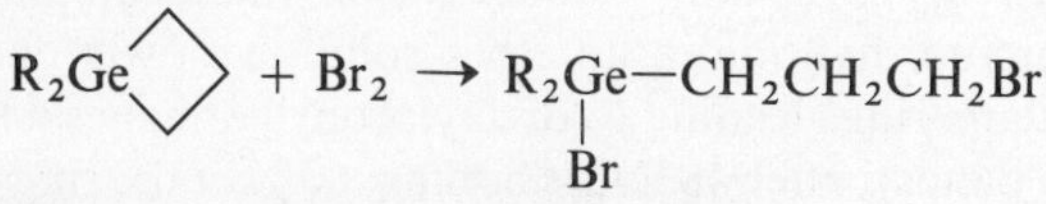

$$R_2Ge\overset{}{<}\!\!\!\rangle + Br_2 \rightarrow R_2Ge\underset{Br}{-}CH_2CH_2CH_2Br$$

The same applies to 4-germaspiro[3,4]octane:

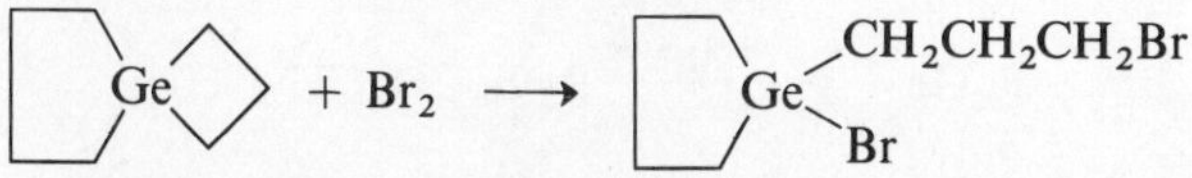

$$\Box Ge\overset{}{<}\!\!\!\rangle + Br_2 \longrightarrow \Box Ge\overset{CH_2CH_2CH_2Br}{\underset{Br}{<}}$$

Iodine also produces an exothermic reaction:

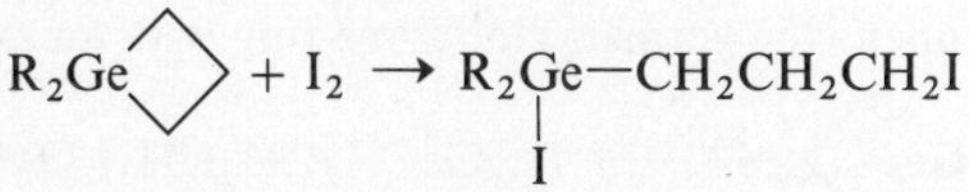

$$R_2Ge\overset{}{<}\!\!\!\rangle + I_2 \rightarrow R_2Ge\underset{I}{-}CH_2CH_2CH_2I$$

With iodine chloride, the chlorine atom becomes linked to the germanium, thus the mono-iodinated derivative may be isolated after alkylation:

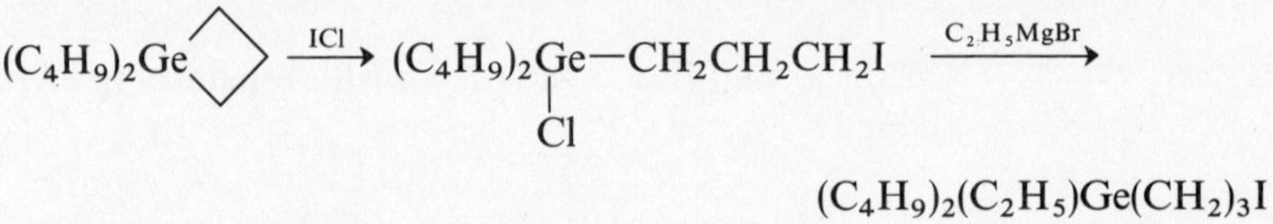

$$(C_4H_9)_2Ge\diamondsuit \xrightarrow{\text{ICl}} (C_4H_9)_2\underset{\underset{Cl}{|}}{Ge}-CH_2CH_2CH_2I \xrightarrow{C_2H_5MgBr}$$

$$(C_4H_9)_2(C_2H_5)Ge(CH_2)_3I$$

Action of protonic acids. Aqueous hydracids do not react with dialkylgermacyclopentanes and dialkylgermacyclohexanes. But, on the other hand, opening of the germacyclobutane ring is quantitative:

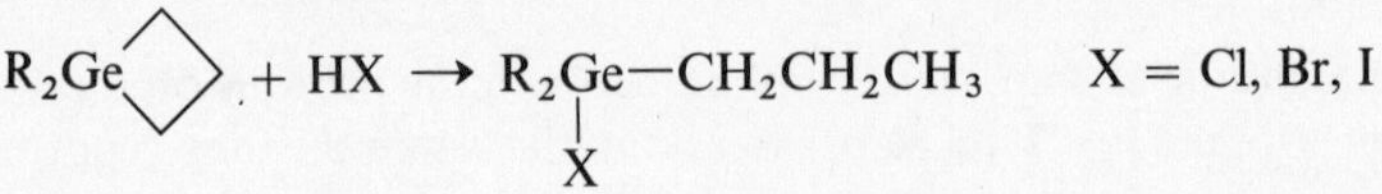

$$R_2Ge\diamondsuit + HX \rightarrow R_2\underset{\underset{X}{|}}{Ge}-CH_2CH_2CH_3 \qquad X = Cl, Br, I$$

Concentrated sulphuric acid reacts slowly with dialkylgermacyclohexanes and somewhat faster with dialkylgermacyclopentanes (3-1-2-1); with dialkylgermacyclobutanes an exothermic reaction immediately occurs:

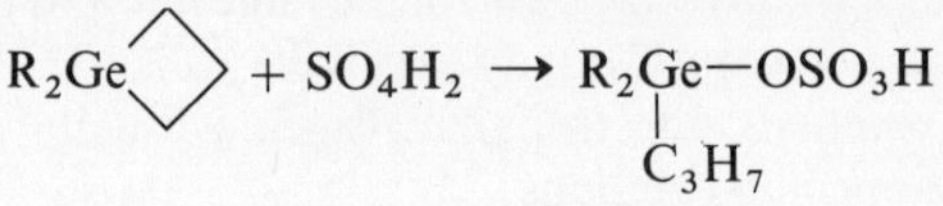

$$R_2Ge\diamondsuit + SO_4H_2 \rightarrow R_2\underset{\underset{C_3H_7}{|}}{Ge}-OSO_3H$$

Organic acids react in a similar way:

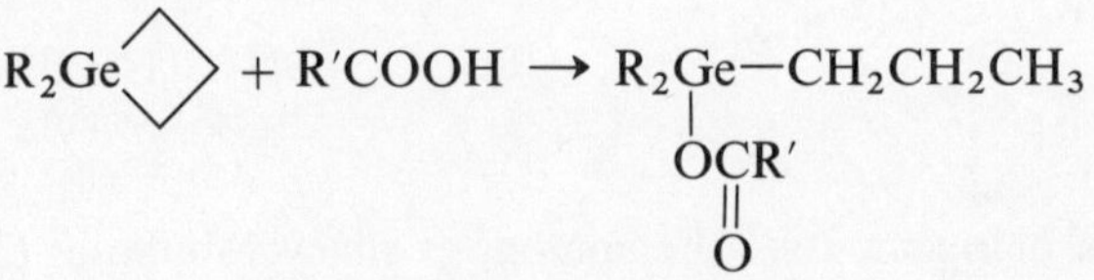

$$R_2Ge\diamondsuit + R'COOH \rightarrow R_2\underset{\underset{\underset{O}{\|}}{\underset{OCR'}{|}}}{Ge}-CH_2CH_2CH_3$$

$$(R' = CH_3-, ClCH_2-)$$

Nucleophilic reagents usually produce slower and incomplete reactions. Lithium aluminium hydride in an ether solution slowly opens dialkylgermacyclobutane rings leading to dialkylpropylgermanes (characteristic IR absorption bands); after 48 hours boiling, 60 % of the ring compounds are still unchanged:

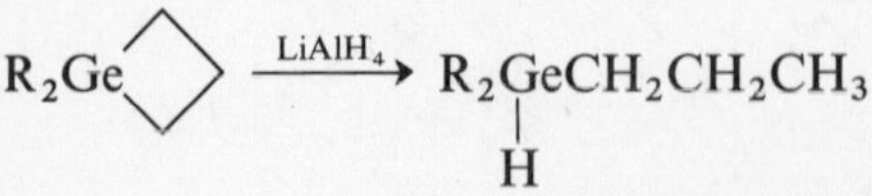

$$R_2Ge\diamondsuit \xrightarrow{\text{LiAlH}_4} R_2\underset{\underset{H}{|}}{Ge}CH_2CH_2CH_3$$

Alcoholic KOH opens the germacyclobutane ring with an average yield:

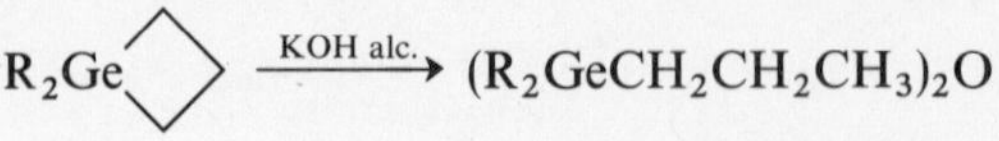

$$R_2Ge\diamondsuit \xrightarrow{\text{KOH alc.}} (R_2GeCH_2CH_2CH_3)_2O$$

alcoholates react easily:

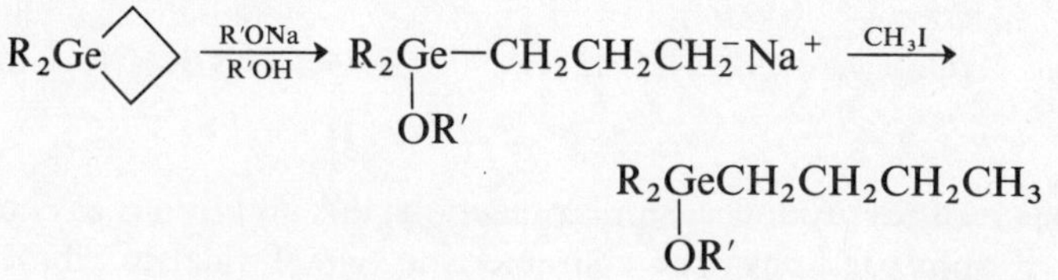

$$R_2Ge\diamond \xrightarrow[\text{R'OH}]{\text{R'ONa}} R_2Ge-CH_2CH_2CH_2^- Na^+ \xrightarrow{\text{CH}_3\text{I}}$$

$$R_2GeCH_2CH_2CH_2CH_3$$
$$|$$
$$OR'$$

$$\begin{cases} R = C_4H_9 \\ R' = CH_3, C_2H_5 \end{cases} \quad \text{(yield: 80\%)}$$

Organolithium compounds in ether are easily added to germacyclobutanes: after hydrolysis, a tetraalkylgermane can be characterized:

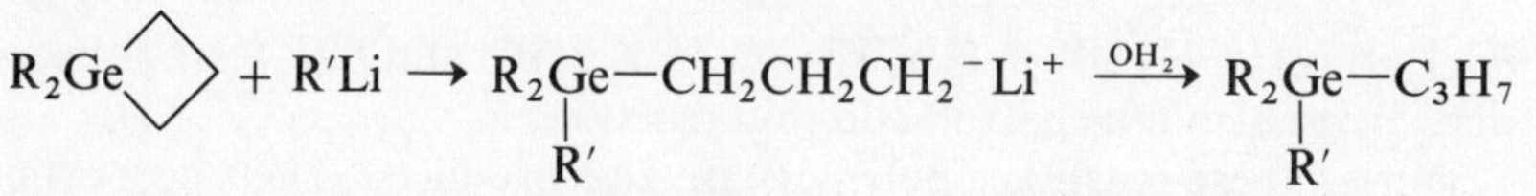

$$R_2Ge\diamond + R'Li \rightarrow R_2Ge-CH_2CH_2CH_2^- Li^+ \xrightarrow{OH_2} R_2Ge-C_3H_7$$
$$\qquad\qquad\qquad\qquad | \qquad\qquad\qquad\qquad\qquad |$$
$$\qquad\qquad\qquad\qquad R' \qquad\qquad\qquad\qquad\qquad R'$$

$$\begin{cases} R = C_2H_5, C_4H_9 \\ R' = C_4H_9, C_6H_5 \end{cases} \quad \text{(yield: 80\%)}$$

Other ring opening reactions. Action of compounds containing a mobile
halogen. Refluxing germanium tetrachloride leads to a tetrachlorinated
digermane:

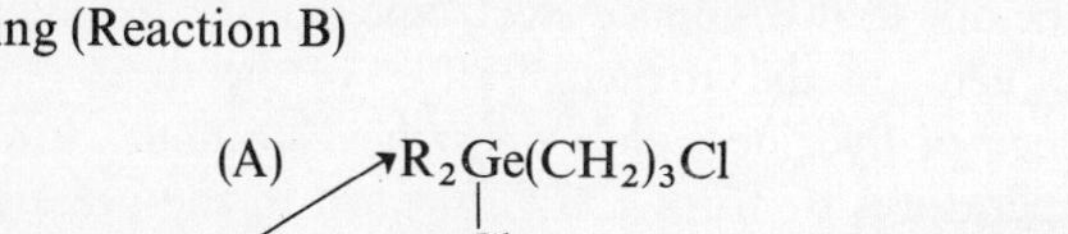

$$R_2Ge\diamond + GeCl_4 \rightarrow R_2Ge-(CH_2)_3GeCl_3$$
$$\qquad\qquad\qquad\qquad\qquad |$$
$$\qquad\qquad\qquad\qquad\qquad Cl$$

Sulphuryl chloride reacts violently. Reaction (A) produces sulphur
dioxide, but substitution reactions release HCl which opens the four-
membered ring (Reaction B)

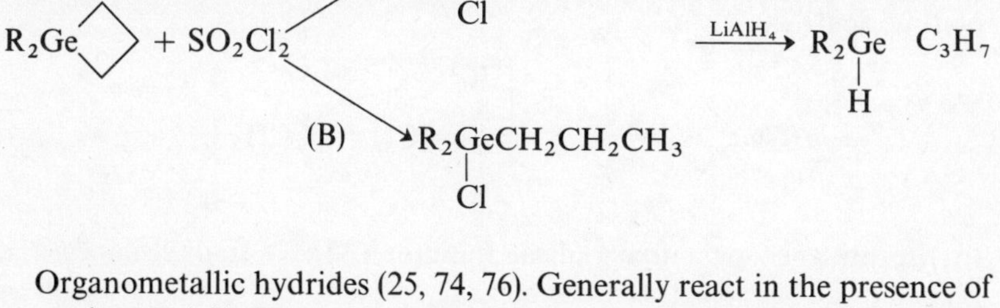

$$\text{(A)} \quad R_2Ge(CH_2)_3Cl$$
$$| \qquad\qquad\qquad$$
$$Cl$$

$$R_2Ge\diamond + SO_2Cl_2 \qquad\qquad \xrightarrow{LiAlH_4} R_2Ge \quad C_3H_7$$
$$\qquad\qquad\qquad\qquad\qquad\qquad\qquad |$$
$$\qquad\qquad\qquad\qquad\qquad\qquad\qquad H$$

$$\text{(B)} \quad R_2GeCH_2CH_2CH_3$$
$$| \qquad\qquad\qquad\quad$$
$$Cl$$

Organometallic hydrides (25, 74, 76). Generally react in the presence of
a catalyst. Trialkylgermanes in the presence of chloroplatinic acid open

the ring with excellent yields:

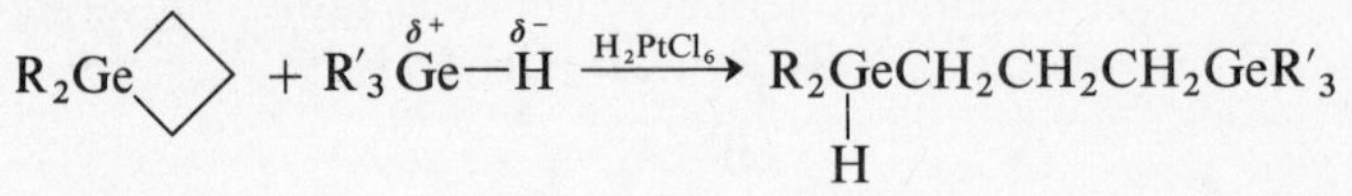

$$R_2Ge\langle\rangle + R'_3\overset{\delta^+}{Ge}-\overset{\delta^-}{H} \xrightarrow{H_2PtCl_6} R_2\underset{\underset{H}{|}}{Ge}CH_2CH_2CH_2GeR'_3$$

Trialkysilanes produce a similar reaction; this addition is easy to observe by IR spectrography: the characteristic Si—H valence vibration band (at $2100\ cm^{-1}$) disappears, giving way to the characteristic Ge—H vibration (at $2010\ cm^{-1}$)

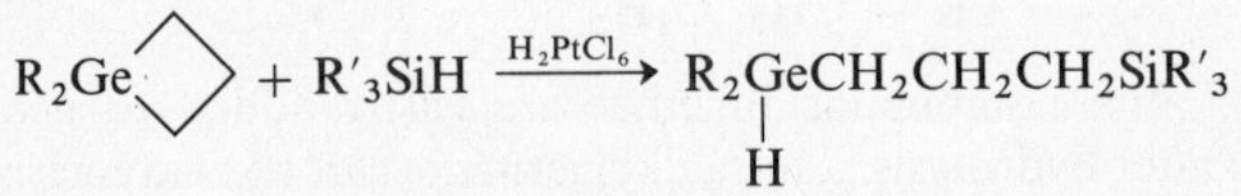

$$R_2Ge\langle\rangle + R'_3SiH \xrightarrow{H_2PtCl_6} R_2\underset{\underset{H}{|}}{Ge}CH_2CH_2CH_2SiR'_3$$

$(C_4H_9)_2Si(H)CH_3$, $Ph_2Si(H)CH_3$, $(C_4H_9)_2SiHCl$ and CH_3SiHCl_2 mixed derivatives also take part in addition reactions.

During these reactions, polymers are also produced. Their proportion depends on the reagent ratios, hydrogenosilane in excess favouring the 1:1 adduct. The reaction mechanism is likely to be that of a polyaddition, the monomeric adduct successively reacting with n cyclobutane derivative molecules, thus leading to a telomeric structure:

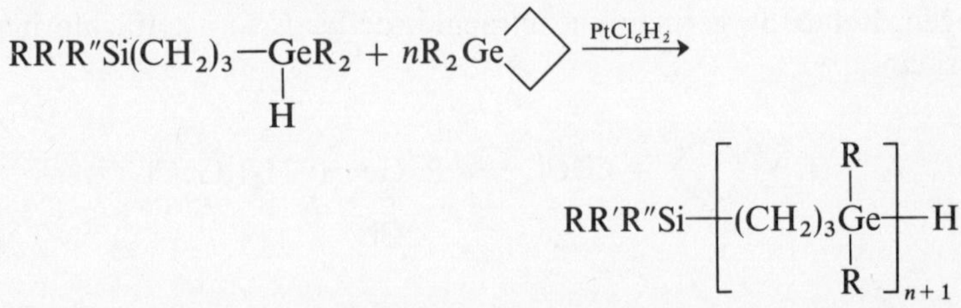

$$RR'R''Si(CH_2)_3-\underset{\underset{H}{|}}{Ge}R_2 + nR_2Ge\langle\rangle \xrightarrow{PtCl_6H_2}$$

$$RR'R''Si\left[(CH_2)_3\underset{\underset{R}{|}}{\overset{\overset{R}{|}}{Ge}}\right]_{n+1}H$$

Trichlorosilane does not lead to the expected addition reaction; this could most likely be due to the reduced electronic density around the hydrogen atom, consequent on the attractive effect of the three chlorine atoms. In the absence of hydrides, dialkylgermacyclobutanes are polymerized at high temperatures or from 150°C upward in the presence of chloroplatinic acid:

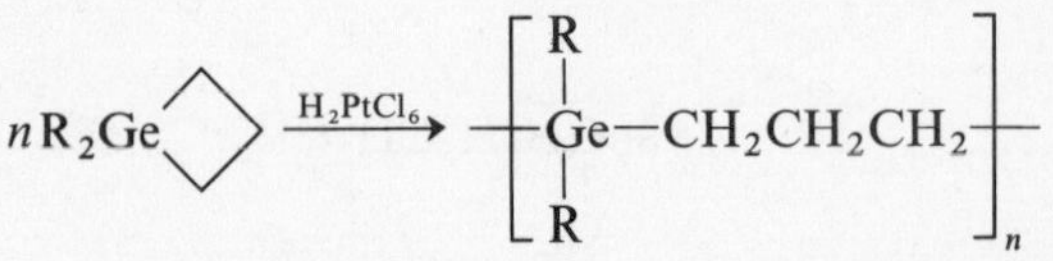

$$nR_2Ge\langle\rangle \xrightarrow{H_2PtCl_6} \left[\underset{\underset{R}{|}}{\overset{\overset{R}{|}}{Ge}}-CH_2CH_2CH_2\right]_n$$

In the presence of a free radical initiator (AIBN) trialkylsilanes and germanes do not react. On the contrary, silanes and germanes such as Cl_3SiH, CH_3Cl_2SiH, $(C_4H_9)_2ClSiH$, $(C_4H_9)_2GeHCl$, the $M^{\delta^+}-H^{\delta^-}$

polarity of which is slight, and which tend to undergo a homolytic scission of this bond, react with dialkylgermacyclobutanes producing a ring opening and forming dialkylpropylchlorogermane most likely resulting from the cleavage of the anti-Farmer adduct:

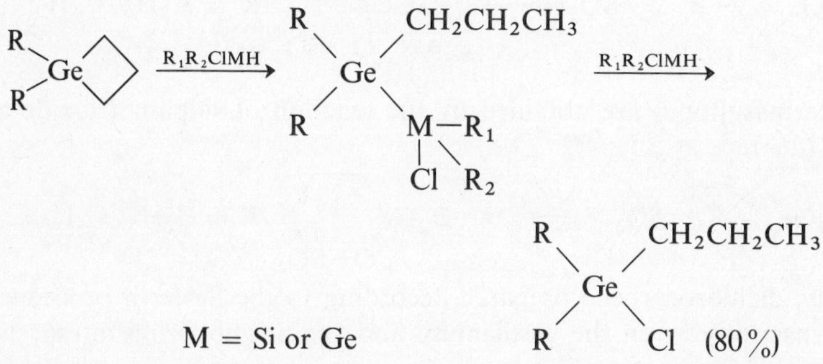

M = Si or Ge   Cl (80%)

Ring expansion of germacyclobutanes. Diazomethane in ether seems to be inert at ordinary temperature. 1,1-Dibutyl-3-methylgermacyclobutane heated at 200°C in the presence of aluminium chloride remains unchanged and isomerization does not occur:

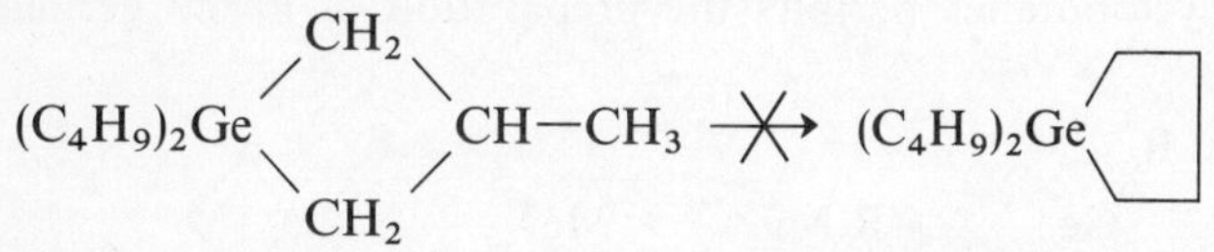

On the other hand, sulphur reacts around 220–230°C, and leads to the germatetrahydrothiophene series (75, 77):

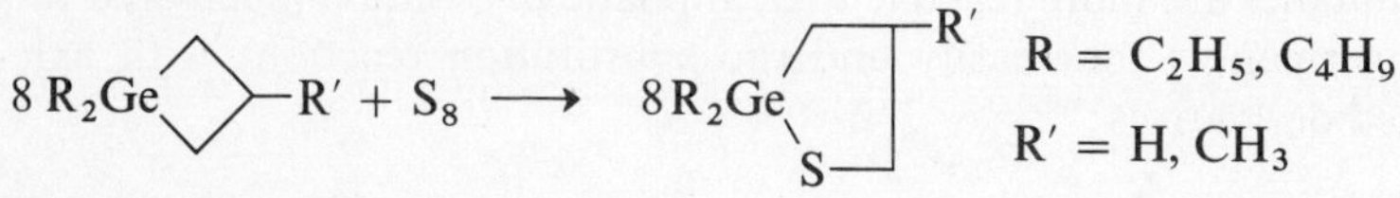

$$R = C_2H_5, C_4H_9$$
$$R' = H, CH_3$$

This reaction is also reported for 3-methylgermacyclobutanes and germaspiranes (77):

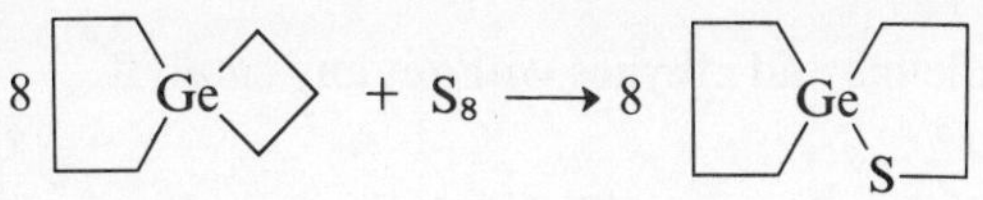

Selenium also produces ring expansions but at higher temperatures (260–270°):

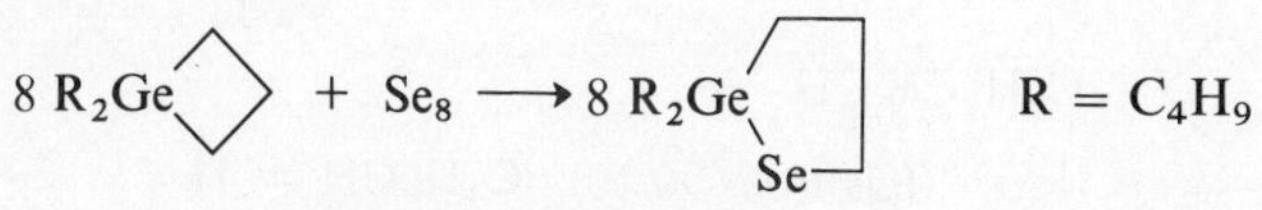

$$R = C_4H_9$$

Under the same conditions, tellurium has no action. Sulphur dioxide reacts around $-10°C$, producing a sulphinate ring (26):

$$\underset{R'}{\overset{R}{\diagdown}}Ge\diagup\hspace{-4pt}\square\hspace{-4pt}\diagdown-R'' + SO_2 \longrightarrow \underset{R'}{\overset{R}{\diagdown}}Ge\diagup\hspace{-4pt}\bigcirc\hspace{-4pt}\underset{O-SO}{} \qquad \begin{aligned} R &= C_2H_5, C_4H_9 \\ R' &= C_2H_5, C_4H_9 \\ R'' &= H, CH_3 \end{aligned}$$

and germasultones are obtained by the reaction of sulphur trioxide at $-70°$ (26a):

$$R_2Ge\diagup\hspace{-4pt}\triangle + SO_3 \xrightarrow{\;CH_2Cl_2\;} R_2Ge\diagup\hspace{-4pt}\bigcirc\hspace{-4pt}\underset{O-SO_2}{} \qquad R = C_2H_5, C_4H_9$$

Finally, dichlorocarbene prepared according to the Seyferth procedure (133) inserts between the germanium and the neighbouring intracyclic carbon (133a):

$$(C_2H_5)_2Ge\diagup\hspace{-4pt}\triangle + CCl_2 \longrightarrow (C_2H_5)_2Ge\diagup\hspace{-4pt}\underset{Cl_2C}{\square}$$

Substitution and addition reactions. The action of Grignard reagents on halogermacyclobutanes permits the preparation of mixed germacyclobutanes:

$$\underset{X}{\overset{R}{\diagdown}}Ge\diagup\hspace{-4pt}\triangle + R'MgX \rightarrow MgX_2 + \underset{R'}{\overset{R}{\diagdown}}Ge\diagup\hspace{-4pt}\triangle$$

Just as Sommer reported for analogous silanes (136), hydrogenogermacyclobutanes are more reactive than aliphatic or 5- and 6-membered ring hydrogermanes: they easily undergo substitution reactions with halogenated derivatives:

$$\underset{H}{\overset{R}{\diagdown}}Ge\diagup\hspace{-4pt}\triangle + R'X \rightarrow \underset{X}{\overset{R}{\diagdown}}Ge\diagup\hspace{-4pt}\triangle + R'H$$

and will add to olefins and alkynes without any catalyst:

$$\underset{H}{\overset{C_4H_9}{\diagdown}}Ge\diagup\hspace{-4pt}\triangle + n\text{-}C_6H_{13}CH{=}CH_2 \rightarrow \underset{n\text{-}C_6H_{13}CH_2CH_2}{\overset{C_4H_9}{\diagdown}}Ge\diagup\hspace{-4pt}\triangle$$

$$\underset{H}{\overset{C_4H_9}{\diagdown}}Ge\diagup\hspace{-4pt}\triangle + C_4H_9C{\equiv}CH \rightarrow \underset{C_4H_9CH{=}CH}{\overset{C_4H_9}{\diagdown}}Ge\diagup\hspace{-4pt}\triangle$$
$$(cis/trans\ 90/10)$$

Reducing properties. Unlike germacyclopentane and germacyclohexane derivatives, germacyclobutane compounds have reducing properties: alcoholic silver nitrate is immediately reduced without heating; mercuric chloride is reduced when heated with germacyclobutanes to the metal. It can be noted that analogous silacyclobutane derivatives also have reducing properties (37–39).

3-1-2-4 Cyclobutanic Polygermanes. Germacyclobutanes containing two or three germanium atoms in the ring are not known*, but a cyclobutane derivative, the ring of which is formed exclusively of germanium atoms, has been isolated by Neumann and coworkers (107, 108) in addition to other germacyclic derivatives from the reaction of diphenylgermane with diethylmercury:

$$4\,(C_6H_5)_2GeH_2 + 4\,(C_2H_5)_2Hg \;\longrightarrow\; 4\,[(C_6H_5)_2Ge\!-\!Hg]_n + 8\,C_2H_6$$

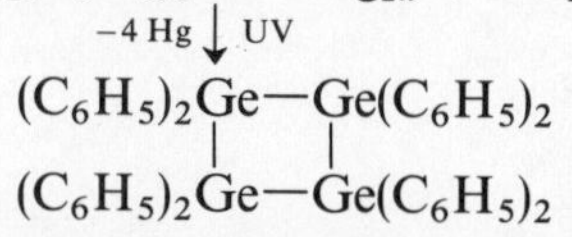

$$\underset{-4\,Hg}{\downarrow}\;UV$$

$$\begin{array}{c}(C_6H_5)_2Ge\!-\!Ge(C_6H_5)_2\\ |\qquad\qquad |\\ (C_6H_5)_2Ge\!-\!Ge(C_6H_5)_2\end{array}$$

On treatment with lithium and bromobenzene in THF, this cyclobutane derivative leads to decaphenyltetragermane or to decaphenylcyclopenta-germane depending on the experimental conditions. Iodine in benzene solution at room temperature opens the ring, producing a linear α,ω-diiodinated tetragermane:

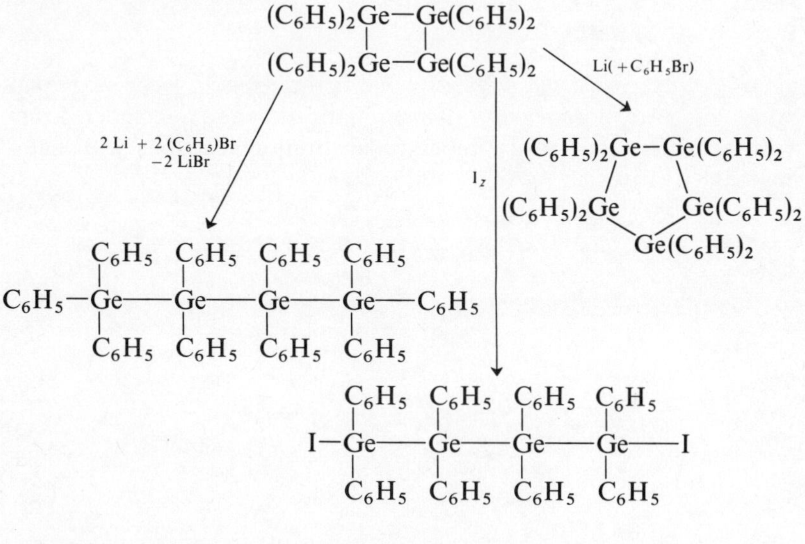

The presence of an $[(i\text{-}C_3H_7)_2Ge]_4$ tetragermane ring has been reported by Carrick and Glockling (17) among the products resulting from the reaction of germanium tetrachloride with isopropylmagnesium chloride. This ring derivative, produced only in very small quantities, was identified by mass spectrometry.

3-1-3 Germacyclobutanes

Table 3-1 showing some physical properties of germacyclobutanes appears on the following pages.

Table 3-1
Germacyclobutanes

COMPOUND	EMPIRICAL FORMULA	M.P. (°C)	B.P. (°C/mm)	n_D^{20}	d_4^{20}	REFERENCES
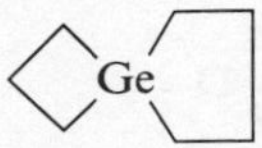 (spiro Ge: cyclopentane + cyclobutane)	$C_7H_{14}Ge$	—	89/38	1.5185	1.2043	75, 77
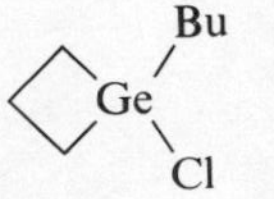 Bu–Ge–Cl (cyclobutane)	$C_7H_{15}GeCl$	—	94/18	1.4900	1.1912	78
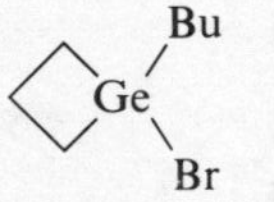 Bu–Ge–Br (cyclobutane)	$C_7H_{15}GeBr$	—	107/18	1.5157	1.4520	78
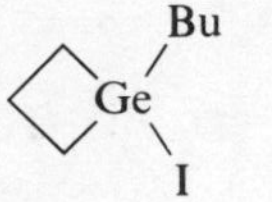 Bu–Ge–I (cyclobutane)	$C_7H_{15}GeI$	—	121/18	1.5558	1.6358	78
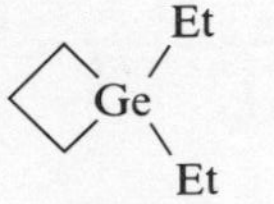 Et–Ge–Et (cyclobutane)	$C_7H_{16}Ge$	—	78/80	1.4738	1.0853	75, 77, 82
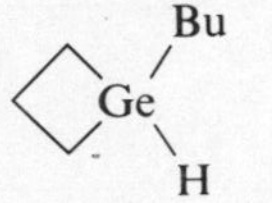 Bu–Ge–H (cyclobutane)	$C_7H_{16}Ge$	—	70/25	1.4781	1.0879	78

Table 3-1—continued

COMPOUND	EMPIRICAL FORMULA	M.P. (°C)	B.P. (°C/mm)	n_D^{20}	d_4^{20}	REFERENCES
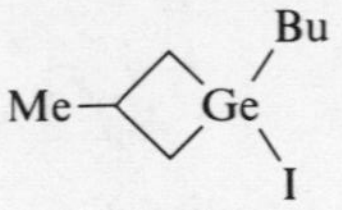 (cyclobutane-Ge ring; Me, Bu, I)	$C_8H_{17}GeI$	—	109/15	1.5410	1.5431	78
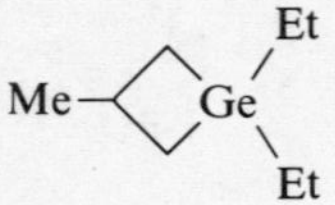 (cyclobutane-Ge ring; Me, Et, Et)	$C_8H_{18}Ge$	—	82/48	1.4682	1.0285	26
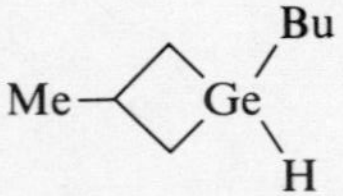 (cyclobutane-Ge ring; Me, Bu, H)	$C_8H_{18}Ge$	—	61/15	1.4701	1.0361	78
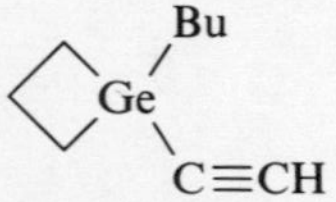 (cyclobutane-Ge ring; Bu, C≡CH)	$C_9H_{16}Ge$	—	67/10	1.4822	1.0795	78
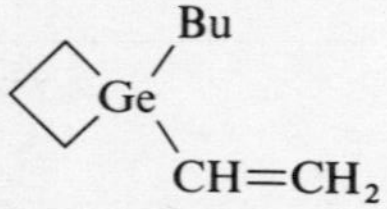 (cyclobutane-Ge ring; Bu, CH=CH$_2$)	$C_9H_{18}Ge$	—	80/18	1.4825	1.0560	78
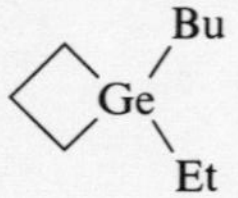 (cyclobutane-Ge ring; Bu, Et)	$C_9H_{20}Ge$	—	75/13	1.4720	1.0401	78

Structure	Formula		bp (°C/mm)	n	d	Ref.
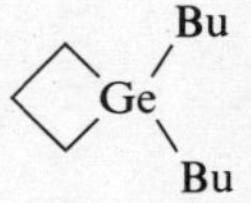 ![cyclobutane Ge(Bu)(Bu)]	$C_{11}H_{24}Ge$	—	112/18	1.4742	1.0163	75, 82
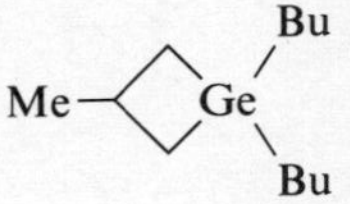 ![Me-cyclobutane Ge(Bu)(Bu)]	$C_{12}H_{26}Ge$	—	113/9	1.4690	0.9937	75
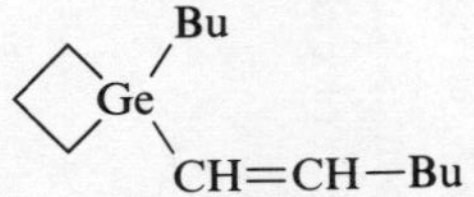 ![cyclobutane Ge(Bu)(CH=CH—Bu)]	$C_{13}H_{26}Ge$	—	75/0.1	1.4828	1.0096	78
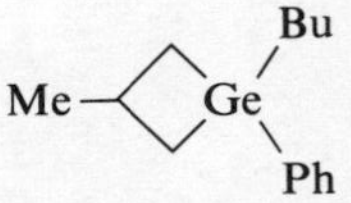 ![Me-cyclobutane Ge(Bu)(Ph)]	$C_{14}H_{22}Ge$	—	82/0.05	1.5355	1.0909	78
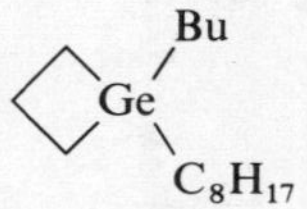 ![cyclobutane Ge(Bu)(C8H17)]	$C_{15}H_{32}Ge$	—	103/0.3	1.4738	0.9799	78
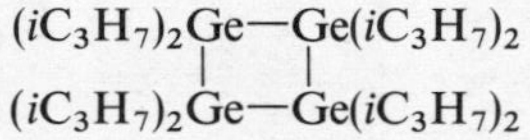 $(iC_3H_7)_2Ge—Ge(iC_3H_7)_2$ $(iC_3H_7)_2Ge—Ge(iC_3H_7)_2$	$C_{24}H_{56}Ge_4$	—	—	—	—	17
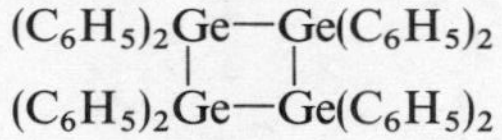 $(C_6H_5)_2Ge—Ge(C_6H_5)_2$ $(C_6H_5)_2Ge—Ge(C_6H_5)_2$	$C_{48}H_{40}Ge_4$	—	F = 238–39 (Kofler)	—	—	52a, 107a, 108, 109

3-2   SATURATED "COMMON" GERMANIUM RING DERIVATIVES

3-2-1   Germacyclanes and substituted derivatives

3-2-1-1   Germacyclanes. In 1932, Schwarz and Reinhardt (132) produced the first germanium ring derivative when treating 1,5-dibromopentane di-Grignard reagent with germanium tetrachloride:

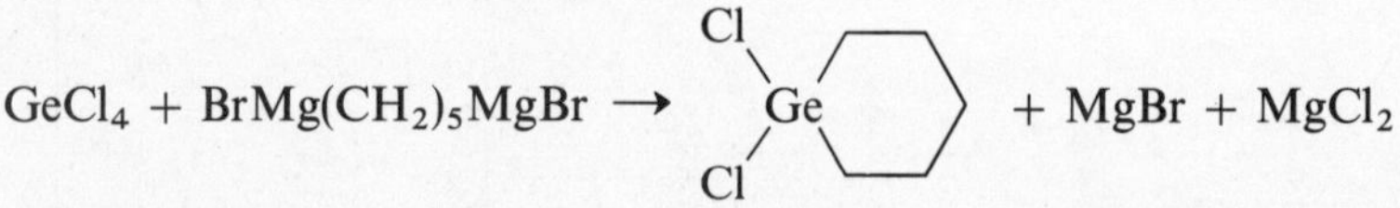

$$GeCl_4 + BrMg(CH_2)_5MgBr \rightarrow \underset{Cl}{\overset{Cl}{Ge}} \langle \rangle + MgBr + MgCl_2 \quad (I)$$

In 1962, Mazerolles (71) investigated this reaction, extending it to the 1,4 dibromobutane and 1,6 dibromohexane di-Grignard reagents. This study pointed out the fact that the action of $GeCl_4$ on aliphatic dimagnesium compounds is quite complex and that the equation proposed by Schwarz only accounts for part of it. The fact is that a dilute ether solution of $GeCl_4$ treated with an equimolar dimagnesium ether solution produces an exothermic reaction leading to:

unreacted germanium tetrachloride;

dichlorogermacyclane resulting from the expected reaction I;

germaspirane resulting from the action of 2 organo dimagnesium molecules with one germanium tetrachloride molecule:

$$(CH_2)_n \begin{array}{c} \text{---MgBr} \\ \\ \text{---MgBr} \end{array} + \underset{Cl}{\overset{Cl}{Ge}}\overset{Cl}{\underset{Cl}{}} + \begin{array}{c} \text{BrMg---} \\ \\ \text{BrMg---} \end{array} (CH_2)_n \rightarrow$$

$$2\,MgBr_2 + 2\,MgCl_2 + (CH_2)_nGe \quad (CH_2)_n;$$

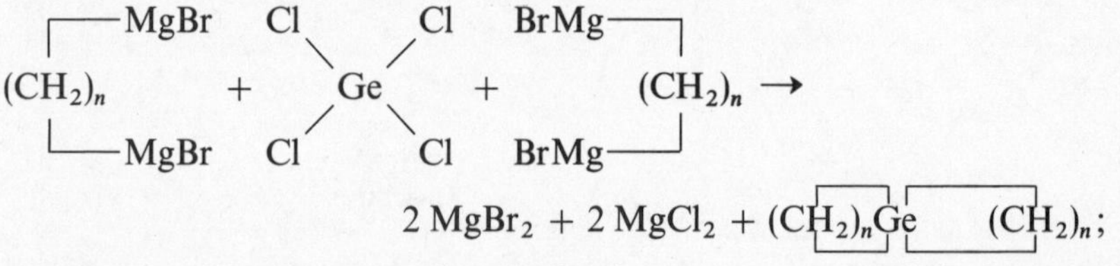

chlorinated polymers of undefined structure containing several germanium atoms. These heavy molecules may result either from the action of two germanium tetrachloride molecules with one dimagnesium molecule:

$$GeCl_4 + BrMg(CH_2)_nMgBr + GeCl_4 \rightarrow$$

$$MgBr_2 + MgCl_2 + Cl_3Ge(CH_2)_nGeCl_3$$

or in a more general manner from the action of N $GeCl_4$ molecules with $(N-1)$ dimagnesium molecules:

$$NGeCl_4 + (N-1)BrMg(CH_2)_nMgBr \rightarrow (N-1)MgBr_2$$

$$+ (N-1)MgCl_2 + Cl_3Ge\text{+}(CH_2)_nGeCl_2]_{N-2}(CH_2)_nGeCl_3$$

Finally, partial ring closures within these chains by the action of one dimagnesium molecule upon two consecutive chlorine atoms is not to be discounted.

Separating the first products is easy after reduction of the mixture by LiH_4Al.

$$GeCl_4 \rightarrow GeH_4 \quad gaseous$$

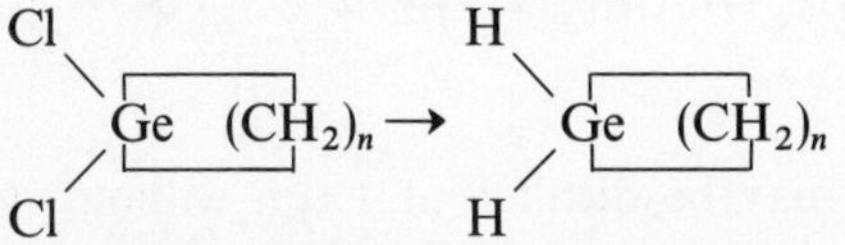

$$\left(\begin{array}{ll} B.P. \quad 91°C & n = 4 \\ \quad\quad 119°C & n = 5 \end{array} \right)$$

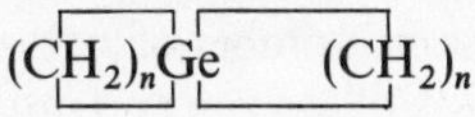

$(CH_2)_n Ge \quad (CH_2)_n$ remains unchanged

$$\left(\begin{array}{ll} B.P. \ 188°C & n = 4 \\ \quad 109°C/17 & n = 5 \end{array} \right)$$

Yields of ring derivatives depend upon the size of the ring; with reference to $GeCl_4$, the yields are about 20% for germacyclopentane and germa-cyclohexane, but close to zero for germacycloheptane. It must be noted that silicon and germanium tetrachlorides react in different ways with aliphatic dimagnesium compounds: in a detailed study of 5-, 6- and 7-membered ring silacyclanes (151), West managed better yields in silacyc-lanic derivatives, but does not seem to have noted that these reactions may have produced silaspiranes. Yields in ring derivatives are better with dialkyldibromo- or diaryldibromogermanes:

$$(C_2H_5)_2GeBr_2 + BrMg(CH_2)_5MgBr \rightarrow 2\,MgBr_2 + \text{[structure]}$$

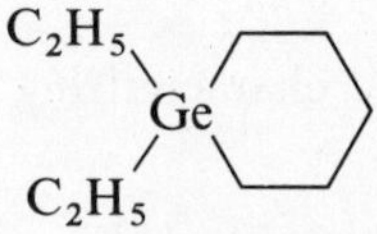

(yield: 59%)

The solvent used also affects the yields greatly: replacing diethylether (in which the dimagnesium compound is not soluble) by benzene favours ring closure:

Ether yield: 45%

$$(C_6H_5)_2GeBr_2 + BrMg(CH_2)_4MgBr \rightarrow 2\,MgBr_2 + \text{[structure]}$$

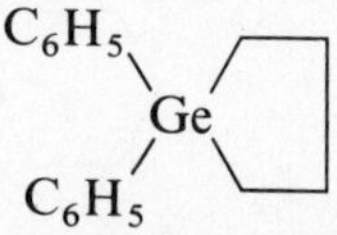

Benzene: 75%

The diphenyl ring derivative can easily be converted into germacyclane, bromine producing the dibrominated derivative, and this can be reduced by $LiAlH_4$:

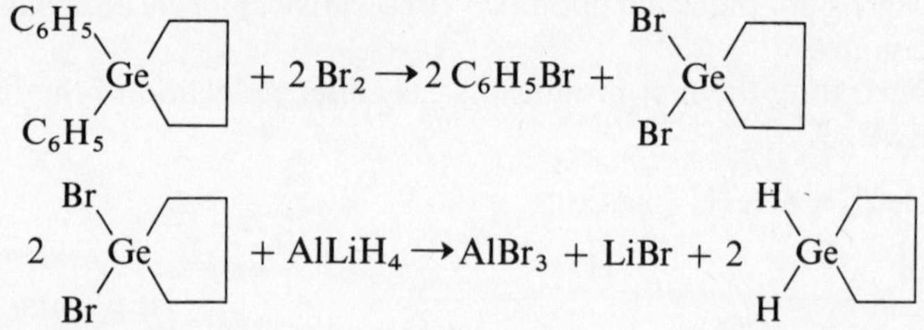

$$C_6H_5\!\!-\!\!Ge\!\!-\!\!\square + 2\,Br_2 \rightarrow 2\,C_6H_5Br + Br\!\!-\!\!Ge\!\!-\!\!\square$$

$$2\,Br\!\!-\!\!Ge\!\!-\!\!\square + AlLiH_4 \rightarrow AlBr_3 + LiBr + 2\,H\!\!-\!\!Ge\!\!-\!\!\square$$

Germacyclanes are liquids that may be distilled at 1 atm without decomposition; their boiling points are notably higher than those of the aliphatic germanium R_2GeH_2 dihydrides with the same number of carbon atoms. Germacyclanes are very sensitive to the action of oxygen and must be distilled and kept under inert gas. Germacyclopentane in contact with air is changed quickly into a polymerized oxide:

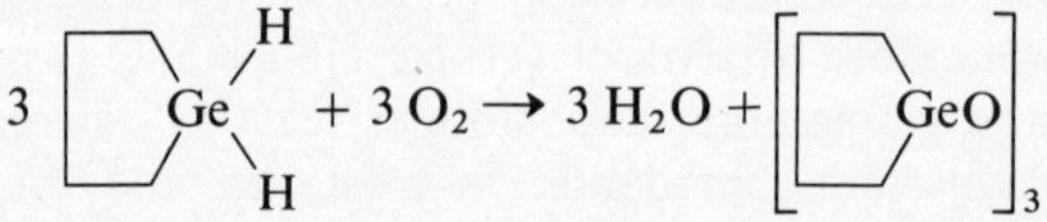

$$3\,\boxed{Ge}\begin{smallmatrix}H\\H\end{smallmatrix} + 3\,O_2 \rightarrow 3\,H_2O + \left[\boxed{GeO}\right]_3$$

Slower oxidation of germacyclohexane takes place in several stages: First one single germanium–hydrogen bond is oxidized: the infrared spectrum of this intermediate derivative shows at $2024\,cm^{-1}$ a strong absorption band characterizing the Ge—H bond (and not the $Ge\!\!<\!\!\begin{smallmatrix}H\\H\end{smallmatrix}$ bond), and another band characterizing O—H (66).

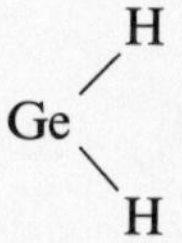

If oxidation is continued, the second Ge—H bond is oxidized, producing a crystallized ring oxide ($M = 640$) for which the infrared spectrum shows no band characteristic of the Ge—H or O—H vibrations. The germacyclohexane oxidation may be written:

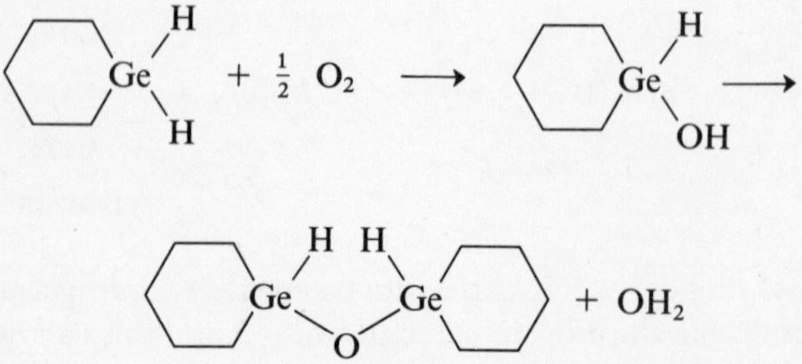

$$\bigcirc\!\!Ge\!\!<\!\!\begin{smallmatrix}H\\H\end{smallmatrix} + \tfrac{1}{2}\,O_2 \longrightarrow \bigcirc\!\!Ge\!\!<\!\!\begin{smallmatrix}H\\OH\end{smallmatrix} \longrightarrow$$

$$\bigcirc\!\!Ge\!\!<\!\!\begin{smallmatrix}H\\\\\end{smallmatrix}\!\!O\!\!-\!\!Ge\!\!<\!\!\begin{smallmatrix}H\\\\\end{smallmatrix}\!\!\bigcirc + OH_2$$

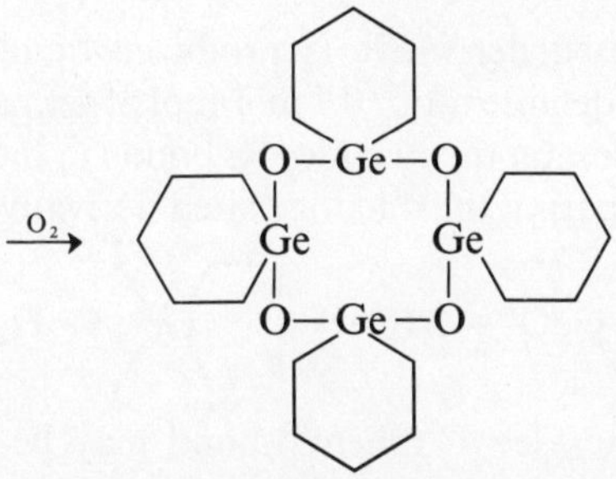

$\xrightarrow{\;O_2\;}$

O—Ge—O / Ge ... Ge / O—Ge—O

3-2-1-2 Halide Derivatives. Substituted halide derivatives may be prepared by taking advantage of the great Ge—H bond reactivity in germacyclanes: when heated, butyl bromide and iodide react with germacyclopentane and germacyclohexane; only the Ge—H bonds are cleaved and the ring is untouched:

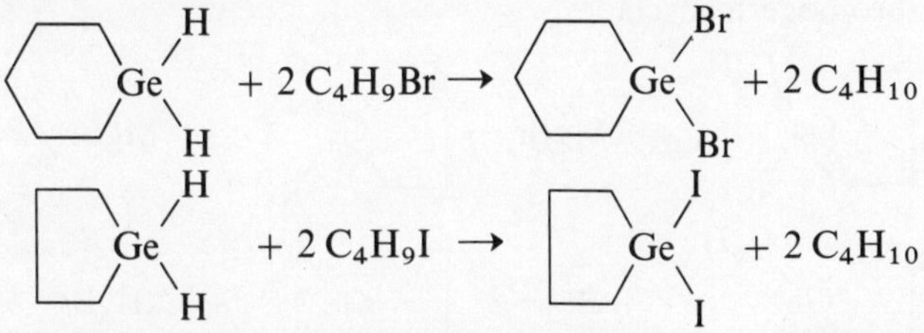

$\text{Ge(H)}_2 + 2\,C_4H_9Br \rightarrow \text{Ge(Br)}_2 + 2\,C_4H_{10}$

$\text{Ge(H)}_2 + 2\,C_4H_9I \rightarrow \text{Ge(I)}_2 + 2\,C_4H_{10}$

As with R_2GeH_2 aliphatic dihydrides, this reaction is complete. Sulphuryl chloride easily gives the dichloride derivative:

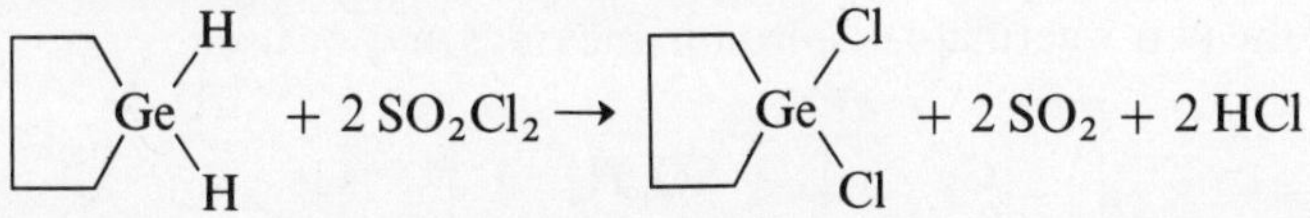

$\text{Ge(H)}_2 + 2\,SO_2Cl_2 \rightarrow \text{Ge(Cl)}_2 + 2\,SO_2 + 2\,HCl$

but it is hard to stop the reaction at its first stage. The milder action of mercuric chloride leads to good yields of the monochlorinated derivative:

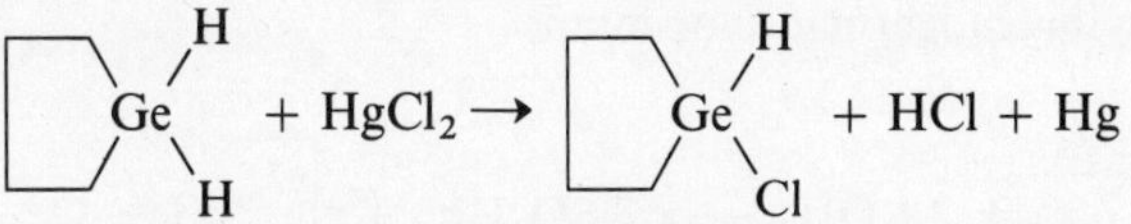

$\text{Ge(H)}_2 + HgCl_2 \rightarrow \text{Ge(H)(Cl)} + HCl + Hg$

Iodine reacts with germacyclanes as with germanium and silicon aliphatic hydrides (6, 7); hydrogen and hydrogen iodide are simultaneously produced, but again the ring remains untouched:

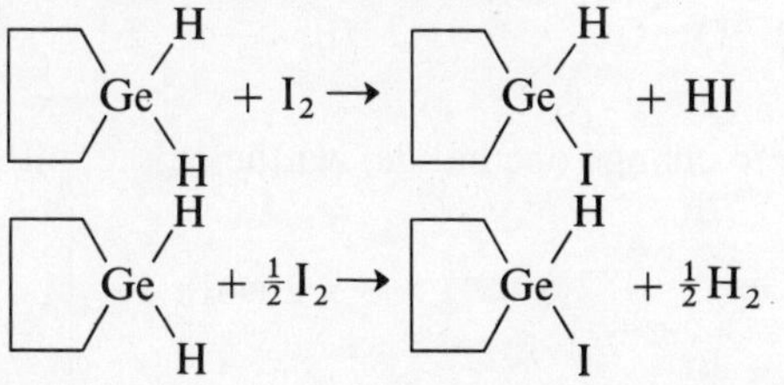

$\text{Ge(H)}_2 + I_2 \rightarrow \text{Ge(H)(I)} + HI$

$\text{Ge(H)}_2 + \tfrac{1}{2}I_2 \rightarrow \text{Ge(H)(I)} + \tfrac{1}{2}H_2$

The monohalogenated derivative is produced quite pure through the action of iodine in a definite ratio : 0.7 mol/mol of germacyclane. Hydrogen halides have no action on the two Ge—C bonds in the ring. This is taken advantage of in preparing the dihalogenated derivatives from the oxides:

$$\text{[Ge ring]}\text{GeO} + 2\,\text{HI} \rightarrow \text{[Ge ring]}\text{Ge}\begin{smallmatrix}I\\I\end{smallmatrix} + H_2O$$

The reactivity of the Ge—C (phenyl) bond may be used in producing phenylbromogermanes:

$$\text{[Ge ring]}\text{Ge}\begin{smallmatrix}C_6H_5\\C_6H_5\end{smallmatrix} + Br_2 \rightarrow \text{[Ge ring]}\text{Ge}\begin{smallmatrix}C_6H_5\\Br\end{smallmatrix} + C_6H_5Br$$

and alkylbromogermacyclanes:

$$\text{[Ge ring]}\text{Ge}\begin{smallmatrix}C_6H_5\\Br\end{smallmatrix} + RMgBr \rightarrow \text{[Ge ring]}\text{Ge}\begin{smallmatrix}C_6H_5\\R\end{smallmatrix} + MgBr_2$$

$$\text{[Ge ring]}\text{Ge}\begin{smallmatrix}C_6H_5\\R\end{smallmatrix} + Br_2 \rightarrow \text{[Ge ring]}\text{Ge}\begin{smallmatrix}Br\\R\end{smallmatrix} + C_6H_5Br$$

With germacyclopentanes, ring opening by concentrated sulphuric acid of one of the two 5-germa[4,4]spirononane rings may be used:

$$\text{[Ge spiro]}\text{Ge} + SO_4H_2 \rightarrow \text{[Ge ring]}\text{Ge}\begin{smallmatrix}C_4H_9\\SO_4H\end{smallmatrix}$$

This derivative, when treated with base and then with hydrobromic acid, leads to butylbromogermacyclopentane:

$$2\,\text{[Ge ring]}\text{Ge}\begin{smallmatrix}C_4H_9\\SO_4H\end{smallmatrix} + 2\,KOH \longrightarrow 2\,SO_4HK + \text{[ring]}\overset{C_4H_9}{\underset{}{Ge}}-O-\overset{C_4H_9}{\underset{}{Ge}}\text{[ring]}$$

$$\text{[ring]}\overset{C_4H_9}{\underset{}{Ge}}-O-\overset{C_4H_9}{\underset{}{Ge}}\text{[ring]} + 2\,HBr \longrightarrow 2\,\text{[ring]}\text{Ge}\begin{smallmatrix}C_4H_9\\Br\end{smallmatrix}$$

It is possible to change one halide, via the oxides, into another halide:

$$\text{[ring]}\text{Ge}\begin{smallmatrix}Br\\Br\end{smallmatrix} + 2\,NaOH \longrightarrow 2\,NaBr + \left[\text{[ring]}\text{GeO}\right]_n$$

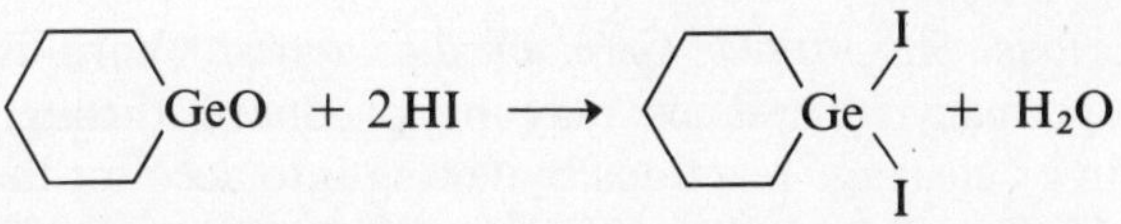

$$\text{(cyclohexyl)}Ge\text{O} + 2\,HI \longrightarrow \text{(cyclohexyl)}Ge\begin{smallmatrix}I\\[-2pt]\\I\end{smallmatrix} + H_2O$$

Action of an appropriate silver salt makes it possible to substitute quantitatively one halogen by another in the order $I \rightarrow Br \rightarrow Cl$

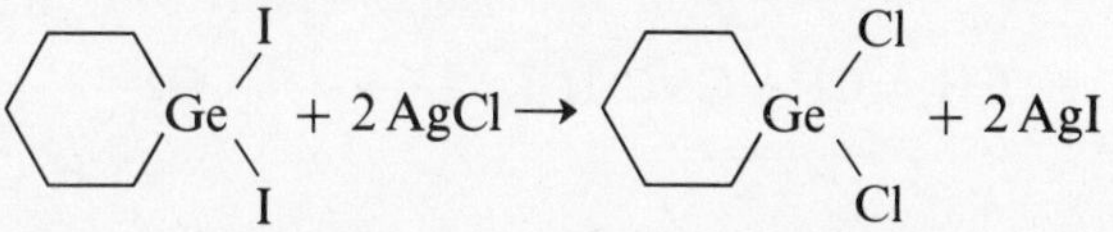

$$\text{(cyclohexyl)}Ge\begin{smallmatrix}I\\[-2pt]\\I\end{smallmatrix} + 2\,AgCl \rightarrow \text{(cyclohexyl)}Ge\begin{smallmatrix}Cl\\[-2pt]\\Cl\end{smallmatrix} + 2\,AgI$$

3-2-1-3   Dialkyl- and Diarylgermacyclanes. Two processes are used in preparing symmetrical derivatives:

Action of a dihalogermacyclane on an organomagnesium compound:

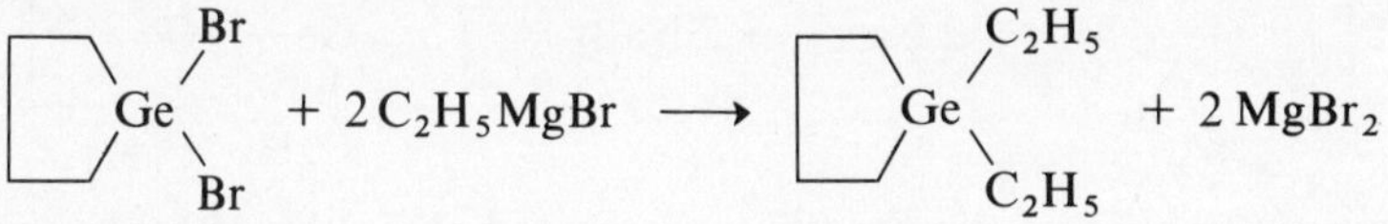

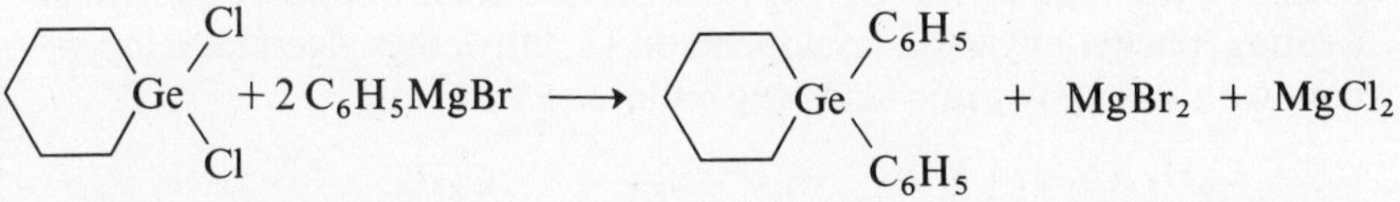

$$\text{(ring)}Ge\begin{smallmatrix}Br\\[-2pt]\\Br\end{smallmatrix} + 2\,C_2H_5MgBr \longrightarrow \text{(ring)}Ge\begin{smallmatrix}C_2H_5\\[-2pt]\\C_2H_5\end{smallmatrix} + 2\,MgBr_2$$

$$\text{(ring)}Ge\begin{smallmatrix}Cl\\[-2pt]\\Cl\end{smallmatrix} + 2\,C_6H_5MgBr \longrightarrow \text{(ring)}Ge\begin{smallmatrix}C_6H_5\\[-2pt]\\C_6H_5\end{smallmatrix} + MgBr_2 + MgCl_2$$

Action of a dialkyl- or diaryldihalogermanes on an aliphatic dimagnesium compound:

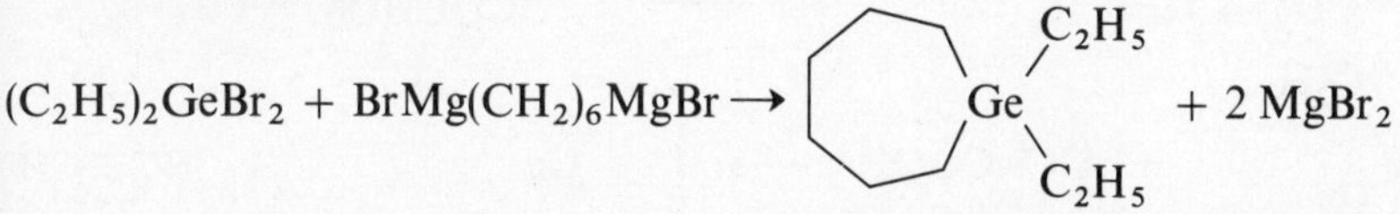

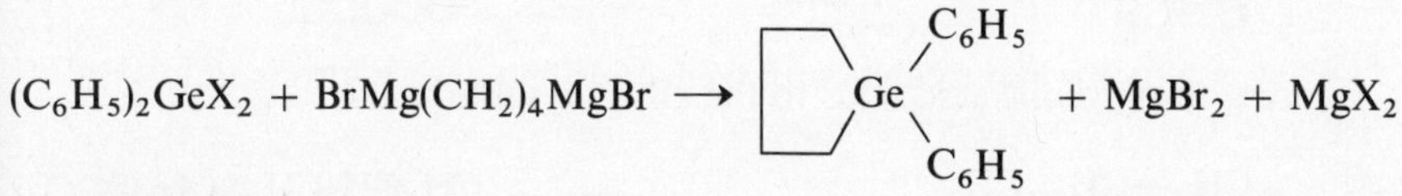

$$(C_2H_5)_2GeBr_2 + BrMg(CH_2)_6MgBr \longrightarrow \text{(ring)}Ge\begin{smallmatrix}C_2H_5\\[-2pt]\\C_2H_5\end{smallmatrix} + 2\,MgBr_2$$

$$(C_6H_5)_2GeX_2 + BrMg(CH_2)_4MgBr \longrightarrow \text{(ring)}Ge\begin{smallmatrix}C_6H_5\\[-2pt]\\C_6H_5\end{smallmatrix} + MgBr_2 + MgX_2$$

Alkylphenyl- or mixed dialkylgermacyclanes are formed from the diphenyl derivative:

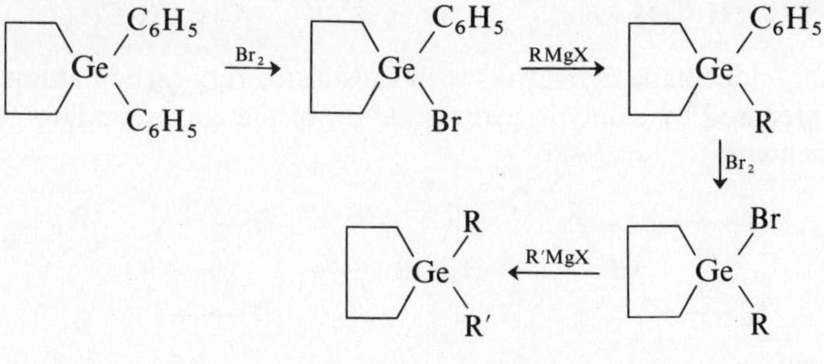

$$\text{(ring)}Ge\begin{smallmatrix}C_6H_5\\[-2pt]\\C_6H_5\end{smallmatrix} \xrightarrow{Br_2} \text{(ring)}Ge\begin{smallmatrix}C_6H_5\\[-2pt]\\Br\end{smallmatrix} \xrightarrow{RMgX} \text{(ring)}Ge\begin{smallmatrix}C_6H_5\\[-2pt]\\R\end{smallmatrix} \xrightarrow{Br_2}$$

$$\text{(ring)}Ge\begin{smallmatrix}R\\[-2pt]\\R'\end{smallmatrix} \xleftarrow{R'MgX} \text{(ring)}Ge\begin{smallmatrix}Br\\[-2pt]\\R\end{smallmatrix}$$

3-2-1-4 Various Substituted Germacyclopentanes. Various types of substituted germacyclopentanes may be produced through addition, condensation or cleavage reactions by taking into account the reactivity of the Ge—H, Ge—X and Ge—C (allyl or ethynyl) bonds (72, 75).

Triethylvinylgermane easily reacts with butylgermacyclopentane

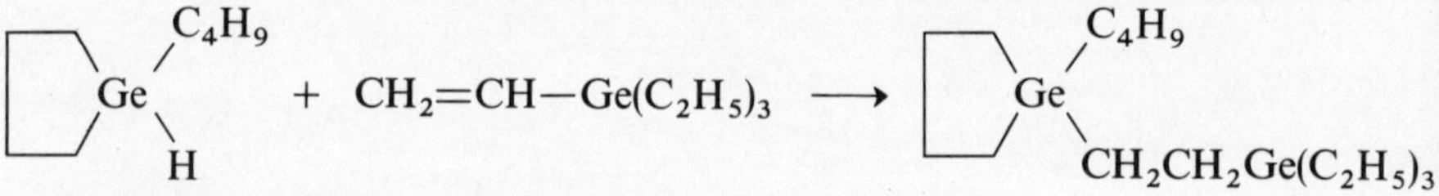

$$\text{Ge} \underset{\text{H}}{\overset{C_4H_9}{<}} + CH_2=CH-Ge(C_2H_5)_3 \longrightarrow \text{Ge} \underset{CH_2CH_2Ge(C_2H_5)_3}{\overset{C_4H_9}{<}}$$

Phenylbromogermacyclopentane condensed by sodium produces an 80 % yield of digermane:

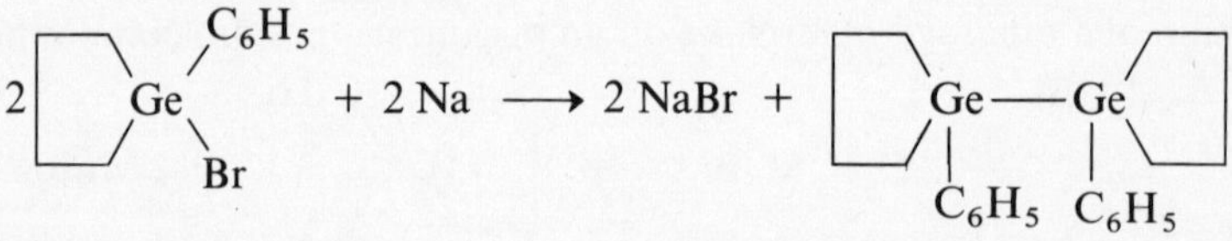

$$2 \; \text{Ge} \underset{Br}{\overset{C_6H_5}{<}} + 2\,Na \longrightarrow 2\,NaBr + \underset{C_6H_5 \;\; C_6H_5}{Ge-Ge}$$

Selective cleavage of the Ge—C (unsaturated chain) bond occurs when treating phenyl-allylgermacyclopentane or ethylethynylgermacyclopentane with strong organic acids (e.g. chloroacetic acids):

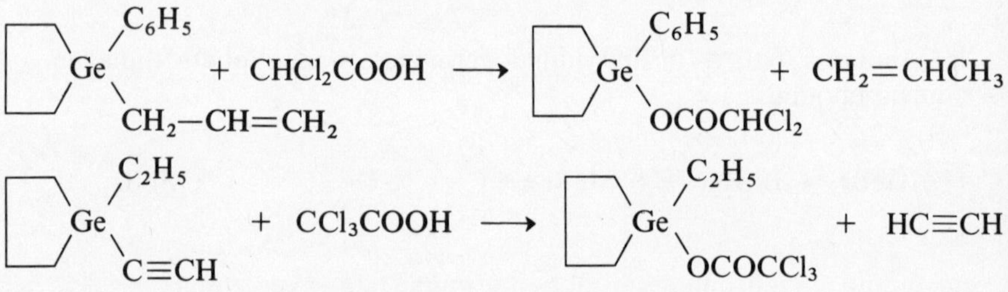

$$\text{Ge} \underset{CH_2-CH=CH_2}{\overset{C_6H_5}{<}} + CHCl_2COOH \longrightarrow \text{Ge} \underset{OCOCHCl_2}{\overset{C_6H_5}{<}} + CH_2=CHCH_3$$

$$\text{Ge} \underset{C\equiv CH}{\overset{C_2H_5}{<}} + CCl_3COOH \longrightarrow \text{Ge} \underset{OCOCCl_3}{\overset{C_2H_5}{<}} + HC\equiv CH$$

whereas thioglycolic acid adds to the allylic double bond:

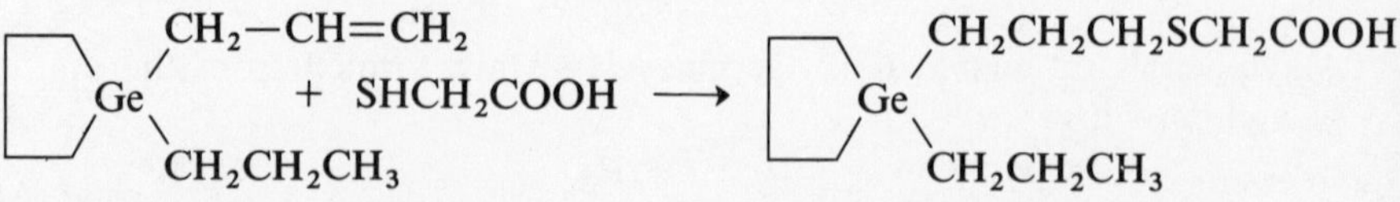

$$\text{Ge} \underset{CH_2CH_2CH_3}{\overset{CH_2-CH=CH_2}{<}} + SHCH_2COOH \longrightarrow \text{Ge} \underset{CH_2CH_2CH_3}{\overset{CH_2CH_2CH_2SCH_2COOH}{<}}$$

Germacyclopentane derivatives with substituted ring carbon atoms have been prepared by catalytic hydrogenation of the corresponding germacyclopentenes:

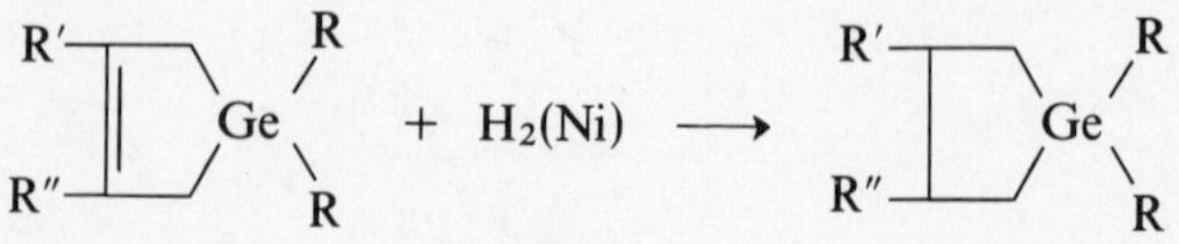

$$R' \underset{R''}{\overset{}{\Vert}} \text{Ge} \underset{R}{\overset{R}{<}} + H_2(Ni) \longrightarrow R' \underset{R''}{\overset{}{\Box}} \text{Ge} \underset{R}{\overset{R}{<}}$$

Hydrogenation of the derivative obtained from butadiene (R = R' = H) is easily carried out under normal pressure. On the other hand, substituted derivatives are hydrogenated only under higher pressures. With the dimethyl derivative (R' = R'' = CH$_3$) a mixture of cyclanic *cis* (75%) and *trans* (25%) isomers is produced. These two derivatives were identified by comparison with the *cis* isomer produced by consecutive hydroboration and protonation of germacyclopentene:

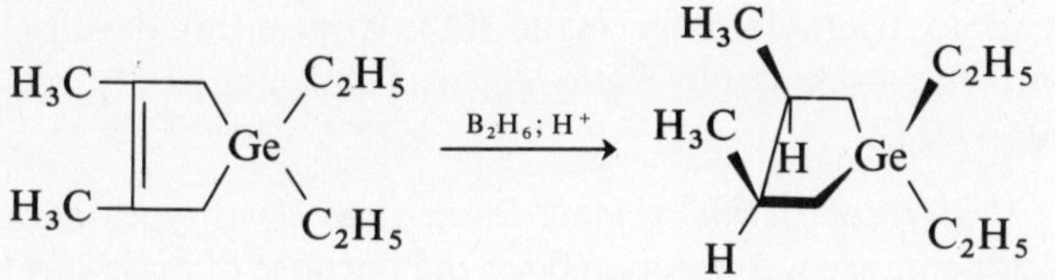

The *cis* dimethyl derivative isomerizes when heating on aluminium chloride:

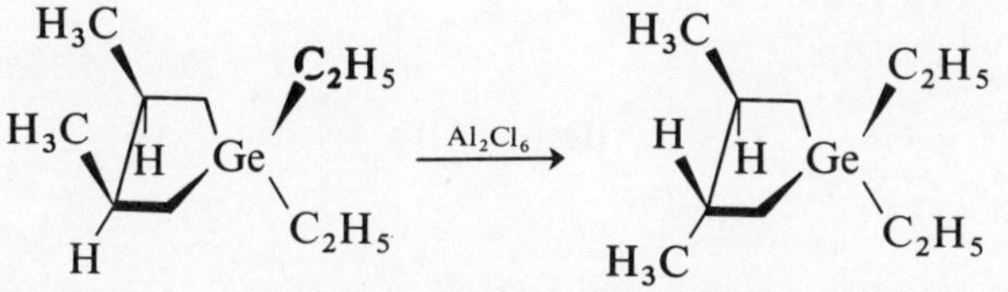

3-2-1-5   Germacyclopentane Ring Stability. The above reactions prove the great stability of the ring in saturated "common ring" germacyclanes. Because there is no ring strain, the Ge—C bonds within the ring resist many reagents: water, oxygen, LiAlH$_4$, alkyl halides, hydrogen halides, etc. A cold ethyl bromide solution of bromine has no reaction on diethylgermacyclopentane, but when this reaction is catalyzed by Al$_2$Br$_6$, the ring is immediately cleaved:

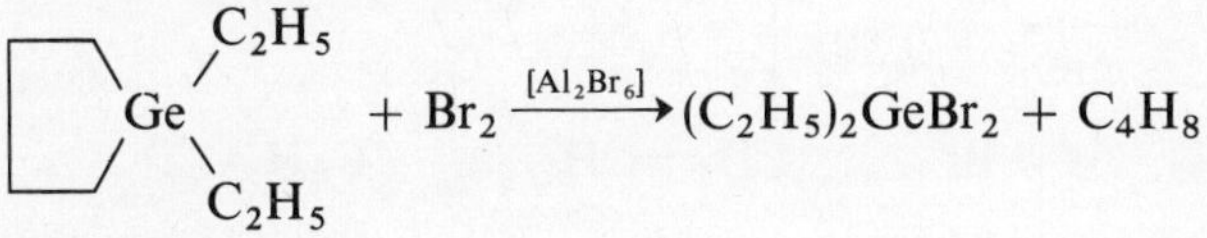

Cold, concentrated sulphuric acid reacts with diethylgermacyclopentane, and in this case the ring is opened:

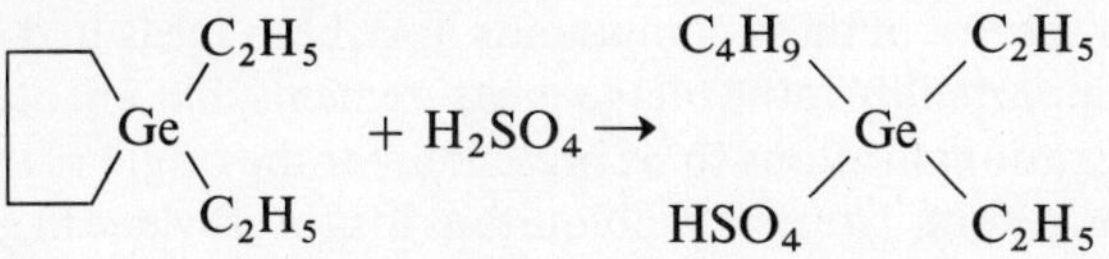

This ring-opening has been proved to occur by the following reactions:

$$\underset{HSO_4}{\overset{C_4H_9}{\diagdown}}Ge\underset{C_2H_5}{\overset{C_2H_5}{\diagup}} \xrightarrow{KOH} [(C_2H_5)_2Ge(C_4H_9)]_2O \xrightarrow{HBr}$$

$$(C_2H_5)_2(C_4H_9)GeBr \xrightarrow{C_2H_5MgBr} (C_2H_5)_3Ge(C_4H_9)$$

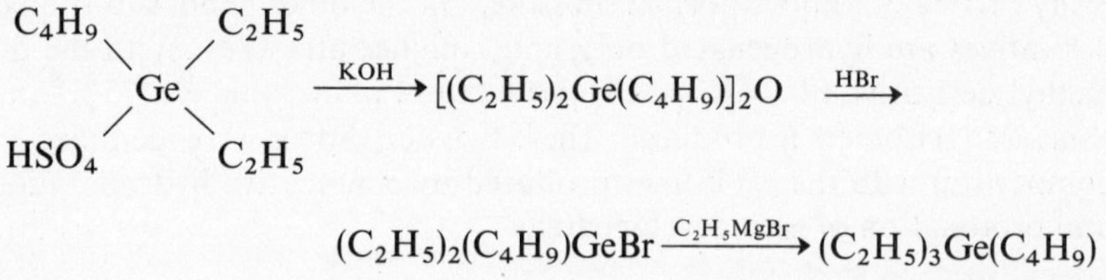

which lead to triethylbutylgermane (60). Concentrated sulphuric acid reacts with the more stable diethylgermacyclohexane only when heated above 30 to 40°C.

3-2-1-6 Deuterated Germacyclanic Derivatives. Two types of deuterated germacyclopentanes were prepared for the purpose of mass spectrometric investigation.

—Ge—D bond derivatives (27, 28). These compounds were obtained by lithium aluminium deuteride reduction of a dihalogenated germacyclopentane:

$$\underset{I}{\overset{I}{\diagdown}}Ge \Big\langle\!\!\!\begin{array}{c}\end{array} + LiAlD_4 \rightarrow AlI_3 + LiI + \underset{D}{\overset{D}{\diagdown}}Ge \Big\langle\!\!\!\begin{array}{c}\end{array}$$

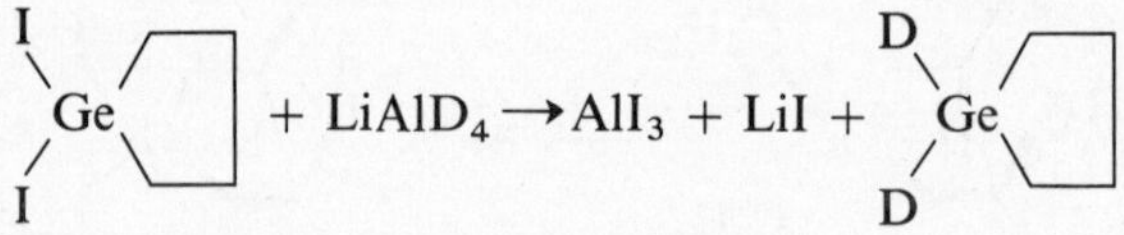

—C—D bond derivatives (28). The following reactions were used in preparing a series of β tetradeuterated germacyclopentanes

$$C_2H_5OCO\text{—}C\equiv C\text{—}COOC_2H_5 \xrightarrow{D_2/Pt}$$

$$C_2H_5OCOCD_2CD_2COOC_2H_5 \xrightarrow[Et_2O]{AlLiH_4} OHCH_2CD_2CD_2CH_2OH$$

$$\underset{CD_2\text{—}CH_2}{\overset{CD_2\text{—}CH_2}{\Big|}}Ge\underset{Br}{\overset{Br}{\diagup}} \xleftarrow{Br_2} \underset{CD_2\text{—}CH_2}{\overset{CD_2\text{—}CH_2}{\Big|}}Ge\underset{C_6H_5}{\overset{C_6H_5}{\diagup}} \xleftarrow[\substack{(1)\ Mg,\ Et_2O\\(2)\ (C_6H_5)_2GeBr_2}]{} $$

$$\downarrow HBr$$

$$BrCH_2CD_2CD_2CH_2Br$$

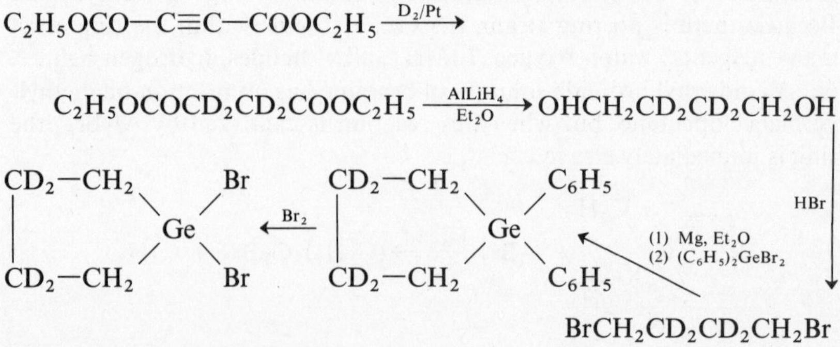

The dibrominated derivative gives the means to form the other deuterated germacyclopentanic derivatives.

The mass spectra of these compounds have been measured. Deuterium labelling, supported by low-voltage measurements, has enabled plausible mechanistic rationalizations to be presented for the origin of the principal ions in these spectra. The nearly ubiquitous loss of ethylene in germacyclopentanes was found to involve only C-2 and C-3

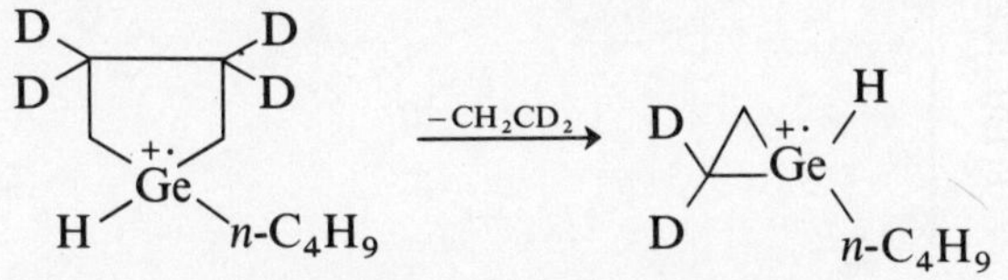

3-2-1-7 Germacyclanes and Substituted Derivatives

Tables 3-2, 3-3, and 3-4 showing some physical properties of germa-cyclopentanes and germacyclohexanes appear on the following pages.

Table 3-2
Germacyclopentanes

COMPOUND	EMPIRICAL FORMULA	M.P. (°C)	B.P. (°C/mm)	n_D^{20}	d_4^{20}	REFERENCES
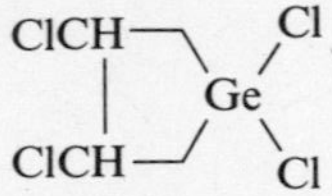	$C_4H_6Cl_4Ge$	—	—	—	—	94
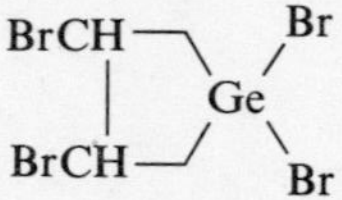	$C_4H_6Br_4Ge$	73	—	—	—	70
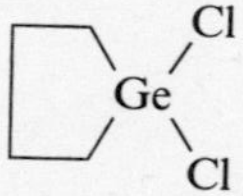	$C_4H_8GeCl_2$	—	94/64	1.5101	1.5239	71, 75
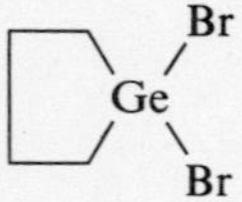	$C_4H_8GeBr_2$	—	97/15 113/35	1.5720	2.0806	71, 75
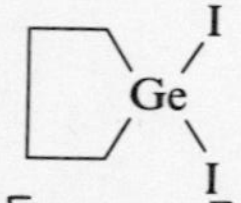	$C_4H_8GeI_2$	—	107/5	1.6770	2.5000	71
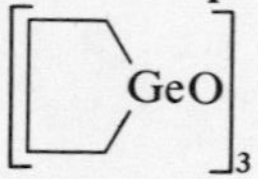	$(C_4H_8GeO)_3$	166	—	—	—	71

Compound	m.p.	b.p.	n_D	d	Ref.
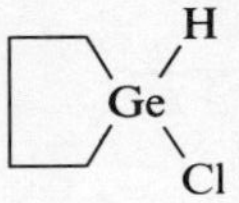 C_4H_9GeCl	—	154/745	1.5078	1.4218	75
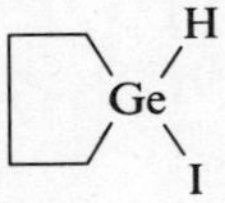 C_4H_9GeI	—	107/48	1.5956	1.9924	71
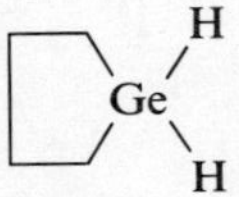 $C_4H_{10}Ge$	—	91–92/760	1.4838	1.2261	71, 75
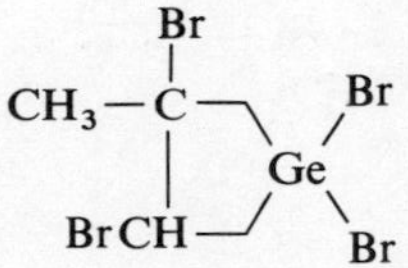 $C_5H_8Br_4Ge$	—	120/0.2	1.6283	2.4994	70
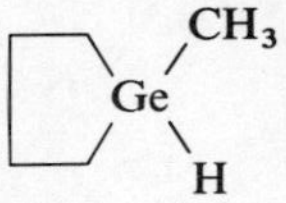 $C_5H_{12}Ge$	—	105/760	1.4700	1.1515	70
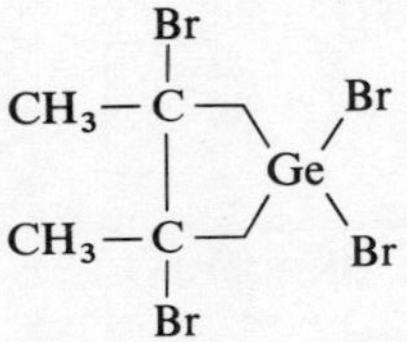 $C_6H_{10}Br_4Ge$	126–27	—	—	—	70
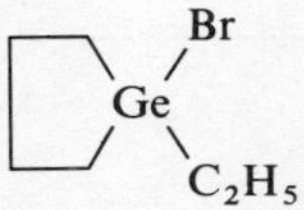 $C_6H_{13}GeBr$	—	75/9	1.5201	1.5261	72

Table 3-2—continued

COMPOUND	EMPIRICAL FORMULA	M.P. (°C)	B.P. (°C/mm)	n_D^{20}	d_4^{20}	REFERENCES
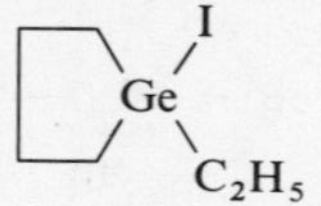 (cyclopentyl)Ge(I)(C$_2$H$_5$)(Cl)	$C_6H_{13}GeI$	—	113/28	1.5631	1.7450	72
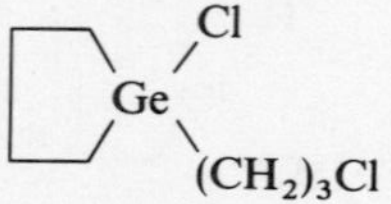 (cyclopentyl)Ge((CH$_2$)$_3$Cl)(Cl)	$C_7H_{14}GeCl_2$	—	136–37/12	1.5158	1.3693	75
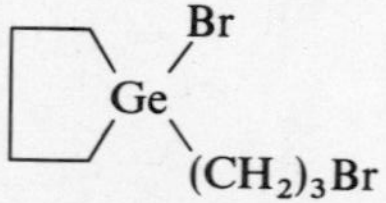 (cyclopentyl)Ge((CH$_2$)$_3$Br)(Br)	$C_7H_{14}GeBr_2$	—	118/1.1	1.5574	1.7933	75
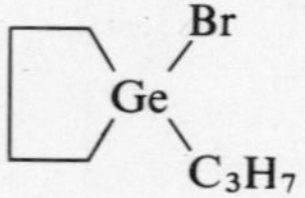 (cyclopentyl)Ge(Br)(C$_3$H$_7$)	$C_7H_{15}GeBr$	—	92.5/10	1.5153	1.4642	72
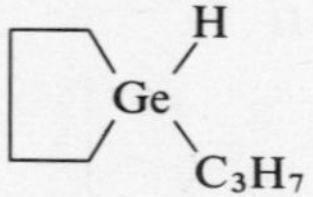 (cyclopentyl)Ge(H)(C$_3$H$_7$)	$C_7H_{16}Ge$	—	89/65	1.4754	1.0995	75
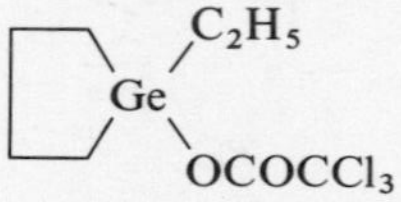 (cyclopentyl)Ge(C$_2$H$_5$)(OCOCCl$_3$)	$C_8H_{13}GeO_2Cl_3$	—	102/1	1.4992	1.4707	70

Structure	Formula		bp	n_D	d	Ref.
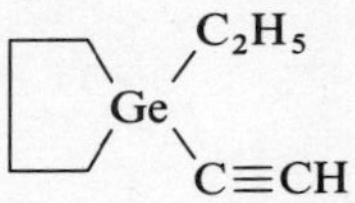 1-ethyl-1-ethynylgermacyclopentane (C_2H_5, $C{\equiv}CH$)	$C_8H_{14}Ge$	—	85/75	1.4862	1.1402	75
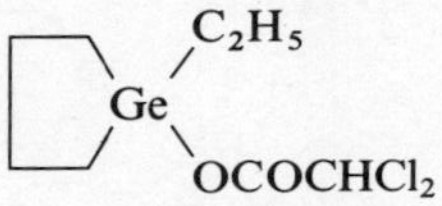 (C_2H_5, $OCOCHCl_2$)	$C_8H_{14}GeO_2Cl_2$	—	106–7/1.4	1.4949	1.4103	70
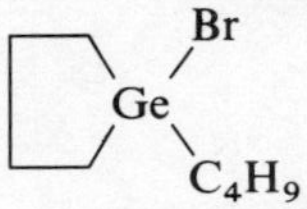 (Br, C_4H_9)	$C_8H_{17}GeBr$	—	82/2.2	1.5120	1.4134	72
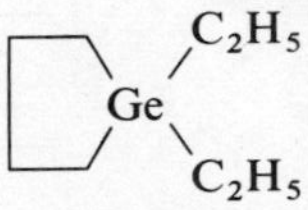 (C_4H_9, C_2H_5)	$C_8H_{18}Ge$	—	173–74/750	1.4725	1.0761	71
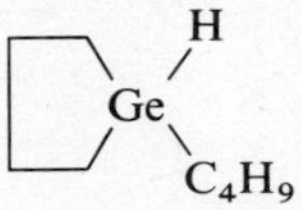 (H, C_4H_9 / C_2H_5)	$C_8H_{18}Ge$	—	179/745	1.4756	1.0694	72
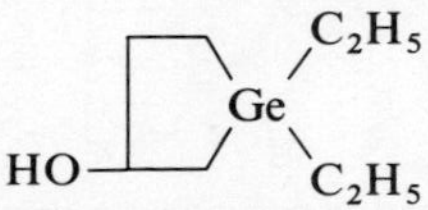 (C_2H_5, C_2H_5; HO, CH_3)	$C_8H_{18}GeO$	—	111/15	1.4944	1.1806	84
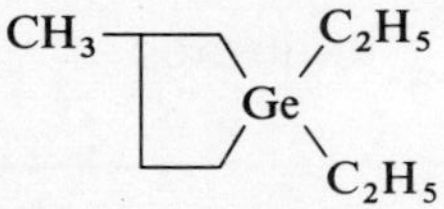 (C_2H_5, C_2H_5; CH_3)	$C_9H_{20}Ge$	—	71.5/14	1.4658	1.0427	84
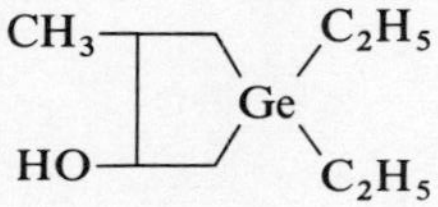 (C_2H_5, C_2H_5; CH_3, HO)	$C_9H_{20}GeO$	—	116.5/16	1.4855	1.1443	84

Table 3-2—continued

COMPOUND	EMPIRICAL FORMULA	M.P. (°C)	B.P. (°C/mm)	n_D^{20}	d_4^{20}	REFERENCES
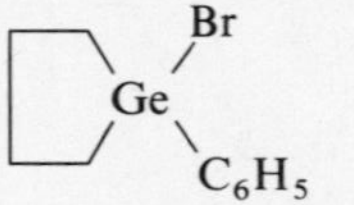 (germacyclopentane) Ge, substituents Br, C$_6$H$_5$	$C_{10}H_{13}GeBr$	—	116/0.9	1.5878	1.5438	71
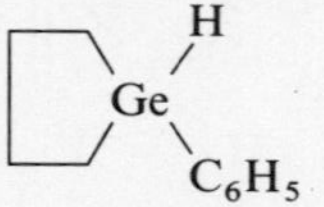 (germacyclopentane) Ge, substituents H, C$_6$H$_5$	$C_{10}H_{14}Ge$	—	127/30	1.5601	1.2215	71
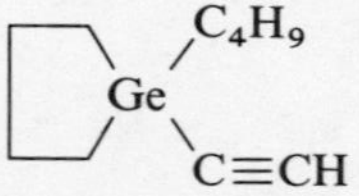 (germacyclopentane) Ge, substituents C$_4$H$_9$, C≡CH	$C_{10}H_{18}Ge$	—	101/28	1.4805	1.0737	75
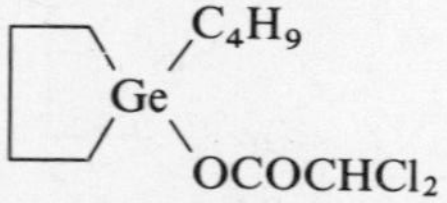 (germacyclopentane) Ge, substituents C$_4$H$_9$, OCOCHCl$_2$	$C_{10}H_{18}GeCl_2O_2$	—	116/0.9	1.4911	1.3240	70
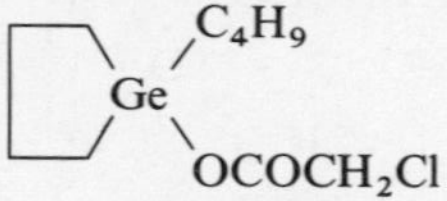 (germacyclopentane) Ge, substituents C$_4$H$_9$, OCOCH$_2$Cl	$C_{10}H_{19}GeClO_2$	—	116/1.4	1.4869	1.2631	70
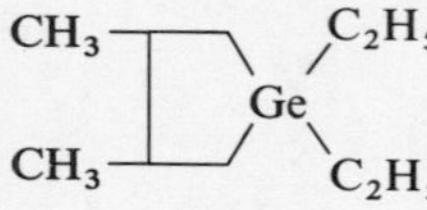 cis (dimethylgermacyclopentane) Ge, substituents C$_2$H$_5$, C$_2$H$_5$; CH$_3$, CH$_3$	$C_{10}H_{22}Ge$	—	90/17	1.4733	1.0455	84

Structure	Formula		b.p./mm	n_D	d	Ref.
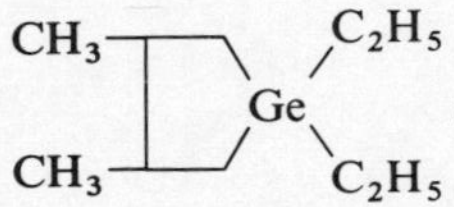 (Ge ring with C_2H_5, C_2H_5, CH_3, CH_3) *trans*	$C_{10}H_{22}GeO$	—	90/17	1.4665	—	84
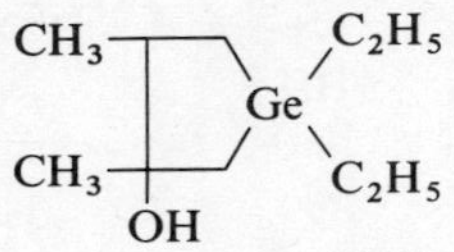 (Ge ring with C_2H_5, C_2H_5, CH_3, CH_3, OH)	$C_{10}H_{22}GeO$	—	112.5/12	1.4889	1.1272	84
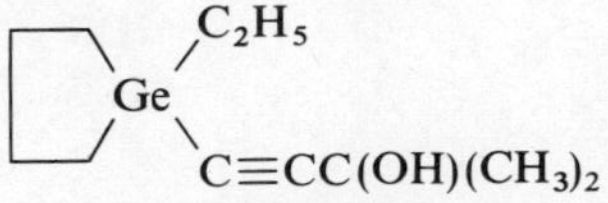 (Ge ring with C_2H_5, C≡CC(OH)(CH$_3$)$_2$)	$C_{11}H_{20}GeO$	—	100/1.4	1.4910	1.0988	70
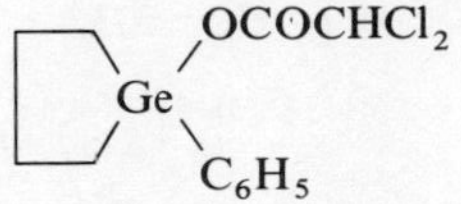 (Ge ring with OCOCHCl$_2$)	$C_{12}H_{14}GeO_2Cl_2$	—	123/0.3	1.5505	1.4290	72
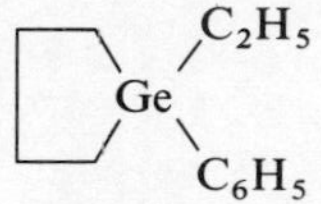 (Ge ring with C_6H_5, C_2H_5)	$C_{12}H_{18}Ge$	—	140/20	1.5441	1.1614	72
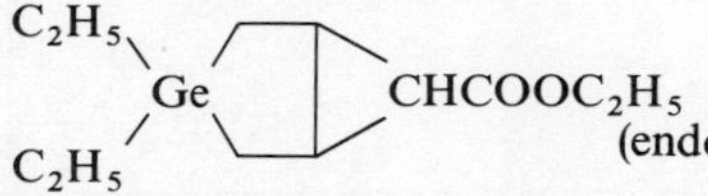 (bicyclic Ge with CHCOOC$_2$H$_5$, C_2H_5, C_2H_5) (endo)	$C_{12}H_{22}GeO_2$	—	102/0.7	1.4899	1.1637	84
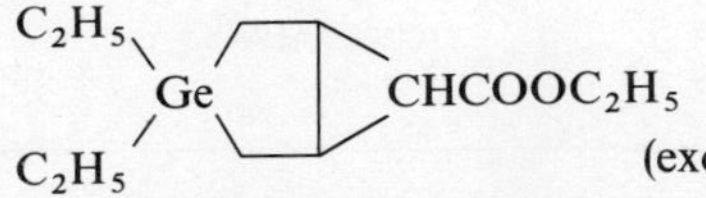 (bicyclic Ge with CHCOOC$_2$H$_5$, C_2H_5, C_2H_5) (exo)	$C_{12}H_{22}GeO_2$	—	102/0.7	1.4899	1.1665	84
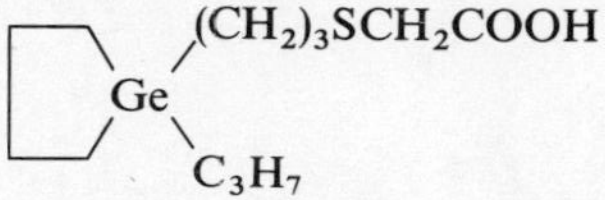 (Ge ring with (CH$_2$)$_3$SCH$_2$COOH, C_3H_7)	$C_{12}H_{24}GeO_2S$	—	158/0.25	1.5201	1.2114	72

Table 3-2—continued

COMPOUND	EMPIRICAL FORMULA	M.P. (°C)	B.P. (°C/mm)	n_D^{20}	d_4^{20}	REFERENCES
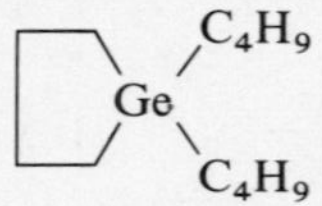	$C_{12}H_{26}Ge$	—	123/18	1.4720	1.0142	70
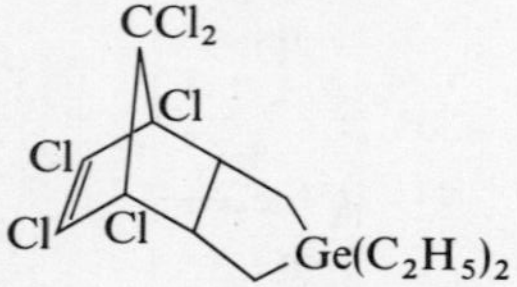	$C_{13}H_{16}Cl_6Ge$	43	—	—	—	84
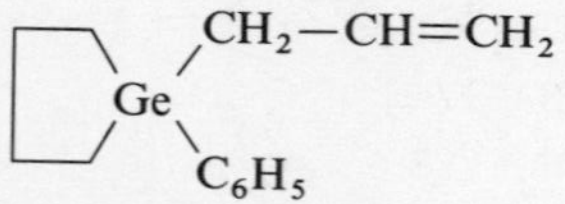	$C_{13}H_{18}Ge$	—	108/1.7	1.5551	1.1575	72, 75
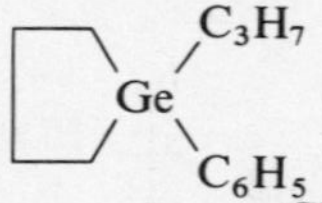	$C_{13}H_{20}Ge$	—	104/0.8	1.5383	1.1354	72
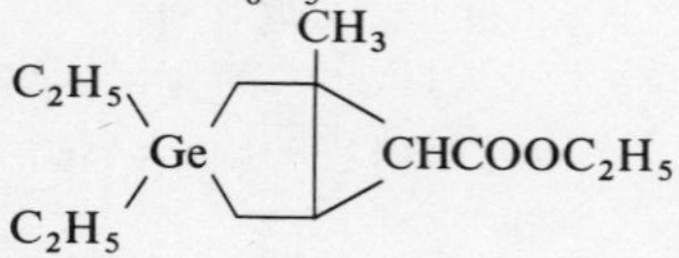	$C_{13}H_{24}GeO_2$	—	92/0.4	1.4850	1.1381	84

Compound	Molecular formula	m.p.	b.p./mm	n	d	References
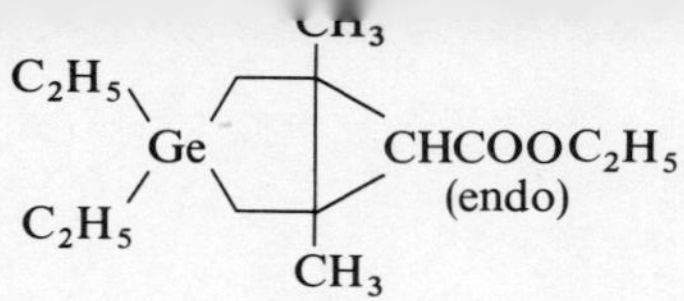 (endo) CHCOOC$_2$H$_5$ structure with Ge, C$_2$H$_5$, C$_2$H$_5$, CH$_3$, CH$_3$	C$_{14}$H$_{26}$GeO$_2$	—	94/0.3	1.4844	1.1092	84
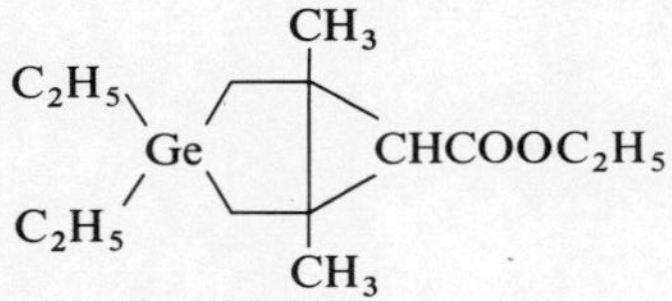 (exo) CHCOOC$_2$H$_5$ structure with Ge, C$_2$H$_5$, C$_2$H$_5$, CH$_3$, CH$_3$	C$_{14}$H$_{26}$GeO$_2$	—	94/0.3	1.4862	1.1213	84
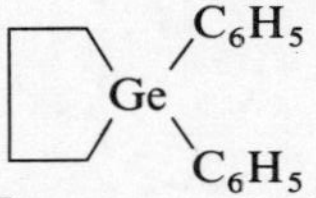 Ge with C$_6$H$_5$, C$_6$H$_5$ (cyclopentane ring)	C$_{16}$H$_{18}$Ge	33	115/0.2 136/0.4	1.5971	1.2189	71, 72, 75
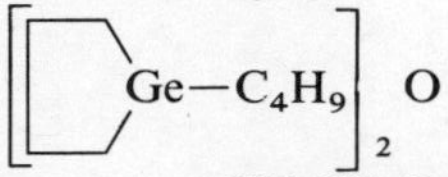 Ge—C$_4$H$_9$, O (bracket]$_2$)	C$_{16}$H$_{34}$Ge$_2$O	—	111/0.2	1.4956	1.1877	72
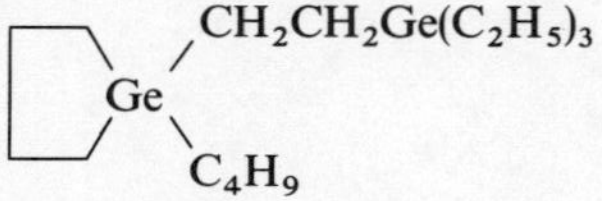 Ge with CH$_2$CH$_2$Ge(C$_2$H$_5$)$_3$, C$_4$H$_9$	C$_{16}$H$_{36}$Ge$_2$	—	131/0.6	1.4940	1.1168	72
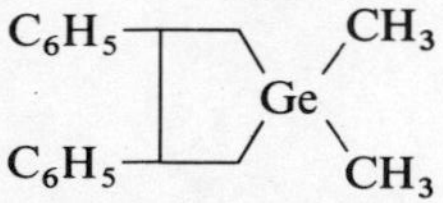 Ge with CH$_3$, CH$_3$, C$_6$H$_5$, C$_6$H$_5$	C$_{18}$H$_{22}$Ge	—	140–41/0.2	1.5920	1.1688	105
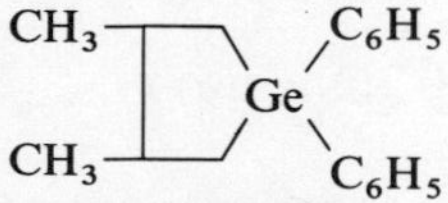 Ge with C$_6$H$_5$, C$_6$H$_5$, CH$_3$, CH$_3$	C$_{18}$H$_{22}$Ge	—	156/0.4	1.5853	1.1710	83
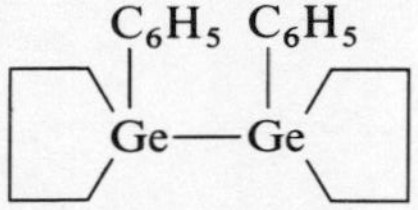 Ge—Ge with C$_6$H$_5$, C$_6$H$_5$, C$_6$H$_5$, C$_6$H$_5$	C$_{20}$H$_{26}$Ge$_2$	—	176/0.6	1.6085	1.3078	72

Table 3-3
Germacyclohexanes

COMPOUND	EMPIRICAL FORMULA	M.P. (°C)	B.P. (°C/mm)	n_D^{20}	d_4^{20}	REFERENCES
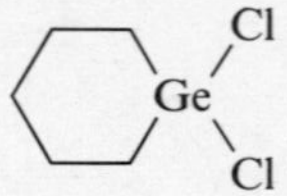	$C_5H_{10}GeCl_2$	—	99/24 55–60/12	1.5107	1.4675	71
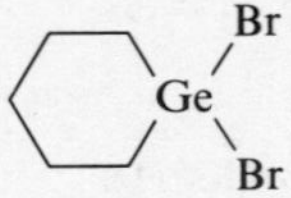	$C_5H_{10}GeBr_2$	—	115/17	1.5659	1.9775	71
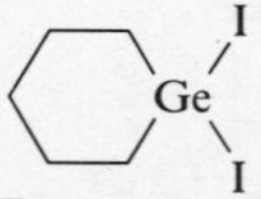	$C_5H_{10}GeI_2$	—	108/2	1.6610	2.3698	71
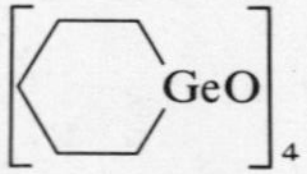	$(C_5H_{10}GeO)_4$	144	—	—	—	71
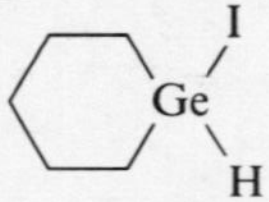	$C_5H_{11}GeI$	—	132/58	1.5835	1.8681	66, 71
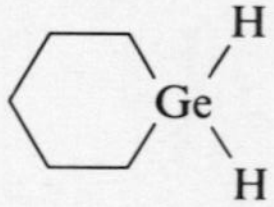	$C_5H_{12}Ge$	—	119–20/760	1.4835	1.1846	66, 71

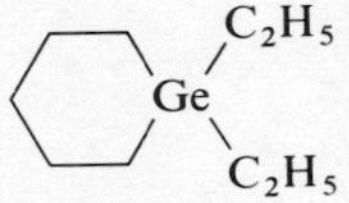	$C_9H_{20}Ge$	—	116–17/72 52/13	1.4742	1.0663	71
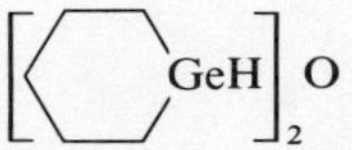	$C_{10}H_{22}Ge_2O$	—	130–31/9	1.5085	—	66, 71
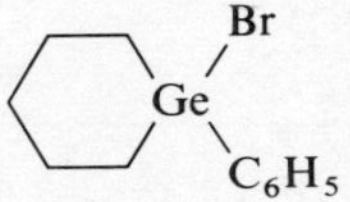	$C_{11}H_{15}GeBr$	—	125/0.8	1.5828	1.5045	71
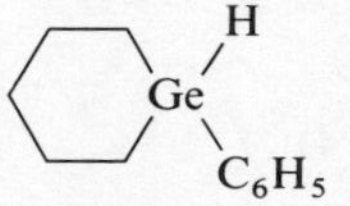	$C_{11}H_{16}Ge$	—	128/14	1.5550	1.1989	66, 71
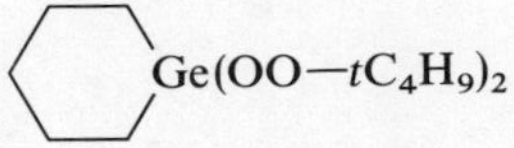	$C_{13}H_{28}GeO_4$	−35	60/0.001 65/0.2	1.4553(26°)	—	24
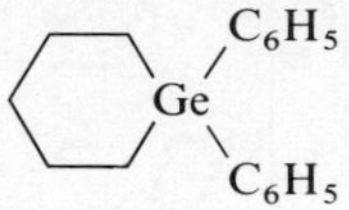	$C_{17}H_{20}Ge$	—	150/1	1.5932	1.2061	71

Table 3-4
Germacycloheptanes

COMPOUND	EMPIRICAL FORMULA	M.P. (°C)	B.P. (°C/mm)	n_D^{20}	d_4^{20}	REFERENCES
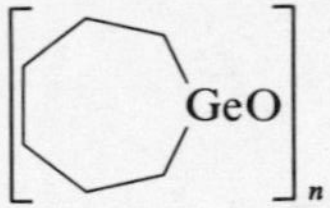	$(C_6H_{12}GeO)_n$	140	—	—	—	70
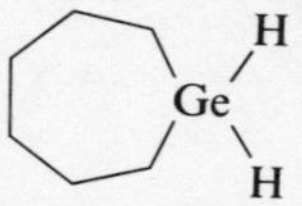	$C_6H_{14}Ge$	—	149–50/760	1.4938	1.1694	70
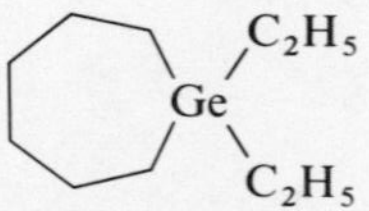	$C_{10}H_{22}Ge$	—	101.5/16	1.4814	1.0641	70

3-2-2 Germaspirans

It was noted (3-2-1-1), that symmetrical germaspirans are formed spontaneously when $GeCl_4$ reacts with an aliphatic dimagnesium compound (13, 71). Mixed germaspiranes may be formed by treating an appropriate dimagnesium aliphatic compound with a dihalogermacyclane:

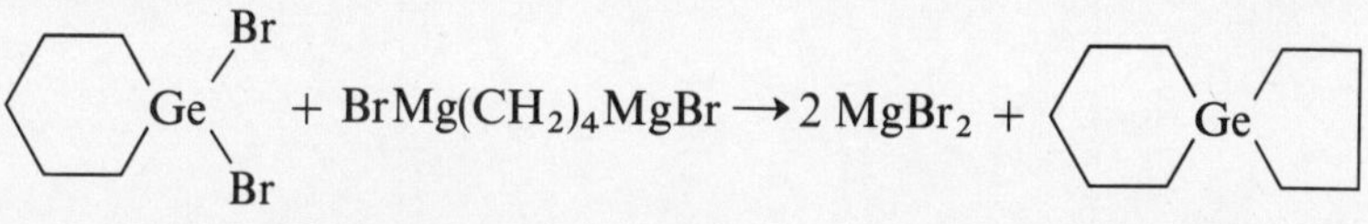

The 4-germa[3.4]spirooctane is prepared in dilute xylene solution by sodium ring closure of 3-chloropropylchlorogermacyclopentane:

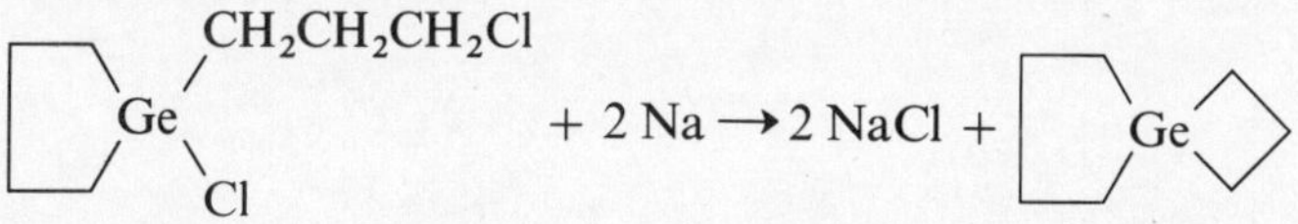

This derivative has the usual reactivity of dialkylgermacyclobutanes: the four-membered ring, being less stable, is easier to open:

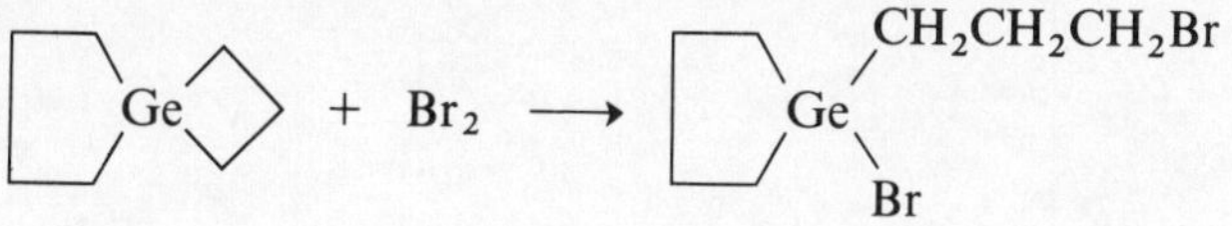

3-2-2-1 Germaspirans

Table 3-5 showing some physical properties of Germaspirans appears on the following page.

Table 3-5
Germaspirans

COMPOUND	EMPIRICAL FORMULA	M.P. (°C)	B.P. (°C/mm)	n_D^{20}	d_4^{20}	REFERENCES
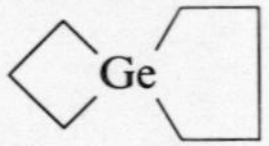	$C_7H_{14}Ge$	—	88–89/38	1.5185	1.2043	75
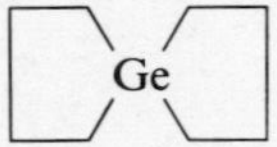	$C_8H_{16}Ge$	—	189/760 70–72/10	1.5118 1.5033 (25°)	1.1837	13, 71, 72
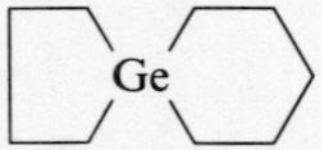	$C_9H_{18}Ge$	—	101/23	1.5058	1.1474	71
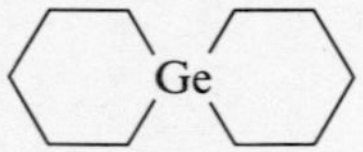	$C_{10}H_{20}Ge$	—	109/17 99–101/10	1.5060 1.5005 (25°)	1.1400	13, 71

3-2-3 Cyclic polygermanes

(a) A germacyclopentane containing two joining Ge atoms within the cycle was produced through the action of a hydrogenogermane on diethylmercury (81):

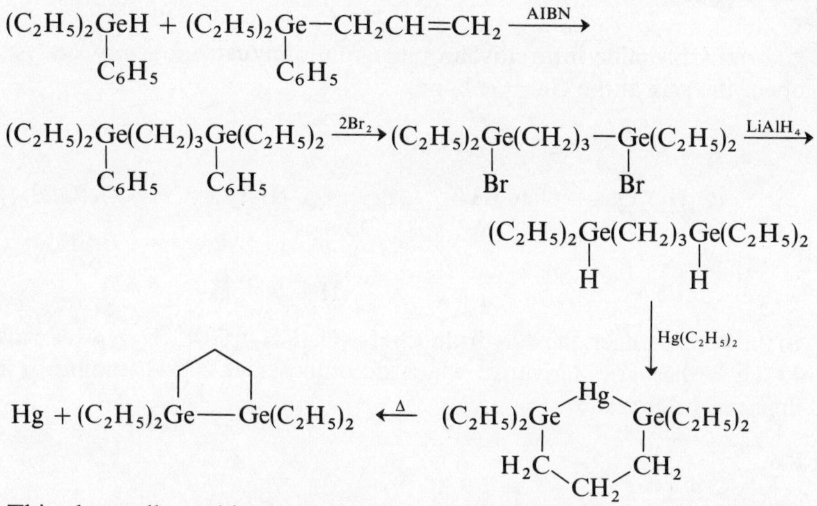

$$(C_2H_5)_2\underset{\underset{C_6H_5}{|}}{Ge}H + (C_2H_5)_2\underset{\underset{C_6H_5}{|}}{Ge}-CH_2CH{=}CH_2 \xrightarrow{\text{AIBN}}$$

$$(C_2H_5)_2\underset{\underset{C_6H_5}{|}}{Ge}(CH_2)_3\underset{\underset{C_6H_5}{|}}{Ge}(C_2H_5)_2 \xrightarrow{2Br_2} (C_2H_5)_2\underset{\underset{Br}{|}}{Ge}(CH_2)_3-\underset{\underset{Br}{|}}{Ge}(C_2H_5)_2 \xrightarrow{\text{LiAlH}_4}$$

$$(C_2H_5)_2\underset{\underset{H}{|}}{Ge}(CH_2)_3\underset{\underset{H}{|}}{Ge}(C_2H_5)_2$$

$$\Big\downarrow {\scriptstyle Hg(C_2H_5)_2}$$

$$Hg + (C_2H_5)_2Ge{-\!\!-\!\!-}Ge(C_2H_5)_2 \xleftarrow{\Delta} (C_2H_5)_2\underset{\underset{H_2C}{|}}{Ge}\overset{Hg}{\underset{CH_2}{\diagdown\diagup}}\underset{\underset{CH_2}{|}}{Ge}(C_2H_5)_2$$

This thermally stable derivative is a colourless liquid. It oxidizes but slowly at room temperature, and quickly at 150°C. In the latter case a mixture of cyclic oxide and linear polymers is formed:

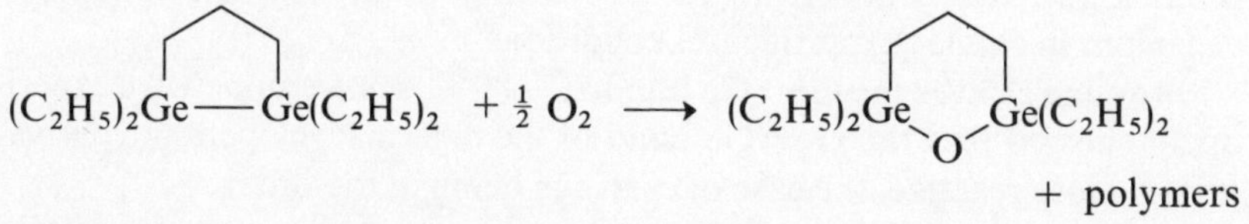

$$(C_2H_5)_2Ge{-\!\!-\!\!-}Ge(C_2H_5)_2 + \tfrac{1}{2}O_2 \longrightarrow (C_2H_5)_2Ge\underset{O}{\diagdown\diagup}Ge(C_2H_5)_2$$
$$+ \text{ polymers}$$

at 180° sulphur adds quantitatively by expanding the ring, leading to a sulphide six-membered ring derivative:

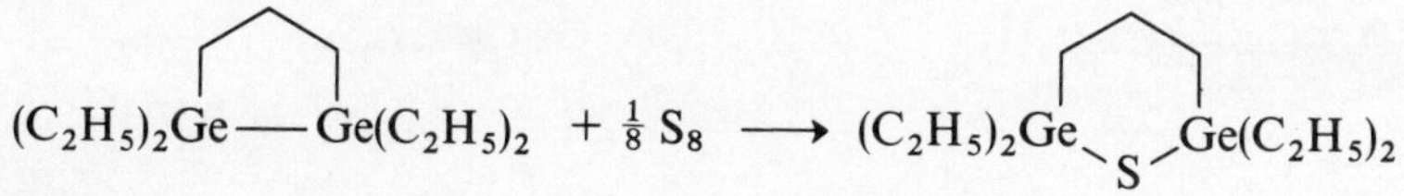

$$(C_2H_5)_2Ge{-\!\!-\!\!-}Ge(C_2H_5)_2 + \tfrac{1}{8}S_8 \longrightarrow (C_2H_5)_2Ge\underset{S}{\diagdown\diagup}Ge(C_2H_5)_2$$

which is identical with the derivative resulting from the action of sodium sulphide on 1,3 bis(diethylbromogermyl)propane:

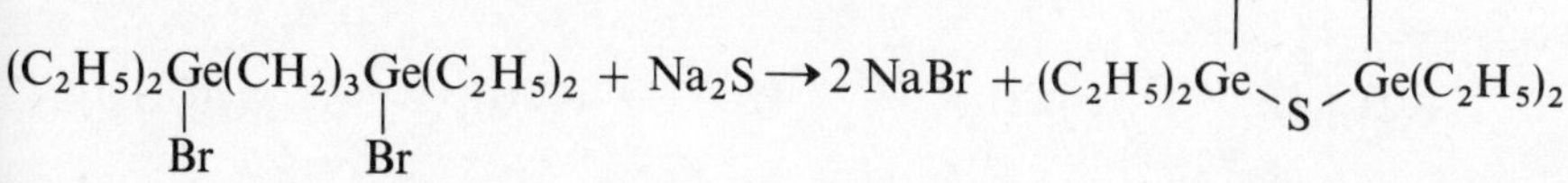

$$(C_2H_5)_2\underset{\underset{Br}{|}}{Ge}(CH_2)_3\underset{\underset{Br}{|}}{Ge}(C_2H_5)_2 + Na_2S \rightarrow 2\,NaBr + (C_2H_5)_2Ge\underset{S}{\diagdown\diagup}Ge(C_2H_5)_2$$

When maintained at 300°C, selenium reacts similarly leading to the analogous selenium ring derivative:

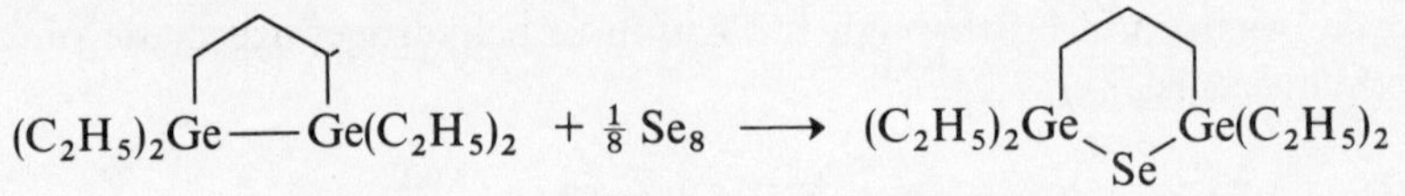

$$(C_2H_5)_2Ge\!\!-\!\!Ge(C_2H_5)_2 + \tfrac{1}{8}Se_8 \longrightarrow (C_2H_5)_2Ge\underset{Se}{\diagdown}Ge(C_2H_5)_2$$

At −80°C bromine, in an ethyl bromide solution, and without any catalyst, opens the ring at the Ge—Ge bond:

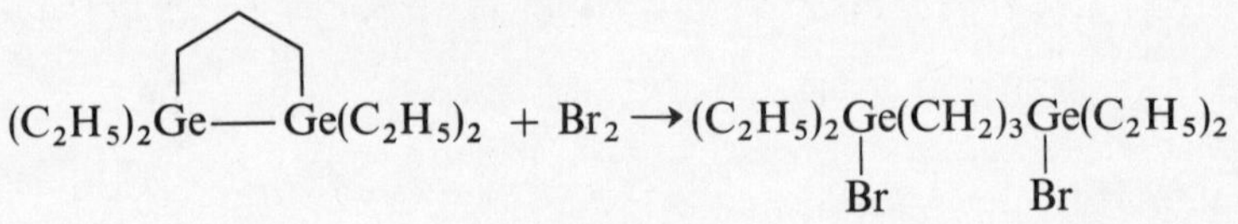

$$(C_2H_5)_2Ge\!\!-\!\!Ge(C_2H_5)_2 + Br_2 \rightarrow \underset{Br}{(C_2H_5)_2Ge(CH_2)_3}\underset{Br}{Ge(C_2H_5)_2}$$

In the same manner, the dihydride $\overset{H}{(C_2H_5)_2Ge}(CH_2)_4\overset{H}{Ge(C_2H_5)_2}$ gives with Et_2Hg a mercuric derivative which decomposes at 200°C leading to a digermacyclohexane:

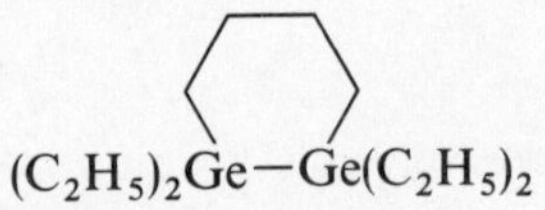

$$(C_2H_5)_2Ge\!-\!Ge(C_2H_5)_2$$

This compound, which is more stable than the digermacyclopentane homologue, reacts not at all or very slowly with oxygen, sulphur and selenium in similar experimental conditions.

Bromine cleaves the Ge—Ge bond at −80°C, but a competitive experiment showed that the Ge—Ge bond of the digermacyclopentane derivative is more reactive, the ease of cleavage being in the order:

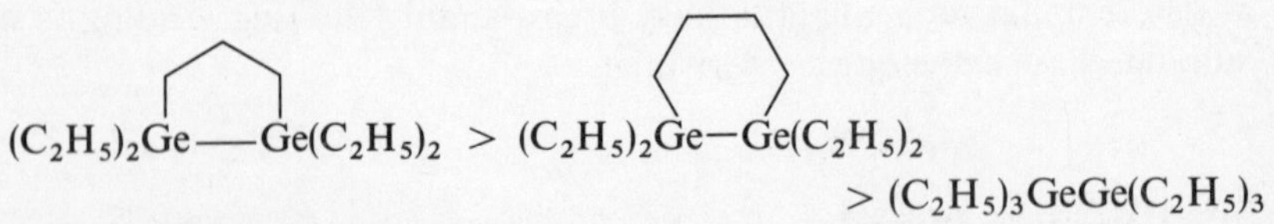

$$(C_2H_5)_2Ge\!\!-\!\!Ge(C_2H_5)_2 > (C_2H_5)_2Ge\!-\!Ge(C_2H_5)_2$$
$$> (C_2H_5)_3GeGe(C_2H_5)_3$$

(b) Germacyclohexanes containing two non-joining germanium atoms have been produced:

by hydrogenating the corresponding cyclenic derivative (22):

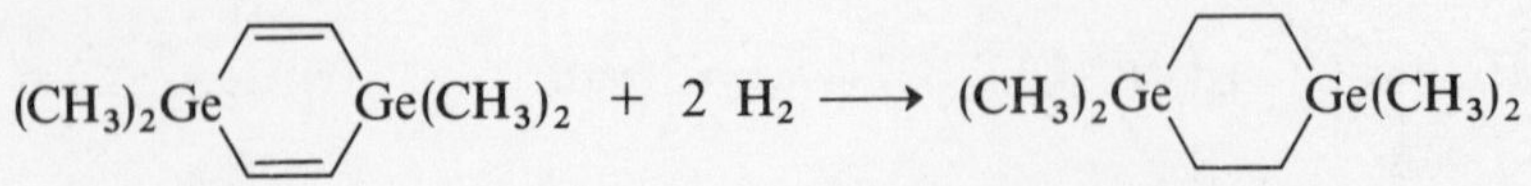

$$(CH_3)_2Ge\diagdown\diagup Ge(CH_3)_2 + 2 H_2 \longrightarrow (CH_3)_2Ge\diagdown\diagup Ge(CH_3)_2$$

and by condensing an equimolar mixture of dialkylgermane and dialkyl-divinylgermane (11):

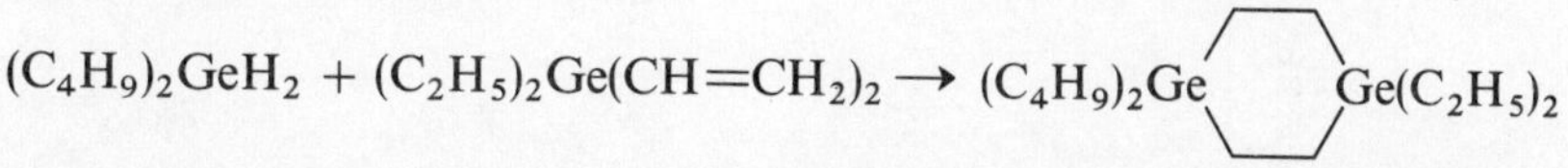

$$(C_4H_9)_2GeH_2 + (C_2H_5)_2Ge(CH=CH_2)_2 \rightarrow (C_4H_9)_2Ge\!\!\bigcirc\!\!Ge(C_2H_5)_2$$

The intermolecular reaction being dominant, the cyclanic derivative yield remains low: 20%.

Action of dibutylgermane on diethyldiallylgermane does not seem to lead to the corresponding cyclanic derivative.

(c) Derivatives containing several germanium atoms within the ring. Mironov and Gar (92) investigated the copper-catalyzed reaction of methylene chloride with germanium: from the mixture of the polychlorinated derivatives formed, they isolated a six-membered ring cyclanic derivative containing 3 germanium atoms:

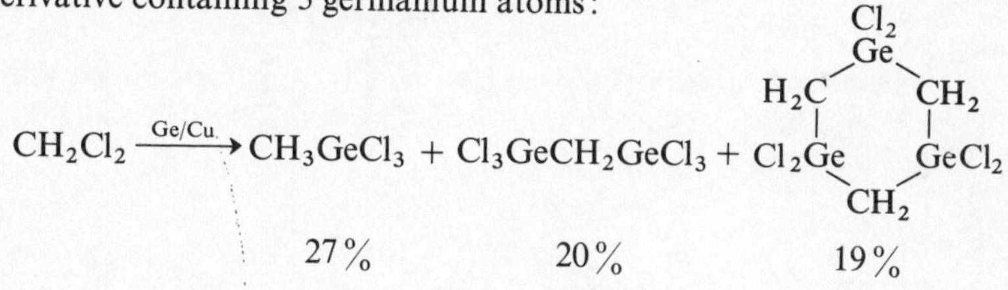

$$CH_2Cl_2 \xrightarrow{\text{Ge/Cu}} CH_3GeCl_3 + Cl_3GeCH_2GeCl_3 + \begin{array}{c} Cl_2 \\ Ge \\ H_2C \quad CH_2 \\ | \qquad | \\ Cl_2Ge \quad GeCl_2 \\ CH_2 \end{array}$$

$$\qquad\qquad\qquad 27\% \qquad\qquad 20\% \qquad\qquad 19\%$$

This derivative reacts with methylmagnesium iodide to produce the corresponding hexamethylated derivative.

Five- and six-membered rings exclusively composed of germanium atoms within the ring have been isolated by Neumann and coworkers (107, 108), in addition to octaphenylcyclotetragermane resulting from the action of diphenyldichlorogermane on sodium naphthalenide in 1,2-dimethoxyethane as in the procedure introduced by Horner and Güsten (48). The yields reported are 37% for decaphenylcyclopentagermane and 7% for the dodecaphenylcyclohexagermane.

3-2-3-1 *Cyclic Polygermanes*

Table 3-6 showing some physical properties of Cyclic Polygermanes appears on the following pages.

Table 3-6
Cyclic Polygermanes

COMPOUND	EMPIRICAL FORMULA	M.P. (°C)	B.P. (°C/mm)	n_D^{20}	d_4^{20}	REFERENCES
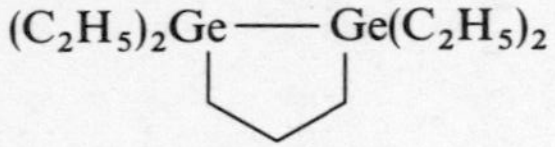 $(C_2H_5)_2Ge\!-\!\!-\!Ge(C_2H_5)_2$	$C_{11}H_{26}Ge_2$	—	128/15	1.5159	1.2088	81
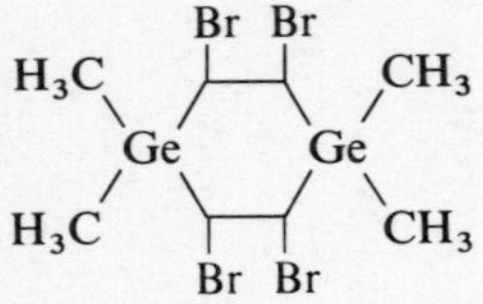 (cyclic dimethyl-dibromo germane)	$C_8H_{16}Br_2Ge_2$	145–49	—	—	—	142
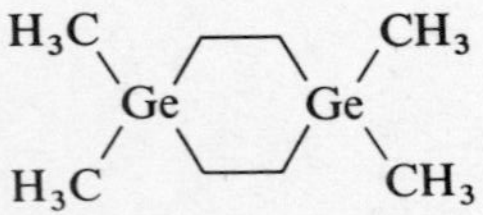 (cyclic dimethyl germane)	$C_8H_{20}Ge_2$	—	70–72/12 79–80/14	1.4861 —	— —	142, 149
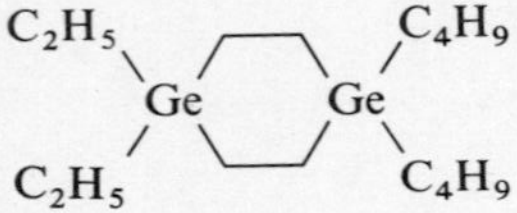 (cyclic dibutyl-diethyl germane)	$C_{16}H_{36}Ge_2$	—	117/0.6	1.4932	1.1198	71
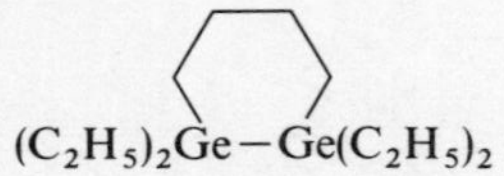 $(C_2H_5)_2Ge\!-\!Ge(C_2H_5)_2$	$C_{12}H_{28}Ge_2$	—	112/2.5	1.5159	1.1974	81

Structure	Formula	mp	bp	n	d	Reference
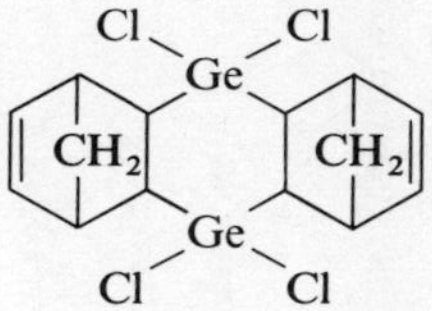 Cl_2Ge(CH$_2$)$Ge(CH_2)$ with Cl, Cl, CH$_2$, CH$_2$ bridges	$C_{14}H_{16}Ge_2Cl_4$	264–65	—	—	—	142
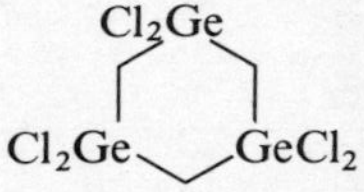 Cl_2Ge—GeCl$_2$ ring	$C_3H_6Ge_3Cl_6$	91–92	150–2/5	—	—	92
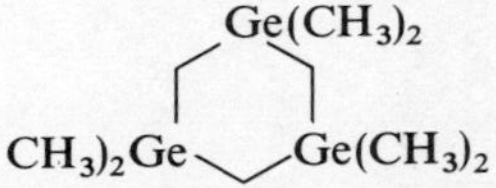 $Ge(CH_3)_2$ ring, $(CH_3)_2Ge$—$Ge(CH_3)_2$	$C_9H_{24}Ge_3$	—	71.3/4	1.4951 1.4935	1.3214 1.3135	92
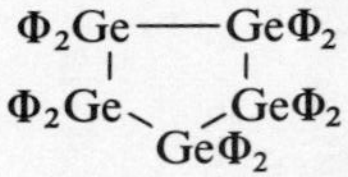 Φ_2Ge—$Ge\Phi_2$, Φ_2Ge—$Ge\Phi_2$ ring	$C_{60}H_{50}Ge_5$	>360	—	—	—	108, 109
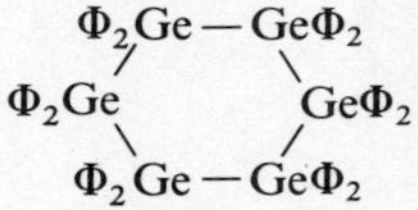 Φ_2Ge—$Ge\Phi_2$ six-membered ring	$C_{72}H_{60}Ge_6$	>360	—	—	—	108, 109

3-3 "MEDIUM" AND "LARGE" RING DERIVATIVES OF GERMANIUM

When the previous synthesis processes are applied to preparing larger germanium rings as in organic chemistry we notice a sudden decrease in yield as soon as the seven-membered ring is reached. In particular, the process using dimagnesium compounds, which is very suitable in preparing germacyclopentanes and germacyclohexanes, produces germacycloheptyl derivatives but in small yields (10%). Action of zinc powder upon the $(C_2H_5)_2Ge[(CH_2)_4Br]_2$ dibrominated derivative in a heptane or alcohol solution gives a mixture of several products. This mixture varies with the conditions of the experiment, but it has not been possible to identify any of the expected nine-membered ring derivative in it.

The same dibrominated derivative, treated with sodium powder in boiling anhydrous xylene, and by using the high dilution technique leads to a mixture of polymers and diethyldibutylgermanium in a 40% yield (79):

$$(C_2H_5)_2Ge[(CH_2)_4Br]_2 \xrightarrow[\text{xylene}]{Na} (C_2H_5)_2Ge(C_4H_9)_2 + \text{polymers}$$

Pyrolysis of heavy metal aliphatic diacid salts enabled Ruzicka (121–124) to produce the first macrocyclic derivatives. By using this process Benkeser and Cunico (14) pyrolyzed 5,5-dimethyl-5-sila-1,9-nonanedioic thorium salt and produced dimethylsilacyclooctanone with poor yield; a similar result was found in organogermanium chemistry. Vacuum calcination of the thorium salt of 6,6-diethyl-6-germa-1,11-undecanedioic acid produced low yields (10%) of ten-membered germacyclanone (79):

$$\{(C_2H_5)_2Ge[(CH_2)_4COO]_2\}_2Th \rightarrow 2\,(C_2H_5)_2Ge\begin{array}{c}\overset{\displaystyle -(CH_2)_4-}{\underset{\displaystyle -(CH_2)_4-}{}}\end{array}C{=}O$$

$$+ ThO_2 + 2\,CO_2$$

The Stoll improved process (137) for condensing diesters by sodium applies very well to germanium derivatives and produces germanium macrorings in good yields (79):

$$R_2Ge[(CH_2)_4COOC_2H_5]_2 + 4\,Na \rightarrow R_2Ge\begin{array}{c}\nearrow (CH_2)_4-\overset{O}{\overset{\|}{C}}-ONa\\ \searrow (CH_2)_4-\overset{O}{\overset{\|}{C}}-ONa\end{array}$$

$$+ 2\,C_2H_5ONa$$

$$R_2Ge \Big\langle {\!\!\begin{array}{l} (CH_2)_4\!-\!\overset{\displaystyle\|}{C}\!-\!ONa \\[1mm] (CH_2)_4\!-\!\overset{\displaystyle\|}{C}\!-\!ONa \end{array}} \quad + 2\,CH_3COOH \rightarrow 2\,CH_3COONa$$

$$+ \; R_2Ge \Big\langle {\!\!\begin{array}{l} (CH_2)_4CH(OH) \\[1mm] \;\;\;\;\;\;\;\;\;| \\[1mm] (CH_2)_4C\!=\!O \end{array}} \qquad (60\%)$$

Germanium-containing thirteen- and fifteen-membered ring acyloïns are prepared in the same way with similar yields. This reaction also produces a little α-diketone which is hard to extract by fractional distillation, but it is very easily extracted by crystallization in pentane at −80°C or by gas chromatography. When the two ester chains are the same, the cyclic germanium acyloïns produced have an uneven number of links. Acyloïns with an even number of links can be produced by using dissymetrical diesters (31, 80):

$$(C_6H_5)_2Ge(C_2H_5)_2 \xrightarrow{\;Br_2\;} (C_2H_5)_2\underset{\overset{\displaystyle|}{Br}}{Ge}C_6H_5 \xrightarrow{\;ClMg(CH_2)_5O-\!\!\bigcirc\!\!O\;}$$

$$(C_2H_5)_2\underset{\overset{\displaystyle|}{C_6H_5}}{Ge}(CH_2)_5O-\!\!\bigcirc\!\!O \xrightarrow{\;HBr\;} (C_2H_5)_2\underset{\overset{\displaystyle|}{Br}}{Ge}(CH_2)_5Br \xrightarrow{\;ClMg(CH_2)_4O-\!\!\bigcirc\!\!O\;}$$

$$(C_2H_5)_2Ge \Big\langle {\!\!\begin{array}{l} (CH_2)_5Br \\[1mm] (CH_2)_4O-\!\!\bigcirc\!\!O \end{array}} \xrightarrow{\;HBr\;} (C_2H_5)_2Ge \Big\langle {\!\!\begin{array}{l} (CH_2)_4Br \\[1mm] (CH_2)_5Br \end{array}} \xrightarrow[\substack{(2)\;CO_2 \\ (3)\;H_2O}]{(1)\;Mg}$$

$$(C_2H_5)_2Ge \Big\langle {\!\!\begin{array}{l} (CH_2)_4COOH \\[1mm] (CH_2)_5COOH \end{array}} \xrightarrow[\;SO_4H_2\;]{C_2H_5OH} (C_2H_5)_2Ge \Big\langle {\!\!\begin{array}{l} (CH_2)_4COOC_2H_5 \\[1mm] (CH_2)_5COOC_2H_5 \end{array}} \xrightarrow[\;xylene\;]{Na}$$

$$(C_2H_5)_2Ge \Big\langle {\!\!\begin{array}{l} (CH_2)_4CH(OH) \\[1mm] \;\;\;\;\;\;\;\;\;| \\[1mm] (CH_2)_5C\!=\!O \end{array}}$$

Germaacyloïns can be used to produce other macrocyclic derivatives of germanium; reduction leads to germacyclic ketones, hydrocarbons or α-glycols depending on the agent used and the experimental conditions (79, 80):

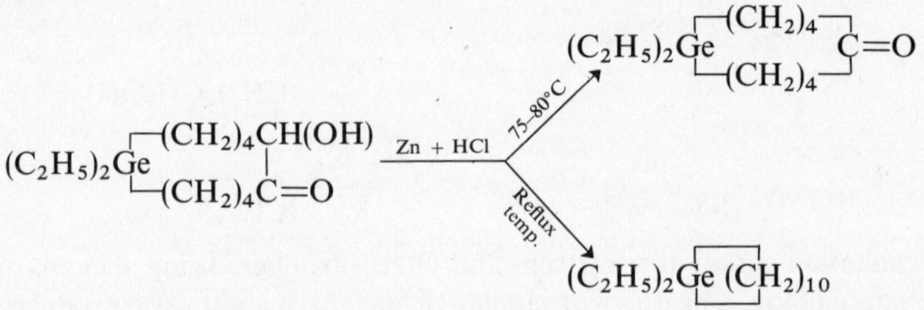

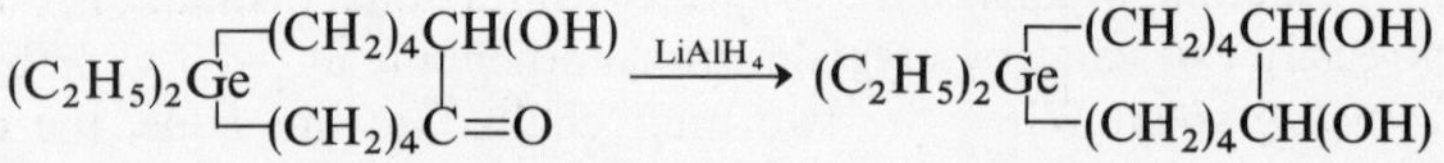

$$(C_2H_5)_2Ge\begin{matrix}(CH_2)_4CH(OH)\\ |\\ (CH_2)_4C=O\end{matrix} \xrightarrow[\text{Reflux temp.}]{\text{75-80°C}} \begin{matrix}(C_2H_5)_2Ge\begin{matrix}(CH_2)_4\\ (CH_2)_4\end{matrix}C=O\\ \\ (C_2H_5)_2Ge\,(CH_2)_{10}\end{matrix}$$

with Zn + HCl

$$(C_2H_5)_2Ge\begin{matrix}(CH_2)_4CH(OH)\\ |\\ (CH_2)_4C=O\end{matrix} \xrightarrow{\text{LiAlH}_4} (C_2H_5)_2Ge\begin{matrix}(CH_2)_4CH(OH)\\ |\\ (CH_2)_4CH(OH)\end{matrix}$$

When acyloïn is oxidized by cupric acetate, an excellent yield of diketones is obtained:

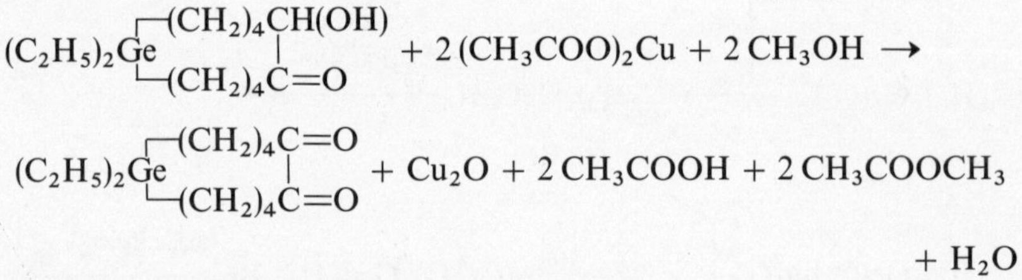

$$(C_2H_5)_2Ge\begin{matrix}(CH_2)_4CH(OH)\\ |\\ (CH_2)_4C=O\end{matrix} + 2\,(CH_3COO)_2Cu + 2\,CH_3OH \rightarrow$$

$$(C_2H_5)_2Ge\begin{matrix}(CH_2)_4C=O\\ |\\ (CH_2)_4C=O\end{matrix} + Cu_2O + 2\,CH_3COOH + 2\,CH_3COOCH_3$$

$$+ H_2O$$

3-3-1   Large ring systems

Table 3-7 showing some physical properties of Large Ring Systems appears on the following pages.

Table 3-7
Large Ring Systems

COMPOUND	EMPIRICAL FORMULA	M.P. (°C)	B.P. (°C/mm)	n_D^{20}	d_4^{20}	REFERENCES
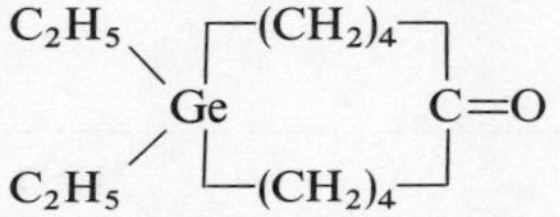 C_2H_5—Ge with $(CH_2)_4$—C=O and $(CH_2)_4$— / C_2H_5	$C_{13}H_{26}GeO$	DNPH 93	116–17/0.6	1.5008	1.0628	31, 79
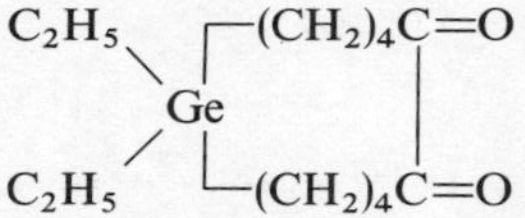 C_2H_5—Ge with $(CH_2)_4C$=O and $(CH_2)_4C$=O / C_2H_5	$C_{14}H_{26}GeO_2$	—	105/0.1	1.4983	1.1521	31, 79
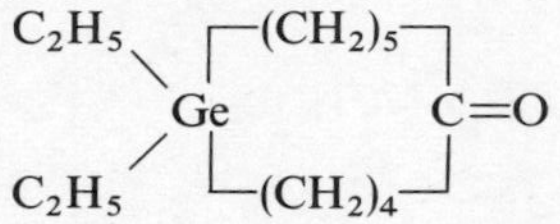 C_2H_5—Ge with $(CH_2)_5$—C=O and $(CH_2)_4$— / C_2H_5	$C_{14}H_{28}GeO$	—	123/1.1	1.5005	1.1096	31, 79
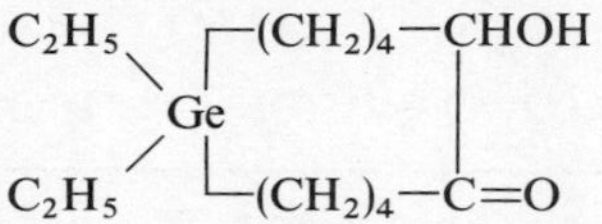 C_2H_5—Ge with $(CH_2)_4$—CHOH and $(CH_2)_4$—C=O / C_2H_5	$C_{14}H_{28}GeO_2$	—	152/0.6	1.5084	1.1644	31, 79
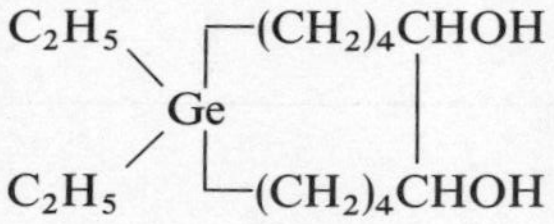 C_2H_5—Ge with $(CH_2)_4CHOH$ and $(CH_2)_4CHOH$ / C_2H_5	$C_{14}H_{30}GeO_2$	—	—	1.5148	1.1529	31, 79

Table 3-7—continued

COMPOUND	EMPIRICAL FORMULA	M.P. (°C)	B.P. (°C/mm)	n_D^{20}	d_4^{20}	REFERENCES
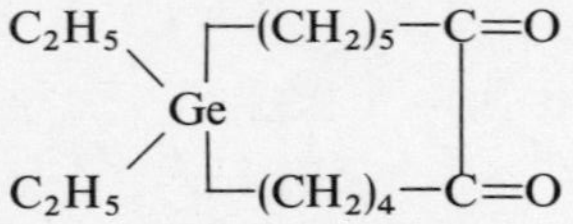 C_2H_5—Ge—$(CH_2)_5$—C=O / C_2H_5—$(CH_2)_4$—C=O	$C_{15}H_{28}GeO$	—	140/1	1.4957	1.1188	31, 80
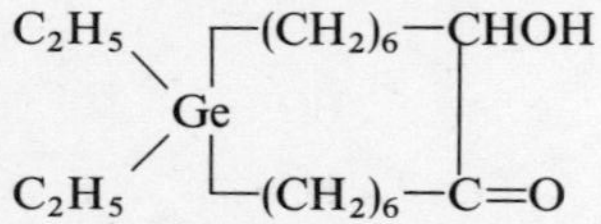 C_2H_5—Ge—$(CH_2)_6$—CHOH / C_2H_5—$(CH_2)_6$—C=O	$C_{18}H_{36}GeO_2$	—	190–95/1.5	—	—	31, 80
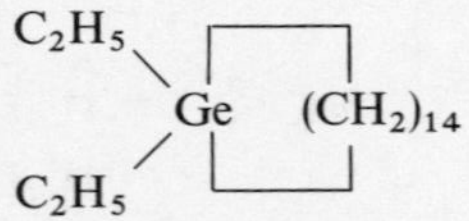 C_2H_5—Ge—$(CH_2)_{14}$ / C_2H_5	$C_{18}H_{38}Ge$	—	135/0.045	1.4918	1.0104	31

3-4 UNSATURATED RING DERIVATIVES OF GERMANIUM

At present we have no knowledge of the existence of the germacyclopropene and germacyclobutene ring systems. On the other hand, many unsaturated five- and six-membered rings containing germanium in the ring have been investigated.

Halogenated germanium ring derivatives are difficult to purify, and therefore dehydrohalogenation of these compounds does not seem the best process for synthesizing germacyclenes. Instead, unsaturated linear derivatives are used as starting materials in the preparations of the unsaturated rings.

3-4-1 Germacyclopentene derivatives

The addition of trichlorosilane to butadiene and other conjugated dienes is exclusively by 1,4 addition (11, 95):

$$CH_2{=}CH{-}CH{=}CH_2 + HSiCl_3 \xrightarrow{\text{Pt/C}} CH_3CH{=}CH{-}CH_2SiCl_3$$

Therefore it was unexpected when the interaction of butadiene with trichlorogermane resulted in a narrow fraction which consisted of a mixture of two compounds, one of which contained two chlorine atoms (90, 91, 102). Gas–liquid chromatography of the products of methylation of the mixture also revealed the presence of two compounds which could be separated by fractional distillation:

$$CH_2{=}CH{-}CH{=}CH_2 \xrightarrow{\text{HGeCl}_3} \begin{cases} Cl_3GeCH_2{-}CH{=}CH{-}CH_3 \\[6pt] Cl_2Ge\begin{array}{c} \diagup CH_2{-}CH \\[-2pt] \quad\quad\; \| \\[-2pt] \diagdown CH_2{-}CH \end{array} \end{cases}$$

$$\xrightarrow{\text{CH}_3\text{MgCl}} \begin{cases} (CH_3)_3GeCH_2{-}CH{=}CH{-}CH_3 \\[6pt] (CH_3)_2Ge\begin{array}{c} \diagup CH_2{-}CH \\[-2pt] \quad\quad\; \| \\[-2pt] \diagdown CH_2{-}CH \end{array} \end{cases}$$

Isoprene also reacts with trichlorogermane to form a mixture of linear and heterocyclic compounds:

$$CH_2{=}\underset{\underset{CH_3}{|}}{C}{-}CH{=}CH_2 \xrightarrow{\text{HGeCl}_3} Cl_3GeCH_2{-}\underset{\underset{CH_3}{|}}{C}{=}CH{-}CH_3 + \underset{\underset{CH_2}{|}}{CH}{=\!\!=}\underset{\underset{CH_2}{|}}{C}{-}CH$$

with the Ge bearing Cl and Cl.

the ratio of the compounds being approximately 3:1.

Under the same conditions, piperylene forms exclusively linear addition products. Nefedov and coworkers (103), who studied the action of trichlorogermane etherate on butadiene at 20°C and under normal atmospheric pressure, achieved a 60 to 65% yield in germacyclopentene derivative and about 30% in polymers. The authors suggested a mechanism for the reaction based upon the similarity between germanium chloride and the analogous dichlorocarbene:

$$2\,(C_2H_5)_2O, HGeCl_3 \rightleftarrows GeCl_2 + 2\,(C_2H_5)_2O, HCl$$

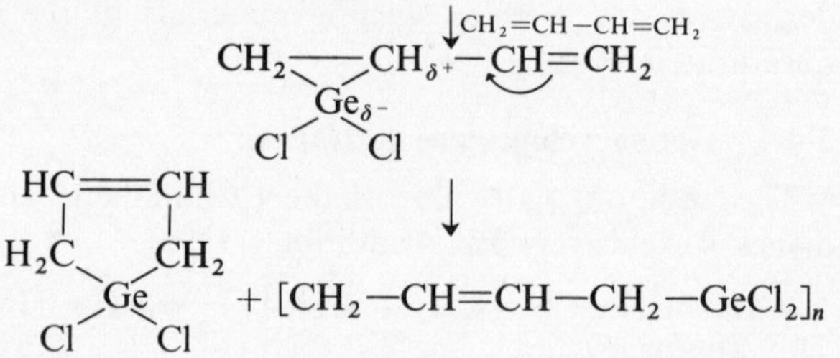

Through the reaction of tribromogermane with butadiene Mironov and Gar (35, 93) produced a mixture of brominated derivatives. Upon treatment with an excess of Grignard reagent these yield both a linear and a cyclopentene derivative

$$HGeBr_3 + CH_2{=}CH{-}CH{=}CH_2 \xrightarrow{MeMgCl}$$

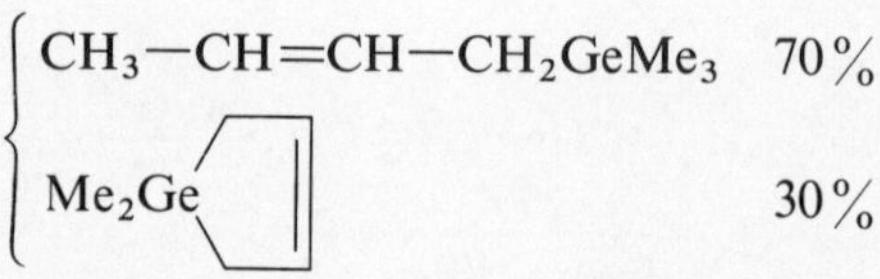

While studying the reaction of germanous iodide with conjugated dienes (64, 83) Mazerolles and Manuel noted that GeI_2 had no action upon piperylene, 1-phenyl butadiene or tetraphenylbutadiene. On the other hand with isoprene or 2,3 dimethylbutadiene the reaction is easily carried out and with the latter it is even exothermic:

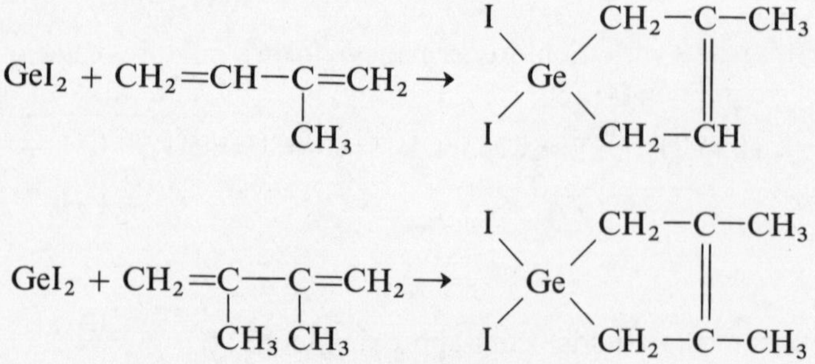

Butadiene has to be heated to about 100°C in an inert solvent (heptane) to react:

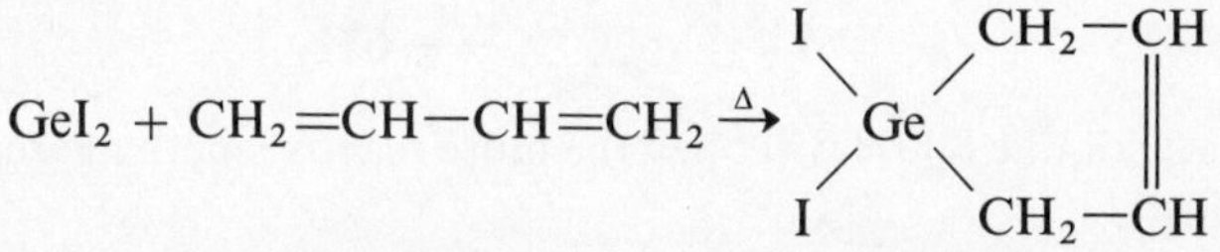

$$GeI_2 + CH_2{=}CH{-}CH{=}CH_2 \xrightarrow{\Delta} $$

Without any solvent, nothing occurs at room temperature and if heated, the butadiene will polymerize (91).

In the three reactions illustrated, the structure of the derivative produced results from a 1,4 addition. Thereby germanous iodide involves a different process from dihalocarbenes which lead to a 1,2 addition (56, 112, 134):

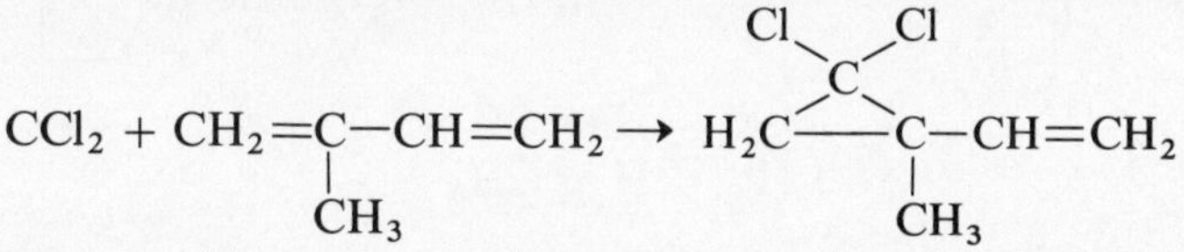

$$CCl_2 + CH_2{=}C{-}CH{=}CH_2 \rightarrow$$

Diiodogermacyclopentenes are thermally unstable and only the butadiene derivative underwent vacuum distillation without decomposition. Both iodine atoms may easily be substituted without destroying the unsaturated ring. The resulting products are generally more stable and may be purified by fractional distillation.

Silver halides react quantitatively in refluxing heptane:

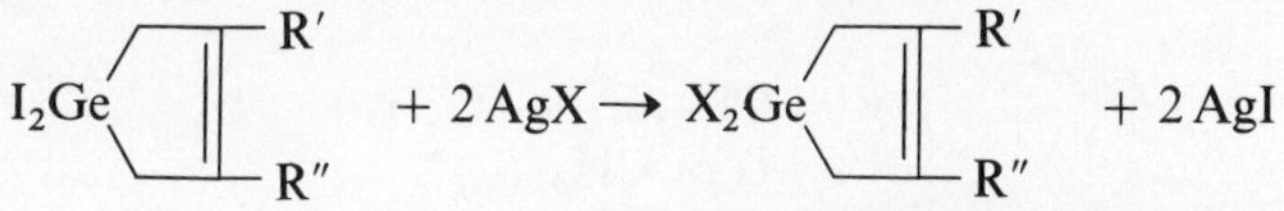

$$X = Cl, Br; R' = R'' = H$$
$$R' = R'' = CH_3$$
$$R' = CH_3; R'' = H$$

Lithium aluminium hydride in ether easily reduces 1,1-diiodo-1-germa-3,4 dimethyl-3-cyclopentene:

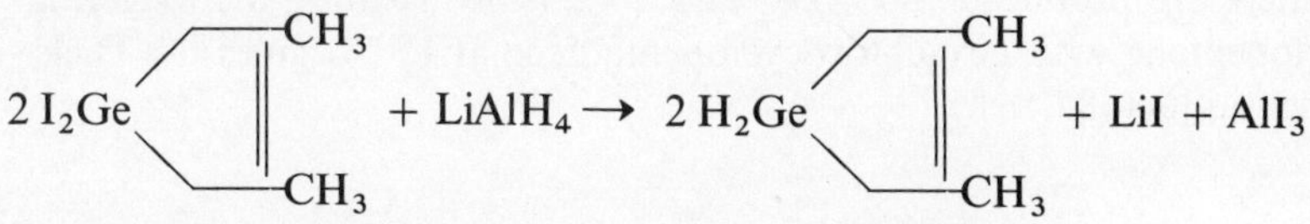

1-Germa-3-methyl-3-cyclopentene, 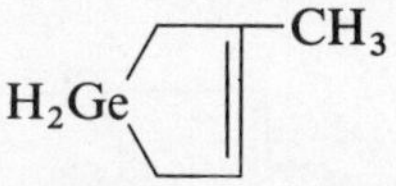 , which is prepared

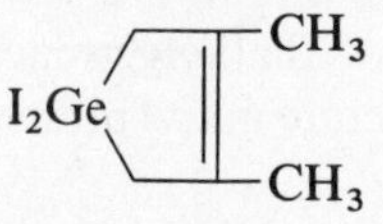

in a similar way, is a colourless very fluid liquid which will quickly polymerize to a transparent solid. Alcoholic KOH transforms into a white oxide-polymer $\left[\begin{array}{c} \text{CH}_3 \\ \text{O}-\text{Ge} \\ \text{CH}_3 \end{array}\right]_n$; under the same

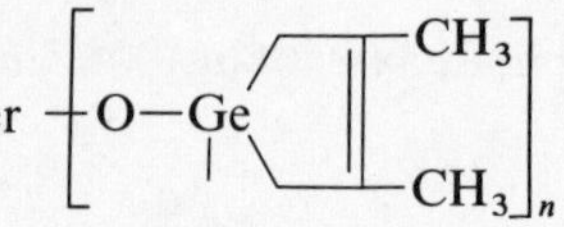

conditions alcoholic KOH will open the more reactive methylgermacyclopentene ring.

Unsaturated germanium diiodide ring derivatives react easily with organomagnesium compounds and produce dialkylgermacyclopentenes in good yields:

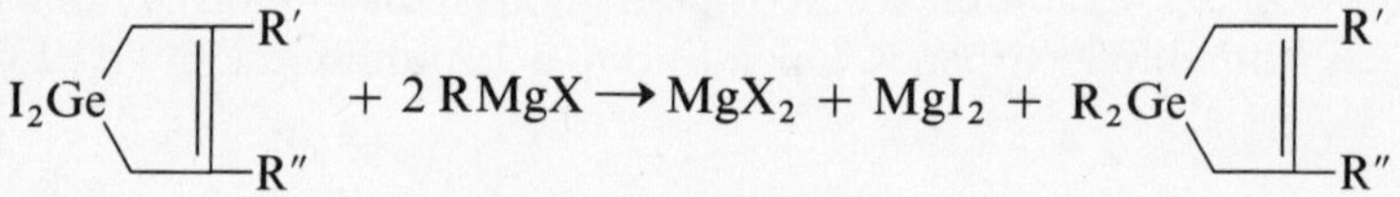

$$I_2Ge\overset{R'}{\underset{R''}{\fbox{}}} + 2\,RMgX \rightarrow MgX_2 + MgI_2 + R_2Ge\overset{R'}{\underset{R''}{\fbox{}}}$$

$$\begin{cases} R' = H \text{ or } CH_3 \\ R'' = H \text{ or } CH_3 \end{cases}$$

Over a Raney nickel catalyst, hydrogen is absorbed by the dialkyl- and diarylcyclopentenes:

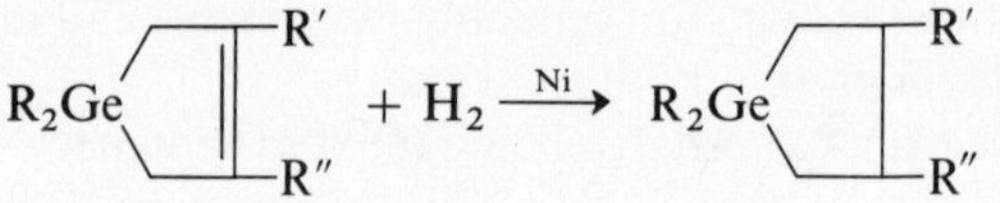

$$R_2Ge\overset{R'}{\underset{R''}{\fbox{}}} + H_2 \xrightarrow{\text{Ni}} R_2Ge\overset{R'}{\underset{R''}{\fbox{}}}$$

$$\begin{cases} R = CH_3, C_2H_5, C_6H_5 \\ R' = H \text{ or } CH_3 \\ R'' = H \text{ or } CH_3 \end{cases}$$

Under atmospheric pressure the butadiene derivative becomes quickly saturated, whereas hydrogenation of the methyl substituted derivatives requires higher pressures. With the dimethyl substituted derivative, two isomers are produced: 80% *cis-* and 20% *trans.* Heating diethylgermacyclopentene with hexachlorocyclopentadiene at 150°C gives the Diels–Alder adduct (84):

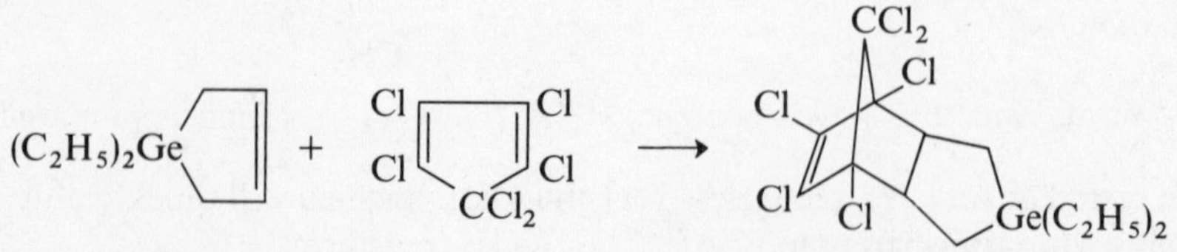

The three different types of dialkylgermacyclopentenes give addition reactions with ethyl diazoacetate in the presence of copper. They lead to a composition of endo and exo derivatives:

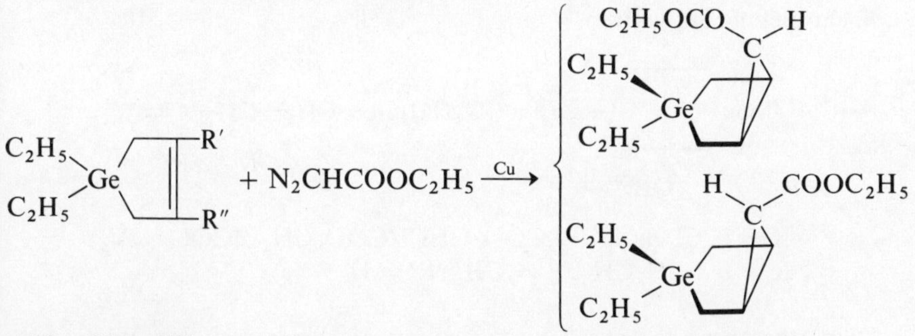

$$\begin{cases} R' = H \text{ or } CH_3 \\ R'' = H \text{ or } CH_3 \end{cases}$$

The methylcyclopentene derivative does not react with phenyllithium in ether solution, but under the same conditions triphenyl-vinylgermane leads to the expected adduct (17a, 133). Triethylgermane which easily reacts with olefinic germanium derivatives (see Chapter 4) will not react with diethylgermamethylcyclopentene in the presence of chloroplatinic acid even at 200°C.

On the other hand, hydroboration easily occurs, at 0°C, in THF and produces the expected germacyclopentanols in 60% yields (84):

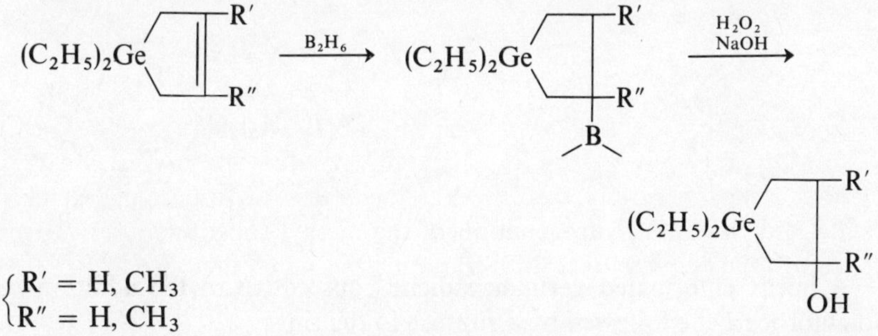

$$\begin{cases} R' = H, CH_3 \\ R'' = H, CH_3 \end{cases}$$

These germacyclanols are liquids distilling without decomposition under reduced pressure. When heated at 260°C on alumina, diethylgermacyclopentanol gives a mixture of germacyclopentene isomers:

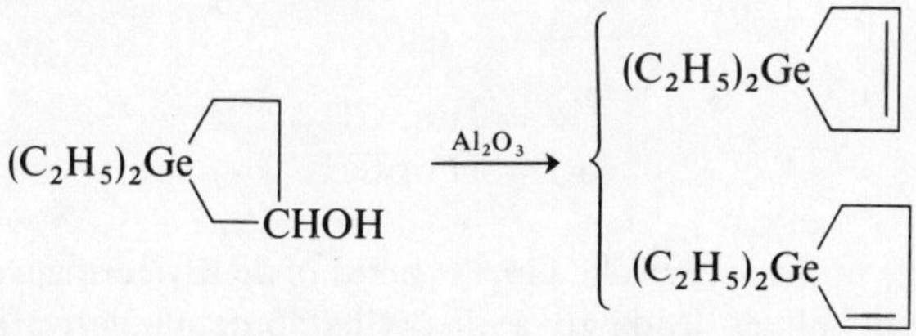

The germacyclopentanols are easily cleaved by protonic acids to give a linear olefinic derivative:

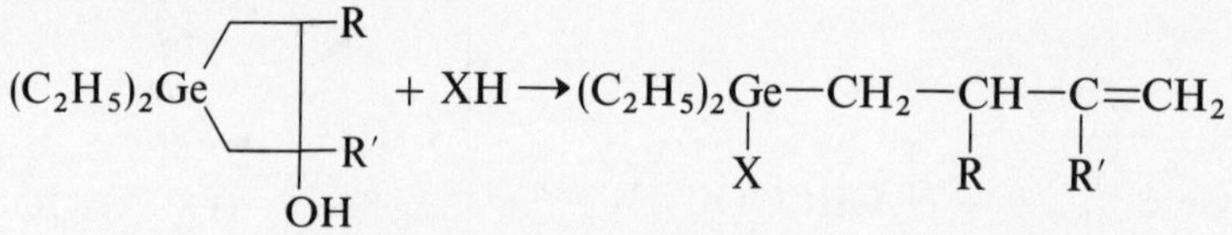

$$(C_2H_5)_2Ge \underset{OH}{\overset{R}{\overbrace{}}} R' + XH \rightarrow (C_2H_5)_2Ge-CH_2-CH-C=CH_2$$
$$ X \quad R \quad R'$$

$$X = F, Cl, Br, I, CH_3COO, CHCl_2COO, CH_2ClCOO, \ldots$$
$$R = R' = H, CH_3 ; R = CH_3, R' = H$$

Dichlorocarbene (prepared by the action of chloroform on potassium tertiobutylate) reacts with dialkylgermacyclopentenes producing two different derivatives:

a germanium cyclohexadiene resulting from a ring expansion reaction similar to the one observed by Parham and Reiff (113, 114) in the indene series:

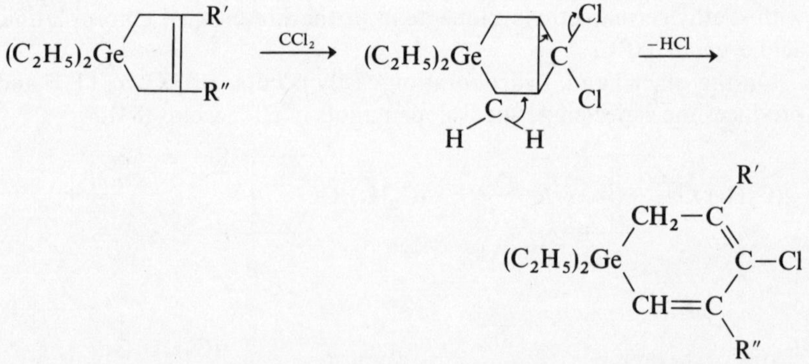

a linear chlorinated germahexadiene consecutive to the addition of dichlorocarbene followed by a rupture in the ring:

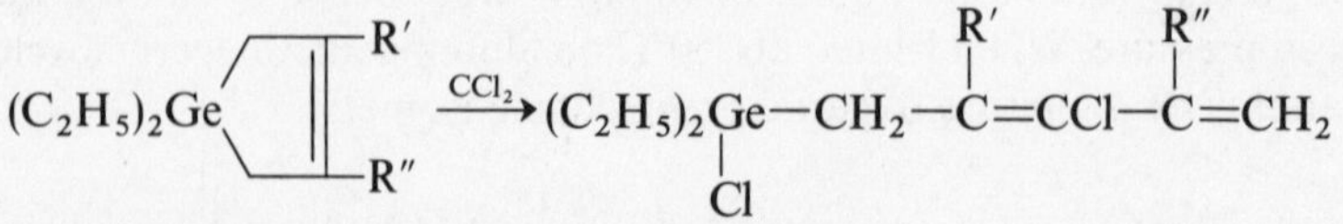

$$\begin{cases} R' = H \text{ or } CH_3 \\ R'' = H \text{ or } CH_3 \end{cases}$$

The cleavage of the intracyclic Ge—C bond of diethylgermacyclopentene by hydrogen halides leads to a diethylhalobutenyl derivative with a terminal double bond:

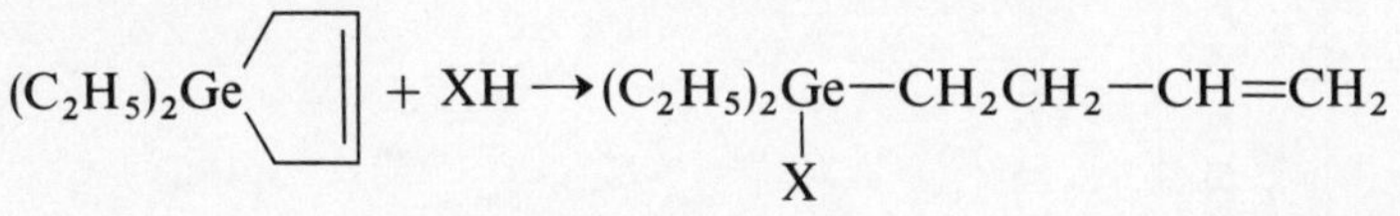

$$(C_2H_5)_2Ge \overset{\diagup}{\diagdown} \| \; + \; XH \rightarrow (C_2H_5)_2\underset{\underset{X}{|}}{Ge}-CH_2CH_2-CH=CH_2$$

$$X = Cl, Br$$

Halogens simultaneously cleave the two intracyclic Ge—C bonds weakened by the double bond in the β position relative to the germanium atom. Iodine reacts at room temperature whereas bromine already cleaves the Ge—C bonds at $-80°C$; in both cases the diethyldihalogermane is formed:

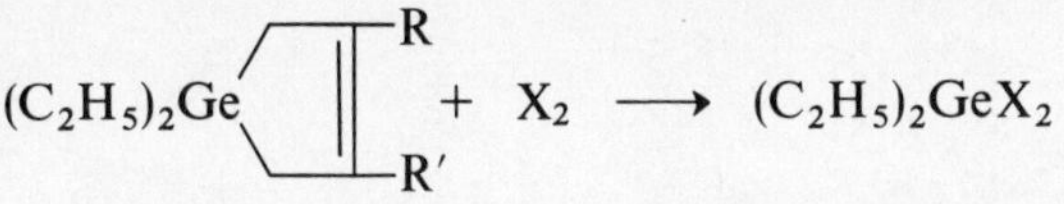

$$(C_2H_5)_2Ge \overset{\diagup R}{\underset{\diagdown R'}{\|}} \; + \; X_2 \; \longrightarrow \; (C_2H_5)_2GeX_2$$

On the other hand, chlorine adds to dichlorogermacyclopentene (92). Bromine does not cleave the Ge—C bond of dibromogermacyclopentenes but gives the normal addition on the double bond (70):

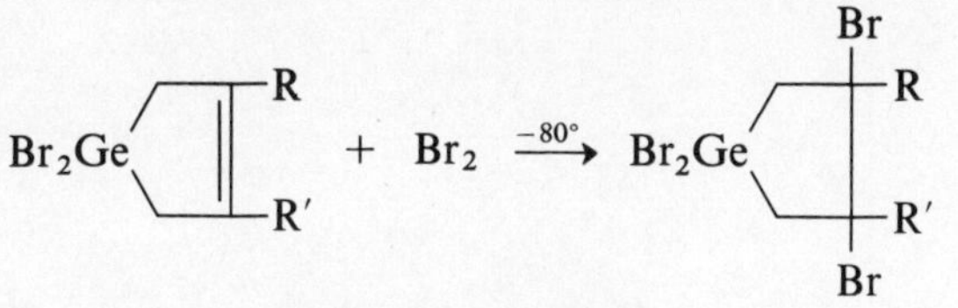

$$Br_2Ge \overset{\diagup R}{\underset{\diagdown R'}{\|}} \; + \; Br_2 \; \overset{-80°}{\longrightarrow} \; Br_2Ge \overset{\diagup\overset{Br}{|}R}{\underset{\diagdown\underset{Br}{|}R'}{}}$$

$$R = R' = H, CH_3$$
$$R = H, R' = CH_3$$

Germacyclopentene Derivatives

Table 3-8 showing some physical properties of germacyclopentene derivatives appears on the following pages.

Table 3-8
Germacyclopentene Derivatives

COMPOUND	EMPIRICAL FORMULA	M.P. (°C)	B.P. (°C/mm)	n_D^{20}	d_4^{20}	REFERENCES
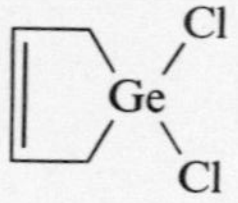 (Ge with Cl, Cl; cyclopentene ring)	$C_4H_6GeCl_2$	—	52/10	1.5223	1.5762	90, 91, 102
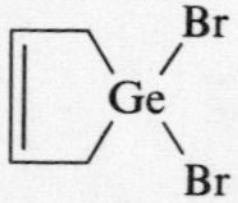 (Ge with Br, Br; cyclopentene ring)	$C_4H_6GeBr_2$	—	95.5/20.5	1.5880	2.1742	35, 93
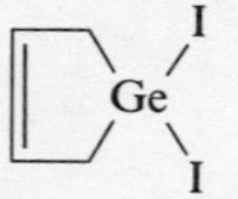 (Ge with I, I; cyclopentene ring)	$C_4H_6GeI_2$	29	77/0.2	1.6890 (30°)	2.5693 (30°)	84
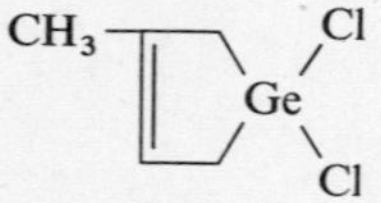 (Ge with Cl, Cl; CH$_3$-cyclopentene ring)	$C_5H_8GeCl_2$	—	98/34	1.5128	1.4694	83
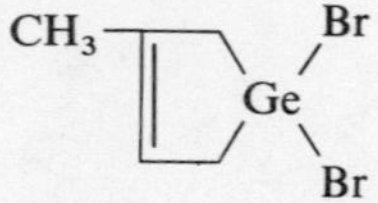 (Ge with Br, Br; CH$_3$-cyclopentene ring)	$C_5H_8GeBr_2$	—	125/25	1.5825	2.059	83
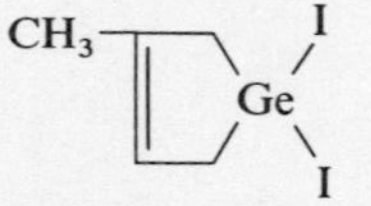 (Ge with I, I; CH$_3$-cyclopentene ring)	$C_5H_8GeI_2$	—	—	1.6515	2.319	83

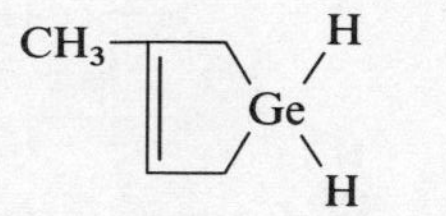 $C_5H_{10}Ge$	—	117/760	1.5212	1.2751	83
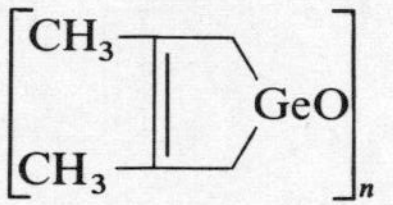 $(C_6H_{10}GeO)_n$	190–91	—	—	—	83
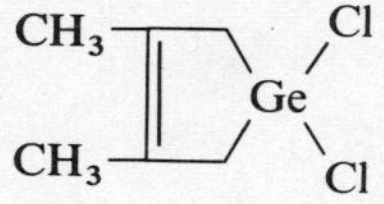 $C_6H_{10}GeCl_2$	—	120/26	1.5178	1.4249	83
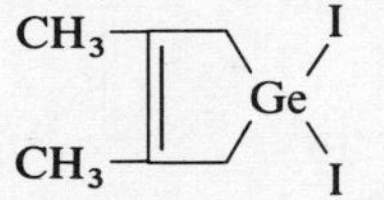 $C_6H_{10}GeI_2$	32–34	—	1.654	2.224	83
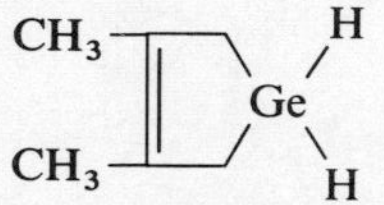 $C_6H_{12}Ge$	—	142/760	1.5025	1.1765	83
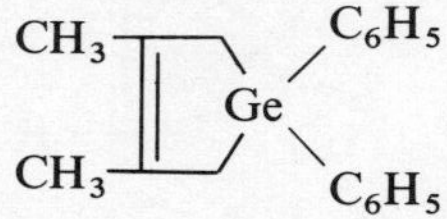 $C_{18}H_{20}Ge$	40–41	142.5/0.25	1.5832 (50°)	1.1743 (50°)	83
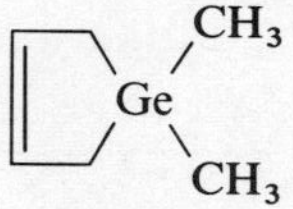 $C_6H_{12}Ge$	—	119/743 121/750	1.4712 1.4723	1.1273 1.1328	90, 91, 94, 102

Table 3-8—continued

COMPOUND	EMPIRICAL FORMULA	M.P. (°C)	B.P. (°C/mm)	n_D^{20}	d_4^{20}	REFERENCES
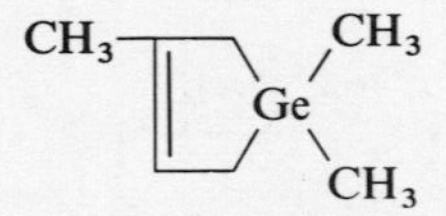 germacyclopentene ring, Ge with CH₃, CH₃	$C_7H_{14}Ge$	—	139/741	1.4740	1.1041	70, 91
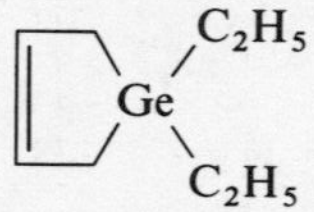 germacyclopentene ring, Ge with CH₃, C₂H₅	$C_8H_{16}Ge$	—	65/17	1.4813	1.1013	84
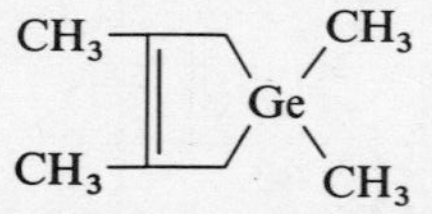 germacyclopentene ring (CH₃, CH₃ on ring), Ge with CH₃, C₂H₅	$C_8H_{16}Ge$	—	71/27	1.4799	1.0902	83
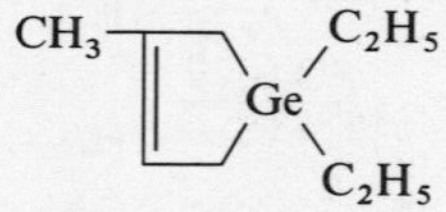 germacyclopentene ring (CH₃, CH₃ on ring), Ge with C₂H₅, C₂H₅	$C_9H_{18}Ge$	—	89/24	1.4805	1.0754	83
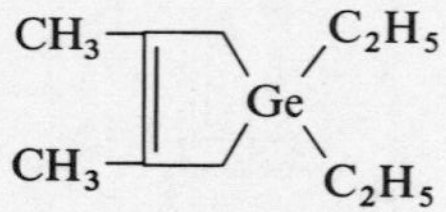 germacyclopentene ring (CH₃, CH₃ on ring), Ge with C₂H₅, C₂H₅	$C_{10}H_{20}Ge$	—	102/26	1.4850	1.0625	83
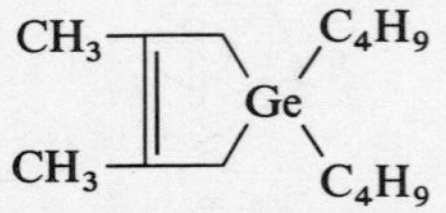 germacyclopentene ring (CH₃, CH₃ on ring), Ge with C₄H₉, C₄H₉	$C_{14}H_{28}Ge$	—	161/36	1.4820	1.0059	83

3-4-2 Germanium pentadiene ring derivatives

At present we have no knowledge of germacyclopentadiene or germa-cyclopentadienes substituted on the germanium atom. Leavitt and co-workers (54) prepared a series of ring substituted heterocyclopentadienes by letting different halides react with 1,4-dilithio-1-2-3-4-tetraphenyl-butadiene. For instance, with a diphenylgermanium dihalide they formed 1,1,2,3,4,5-hexaphenyl-1-germa-2,4-cyclopentadiene:

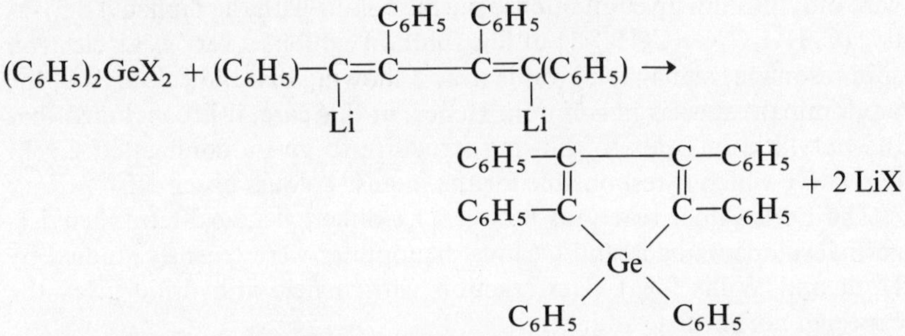

$$(C_6H_5)_2GeX_2 + (C_6H_5)\!-\!\overset{\overset{\displaystyle C_6H_5}{|}}{C}\!=\!\overset{\overset{\displaystyle C_6H_5}{|}}{C}\!-\!\!-\!\!-\!C\!=\!\underset{\underset{\displaystyle Li}{|}}{C}(C_6H_5) \rightarrow$$

with Li substituents shown below the carbons.

$$
\begin{array}{c}
C_6H_5-C \!\!\!-\!\!\!-\!\!\!-\!\!\! C-C_6H_5 \\
\| \qquad\qquad \| \\
C_6H_5-C \qquad\quad C-C_6H_5 \\
\diagdown \quad \diagup \\
Ge \\
\diagup \quad \diagdown \\
C_6H_5 \qquad C_6H_5
\end{array}
\quad + \; 2\,LiX
$$

whereas with germanium tetrachloride the corresponding spiran is formed. Gilman and Gorsich (40) prepared another type of spiran by the reactional 2,2′-diphenylenedilithium with germanium tetrachloride

$$2\;\;\underset{\text{Li}\quad\text{Li}}{\text{(biphenylene)}} + \; GeCl_4 \longrightarrow 2\,LiCl \; + \; \text{(dibenzogermole spiran)}$$

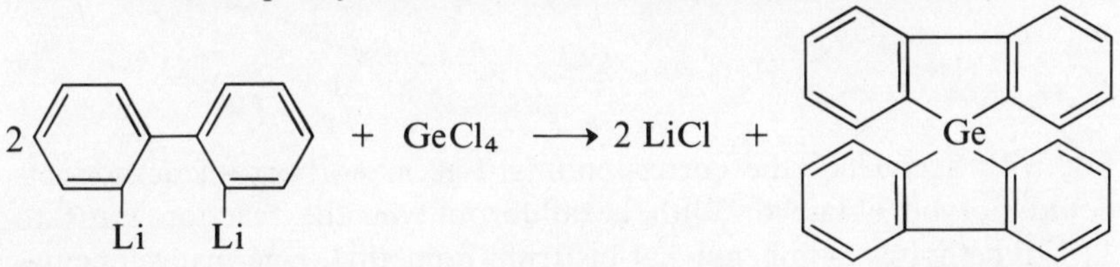

(yield: 29 %)

Although attempts to prepare halogen-substituted germoles by the reaction of $(C_6H_5)_4C_4Li_2$ with chlorogermanes have been reported to be unsuccessful (55), Curtis (21) has found that reverse addition of a diethyl ether suspension of 1,4-dilithiotetraphenylbutadiene to a diethyl ether or tetrahydrofuran solution of $RGeCl_3$ or $GeCl_4$ gives in 50–70 % yield the corresponding heterocycles:

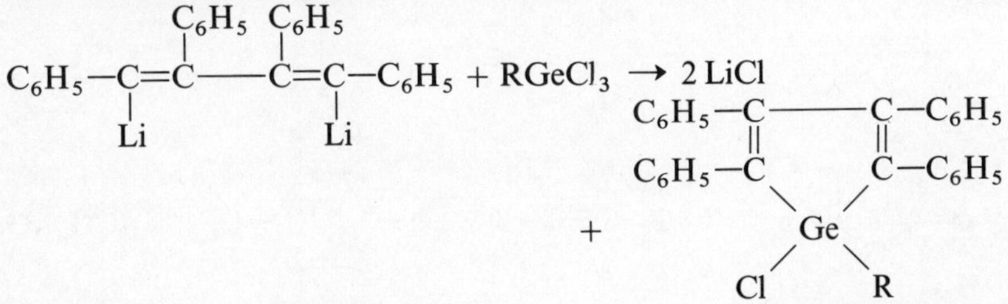

$$C_6H_5-\overset{\overset{\displaystyle C_6H_5}{|}}{C}\!=\!\overset{\overset{\displaystyle C_6H_5}{|}}{\underset{\underset{\displaystyle Li}{|}}{C}}\!-\!\!-\!\!-\!C\!=\!\underset{\underset{\displaystyle Li}{|}}{C}\!-\!C_6H_5 \; + \; RGeCl_3 \; \rightarrow \; 2\,LiCl$$

$$
\begin{array}{c}
C_6H_5-C \!\!\!-\!\!\!-\!\!\!-\!\!\! C-C_6H_5 \\
\| \qquad\qquad \| \\
C_6H_5-C \qquad\quad C-C_6H_5 \\
\diagdown \quad \diagup \\
Ge \\
\diagup \quad \diagdown \\
Cl \qquad R
\end{array}
$$

with a leading $+$ sign.

$$R = Cl, C_6H_5$$

The reduction of these compounds by $LiAlH_4$ gives the corresponding hydrides $(C_6H_5)_4C_4GeH_2$ and $(C_6H_5)_4C_4GeH(C_6H_5)$ in 70–90 % yield. The last of these, when treated by n-butyllithium in THF at $-78°C$, gives a bright-red solution from which is regenerated the starting hydride by addition of water. Quenching with trimethylchlorosilane gives the silyl derivative, $(C_6H_5)_4C_4Ge(C_6H_5)Si(CH_3)_3$. Thus the bright-red colour is ascribed to the germyllithium compound, $(C_6H_5)_4C_4Ge(Li)C_6H_5$. Like-wise butyllithium gives an intense purple colour with the diphenyl deriva-tive $(C_6H_5)_4C_4Ge(C_6H_5)_2$; but this solution exhibits a very weak electron spin resonance signal at a g value near 2 showing that radicals are not the predominant species in solution. Hence, in this case, it is concluded that the butyllithium adds to the ring structure to give a conjugated C—Li derivative which is responsible for the intense colours observed.

The Diels–Alder reactions between 1,1-dimethyl-2,3,4,5-tetraphenyl-1-germacyclopentadiene and various dienophiles were recently studied by Hota and Willis (48a). The reaction with maleic anhydride gives the expected norbornene adduct:

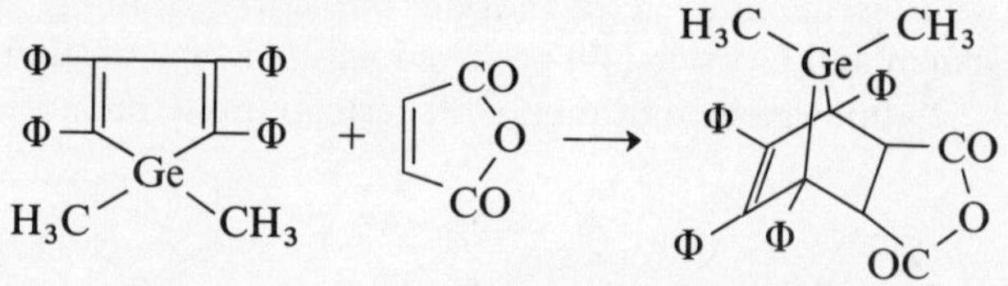

but with acetylenes the corresponding 7-germanorbornadiene adducts could not be obtained. With hexafluorobutyne the reaction leads to 1, 1-dimethyl-2, 5-diphenyl-3, 4-bis(trifluoromethyl)-1-germacyclopenta-diene and bis (trifluoromethyl)tetraphenylbenzene, presumably arising from elimination of dimethylgermane from the initial bicyclic adduct:

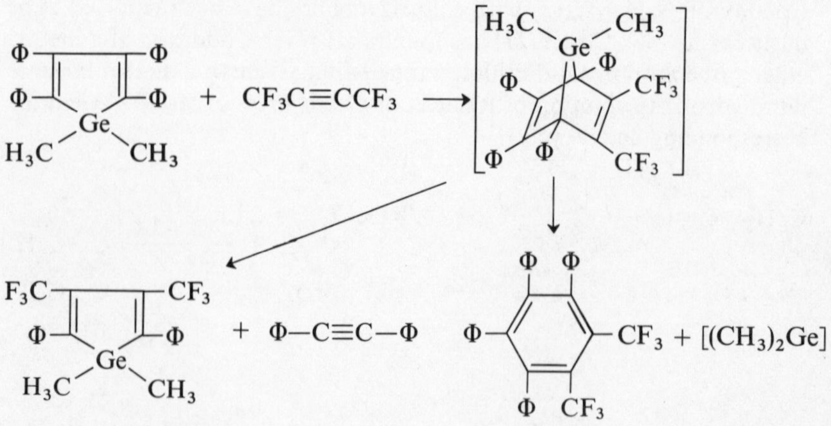

With phenylacetylene (48a, 104a) the reaction occurs rapidly under mild conditions but the adduct decomposes giving pentaphenylbenzene, the germanium being incorporated into a polymeric residue. Although the silole with dimethylacetylene dicarboxylate gives a stable silanorbornadiene (39a), the reaction of the germole with this acetylene did not give the expected adduct (48a). Instead, the main product identified was dimethyl tetraphenylphthalate arising through the decomposition of the germanorbornadiene; the dimethylgermene fragment apparently reacted with further acetylene to give a five-membered heterocycle:

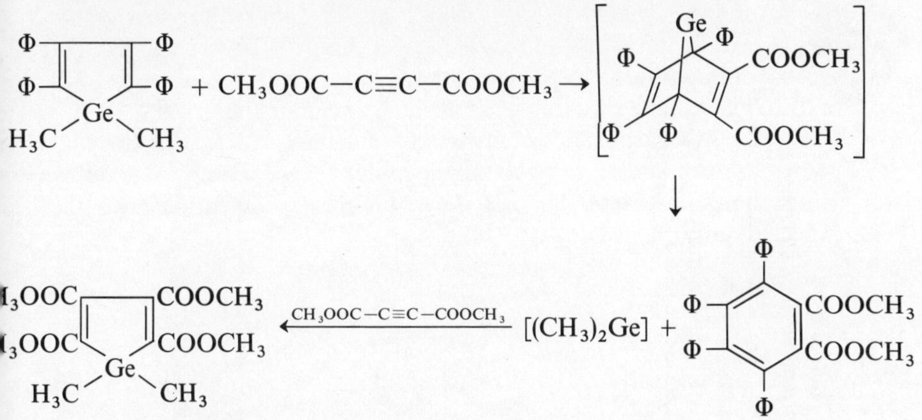

Hota and Willis explain the difference in stability between the 7-germanorbornadiene which they were unable to isolate and the readily prepared silicon analogue by the slight difference between the length of the Si—C (I, 89 Å) and Ge—C (I, 98 Å) bonds, which is sufficient to distort the ring system and cause elimination of dimethylgermene and other fragments.

3-4-2-1 Germanium Pentadiene Ring Derivatives

Table 3-9 showing some physical properties of germanium pentadiene ring derivatives appears on the following pages.

Table 3-9
Germanium Pentadiene Ring Derivatives

COMPOUND	EMPIRICAL FORMULA	M.P. (°C)	B.P. (°C/mm)	n_D^{20}	d_4^{20}	REFERENCES
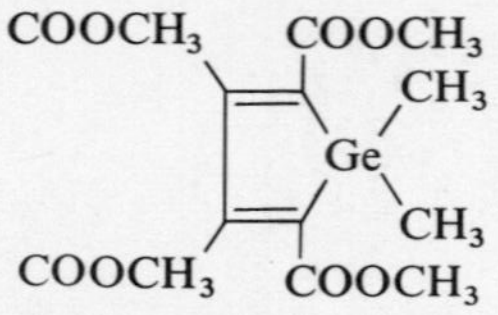	$C_{14}H_{18}O_8Ge$	—	—	—	—	48a
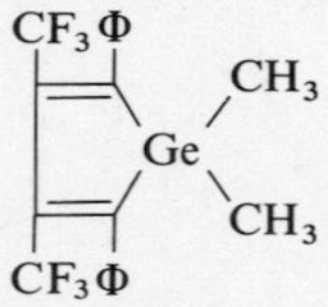	$C_{20}H_{16}F_6Ge$	—	—	—	—	48a
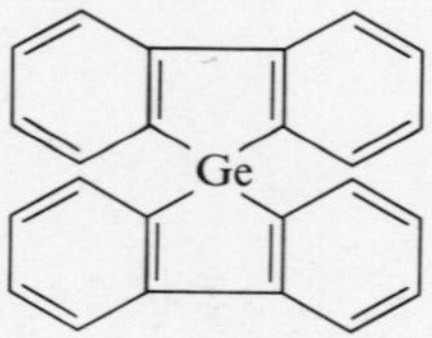	$C_{24}H_{16}Ge$	244.5–46	—	—	—	40
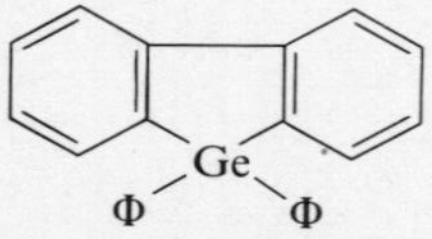	$C_{24}H_{18}Ge$	152–53	—	—	—	42

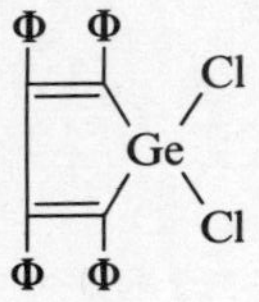	$C_{28}H_{20}GeCl_2$	197–99	—	—	—	21
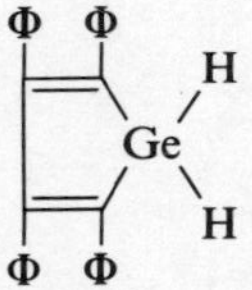	$C_{28}H_{22}Ge$	192–93	—	—	—	21
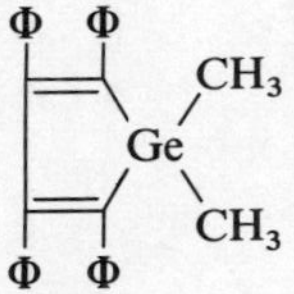	$C_{30}H_{26}Ge$	179–81	—	—	—	48a
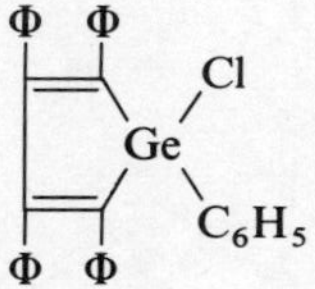	$C_{34}H_{25}GeCl$	210–11	—	—	—	21
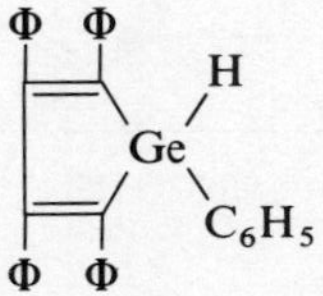	$C_{34}H_{26}Ge$	187–88	—	—	—	21

Table 3-9—continued

COMPOUND	EMPIRICAL FORMULA	M.P. (°C)	B.P. (°C/mm)	n_D^{20}	d_4^{20}	REFERENCES
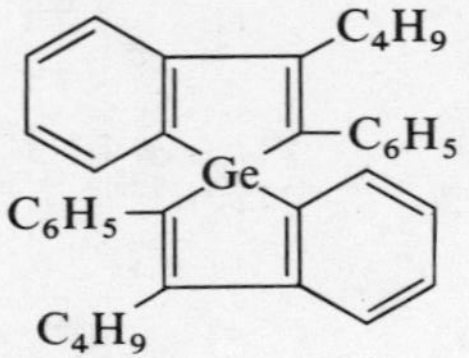	$C_{36}H_{36}Ge$	—	—	—	—	115
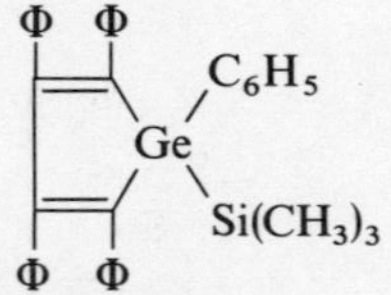	$C_{37}H_{34}SiGe$	178–80	—	—	—	21
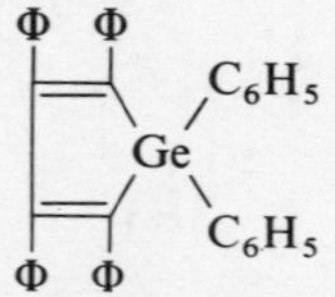	$C_{40}H_{30}Ge$	198–99	—	—	—	21
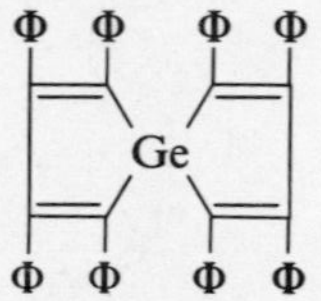	$C_{56}H_{40}Ge$	258–60	—	—	—	54, 55

3-4-3   Digermacyclohexadienes

We noted (3-1-1) that Vol'pin and coworkers through the action of germanous iodide on acetylenic derivatives produced six-membered cyclenic compounds containing two germanium atoms within the ring (145).

The reaction of GeI_2 with acetylene is carried out at 130 to 140°C under a pressure of 10 atm in benzene and leads to a mixture of the cyclic germanium derivative (44 % yield) and of polymers:

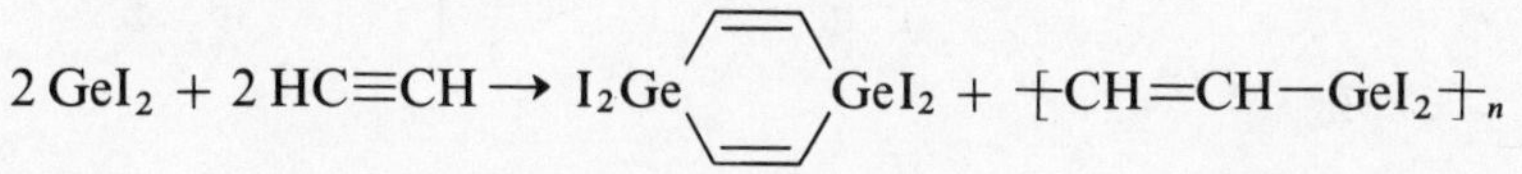

$$2\,GeI_2 + 2\,HC{\equiv}CH \rightarrow I_2Ge\diagup\!\!\!\!\diagdown GeI_2 + \left[CH{=}CH{-}GeI_2\right]_n$$

When recrystallized, the ring derivative appears as colourless crystals insoluble in ether and saturated hydrocarbons and moderately soluble in benzene, chloroform or carbon tetrachloride, M.P. = 225–226°C.

In contrast to the highly reactive dihalocarbenes, germanium diiodide does not noticeably react with diphenylacetylene either at room temperature or on being moderately heated, but above 200°C an addition reaction occurs:

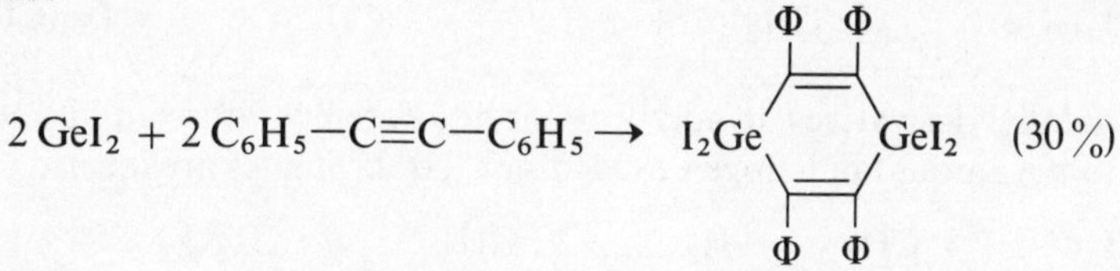

$$2\,GeI_2 + 2\,C_6H_5{-}C{\equiv}C{-}C_6H_5 \rightarrow \begin{matrix}\Phi \quad \Phi \\ I_2Ge\diagup\!\!\!\!\diagdown GeI_2 \\ \Phi \quad \Phi\end{matrix} \quad (30\%)$$

Simultaneously, trimerization reaction of diphenylacetylene to hexaphenylbenzene (GeI_2 catalyst) takes place.

The digermacyclohexadiene ring is stable and will resist many reagents:
silver nitrate reacts immediately with the two germanium-linked iodine atoms;
organomagnesium compounds quantitatively transform the halogenated compounds into alkyl derivatives.

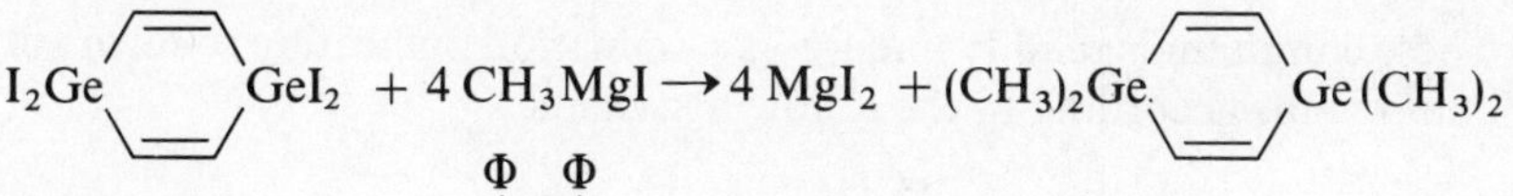

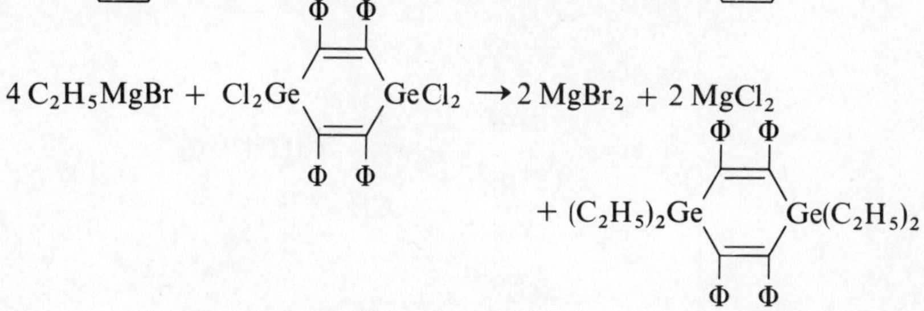

$$I_2Ge\diagup\!\!\!\!\diagdown GeI_2 + 4\,CH_3MgI \rightarrow 4\,MgI_2 + (CH_3)_2Ge\diagup\!\!\!\!\diagdown Ge(CH_3)_2$$

$$4\,C_2H_5MgBr + \begin{matrix}\Phi \quad \Phi \\ Cl_2Ge\diagup\!\!\!\!\diagdown GeCl_2 \\ \Phi \quad \Phi\end{matrix} \rightarrow 2\,MgBr_2 + 2\,MgCl_2$$

$$+ \begin{matrix}\Phi \quad \Phi \\ (C_2H_5)_2Ge\diagup\!\!\!\!\diagdown Ge(C_2H_5)_2 \\ \Phi \quad \Phi\end{matrix}$$

NaOH attacks the germanium-linked halogen atoms without cleaving the ring. The hydroxide produced (intense absorption at $3322\,cm^{-1}$) is gradually dehydrated on heating to form a polymeric oxide with a mean molecular weight of about 25 000:

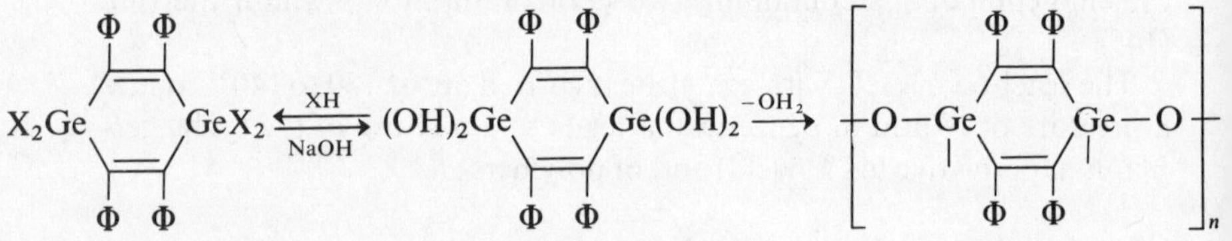

X = Cl, Br, I

Treatment of the hydroxide or oxide with hydrogen chloride, bromide or iodide leads respectively to the chloride, bromide or iodide derivative. The iodide derivative from acetylene also shows the same reactions with NaOH, silver nitrate and organomagnesium compounds. Dialkyl derivatives may be hydrogenated in the presence of Raney nickel (142, 149) forming the corresponding saturated 1,4-digermacyclohexanes:

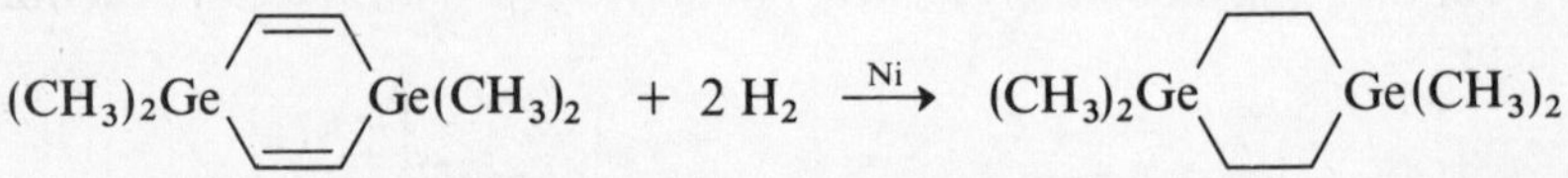

The methyl derivative can add bromine at low temperatures, and, according to the amount of halogen used, di- or tetrabromides are formed (142):

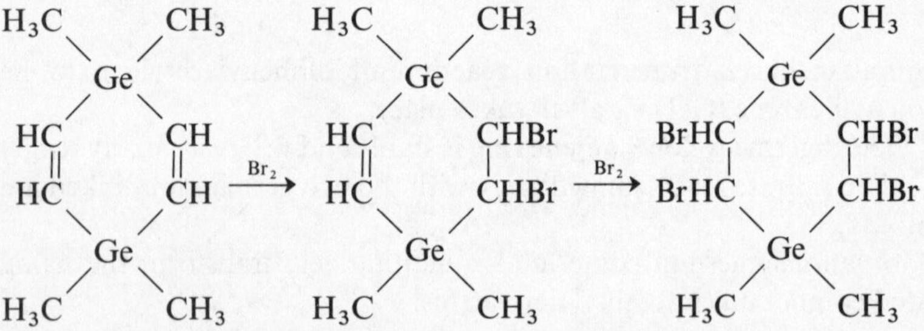

These compounds readily undergo β elimination on heating to form vinyl compounds according to the following scheme:

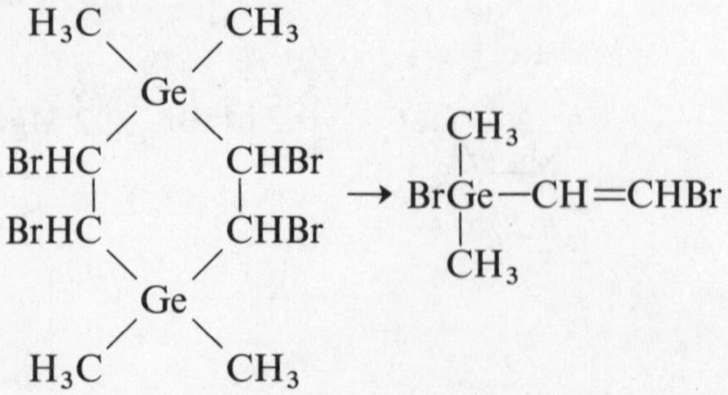

An identical ring cleavage was observed by Johnson and coworkers with the substituted derivative prepared from tolane. Similar to 1,1,4,4-tetra-methylcyclohexadiene-2,5 (148), the methyl derivative undergoes hydroxylation (Prévost reaction). Interaction of this derivative with the same amount of a complex of silver benzoate and iodine gives rise to an adduct with one double bond. The dichloride compound, which has dienophilic properties, gives adducts with cyclopentadiene (142); this reaction proceeds stepwise and, depending on the amount of cyclopentadiene, it is possible to isolate either the monoadduct or the diadduct:

Neither the methyl nor the phenyl derivatives give such adducts, even on prolonged heating up to 160°C.

3-4-3-1 Physical Studies and Structure. Because of the controversy concerning the structure of these derivatives, a great many physical investigations have been carried out on these compounds with the aim of establishing their correct formulae.

IR and Raman spectra of all compounds $R_2Ge\diagup\diagdown GeR_2$ (50, 58,

141–143, 148) reveal an absorption band in the $500-600\ cm^{-1}$ region characteristic of the valence vibration of the Ge—C bond; for the methyl derivative ($R=CH_3$), the frequency ($535\ cm^{-1}$) is smaller than in usual methylgermanes ($572\ cm^{-1}$ for Me_4Ge). These cyclenic compounds have a similar vibration frequency of the double bond in the Raman spectra ($1557\ cm^{-1}$ for $R=CH_3$ and $1550\ cm^{-1}$ for $R=Cl$) than for the linear corresponding olefinic digermanes: $1560\ cm^{-1}$ and $1550\ cm^{-1}$ for 1,2-bis(trimethylgermyl)- and bis(trichlorogermyl)ethylene respectively. The low frequency values are due to the influence of two germanium atoms in the α position with respect to the double bond.

The absorption maximum of the methyl compound ($R=CH_3$) in the UV spectrum (206 nm) undergoes a bathochromic shift compared to 1,2-bis(trimethylgermyl)ethylene (195 nm) (142).

The NMR spectrum of the methyl derivative ($R=CH_3$) (141, 142), which has two types of hydrogen atoms, contains only two resonances of hydrogen having relative areas of $3:1$. For the methyl ($R=CH_3$), the iodo- ($R=I$) and the chloro- ($R=Cl$) derivatives, the chemical shifts of protons at the double bond are 6.95, 7.02 and 7.32 ppm respectively. These values are in a lower field than with the corresponding open-chain derivatives (6.1 ppm for $Me_3GeCH=CHGeMe_3$ and 6.95 ppm for $Cl_3GeCH=CHGeCl_3$).

An X-ray investigation (149) on a monocrystal of the chloride

$$Cl_2Ge\diagdown\!\!\diagup\!\!\diagdown\!\!\diagup\!GeCl_2$$ showed that this molecule consists of a planar

centrosymmetrical six-membered ring with the following bond lengths (Å): $C=C$ 1.35; $Ge—C$ 1.98; $Ge—Cl$ 2.15 and the following bond angles: $C—Ge—C$ and $C—C—Ge$ about $120°$, $Cl—Ge—Cl$ $103°$. In the iodide compound, $Ge—I = 2.48$ Å. Unit cells and space groups for a series of 1,4-digermacyclohexadienes were determined by Bokii and coworkers (14a).

An electron diffraction study (140) of the structure of the dichloride compound in the vapour phase has also shown that the molecule has the structure of a six-membered ring, planar or nearly so, with the basic

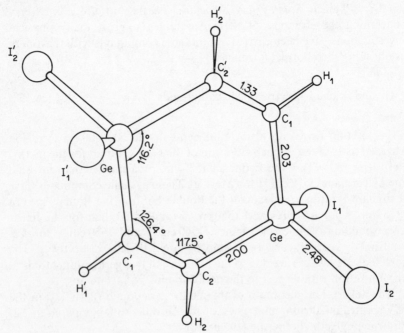

FIG. 3-1   Structure of 1,1,4,4-tetraiodo-1,4-digerma-2,5-cyclohexadiene (142)
(Reproduced by courtesy of the authors)

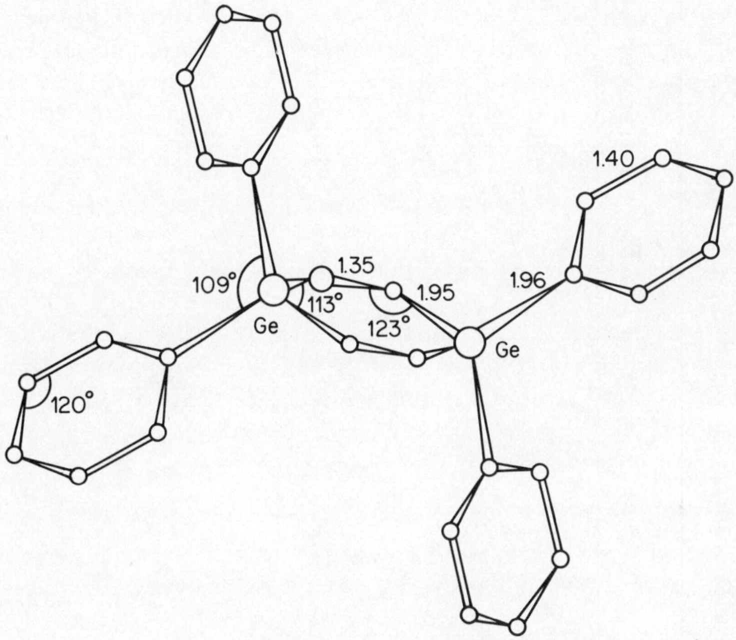

FIG. 3-2   Structure of 1,1,4,4-tetraphenyl-1,4-digerma-2,5-cyclohexadiene (142)
(Reproduced by courtesy of the authors)

parameters: Ge—Cl 2.143 ± 0.01 Å; Ge—C 1.932 ± 0.02 Å; C=C 1.35 ± 0.05 Å; Cl—Ge—Cl $94° \pm 5°$; Cl—Ge—C $109° \pm 5°$; C—Ge—C $117°20' \pm 5°$; Ge—C—C $121°28' \pm 3°$. Figures 3-1 and 3-2 (142) show the structures of the 1,1,4,4-tetraiodo- and the 1,1,4,4-tetraphenyl-1,4-digerma-2,5-cyclohexadienes.

Mass spectra

Johnson and coworkers (49, 50) carried out a detailed mass spectrographic study of these derivatives. Their findings prove that without doubt the compounds resulting from the action of germanous diiodide on acetylenic derivatives (acetylene, tolane) display a cyclodienic structure the two germanium atoms being in the 1,4- position:

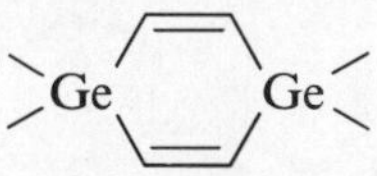

3-4-3-2   Digermahexadiene Derivatives

Table 3-10 showing some physical properties of digermahexadiene derivatives appears on the following pages.

Table 3-10
Digermahexadiene Derivatives

COMPOUND	EMPIRICAL FORMULA	M.P. (°C)	B.P. (°C/mm)	n_D^{20}	d_4^{20}	REFERENCES
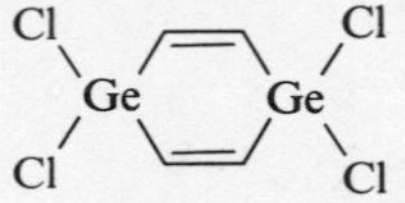	$C_4H_4Ge_2Cl_4$	145–46	—	—	—	141, 142
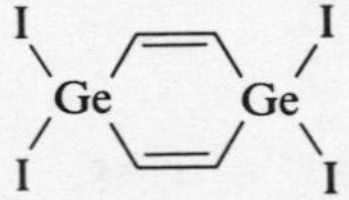	$C_4H_4Ge_2I_4$	225–26	—	—	—	141, 142
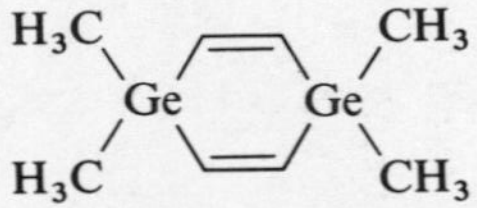	$C_8H_{16}Ge_2$	—	67/14	1.5009	1.2850	141, 142
	$C_8H_{16}Ge_2Br_2$	72–74	—	—	—	142
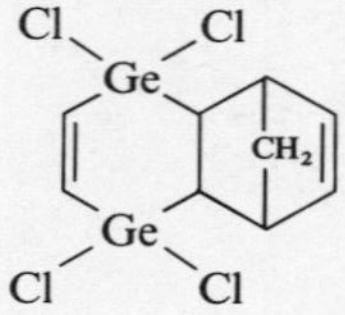	$C_9H_{10}Ge_2Cl_4$	175–77	—	—	—	142

Structure	Formula					Ref.
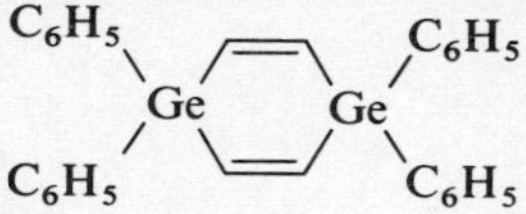	$C_{28}H_{24}Ge_2$	149–50	—	—	—	142
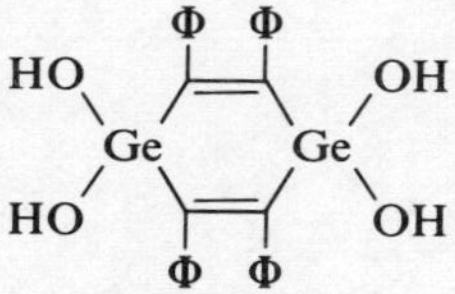	$C_{28}H_{24}Ge_2O_4$	—	—	—	—	148
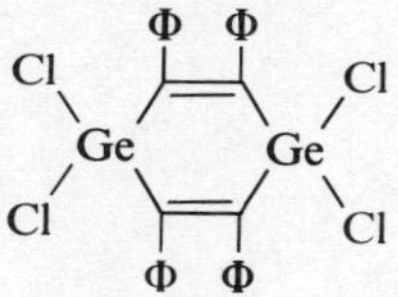	$C_{28}H_{20}Ge_2Cl_4$	290–92	—	—	—	148
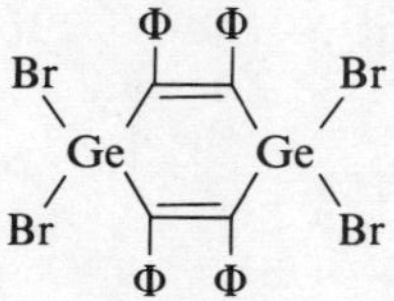	$C_{28}H_{20}Ge_2Br_4$	317–18	—	—	—	148
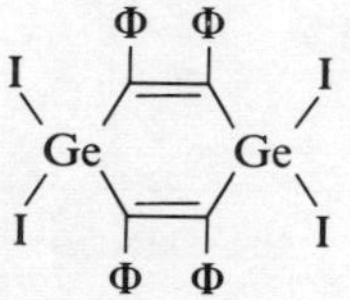	$C_{28}H_{20}Ge_2I_4$	301–3	—	—	—	145, 148
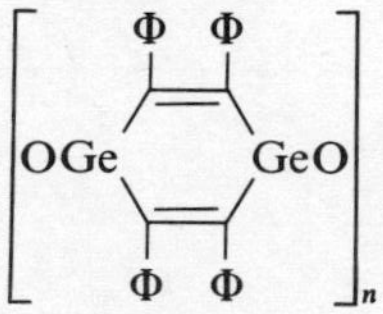	$(C_{28}H_{20}Ge_2O_2)_n$	>600	—	—	—	148

Table 3-10—continued

COMPOUND	EMPIRICAL FORMULA	M.P. (°C)	B.P. (°C/mm)	n_D^{20}	d_4^{20}	REFERENCES
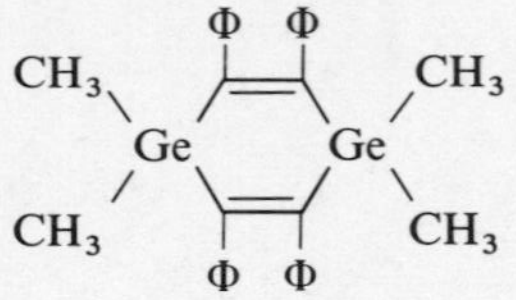	$C_{32}H_{32}Ge_2$	295	—	—	—	148
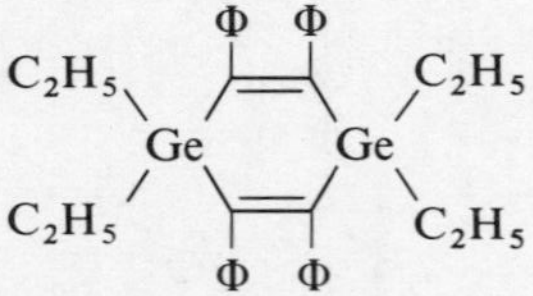	$C_{36}H_{40}Ge_2$	239–40	—	—	—	148
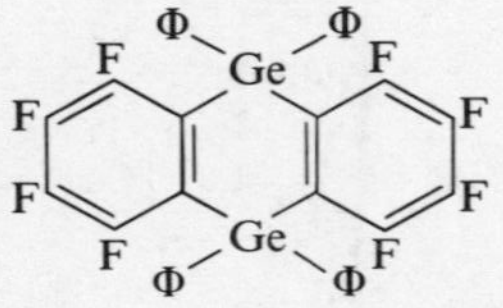	$C_{36}H_{20}Ge_2F_8$	278–82.5	—	—	—	19

3-5   HETEROCYCLIC COMPOUNDS

3-5-1   Synthesis

Germanium heterocycles can be obtained by the following reactions:

3-5-1-1   Intramolecular Cyclization. On being heated in the presence of a catalyst, some types of enoxygermanes lead to five-membered-ring heterocycles (63)

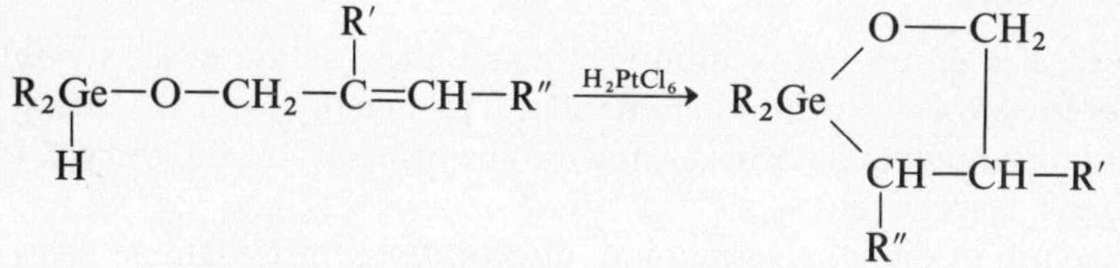

$$R_2Ge-O-CH_2-\underset{\underset{H}{|}}{\overset{\overset{R'}{|}}{C}}=CH-R'' \xrightarrow{H_2PtCl_6} R_2Ge\underset{\underset{R''}{|}}{\overset{O-CH_2}{\diagdown}}\overset{|}{\underset{CH-CH-R'}{}}$$

Another intramolecular cyclization was observed when heating γ-ketohydrogenogermane in the presence of catalyst (126)

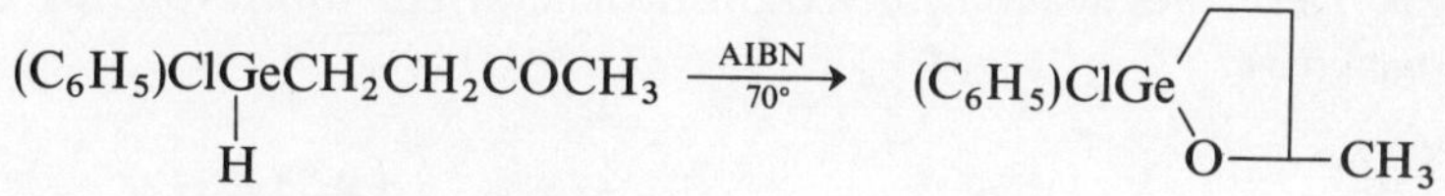

$$(C_6H_5)Cl\underset{\underset{H}{|}}{Ge}CH_2CH_2COCH_3 \xrightarrow[70°]{AIBN} (C_6H_5)ClGe\underset{O-CH_3}{\diagdown}$$

3-5-1-2   Intermolecular Cyclization

(a) Cyclizing polymerization of $\diagdown GeO$, $\diagdown GeS$ and $\diagdown GeNR$ derivatives

Dialkylgermoxanes formed during the hydrolysis of dialkyldihalo-germanes polymerize spontaneously, leading generally to a cyclic trimer (7):

$$3 R_2GeX_2 + 6 NaOH \rightarrow 6 NaX + 3 OH_2 + (R_2GeO)_3$$

Di, *n*- and *sec*-alkoxydibutylgermanes are readily and almost quantitatively hydrolysed to trimeric dibutylgermanium oxide by water at room temperature (67):

$$(C_4H_9)_2Ge(OR)_2 + OH_2 \rightarrow (C_4H_9)_2Ge\overset{OH}{\underset{OR}{\diagup\diagdown}} + ROH$$

$$3(C_4H_9)_2Ge\overset{OH}{\underset{OR}{\diagup\diagdown}} \rightarrow [(C_4H_9)_2GeO]_3 + 3 ROH \qquad R = C_2H_5, i\text{-}C_3H_7$$

Di-tertiobutoxydibutylgermane gives the same reaction but only in the presence of *p*-toluenesulphonic acid. In the same manner, dimeric dibutylgermanium sulphide is obtained by passing hydrogen sulphide gas into a solution of di-n- and sec-alkoxydibutylgermane in the parent alcohol (67):

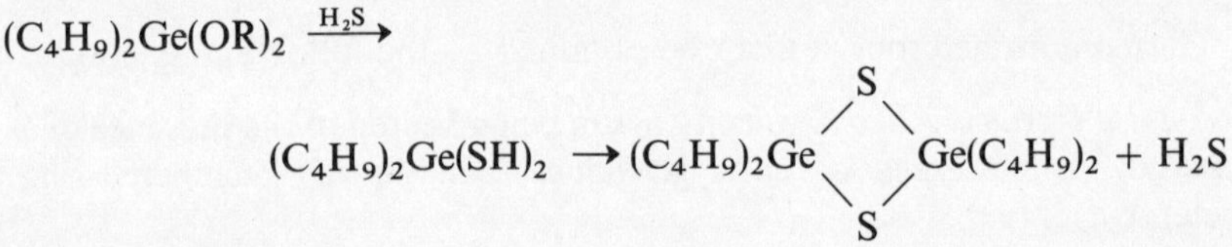

$$(C_4H_9)_2Ge(OR)_2 \xrightarrow{H_2S}$$

$$(C_4H_9)_2Ge(SH)_2 \rightarrow (C_4H_9)_2Ge \diagdown \overset{S}{\underset{S}{\diagup\diagdown}} Ge(C_4H_9)_2 + H_2S$$

In the case of di-tert-butoxydibutylgermane, the reaction proceeds only in the presence of p-toluenesulphonic acid. The dimethyl- and diisopropyl-germanium sulphides have been shown to be trimeric (16) and dimeric (5) in benzene, respectively.

The action of dimethylgermanium dichloride on methylamine gives a six-membered Ge—N ring compound (120a, 131).

(b) Mixed bimolecular cyclization

Dialkylgermanes react with tetramethylene glycol with evolution of hydrogen (63):

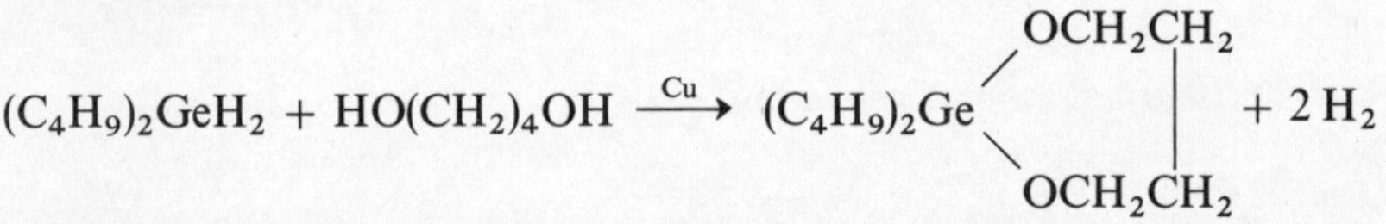

$$(C_4H_9)_2GeH_2 + HO(CH_2)_4OH \xrightarrow{Cu} (C_4H_9)_2Ge \overset{OCH_2CH_2}{\underset{OCH_2CH_2}{\diagup\diagdown|}} + 2H_2$$

A general procedure for preparing germanium heterocycles with Ge—O, Ge—S and Ge—N bonds involves the reaction of germanium dihalides with difunctional compounds (—OH, —SH, —NH₂) in the presence of a base (8, 22, 152, 154–156):

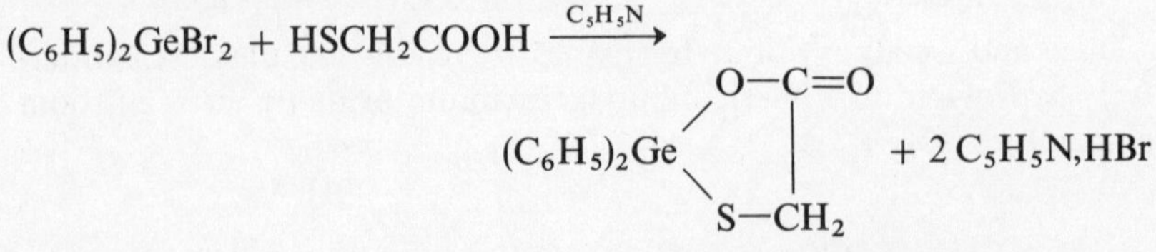

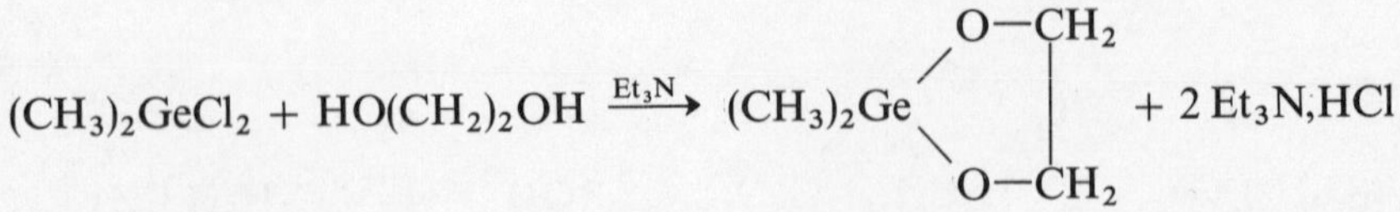

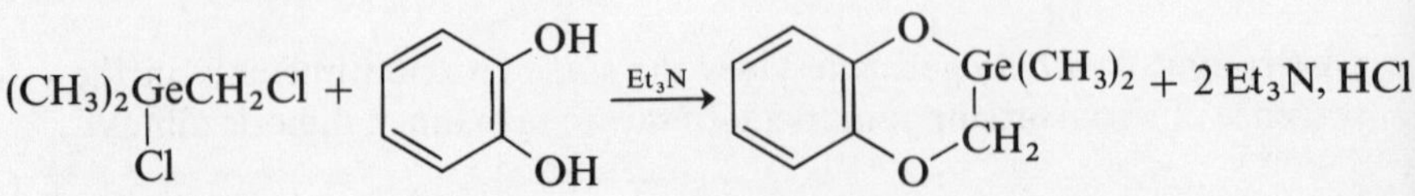

$$(C_6H_5)_2GeBr_2 + HSCH_2COOH \xrightarrow{C_5H_5N}$$

$$(C_6H_5)_2Ge \overset{O—C=O}{\underset{S—CH_2}{\diagup\diagdown|}} + 2C_5H_5N,HBr$$

$$(CH_3)_2GeCl_2 + HO(CH_2)_2OH \xrightarrow{Et_3N} (CH_3)_2Ge \overset{O—CH_2}{\underset{O—CH_2}{\diagup\diagdown|}} + 2Et_3N,HCl$$

$$(CH_3)_2GeCH_2Cl \underset{Cl}{|} + \underset{\text{(benzene ring with two OH)}}{} \xrightarrow{Et_3N} \underset{\text{(benzene ring)}}{} \overset{O}{\underset{O}{\diagdown\diagup}} \overset{Ge(CH_3)_2}{\underset{CH_2}{}} + 2Et_3N, HCl$$

With substituted diphenol, a mixture of position isomers is obtained:

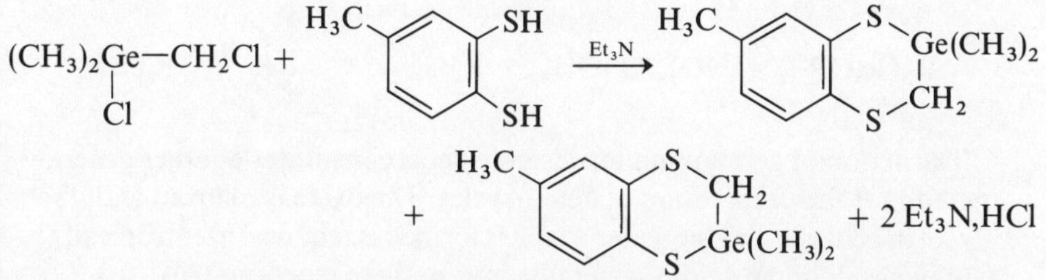

$$(CH_3)_2Ge-CH_2Cl + \overset{H_3C}{\underset{SH}{\bigodot}} \xrightarrow{Et_3N} \cdots$$

The reaction of chloromethyldimethylchlorogermane with ortho-amino-phenol gives only the Ge—O bond derivatives (152):

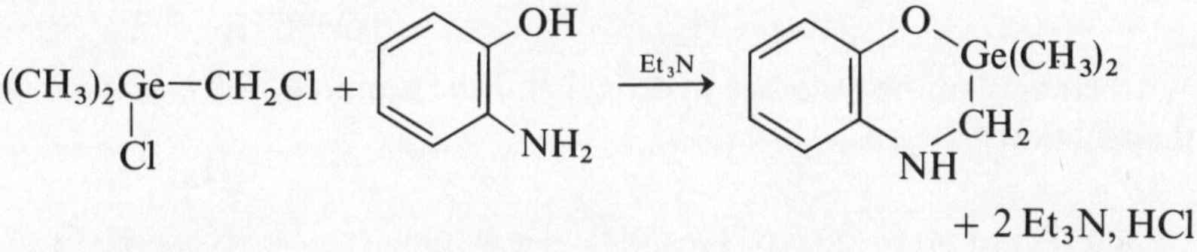

Chloromethyldimethylchlorogermane reacts with H_2O, H_2S, and NH_3 under mild conditions with substitution of the Ge—Cl to give 1,3-bis-(chloromethyl)tetramethyldigermoxane, -digermathiane, and -digerma-zane, respectively. In a second step, the two carbon-bound chlorine atoms may be substituted by sulphur yielding germanium heterocycles (159). A large number of cyclic glycoxygermanes have been synthesized by Mathur and coworkers (68, 87) by the reaction of dialkyldialkoxygermanes with various glycols in the presence of an acid catalyst such as p-toluene-sulphonic acid:

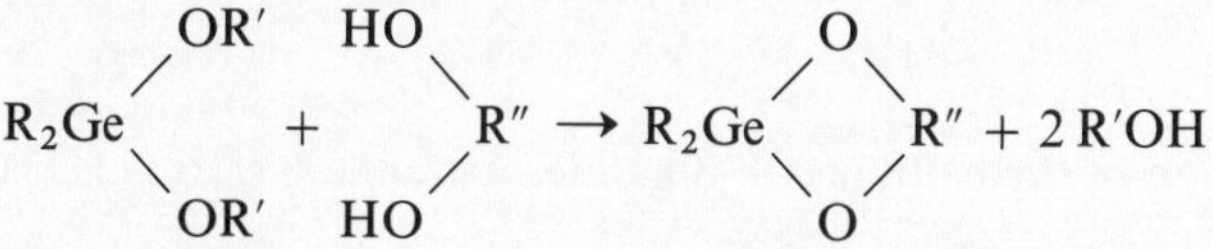

$R = C_4H_9, C_6H_5;$

$R' = C_2H_5, i\text{-}C_3H_7;$

$R'' = -(CH_2)_2, -(CH_2)_3-, -CH_2CH_2CH(CH_3), -(CH_2)_2O(CH_2)_2-$
etc.

With ethylene glycol, the reaction proceeds even in absence of the catalyst.

With hydroxycarboxylic acids (salicylic and mandelic acids) the hydroxyl group takes part in the reaction along with the carboxyl group (69):

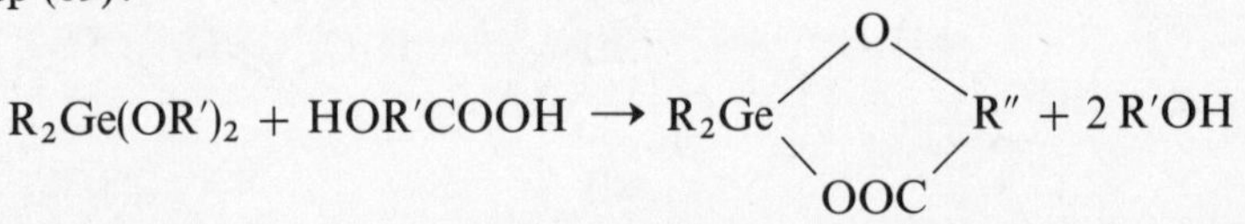

$$R_2Ge(OR')_2 + HOR'COOH \rightarrow R_2Ge\underset{OOC}{\overset{O}{<}}R'' + 2\,R'OH$$

The action of germanium oxides on glycols constitutes another general method for the preparation of heterocycles. The oxide is refluxed with the glycol in benzene and the water which is formed is removed azeotropically: p-toluenesulphonic acid is generally used in these reactions (86).

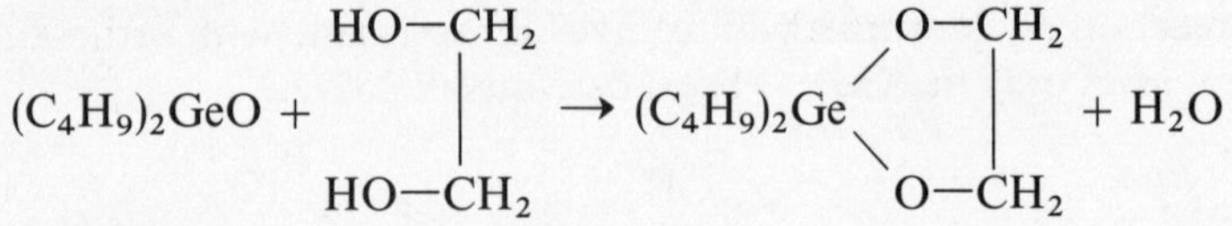

$$(C_4H_9)_2GeO + \begin{array}{c} HO-CH_2 \\ | \\ HO-CH_2 \end{array} \rightarrow (C_4H_9)_2Ge\overset{O-CH_2}{\underset{O-CH_2}{<}}\begin{array}{c} | \\ | \end{array} + H_2O$$

With ethanolamine both the hydroxyl and the amino group take part in the reaction:

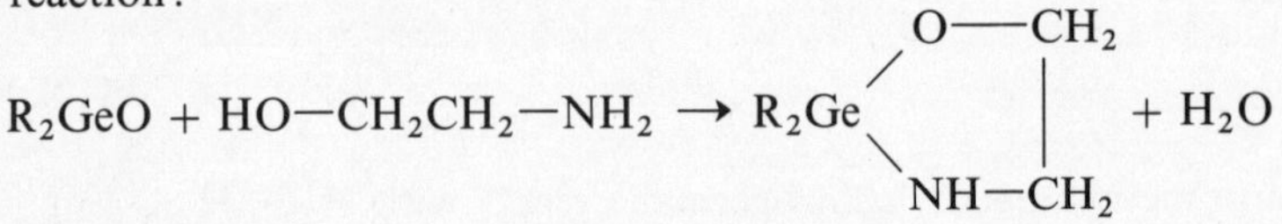

$$R_2GeO + HO-CH_2CH_2-NH_2 \rightarrow R_2Ge\overset{O--CH_2}{\underset{NH-CH_2}{<}}\begin{array}{c} | \\ | \end{array} + H_2O$$

$$R = C_4H_9, C_6H_5$$

Germanium dioxide itself reacts with α and β glycols in a Dean–Stark apparatus giving heterocyclic spirans (61):

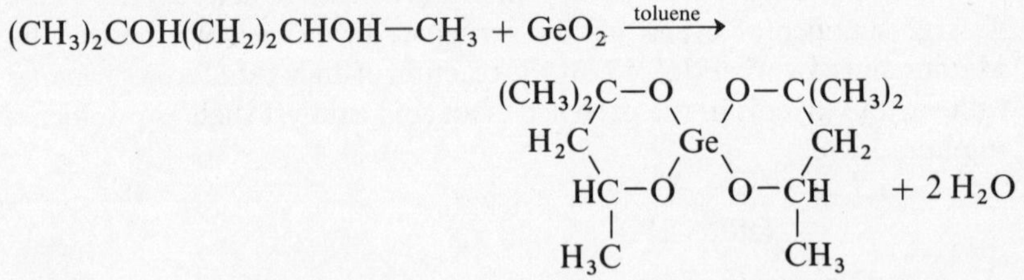

$$(CH_3)_2COH(CH_2)_2CHOH-CH_3 + GeO_2 \xrightarrow{\text{toluene}}$$

$$\begin{array}{cc} (CH_3)_2C-O & O-C(CH_3)_2 \\ H_2C & Ge & CH_2 \\ HC-O & O-CH \\ | & | \\ H_3C & CH_3 \end{array} + 2\,H_2O$$

Reaction of germanium α,ω-dihalides with alkali salts (73, 77):

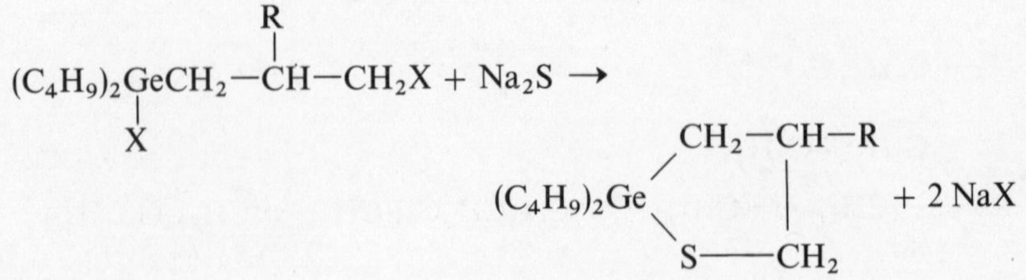

$$\begin{array}{c} R \\ | \\ (C_4H_9)_2GeCH_2-CH-CH_2X + Na_2S \rightarrow \\ | \\ X \end{array}$$

$$(C_4H_9)_2Ge\overset{CH_2-CH-R}{\underset{S---CH_2}{<}}\begin{array}{c} | \\ | \end{array} + 2\,NaX$$

$$\begin{cases} X = Cl, Br \\ R' = H, CH_3 \end{cases}$$

This reaction also occurs with appropriate halodigermanes (70, 81):

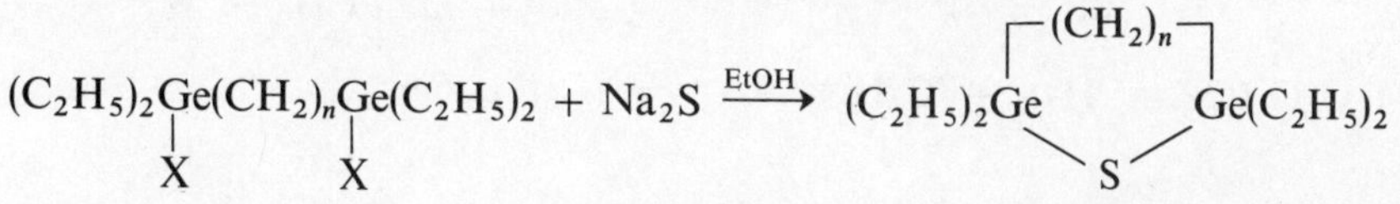

$$(C_2H_5)_2Ge(CH_2)_nGe(C_2H_5)_2 + Na_2S \xrightarrow{EtOH} (C_2H_5)_2Ge\underset{S}{\overset{(CH_2)_n}{\diagup\diagdown}}Ge(C_2H_5)_2$$

$$n = 2, 3, 4$$

Likewise, alcoholic sodium hydroxide gives the corresponding cyclic oxide:

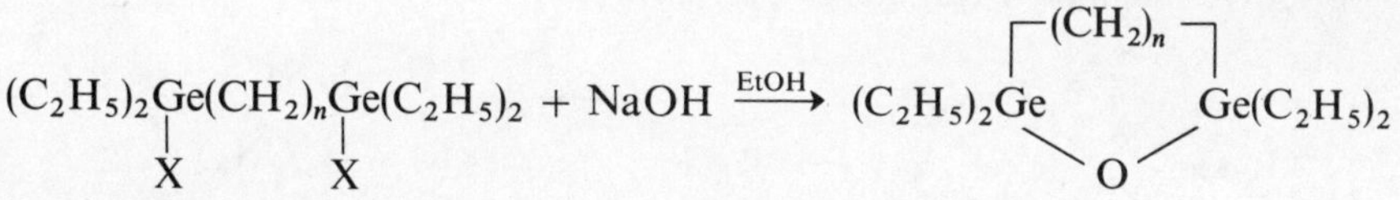

$$(C_2H_5)_2Ge(CH_2)_nGe(C_2H_5)_2 + NaOH \xrightarrow{EtOH} (C_2H_5)_2Ge\underset{O}{\overset{(CH_2)_n}{\diagup\diagdown}}Ge(C_2H_5)_2$$

$$n = 2, 3, 4$$

Silver oxide in tetrahydrofuran gives the same oxide ($n = 2$).

Lithium derivatives have been used for the preparation of heterocyclic compounds with Ge—N bonds (36):

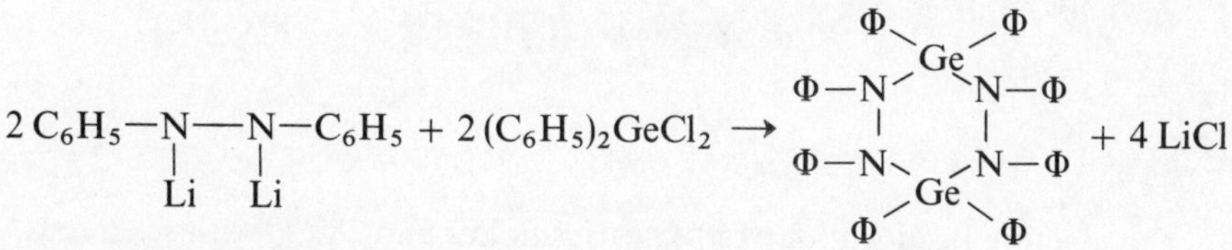

$$2\ C_6H_5{-}N{-}N{-}C_6H_5 + 2\ (C_6H_5)_2GeCl_2 \rightarrow \begin{array}{c}\text{ring structure}\end{array} + 4\ LiCl$$

(with Li, Li on the nitrogens)

Ge—Si bonds (44):

$$\Phi_2GeCl_2 + LiSi\Phi_2(Si\Phi_2)_2Si\Phi_2Li \rightarrow \begin{array}{c}\text{ring structure}\end{array} + 2\ LiCl$$

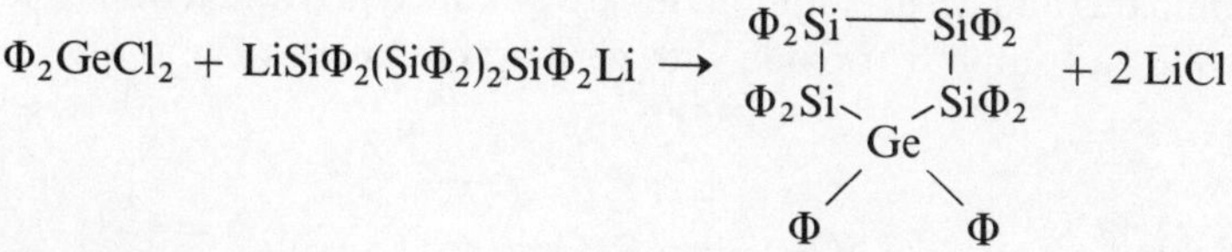

and Ge—C bonds (19):

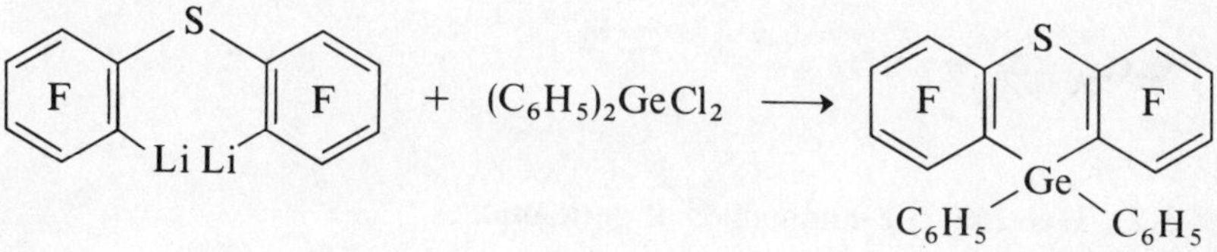

With germanium tetrachloride the spiro-compound is obtained (M.P. 230°C). A six-membered-ring germanium derivative containing a tin atom in the ring was obtained by Henry and Noltes (47) by condensation of diphenyltin dihydride on diphenyldivinylgermane (yield 17%):

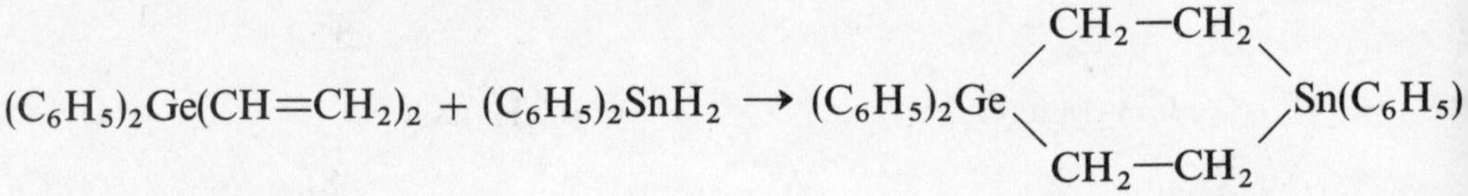

$$(C_6H_5)_2Ge(CH{=}CH_2)_2 + (C_6H_5)_2SnH_2 \rightarrow (C_6H_5)_2Ge\underset{CH_2{-}CH_2}{\overset{CH_2{-}CH_2}{<\,\,>}}Sn(C_6H_5)$$

3-5-1-3 Ring Expansion

Germacyclobutanes react above 250°C with sulphur and selenium to give five-membered-ring heterocycles (75, 77):

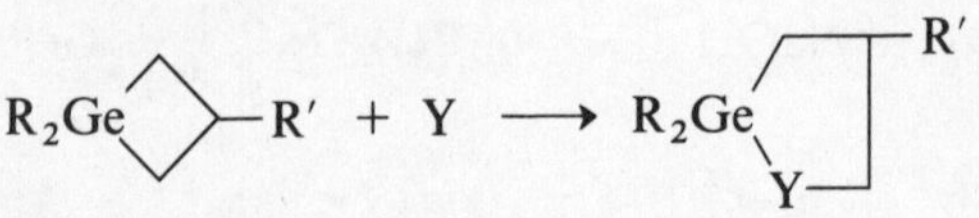

$$R_2Ge\,\triangleright{-}R' + Y \longrightarrow R_2Ge \quad \begin{cases} Y = S, Se \\ R' = H, CH_3 \\ R = C_2H_5, C_4H_9 \end{cases}$$

Digermacyclopentanes give a ring expansion with oxygen, sulphur and selenium (81)

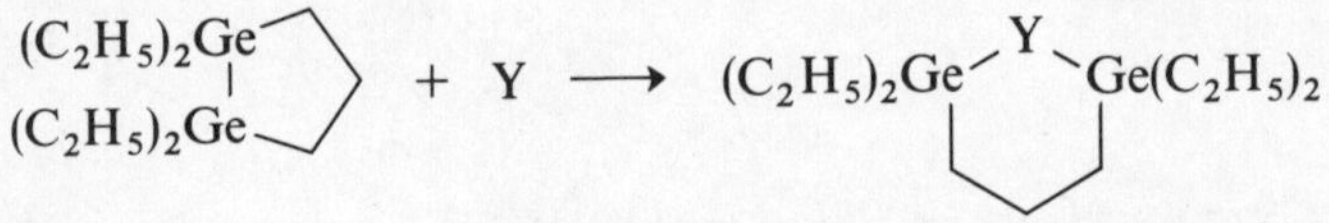

$$\begin{matrix}(C_2H_5)_2Ge\\|\\(C_2H_5)_2Ge\end{matrix}\,\rangle + Y \longrightarrow (C_2H_5)_2Ge^{\diagup Y \diagdown}Ge(C_2H_5)_2$$

$$Y = O, S, Se$$

The reaction occurs about 150°C with sulphur and 280°C with selenium; with oxygen polygermoxanes are also formed. The six-membered germa-sulphinates and germasulphonates are obtained in the reaction of sulphur dioxide and sulphur trioxide with germacyclobutanes (26, 26a):

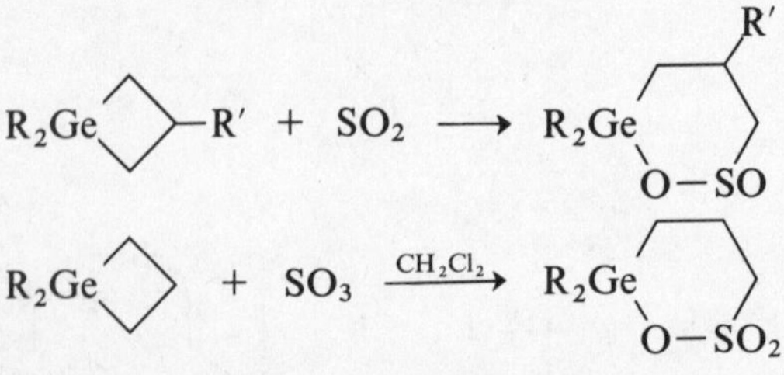

$$R_2Ge\,\square{-}R' + SO_2 \longrightarrow R_2Ge \underset{O{-}SO}{\overset{R'}{<\,\,>}} \quad \begin{cases} R = C_2H_5, C_4H_9 \\ R' = H, CH_3 \end{cases}$$

$$R_2Ge\,\square + SO_3 \xrightarrow{CH_2Cl_2} R_2Ge\underset{O{-}SO_2}{<\,\,>} \quad R = C_2H_5, C_4H_9$$

3-5-2 Heterocyclic compounds of germanium

Table 3-11 (a–e) showing some physical properties of heterocyclic compounds of germanium appears on the following pages.

Table 3-11
Heterocyclic Compounds of Germanium
(a) Heterocyclic oxygen compounds

COMPOUND	EMPIRICAL FORMULA	M.P. (°C)	B.P. (°C/mm)	n_D^{20}	d_4^{20}	REFERENCES
[(CH₃)₂GeO]ₙ	(C₂H₆GeO)ₙ n = 3, 4	91–92	86–96/1	—	—	16, 97, 117, 118, 119, 165, 166
	(C₄H₈GeO)₃	166	—	—	—	71
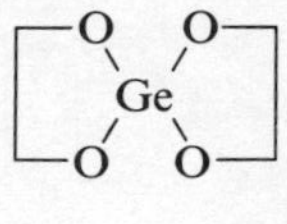	C₄H₈GeO₄	61	—	—	—	61
(n–C₄H₉GeO)₄ H	(C₄H₁₀GeO)₄	—	144/0.2	1.4810	1.3621	65
[(C₂H₅)₂GeO]ₙ	(C₄H₁₀GeO)ₙ n = 3 n = 4	18 29	— 128–29/3	— 1.4711	— 1.3582 (31°)	33, 138 2
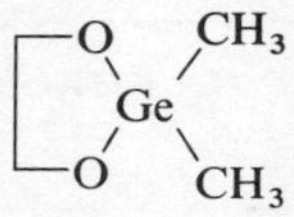	C₄H₁₀GeO₂	—	73–74/8	1.4785 (30°)	1.4527 (30°)	153, 165

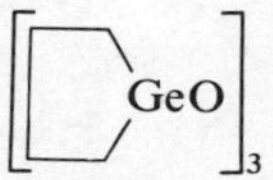

Table 3-11(a)—continued

COMPOUND	EMPIRICAL FORMULA	M.P. (°C)	B.P. (°C/mm)	n_D^{20}	d_4^{20}	REFERENCES
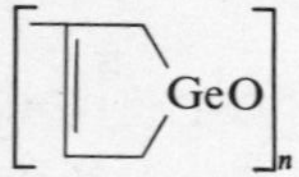 $\left[\text{GeO}\right]_n$	$(C_5H_8GeO)_n$	—	—	—	—	83
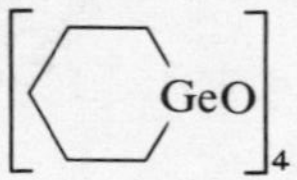 $\left[\text{GeO}\right]_4$	$(C_5H_{10}GeO)_4$	144	—	—	—	71
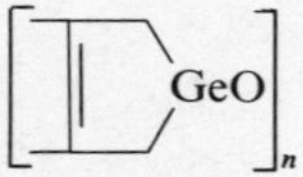 $\left[\text{GeO}\right]_n$	$(C_6H_{10}GeO)_n$	190–91	—	—	—	83
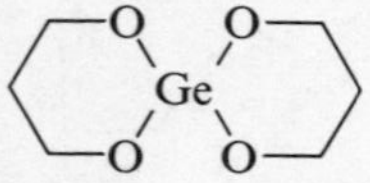	$C_6H_{12}GeO_4$	—	140/1	—	—	85
$[(n\text{-}C_3H_7)_2GeO]_3$ $[(i\text{-}C_3H_7)_2GeO]_3$	$(C_6H_{14}GeO)_3$ $(C_6H_{14}GeO)_3$	5, 8 44	320/760 321/760	1.4730 —	1.240 —	3 4
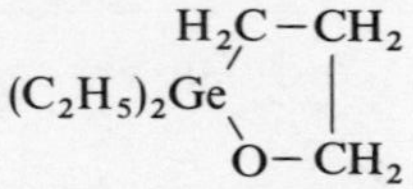 H_2C-CH_2 $(C_2H_5)_2Ge\ \ \ \ \ \ \ \begin{smallmatrix}\\O-CH_2\end{smallmatrix}$	$C_7H_{16}GeO$	—	72/12	1.4799	1.1849	125
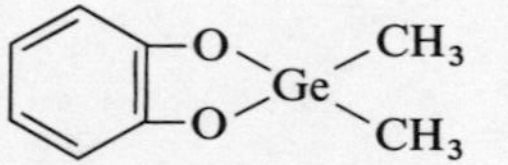	$C_8H_{10}GeO_2$	198	—	—	—	153

Structure	Formula	m.p.	b.p./press.	n	d	Ref.
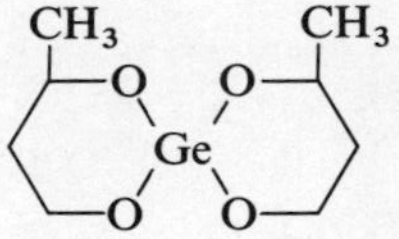 spiro dioxastannane with two CH₃ groups (Ge center)	$C_8H_{16}GeO_4$	—	93/0.4	—	—	85
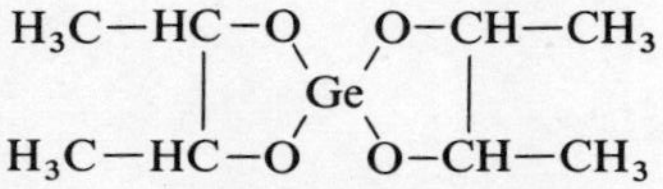 $H_3C-HC-O$ $O-CH-CH_3$ / Ge / $H_3C-HC-O$ $O-CH-CH_3$	$C_8H_{16}GeO_4$	—	sublimes at 130–40/1	—	—	85
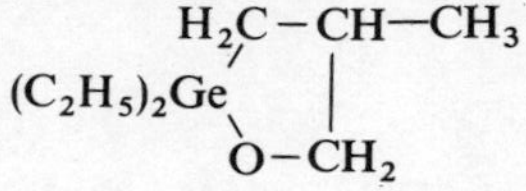 $(C_2H_5)_2Ge$ with $H_2C-CH-CH_3$, $O-CH_2$	$C_8H_{18}GeO$	—	75/15′	1.4698	1.1393	125
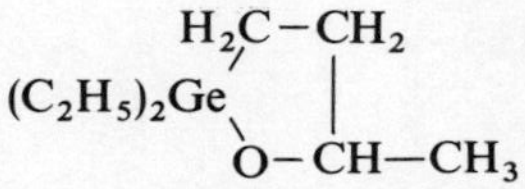 $(C_2H_5)_2Ge$ with H_2C-CH_2, $O-CH-CH_3$	$C_8H_{18}GeO$	—	80/15	1.4646	1.1310	125
$[(n-C_4H_9)_2GeO]_3$	$(C_8H_{18}GeO)_3$	–17	180–82/1	1.4712	1.161	7, 116
$[(t-C_4H_9)_2GeO]_3$	$(C_8H_{18}GeO)_3$	—	180–82/1	—	—	67
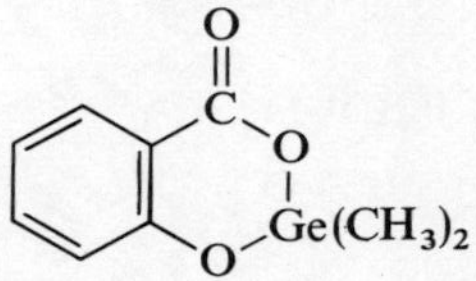 $O=C-O-Ge(CH_3)_2$ fused to benzene ring, O	$C_9H_{10}GeO_3$	156–62	160/3	—	—	153
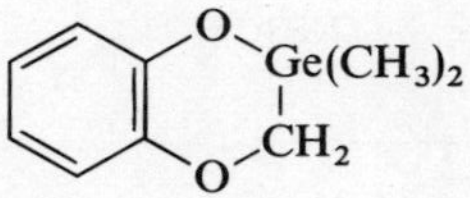 $O-Ge(CH_3)_2$ / CH_2 fused to benzene ring, O	$C_9H_{12}GeO_2$	95–96/1	—	—	—	152

Table 3-11(a)—continued

COMPOUND	EMPIRICAL FORMULA	M.P. (°C)	B.P. (°C/mm)	n_D^{20}	d_4^{20}	REFERENCES
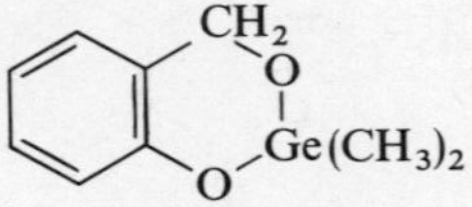 (structure)	$C_9H_{12}GeO_2$	66	118/10	—	—	154
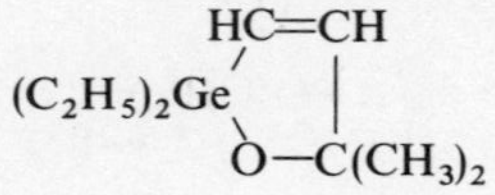 (structure)	$C_9H_{18}GeO$	—	78/17	1.4597	1.093	125
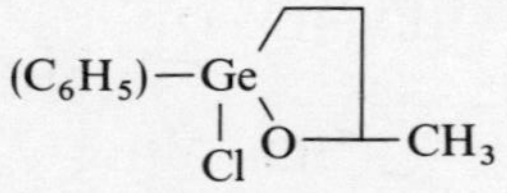 (structure)	$C_{10}H_{13}GeClO$	—	77/0.2	1.5400	1.2058	126
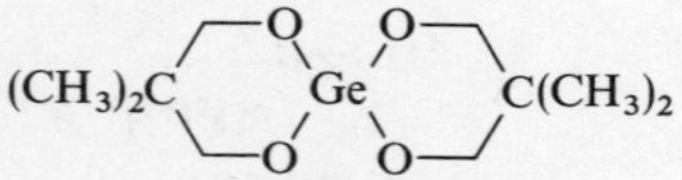 (structure)	$C_{10}H_{20}GeO_4$	—	90–95/0.8	—	—	85
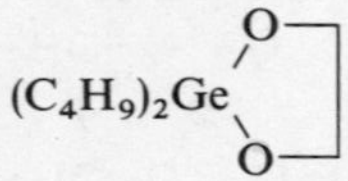 (structure)	$C_{10}H_{22}GeO_2$	—	98–103/2.5–3 103–6/3.5 108/4	1.4653 1.4712	—	68, 86
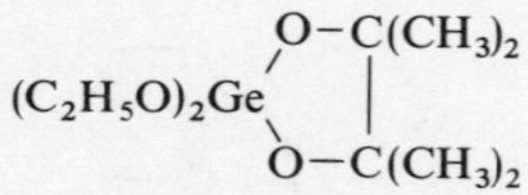 (structure)	$C_{10}H_{22}GeO_4$	—	100–10/5.5–6.5	—	—	85

Structure	Molecular formula	mp	bp	n	d	Ref.
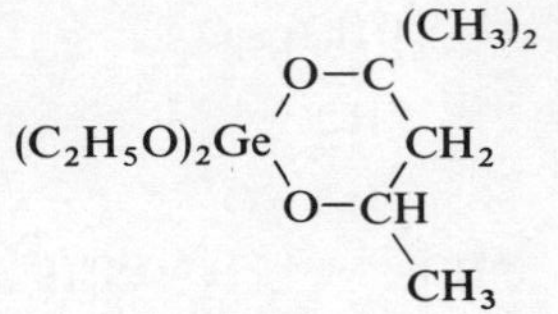 $(C_2H_5O)_2Ge$ ring with $(CH_3)_2$, CH_2, CH—CH_3	$C_{10}H_{22}GeO_4$	—	114–18/7	—	—	85
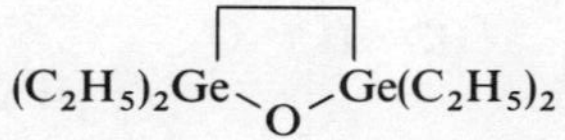 $(C_2H_5)_2Ge$—O—$Ge(C_2H_5)_2$	$C_{10}H_{24}Ge_2O$	—	86/1.5	1.4971	1.2759	70
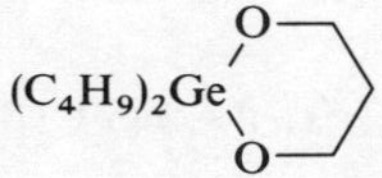 $(C_4H_9)_2Ge$ dioxane ring	$C_{11}H_{24}GeO_2$	—	104–8/4	1.4705	—	68
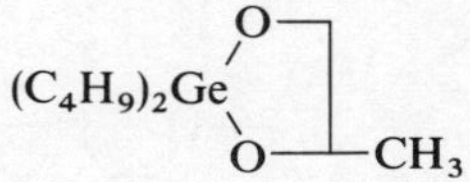 $(C_4H_9)_2Ge$ dioxolane ring with CH_3	$C_{11}H_{24}GeO_2$	—	106–9/3.5	1.4540	—	86
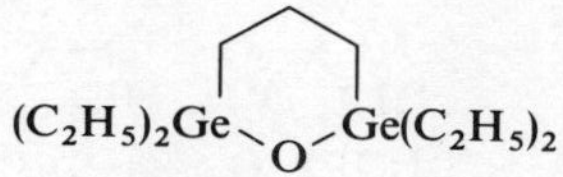 $(C_2H_5)_2Ge$—O—$Ge(C_2H_5)_2$	$C_{11}H_{26}Ge_2O$	—	77/0.6	1.4820	1.2373	81
$[(C_6F_5)_2GeO]_4$	$(C_{12}GeOF_{10})_4$	238–48	—	—	—	32
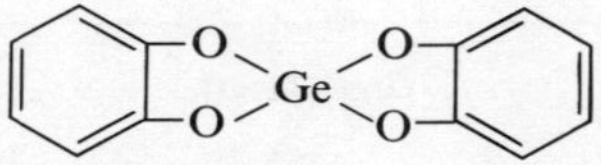 catechol germanium spiro structure	$C_{12}H_8GeO_4$	215–35 dec	—	—	—	163
$[(C_6H_5)_2GeO]_3$	$(C_{12}H_{10}GeO)_3$	149	300/2	—	—	89
$[(C_6H_5)_2GeO]_4$	$(C_{12}H_{10}GeO)_4$	218	—	—	—	89, 99
$[(C_6H_5)_2GeO]_n$	$(C_{12}H_{10}GeO)_n$	298	—	—	—	29, 88

Table 3-11(a)—continued

COMPOUND	EMPIRICAL FORMULA	M.P. (°C)	B.P. (°C/mm)	n_D^{20}	d_4^{20}	REFERENCES
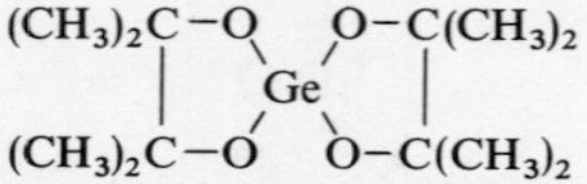 (CH₃)₂C−O, O−C(CH₃)₂ … Ge … (CH₃)₂C−O, O−C(CH₃)₂	$C_{12}H_{24}GeO_4$	—	sublimes at 90–100/0.3	—	—	85
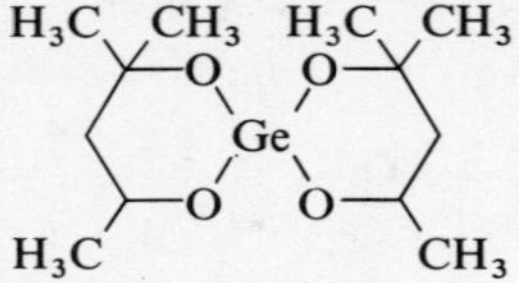 H₃C CH₃ / H₃C CH₃ ring structure with Ge and O	$C_{12}H_{24}GeO_4$	—	126/10 78/0.05	— —	— —	61, 85
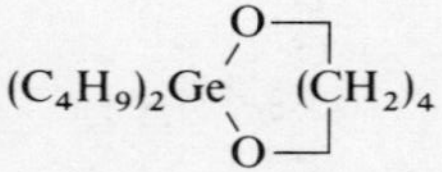 (C₄H₉)₂Ge with (CH₂)₄ (O−(CH₂)₄−O)	$C_{12}H_{26}GeO_2$	—	135/10 107–11/3.5	1.4648 1.4725	1.1025 —	63, 68
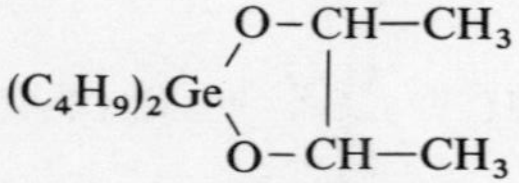 (C₄H₉)₂Ge O−CH−CH₃ / O−CH−CH₃	$C_{12}H_{26}GeO_2$	—	90–94/1.5	1.4555	—	68
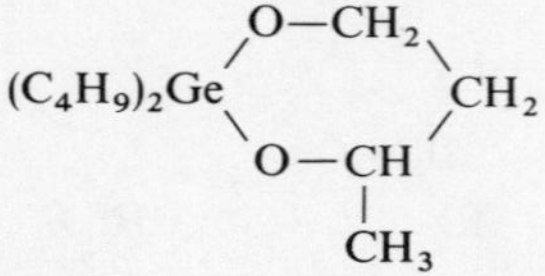 (C₄H₉)₂Ge O−CH₂ / O−CH−CH₃ , CH₂	$C_{12}H_{26}GeO_2$	—	101–5/3	1.4550	—	68
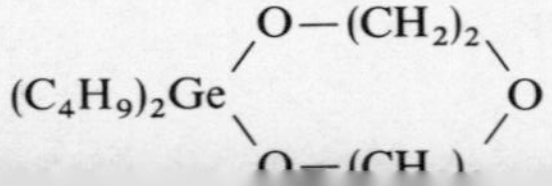 (C₄H₉)₂Ge O−(CH₂)₂−O / O−(CH₂)	$C_{12}H_{26}GeO_3$	—	120–22/2.7	1.4695	—	68

Structure	Molecular formula	m.p.	b.p./mm	n_D	d	Ref.
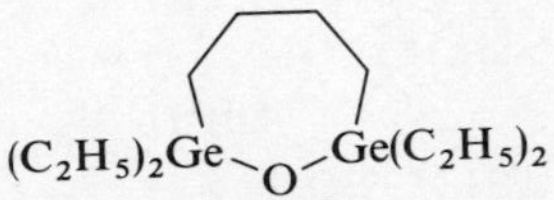 $(C_2H_5)_2Ge$–O–$Ge(C_2H_5)_2$ (cyclic)	$C_{12}H_{28}Ge_2O$	—	93/0.9	1.4862	1.2253	81
$[(m\text{-}CH_3C_6H_4)_2GeO]_3$	$(C_{14}H_{14}GeO)_3$	83–84	—	—	—	30
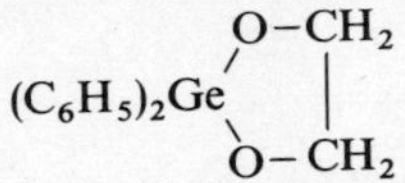 $(C_6H_5)_2Ge$(O–CH_2…CH_2–O)	$C_{14}H_{14}GeO_2$	—	140–42/0.1 140/0.6	—	—	86, 87
$[(p\text{-}CH_3OC_6H_4)_2GeO]_n$	$(C_{14}H_{14}GeO_3)_n$	281–301	—	—	—	30
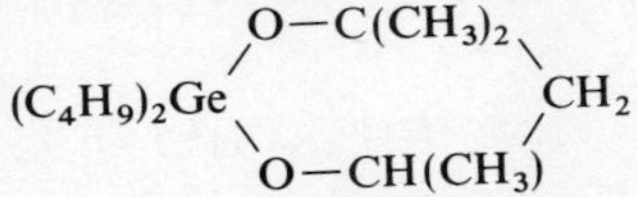 $(C_4H_9)_2Ge$ [O–$C(CH_3)_2$–CH_2 / O–$CH(CH_3)$]	$C_{14}H_{30}GeO_2$	—	116–18/4 105–8/2.5	1.4500	—	68
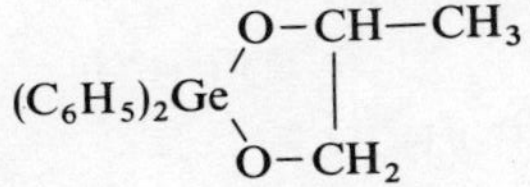 $(C_6H_5)_2Ge$ [O–CH–CH_3 / O–CH_2–CH_2]	$C_{15}H_{16}GeO_2$	—	127/0.6	—	—	87
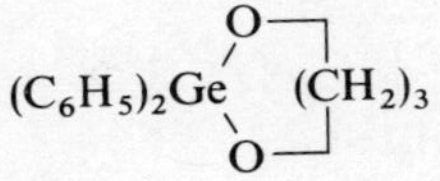 $(C_6H_5)_2Ge$ [$(CH_2)_3$]	$C_{15}H_{16}GeO_2$	—	150/0.6	—	—	87
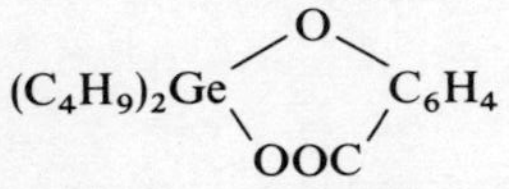 $(C_4H_9)_2Ge$ [O–C_6H_4–OOC]	$C_{15}H_{22}GeO_3$	—	158/0.5	1.5395	—	69
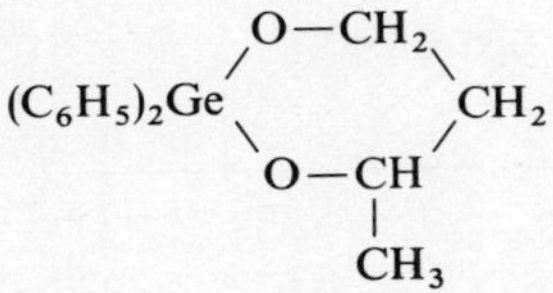 $(C_6H_5)_2Ge$ [O–CH_2–CH_2 / O–CH…CH_3]	$C_{16}H_{18}GeO_2$	—	135/0.8	1.5665	—	87

Table 3-11(a)—continued

COMPOUND	EMPIRICAL FORMULA	M.P. (°C)	B.P. (°C/mm)	n_D^{20}	d_4^{20}	REFERENCES
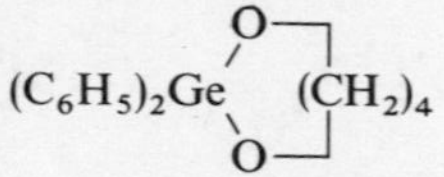 $(C_6H_5)_2Ge$, O—(CH$_2$)$_4$—O	$C_{16}H_{18}GeO_2$	—	142/0.2	—	—	87
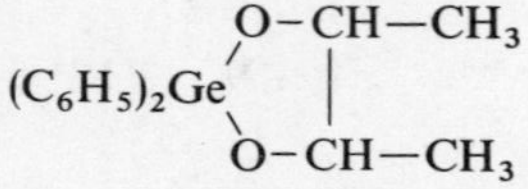 $(C_6H_5)_2Ge$, O—CH—CH$_3$ / O—CH—CH$_3$	$C_{16}H_{18}GeO_2$	—	127/0.6	1.5680	—	87
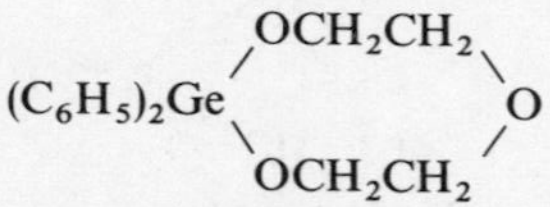 $(C_6H_5)_2Ge$, OCH$_2$CH$_2$—O—OCH$_2$CH$_2$	$C_{16}H_{18}GeO_3$	—	155/0.2	—	—	87
$[(p\text{-}C_2H_5OC_6H_4)_2GeO_3]_4$	$(C_{16}H_{18}GeO_3)_4$	177–79	—	—	—	30
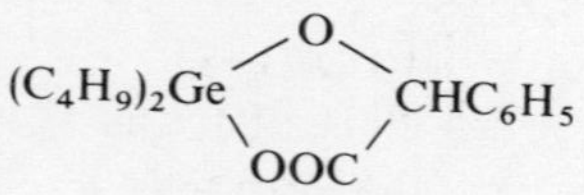 $(C_4H_9)_2Ge$, O—CHC$_6$H$_5$ / OOC	$C_{16}H_{24}GeO_3$	—	150–60/0.4 sublimes	—	—	69
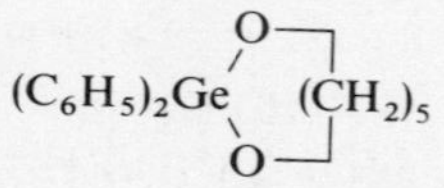 $(C_6H_5)_2Ge$, O—(CH$_2$)$_5$—O	$C_{17}H_{20}GeO_2$	—	138/0.2	—	—	87
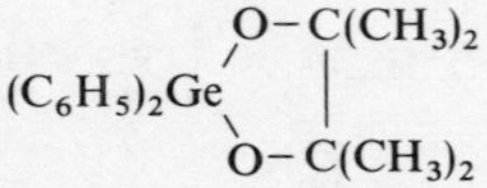 $(C_6H_5)_2Ge$, O—C(CH$_3$)$_2$ / O—C(CH$_3$)$_2$	$C_{18}H_{22}GeO_2$	—	133/0.3	1.5595	—	87

Structure	Formula	m.p.	b.p.	n_D		Ref.
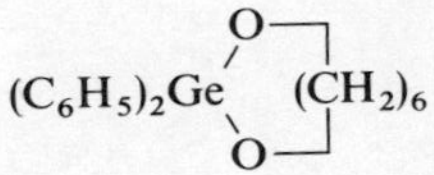 $(C_6H_5)_2Ge$ O—$(CH_2)_6$—O	$C_{18}H_{22}GeO_2$	—	129/0.2	1.5570	—	87
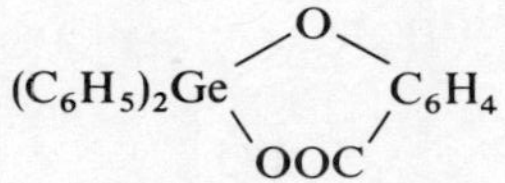 $(C_6H_5)_2Ge$ O—C_6H_4—OOC	$C_{19}H_{14}GeO_3$	—	—	—	—	69
$[(\beta\text{-}C_{10}H_7)_2GeO]_n$	$(C_{20}H_{14}GeO)_n$	211–12.5	—	—	—	29, 30
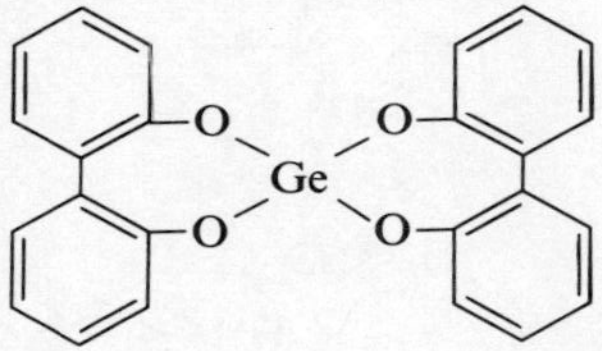	$C_{24}H_{16}GeO_4$	340	—	—	—	100, 164
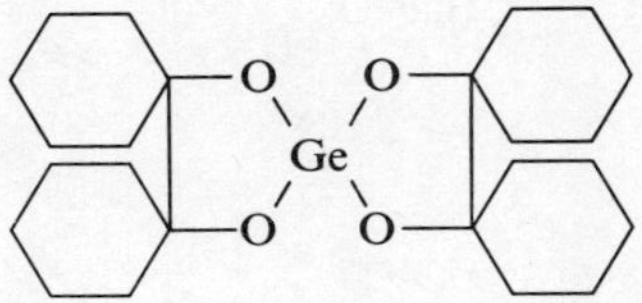	$C_{24}H_{40}GeO_4$	279	—	—	—	100
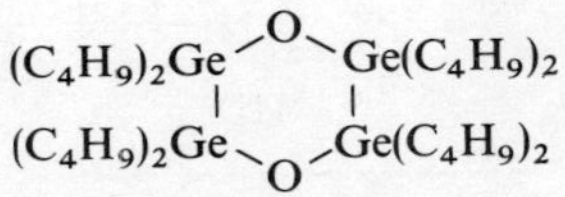 $(C_4H_9)_2Ge$—O—$Ge(C_4H_9)_2$ $(C_4H_9)_2Ge$—O—$Ge(C_4H_9)_2$	$C_{32}H_{72}Ge_4O_2$	—	130–32/5.10^{-5}	1.4993	—	16a
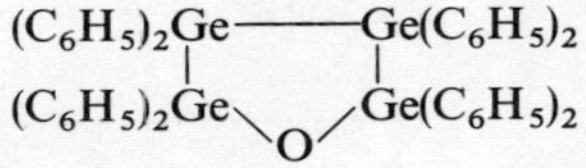 $(C_6H_5)_2Ge$—$Ge(C_6H_5)_2$ $(C_6H_5)_2Ge$—O—$Ge(C_6H_5)_2$	$C_{48}H_{40}Ge_4O$	206–8	—	—	—	109

Table 3-11

(b) Heterocyclic nitrogen compounds

COMPOUND	EMPIRICAL FORMULA	M.P. (°C)	B.P. (°C/mm)	n_D^{20}	d_4^{20}	REFERENCES
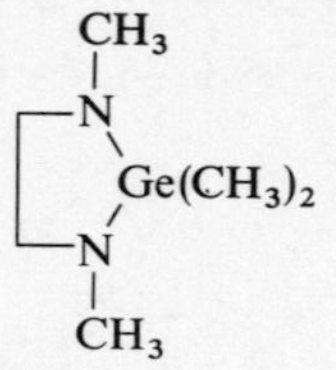	$C_6H_{16}GeN_2$	—	152–53/742	—	—	161
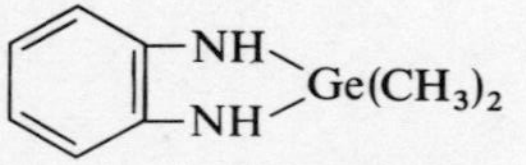	$C_8H_{12}GeN_2$	—	144/5	—	—	153
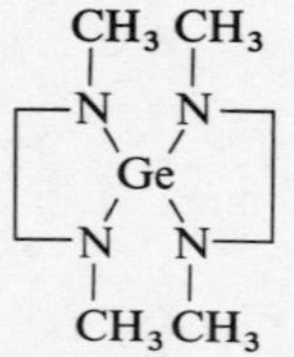	$C_8H_{20}GeN_4$	—	106/16	1.4859 (28°)	—	160
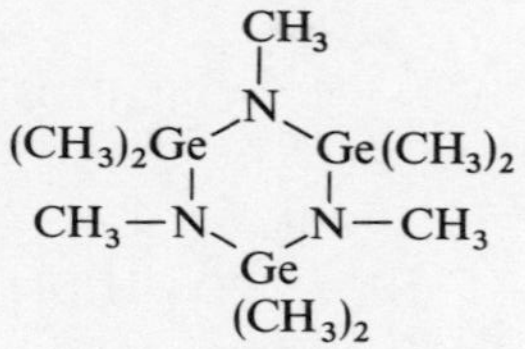	$C_9H_{27}Ge_3N_3$	—	80/2	—	—	120a, 131

Structure	Empirical formula	m.p.	b.p.	n_D	d	Ref.
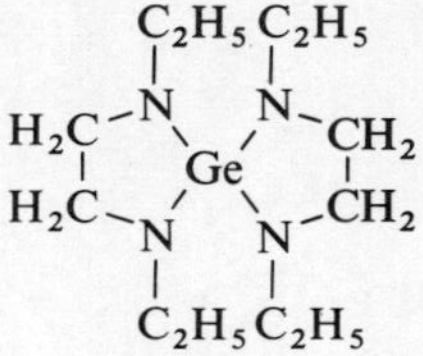	$C_{12}H_{28}GeN_4$	—	140–42/9	1.4904 (28°)	—	160
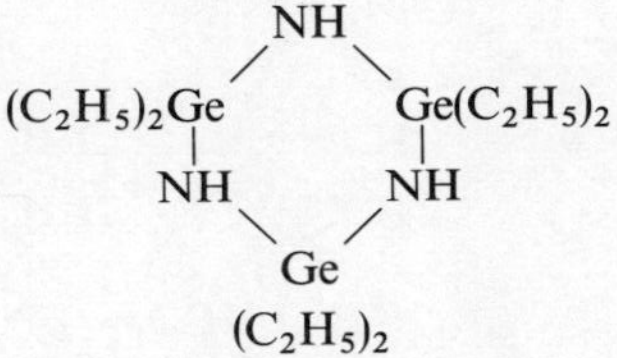	$C_{12}H_{33}Ge_3N_3$	—	137/3	1.4891	1.3235	95a
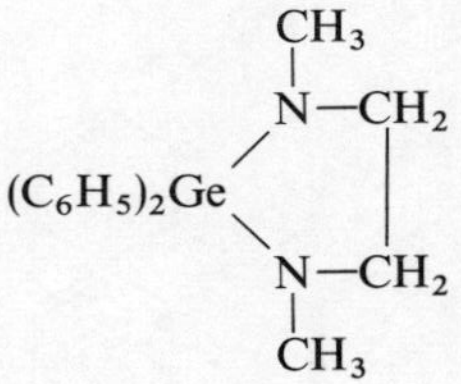	$C_{16}H_{20}GeN_2$	—	172/13	—	—	162
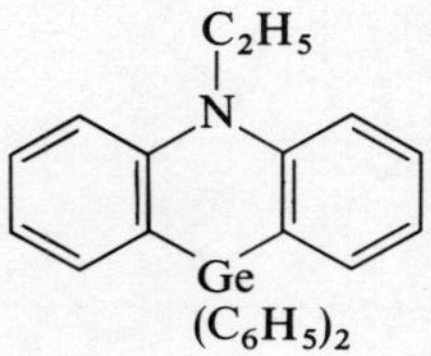	$C_{26}H_{23}GeN$	120–21	—	—	—	41

Table 3-11(b)—continued

COMPOUND	EMPIRICAL FORMULA	M.P. (°C)	B.P. (°C/mm)	n_D^{20}	d_4^{20}	REFERENCES
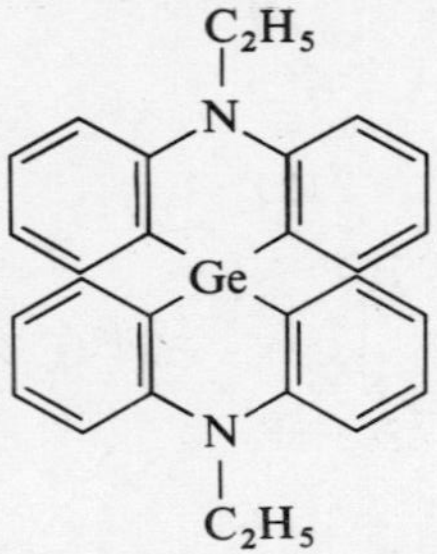	$C_{28}H_{26}GeN_2$	212–14	—	—	—	41
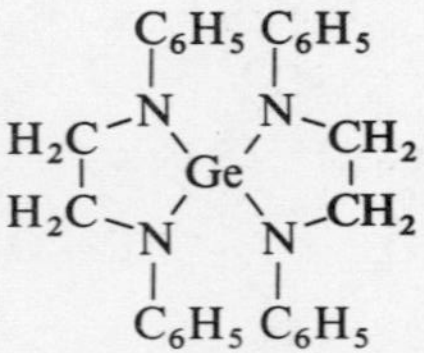	$C_{28}H_{28}GeN_4$	305–10	—	—	—	160
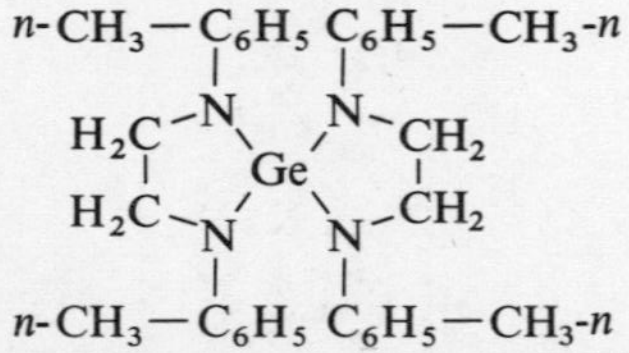	$C_{32}H_{40}GeN_4$	317–20	—	—	—	160

Table 3-11(b)—continued

COMPOUND	EMPIRICAL FORMULA	M.P. (°C)	B.P. (°C/mm)	n_D^{20}	d_4^{20}	REFERENCES
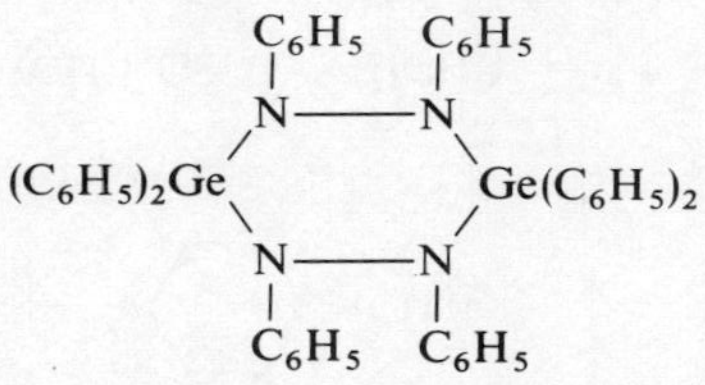 (structure)	$C_{48}H_{40}Ge_2N_4$	306–7	—	—	—	36

(structure)

C_6H_5 C_6H_5
$-N$ $Ge(C_6H_5)_2$ $N-$ C_6H_5

$-N$ $N-$ C_6H_5
$(C_6H_5)_2Ge$ C_6H_5

Table 3-11

(c) *Heterocyclic sulphur compounds*

COMPOUND	EMPIRICAL FORMULA	M.P. (°C)	B.P. (°C/mm)	n_D^{20}	d_4^{20}	REFERENCES
[(CH₃)₂GeS]₃	(C₂H₆GeS)₃	55.5	302 / 110/1	— / —	— / —	117, 120, 166
(structure)	C₄H₈GeS₄	165	—	—	—	22
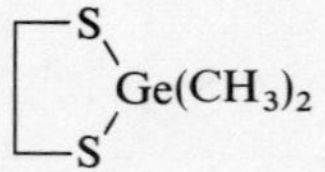 (structure)	C₄H₁₀GeS₂	13	87–93/12	—	—	153
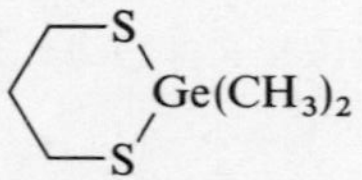 (structure)	C₅H₁₂GeS₂	24–27	103–5/10	—	—	155
[(i-C₃H₇)₂GeS]₂	(C₆H₁₄GeS)₂	—	117–21/1 / 312	1.5510 / —	1.3270 / —	5
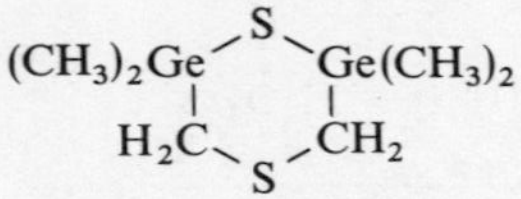 (structure)	C₆H₁₆Ge₂S₂	76	72/0.1	—	—	159
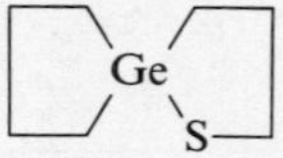 (structure)	C₇H₁₄GeS	—	120/18	1.5669	1.3198	77

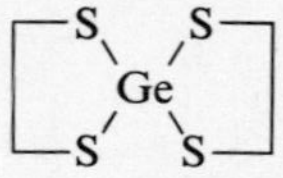

Structure	Formula	M.p.	B.p./mm	n_D	d	Ref.
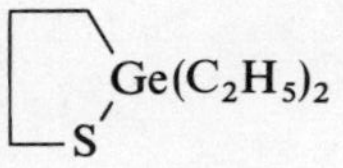 cyclo-$\mathrm{Ge(C_2H_5)_2}$, S	$\mathrm{C_7H_{16}GeS}$	—	107/23	1.5241	1.2102	77
$[(\mathrm{C_4H_9})_2\mathrm{GeS}]_2$ $[(\mathrm{C_4H_9})_2\mathrm{GeS}]_3$	$(\mathrm{C_8H_{18}GeS})_2$ $(\mathrm{C_8H_{18}GeS})_3$	— —	176–80/0.8 222–25/1	— —	— —	67 130
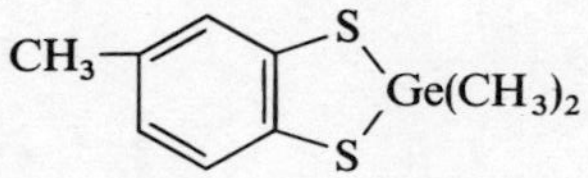 benzene-$\mathrm{S-Ge(CH_3)_2-S}$, $\mathrm{CH_3}$	$\mathrm{C_9H_{12}GeS_2}$	—	120/2	1.6381 (25°)	—	156
benzene-$\mathrm{S-CH_2-Ge(CH_3)_2-S}$, $\mathrm{CH_3}$	$\mathrm{C_{10}H_{14}GeS_2}$	—	132/2	—	—	152
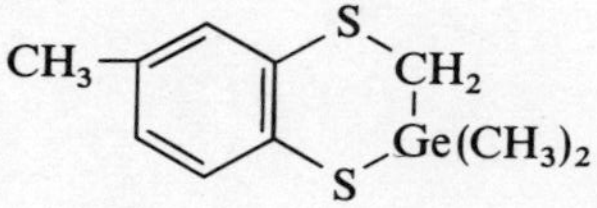 benzene-$\mathrm{S-Ge(CH_3)_2-CH_2-S}$, $\mathrm{CH_3}$	$\mathrm{C_{10}H_{14}GeS_2}$	—	132/2	—	—	152
$(\mathrm{C_2H_5})_2\mathrm{Ge-S-Ge(C_2H_5)_2}$	$\mathrm{C_{10}H_{24}Ge_2S}$	—	95/0.1	1.5276	1.2805	70
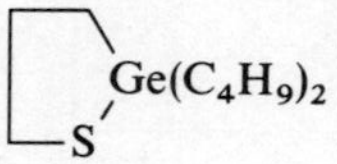 cyclo-$\mathrm{Ge(C_4H_9)_2}$, S	$\mathrm{C_{11}H_{24}GeS}$	—	130/4 94/0.4	1.5089	1.1043	75, 77
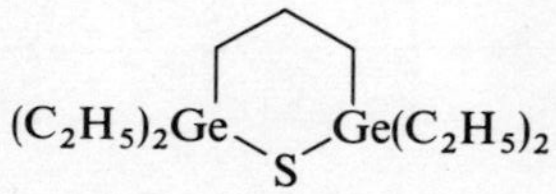 $(\mathrm{C_2H_5})_2\mathrm{Ge-S-Ge(C_2H_5)_2}$	$\mathrm{C_{11}H_{26}Ge_2S}$	—	98/0.5	1.5316	1.2679	81
$[(\mathrm{C_6H_5})_2\mathrm{GeS}]_2$ $[(\mathrm{C_6H_5})_2\mathrm{GeS}]_3$	$(\mathrm{C_{12}H_{10}GeS})_2$ $(\mathrm{C_{12}H_{10}GeS})_3$	203 170	— —	— —	— —	45 45

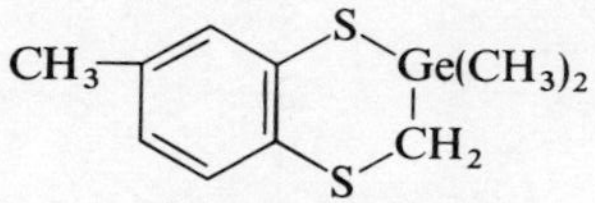

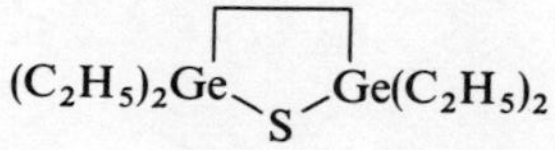

Table 3-11(c)—continued

COMPOUND	EMPIRICAL FORMULA	M.P. (°C)	B.P. (°C/mm)	n_D^{20}	d_4^{20}	REFERENCES
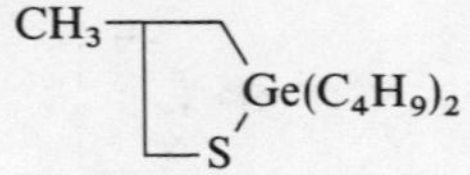 $CH_3\text{—...—Ge}(C_4H_9)_2\text{—S}$	$C_{12}H_{26}GeS$	—	103/0.7 90/0.2	1.5048	1.0836	77
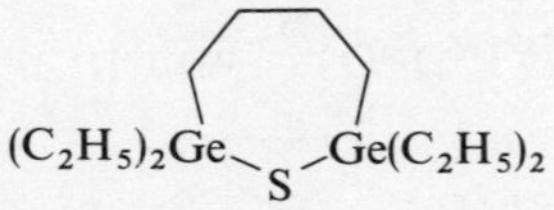 $(C_2H_5)_2Ge\text{—S—}...\text{—S—Ge}(C_2H_5)_2$	$C_{12}H_{28}Ge_2S$	—	125/1	1.5350	1.2602	81
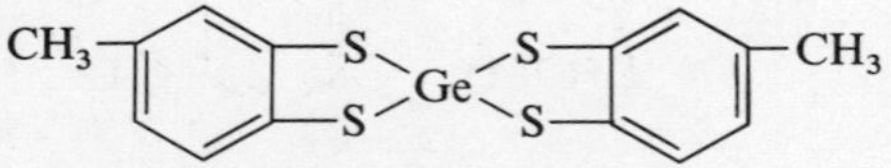 (benzothiophene-type Ge structure with CH_3 groups)	$C_{14}H_{12}GeS_4$	—	—	—	—	32a
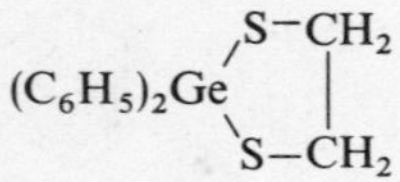 $(C_6H_5)_2Ge(S\text{—}CH_2\text{—}CH_2\text{—}S)$	$C_{14}H_{14}GeS_2$	87–88	—	—	—	22
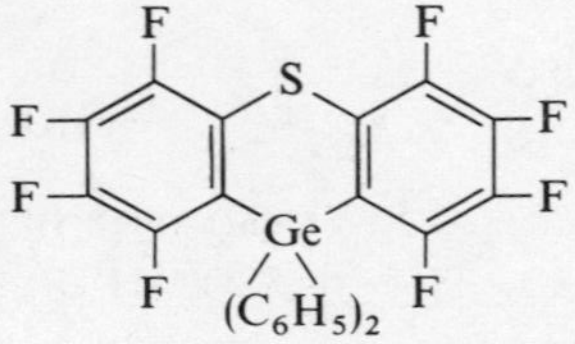 (perfluorinated Ge–S structure with $(C_6H_5)_2$)	$C_{24}H_{10}GeSF_8$	140–43	—	—	—	19

Table 3-11(c)—continued

COMPOUND	EMPIRICAL FORMULA	M.P. (°C)	B.P. (°C/mm)	n_D^{20}	d_4^{20}	REFERENCES
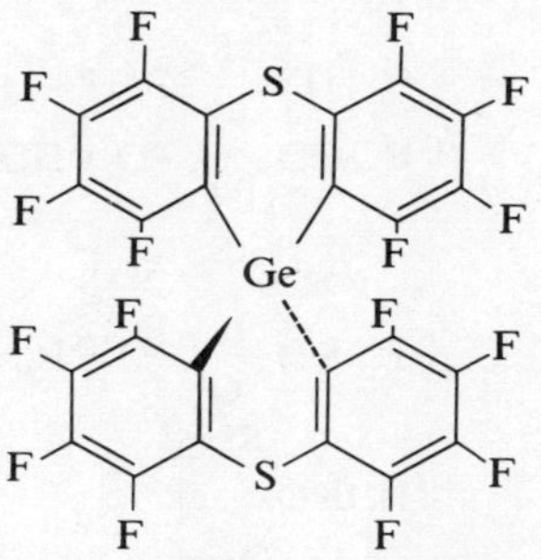	$C_{24}F_{16}GeS_2$	229.5-31.5	—	—	—	19

Table 3-11

(d) Heterocyclic selenium compounds

COMPOUND	EMPIRICAL FORMULA	M.P. (°C)	B.P. (°C/mm)	n_D^{20}	d_4^{20}	REFERENCES
$[(CH_3)_2GeSe]_3$	$(C_2H_6GeSe)_3$	53	—	—	—	128, 129
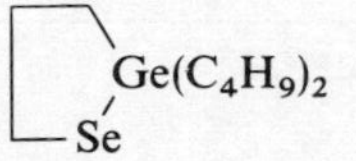	$C_{11}H_{24}GeSe$	—	98/0.3	1.5252	1.2754	77
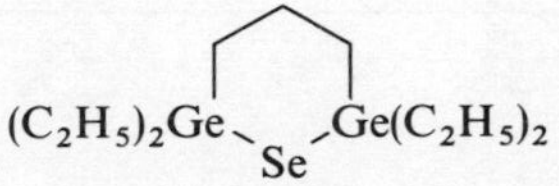	$C_{11}H_{26}Ge_2Se$	—	112/0.5	1.5545	1.4440	81

Table 3-11

(e) Mixed heterocycles

COMPOUND	EMPIRICAL FORMULA	M.P. (°C)	B.P. (°C/mm)	n_D^{20}	d_4^{20}	REFERENCES
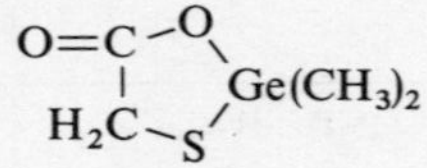	$C_4H_8GeO_2S$	129–32	155/5 sublimes	—	—	153
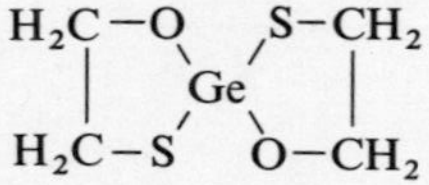	$C_4H_8GeO_2S_2$	181	—	—	—	22
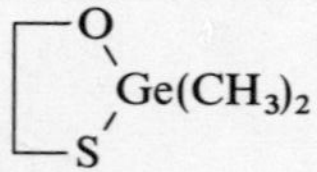	$C_4H_{10}GeOS$	−2	64/12	—	—	153
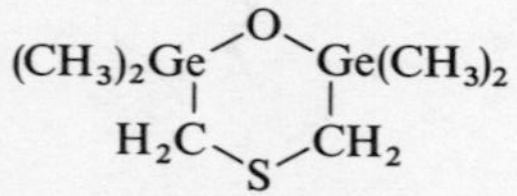	$C_6H_{16}Ge_2OS$	—	79–81/0.1	—	—	159
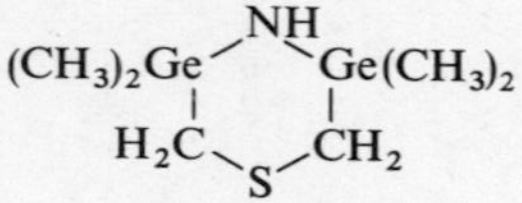	$C_6H_{17}Ge_2NS$	170–72	—	—	—	159
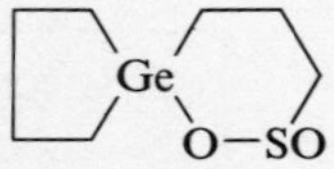	$C_7H_{14}GeO_2S$	71	—	—	—	70
$CH_3Ge(OCH_2CH_2)_3N$	$C_7H_{15}GeO_3N$	158–59	—	—	—	150

Structure	Formula	M.p.	B.p.	n_D	d	Ref.
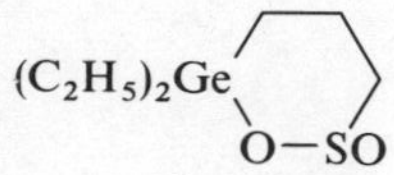 $(C_2H_5)_2Ge$ ring with SO	$C_7H_{16}GeO_2S$	—	103/0.5	1.5111	1.3505	26
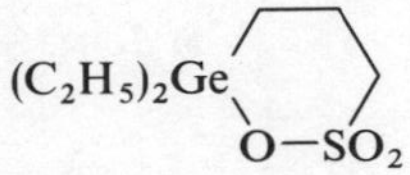 $(C_2H_5)_2Ge$ ring with SO_2	$C_7H_{16}GeO_3S$	51–53	170/0.2	—	—	26a
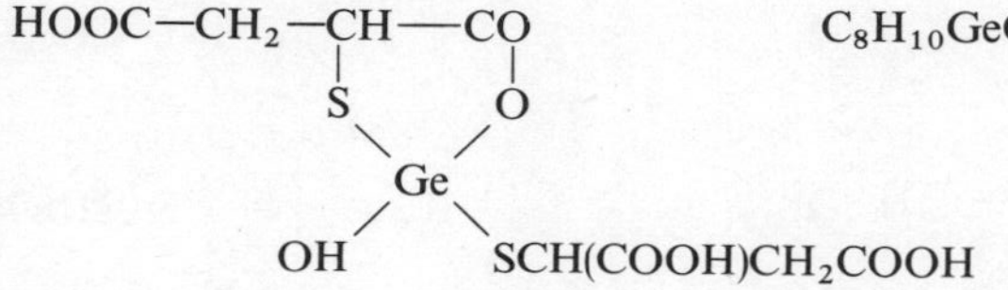 $HOOC-CH_2-CH-CO$ ring (S, Ge, O), Ge–SCH(COOH)CH$_2$COOH, OH	$C_8H_{10}GeO_9S_2$	—	—	—	—	18
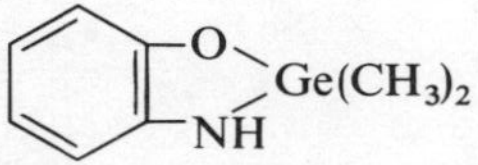 $O-Ge(CH_3)_2$ / NH (benzene ring)	$C_8H_{11}GeON$	—	142–46/3	—	—	153
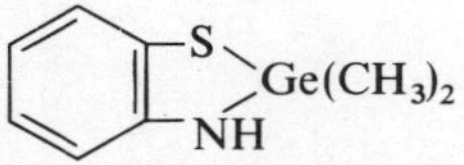 $S-Ge(CH_3)_2$ / NH (benzene ring)	$C_8H_{11}GeSN$	121–23	—	—	—	153
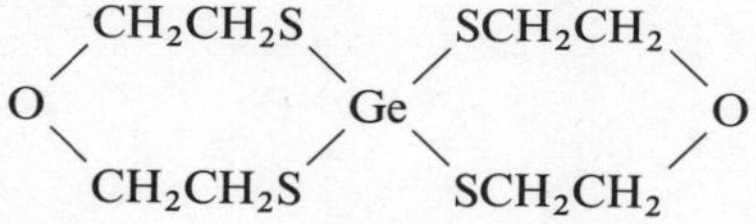 Ge with SCH_2CH_2, SCH_2CH_2, CH_2CH_2S, CH_2CH_2S ring	$C_8H_{16}GeO_2S_4$	159	—	—	—	9, 10
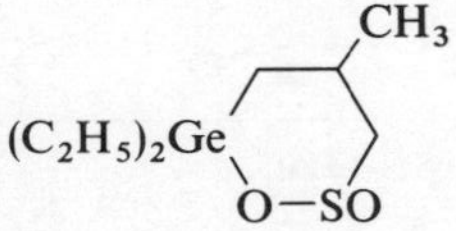 $(C_2H_5)_2Ge$ CH_3 ring with SO	$C_8H_{18}GeO_2S$	—	104/0.2	1.5026	1.2799	26

Table 3-11(e)—continued

COMPOUND	EMPIRICAL FORMULA	M.P. (°C)	B.P. (°C/mm)	n_D^{20}	d_4^{20}	REFERENCES
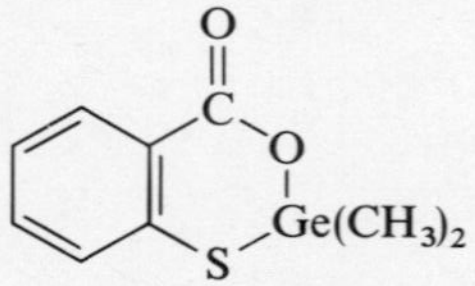	$C_9H_{10}GeO_2S$	166–69	160/3 sublimes	—	—	153
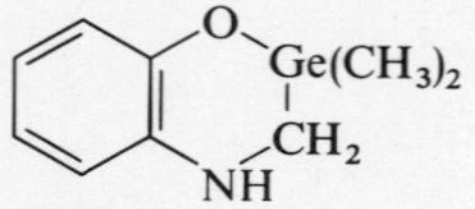	$C_9H_{13}GeON$	—	122/1	—	—	152
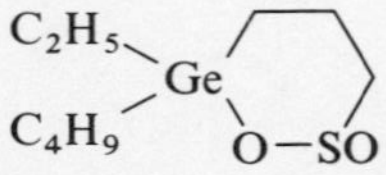	$C_9H_{20}GeO_2S$	—	115/0.1	1.5050	1.2664	26
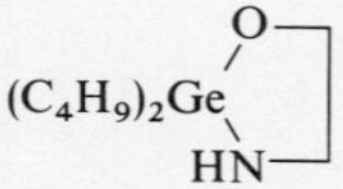	$C_{10}H_{23}GeON$	—	121–22/2	1.4650	—	86
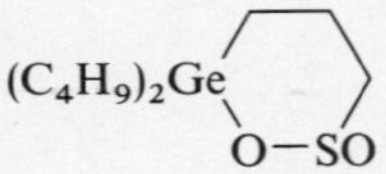	$C_{11}H_{24}GeO_2S$	—	111/0.03	1.4980	1.1990	26
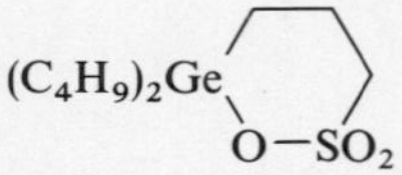	$C_{11}H_{24}GeO_3S$	—	160/0.02	1.4908	1.2534	26a
$C_6H_5Ge(OCH_2CH_2)_3N$	$C_{12}H_{17}GeO_3N$	232–32.5	—	—	—	150

Structure	Formula	m.p. (°C)	b.p. (°C/mm)	n_D	d	Ref.
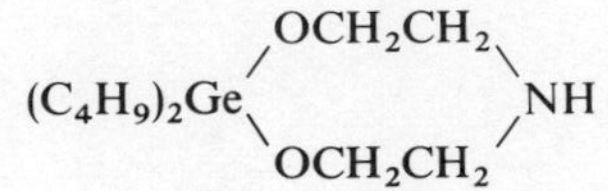 $(C_4H_9)_2Ge$ with OCH_2CH_2—NH—OCH_2CH_2	$C_{12}H_{27}GeO_2N$	—	127–30/0.5–0.7	1.4750	—	84
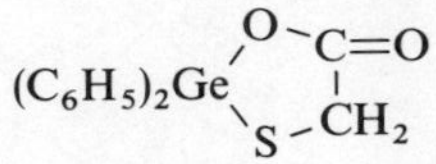 $(C_6H_5)_2Ge$ with O—$C{=}O$ / S—CH_2	$C_{14}H_{12}GeO_2S$	143–45	—	—	—	22
$(C_6H_5)_2Ge$ with O—CH_2 / S—CH_2	$C_{14}H_{14}GeOS$	80–81	—	—	—	22
$(C_6H_5)_2Ge$ with O—CH_2 / NH—CH_2	$C_{14}H_{15}GeON$	—	—	—	—	86
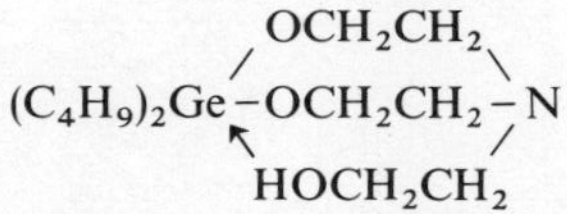 $(C_4H_9)_2Ge$ with OCH_2CH_2—N—OCH_2CH_2—$HOCH_2CH_2$	$C_{14}H_{31}GeO_3N$	—	136/0.1	1.4810	—	86
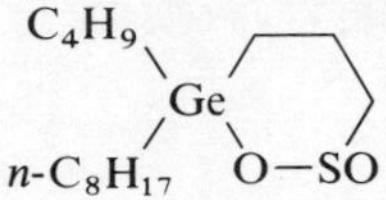 C_4H_9, $n\text{-}C_8H_{17}$ Ge–O–SO	$C_{15}H_{32}GeO_2S$	—	165/0.1	1.4902	1.1184	26
$(C_2H_5)_3SiN$ $(CH_3)_2$ Ge, $NSi(C_2H_5)_3$ Sn $(CH_3)_2$	$C_{16}H_{42}GeSi_2SnN_2$	—	124–28/0.2	—	—	127
$(C_6H_5)_2Sn$ $Ge(C_6H_5)_2$	$C_{28}H_{28}GeSn$	124–25	—	—	—	47, 111

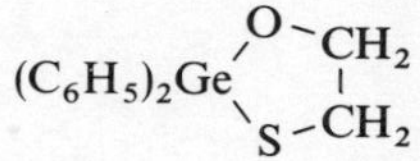

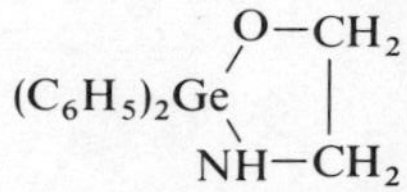

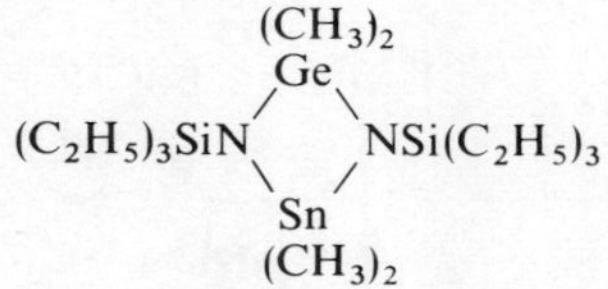

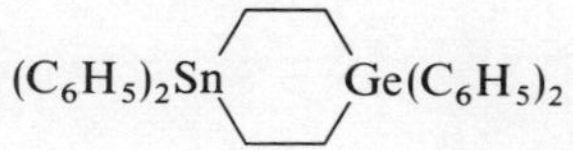

Table 3-11(e)—continued

COMPOUND	EMPIRICAL FORMULA	M.P. (°C)	B.P. (°C/mm)	n_D^{20}	d_4^{20}	REFERENCES
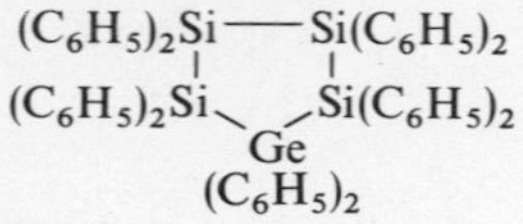 $(C_6H_5)_2Si$—$Si(C_6H_5)_2$ $(C_6H_5)_2Si$ $Si(C_6H_5)_2$ Ge $(C_6H_5)_2$	$C_{60}H_{50}GeSi_4$	316	—	—	—	44

3-6 TABLE: OTHER CYCLIC DERIVATIVES

Table 3-12
Other Cyclic Derivatives

COMPOUND	EMPIRICAL FORMULA	M.P. (°C)	B.P. (°C/mm)	n_D^{20}	d_4^{20}	REFERENCES
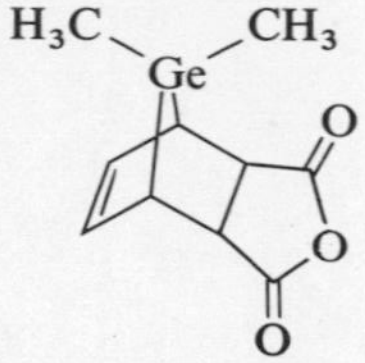	$C_{34}H_{28}O_3Ge$	—	—	—	—	48a
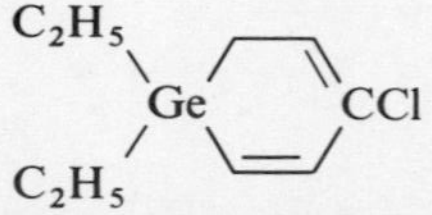	$C_9H_{15}GeCl$	—	76/0.5	1.5167	1.2359	70

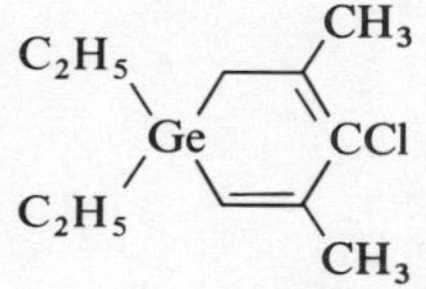 $C_{11}H_{19}GeCl$	—	100/0.7	1.5211	1.1878	70
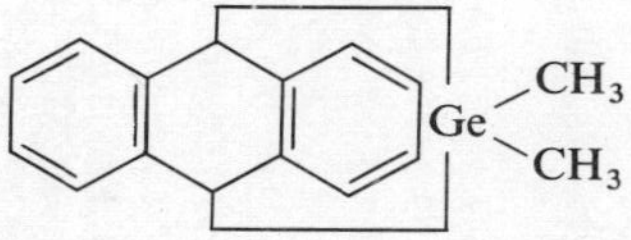 $C_{16}H_{16}Ge$	—	—	—	—	167
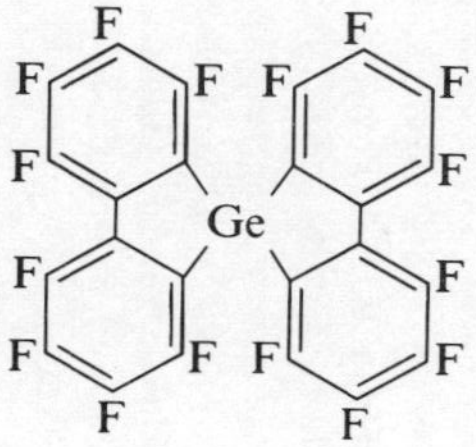 $C_{24}F_{16}Ge$	230.5–32.0	—	—	—	19a

REFERENCES

1. Abel, E. W., and D. A. Armitage, *Advan. Organometal. Chem.*, **5**, 1–92 (1967).
2. Anderson, H. H., *J. Am. Chem. Soc.*, **72**, 2089 (1950).
3. Anderson, H. H., *J. Am. Chem. Soc.*, **74**, 2370 (1952).
4. Anderson, H. H., *J. Am. Chem. Soc.*, **75**, 814 (1953).
5. Anderson, H. H., *J. Am. Chem. Soc.*, **78**, 1692 (1956).
6. Anderson, H. H., *J. Am. Chem. Soc.*, **80**, 5083 (1958).
7. Anderson, H. H., *J. Am. Chem. Soc.*, **83**, 547 (1961).
8. Backer, H. I., and W. Drenth, *Rec. Trav. Chim.*, **70**, 559 (1951).
9. Backer, H. I., and F. Stienstra, *Rec. Trav. Chim.*, **52**, 1033 (1933).
10. Backer, H. I., and F. Stienstra, *Rec. Trav. Chim.*, **54**, 607 (1935).
11. Bailey, D. J., and A. N. Pines, *Ind. Eng. Chem.*, **46**, 2363 (1954).
12. Bajer, F. J., and H. W. Post, *J. Org. Chem.*, **27** (4), 1422 (1962).
13. Bajer, F., and H. W. Post, *J. Organometal. Chem.*, **11**, 187 (1968).
14. Benkeser, R. A., and R. F. Cunico, *J. Org. Chem.*, **32**(2), 395 (1967).
14a. Bokii, N. C., G. N. Zakharova, and Yu. T. Struchkov, *Zh. Strukt. Chem.*, **6**(3), 476 (1965).
15. Bouveault, L., and R. Locquin, *Bull. Soc. Chim. France*, **35**(3), 629 (1906).
16. Brown, M. P., and E. G. Rochow, *J. Am. Chem. Soc.*, **82**, 4166 (1960).
16a. Bulten, E. J., and J. G. Noltes, *Tetrahedron Letters*, **29**, 3471 (1966).
17. Carrick, A., and F. Glockling, *J. Chem. Soc.*, A (6), 623 (1966).
17a. Cason, L. F., and H. G. Brooks, *J. Am. Chem. Soc.*, **74**, 4582 (1952).
18. Clark, E. R., *J. Inorg. Nucl. Chem.*, **25**, 353 (1963).
19. Cohen, S. C., and A. G. Massey, *J. Organometal. Chem.*, **12**, 341 (1968).
19a. Cohen, S. C., M. L. N. Reddy, and A. G. Massey, *Chem. Commun.*, (9), 451, 1967.
20. Coulson, C. A., and W. E. Moffitt, *J. Chem. Phys.*, **15**, 151 (1947).
21. Curtis, M. D., *J. Am. Chem. Soc.*, **89** (16), 4241 (1967).
22. Davidson, W. E., K. Hills, and M. C. Henry, *J. Organometal. Chem.*, **3**, 285 (1965).
23. Davies, A. G., and C. D. Hall, *Chem. Ind. (London)*, 1695 (1958).
24. Davies, A. G., and C. D. Hall, *J. Chem. Soc.*, 3835 (1959).
25. Dubac, J., and P. Mazerolles, *Bull. Soc. Chim. France*, 4027 (1967).
26. Dubac, J., and P. Mazerolles, *Compt. Rend.*, **267**, 411 (1968).
26a. Dubac, J., and P. Mazerolles, *Bull. Soc. Chim. France*, 3608 (1969).
27. Duffield, A. M., H. Budzikiewicz, and C. Djerassi, *J. Am. Chem. Soc.*, **87**, 2920 (1965).
28. Duffield, A. M., C. Djerassi, P. Mazerolles, J. Dubac, and G. Manuel, *J. Organometal. Chem.*, **12**, 123 (1968).
29. Emelyanova, L. I., and L. G. Makarova, *Izv. Akad. Nauk SSSR Otd. Khim. Nauk*, 2067 (1961).
30. Emelyanova, L. I., V. N. Vinogradova, L. G. Makarova, and A. N. Nesmeyanov, *Izv. Akad. Nauk SSSR Otd. Khim. Nauk* 53 (1962).
31. Faucher, A., and P. Mazerolles, *2° Symposium Intern. du silicium*, Bordeaux, 1968.
32. Fenton, D. E., A. G. Massey, and D. S. Urch, *J. Organometal. Chem.*, **6**, 352 (1966).
32a. Fink, F. H., J. A. Turner, and D. A. Payne, *J. Am. Chem. Soc.*, **88**, 1571 (1966).
33. Flood, E. A., *J. Am. Chem. Soc.*, **54**, 1663 (1932).
34. Fritz, G., and J. Grobe, *Z. Anorg. Allgem. Chem.*, **315**, 167 (1962).
35. Gar, T. K., and V. F. Mironov, *Izv. Akad. Nauk SSSR Otd. Khim. Nauk*, (5), 855 (1965).
36. George, M. V., P. B. Talukdar, and H. Gilman, *J. Organometal. Chem.*, **5**, 397 (1966).
37. Gilman, H., and W. H. Atwell, *J. Org. Chem.*, **28**, 2905 (1963).
38. Gilman, H., and W. H. Atwell, *J. Org. Chem.*, **2**, 277 (1964).

39. Gilman, H., and W. H. Atwell, *J. Am. Chem. Soc.*, **86**, 2687 (1964).
39a. Gilman, H., S. G. Cottis, and W. H. Atwell, *J. Am. Chem. Soc.*, **86**, 1596 (1964).
40. Gilman, H., and R. D. Gorsich, *J. Am. Chem. Soc.*, **80**, 1883 (1958).
41. Gilman, H., and E. Zuech, *J. Am. Chem. Soc.*, **82**, 2522 (1960).
42. Gverstsiteli, I. M., T. P. Doksopulo, M. M. Menteshashvili, and I. I. Agkhazava, *Soobshch. Akad. Nauk Gruz.*, **40**(2), 333 (1965).
43. Hansley, W. L., *J. Am. Chem. Soc.*, **57**, 2303 (1935).
44. Hengge, E., and U. Brychcy, *Monatsh. Chem.*, **97**(5), 1309 (1966).
45. Henry, M. C., and W. E. Davidson, *Can. J. Chem.*, **41**, 1276 (1963).
46. Henry, M. C., and J. G. Noltes, *J. Am. Chem. Soc.*, **82**, 558 (1960).
47. Henry, M. C., and J. G. Noltes, *J. Am. Chem. Soc.*, **82**, 561 (1960).
48. Horner, L., and H. Güsten, *Ann. Chem.*, **652**, 99 (1962).
48a. Hota, N. K., and C. J. Willis, *J. Organometal. Chem.*, **15**, 89 (1968).
49. Johnson, F., and R. S. Gohlke, *Tetrahedron*, (26), 1291 (1962).
50. Johnson, F., R. S. Gohlke, and W. A. Nasutavicus, *J. Organometal. Chem.*, **3**, 233 (1965).
51. Knoth, W. H., and R. V. Lindsey, *J. Org. Chem.*, **23**, 1392 (1958).
52. Kolesnikov, S. P., and V. I. Shiryaev, and O. M. Nefedov, *Izv. Akad. Nauk SSSR*, (3), 584 (1966).
52a. Kraus, C. A., and C. L. Brown, *J. Am. Chem. Soc.*, **52**, 4031 (1930).
53. Kursanov, D. N., and M. E. Vol'pin, *Zh. Vses. Khim. Obshch. im. D.I. Mendeleeva*, **7**, 282 (1962).
54. Leavitt, F. C., T. A. Manuel, and F. Johnson, *J. Am. Chem. Soc.*, **81**, 3163 (1959).
55. Leavitt, F. C., T. A. Manuel, F. Johnson, L. U. Matternas, and D. S. Lehman, *J. Am. Chem. Soc.*, **82** (19), 5099 (1960).
56. Ledwith, A., and R. M. Bell, *Chem. Ind. (London)*, 459 (1959).
57. Leites, L. A., *Izv. Akad. Nauk SSSR Otd. Khim. Nauk*, (8), 1525 (1963).
58. Leites, L. A., V. G. Dulova, and M. E. Vol'pin, *Izv. Akad. Nauk SSSR Otd. Khim. Nauk*, (4), 731 (1963).
59. Leites, L. A., T. K. Gar, and V. F. Mironov, *Dokl. Akad. Nauk SSSR*, **158**(2), 400 (1964).
60. Lesbre, M., and P. Mazerolles, *Compt. Rend.*, **246**, 1708 (1958).
61. Lesbre, M., and R. W. Russell, *Bull. Soc. Chim.*, *France*, **4**, 566 (1959).
62. Lesbre, M., and J. Satgé, *Compt. Rend.*, **252**, 1976 (1961).
63. Lesbre, M., and J. Satgé, *Compt. Rend.*, **254**, 4051 (1962).
64. Manuel, G., and P. Mazerolles, *Bull. Soc. Chim.*, *France*, 2447 (1965).
65. Massol, M., and J. Satgé, *Bull. Soc. Chim. France*, **9**, 2737 (1966).
66. Mathis, R., and P. Mazerolles, *Bull. Soc. Chim. France*, 1913 (1962).
67. Mathur, S., G. Chandra, A. K. Rai, and R. C. Mehrotra, *J. Organometal. Chem.*, **4**, 294 (1965).
68. Mathur, S., G. Chandra, A. K. Rai, and R. C. Mehrotra, *J. Organometal. Chem.*, **4**, 371 (1965).
69. Mathur, S., and R. C. Mehrotra, *J. Organometal. Chem.*, **7**, 227 (1967).
70. Mazerolles, P., *et al.* Unpublished work.
71. Mazerolles, P., *Bull. Soc. Chim. France*, 1907 (1962).
72. Mazerolles, P., and J. Dubac, *Compt. Rend.*, **257**, 1103 (1963).
73. Mazerolles, P., and J. Dubac, *3rd International Symposium on Organometallic Chemistry*, Munich, 1967.
74. Mazerolles, P., and J. Dubac, *Compt. Rend.*, **265**, 403 (1967).
75. Mazerolles, P., J. Dubac, and M. Lesbre, *J. Organometal. Chem.*, **5**, 35 (1966).

76. Mazerolles, P., J. Dubac, and M. Lesbre, *Tetrahedron Letters*, (3), 255 (1967).
77. Mazerolles, P., J. Dubac, and M. Lesbre, *J. Organometal. Chem.*, 12, 143 (1968).
78. Mazerolles, P., J. Dubac, and M. Lesbre, *Compt. Rend.*, 266, 1794 (1968).
79. Mazerolles, P., and A. Faucher, *Bull. Soc. Chim. France*, (6), 2134 (1967).
80. Mazerolles, P., A. Faucher, and A. Laporterie, *Bull. Soc. Chim. France*, (3), 887 (1969).
81. Mazerolles, P., M. Lesbre, and M. Joanny, *J. Organometal. Chem.*, 16, 227 (1969).
82. Mazerolles, P., M. Lesbre, and J. Dubac, *Compt. Rend.*, 260, 2255 (1965).
83. Mazerolles, P., and G. Manuel, *Bull. Soc. Chim. France*, (1), 327 (1966).
84. Mazerolles, P., G. Manuel, and F. Thoumas, *Compt. Rend.*, 267, 619 (1968).
85. Mehrotra, R. C., and G. Chandra, *J. Chem. Soc.*, (5), 2804 (1963).
86. Mehrotra, R. C., and S. Mathur, *J. Organometal. Chem.*, 6, 11 (1966).
87. Mehrotra, R. C., and S. Mathur, *J. Organometal. Chem.*, 6, 425 (1966).
88. Metlesics, W., and H. Zeiss, *J. Am. Chem. Soc.*, 82, 3321 (1960).
89. Metlesics, W., and H. Zeiss, *J. Am. Chem. Soc.*, 82, 3324 (1960).
90. Mironov, V. F., and T. K. Gar, *Izv. Akad. Nauk SSSR, Otd. Khim. Nauk*, (3), 578 (1963).
91. Mironov, V. F., and T. K. Gar, *Dokl. Akad. Nauk SSSR*, 152 (5), 1111, (1963).
92. Mironov, V. F., and T. K. Gar, *Izv. Akad. Nauk SSSR*, (10), 1887 (1964).
93. Mironov, V. F., and T. K. Gar, *Izv. Akad. Nauk SSSR, Otd. Khim. Nauk*, (4), 755 (1965).
94. Mironov, V. F., and T. K. Gar, *Izv. Akad. Nauk SSSR Ser. Khim.*, (3), 482 (1966).
94a. Mironov, V. F., T. K. Gar and S. A. Mikhailyants, *Dokl. Akad. Nauk SSSR*, 188 (I), 120 (1969).
95. Mironov, V. F., and V. V. Nepomnina, *Izv. Akad. Nauk SSSR, Otd. Khim. Nauk*, (8), 1419 (1960).
95a. Mironov, V. F., E. S. Sobolev, and L. M. Antipin, *Zh. Obshch. Khim.*, 37(11), 2573 (1967).
96. Moedritzer, K., *J. Organometal. Chem.*, 5, 254 (1966).
97. Moedritzer, K., and J. R. van Wazer, *Inorg. Chem.*, 4, 1753(1965).
98. Morgan, G. T., and H. D. K. Drew, *J. Chem. Soc.*, 125, 1261 (1924).
99. Morgan, G. T., and H. D. K. Drew, *J. Chem. Soc.*, 127, 1760 (1925).
100. Müller, R., and L. Heinrich, *Chem. Ber.*, 95, 2276 (1962).
101. Müller, R., R. Kohne, and H. Beyer, *Chem. Ber.*, 95, 3030 (1962).
102. Nefedov, O. M., S. P. Kolesnikov, A. S. Khachaturov, and A. D. Petrov, *Dokl. Akad. Nauk SSSR*, 154 (6), 1389 (1964).
103. Nefedov, O. M., S. P. Kolesnikov, and V. I. Sheichenko, *Angew. Chem.*, 76, 498 (1964).
104. Nefedov, O. M., and M. N. Manakov, *Izv. Akad. Nauk SSSR Otd. Khim. Nauk*, (4), 769 (1963).
104a. Nefedov, O. M., and M. N. Manakov, *Angew. Chem. Intern. Ed. Engl.*, 5, 1021 (1966).
105. Nefedov, O. M., M. N. Manakov, and A. D. Petrov, *Dokl. Akad. Nauk SSSR*, 147, 1376 (1962).
106: Nefedov, O. M., V. I. Shiryaev, R. A. Strazdynya, and M. N. Manakov, *USSR, avt. svid.* 187796 (1966) *Byull. Izobr.*, (21), 42 (1966).
107. Neumann, W. P., *Angew. Chem.*, 75(14), 679 (1963).
107a. Neumann, W. P., *Angew. Chem. Intern. Ed.*, 2, 555 (1963).
108. Neumann, W. P., and K. Kühlein, *Tetrahedron Letters*, (23), 1541 (1963).
109. Neumann, W. P., and K. Kühlein, *Ann. Chem.*, 683, 1 (1965).
110. Neumann, W. P., and K. Kühlein, *Ann. Chem.*, 702, 13 (1967).
111. Noltes, J. G., G. J. M. van der Kerk, *Chimia (Switz.)*, 16, 122 (1962).
112. Orchin, M., and E. C. Herrick, *J. Org. Chem.*, 24, 139 (1959).

113. Parham, W. E., and H. E. Reiff, *J. Am. Chem. Soc.*, **77**, 1177 (1955).
114. Parham, W. E., H. E. Reiff, and P. Swartzentruber, *J. Am. Chem. Soc.*, **78**, 1437 (1956).
115. Rausch, M. D., and L. P. Klemann, *J. Am. Chem. Soc.*, **89**(22), 5732 (1967).
116. Rijkens, F., and G. J. M. van der Kerk, *Investigations in the Field of Organogermanium Chemistry*, Germanium Research Committee, Utrecht, 1964.
117. Rochow, E. G., *J. Am. Chem. Soc.*, **70**, 1801 (1948).
118. Rochow, E. G., and A. L. Allred, *J. Am. Chem. Soc.*, **77**, 4489 (1955).
119. Ruidisch, J., and M. Schmidt, *Chem. Ber.*, **96**, 821 (1963).
120. Ruidisch, J., and M. Schmidt, *Chem. Ber.*, **96**, 1424 (1963).
120a. Ruidisch, J., and M. Schmidt, *Angew. Chem.*, **76**, 686 (1964).
121. Ruzicka, L., H. A. Boekenoogen, M. Hurtin, and G. A. Klinkenberg, *Helv. Chim. Acta*, **14**, 1319 (1931).
122. Ruzicka, L., W. Brugger, M. Pfeiffer, M. Schinz, and M. Stoll, *Helv. Chim. Acta*, **9**, 339 (1926).
123. Ruzicka, L., W. Brugger, M. Pfeiffer, H. Schinz, and M. Stoll, *Helv. Chim. Acta*, **9**, 499 (1926).
124. Ruzicka, L., M. Stoll, and H. C. Schinz, *Helv. Chim. Acta*, **11**, 670 (1928).
125. Satgé, J., *et al.* Unpublished work.
126. Satgé, J., P. Rivière, and M. Lesbre, *Compt. Rend.*, **265**, 494 (1967).
127. Scherer, O. J., and D. Biller, *Z. Naturforsch.*, **B22**(10), 1079 (1967).
128. Schmidt, M., and H. Ruf, *Angew. Chem.*, **73**, 64 (1961).
129. Schmidt, M., and H. Ruf, *J. Inorg. Nucl. Chem.*, **25**, 557 (1963).
130. Schmidt, M., and H. Schumann, *Z. Anorg. Allgem. Chem.*, **325**, 130 (1963).
131. Schumann-Ruidisch, I., and B. Jutzi-Mebert, *J. Organometal. Chem.*, **11**(1), 77–83 (1968).
132. Schwarz, R., and W. Reinhardt, *Chem. Ber.*, **65**, 1743 (1932).
133. Seyferth, D., and M. A. Weiner, *J. Am. Chem. Soc.*, **84**, 361 (1962).
134. Shono, T., and R. Oda, *J. Chem. Soc. Japan*, **80**, 1200 (1959).
135. Sommer, L. H., and G. A. Baum, *J. Am. Chem. Soc.*, **76**, 5002 (1954).
136. Sommer, L. H., U. F. Bennet, P. G. Campbell, and D. R. Weyenberg, *J. Am. Chem. Soc.*, **79**, 3295 (1957).
137. Stoll, H., and J. Hulstkampp, *Helv. Chim. Acta*, **30**, 1815 (1947).
138. Trautman, C. E., and H. A. Ambrose, U.S. Pat. 2 416 360 (1947).
139. Vilkov, L. V., L. N. Gorokhov, V. S. Mastryukov, and A. D. Rusin, *Zh. Fiz. Khim.*, **38**(11), 2674 (1964).
140. Vilkov, L. V., and V. S. Mastryukov, *Zh. Strukt. Khim.*, **6**(6), 811 (1965).
141. Vol'pin, M. E., V. G. Dulova, and D. N. Kursanov, *Izv. Akad. Nauk SSSR Otd. Khim. Nauk*, 727 (1963).
142. Vol'pin, M. E., V. G. Dulova, Yu. T. Struchkov, N. K. Bokiy, and D. N. Kursanov, *J. Organometal. Chem.*, **8**, 87 (1967).
143. Vol'pin, M. E., Yu. D. Koreshkov, V. G. Dulova, and D. N. Kursanov, *Tetrahedron*, **18**, 107 (1962).
144. Vol'pin, M. E., Yu. D. Koreshkov, and D. N. Kursanov, *Izv. Akad. Nauk SSSR*, 1355 (1961).
145. Vol'pin, M. E., and D. N. Kursanov, *Izv. Akad. Nauk SSSR*, 1903 (1960).
146. Vol'pin, M. E., and D. N. Kursanov, *Zh. Obshch. Khim.*, **32**, 1137 (1962).
147. Vol'pin, M. E., and D. N. Kursanov, *Zh. Obshch. Khim.*, **32**, 1142 (1962).
148. Vol'pin, M. E., and D. N. Kursanov, *Zh. Obshch. Khim.*, **32**, 1455 (1962).
149. Vol'pin, M. E., Yu. T. Struchkov, L. V. Vilkov, V. S. Mastryukov, V. G. Dulova, and D. N. Kursanov, *Izv. Akad. Nauk SSSR Otd. Khim. Nauk*, (11), 2067 (1963).

150. Voronkov, M. G., G. I. Zelchev, and V. F. Mironov, *Avt. Svid.* (190897), (1965), *Byll. Izobr.*, (3), 26 (1967).

151. West, R., *J. Am. Chem. Soc.*, **76**, 6012 (1954).

152. Wieber, M., and C. D. Frohning, *J. Organometal. Chem.*, **8**(3), 459 (1967).

153. Wieber, M., and M. Schmidt, *Z. Naturforsch.*, **18B**, 846–49 (1963).

154. Wieber, M., and M. Schmidt, *J. Organometal. Chem.*, **1**, 93 (1964).

155. Wieber, M., and M. Schmidt, *J. Organometal. Chem.*, **1**, 336 (1964).

156. Wieber, M., and M. Schmidt, *J. Organometal. Chem.*, **2**, 129 (1964).

157. Wieber, M., and M. Schmidt, *Angew. Chem. Intern. Ed. Engl.*, **3**, 153 (1964).

158. Wieber, M., and M. Schmidt, *Angew. Chem. Intern. Ed. Engl.*, **3**, 657 (1964).

159. Wieber, M., and G. Schwarzmann, *Monatsh. Chem.*, **99**(1), 255 (1968).

160. Yoder, C. H., and J. J. Zuckerman, *J. Inorg. Chem.*, **3**, 1329 (1964).

161. Yoder, C. H., and J. J. Zuckerman, *J. Am. Chem. Soc.*, **88**, 2170 (1966).

162. Yoder, C. H., and J. J. Zuckerman, *J. Am. Chem. Soc.*, **88**, 4831 (1966).

163. Yoder, C. H., and J. J. Zuckerman, *Inorg. Chem.*, **6**(1), 163 (1967).

164. Zuckerman, J. J., *J. Chem. Soc.*, 1322 (1963).

165. Zueva, G. Ya., I. F. Manucharova, I. P. Yakovlev, and V. A. Ponomarenko, *Izv. Akad. Nauk SSSR Neorg. Matern*, **2**, 229 (1966).

166. Brit. Pat. 654 571 (1951); *CA*, **46**, 4561 b (1952).

167. Fr. Pat. 1 467 549; *CA*, **68**, 49769 a (1968).

168. U.S. Pat. 2 506 386; *CA*, **44**, 7344 d (1950).

169. U.S. Pat. 3 334 978; *CA*, **68**, 68 668 h (1968).

4 Germanium–hydrogen bond

4-1 CHEMISTRY OF TRICHLOROGERMANE

Reactions of trichlorogermane, $HGeCl_3$, are of particular interest. Trichlorogermane is believed to be extensively ionized in water, even in the presence of high concentrations of mineral acids (3, 7):

$$HGeCl_3 \rightleftarrows H^+ + GeCl_3^-$$

In NMR spectroscopy, the proton signals of trichlorogermane and chloroform lie at appreciably lower field than does the signal from trichlorosilane, which is in accord with the degree of protonization of the hydrogen as indicated in the sequence:

$$HGeCl_3 > HCCl_3 > HSiCl_3 \qquad (32)$$

On the other hand, trichlorogermane exhibits great instability and dissociates easily into germanium dichloride and HCl. Careful low-temperature vacuum distillation gives $GeCl_2$ in polymeric form (23, 41).

$$HGeCl_3 \rightarrow HCl + (GeCl_2)_n$$

The alcoholic solutions are also unstable and seem to liberate progressively hydrogen and germanium dichloride (1).

4-1-1 Reducing properties

In 1886, Winckler noted the ability of trichlorogermane to reduce inorganic compounds such as mercuric chloride and chromic acid (43). Nefedov and others (30) showed that trichlorogermane easily reduces aromatic nitro-compounds to the corresponding amines at room temperature (yields 80–90%); benzophenone is partially reduced to diphenylcarbinol (33). The trichlorogermane etherates, for example $2 Et_2O \cdot HGeCl_3$, exhibit higher reducing power and reduce $FeCl_3$ at about 20°C and also nitrocompounds in quantitative yield. The etherates have a greater tendency to generate germanium dichloride than does $HGeCl_3$ itself. During vacuum distillation they decompose according to the scheme:

$$2 Et_2O \cdot HGeCl_3 \rightarrow 2 Et_2O \cdot HCl + (GeCl_2)_x$$

and act as a source of $GeCl_2$ (25–27).

4-1-2   Additions to unsaturated compounds

Trichlorogermane is considerably more reactive than trichlorosilane. As with the latter, the first addition reactions to olefins were carried out (4, 38, 40) from the azeotropic mixture (70% GeHCl$_3$, 30% GeCl$_4$) in a sealed tube and in the presence of peroxides at a temperature below 100°C (11, 13, 14). Later, Mironov and coworkers (15–21) established that trichlorogermane readily adds across the multiple bond of practically all unsaturated compounds without any need for catalyst or initiator, giving alkylgermanium trichlorides (Table 4-1).

$$Cl_3GeH + \quad \diagdown \!\!\! C = C \!\!\! \diagup \quad \rightarrow \quad Cl_3Ge - \overset{|}{\underset{|}{C}} - CH \diagup$$

Trichlorogermane adds smoothly at room temperature, even to unactivated double bonds such as those in ethylene while additions of trichlorosilane mainly occur only in the presence of a catalyst, and under more drastic conditions at temperatures above 200°C (36). Because of this, the thermal reactions of hydrogermylation by the trichlorogermane are more numerous than the hydrosilylation reactions. For example, $GeHCl_3$ adds vigorously to cyclopentadiene, while $SiHCl_3$ remains inactive even in the presence of a catalyst (15).

It is interesting to note that the uncatalysed yields of the germyl adducts were found to be much higher than those obtained in the presence of catalysts or initiator: thus, with ethylene the yield of $EtGeCl_3$ is 55% without catalyst (37) and only 25% in the presence of H_2PtCl_6 (38). With octene-1, the yields of adduct are respectively 55% in the absence and 18% in the presence of peroxide (37, 40). Trichlorogermane adds more easily than hydrosilanes to halo-olefins with the halogen atom at the double bond:

$$HGeCl_3 + CH_2{=}CHCl \rightarrow ClCH_2CH_2GeCl_3$$

$$\beta \text{ adduct}$$

in contrast to the corresponding $HSiCl_3$ addition, which gives the addition product in only low yield.

Addition of $HGeCl_3$ to more highly chlorinated ethylene derivatives often requires higher temperature (2, 13, 37, 39)

$$CH_2{=}CCl_2 \rightarrow Cl_2CHCH_2GeCl_3 \qquad (62\% \text{ yield})$$

$$CHCl{=}CHCl \rightarrow ClCH_2CHClGeCl_3$$

$$CH_2{=}CF_2 \rightarrow F_2CHCH_2GeCl_3 \qquad (\text{exothermic})$$

$$\underset{\underset{Cl}{|}}{CH_2}{=}\underset{\underset{Cl}{|}}{C}{-}CH_3 \rightarrow CH_3{-}\underset{\underset{Cl}{|}}{\overset{\overset{H}{|}}{C}}{-}CH_2GeCl_3$$

The trichloro(2-chloroalkyl)germanes undergo β-elimination with Grignard reagents:

$$Cl_3GeCH_2CH_2Cl \xrightarrow[CH_3MgX]{} (CH_3)_4Ge + CH_2{=}CH_2$$

and with alcoholic alkali; the four chlorine atoms can be titrated with 0.1N NaOH in ethanol (16).

$$Cl_3GeCH_2CH_2Cl + 4\,NaOH \rightarrow C_2H_4 + Ge(OH)_4 + 4\,NaCl$$

With allyl chloride, the monoadduct γ chloropropyltrichlorogermane is obtained in 77% yield (14, 37).

With 1,3-butadiene, two concomitant reactions were observed (2, 17). In addition to the normal 1,4 adduct:

$$HGeCl_3 + CH_2{=}CHCH{=}CH_2 \rightarrow CH_3CH{=}CHCH_2GeCl_3$$

a cyclic adduct, 1,1 dichloro-1 germacyclopentene-3, was obtained:

$$(1) \qquad\qquad HGeCl_3 \rightleftarrows HCl + GeCl_2$$

$$(2) \quad GeCl_2 + CH_2{=}CHCH{=}CH_2 \rightarrow \underset{Cl\diagup Ge\diagdown Cl}{\overset{HC{=}CH}{\underset{\underset{}{}}{H_2C\diagdown\quad\diagup CH_2}}}$$

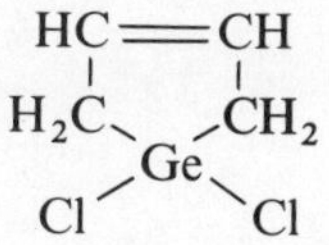

Isoprene also reacts with $HGeCl_3$ with the formation of a heterocyclic compound (17, 18).

The influence of the solvent in these reactions is particularly evident in the following cases:

(a) With acetylene, trichlorogermane adds very easily without catalyst and solvent in an exothermic reaction to give 1,2-bis(trichlorogermyl)-

ethane in 90% yield (37).

$$CH\equiv CH + 2(HGeCl_3) \rightarrow Cl_3GeCH_2CH_2GeCl_3$$

With a saturated solution of acetylene in hexane, the mono addition product vinyltrichlorogermane $CH_2=CH-GeCl_3$ was also obtained in 13% yield (Mironov and Gar (20)). In ethereal solution, only polymers were formed (21).

(b) With ethylene, reaction can take two courses which are quite different in the presence or absence of ether.

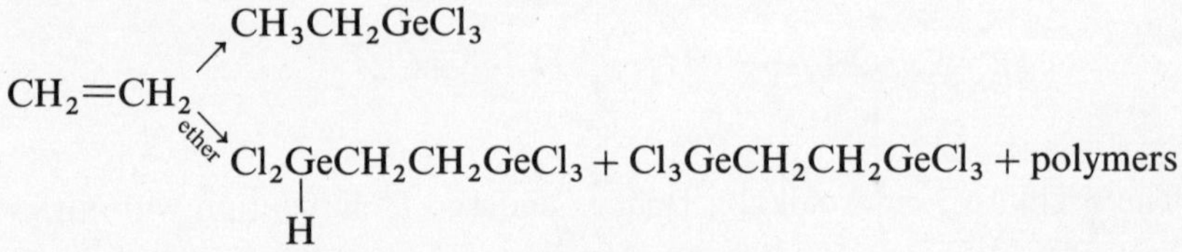

$$CH_2=CH_2 \nearrow CH_3CH_2GeCl_3$$
$$\searrow_{ether} Cl_2GeCH_2CH_2GeCl_3 + Cl_3GeCH_2CH_2GeCl_3 + polymers$$
$$\overset{|}{H}$$

The different courses of reaction referred to above may be explained by the double equilibrium:

$$GeHCl_3 \rightleftarrows H^+(GeCl_3)^- \rightleftarrows HCl + GeCl_2$$

In ether this equilibrium is displaced to the right, and $GeCl_2$ adds across the double bond, giving an unstable three-membered heterocycle (5):

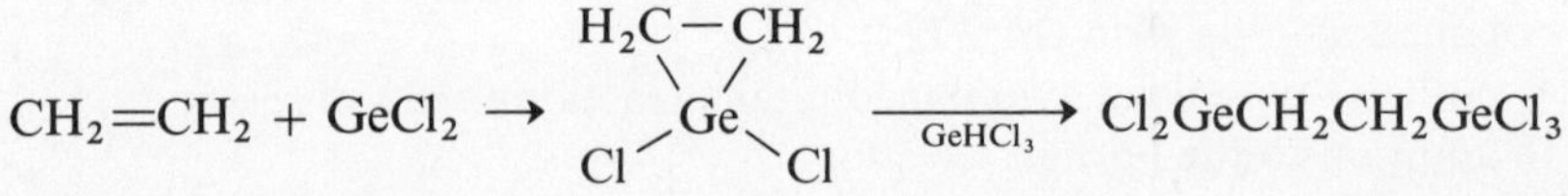

$$CH_2=CH_2 + GeCl_2 \rightarrow \begin{matrix} H_2C-CH_2 \\ \diagdown / \\ Ge \\ Cl \diagup \diagdown Cl \end{matrix} \xrightarrow[GeHCl_3]{} Cl_2GeCH_2CH_2GeCl_3$$

The reactions of dietherates $HGeCl_3$, $2 Et_2O$, and $HGeCl_3 \cdot 2$ THF with unsaturated compounds can also be interpreted as addition reactions of germanium dichloride (27):

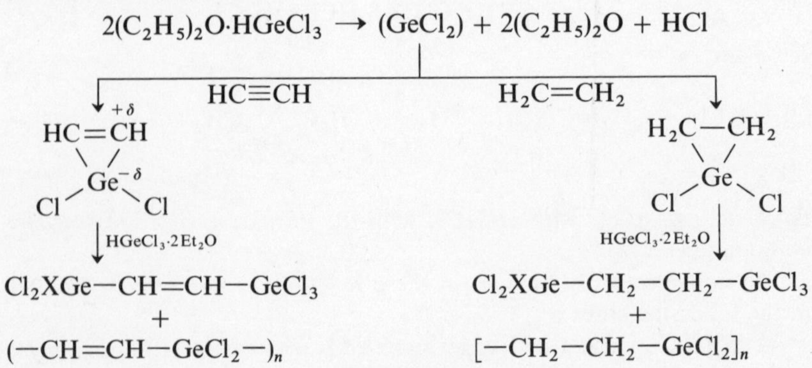

$$2(C_2H_5)_2O \cdot HGeCl_3 \rightarrow (GeCl_2) + 2(C_2H_5)_2O + HCl$$

$$\xleftarrow{HC\equiv CH} \quad \xrightarrow{H_2C=CH_2}$$

$$\begin{matrix} HC=CH \\ \diagdown / \\ Ge^{-\delta} \\ Cl \diagup \diagdown Cl \end{matrix} \qquad \begin{matrix} H_2C-CH_2 \\ \diagdown / \\ Ge \\ Cl \diagup \diagdown Cl \end{matrix}$$

$$\downarrow HGeCl_3 \cdot 2 Et_2O \qquad\qquad HGeCl_3 \cdot 2 Et_2O \downarrow$$

$$Cl_2XGe-CH=CH-GeCl_3 \qquad Cl_2XGe-CH_2-CH_2-GeCl_3$$
$$+ \qquad\qquad +$$
$$(-CH=CH-GeCl_2-)_n \qquad [-CH_2-CH_2-GeCl_2]_n$$

$$(X = H, Cl)$$

A similar reaction reported by Nefedov and Kolesnikov is the alkylation of dietherate $HGeCl_3 \cdot 2\,Et_2O$ by methyllithium or methylmagnesium bromide which leads to the telomers $Me(GeMe_2)_nMe$, $n > 2$, as well as some cyclic $(Me_2Ge)_m$, $m = 4$ and 6 (25, 26). With acrylic acid, dietherates give $Cl_3GeCH_2CH_2CO_2H$ in good yield (Mironov and Gar (23)).

Some of the above reactions are summarized in Table 4-1.

Table 4-1

Addition of Trichlorogermane to Unsaturated Compounds

C_n	STARTING COMPOUND	CATALYST	PRODUCT	YIELD (%)	B.P. (°C/mm)	n_D^{20}	REFERENCES
C_2	CH≡CH	None	$Cl_3GeCH_2CH_2GeCl_3$	90	130/12	—	32
	$CH_2=CH_2$	None	$C_2H_5GeCl_3$	55	142/761	—	32
	$CH_2=CHCl$	None	$ClCH_2CH_2GeCl_3$	33	75/15	1.5092	10, 11
	$Cl_2C=CH_2$	None	$Cl_3GeCH_2CHCl_2$	58	74/10	1.5176	10, 11
	trans-ClCH=CHCl	None	$ClCH_2CHClGeCl_3$	62	88/12	1.5240	10, 11
	$Cl_2C=CHCl$	None	$Cl_3GeC_2H_2Cl_3$	55	92/9	1.5341	10, 11
	$Cl_2C=CCl_2$	None	$Cl_2CHCl_2CGeCl_3$	48	123/15	1.5378	10, 11
	$Cl_3SiCH=CH_2$	None	$Cl_3SiCH_2CH_2GeCl_3$	83	120/15	—	32
C_3	$CH_2=CHCH_3$	None	$Cl_3GeCH_2CH_2CH_3$	85	163.5/756	1.4749	12
	$CH_2=C(Cl)CH_3$	None	$Cl_3GeCH_2CH(Cl)CH_3$	54	98/32	—	13
	$CH_2=CHCN$	None	$Cl_3GeCH_2CH_2CN$	53	135/22	—	32
	$CH_3CH=CHCl$	None	$Cl_3GeCH(CH_3)CH_2Cl$	30	69/7	—	13
	$CH_2=CHCH_2Cl$	None	$Cl_3Ge(CH_2)_3Cl$	77	105/20	—	32
	$Cl_3GeCH_2CH=CH_2$	None	$Cl_3GeCH_2CH_2CH_2GeCl_3$	65	123.5/5	1.5314	12
C_4	$CH_3CH_2CH=CH_2$	None	$Cl_3GeCH_2CH_2CH_2CH_3$	75	182/737	1.4520	12
	$CH_2=C(CH_3)_2$	None	$Cl_3GeCH_2CH(CH_3)_2$	70	108/25	—	13
	$CH_2=CHCH=CH_2$	None	$Cl_3GeCH_2—CH=CHCH_3$	61	177/759	1.5080	12
	$CH_2=CHOCOCH_3$	None	$Cl_3GeCH_2CH_2OCOCH_3$	67	97/15	1.4868	12
	$CH_2=C(CH_3)—CH_2Cl$	None	$Cl_3GeCH_2CH(CH_3)CH_2Cl$	58	91/11	—	12

C$_5$	Pentene-1	Peroxide	Cl$_3$Ge(CH$_2$)$_4$CH$_3$	21	102/40	—	35
	4-Chloropentene-1	Peroxide	Cl$_3$Ge(CH$_2$)$_3$CHClCH$_3$	10	80/0.4	—	35
	Allyl acetate	None	Cl$_3$Ge(CH$_2$)$_3$CO$_2$CH$_3$	41	127/12	—	32
C$_6$	Hexene-1	Peroxide	CH$_3$(CH$_2$)$_5$GeCl$_3$	22	97/14	1.4719 (25°)	4
	Cyclohexene	None	C$_6$H$_{11}$GeCl$_3$	85	79/3	—	13
C$_7$	Heptene-1	Peroxide	Cl$_3$Ge(CH$_2$)$_6$CH$_3$	18	102/10	—	35
C$_8$	Octene-1	Peroxide	CH$_3$(CH$_2$)$_7$GeCl$_3$	18	80/0.6	—	12
	Phenylacetylene	None	Cl$_3$GeCH=CHC$_6$H$_5$	43.5	146/16	1.5833	12
	Styrene	None	Cl$_3$GeCH$_2$CH$_2$C$_6$H$_5$	30	132/10	1.5550	12

4-1-3   Additions to a three-membered ring

Trichlorogermane and its etherate add very easily to aryl substituted cyclopropanes, giving in good yields the corresponding organotrichlorogermanes (electrophilic addition):

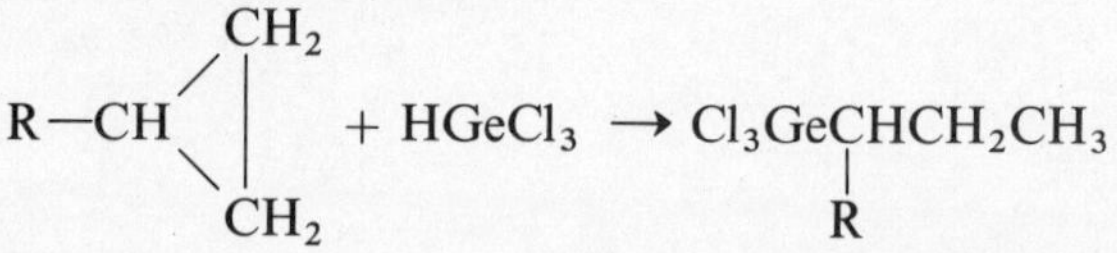

$$R\!-\!CH\!\!\begin{array}{c} \diagup CH_2 \\ | \\ \diagdown CH_2 \end{array} + HGeCl_3 \ \rightarrow \ Cl_3Ge\underset{\underset{R}{|}}{C}HCH_2CH_3$$

For example, phenylcyclopropane leads to the formation of 1-phenyl-(trichlorogermyl)propane (85% yield) while *n*-amyl cyclopropane gives a mixture of the two isomeric octyl trichlorogermanes (31). In contrast chlorocyclopropanes and cyclopropane itself do not react with $HGeCl_3$ or its etherates.

With ethylene oxide, $HGeCl_3$ reacts in a different way than does $HSiCl_3$:

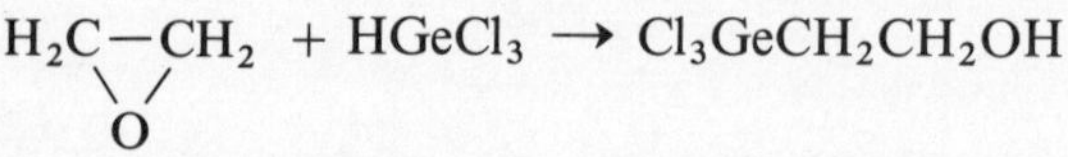

$$H_2C\!-\!CH_2 + HGeCl_3 \ \rightarrow \ Cl_3GeCH_2CH_2OH$$
$$\diagdown \!\!\!\! \underset{O}{\diagup}$$

The alcohol was isolated after alkylation with a Grignard reagent (23).

4-1-4   Reactions with ketones

Mironov and coworkers showed that trichlorogermane adds at the carbonyl group of formaldehyde with formation of a germanium–carbon bond (23):

$$HCHO + HGeCl_3 \ \rightarrow \ Cl_3GeCH_2OH$$

Under the influence of $HGeCl_3$, ketones of type $RCOCH_2R'$ condense to given unsaturated ketones which readily add trichlorogermane with formation of trichlorogermyl ketones (Nefedov, Kolesnikov and Perlmuther, (33)):

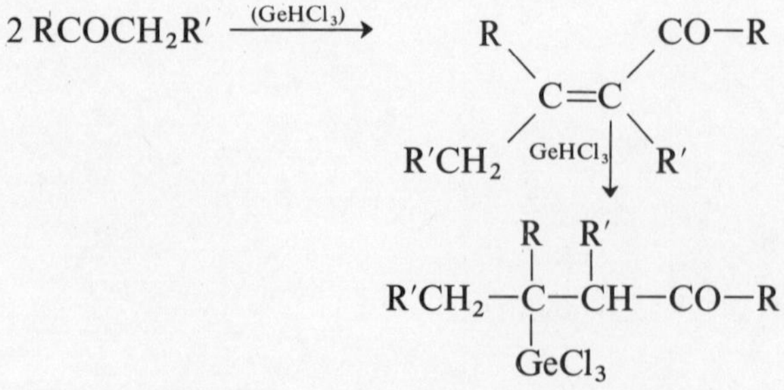

$$2\,RCOCH_2R' \xrightarrow{\text{(GeHCl}_3)} \ \begin{array}{c} R \\ \diagdown \\ R'CH_2 \diagup \end{array} C\!=\!C \begin{array}{c} CO\!-\!R \\ \diagup \\ \diagdown R' \end{array}$$

$$\downarrow \text{GeHCl}_3$$

$$R'CH_2\!-\!\underset{\underset{GeCl_3}{|}}{\overset{\overset{R}{|}}{C}}\!-\!\overset{\overset{R'}{|}}{CH}\!-\!CO\!-\!R$$

(R = alkyl or aryl, R' = H, alkyl or aryl)

Acetone reacts vigorously, being first condensed to mesityl oxide (6); cyclohexanone affords a 97% yield of 2-(1'-trichlorogermylcyclohexyl) cyclohexanone:

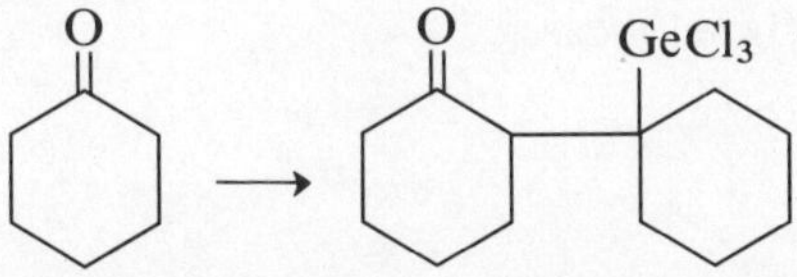

which is also formed on treatment of $HGeCl_3$ with 1-methoxycyclohexene or with 2-cyclohexylidene cyclohexanone. Cycloheptanone gives a 98% yield of 2-(1'trichlorogermylcyclohexyl) cycloheptanone. Benzophenone reacts quite differently: the strong reducing action of $HGeCl_3$ leads to the alcohol (diphenylcarbinol), which condenses to give the ether, and decomposition gives alcohol and aryltrichlorogermane (33):

$$(C_6H_5)_2CO \xrightarrow{\text{(GeHCl}_3)} (C_6H_5)_2CHOH \xrightarrow[-H_2O]{\text{(GeHCl}_3)}$$

$$(C_6H_5)_2CH-O-CH(C_6H_5)_2$$
$$\downarrow \text{GeHCl}_3$$
$$(C_6H_5)_2CH-GeCl_3 + (C_6H_5)_2CHOH$$

Other alcohols react similarly with trichlorogermane. Thus, methanol gives CH_3GeCl_3 (47% yield).

4-1-5 Hydrogermylation of aromatic compounds

In contrast with trichlorosilane, the electrophilic addition of $HGeCl_3$ to aromatic double bonds occurs easily without catalyst at moderate temperatures. Hydrogermylations of naphthalene, dimethylnaphthalene, anthracene and phenanthrene were carried out by heating an equimolar mixture of $HGeCl_3$ and the aromatic at 100–130°C for 30 minutes. The reaction mixture was then methylated with excess MeMgBr in ether and the products separated by vacuum distillation. The trimethylgermyl groups are located in the tetrahydrobenzoidal ring, as confirmed by NMR spectrum (Kolesnikov and Nefedov (9)):

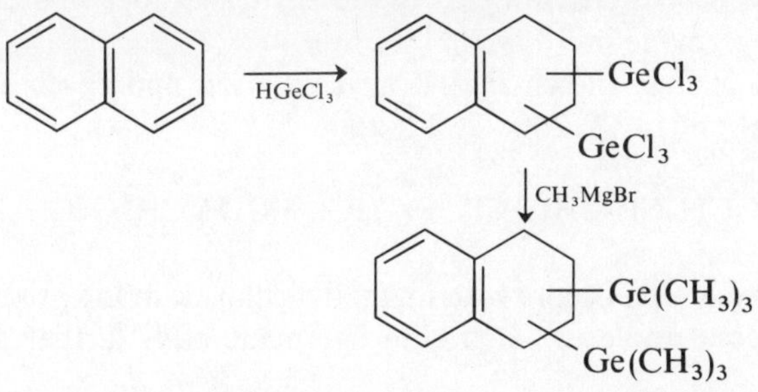

No reaction with benzene, toluene or the xylenes was observed under these conditions but alkoxybenzenes do yield adducts. For example, anisole gave, after subsequent methylation, a mixture of isomeric methoxy tris(trimethylgermyl)cyclohexanes.

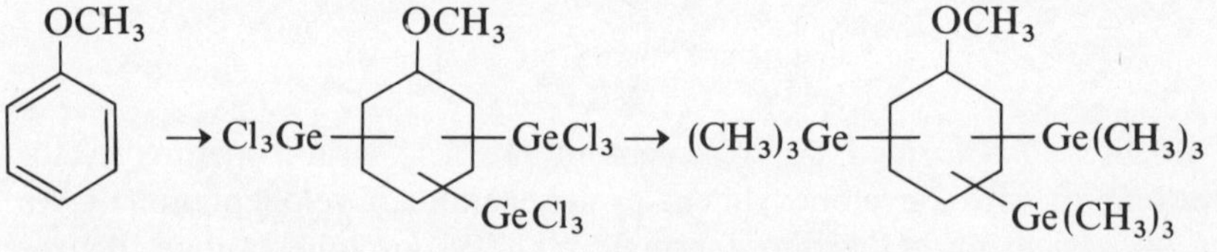

Trichlorogermane adds to heterocyclic compounds like thiophene when cooled to $-70°C$ (exothermic reaction). To determine the reactivity of $HGeCl_3$ in relation to the basicity of aromatic hydrocarbons, the reaction of deuterium exchange between $DGeCl_3$ and some hydrocarbons was examined (8).

4-1-6 Condensation–elimination reactions

A migration of the hydrogen proton can be observed from the $HGeCl_3$ molecule when one mixes, in the presence of ether, trichlorogermane with mobile-type halogen compounds such as *tert*-butyl chloride, chlorotriphenylmethane, allylic halides, etc.

$$Me_3CCl + HGeCl_3 \xrightarrow{\text{ether}} HCl + Me_3C-GeCl_3 \quad 80\%$$

$$Ph_3CCl + HGeCl_3 \xrightarrow{\text{ether}} HCl + Ph_3C-GeCl_3 \quad 77\%$$

$$H_2C=CHCH_2Cl + HGeCl_3 \longrightarrow HCl + Cl_3GeCH_2CH=CH_2$$

In the absence of ether, the last reaction is replaced by simple addition of the trichlorogermane across the double bond of allyl chloride (11, 12). Allyl bromide undergoes the same condensation reaction with elimination of HBr, which reacts with part of the allyltrichlorogermane to give allyltribromogermane in 9% yield (38).

It was further shown that benzyl chloride undergoes an analogous condensation (14, 27, 28):

$$C_6H_5CH_2Cl + HGeCl_3 \rightarrow HCl + C_6H_5CH_2GeCl_3 \quad 45\%$$

A similar reaction occurs with methallyl chloride in the presence of ether, but a second molecule of trichlorogermane adds further to the initial product of the condensation reaction (11):

$$H_2C{=}\underset{\underset{CH_3}{|}}{C}{-}CH_2Cl + HGeCl_3 \rightarrow HCl + H_2C{=}\underset{\underset{CH_3}{|}}{C}{-}CH_2GeCl_3$$

$$H_2C{=}\underset{\underset{CH_3}{|}}{C}{-}CH_2GeCl_3 + HGeCl_3 \rightarrow Cl_3GeCH_2\underset{\underset{CH_3}{|}}{C}HCH_2GeCl_3$$

In the absence of ether, only the addition reaction occurs, giving (15):

$$Cl_3GeCH_2\underset{\underset{CH_3}{|}}{C}HCH_2Cl$$

With propargyl chloride, the condensation reaction is also followed by the addition of two molecules of $HGeCl_3$ to the triple bond:

$$HC{\equiv}CCH_2Cl \rightarrow HC{\equiv}CCH_2GeCl_3 \rightarrow (Cl_3Ge)_2CHCH_2CH_2GeCl_3$$

The trichlorogermane etherate reacts more vigorously than trichlorogermane itself (5, 9, 20):

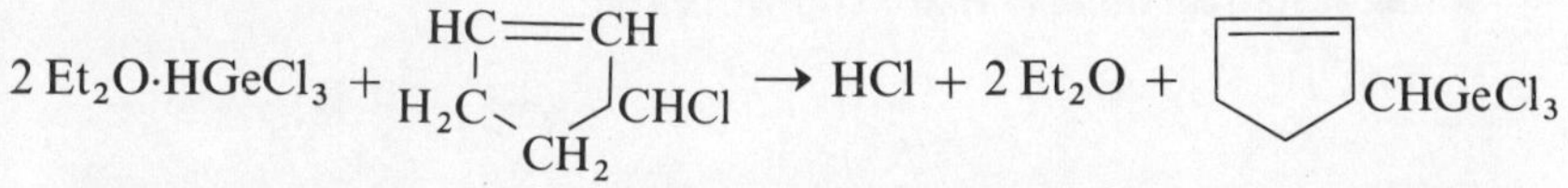

$$2\,Et_2O{\cdot}HGeCl_3 + \underset{\underset{CH_2}{|}}{\overset{HC{=\!=\!}CH}{H_2C\diagdown\diagup}}CHCl \rightarrow HCl + 2\,Et_2O + \bigtriangleup\!\!\square CHGeCl_3$$

This alkenyltrichlorogermane, in which the Ge atom is located in the α position relative to the $C{=}C$ bond, exhibits a high reactivity in polar addition reactions and an abnormally high intensity of the Raman lines corresponding to double bond vibration (15). Nesmeyanov explains

Table 4-2
Condensation Reactions with Allyl Halides

ALLYL HALIDES	REACTION PRODUCTS	YIELD %
$ClCH_2CH{=}CH_2$ (ether)	$ClGeCH_2CH{=}CH_2$	27
	$ClCH_2{-}\underset{\underset{GeCl_3}{|}}{C}H{-}CH_2GeCl_3$	57
$BrCH_2CH{=}CH_2$	$\begin{cases} ClGeCH_2CH{=}CH_2 \\ Br_3GeCH_2CH{=}CH_2 \end{cases}$	49.5 9
$ICH_2CH{=}CH_2$	$Cl_3GeCH_2CH{=}CH_2$	38
$ClCH_2{-}\underset{\underset{CH_3}{|}}{C}{=}CH_2$ (ether)	$Cl_3GeCH_2{-}\underset{\underset{CH_3}{|}}{C}{=}CH_2$	60

these facts by σ–π conjugation of bonds 1.2 and 3.4 in the system

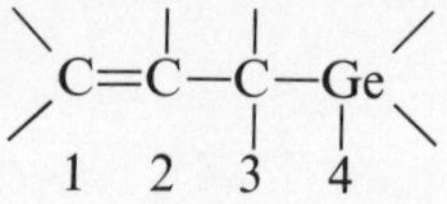

$$
\begin{array}{c}
\diagdown \qquad | \quad | \qquad \diagup \\
{}_{\diagup}C{=}\underset{\;}{C}{-}\underset{\;}{C}{-}Ge_{\diagdown} \qquad (34) \\
\diagup \quad | \quad | \quad \diagdown \\
1 \quad 2 \quad 3 \quad 4
\end{array}
$$

In the aromatic series, Shostakowskii reported a condensation reaction of trichlorogermane with p-dichlorobenzene, giving p-chlorophenyl-trichlorogermane (42). These condensation reactions have no close analogies in the chemistry of other elements of Group IV B.

4-1-7   Reactions of tribromogermane

Gar and Mironov investigated the synthesis and transformations of tribromogermane (5), which dissociates in the neighbourhood of 15°C. As with trichlorogermane, without catalysts this compound adds exo-thermally to unsaturated compounds, and undergoes similar condensation reactions (28).

Some of its reactions (44) are shown below:

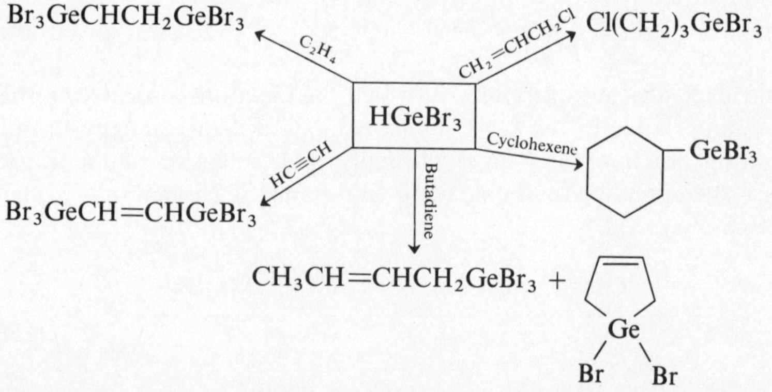

The ether complex of tribromogermane by reactions with formaldehyde forms $C_2H_5OCH_2GeBr_3$, possibly as a result of the condensation of Br_3GeCH_2OH with ethanol, which is one of the products of the decomposition of the ether complex (23).

References (4-1)

1. Delwaulle, M. L., *J. Phys. Chem.*, **56**, 355, 1952.
2. Dzurinskaya, N. G., V. F. Mironov, and A. D. Petrov, *Dokl. Akad. Nauk SSSR*, **138**, 1107, 1961.
3. Everest, A., *Res. Correspondence*, **8**, 61, 1955.
4. Fischer, A. K., R. C. West, and E. G. Rochow, *J. Am. Chem. Soc.*, **76**, 5878, 1954.
5. Gar, T. K., and V. F. Mironov, *Izv. Akad. Nauk SSSR*, 365, 1965.

6. Gar, T. K., and V. F. Mironov, *Zh. Obshch. Khim.*, **36**, 1709, 1966.
7. Gilman, H., and W. Melvin, *J. Am. Chem. Soc.*, **71**, 4050, 1949.
8. Kolesnikov, S. P., O. M. Nefedov, and V. I. Sheichenko, *Izv. Akad. Nauk SSSR*, 443, 1963.
9. Kolesnikov, S. P., and O. M. Nefedov, *Angew. Chem.*, **4**, 352, 1965.
10. Leites, A., T. K. Gar, and V. F. Mironov, *Izv. Akad. Nauk SSSR*, 400, 1964.
11. Mironov, V. F., N. G. Dzyrinskaya, and A. D. Petrov, *Dokl. Akad. Nauk SSSR*, **131**, 98, 1960.
12. Mironov, V. F., N. G. Dzyrinskaya, and A. D. Petrov, *Izv. SSSR*, 2066, 1960.
13. Mironov, V. F., and N. G. Dzyrinskaya, *Dokl. Akad. Nauk SSSR*, **138**, 1107, 1961.
14. Mironov, V. F., N. G. Dzyrinskaya, T. K. Gar, and A. D. Petrov, *Izv. Akad. Nauk SSSR*, 460, 1962.
15. Mironov, V. F., T. K. Gar, and A. Leites, *Izv. Akad. Nauk SSSR*, 1387, 1962.
16. Mironov, V. F., and N. G. Dzyrinskaya, *Izv. Akad. Nauk SSSR*, 75, 1963.
17. Mironov, V. F., and T. K. Gar, *Izv. Akad. Nauk SSSR*, 578, 1963.
18. Mironov, V. F., and T. K. Gar, *Dokl. Akad. Nauk SSSR*, **152**, 1112, 1963.
19. Mironov, V. F., and A. L. Kravchenko, *Dokl. Akad. Nauk SSSR*, **155**, 843, 1964.
20. Mironov, V. F., and T. K. Gar, *Izv. Nauk SSSR*, 1420, 1964.
21. Mironov, V. F., and T. K. Gar, *Izv. Nauk SSSR*, 1515, 1964.
22. Mironov, V. F., and T. K. Gar, *Izv. Akad. Nauk SSSR*, 482, 1966.
23. Mironov, V. F., E. M. Berliner, and T. K. Gar, *Zh. Obshch. Khim.*, **37**, 962, 1967.
24. Moulton, W., and G. Miller, *J. Am. Chem. Soc.*, **78**, 2702, 1956.
25. Nefedov, O. M., and S. P. Kolesnikov, *Izv. Akad. Nauk SSSR*, 2068, 1963.
26. Nefedov, O. M., and S. P. Kolesnikov, *Izv. Akad. Nauk SSSR*, 773, 1964.
27. Nefedov, O. M., S. P. Kolesnikov, and L. Scheinchenko, *Angew. Chem.*, **76**, 498, 1964.
28. Nefedov, O. M., S. P. Kolesnikov, and L. Scheinchenko, *Angew. Chem.*, **3**, 508, 1964.
29. Nefedov, O. M., S. P. Kolesnikov, A. S. Chachakorov, and A. D. Petrov, *Dokl. Akad. Nauk SSSR*, **154**, 1389, 1964.
30. Nefedov, O. M., S. P. Kolesnikov, and N. Mahkova, *Izv. Akad. Nauk SSSR*, 2224, 1964.
31. Nefedov, O. M., S. P. Kolesnikov, and N. Novitskaya, *Izv. Akad. Nauk SSSR*, 579, 1965.
32. Nefedov, O. M., S. P. Kolesnikov, L. Scheinchenko, and N. Scheinker, *Dokl. Akad. Nauk SSSR*, **162**, 590, 1965.
33. Nefedov, O. M., S. P. Kolesnikov, and B. L. Perlmuther, *Angew. Chem.*, *Intern. Ed.*, **6**, 628, 1967.
34. Nesmeyanov, A. N., and M. I. Kabachnick, *Zh. Obshch. Khim.*, **25**, 41, 1955.
35. Petrov, A. D., V. F. Mironov, and E. Golpy, *Izv. Akad. Nauk SSSR*, 1146, 1956.
36. Petrov, A. D., V. F. Mironov, and A. Chernyshev, *Progr. Chem.*, **26**, 292, 1957.
37. Petrov, A. D., V. F. Mironov, and N. G. Dzyrinskaya, *Dokl. Akad. Nauk SSSR*, **128**, 302, 1959.
38. Ponomarenko, V. A., Y. Vzenkova, and P. Egorov, *Dokl. Akad. Nauk SSSR*, **112**, 405, 1958.
39. Ponomarenko, V. A., Y. A. Zueva, and P. S. Andrew, *Izv. Akad. Nauk SSSR*, 1758, 1961.
40. Riemschneider, R., V. Menge, and P. Z. Klang, *Z. Naturforsch.*, **11**, 115, 1956.
41. Schumb, W. C., and M. Smyth, *J. Am. Chem. Soc.*, **77**, 3003, 1955.
42. Shostakowskii, M. F., A. Sokolov, and T. Erurakova, *Zh. Obshch. Khim.*, **32**, 1714, 1962.
43. Winckler, C., *J. Prakt. Chem.*, **34**, 177, 1886.
44. *Annual Surveys of Organometallic Chemistry*, **2**, 151, Elsevier, 1966.

4-2 ORGANOHALOGERMANIUM HYDRIDES

4-2-1 Preparations

$R_2(X)GeH$, $R(X)_2GeH$ and $R(X)GeH_2$ are intermediate mixed hydride derivatives between R_3GeH trialkyl- and R_2GeH_2 dialkylgermanes on the one hand and X_3GeH and X_2GeH_2 halogenogermanes on other hand.

The first organohalogermanes, n-$Bu_2(X)GeH$ and n-$Bu(X)GeH_2$ (X = Cl, Br, I), described in 1959 by H. H. Anderson (10), were isolated after partial cleavage of the butylgermane Ge—H bonds in n-$BuGeH_3$ and n-Bu_2GeH_2 by $HgCl_2$, $HgBr_2$ and by iodine. At about the same time, Lesbre and Satgé succeeded in developing new dialkylhalogermane syntheses as they studied hydrogen–halogen exchange reactions between organogermanium hydrides and organic halides (96, 142):

$$R_2GeH_2 + RX \longrightarrow R_2(X)GeH + RH$$

Action of gaseous hydrogen chloride on CH_3GeH_3 and $(CH_3)_2GeH_2$ in the presence of $AlCl_3$ enabled Amberger and Boeters (3) to isolate $(CH_3)(Cl)GeH_2$, $(CH_3)(Cl)_2GeH$ and $(CH_3)_2(Cl)GeH$. On their part, Griffiths and Onyszchuk noted that the action of gaseous HCl and HBr on CH_3GeH_3 over $AlCl_3$ or $AlBr_3$ leads exclusively to $(CH_3)(Cl)_2GeH$ and $(CH_3)(Br)GeH_2$ (64).

Cleavage of dipropylgermane by HgI_2 leads to dipropyliodogermane, n-Pr_2IGeH (14). Diethyl and diphenylchlorogermane formation is recorded in the hydrogen–halogen exchange reaction between R_2GeH_2 dihydrides and R_2GeCl_2 dichlorides over $AlCl_3$ (90, 96, 142):

$$R_2GeH_2 + R_2GeCl_2 \xrightarrow{\text{AlCl}_3} R_2(Cl)GeH$$

$(C_6H_5)_2(X)GeH$ and $(C_6H_5)(X)GeH_2$ phenylhalogermanes (X = Cl, Br, I) are also produced through cleavage of one phenylgermane Ge—H bond in $(C_6H_5)_2GeH_2$ and $(C_6H_5)GeH_3$ by different organic or inorganic halogenating agents (140, 150). These may be chloromethyl ether, $HgCl_2$, N-chloro- or bromo succinimide, $HgBr_2$ and N-iodosuccinimide.

$(C_6H_5)_2GeH_2$ or $C_6H_5GeH_3$ + $ClCH_2OCH_3$ $\xrightarrow{\text{AlCl}_3}$

$(C_6H_5)_2(Cl)GeH$ or $(C_6H_5)(Cl)GeH_2$ + CH_3OCH_3

$(C_6H_5)_2GeH_2$ or $C_6H_5GeH_3$ + $(CH_2CO)_2NBr$ $\longrightarrow$

$(C_6H_5)_2(Br)GeH$ or $(C_6H_5)(Br)GeH_2$ + $(CH_2CO)_2NH$

$R(X)_2GeH$ organodihalogermane synthesis is harder to achieve and only the dichlorides are produced through direct cleavage of $RGeH_3$

trihydrides (105, 106, 140, 150).

$$RGeH_3 \xrightarrow[\text{AlCl}_3]{\text{ClCH}_2\text{COCH}_3} R(Cl)GeH_2 \xrightarrow[\text{AlCl}_3]{\text{ClCH}_2\text{OCH}_3} R(Cl)_2GeH$$

$$(R = Et, Bu, Ph)$$

More recently, Mironov and Kravchenko have described a new synthesis of these dichlorides through partial alkylation of trichlorogermane etherate using tetraalkyltin and lead compounds (122, 123).

$$Cl_3GeH + R_4M \xrightarrow{\text{ether}} R(Cl)_2GeH + R_3MCl$$

$$\begin{cases} M = Sn \quad R = Me \text{ or } Et \\ M = Pb \quad R = Et \end{cases}$$

Organodibromo- and -diiodogermanes and $R_2(F)GeH$, $R(F)GeH_2$ and $R(F)_2GeH$ type organofluorogermanes are isolated by means of an indirect synthesis. The ammoniacal hydrolysis of $R_2(Cl)GeH$, $R(Cl)GeH_2$ and $R(Cl)_2GeH$ organochlorogermanes leads to the corresponding organohydro-germanium oxides: $[R_2(H)Ge]_2O$, $[R(H)_2Ge]_2O$ and $[R(H)GeO]_n$.

Hydrohalic acid cleavage of these oxides leads to the different types of organohalogermanes in good yields (105, 106, 140, 150).

$$[R_2(H)Ge]_2O + 2HX \rightarrow R_2(X)GeH + H_2O \quad \begin{cases} R = Et, Bu, C_6H_5 \\ X = F, Cl, Br, I \end{cases}$$

$$[R(H)_2Ge]_2O + 2HX \rightarrow R(X)GeH_2 + H_2O \quad \begin{cases} R = Et, Bu, C_6H_5 \\ X = F, Cl, Br, I \end{cases}$$

$$[R(H)GeO]_n + 2nHX \rightarrow nR(X)_2GeH + nH_2O \quad \begin{cases} R = C_6H_5 \\ X = F, Cl, Br, I \end{cases}$$

4-2-2 Stability of organohalogermanium hydrides

The stability of the alkylhalogermanes belonging to these three classes decreases following the order $X = Cl > Br > I > F$. Moreover the stability will further decrease as the degree of substituting halogens increases (105, 106):

$$R_2(X)GeH \rightarrow R(X)GeH_2 \rightarrow R(X)_2GeH$$

Ethylfluorogermane is readily decomposed at room temperature (105, 106):

$$2\,Et(F)GeH_2 \rightarrow EtGeH_3 + (Et(F)_2GeH)$$
$$\text{unstable}$$

Diethyl- and dibutylfluorogermanes will disproportionate similarly, but at a slower rate (105, 106):

$$2\,R_2(F)GeH \rightarrow R_2GeH_2 + R_2GeF_2$$

Diethylchlorogermane, either when heated or passed over an $AlCl_3$ catalyst, gives the following equilibrium (105):

$$Et_2(Cl)GeH \rightleftarrows Et_2GeCl_2 + Et_2GeH_2$$

Phenylhalogermanes are usually less stable than the corresponding alkylgermane. But in contrast to dialkylfluorogermanes, diphenyl-fluorogermane is the most stable among the diphenylhalogermane series for which the stability order is (140):

$$(C_6H_5)_2(F)GeH > (C_6H_5)_2(Cl)GeH > (C_6H_5)_2(Br)GeH > (C_6H_5)_2(I)GeH$$

Thermal decomposition of $(C_6H_5)_2ClGeH$ leads to three main products:

$$(C_6H_5)_2GeH_2;\quad (C_6H_5)_2GeCl_2 \quad\text{and}\quad (C_6H_5)_3GeCl.$$

Similar thermal dismutations are recorded in the $(C_6H_5)_2(X)GeH$ diphenylhalogermane series (140). But dismutation of $Ph_2(Cl)GeH$ over $AlCl_3$ is entirely different (152)

$$4Ph_2(Cl)GeH \xrightarrow[25°]{AlCl_3} GeH_4 + Ph_2GeCl_2 + 2\,Ph_3GeCl$$

$(C_6H_5)(X)GeH_2$ monophenylmonohalogermanes are less stable and have the following stability order: $Cl > Br > I > F$. Their thermal decomposition is quite complex. After being heated for 4 h at 150°C $Ph(Cl)GeH_2$ is 50% decomposed, giving many products:

$$GeH_4, C_6H_5GeH_3, (C_6H_5)_2(Cl)GeH \text{ and } C_6H_5(Cl)_2GeH,$$

$$C_6H_5GeCl_3, (C_6H_5)_2GeCl_2 \quad (140).$$

Over $AlCl_3$, the dismutation reaction occurs immediately and is much simpler:

$$2\,(C_6H_5)(Cl)GeH_2 \xrightarrow[25°]{AlCl_3} GeH_4 + (C_6H_5)_2GeCl_2$$

$(C_6H_5)(X)_2GeH$ phenyldihalogermanes are the least stable of all the phenylhalogermanes.

Phenyldiiodo- and phenyldifluorogermane are notably unstable. After two hours heating at 100°C, phenyldichlorogermane is entirely decomposed into $H_2, C_6H_6, C_6H_5GeCl_3$, polymers and germanium metal (140).

4-2-3 Reactivity of organohalogermanium hydrides

Alkyl- and phenylhalogermanes display great reactivity towards unsaturated carbon–carbon bonds. Addition reactions to olefins and alkynes usually occur in high yields and without the use of a catalyst. Their reactivity is often similar to the reactivity of X_3GeH trihalogenogermanes (X = Cl, Br), and in any case greater than that of R_3GeH trialkylgermanes which require platinum catalysts (H_2PtCl_6 or Pt-asbestos) for several of these addition reactions (142).

Addition to 1-alkenes always proceeds in an antiMarkovnikov fashion: the germanium group is linked to the terminal carbon of the double bond and produces linear derivatives (99, 101, 105, 140):

$$R_2(X)GeH + H_2C{=}CH{-}R' \rightarrow R_2(X)GeCH_2CH_2R'$$

(R = alkyl or phenyl)

Inhibition of these reactions by hydroquinone or "galvinoxyl" and their notable acceleration under action of radical reaction initiators such as azobis isobutyronitrile or UV radiation suggest that the addition mechanism is of a radical type:

$$R_2(X)GeH \xrightarrow[\text{or UV}]{\text{AIBN}} R_2(X)Ge^{\cdot} + H^{\cdot}$$

$$R_2(X)Ge^{\cdot} + \overset{\diagdown}{\underset{\diagup}{C}}{=}\overset{\diagup}{\underset{\diagdown}{C}} \rightarrow R_2(X)Ge{-}\overset{|}{\underset{|}{C}}{-}\overset{\diagup}{\underset{\diagdown}{C^{\cdot}}}$$

$$R_2(X)Ge{-}\overset{|}{\underset{|}{C}}{-}\overset{\diagup}{\underset{\diagdown}{C^{\cdot}}} + R_2(X)GeH \rightarrow R_2(X)Ge{-}\overset{|}{\underset{|}{C}}{-}\overset{\diagup}{\underset{\diagdown}{CH}} + R_2(X)Ge^{\cdot}$$

The alkylhalogermanes also add very easily in a simple thermal reaction to several functionally substituted alkenes. The germanium group is taken up by the terminal carbon of the double bond (99–101, 105):

$$R_2(X)GeH + CH_2{=}CH{-}A \rightarrow R_2(X)GeCH_2CH_2A$$

$$A = -CH_2Cl, -OC_4H_9, -CH_2OH, -CN, -CH_2CN, -COOCH_3,$$

$$-COOC_2H_5, -CO{-}CH_3, -CH(OC_2H_5)_2, -O{-}COCH_3,$$

$$-CH_2{-}O{-}COCH_3$$

Similar reactions are recorded with branched olefins such as methyl methacrylate, methacroleine, methallyl chloride, isopropenyl acetate and also with cyclohexene.

Several syntheses were carried out with these adducts and led to new organogermanium derivatives with substituted functions (99–101, 105).

$$R_2(X)GeCH_2CH_2CH_2Cl \xrightarrow{RMgX} R_3GeCH_2CH_2CH_2Cl \xrightarrow{NH_3}$$

$$R_3GeCH_2CH_2CH_2NH_2$$

$$R_3GeCH_2CH_2CH_2Cl \xrightarrow{Mg}$$

$$R_3GeCH_2CH_2CH_2MgCl \underset{O_2}{\overset{C_2H_4O}{<}} \begin{matrix} R_3GeCH_2CH_2CH_2CH_2CH_2OH \\ R_3GeCH_2CH_2CH_2OH \end{matrix}$$

$$R_2(Cl)Ge(CH_2)_n-CN \xrightarrow{LiAlH_4} \underset{\underset{H}{|}}{R_2Ge}-(CH_2)_nCH_2NH_2 \qquad (n = 2, 3)$$

$$R_2(Cl)GeCH_2CH_2CH_2CN \xrightarrow{RMgBr} R_3GeCH_2CH_2CH_2\underset{\underset{O}{\|}}{C}-R$$

$$R_2(X)GeCH_2CH_2COOCH_3 \xrightarrow{LiAlH_4} R_2-\underset{\underset{H}{|}}{Ge}-CH_2CH_2CH_2OH$$

$$R_2(X)GeCH_2CH_2COOC_2H_5 \xrightarrow{RMgBr} R_3GeCH_2CH_2C(OH)R_2$$

$$R_2(X)GeCH_2-\underset{\underset{CH_3}{|}}{CH}-COOCH_3 \xrightarrow{LiAlH_4} R_2Ge-CH_2-\underset{\underset{CH_3}{|}}{CH}-CH_2OH$$
$$\phantom{R_2(X)GeCH_2-CH-COOCH_3 \xrightarrow{LiAlH_4} R_2Ge-CH_2-}\underset{H}{|}$$

$$R_2(Cl)GeCH_2CH_2-O-COCH_3 \xrightarrow{RMgBr} R_3GeCH_2CH_2OH$$

Germanium–hydrogen bond reactivity is always reported to rise according to the following order in monohydride series:

$$R_3GeH < R_2(X)GeH < R(X)_2GeH$$

It is the hydride character of the hydrogen atom in R_3GeH (R = alkyl) that favours addition and polar substitution mechanisms (142). This character decreases as the alkyl groups are progressively replaced by halogens (105):

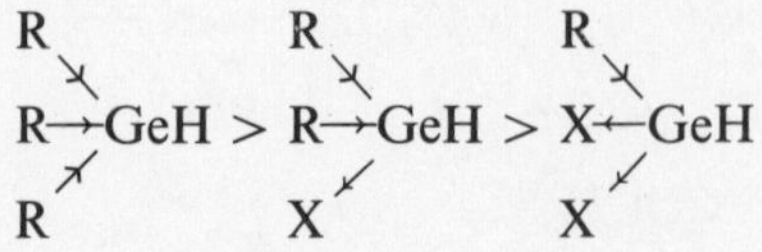

$$\underset{R}{\overset{R}{R{\to}GeH}} > \underset{X}{\overset{R}{R{\to}GeH}} > \underset{X}{\overset{R}{X{\leftarrow}GeH}}$$

Especially with $R(X)_2GeH$ the decrease of the Ge—H bond polarity and energy promotes its homolytic rupture $\geqslant Ge^{\cdot} + H^{\cdot}$. Comparative reactions of R_3GeH, $R_2(Cl)GeH$, $R(Cl)_2GeH$ monohydrides with allyl chloride illustrate this polarity and reactivity:

Σ_3GeH	REACTION TEMP. ($^\circ$C)	RELATIVE% I (ADDITION)	II (REDUCTION)	REFERENCES
Et_3GeH	80	0	100	105–142
$Et_2(Cl)GeH$	150	90	10	105
$Et(Cl)_2GeH$	25	100	0	105

I addition: $\Sigma_3GeH + CH_2{=}CHCH_2Cl \rightarrow \Sigma_3GeCH_2CH_2CH_2Cl$

II reduction: $\Sigma_3GeH + CH_2{=}CHCH_2Cl \rightarrow \Sigma_3GeCl +$

$$CH_3CH{=}CH_2$$

The existence of a low residual negative electric charge on the hydrogen atom in $R_2(X)GeH$ may explain the slight percentage of reduction next to the predominant addition with $Et_2(Cl)GeH$. This addition can be compared with the reduction by Et_3GeH. The single mode of addition of $Et(Cl)_2GeH$ indicates that the mechanism is entirely radical in nature.

An addition of alkylhalogermanes to acetylene leads to a saturated digermane after a second hydride molecule is added to the intermediate vinylic derivative (99, 105):

$$R_2(X)GeH + HC{\equiv}CH \rightarrow R_2(X)GeCH{=}CH_2$$

$$+ R_2(X)GeH \rightarrow R_2(X)GeCH_2CH_2Ge(X)R_2$$

With 1-hexyne and phenylacetylene addition will lead to linear adducts (*trans* and *cis* forms)

$$R_2(X)GeH + HC{\equiv}C{-}R' \rightarrow R_2(X)GeCH{=}CH{-}R'$$

Such derivatives, on being treated with an acetylenic Grignard reagent, lead to germanium enynes (100, 105):

$$R_2(X)GeCH{=}CH{-}R' + R''{-}C{\equiv}C{-}MgX \rightarrow R_2Ge\diagup^{CH=CH-R'}_{\diagdown C{\equiv}C-R''}$$

Next to the $Et_2(Cl)GeCH{=}CHCH_2Cl$ (*cis* and *trans*) isomeric adducts, addition of diethylchlorogermane to propargyl chloride also reveals a slight percentage of a $R_2(Cl)Ge-\underset{\underset{CH_2}{\|}}{C}-CH_2Cl$ branched adduct and a

high percentage of $Et_2(Cl)GeCH=C(CH_3)Cl$ *cis* and *trans* isomers and $Et_2(Cl)GeCH_2-C(Cl)=CH_2$. The formation of two latter addition derivatives besides the normal adduct can be explained by the following scheme (108):

$$\Sigma_3Ge^{\cdot} + CH\equiv C-CH_2Cl \rightarrow \Sigma_3GeCH=\overset{\cdot}{C}-CH_2Cl$$

$$\Sigma_3GeCH=\overset{\cdot}{C}CH_2Cl + \Sigma_3GeH \rightarrow \Sigma_3GeCH=CH-CH_2Cl + \Sigma_3Ge^{\cdot}$$

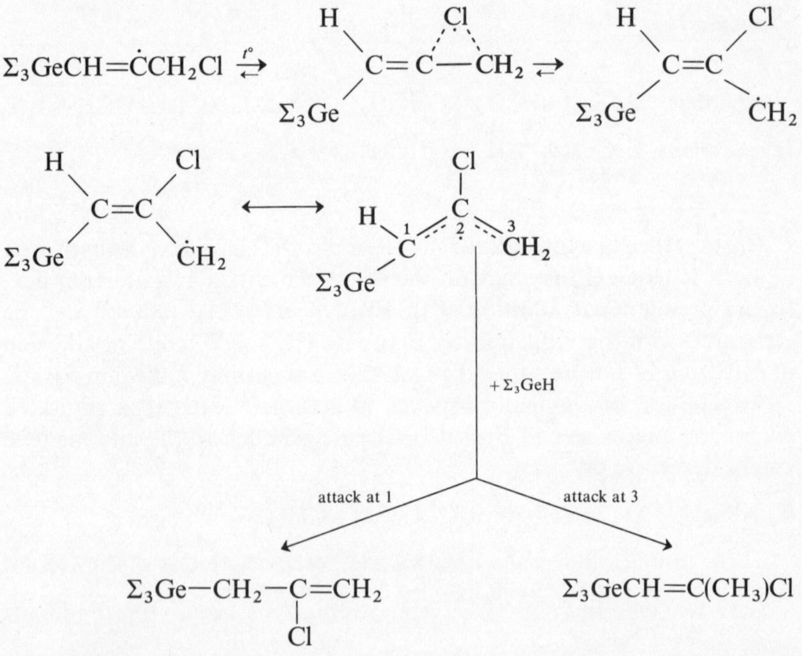

$$\Sigma_3GeCH=\overset{\cdot}{C}CH_2Cl \underset{}{\overset{r^{\circ}}{\rightleftarrows}}
\begin{array}{c}
\underset{\Sigma_3Ge}{\overset{H}{\diagdown}}C=\underset{}{\overset{Cl}{C}}-CH_2 \rightleftarrows
\end{array}
\begin{array}{c}
\underset{\Sigma_3Ge}{\overset{H}{\diagdown}}C=C\underset{\overset{\cdot}{C}H_2}{\overset{Cl}{\diagup}}
\end{array}$$

$$\begin{array}{c}
\underset{\Sigma_3Ge}{\overset{H}{\diagdown}}C=C\underset{\overset{\cdot}{C}H_2}{\overset{Cl}{\diagup}} \longleftrightarrow
\underset{\Sigma_3Ge}{\overset{H}{\diagdown}}\overset{\overset{Cl}{|}}{\underset{1}{C}}\overset{2}{C}\overset{3}{\cdots}CH_2
\end{array}$$

$+\Sigma_3GeH$

attack at 1   attack at 3

$$\Sigma_3Ge-CH_2-\underset{\underset{Cl}{|}}{C}=CH_2 \qquad\qquad \Sigma_3GeCH=C(CH_3)Cl$$

The ratio of reduction of propargyl chloride to propyne, which is 25% in this case, rises to 55% when the same reaction is carried out in a polar solvent such as acetonitrile (105, 107, 108). On the other hand, $Et(Cl)_2GeH$, ethyldichlorogermane produces only the linear adduct.

$$(C_2H_5)(Cl)_2GeH + CH\equiv C-CH_2Cl \rightarrow$$

$$(C_2H_5)(Cl)_2GeCH=CH-CH_2Cl \text{ (cis and trans)}$$

There is also lower percentage (10%) of the $Et(Cl)_2GeCH_2C\equiv CH$ condensation derivative which is identified by its di- and trigermanium addition derivatives:

$Et(Cl)_2GeCH_2CH=CH-Ge(Cl)_2Et$ and

$$Et(Cl)_2GeCH_2CH_2CH[Ge(Cl)_2Et]_2$$

This side reaction is amplified in acetonitrile (105, 108):

$\Sigma_3GeH + CH\equiv CCH_2Cl$ (WITHOUT SOLVENT)	t ($^\circ$C)	I	II (RELATIVE%)	III
$Et_3GeH\ (H_2PtCl_6)$	25	65	35	0
$Et_2(Cl)GeH$	150	75	25	0
$Et(Cl)_2GeH$	25	90	0	10

$\Sigma_3GeH + CH\equiv CCH_2Cl$ (SOLVENT CH_3CN)	t ($^\circ$C)	I	II (RELATIVE%)	III
$Et_3GeH\ (H_2PtCl_6)$	80	20	80	0
$Et_2(Cl)GeH$	80	45	55	0
$Et(Cl)_2GeH$	80	75	0	25

I addition $\Sigma_3GeH + HC\equiv C-CH_2Cl \rightarrow \Sigma_3GeCH=CH-CH_2Cl$ + secondary adducts
II reduction $\Sigma_3GeH + HC\equiv C-CH_2Cl \rightarrow \Sigma_3GeCl + CH\equiv CH-CH_3$
III condensation $\Sigma_3GeH + HC\equiv CCH_2Cl \rightarrow \Sigma_3GeCH_2C\equiv CH + HCl$

These results confirm the polarity and reactivity trends of the Ge—H bond in the R_3GeH, $R_2(Cl)GeH$, $R(Cl)_2GeH$ hydride series.

To the simultaneous radical and ionic mechanisms noted with $Et_2(Cl)GeH$ is opposed an essentially radical addition mechanism with $Et(Cl)_2GeH$. However, in this case there is a slight percentage of condensation–elimination.

The progressive decrease in the reducing character of the Ge—H hydrogen and the appearance of a slight condensation–elimination with $Et(Cl)_2GeH$ suggest that the GeH polarity inversion reported in the $\overset{\delta+}{Et_3}\overset{\delta-}{GeH}$, $Et_2(Cl)GeH$, $Et(Cl)_2GeH$, $\overset{\delta-}{Cl_3}\overset{\delta+}{GeH}$ series may occur at the $Et(Cl)_2GeH$ stage (105).

Electronegativity calculations for the Σ_3Ge groups with the use of a bond-by-bond iteration method for R_3GeH, $R_2(Cl)GeH$, $R(Cl_2)GeH$ and Cl_3GeH, as well as the polarity inversion observed in these series for the

Ge—H bond, suggest for the germanium atom the following electro-negativities (91):

$$\chi^\circ_{Ge} = 2.04\text{--}2.05 \quad \text{when} \quad \gamma^\circ_H = 2.10$$

$$\chi^\circ_{Ge} = 2.08\text{--}2.09 \quad \text{when} \quad \gamma^\circ_H = 2.13$$

Alkyhalogermane additions upon the conjugated structures of α-ethylenic aldehydes, ketones, esters and nitriles occurs in the 1,2-carbon–carbon double bond position. It is of the radical type and produces only linear adducts (101, 105, 148). These additions, without any catalyst, produce excellent yields. R_3GeH and R_2GeH_2 alkylgermanes usually display a lesser reactivity in similar reactions (142).

In contrast, addition to conjugated dienes such as 1,3-butadiene and 2,3-dimethyl-1,3-butadiene occurs in a 1,4 manner (105, 147, 148):

$$R_2(X)GeH + CH_2=CH-CH=CH_2 \rightarrow$$

$$R_2(X)GeCH_2CH=CH-CH_3 \; (cis \text{ and } trans)$$

$$(R = Et, X = Cl, Br, I)$$

$$R(X)_2GeH + CH_2=\underset{CH_3}{\overset{}{C}}-\underset{CH_3}{\overset{}{C}}=CH_2 \rightarrow R(X)_2GeCH_2-\underset{CH_3}{\overset{|}{C}}=\underset{CH_3}{\overset{|}{C}}-CH_3$$

$$(R = Bu, X = Cl)$$

An addition of the same type is reported with triethylgermane and 1,3-butadiene in the presence of chloroplatinic acid (148):

$$Et_3GeH + CH_2=CH-CH=CH_2 \xrightarrow{H_2PtCl_6}$$

$$Et_3GeCH_2CH=CH-CH_3 \xleftarrow{EtMgX} Et_2\underset{X}{\overset{|}{Ge}}-CH_2CH=CH-CH_3$$

(cis and trans)

Et_3GeH, Et_2ClGeH, $EtCl_2GeH$ and Cl_3GeH were condensed on iso-propenylacetylene. In the presence of Pt (from reduction of H_2PtCl_6) addition of Et_3GeH and $Et_2(Cl)GeH$ is oriented exclusively towards the formation of germanium 1,3-dienes:

$$R_3GeCH=CH-C(CH_3)=CH_2$$

Only ethylchlorogermanes react on radical catalysis (azobisisobutyroni-trile) yielding germanium dienes, alkynes and allenes. The reaction mechanisms and the structure of the isomers, and their percentages relative to the type of Ge-bonded substituents are discussed (106b).

$(C_6H_5)_2(Cl)GeH$ and $C_6H_5(Cl)_2GeH$ phenylchlorogermanes are con-densed with the carbonyl group of saturated ketones and aldehydes such as cyclohexanone, methylisobutylketone, isobutyraldehyde, benzalde-hyde. This requires radical initiators which may be UV light or azobis-isobutyronitrile (140):

$$\Sigma_3GeH \xrightarrow{UV} \Sigma_3Ge^{\cdot} + H^{\cdot}$$

$$\Sigma_3Ge^{\cdot} + O{=}C\overset{\diagup}{\diagdown} \rightarrow \Sigma_3Ge{-}O{-}C^{\cdot}\overset{\diagup}{\diagdown}$$

$$\Sigma_3Ge{-}O{-}C^{\cdot}\overset{\diagup}{\diagdown} + \Sigma_3GeH \rightarrow \Sigma_3Ge{-}O{-}\overset{\diagup}{C}H + \Sigma_3Ge^{\cdot}$$

However with the addition of phenyldichlorogermane a notable quantity of alcohol is produced by reduction, i.e. by hydrogenolysis of the Ge–O bond of the adduct:

$$(C_6H_5)(Cl)_2Ge{-}O{-}\overset{\diagup}{C}H + (C_6H_5)(Cl)_2GeH \rightarrow$$

$$(C_6H_5)(Cl)_2Ge{-}Ge(Cl)_2(C_6H_5) + HO\overset{\diagup}{C}H$$

Hydrogenolysis of the Ge–O bond has not been reported either with diphenylchlorogermane or triphenylgermane.

The electrophilic activity of the Ge–H hydrogen seems much greater with $C_6H_5(Cl)_2GeH$. It may be explained by the cumulative attraction effects of the phenyl group and two chlorine atoms bound to the ger-manium.

$Ph_2(Cl)GeH$ and $Ph(Cl)_2GeH$ phenylchlorogermanes preferentially add to the double bonds of α- or γ-olefinic ketones such as vinyl methyl ketone and allylacetone, forming linear adducts (151, 153):

$$(C_6H_5)_2(Cl)GeH + CH_2{=}CHCOCH_3 \xrightarrow{UV}$$

$$(C_6H_5)_2(Cl)GeCH_2CH_2COCH_3$$

$$(C_6H_5)(Cl)_2GeH + CH_2{=}CHCH_2CH_2COCH_3 \xrightarrow{AIBN}$$

$$(C_6H_5)(Cl)_2Ge(CH_2)_4COCH_3$$

But with 2-methyl-2-heptene-6-one, steric hindrance at the double bond directs the germanium group to the 3-carbon (151, 153):

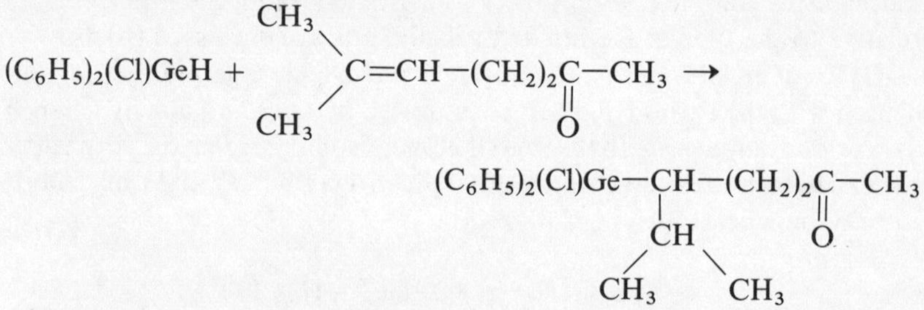

$$(C_6H_5)_2(Cl)GeH + \underset{CH_3}{\overset{CH_3}{>}}C{=}CH{-}(CH_2)_2\underset{\underset{O}{\|}}{C}{-}CH_3 \rightarrow$$

$$(C_6H_5)_2(Cl)Ge{-}\underset{\underset{\underset{CH_3\ \ \ CH_3}{/\ \ \backslash}}{CH}}{CH}{-}(CH_2)_2\underset{\underset{O}{\|}}{C}{-}CH_3$$

Triphenylgermane with larger germanium substituent groups adds to the carbonyl group of this ketone.

Without any catalyst, phenyldichlorogermane adds to the mesityl oxide olefinic C=C bond. But NMR data on the successive steps of the reaction proves that this is a 1,4 addition by way of a transient enolic form (151, 153):

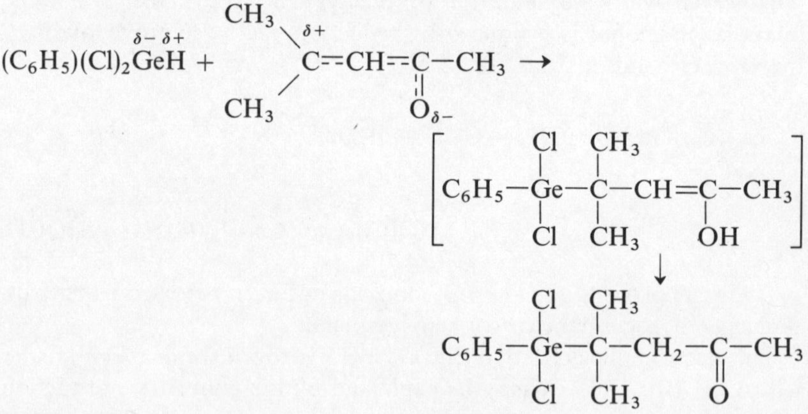

$$(C_6H_5)(Cl)_2\overset{\delta-\ \delta+}{GeH} + \underset{CH_3}{\overset{CH_3}{>}}\overset{\delta+}{C}{=}CH{=}\underset{\underset{O_{\delta-}}{\|}}{C}{-}CH_3 \rightarrow$$

$$\left[C_6H_5{-}\underset{\underset{Cl}{|}}{\overset{\overset{Cl}{|}}{Ge}}{-}\underset{\underset{CH_3}{|}}{\overset{\overset{CH_3}{|}}{C}}{-}CH{=}\underset{\underset{OH}{|}}{C}{-}CH_3 \right]$$

$$\downarrow$$

$$C_6H_5{-}\underset{\underset{Cl}{|}}{\overset{\overset{Cl}{|}}{Ge}}{-}\underset{\underset{CH_3}{|}}{\overset{\overset{CH_3}{|}}{C}}{-}CH_2{-}\underset{\underset{O}{\|}}{C}{-}CH_3$$

Cl_3GeH, trichlorogermane, reacts in the same way with mesityl oxide which would confirm the marked $\overset{\delta-\ \ \ \delta+}{Ge{-}H}$ polarity in phenyldichlorogermane. We may recall that use of triphenylgermane leads to a reversed 1,4 radical addition forming *cis* and *trans* $Ph_3Ge{-}O{-}\underset{\underset{CH_3}{|}}{C}{=}CCH{+}(CH_3)_2$

enoxygermanes. Steric hindrance presumably orients the triphenylgermyl radical towards the oxygen (151, 153). Action of hydrides such as $(C_6H_5)_2GeH_2$ diphenylgermane and $(C_6H_5)(Cl)GeH_2$ phenylchlorogermane with vinyl methyl ketone and allylacetone, apart from to the

normal mono- and di-addition derivatives also leads to polymers and
to ring derivatives with a Ge—O—C bond (151, 153):

$$C_6H_5(Cl)GeH_2 + CH_2{=}CH(CH_2)_nCOCH_3 \rightarrow$$

$(n = 0 \text{ or } 2)$

$$\begin{cases} C_6H_5(Cl)\underset{\underset{H}{|}}{Ge}CH_2CH_2(CH_2)_nCOCH_3 \quad\quad (n = 0 \text{ or } 2) \\[1em] C_6H_5(Cl)Ge[CH_2CH_2(CH_2)_nCOCH_3]_2 \\ C_6H_5(Cl)Ge{-}CH_2CH_2(CH_2)_n\underset{\diagdown\quad\quad\diagup}{C}H{-}CH_3 \\ \quad\quad\quad\quad\quad\quad\quad\quad\quad O \\ \text{polymers} \end{cases}$$

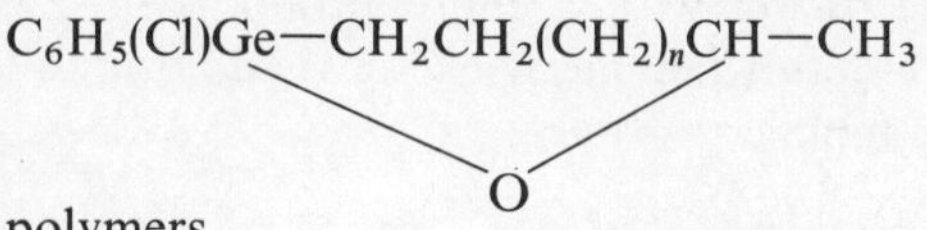

These heterocyclic derivatives result from intramolecular addition of
Ge—H to the carbonyl group in the monoaddition derivative. Monoaddition derivatives previously isolated and heated on AIBN do lead to
these heterocycles with average yields of 70 % (151, 153):

$$C_6H_5(Cl)\underset{\underset{H}{|}}{Ge}CH_2CH_2(CH_2)_nCOCH_3 \xrightarrow[70°C]{AIBN}$$

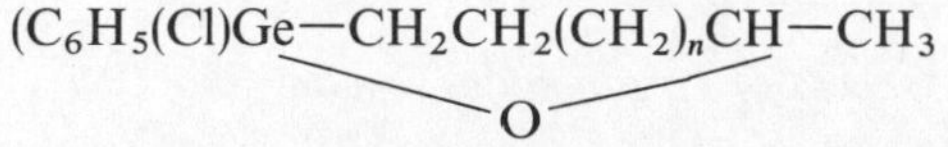

$$(C_6H_5(Cl)Ge\underset{\diagdown\quad\quad\quad\quad\quad\diagup}{-}CH_2CH_2(CH_2)_nCH{-}CH_3 \quad\quad (n = 0 \text{ or } 2)$$
$$\quad\quad\quad\quad\quad\quad\quad\quad O$$

The ionic addition of phenyldichlorogermane to methyl ethynyl
ketone $CH{\equiv}C{-}COCH_3$ leads to mono- and di-addition derivatives
$PhCl_2GeCH{=}CHCOCH_3$ (*cis* and *trans*) and

$$(PhCl_2Ge)_2CH{-}CH_2COCH_3 \quad (151).$$

The trend of the condensation reactions of the organogermanium
hydrides with diphenylketene is intimately linked with the polarity of
the Ge—H bond of these hydrides and the steric effects (141, 151). Triethylgermane with $\overset{\delta+\quad\delta-}{Ge{-}H}$ polarity and triphenylgermane, mainly for
reasons of hindrance, add to the carbonyl group of diphenylketene with
formation of vinyloxygermanes:

$$\Sigma_3GeH + Ph_2C{=}C{=}O \rightarrow \Sigma_3GeO{-}CH{=}CPh_2$$

Phenyldichlorogermane with $\overset{\delta-\quad\delta+}{Ge{-}H}$ polarity also adds on the carbonyl group but the germyl group is fixed on the central positive carbon

of the diphenylketene with formation of an enol which rearranges thermally in α-germylketone:

$$C_6H_5(Cl)_2GeH + Ph_2C=C=O \rightarrow C_6H_5(Cl)_2Ge-\underset{OH}{\overset{|}{C}}=CPh_2$$

$$\downarrow 90°$$

$$C_6H_5(Cl)_2Ge-\underset{\underset{O}{\parallel}}{C}-CHPh_2$$

Trichlorogermane with a more pronounced $\overset{\delta-}{Ge}-\overset{\delta+}{H}$ polarity gives directly the α-germylketone by classical addition to the ketenes of compounds with mobile hydrogen.

$$Cl_3GeH + Ph_2C=C=O \rightarrow Cl_3Ge-\underset{\underset{O}{\parallel}}{C}-CHPh_2$$

The reduction of these α-germylketones by $LiAlH_4$ leads to the α-germanium secondary alcohols with $Ge-H_2$ or $Ge-H_3$ groups:

$$R(Cl)_2Ge-\underset{\underset{O}{\parallel}}{C}-CHPh_2 \xrightarrow{\text{LiAlH}_4} R(H)_2GeCHOH-CHPh_2$$

These α-germanium alcohols are not very stable. Their thermal decomposition is observed in the range of 120–140°C.

4-2-3-1   Other Aspects of Organohalogermane Reactivity. The action of sodium hydroxide on $(C_4H_9)(Cl)GeH_2$ (10) or of silver carbonate on $(CH_3)(Br)GeH_2$ (64) or $(C_4H_9)_2(I)GeH$ (142), leads, with simultaneous hydrolysis of the Ge—X and Ge—H bonds, to the corresponding alkylgermanium oxides. On the other hand, ammoniacal hydrolysis of alkyl- and phenylhalogermanes leads to the expected organohydrogermanium oxides (105, 106, 140, 150):

$$[R_2(H)Ge]_2O, [R(H)_2Ge]_2O \text{ and } [R(H)GeO]_n \qquad R = Et, Bu, phenyl$$

In the same way ammonolysis of alkylchlorogermanes gives alkylhydrogermylamines. The unstable $R_nH_{3-n}GeNH_2$ ($n = 1, 2$) alkylhydrogermylamines first produced condense to the secondary or tertiary alkylhydrogermylamines (106):

$$Et_2(Cl)GeH \xrightarrow{NH_3} Et_2(H)Ge-\underset{H}{\overset{|}{N}}-Ge(H)Et_2 + [Et_2(H)Ge]_3N$$

$[Et(H)_2Ge]_3N$ tris(ethyldihydrogermyl)amine is the only one isolated after ammonolyzing $Et(H)_2GeCl$. Action of diethylchlorogermane on lithium nitride also leads to tris(diethylhydrogermyl)amine (106):

$$3 Et_2(H)GeCl + Li_3N \longrightarrow [Et_2(H)Ge]_3N + 3 LiCl$$

Owing to its instability, it has not been possible to isolate methyl-methoxygermane after reacting sodium methylate with methylbromo-germane (64):

$$CH_3(Br)GeH_2 + CH_3ONa \longrightarrow CH_3(CH_3O)GeH_2 \longrightarrow$$

$$CH_3OH + (CH_3GeH)_n$$

By the same method from dialkylchlorogermanes the dialkyl(methoxy)-germanes can be isolated (106a).

On the other hand, action of $Pb(SEt)_2$ on diethylchlorogermane leads to thiogermane containing both a Ge—H and a Ge—S bond (145):

$$2 Et_2-\underset{\underset{H}{|}}{Ge}-Cl + Pb(SEt)_2 \xrightarrow{C_6H_6} 2 Et_2\underset{\underset{H}{|}}{Ge}-SEt + PbCl_2$$

Dialkylchlorogermanes react with mercuridiacetic esters. (Dialkyl-chlorogermyl) acetic esters are formed (12):

$$R_2Ge\overset{\diagup Cl}{\diagdown H} + Hg(CH_2COOCH_3)_2 \longrightarrow R_2Ge\overset{\diagup Cl}{\diagdown CH_2COOCH_3}$$

$$+ Hg + CH_3COOCH_3$$

Dialkyliodogermanes also react with mercuridiacetic esters, but in this case condensation occurs on the germanium–iodine bond:

$$2 R_2Ge\overset{\diagup I}{\diagdown H} + Hg(CH_2COOCH_3)_2 \longrightarrow 2 R_2\underset{\underset{H}{|}}{Ge}-CH_2COOCH_3 + HgI_2$$

Alkylation of $R_2(X)GeH$ and $R(X)_2GeH$ organohalogermanes by Grignard reagents leads to mixed triorganogermanes: $R_2R'GeH$ and RR'_2GeH (122, 150):

$$R(X)_2GeH \xrightarrow{R'MgBr} RR'_2GeH$$

4-2-4   Addition reactions of organohalogermanium hydrides

Table 4-3 (a–c) showing some addition reactions of organohalogermanium hydrides appears on the following pages.

Table 4-3

Addition Reactions of Organohalogermanium Hydrides

(a) To ethylenic compounds

UNSATURATED COMPOUND	GERMANE	CATALYST	REACTION TEMP. (°C)	TIME (h)	REACTION PRODUCTS	B.P. OR (M.P.) (°C/mm)	YIELD (%)	REFERENCES
$CH_2=CH_2$	$CH_3(Cl)_2GeH$	none	Reflux	6	$(C_2H_5)(CH_3)GeCl_2$	149/750	90	122
$CH_2=CH-CN$	$(C_4H_9)_2(Cl)GeH$	none	100	15	$(C_4H_9)_2(Cl)GeCH_2CH_2CN$	168.5/10	70	99, 148
$CH_2=CHCH_2Cl$	$(C_2H_5)_2(Cl)GeH$	none	140	15	$(C_2H_5)_2(Cl)GeCH_2CH_2CH_2Cl$	109/10	90	99, 105
	$(C_4H_9)_2(Cl)GeH$	none	150	15	$(C_4H_9)_2(Cl)GeCH_2CH_2CH_2Cl$	148/8	90	99, 105
$CH_2=CH-CH_2Cl$	$(C_2H_5)(Cl)_2GeH$	none	100	2	$(C_2H_5)(Cl)_2GeCH_2CH_2CH_2Cl$	112.5/10	95	105
$CH_2=CH-CH_2OH$	$(C_4H_9)_2(Cl)GeH$	none	150	20	$(C_4H_9)_2(Cl)GeCH_2CH_2CH_2OH$	116/0.5	65	99, 105
$CH_2=CH-COOH$	$(C_2H_5)_2(Cl)GeH$	none	90	20	$(C_2H_5)_2(Cl)GeCH_2CH_2COOH$	132/0.5	45	105
$CH_2=CH-CH=CH_2$	$(C_2H_5)_2(Cl)GeH$	none	110	15	$(C_2H_5)_2(Cl)GeCH_2CH=CHCH_3$ 90% cis- 10% trans	99–100/22	31	105, 147, 148
	$(C_2H_5)_2(Br)GeH$	none	110	15	$(C_2H_5)_2(Br)GeCH_2CH=CHCH_3$ 90% cis- 10% trans	112/22	49	105, 147, 148
	$(C_2H_5)_2(I)GeH$	none	110	15	$(C_2H_5)_2(I)GeCH_2CH=CHCH_3$ 90% cis- 10% trans	126/22	83	105, 147, 148
$CH_2=CH-CH_2CN$	$(C_2H_5)_2(Cl)GeH$	none	150	15	$(C_2H_5)_2(Cl)GeCH_2CH_2CH_2CN$	156/13	79	100, 105
$CH_2=C-CH_2Cl$ (CH_3)	$(C_4H_9)_2(Cl)GeH$	none	150	15	$(C_4H_9)_2(Cl)GeCH_2-CH-CH_2Cl$ (CH_3)	161/10	78	100, 105
$CH_2=CH-COCH_3$	$(C_2H_5)_2(Cl)GeH$	none	reflux 100°	1	$(C_2H_5)_2(Cl)GeCH_2CH_2COCH_3$	137/18	70	105, 148
$CH_2=CH-COCH_3$	$(C_2H_5)_2(Cl)GeH$	H_2PtCl_6	60	3	$(C_2H_5)_2(Cl)GeCH_2CH_2CH_2COCH_3$ (75%) $(C_2H_5)_2(Cl)GeO-C(CH_3)=CHCH_3$ (25%)	—	—	105
$CH_2=CH-COCH_3$	$(C_2H_5)_2(Cl)GeH$	UV	30	4	$(C_2H_5)_2(Cl)GeCH_2CH_2COCH_3$	$140/3.10^{-3}$	66	151, 153
$CH_2=CH-COCH_3$	$(C_6H_5)_2(Cl)_2GeH$	none	exothermic	1½	$(C_6H_5)_2(Cl)GeCH_2CH_2COCH_3$	$108/10^{-2}$	70	151, 153
$CH_2=CH-COCH_3$	$(C_6H_5)(Cl)GeH_2$	AIBN	instantaneous exothermic		$(C_6H_5)(Cl)Ge(CH_2)_2CHCH_3$ (35%) (with O ring) $(C_6H_5)(Cl)Ge[(CH_2)_2COCH_3]_2$ (60%) Polymers (5%)	$90/4.10^{-2}$ (108)	—	151

Olefin	Germane	Catalyst	Temp. (°C)	Time (hr)	Product	b.p. (°C/mm)	Yield (%)	Ref.
$CH_2{=}C(CH_3)CHO$	$(C_4H_9)_2(Cl)GeH$	H_2PtCl_6 hydroquinone	140	20	$(C_4H_9)_2(Cl)GeCH_2CH(CH_3)CHO$	151/13	40	101, 105
$CH_2{=}CH{-}COOCH_3$	$(C_4H_9)_2(Cl)GeH$	none	150	15	$(C_4H_9)_2(Cl)GeCH_2CH_2COOCH_3$	164/12	65	101, 105, 148
$H_2C{=}CH{-}COOCH_3$	$(C_2H_5)_2(Cl)GeH$	none	exothermic	1/6	$(C_2H_5)_2(Cl)GeCH_2CH_2COOCH_3$	132/15	~100	101, 105
$H_2C{=}CH{-}COOC_2H_5$	$(C_2H_5)_2(Cl)GeH$	none	140	15	$(C_2H_5)_2(Cl)GeCH_2CH_2COOC_2H_5$	145/20	68	101, 105, 148
$H_2C{=}C(CH_3){-}COOCH_3$	$(C_4H_9)_2(Cl)GeH$	none	150	15	$(C_4H_9)_2(Cl)GeCH_2{-}CH(CH_3)COOCH_3$	173/18	65	101, 105
$H_2C{=}C(CH_3){-}OCOCH_3$	$(C_2H_5)_2(Cl)GeH$	none	exothermic, 80	1	$(C_2H_5)_2(Cl)GeCH_2CH(CH_3)OCOCH_3$	130/18	85	101, 105
$H_2C{=}C(CH_3){-}C(CH_3){=}CH_2$	$(C_2H_5)_2(Cl)GeH$	none	160	18	$(C_2H_5)_2(Cl)GeCH_2{-}C(CH_3){=}C(CH_3){-}CH_3$	111/10	52	105, 147, 148
$H_2C{=}C(CH_3){-}C(CH_3){=}CH_2$	$(C_4H_9)_2(Cl)GeH$	none	160	18	$(C_4H_9)_2(Cl)GeCH_2{-}C(CH_3){=}C(CH_3){-}CH_3$	159/13	41	105, 147, 148
$H_2C{=}C(CH_3){-}C(CH_3){=}CH_2$	$(C_4H_9)(Cl)_2GeH$	none	160	18	$(C_4H_9)(Cl)_2GeCH_2{-}C(CH_3){=}C(CH_3){-}CH_3$	141/17	77	105, 147, 148
$CH_2{=}CH(CH_2)_3CH_3$	$(C_6H_5)_2(F)GeH$	AIBN	70	1/2	$(C_6H_5)_2(F)Ge(CH_2)_5CH_3$	124/0.2	92	140
$CH_2{=}CH(CH_2)_3CH_3$	$(C_6H_5)_2(Cl)GeH$	none	exothermic	3/4	$(C_6H_5)_2(Cl)Ge(CH_2)_5CH_3$	148/0.1	81	140
$CH_2{=}CH(CH_2)_3CH_3$	$(C_6H_5)_2(Br)GeH$	none	exothermic	1	$(C_6H_5)_2(Br)Ge(CH_2)_5CH_3$	$128/7 \cdot 10^{-2}$	91	140
$CH_2{=}CH(CH_2)_3CH_3$	$(C_6H_5)(I)GeH$	AIBN	70	1/2	$(C_6H_5)(I)Ge(CH_2)_5CH_3$	$138/5 \cdot 10^{-2}$	85	140
$CH_2{=}CH(CH_2)_3CH_3$	$(C_6H_5)(F)_2GeH$	none	45	instantaneous	$(C_6H_5)(F)_2Ge(CH_2)_5CH_3$	86/0.2	81	140
$CH_2{=}CH(CH_2)_3CH_3$	$(C_6H_5)(Cl)_2GeH$	none	exothermic	instantaneous	$(C_6H_5)(Cl)_2Ge(CH_2)_5CH_3$	94/0.1	84	140
$CH_2{=}CH(CH_2)_3CH_3$	$(C_6H_5)(Br)_2GeH$	none	exothermic	instantaneous	$(C_6H_5)(Br)_2Ge(CH_2)_5CH_3$	$105/2 \cdot 10^{-2}$	89	140
$CH_2{=}CH(CH_2)_3CH_3$	$(C_6H_5)(I)_2GeH$	none	exothermic	instantaneous	$(C_6H_5)(I)_2Ge(CH_2)_5CH_3$	$128/6 \cdot 10^{-3}$	92	140

Table 4-3(a)—continued

UNSATURATED COMPOUND	GERMANE	CATALYST	REACTION TEMP. (°C)	TIME (h)	REACTION PRODUCTS	B.P. OR (M.P.) (°C/mm)	YIELD (%)	REFERENCES
$CH_2=CH-CH_2OCOCH_3$	$(C_4H_9)_2(Cl)GeH$	none	120	10	$(C_4H_9)_2(Cl)Ge(CH_2)_3OCOCH_3$	186/18	80	101, 105
(cyclohexene)	$(C_4H_9)_2(Cl)GeH$	none	150	24	$Ge(Cl)(C_4H_9)_2$ (cyclohexyl)	163.5/13	70	99
$C_4H_9OCH=CH_2$	$(C_4H_9)_2(Cl)GeH$	none	140	5	$(C_4H_9)_2(Cl)GeCH_2CH_2OC_4H_9$	164/10	80	99, 105
$CH_2=CH-(CH_2)_2C-CH_3$ $\overset{\parallel}{O}$	$(C_6H_5)_2(Cl)GeH$	UV	exothermic	4	$(C_6H_5)_2(Cl)Ge(CH_2)_4C-CH_3$ $\overset{\parallel}{O}$	155/5.10^{-3}	90	151, 153
$CH_2=CH-(CH_2)_2C=O$ $-CH_3$	$(C_6H_5)(Cl)_2GeH$	UV	exothermic	instantaneous	$(C_6H_5)(Cl)_2Ge(CH_2)_4C-CH_3$ $\overset{\parallel}{O}$	124/10^{-3}	95	151, 153
$CH_2=CH(CH_2)_2COCH_3$	$(C_6H_5)(Cl)GeH_2$	none	exothermic	½	$(C_6H_5)(Cl)Ge(H)[(CH_2)_2COCH_3]$	120/2.10^{-2}	70	151
					$(C_6H_5)(Cl)Ge[(CH_2)_4COCH_3]_2$	170/2.10^{-2}	15	151
$\underset{CH_3}{\overset{CH_3}{>}}C=CH-C-CH_3$ $\overset{\parallel}{O}$	$(C_6H_5)(Cl)_2GeH$	none	exothermic (45)	3	$(C_6H_5)(Cl)_2Ge-CH-CH_2-\underset{CH_3}{\overset{\parallel O}{C}}-C-CH_3$	(67)	80	151, 153
$CH_2=CH(CH_2)_5CH_3$	$(C_4H_9)_2(Cl)GeH$	none	150	15	$(C_4H_9)_2(Cl)GeCH_2(CH_2)_6CH_3$	134.5/0.6	78	99, 105
	$(C_4H_9)_2(Br)GeH$	none	150	15	$(C_4H_9)_2(Br)GeCH_2(CH_2)_6CH_3$	123.5/0.3	85	99, 105
	$(C_4H_9)_2(I)GeH$	none	150	15	$(C_4H_9)_2(I)GeCH_2(CH_2)_6CH_3$	143.5/0.5	90	99, 105
(cyclohexene)	$(C_2H_5)_2(Cl)GeH$	none	110	15	$(C_2H_5)_2(Cl)GeCH_2CH_2$ (cyclohexenyl)	162/18	44	105, 148
$\underset{CH_3}{\overset{CH_3}{>}}C=CH(CH_2)_2-\overset{\parallel}{C}-CH_3$ O	$(C_6H_5)_2(Cl)GeH$	UV	exothermic	2	$(C_6H_5)_2(Cl)Ge-CH-(CH_2)_3\overset{\parallel O}{C}-CH_3$ $\underset{CH_3}{\overset{|}{CH}}$ CH_3	148/5.10^{-3}	78	151, 153
$CH_3-\underset{CH_3}{\overset{|}{C}}=CH-CH_2-CH_2C-CH_3$ O	$(C_6H_5)(Cl)_2GeH$	none	exothermic	instantaneous	$(C_6H_5)(Cl)_2Ge-\overset{|}{CH}-CH_2CH_2CCH_3$ $\overset{\parallel}{O}$ $\underset{CH}{}$	122/10^{-2}	80	151, 153

(b) To acetylenic compounds

UNSATURATED COMPOUND	GERMANE	CATALYST	REACTION TEMP. (°C)	TIME (h)	REACTION PRODUCTS	B.P. OR (M.P.) (°c/mm)	YIELD (%)	REFERENCES
HC≡CH	$(C_2H_5)_2(Cl)GeH$	none	110	10	$(C_2H_5)_2(Cl)GeCH_2CH_2Ge(Cl)(C_2H_5)_2$	163/10	52	99, 105
HC≡C—CH₂Cl	$(C_2H_5)_2(Cl)GeH$	none	150	15	$(C_2H_5)_2(Cl)GeCH=CHCH_2Cl$ 30% *cis*- 15% *trans*			
					$(C_2H_5)_2(Cl)GeC-CH_2Cl$ (10%) [‖CH₂]			
					$(C_2H_5)_2(Cl)GeCH_2-C=CH_2$ [Cl] ; $(C_2H_5)_2(Cl)Ge-CH=C-CH_3$ [Cl] (45%)	108–13/13	47	105, 107, 108
HC≡C—CH₂Cl	$(C_2H_5)_2(Cl)GeH$	AIBN or UV	80	15	$(C_2H_5)_2GeCl_2$		16	
					$(C_2H_5)_2(Cl)GeCH=CHCH_2Cl$ 30% *cis*- 20% *trans*			
					$(C_2H_5)_2(Cl)GeC-CH_2Cl$ (10%) [‖CH₂]			
					$(C_2H_5)_2(Cl)GeCH_2-C=CH_2$ [Cl] ; $(C_2H_5)_2(Cl)GeCH=C-CH_3$ [Cl] (40%)	—	72	105, 107, 108
HC≡CCH₂Cl	$(C_2H_5)(Cl)_2GeH$	none	exothermic	1–1¼	$(C_2H_5)_2GeCl_2$; $(C_2H_5)(Cl)_2GeCH=CHCH_2Cl$ 60% *cis*- 40% *trans*	113–15/15	13 / 80	105, 107
HC≡CCH₂Cl	$(C_2H_5)(Cl)_2GeH$	solvent CH₃CN	80	5	$(C_2H_5)(Cl)_2GeCH=CHCH_2Cl$ 65% *cis*- 35% *trans* ; $(C_2H_5)(Cl)_2GeCH_2CH_2CH=CHGe(Cl)_2(C_2H_5)$; $[(C_2H_5)(Cl)_2GeCH_2CH(GeCl)_2(C_2H_5)]_2$	113–15/15	53	105, 107
HC≡C—COCH₃	$(C_6H_5)(Cl)_2GeH$	none	exothermic	½	$(C_6H_5)(Cl)_2GeCH=CH-COCH_3$ (8%) ; $[(C_6H_5)(Cl)_2Ge]_2CH-CH_2COCH_3$ (92%)	—	54	151

Table 4-3(b)—continued

UNSATURATED COMPOUND	GERMANE	CATALYST	REACTION TEMP. (°C)	TIME (h)	REACTION PRODUCTS	B.P. OR (M.P.) (°C/mm)	YIELD (%)	REFERENCES
HC≡C—COCH₃	(C₆H₅)(Cl)₂GeH	H₂PtCl₆	0	$\frac{7}{12}$	(C₆H₅)(Cl)₂GeCH=CHCOCH₃ (40%) 66% *cis*- 34% *trans*	—	70	151
CH≡C—C=CH₂ / CH₃	(C₂H₅)₂(Cl)GeH	H₂PtCl₆	100 autoclave	15	[(C₆H₅)(Cl)₂Ge]₂CH—CH₂COCH₃ (60%) (C₂H₅)₂(Cl)GeCH=CH—C=CH₂ / CH₃ (*trans*)	105/16	60	106b
CH≡C—C=CH₂ / CH₃	(C₂H₅)₂(Cl)GeH	AIBN	100 autoclave	15	(C₂H₅)₂(Cl)GeCH=CH—C=CH₂ / CH₃ (30%) (*cis*) (C₂H₅)₂(Cl)GeCH₂CH—C≡CH / CH₃ (24%) (C₂H₅)₂(Cl)GeCH₂—C=C=CH₂ / CH₃ (46%)	—	35	106b
CH≡C—C=CH₂ / CH₃	(C₂H₅)(Cl)₂GeH	none	100 autoclave	15	(C₂H₅)(Cl)₂GeCH=CH—C=CH₂ / CH₃ (25%) (*cis*) (C₂H₅)(Cl)₂GeCH₂—CH—C≡CH / CH₃ (15%) (C₂H₅)(Cl)₂GeCH₂—C=C=CH₂ / CH₃ (60%)	101/15	35	106b
HC≡C(CH₂)₃CH₃	(C₄H₉)₂(Cl)GeH	none	100	2	(C₄H₉)₂(Cl)GeCH=CH(CH₂)₃CH₃	156.5/12	96	105, 107
HC≡C—C₆H₅	(C₂H₅)₂(Cl)GeH	none	80	2	(C₂H₅)₂(Cl)GeCH=CH(C₆H₅) 90% *cis*- 10% *trans*	149–50/10	54	105
HC≡C—C₆H₅	(C₂H₅)(Cl)₂GeH	none	80	1	(C₂H₅)(Cl)₂GeCH=CHC₆H₅ 60% *cis*- 40% *trans*	158/15	63	105
HC≡C—C₆H₅	(C₄H₉)₂(Cl)GeH	none	120	6	(C₄H₉)₂(Cl)GeCH=CH(C₆H₅) 100% *trans*	140/0.4	75	99, 105

(c) *To carbonyl compounds*

UNSATURATED COMPOUND	GERMANE	CATALYST	REACTION TEMP. (°C)	TIME (h)	REACTION PRODUCTS	B.P. OR (M.P.) (°C/mm)	YIELD (%)	REFERENCES
CH_3—CH—CHO with CH_3	$(C_6H_5)_2(Cl)GeH$	UV	30	2	$(C_6H_5)_2GeOCH_2$—CH(CH_3)_2, Cl	$106/4.10^{-2}$	76	140
CH_2—CH_2 CH_2—CH_2 C=O	$(C_6H_5)_2(Cl)GeH$	UV	30	3	$(C_6H_5)_2Ge$—OC_6H_{11}, Cl	$136/3.10^{-3}$	80	140
$C_6H_5C≡C=O$	$(C_6H_5)(Cl)_2GeH$	none	60	2	$(C_6H_5)(Cl)_2Ge$—$C=C(C_6H_5)_2$, OH		90	141, 151
$(C_6H_5)(Cl)Ge$—$(CH_2)_4COCH_3$, H	—	UV	30	16	$\xrightarrow{90°}$ $(C_6H_5)(Cl)_2GeCOCH(C_6H_5)_2$ $(C_2H_5)(Cl)Ge(CH_2)_4CHCH_3$, O	$108/3.10^{-2}$	76	151
					$\left[\begin{array}{c}C_6H_5\\-Ge—(CH_2)_4\\Cl\end{array}\;\begin{array}{c}CH_3\\-C—O\\H\end{array}\right]_n$		18	

4-2-5   Table: Organohalogermanium hydrides

Table 4-4
Organohalogermanium Hydrides

(a) R$_2$XGeH

	COMPOUND	M.P. (°C)	B.P. (°C/mm)	n_D^{20}	d_4^{20}	REFERENCES
C$_2$H$_7$GeCl	(CH$_3$)$_2$(Cl)GeH	—	89/760	—	—	3
C$_4$H$_{11}$GeF	(C$_2$H$_5$)$_2$(F)GeH	—	112/760	1.4132	1.2158	106
C$_4$H$_{11}$GeCl	(C$_2$H$_5$)$_2$(Cl)GeH	—	137/760	1.4572	1.2409	96, 142
C$_4$H$_{11}$GeBr	(C$_2$H$_5$)$_2$(Br)GeH	—	153/760	1.4888	1.5340	96, 142
C$_4$H$_{11}$GeI	(C$_2$H$_5$)$_2$(I)GeH	—	70/23	1.5382	1.7717	96, 142
C$_6$H$_{15}$GeCl	(n-C$_3$H$_7$)$_2$(Cl)GeH	—	55/7	1.4624	1.1622	12
C$_6$H$_{15}$GeI	(n-C$_3$H$_7$)$_2$(I)GeH	—	64/2.5	1.5226	1.5896	14
C$_8$H$_{19}$GeF	(n-C$_4$H$_9$)$_2$(F)GeH	—	114/60	1.4344	1.0848	106
C$_8$H$_{19}$GeCl	(n-C$_4$H$_9$)$_2$ClGeH	—	105/17	1.4620	1.1100	10, 142
C$_8$H$_{19}$GeBr	(n-C$_4$H$_9$)$_2$(Br)GeH	—	113/14	1.4832	1.3050	10, 142
C$_8$H$_{19}$GeI	(n-C$_4$H$_9$)$_2$(I)GeH	—	74/1	1.5148	1.4700	10, 142
C$_{12}$H$_{11}$GeF	(C$_6$H$_5$)$_2$(F)GeH	—	66/2.10^{-3}	1.5790	1.3436	140
C$_{12}$H$_{11}$GeCl	(C$_6$H$_5$)$_2$(Cl)GeH	—	102/0.4	1.6010	1.3514	140, 150
C$_{12}$H$_{11}$GeBr	(C$_6$H$_5$)$_2$(Br)GeH	—	90/10^{-3}	1.6186	1.5340	140, 150
C$_{12}$H$_{11}$GeI	(C$_6$H$_5$)$_2$(I)GeH	27	—	—	—	140, 150

(*b*) $RXGeH_2$

COMPOUND		M.P. (°C)	B.P. (°C/mm)	n_D^{20}	d_4^{20}	REFERENCES
$CH_3(Cl)GeH_2$	CH_5GeCl	—	71/760	—	—	3
$CH_3(Br)GeH_2$	CH_5GeBr	−89.2	80.5/760	—	—	64
$C_2H_5(F)GeH_2$	C_2H_7GeF	—	65/760	1.3998	—	105, 106
$C_2H_5(Cl)GeH_2$	C_2H_7GeCl	—	90.5/760	1.4547	1.3702	105, 106
$C_2H_5(Br)GeH_2$	C_2H_7GeBr	—	113/760	1.4962	1.7341	105, 106
$C_2H_5(I)GeH_2$	C_2H_7GeI	—	138/760	1.5612	2.0277	105, 106
$(C_3H_7)(Cl)GeH_2$	C_3H_9GeCl	—	—	—	—	41
$(n\text{-}C_4H_9)(Cl)GeH_2$	$C_4H_{11}GeCl$	—	140/760	1.4598	1.2460	10, 106, 172
$(n\text{-}C_4H_9)(Br)GeH_2$	$C_4H_{11}GeBr$	—	159/760	1.4910	1.5360	10, 106, 172
$(n\text{-}C_4H_9)(I)GeH_2$	$C_4H_{11}GeI$	—	85/27	1.5412	1.7760	10, 172
$(n\text{-}C_6H_{13})(Cl)GeH_2$	$C_6H_{15}GeCl$	—	86/27	1.4623	1.1671	105
$C_6H_5(F)GeH_2$	C_6H_7GeF	—	76/18	1.5329	1.4422	140
$C_6H_5(Cl)GeH_2$	C_6H_7GeCl	—	80/13	1.5650	1.4364	140, 150
$C_6H_5(Br)GeH_2$	C_6H_7GeBr	—	38/0.2	1.5961	1.7322	140, 150
$C_6H_5(I)GeH_2$	C_6H_7GeI	29	—	—	—	140, 150

(c) RX_2GeH

COMPOUND		M.P. (°C)	B.P. (°C/mm)	n_D^{20}	d_4^{20}	REFERENCES
CH_4GeCl_2	$(CH_3)(Cl)_2GeH$	—	101.5/750	1.4701	1.6356	3, 64, 123
$C_2H_6GeCl_2$	$(C_2H_5)(Cl)_2GeH$	—	130/760	1.4747	1.5339	105, 106, 123
$C_2H_6GeBr_2$	$(C_2H_5)(Br_2)GeH$	—	—	—	—	105
$C_2H_6GeI_2$	$(C_2H_5)(I_2)GeH$	—	—	—	—	105
$C_4H_{10}GeCl_2$	$(n\text{-}C_4H_9)(Cl_2)GeH$	—	74.5/25	1.4738	1.3722	106, 147
$C_6H_{14}GeCl_2$	$(n\text{-}C_6H_{13})(Cl_2)GeH$	—	106/20	1.4723	1.2931	105
$C_6H_6GeF_2$	$(C_6H_5)(F_2)GeH$	38	—	—	—	140
$C_6H_6GeCl_2$	$(C_6H_5)(Cl_2)GeH$	—	94/11	1.5642	1.5435	140, 150
$C_6H_6GeBr_2$	$(C_6H_5)(Br_2)GeH$	—	$38/10^{-3}$	1.6190	2.0469	140
$C_6H_6GeI_2$	$(C_6H_5)(I_2)GeH$	—	$80/2.10^{-2}$	—	2.4	140

4-3   ORGANOGERMANIUM HYDRIDES

4-3-1   Preparations

Several methods of preparation have been described in the literature.

4-3-1-1   From Organogermyllithium or Organogermylsodium Derivatives.
Triphenylgermane, $(C_6H_5)_3GeH$, is formed quantitatively when triphenylgermylsodium, $(C_6H_5)_3GeNa$, reacts with ammonium bromide in liquid
ammonia or by treating a benzene solution of the same compound with
water (87).
Triethylgermane, $(C_2H_5)_3GeH$, is prepared by a similar method:
ammonolysis of triethylgermyllithium (86):

$$(C_2H_5)_3Ge—Ge(C_2H_5)_3 \xrightarrow[\text{EtNH}_2]{\text{Li}} (C_2H_5)_3GeLi$$

$$(C_2H_5)_3GeLi + NH_3 \longrightarrow (C_2H_5)_3GeH + LiNH_2$$

Methyl- ethyl- and propylgermanes, CH_3GeH_3, $C_2H_5GeH_3$ and n-
$C_3H_7GeH_3$, have been prepared by the reaction of sodium germanyl,
H_3GeNa, with the appropriate alkyl halides in liquid ammonia (132, 171):

$$NaGeH_3 + RX \rightarrow RGeH_3 + NaX$$

Recently Cradock and coworkers prepared with good yields CH_3GeH_3,
$C_2H_5GeH_3$ and $CH_3OCH_2GeH_3$ by the action of CH_3I, C_2H_5Br and
CH_3OCH_2Cl on $KGeH_3$ in hexamethylphosphoramide (25), and
$CH_3OCH_2GeH_3$ is also prepared by the action of CH_3OCH_2Cl on
freshly prepared solid $NaGeH_3$ (50).
The compounds CH_3GeH_3 and $(CH_3)_3SiGeH_3$ are formed by a
coupling reaction between GeH_3K and CH_3I and $(CH_3)_3SiCl$ respectively
in monoglyme or diglyme (4).
Methyl(germyl)silanes $(CH_3)_nSi(GeH_3)_{4-n}$ in which $n = 0$, 1, 2, or 3
have been synthesized by the reaction of sodium germyl, $NaGeH_3$,
with the appropriate methylchlorosilane. The 1,1,1-trimethyldigermane
has been prepared by the reaction of sodium germyl with trimethylfluorogermane (34a).
Ethylisoamyl- and diethylisoamylgermane have been prepared by reaction of isoamyl bromide or ethyl iodide with lithium germanyl salts in
ethylamine (59):

$$C_2H_5GeLiH_2 + i\text{-}C_5H_{11}Br \rightarrow (C_2H_5)(i\text{-}C_5H_{11})GeH_2$$

$$(C_2H_5)(i\text{-}C_5H_{11})GeLiH + C_2H_5I \rightarrow (C_2H_5)_2(i\text{-}C_5H_{11})GeH$$

Ethylmethylgermane, $(C_2H_5)(CH_3)GeH_2$, was prepared by treatment of the sodium salt of ethylgermane with methyl chloride in liquid ammonia (16).

The hydrolysis of tribenzylgermyllithium leads to tribenzylgermane and the action of D_2O to deuteride (27):

$$(PhCH_2)_3GeLi \xrightarrow{H_2O} (PhCH_2)_3GeH$$

$$(PhCH_2)_3GeLi \xrightarrow{D_2O} (PhCH_2)_3GeD$$

4-3-1-2 Reduction of Germanium Halides and Sulphides by Zinc Amalgam and Aqueous HCl. Organogermanium halides such as Ph_3GeBr or sulphides such as (Me_2GeS) can be reduced to the corresponding hydrides by zinc and hydrochloric acid (193):

$$(C_6H_5)_3GeBr \xrightarrow{Zn+HCl} (C_6H_5)_3GeH$$

$$((CH_3)_2GeS) \xrightarrow{Zn+HCl} (CH_3)_2GeH_2$$

However, under similar conditions triphenyltin chloride or iodide gave no material showing Sn—H absorption and diphenyldichlorosilane and triphenylchlorosilane gave no products showing Si—H absorption.

The reason for the anomalously easy formation of Ge—H bonds in reduction of this type is consistent with recent conclusions that germanium is more electronegative than either silicon or tin (2, 91). However, these low yield reactions cannot be adopted as synthetic methods for the preparation of organogermanium hydrides.

4-3-1-3 Reduction of Organogermanium Halides, Oxides, Hydroxides, Alkoxides, Sulphides, etc. by Lithium Aluminium Hydride. In 1947, Finholt and coworkers reported a facile synthesis of the organic hydrides of tin and silicon by reduction of the corresponding halides by lithium aluminium hydride, $LiAlH_4$ (43). This new reduction procedure was successfully used in the preparation of the organic hydrides of germanium from halides. Triphenyl- and tricyclohexylgermane were prepared in this way (76, 79):

$$4 R_3GeCl + LiAlH_4 \rightarrow 4 R_3GeH + LiCl + AlCl_3$$

This method has been generalized later for the synthesis of polyhydrides and it has been used by many authors (7, 8, 10, 19, 21, 22, 32, 33, 39, 47, 75, 77, 142, 149, 192):

$$R_nGeX_{4-n} \xrightarrow{LiAlH_4} R_nGeH_{4-n}$$

Solvents generally used are diethyl- and di-*n*-butyl ether.

Tricyclohexylgermanol, $(C_6H_{11})_3GeOH$, is reduced to tricyclohexyl-germane by $LiAlH_4$ (79):

$$4\,(C_6H_{11})_3GeOH + LiAlH_4 \rightarrow 4\,(C_6H_{11})_3GeH + LiOH + Al(OH)_3$$

Bis(triphenylgermanium)oxide, $[(C_6H_5)_3Ge]_2O$, is also reduced to triphenylgermane by the same reagent (79). Alkoxygermanes, R_3GeOR', (97) and alkylthiogermanes, R_3GeSR', (145) are reduced by $LiAlH_4$ with formation of trialkylgermanes. The germanium–phosphorus bond in Et_3GePPh_2 is also cleaved with formation of Et_3GeH and diphenyl-phosphine (146).

4-3-1-4 Reduction using Lithium Hydride, Lithium Deuteride or Sodium Hydride. Organogermanium hydrides and deuterides can be prepared from the corresponding chlorides, by reacting them with lithium hydride or deuteride in dioxane or diisoamyl ether (136, 137) with similar or slightly lower yields than with lithium aluminium hydride. Organo-germanium halides can also be reduced by NaH in the presence of alkylaluminium halides (15). Dimethylgermane was prepared by reaction of dimethyldichlorogermane with a suspension of NaH in mineral oil in the presence of triphenylborane at 110°C over 3 h (15a).

4-3-1-5 Reduction by Lithium Aluminium Hydride tri-tert-butoxide. Sujishi and Keith had obtained excellent yields of monogermane by the reduction of $GeCl_4$ with $LiAlH(O\text{-}tert\text{-}C_4H_9)_3$ (169). The three methyl-germanes, CH_3GeH_3, $(CH_3)_2GeH_2$ and $(CH_3)_3GeH$, have been prepared by the same method in anhydrous dioxane with 60–70 % yields (176).

4-3-1-6 Reduction by Sodium Borohydride. The reduction of trialkyl-germanium halides by sodium borohydride in tetrahydrofuran can be carried out in almost quantitative yield. The reaction takes place in two phases (142):

$$R_3GeX + NaBH_4 \rightarrow R_3Ge(BH_4) + NaX$$

$$R_3Ge(BH_4) + 3H_2O \rightarrow R_3GeH + B(OH)_3 + 3H_2$$

The reduction of dialkylgermanium dihalides under the same experi-mental conditions leads to dialkylgermanes with 70 % yields (142). Germanium tetrachloride is reduced to GeH_4 in the same solvent with a 43.7 % yield (104).

In 1963, Griffiths (63) obtained germane, GeH_4, and the methylger-manes CH_3GeH_3, CD_3GeH_3, $(CH_3)_2GeH_2$, $(CD_3)_2GeH_2$ and $(CH_3)_3GeH$ by the same hydroborate method after addition of an aqueous solution of sodium hydroborate to acidic solutions of GeO_2, CH_3GeBr_3, CD_3GeBr_3, $(CH_3)_2GeBr_2$, $(CD_3)_2GeBr_2$ and $(CH_3)_3GeBr$. At reaction temperatures of 30–55°C, yields of 90–100% were obtained for the alkyl-germanes. The yield of germane was virtually quantitative and temperature-independent in the 0–80°C range (63).

Ethylgermanes, $C_2H_5GeH_3$ and $(C_2H_5)_2GeH_2$, were prepared by the same method by Birchall and Jolly (16).

4-3-1-7 Action of Grignard Reagents on Germanium Tetrachloride. In the action of germanium tetrachloride on isopropylmagnesium bromide tri-isopropylgermane, i-Pr_3GeH, is the main compound formed (111).

In 1962, Glockling and Hooton showed that the arylmagnesium halides react with germanium tetrachloride with formation of hexa-aryldigermanes, Ar_6Ge_2, tetrasubstituted germanes, R_4Ge, and also of triarylgermanes, Ar_3GeH (in the case of o-, m- and p-$CH_3C_6H_4MgX$). The hypothesis of an intermediate germyl Grignard reagent, Ar_3GeMgX, whose hydrolysis leads to the germane, Ar_3GeH, has been advanced (60).

The reaction of cyclohexylmagnesium chloride with $GeCl_4$ also gives rise to the development of important amounts of tricyclohexylgermane besides the mono-, di-, and tri-substituted products. The intermediacy of a germyl Grignard reagent, $(C_6H_{11})_3GeMgX$, is evidenced by the action of D_2O which leads to the corresponding deuteride (119):

$$(C_6H_{11})_3GeMgX \xrightarrow{D_2O} (C_6H_{11})_3GeD$$

Seyferth (157) explained the presence of hexavinyldigermane during the action of vinylmagnesium bromide on germanium tetrachloride by the formation of a transitory compound, $(CH_2{=}CH)_3GeMgBr$, formed from germanium dichloride:

$$GeCl_4 + 2\,RMgX \rightarrow GeCl_2 + R{-}R + 2\,MgXCl$$

$$GeCl_2 + 2\,RMgX \rightarrow GeR_2 + 2\,MgXCl$$

$$GeR_2 + RMgX \rightarrow R_3GeMgX \qquad [R = CH_2{=}CH{-}]$$

This reaction scheme permits the interpretation of the results of Valade and coworkers (118, 119) in the reaction of the sterically hindered Grignard reagents with germanium tetrachloride:

$$GeCl_4 \xrightarrow{RMgCl} \begin{cases} R_nGeCl_{4-n} \qquad (n = 2, 3, 4) \\ \\ GeCl_2 + R({+}H) + R({-}H) \\ \quad \xrightarrow{RMgCl} R_3GeMgCl \xrightarrow{H_2O} R_3GeH \end{cases}$$

The hypothesis of the transitory passage through germanium dichloride has been corroborated by the experiments of Glockling and Hooton (61) who showed that germanium diiodide, GeI_2, reacts on the phenyllithium or the mesitylmagnesium bromide to yield the corresponding triaryl-germanes:

$$GeI_2 \xrightarrow{C_6H_5Li} (C_6H_5)_3GeLi$$

$$(C_6H_5)_3GeLi \xrightarrow{H_2O} (C_6H_5)_3GeH$$

4-3-1-8 Formation of a Germanium–Hydrogen Bond in the Cleavage Reactions of Germacyclobutanes by Hydrogenosilanes and Hydrogeno-germanes. Silanes such as Et_3SiH, Bu_2MeSiH, Ph_2MeSiH, Bu_2ClSiH, $MeCl_2SiH$ cleave the Ge—C bond of the dialkylgermacyclobutanes with formation of a Ge—H bond (113, 114):

$$\underset{/}{\overset{\backslash}{-}}Si{-}H + \underset{/}{\overset{R}{\underset{R}{\overset{\backslash}{Ge}}}}\diamond \xrightarrow{H_2PtCl_6} \underset{/}{\overset{\backslash}{-}}SiCH_2CH_2CH_2{-}\underset{\underset{H}{|}}{\overset{R}{\underset{\backslash}{Ge}}}\overset{R}{\diagup}$$

A similar reaction is observed with Et_3GeH, always in the presence of chloroplatinic acid.

4-3-1-9 Other Modes of Formation of Ge—H Bond. Bis(triethylgermyl)-cadmium reacts with water, propanol, acetic acid and triethyltin hydride with the formation of triethylgermane (186):

$$[(C_2H_5)_3Ge]_2Cd + RH \rightarrow (C_2H_5)_3GeH + (C_2H_5)_3GeR + Cd$$

$$R = OH, n\text{-}C_3H_7O{-}, CH_3COO{-}$$

$$[(C_2H_5)_3Ge]_2Cd + 2(C_2H_5)_3SnH \rightarrow$$

$$2(C_2H_5)_3GeH + [(C_2H_5)_3Sn]_2Cd$$
$$\downarrow$$
$$Cd + (C_2H_5)_6Sn_2$$

Triethylgermane and $[(C_2H_5)_3Sn]_3Sb$ or $[(C_2H_5)_3Sn]_3Bi$ are formed in the interaction of triethyltin hydride with $[(C_2H_5)_3Ge]_3Sb$ (183) or $[(C_2H_5)_3Ge]_3Bi$ (88) respectively.

Reaction of tris(triethylgermyl)thallium with triethyltin hydride does not lead to tris(triethyltin)thallium, $[(C_2H_5)_3Sn]_3Tl$, but to its decomposition products. The formation of triethylgermane in 77% yield is also noted (89):

$$6(C_2H_5)_3SnH + 2[(C_2H_5)_3Ge]_3Tl \rightarrow$$

$$6(C_2H_5)_3GeH + 2Tl + 3(C_2H_5)_6Sn_2$$

Glacial acetic acid reacts analogously with bis-(triethylgermyl)mercury (88) and tris-(triethylgermyl)bismuth (191):

$$CH_3COOH + (C_2H_5)_3GeHgGe(C_2H_5)_3 \rightarrow$$

$$Hg + (C_2H_5)_3GeH + (C_2H_5)_3GeOCOCH_3$$

$$3\,CH_3COOH + 2\,[(C_2H_5)_3Ge]_3Bi \rightarrow$$

$$2\,Bi + 3\,(C_2H_5)_3GeH + 3\,(C_2H_5)_3GeOCOCH_3$$

The deprotonation–alkylation method is described for the synthesis of methylgermane in 1,2-dimethoxyethane as solvent (141b):

$$2\,KOH + GeH_4 \rightarrow K^+ + GeH_3^- + KOH\cdot H_2O$$

$$CH_3I + GeH_3^- \rightarrow CH_3GeH_3 + I^-$$

Polarographic reduction of Ph_2MX_2 leads to $Ph_2MH_2(M = Si, Ge)$ (34).

4-3-2   Thermal stability of organogermanium hydrides

Johnson and Harris have been the first to point out the thermal disproportionation of triphenylgermane to tetraphenylgermane and diphenylgermane (76). The same authors also showed the instability of $(C_6H_5)_2GeH_2$, which proceeds slowly at a normal temperature, giving various oxidation and redistribution products, one of which is the tetraphenylgermane (75).

Rivière and Satgé found at 200°C in 12 hours a partial thermal decomposition of phenylgermane (140):

$$C_6H_5GeH_3 \xrightarrow{200°C} (C_6H_5)_2GeH_2 + GeH_4(Ge + 2H_2)$$

In the presence of $AlCl_3$ this decomposition is instantaneous even at ambient temperature.

The alkylgermanes are more stable than the arylgermanes; their stability is higher than that of the stannanes and it can be compared with that of the analogous silanes; it increases in the order $RGeH_3 < R_2GeH_2 < R_3GeH$ (142).

The decomposition of tricyclohexylgermane at a low temperature should give the primary reaction products (134):

$$(C_6H_{11})_3GeH \rightarrow 3\,C_6H_{10} + Ge + 2H_2$$

It was found that the thermal decomposition of tricyclohexylgermane begins at a temperature as low as $\sim 360°C$. The main decomposition products are hydrogen, cyclohexane, benzene, germanium, cyclohexene and highly condensed compounds containing cyclohexyl rings. The formation of benzene and cyclohexane can be explained by the disproportionation of the cyclohexene (134).

4-3-3 Reactivity of the germanium–hydrogen bond in organogermanium hydrides

4-3-3-1 General Reactivity

(a) Oxidation reactions

The organogermanium hydrides are much less oxidizable than the analogous alkylstannanes. Their sensitiveness to oxidizing agents can rather be compared with that of the silanes. The three types of hydrides, R_3GeH, R_2GeH_2 and $RGeH_3$, slowly oxidize in air. They are transformed into oxides with decreasing basicity $(R_3Ge)_2O$, (R_2GeO) and $(RGeO)_2O$ (142).

Tricyclohexylgermane is readily oxidized to the corresponding germanol, $(C_6H_{11})_3GeOH$, on standing in the air for a short time (79). Potassium permanganate in acetone solution partially converts triethylgermane into the oxide, $(Et_3Ge)_2O$ (8).

The action of gaseous oxygen on tetraethyldihydrodigermoxane gives through the oxidation of the Ge—H bonds the corresponding oxide (106):

$$(C_2H_5)_2\underset{\overset{|}{H}}{Ge}-O-\underset{\overset{|}{H}}{Ge}-(C_2H_5)_2 \xrightarrow{O_2} [(C_2H_5)_2GeO]_n$$

(b) Action of acids

Hydrohalic acids. Anderson reported that triethylgermane is readily converted to triethylchlorogermane by refluxing hydrochloric acid (8).

Triphenylgermane is also transformed into the respective triphenylgermanium halide by the action of such acids (87):

$$Ph_3GeH + HX \rightarrow Ph_3GeX + H_2$$

Anhydrous hydrochloric acid only reacts with amylgermane and dibutylgermane in presence of a catalyst ($AlCl_3$) (142). The cleavage of the GeH bonds of methylgermane, CH_3GeH_3, by gaseous hydrochloric acid in the presence of $AlCl_3$ allowed Amberger and Boeters (3) on the one hand and Griffiths and Onyszchuk (64) on the other to isolate mono and dichloromethylgermane, $(CH_3)(Cl)GeH_2$ and $(CH_3)(Cl)_2GeH$. The action of gaseous HBr (catalyst $AlBr_3$) leads exclusively to methylbromogermane $(CH_3)BrGeH_2$ (64).

Oxyacids. Concentrated sulphuric acid (36N) reacts slowly in the cold with the alkylgermanes with hydrogen release (8):

$$2 R_3GeH + H_2SO_4 \rightarrow (R_3Ge)_2SO_4 + 2 H_2 \text{ (yield 70\%)}$$

The same acid reacts more readily with dialkylgermanes (142). Fuming nitric acid reacts violently with the organogermanium hydrides and the action of hot sulphuric–nitric mixture results in quantitative oxidation to GeO_2. Tri-1-naphthylgermane shows little reaction with cold basic solutions. However, this compound gives a steady stream of hydrogen when warmed with KOH in moist piperidine (192).

Benzenesulphonic acid reacts slowly with triethylgermane in toluene solution (142):

$$(C_2H_5)_3GeH + C_6H_5SO_3H \rightarrow (C_2H_5)_3GeSO_3C_6H_5 + H_2$$

$$(yield\ 70\%)$$

Boric acid reacts with triethylgermane in the presence of copper powder to give triethylgermanium borate (98):

$$3(C_2H_5)_3GeH + B(OH)_3 \rightarrow [(C_2H_5)_3GeO]_3B + 3H_2 \qquad (yield\ 50\%)$$

Organic acids. No reaction is observed between alkylgermanes and refluxing acetic or formic acids (8). However, in the presence of copper powder, acetic acid and tributylgermane react quantitatively (98).

Triethylgermane adds to the $C=C$ bond of acrylic acid in the absence of a catalyst (93, 142); in the presence of copper powder, however, the condensation reaction occurs (98):

$$(C_2H_5)_3GeH + CH_2=CH-COOH \rightarrow$$

$$CH_2=CHCOOGe(C_2H_5)_3 + H_2$$

The strong acids C_3F_7COOH and C_2F_5COOH react quantitatively with triethylgermane (8).

Fluoroacetic acids, CF_3COOH, CHF_2COOH and CH_2FCOOH, convert $(C_2H_5)_3GeH$ to the corresponding esters with yields of 95, 60 and 20%, respectively. On the other hand, trichloroacetic, tribromoacetic and monoiodoacetic acids are reduced by Et_3GeH to acetic acid (yield 80–94%) (8). The different behaviour of trifluoroacetic acid on the one hand, and trichloro-, tribromo- and monoiodoacetic acids on the other, may be explained by the comparative values of the bond energy: $C-F$: 107, $C-H$: 87, $C-Cl$: 67, $C-Br$: 54, $C-I$: 46 kcal/mol.

(c) Action of bases

Fuchs and Gilman (47) reported that the trihexylgermane did not evolve hydrogen at a noticeable rate when treated with dilute alcoholic potassium hydroxide solution although triphenylgermane reacted readily. In a study on the alkaline solvolysis of the trialkylsilanes, germanes and stannanes, Schott and Harzdorf (155) report that the Sn—H bond is very sensitive, the Si—H bond rather sensitive and the Ge—H bond extremely resistant to alkaline hydrolysis. The alkylgermanes R_3GeH, R_2GeH_2 and $RGeH_3$ do not react noticeably with 20% aqueous sodium hydroxide. On the other hand, a monoalkylgermane, $n\text{-}C_6H_{13}GeH_3$, on being treated with 30% alcoholic potassium at 80°C did react with slow hydrogen release (142):

$$2\,RGeH_3 + 6\,OH^- + 6\,C_2H_5OH(H_2O) \rightarrow$$

$$(RGeO)_2O + 6\,H_2 + 6\,C_2H_5O^-(OH^-) + 3\,H_2O$$

The same reaction is slower with the dihydrides, R_2GeH_2.

(d) Action of halogens

Organogermanium hydrides react immediately with halogens to form corresponding organogermanium halides (75, 76, 79, 86, 87, 142, 192):

$$R_nGeH_{4-n} + X_{2(4-n)} \rightarrow$$

$$R_nGeX_{4-n} + HX_{4-n} \qquad (n = 1, 2, 3) \quad (X = Cl, Br, I)$$

The action of iodine on an excess of dibutylgermane $(C_4H_9)_2GeH_2$ or butylgermane $(C_4H_9)GeH_3$ at 0°C leads to the butyliodogermanes of type $RIGeH_2$ and R_2IGeH (10).

(e) Action of inorganic halides

Triethylgermane reduces salts of some transition metals either to a lower oxidation state or sometimes to the free metal (8). These reductions apparently occur with transition elements in which the oxidation potential, $\Delta E°$, for the highest oxidation state present has a value between approximately -0.06 and -2.0 volt. Et_3GeH reduces the salts of Pt, Pd, Au, Hg to the corresponding metal, Cu^{II} to Cu^{I}, Ti^{IV} to Ti^{III} or Ti^{II}, Cr^{VI} to Cr^{III}, V^V to V^{IV} or V^{III}; as expected no reduction of $CdCl_2$ occurs since $\Delta E°$ for Cd is 0.40 volt.

Germanium tetrachloride is reduced in two stages (to Ge^{II} and then to Ge^0) by triethylgermane (142).

Trichlorogermane etherate, $HGeCl_3 \cdot 2\,Et_2O$, has been prepared by partial reduction of $GeCl_4$ by means of Et_3GeH in ether (106):

$$GeCl_4 + Et_3GeH \xrightarrow[35°C]{ether} HGeCl_3 \cdot 2\,Et_2O + Et_3GeCl \qquad \text{(yield 80%)}$$

Aluminium chloride is less easily reduced to metallic aluminium:

$$3\,(C_4H_9)_3GeH + AlCl_3 \rightarrow 3\,(C_4H_9)_3GeCl + Al + 3/2\,H_2$$

Tributylgermane at 220°C reduces carbon tetrachloride with almost quantitative formation of Bu_3GeCl and chloroform (142).

Sulphuryl chloride reacts vigorously with triethylgermane at room temperature. The halogenation is quantitative:

$$2(C_2H_5)_3GeH + SO_2Cl_2 \rightarrow 2(C_2H_5)_3GeCl + SO_2 + H_2$$

The action of a deficiency of mercuric halides HgX_2 (X = Cl, Br) on mono- and dialkylgermanes $RGeH_3$ and R_2GeH_2 allows preparation of the organohalogermanium hydrides, $RXGeH_2$ and R_2XGeH (10, 105, 106). The iodides R_2IGeH can be synthesized by the action of HgI_2 (14) or iodine (10). The same reactions are observed with the phenylgermanes, $PhGeH_3$ and Ph_2GeH_2 (140).

Gradual addition of a deficiency of $Hg(CN)_2$ or $Hg(SCN)_2$ to di-n-butylgermane, furnishes approximately 74 % yields of $(n\text{-}C_4H_9)_2Ge(H)CN$ or $(n\text{-}C_4H_9)_2Ge(H)NCS$ respectively (6).

(f) Action of halogenated organic derivatives

The reduction of halogenated organic derivatives by alkylgermanes can be explained in terms of a free-radical chain process.

The alkylgermanes react without a catalyst with alkyl halides to give the corresponding hydrocarbon (96, 142):

$$R_3GeH + R'X \rightarrow R_3GeX + R'H$$

$$RGeH_3 + 3R'X \rightarrow RGeX_3 + 3R'H$$

Iodides react more readily than bromides and the latter more readily than chlorides.

The reactivity of the trihydrides $RGeH_3$ is greater than that of the dihydrides, R_2GeH_2, and the monohydrides, R_3GeH in this type of reaction (96, 142).

The partial cleavage of the Ge—H bonds of a dialkylgermane, R_2GeH_2, by a deficiency of organic halide leads to the formation of dialkylhalogermanes R_2GeHX (X = Br, I) (96, 142).

Allyl halides (bromide and chloride) are reduced to propene with the formation of R_3GeX (105, 142). In the presence of chloroplatinic acid, the yield of the reduction is lower. The C=C addition product and the condensation derivative are also formed (105, 121):

$$CH_2{=}CHCH_2Cl + HGe(C_2H_5)_3 \xrightarrow{\text{H}_2\text{PtCl}_6}$$

$$\nearrow CH_2{=}CH{-}CH_2Ge(C_2H_5)_3 \quad (26\%)$$

$$\rightarrow ClCH_2CH_2CH_2Ge(C_2H_5)_3 \quad (5\%)$$

$$\searrow CH_2{=}CH{-}CH_3 + ClGe(C_2H_5)_3 \quad (55\%)$$

In the presence of the same catalyst, allyl bromide and iodide react with Et_3GeH to give reduction and condensation products.

With propargyl chloride, the reducing character of Et_3GeH accounts for the large formation of Et_3GeCl in the absence of solvent or in acetonitrile. On the other hand, in the presence of chloroplatinic acid, the addition reaction preponderates over the reduction reaction (relative percentages 65 and 35%) (cf addition–reaction Tables). The aryl halides react with more difficulty. The reduction of iodobenzene to benzene by Bu_2GeH_2 (40% yield) is effected only after long refluxing (142).

Organic acid chlorides such as benzoyl chloride are reduced by R_3GeH to benzaldehyde (142). Triphenylgermane reacts with benzoyl chloride to give triphenylchlorogermane (76). Chloromethyl ether is reduced quantitatively to dimethyl ether by Et_3GeH. The same reagent cleaves selectively, in the presence of traces of $AlCl_3$, one or two $Ge-H$ bonds of the alkyl- and phenylgermanes to give alkylchlorogermanes (105, 106, 142) and phenylchlorogermanes (140, 150) of type R_2ClGeH, $RClGeH_2$ and RCl_2GeH.

The action of N-bromosuccinimide on Et_3GeH in CCl_4 leads to the substitution of the $Ge-H$ hydrogen and the formation of Et_3GeBr (48). Partial substitution of the hydrogen atoms in phenylsilane, $PhSiH_3$, gives phenylbromosilane, $PhSiH_2Br$ (48). In the same way, the action of N-chloro-, N-bromo- and N-iodosuccinimide on $PhGeH_3$ and Ph_2GeH_2 in petroleum ether leads to the chloro-, bromo- and iodophenylgermanes, $C_6H_5(X)GeH_2$ and $(C_6H_5)_2XGeH$ (140).

Another hydrogen–halogen exchange reaction is observed in the reduction of dichloro- or dibromodialkylgermanes by trialkylgermanes in the presence of aluminium chloride or bromide (142):

$$R_2GeX_2 + R_3GeH \xrightarrow{AlX_3} R_2GeHX + R_3GeX \qquad (X = Cl \text{ or } Br)$$

Group IV hydrides can be used as radical-generating and trapping agents and also to effect some radical cyclization reaction (82a):

$$ICH_2CH_2CH_2I \xrightarrow{R_3MH} ICH_2CH_2CH_2 \cdot \rightarrow \overline{CH_2CH_2CH_2}$$

The reaction of 1,3-di-iodopentane with Ph_3GeH between 71°C and 225°C leads to cyclization and reduction products. The percentage of cyclization increases with the reaction temperature.

The reaction of 1,5-di-iodopentane with Ph_3GeH results only in reduction (>99.5%) (82a).

(g) Other reduction reactions

Polar mechanisms in germanium and tin hydride reductions are noted by Carey and Tremper (23a). In the hydride transfer from triphenylgermane, silane and stannane to the 4-t-butyl-1-phenylcyclohexyl cation and the diphenylcyclopropylmethyl cation, tin and germanium hydrides are both better donors than silicon hydrides.

4-3-3-2   Reactions with Sulphur, Selenium, Tellurium and their Organic Derivatives. In a preliminary paper (178), Vyazankin and coworkers reported in 1966 that compounds of the type $[(C_2H_5)_3M]_2M'$ in which M = Sn or Ge and M' = S, Se, Te can be synthesized by heating elementary sulphur, selenium, tellurium, with triethyltin hydride and triethylgermane. By this method, bis(triethylgermanium)sulphide, $(Et_3Ge)_2S$, bis(triethylgermanium) selenide, $(Et_3Ge)_2Se$, and bis(triethylgermanium) telluride were obtained in yields of 20, 18 and 60% respectively.

Bis(triethylgermanium) selenide is also obtained in 45% yield by heating triethylgermane with diethyl selenide (20 hours at 140°C) (178, 180):

$$2(C_2H_5)_3GeH + (C_2H_5)_2Se \rightarrow 2C_2H_6 + [(C_2H_5)_3Ge]_2Se$$

The same authors reported (179, 181) the reaction of triethylgermane with diethyl telluride (7 hours at 140°C):

$$(C_2H_5)_2Te \xrightarrow{(C_2H_5)_3GeH}$$

$$(C_2H_5)_3GeTeC_2H_5 \xrightarrow{(C_2H_5)_3GeH} [(C_2H_5)_3Ge]_2Te$$

The action of triethylgermane on sulphur and selenium gives as final (and not intermediate) products triethylmercaptogermane $(C_2H_5)_3GeSH$, and triethylselenylgermane, $(C_2H_5)_3GeSeH$ in yields of 49 and 63% respectively. Simultaneously bis(triethylgermyl) sulphide and *bis*(triethylgermyl) selenide, respectively, also were formed (179, 180).

A different type of compound is obtained by heating selenium with trialkylgermanes up to 200°C (181):

$$R_3GeH + Se \rightarrow R_3GeSeH   (R = i\text{-}C_3H_7, cyclo\text{-}C_6H_{11})$$

This reaction may be accompanied by the formation of symmetrical compounds:

$$2R_3GeSeH \rightarrow H_2Se + R_3GeSeGeR_3$$

$$R_3GeSeH + R_3GeH \rightarrow H_2 + R_3GeSeGeR_3$$

The reaction between triethylgermane and tellurium at 190–200°C leads only to *bis*(triethylgermyl) telluride in 75 % yield. No unsymmetrical products have been isolated in this reaction. Triethylsilane does not react with tellurium when heated to 240°C. On the other hand, the reaction of triethyltin hydride takes place at 130°C (181). It has also been shown that, in contrast to triethylgermane or triethyltin hydride, triethylsilane does not react with sulphur or selenium even at 215°C for 6 h (180). Hence the stability of the hydride bond to the action of sulphur, selenium and tellurium diminishes in the series Si—H > Ge—H > Sn—H.

4-3-3-3   Hydrogenolysis of Metal–Nitrogen and Metal–Oxygen Bonds by Organogermanium Hydrides

(a) Cleavage of metal–nitrogen bonds

The synthesis of linear and branched compounds with Ge—Sn bonds involves hydrogenolytic fission of Ge—N and Sn—N bonds with organotin- and germanium hydrides (26).

Both triethyl- and triphenyltin diethylamine react with triphenyl-germane under relatively mild conditions:

$$R_3SnN(C_2H_5)_2 + (C_6H_5)_3GeH \rightarrow$$

$$R_3Sn—Ge(C_6H_5)_3 + (C_2H_5)_2NH \qquad R = C_2H_5, C_6H_5$$

The reaction with alkylgermanium hydrides does not proceed under comparable conditions. The observed solvent and substituent effects suggest that electrophilic attack of the organogermanium-hydride hydrogen on nitrogen is involved.

The reaction of a bis(dialkylamino)tin derivative with triphenyl-germane in a 1:1 ratio affords an interesting intermediate containing a Ge—Sn and a Sn—N bond (26):

$$(C_2H_5)_2Sn[N(C_2H_5)_2]_2 + (C_6H_5)_3GeH \rightarrow$$

$$(C_6H_5)_3Ge—Sn(C_2H_5)_2—N(C_2H_5)_2 + (C_2H_5)_2NH$$

Hydrogenolysis reactions of this compound with triphenyltin hydride and triphenylgermane led to compounds with three catenated metal atoms:

$$(C_6H_5)_3Ge—Sn(C_2H_5)_2—N(C_2H_5)_2 + (C_6H_5)_3GeH \rightarrow$$

$$(C_6H_5)_3Ge—Sn(C_2H_5)_2—Ge(C_6H_5)_3 + (C_2H_5)_2NH$$

In the reaction between ethyltris(diethylamino)tin and triphenyl-germane one or two diethylamino groups are readily split off; the third one is cleaved only with difficulty and tris(triphenylgermyl)ethyltin was isolated in 3 % yield only:

$$(C_2H_5)Sn[N(C_2H_5)_2]_3 + 3(C_6H_5)_3GeH \rightarrow$$

$$(C_2H_5)Sn[Ge(C_6H_5)_3]_3 + 3(C_2H_5)_2NH$$

1:1 or 1:2 reactions between triphenylgermane and ethyltris(diethyl-amino)tin afforded the intermediates:

$$\underset{\overset{|}{N(C_2H_5)_2}}{\overset{N(C_2H_5)_2}{(C_2H_5)-Sn-Ge(C_6H_5)_3}} \quad \text{and} \quad \underset{\overset{|}{Ge(C_6H_5)_3}}{\overset{Ge(C_6H_5)_3}{(C_2H_5)-Sn-N(C_2H_5)_2}}$$

Diphenylgermane, but not dibutylgermane, is capable of effecting the hydrogenolytic fission of the tin–nitrogen bond. The reaction of tri-organotin diethylamines with diphenylgermane in a 2:1 ratio afforded tri-metal derivatives in good yield:

$$2R_3SnN(C_2H_5)_2 + (C_6H_5)_2GeH_2 \rightarrow$$

$$R_3Sn-Ge(C_6H_5)_2-SnR_3 + 2(C_2H_5)_2NH \qquad (R = C_2H_5, C_6H_5)$$

Polymeric products containing alternating germanium and tin atoms in the chain have been synthesized starting with bifunctional reagents like $R_2Sn(NR'_2)_2$ and $(C_6H_5)_2GeH_2$ (26):

$$nR_2Sn(NR'_2)_2 + n(C_6H_5)_2GeH_2 \rightarrow \pleft[Sn(R_2)Ge(C_6H_5)_2\pright]_n + 2nR'_2NH$$

Neumann and coworkers also recorded the synthesis of derivatives with Ge—Sn (129, 165) and Ge—Pb (128) bonds by cleavage of the Sn—N bonds of the stannyldialkylamines and Pb—N bonds of the plumbyldialkylamines:

$$(C_6H_5)_3GeH + R_3PbN(C_2H_5)_2 \rightarrow (C_6H_5)_3Ge-PbR_3 + (C_2H_5)_2NH$$

The action of diphenylgermane, $(C_6H_5)_2GeH_2$, on $(C_6H_5)_2Pb(NEt_2)_2$ did not give a cyclic compound, but rather the linear derivative $(C_6H_5)_3Pb-Ge(C_6H_5)_2-Pb(C_6H_5)_3$ with two Ge—Pb bonds (128).

(b) Cleavage of metal–oxygen bonds

The cleavage of the tin–oxygen bonds by triethylgermane (or triethyl-silane) is observed under severe conditions (about 180°C) (182). In this case, no formation of derivatives with Ge—Sn bonds takes place.

$$R_3GeH + 2(C_2H_5)_3SnOCH_3 \rightarrow$$

$$R_3GeOCH_3 + CH_3OH + (C_2H_5)_6Sn_2$$

The formation of hexaethylditin and other products can be explained on the assumption that the initial act in the reaction is the replacement of hydride hydrogen by a methoxy group:

$$R_3GeH + (C_2H_5)_3SnOCH_3 \rightarrow R_3GeOCH_3 + (C_2H_5)_3SnH$$

The triethyltin hydride reacts further with excess on triethylmethoxytin with the formation of methanol and hexaethylditin:

$$(C_2H_5)_3SnH + (C_2H_5)_3SnOCH_3 \rightarrow$$

$$CH_3OH + (C_2H_5)_3Sn{-}Sn(C_2H_5)_3$$

4-3-4 Reactions of organogermanium hydrides with organometallic derivatives

4-3-4-1 With Organolithium Compounds and Grignard Reagents

(a) Action of organolithium compounds

Johnson and Harris (76) reported that when triphenylgermane was added to a refluxing ether solution of an excess of phenyllithium, followed by refluxing for 12 hours, there was isolated tetraphenylgermane in 70% yield:

$$Ph_3GeH + PhLi \rightarrow Ph_4Ge + LiH$$

The reverse addition of an excess of phenyllithium to refluxing triphenylgermane gave 50–60% yields of hexaphenyldigermane along with some tetraphenylgermane. The formation of hexaphenyldigermane may indicate the presence of triphenylgermyllithium. A possible mechanism to account for the formation of the hexaphenyldigermane may be:

$$Ph_3GeH + PhLi \rightarrow Ph_3GeLi + PhH$$

$$Ph_3GeH + Ph_3GeLi \rightarrow Ph_3GeGePh_3 + LiH$$

Gilman and Gerow (53) have found that triphenylgermane reacts with phenyllithium in ether solution to give triphenylgermyllithium in good yield:

$$(C_6H_5)_3GeH + PhLi \rightarrow$$

$$PhH + (C_6H_5)_3GeLi \xrightarrow[H_2O]{CO_2} (C_6H_5)_3GeCOOH$$

Under the same conditions, butyllithium gave a quantitative yield of triphenylgermyllithium. This reaction affords an excellent method for the preparation of triphenylgermyllithium. Methyllithium gave, after refluxing 24 hours, 70–80% yields of triphenylgermyllithium (and after carbonation, triphenylgermylcarboxylic acid) along with 16% of triphenylmethylgermane.

The reaction between triphenylgermyllithium and triphenylgermane gave hexaphenyldigermane in only 12% yield.

It has been reported that both triphenylsilane (56) and triethylsilane (55) react with RLi compounds to give the tetrasubstituted silane and lithium hydride:

$$R_3SiH + R'Li \rightarrow R_3SiR' + LiH$$

Triethylgermane reacts with butyllithium or phenyllithium in a very different way. There is no alkylation reaction, but formation of $(C_2H_5)_3GeLi$ with a yield below 10% (74).

The reactions of organogermanium hydrides with organolithium reagents to give organogermyllithium compounds are consistent with the concept of a higher electronegativity for germanium compared with that of silicon (2, 91).

Diphenylgermane in ether, when treated at $-10°C$ with n-butyllithium in pentane, gave butyldiphenylgermane (36%) and digermanes $Ph_2Ge(H)Ge(H)Ph_2$ and $Ph_2Ge(H)GeBuPh_2$ formed by reactions of the type (28):

$$Ph_2Ge(H)Li + Ph_2GeH_2 \rightarrow LiH + Ph_2Ge(H)Ge(H)Ph_2$$

In contrast, when n-butyllithium is added to $(C_6H_5)_4C_4Ge(H)C_6H_5$ in THF at $-78°C$ a bright red colour is produced. The addition of water to this red solution regenerates the starting hydride. If trimethylchlorosilane is added in place of water, the silyl derivative $(C_6H_5)_4C_4Ge(C_6H_5)Si(CH_3)_3$ is formed in 90% yield:

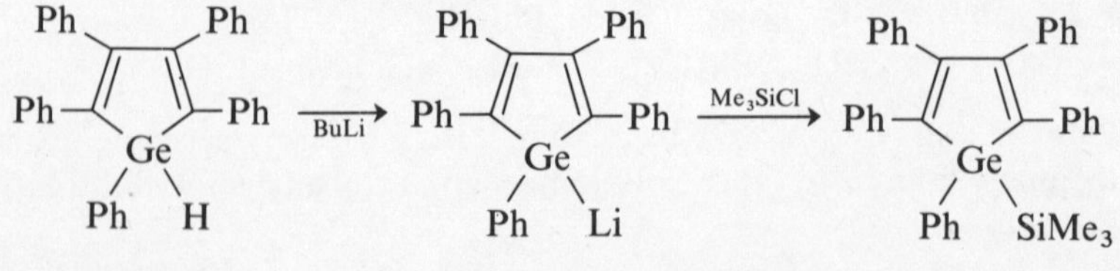

The bright red colour is ascribed to the germyllithium compound $(C_6H_5)_4C_4Ge(Li)C_6H_5$ (33). The germanium hydride

$$(C_6H_5)_4C_4Ge(H)C_6H_5$$

thus does form the germyllithium derivative $(C_6H_5)_4C_4Ge(Li)C_6H_5$ and triphenylgermane also forms triphenylgermyllithium under the same conditions. It thus appears that the ring structure does not confer

enhanced acidity on the hydride

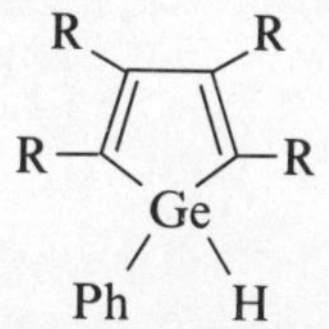

$$
\begin{array}{c}
R \qquad R \\
R \!-\!\!\overset{\displaystyle\big|}{\underset{\displaystyle\diagup\,\big\backslash}{Ge}}\!\!-\! R \\
Ph \qquad H
\end{array}
$$

Vyazankin, Razuvaev and coworkers (189) reported that triethyl-germyllithium reacts with triethylsilane, -germane and -stannane to give compounds of the $(C_2H_5)_3GeM(C_2H_5)_3$ type:

$$(C_2H_5)_3GeLi + (C_2H_5)_3GeH \;\rightarrow\; LiH + (C_2H_5)_3GeGe(C_2H_5)_3 \quad (42\%)$$

Tris(triphenylgermyl)germane $(Ph_3Ge)_3GeH$ (61) and α-naphthylphenyl methylgermane (22) are all cleared smoothly and in high yield by butyl-lithium:

$$R_3GeH + R'Li \;\rightarrow\; R_3GeLi + R'H$$

In the reaction between tribenzylgermane and butyllithium are observed alkylation, hydrogen–metal exchange, cleavage of $Ge-CH_2Ph$ bonds and coupling to hexabenzyldigermane:

$$(PhCH_2)_3GeH + BuLi \;\rightarrow\;$$

$$(PhCH_2)_3GeLi \!-\!\!\!\left[\begin{array}{l} \xrightarrow{\text{MeI}} (PhCH_2)_3GeMe \\[12pt] \xrightarrow[\text{(PhCH$_2$)$_3$GeH}]{} (PhCH_2)_6Ge_2 + LiH \end{array}\right.$$

Formation of $(PhCH_2)_3GeBu$ may be interpreted in terms of nucleophilic attack by BuLi on the polar hydride $(PhCH_2)_3\overset{\delta+}{Ge}-\overset{\delta-}{H}$ with displacement of H^-:

$$(PhCH_2)_3GeH + BuLi \;\rightarrow\; (PhCH_2)_3GeBu + LiH$$

Displacement of $PhCH_2$ leads to the formation of dibenzylbutylgermane and tetrabenzylgermane:

$$(PhCH_2)_3GeH + BuLi \;\rightarrow\; (PhCH_2)_2Ge(H)Bu + PhCH_2Li$$

$$(PhCH_2)_3GeH + PhCH_2Li \;\rightarrow\; (PhCH_2)_4Ge + LiH$$

An independent experiment confirmed that tribenzylgermane and benzyl-lithium give, among other products, tetrabenzylgermane.

Dibenzylbutylgermane retains the polar character of the original hydride and is therefore susceptible to further attack in the presence of an excess of butyllithium giving dibenzyldibutylgermane and benzyl-dibutylgermane (27):

$$(PhCH_2)_2Ge(H)Bu + BuLi \nearrow (PhCH_2)_2GeBu_2 + LiH$$
$$\searrow PhCH_2Ge(Bu)_2H$$

For the same reason benzyldibutylgermane will react with butyllithium giving, finally, benzyltributylgermane and tetrabutylgermane:

$$PhCH_2Ge(Bu)_2H + BuLi \nearrow PhCH_2GeBu_3$$
$$\searrow Bu_3GeH \xrightarrow{BuLi} Bu_4Ge + LiH$$

Experiments have been carried out by the same authors (27) on the reaction between diphenylgermane and butyllithium with a view to obtaining evidence for a dilithioderivative and to examining the selectivity of the reaction. With two equivalents of butyllithium and subsequent reaction with ethyl bromide there was evidence for the formation of Ph_2GeLi_2. Cleavage of phenyl groups was not observed but otherwise the behaviour was similar to that of benzylgermanes, metalation, alkylation and coupling reactions taking place:

$$Ph_2GeH_2 + 2\,BuLi$$
$$\downarrow$$
$$Ph_2GeLi_2 + Ph_2GeBuLi + Ph_2GeBu_2 + Ph_2LiGe\text{—}GeLiPh_2$$
$$\xrightarrow[EtBr]{}\Big\downarrow$$
$$Ph_2GeEt_2 + Ph_2GeBuEt + Ph_2GeBu_2 \qquad Ph_2EtGe\text{—}GeEtPh_2$$
$$\quad 2\% \qquad\qquad 20\% \qquad\qquad 12\% \qquad\qquad 28\%$$

(d) Action of Grignard reagents

Gilman and Zuech (57) have found that certain Grignard reagents will metalate triphenylgermane in refluxing tetrahydrofuran. Reaction of triphenylgermane with allylmagnesium chloride in refluxing tetrahydrofuran gave triphenylgermylmagnesium chloride, $Ph_3GeMgCl$.

The presence and structure of the Grignard reagent were confirmed by carbonation, which afforded the triphenylgermanecarboxylic acid, and by identification of the tetrahydrofuran cleavage product 4-hydroxy-butyltriphenylgermane:

$$(C_6H_5)_3GeH + CH_2\text{=}CH\text{—}CH_2MgCl \rightarrow (C_6H_5)_3GeMgCl$$
$$\underset{(H_2O)}{\overset{CO_2}{\swarrow}} \qquad\qquad\qquad \underset{(H_2O)}{\overset{THF}{\searrow}}$$
$$(C_6H_5)_3GeCOOH \qquad\qquad (C_6H_5)_3Ge(CH_2)_4OH$$

Triphenylallylgermane is also formed in low yield (3.9%). Allyl-magnesium bromide was also found to metalate triphenylgermane, as did phenylmagnesium bromide. n-Butylmagnesium bromide, on the other hand, appeared to be completely unreactive. Allylmagnesium chloride reacts in the THF with dibutylgermane according to the following alkylation reaction (142):

$$(C_4H_9)_2GeH_2 + CH_2{=}CH{-}CH_2MgBr \rightarrow$$

$$(C_4H_9)_2Ge(H)CH_2CH{=}CH_2 \quad (35\%)$$

Benzylmagnesium chloride also reacts in the same solvent with dibutyl-germane and heptylgermane, respectively, with formation of dibutyl-benzylgermane (yield, 28%), heptylbenzylgermane and heptyldibenzyl-germane (yield, 15%) (142). The action of phenylmagnesium bromide on dibutylgermane under the same experimental conditions leads to dibutylphenylgermane in 20% yield (142).

4-3-4-2 With Miscellaneous Organometallic Compounds. Vyazankin, Razuvaev and coworkers described in 1963 the first organogermanium compound of mercury, $(C_2H_5)_3GeHgGe(C_2H_5)_3$ (182a, 187, 188). This derivative was obtained by interaction between diethylmercury and triethylgermane in the absence of atmospheric oxygen at 100–120°C:

$$(C_2H_5)_3GeH + (C_2H_5)_2Hg \rightarrow C_2H_6 + (C_2H_5)_3GeHgC_2H_5$$

$$2\,(C_2H_5)_3GeH + (C_2H_5)_2Hg \rightarrow 2\,C_2H_6 + (C_2H_5)_3GeHgGe(C_2H_5)_3$$

Bis(trimethylgermyl)mercury (36) and bis(triphenylgermyl)mercury (181a) can be prepared in the same way.

In the action of diphenylgermane on diethylmercury, Neumann (126, 127) reports the formation of a crystalline cyclic tetramer of diphenyl-germanium:

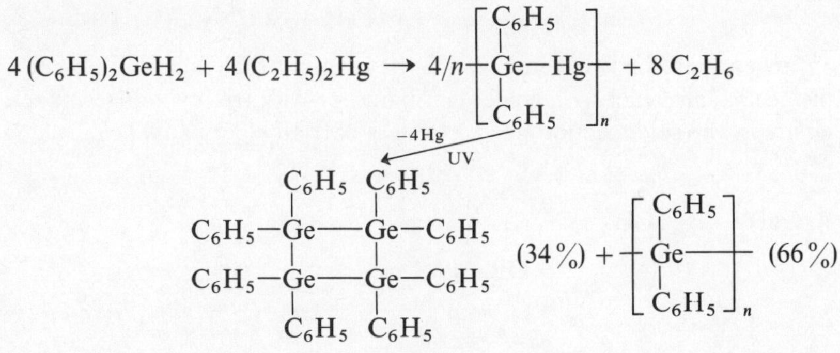

The reaction of diethylzinc with triethylgermane proceeds with the formation of compounds containing Ge—Zn—Ge group but also tetra-ethylgermane and hexaethyldigermane in accordance with the following equations (190):

$$(C_2H_5)_3GeH + (C_2H_5)_2Zn \rightarrow C_2H_6 + Zn + (C_2H_5)_4Ge$$

$$2(C_2H_5)_3GeH + (C_2H_5)_2Zn \rightarrow 2C_2H_6 + Zn + (C_2H_5)_6Ge_2$$

Triethylgermane reacts with diethylcadmium under mild conditions (3 h at 80–85°C) with the formation of bis(triethylgermyl)cadmium and ethane in yields of 78.6 and 90.3 % (185):

$$2(C_2H_5)_3GeH + (C_2H_5)_2Cd \rightarrow 2C_2H_6 + (C_2H_5)_3GeCdGe(C_2H_5)_3$$

By heating triethylgermane with triethylthallium, Vyazankin, Razuvaev and coworkers obtained tris(triethylgermyl)thallium and ethane in yields of 91 and 100 % respectively (89, 184).

Organometallic compounds with covalent bonds M—Sb and M—Bi (M = Si, Ge, Sn) can readily be obtained according to the general equation (88, 183, 191):

$$3(C_2H_5)_3MH + (C_2H_5)_3M' \rightarrow 3C_2H_6 + [(C_2H_5)_3M]_3M'$$

$$M = Si, Ge, Sn \qquad M' = Sb, Bi$$

In this type of reaction, triethylbismuth is more reactive than tri-ethylantimony, while the activity of the hydrides decreases in the series

$$(C_2H_5)_3SnH > (C_2H_5)_3GeH > (C_2H_5)_3SiH$$

The correctness of the series of hydride reactivity indicated above is also confirmed by the following reactions:

$$3(C_2H_5)_3GeH + [(C_2H_5)_3Si]_3M' \rightarrow$$

$$3(C_2H_5)_3SiH + [(C_2H_5)_3Ge]_3M' \qquad (M' = Bi, Sb)$$

$$6(C_2H_5)_3SnH + 2[(C_2H_5)_3Ge]_3Bi \rightarrow$$

$$6(C_2H_5)_3GeH + 2[(C_2H_5)_3Sn]_3Bi$$

Lutsenko and coworkers (13, 102) have found that esters of mercury-bis-acetic acid react with trialkylgermanes or with trialkyliodogermanes, giving as the sole reaction product esters of trialkylgermanyl acetic acid:

$$R_3GeH + Hg\left(CH_2-C\begin{matrix} \nearrow O \\ \searrow OR' \end{matrix}\right)_2 \rightarrow$$

$$R_3GeCH_2COOR' + Hg + CH_3COOR'$$

The structure of the latter has been established by reduction with lithium aluminium hydride, the corresponding germanium-containing alcohols being obtained. This represents a new method of preparation of esters containing Ge in the α-position to the ester group. However, trialkyliodogermanes were found more convenient for the preparation of organogermanium esters.

This reaction can be used to synthesize esters of dialkylgermyl-bis-acetic acid and dialkylgermyl acetic acid. By the reaction of dialkylgermanes with methylmercury-bis-acetate it proved possible, depending on the ratio of the reactants, to replace one or two of the hydrogen atoms by the organic residue (14):

$$R_2GeH_2 + Hg(CH_2COOCH_3)_2 \rightarrow$$

$$R_2Ge(H)CH_2COOCH_3 + Hg + CH_3COOCH_3 \quad \text{(Yield, 60\%)}$$

$$R_2GeH_2 + 2\,Hg(CH_2COOCH_3)_2 \rightarrow$$

$$R_2Ge(CH_2COOCH_3)_2 + 2\,Hg + 2\,CH_3COOCH_3 \quad \text{(Yield, 52\%)}$$

$$R = n\text{-}C_3H_7,\ n\text{-}C_4H_9$$

Di-n-butylethynylstibine reacts with tri-n-butylgermane by exchanging its ethynyl group for the hydrogen bonded to the germanium atom (125):

$$(C_4H_9)_2SbC \equiv CH + (C_4H_9)_3GeH \rightarrow$$

$$(C_4H_9)_3GeC \equiv CH + (C_4H_9)_2SbH$$

The reaction of triethylgermane with tetraethyllead proceeds with gradual replacement of ethyl groups in $(C_2H_5)_4Pb$ by triethylgermyl groups:

$$n(C_2H_5)_3GeH + (C_2H_5)_4Pb \rightarrow$$

$$[(C_2H_5)_3Ge]_nPb(C_2H_5)_{4-n} + nC_2H_6 \quad \text{(with } n = 1\text{–}4\text{)}$$

The side processes accompanying this reaction with formation of lead, tetraethylgermane, hexaethyldigermane have been discussed (182b).

Excess of dimethylgermane reacts with triiron dodecacarbonyl to give $[(CH_3)_2Ge]_3Fe_2(CO)_6$.

Diphenylgermane reacts also with the same compound.

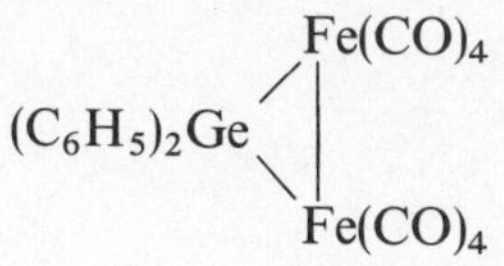

$$(C_6H_5)_2Ge \underset{\diagdown}{\overset{\diagup}{}} \begin{matrix} Fe(CO)_4 \\ | \\ Fe(CO)_4 \end{matrix}$$

was isolated (20a).

4-3-5   Addition reactions of organogermanium hydrides

4-3-5-1   *With Ethylenic and Acetylenic Derivatives*

(a) Arylgermanes additions

Very many addition reactions of organogermanium hydrides to unsaturated ethylenic or acetylenic substrates have been described in the literature. The Tables appended indicate a large variety of them. In 1957, Gilman and coworkers described the first addition reactions of triphenylgermane on various ethylenic compounds, such as octene-1 and cyclohexene (46), octadecene, triphenylallylgermane (54) and triphenylallylsilane (117). All these reactions are initiated by the peroxides and ultraviolet irradiation and proceed *via* a radical mechanism. Peroxide initiation fails to cause addition of triphenylgermane to 1,1-diphenylethylene (46).

However, triphenylgermane adds to octene-1 and cyclohexene in the absence of catalyst in yields of 29 and 39 % below those observed in the presence of a radical catalyst (73). In the absence of solvent and a catalyst, triphenylgermane also adds to a large range of organic compounds containing olefinic double bonds with and without activating functional groups, according to the general reaction (73):

$$(C_6H_5)_3GeH + CH_2{=}CH{-}Y \rightarrow (C_6H_5)_3GeCH_2CH_2Y$$

$$Y = C_6H_5, -OCOCH_3, -CN, -CONH_2, -COCH_3,$$

$$-COOCH_3, -(CH_2)_2CH_2OH$$

Triphenylgermane additions to phenylacetylene and 2-methyl-3-butyn-2-ol are also observed without catalyst, but with low or medium yields (12 and 49 % respectively) (73). No reducing properties of the triphenylgermane towards the functional group were observed.

The study of the additions of the phenyl- and phenylhalogermanes to hexene-1 served to establish beyond doubt the radical mechanism of the triphenylgermane addition on the ethylenic double bond (140). An important catalytic action of UV irradiation and azobisisobutyronitrile (AIBN) and the complete inhibition of these reactions by "galvinoxyl" were observed. The addition mechanism can be indicated schematically as follows:

$$(C_6H_5)_3GeH \xrightarrow[\text{AIBN}]{\text{UV}} (C_6H_5)_3Ge^{\cdot} + H^{\cdot}$$

$$(C_6H_5)_3Ge^{\cdot} + CH_2{=}CH{-}R \rightarrow$$

$$(C_6H_5)_3GeCH_2{-}\dot{C}H{-}R$$

$$(C_6H_5)_3GeCH_2{-}\dot{C}H{-}R + (C_6H_5)_3GeH \rightarrow$$

$$(C_6H_5)_3GeCH_2CH_2R + (C_6H_5)_3Ge^{\cdot}$$

Owing to steric hindrance and the stability of intermediate radicals, the germanium group adds preferentially at the end of the chain with the formation of a more stable secondary radical. This is confirmed by the complete absence of a branched addition derivative of the type $\geqslant$Ge—CH—R. Apart from this, the intermediary radicals
$\quad\quad\quad\quad\;|$
$\quad\quad\quad\quad CH_3$
$\geqslant$Ge—CH$_2$ĊH—R which are formed seem to initiate the sequence of the reaction and it appears that a chain reaction mechanism is involved. This seems to be confirmed by the fact that no recombinations of radicals $\geqslant$Ge' and H' with formation of hydrogen or digermane $\geqslant$Ge—Ge$\leqslant$ have ever been observed (140).

The competition of the addition reactions of triphenylgermane (and the phenylchlorogermanes) on the double olefinic bond and the carbonyl group has been studied on some α- and γ-ethylenic ketones (151, 153). Triphenylgermane adds by a radical mechanism to the allylacetone and vinyl methyl ketone carbon–carbon double bond. But with 2-methyl-2-heptene-6-one, steric hindrance at the double bond directs the triphenylgermanium group to the carbonyl group (cf. addition Table, arylgermanium hydrides). A radical 1-4 addition of triphenylgermane to mesityl oxide is also observed (151, 153):

$$
\begin{array}{c}
CH_3 \\
\diagdown \\
C{=}CH{-}C{=}O + Ph_3GeH \\
\diagup | \\
CH_3 CH_3
\end{array}
\xrightarrow{\text{AIBN}}
\begin{array}{c}
CH_3 \\
\diagdown \\
CH{-}CH{=}C{-}OGePh_3 \\
\diagup | \\
CH_3 CH_3
\end{array}
$$

The action of diphenylgermane on methyl vinyl ketone and allylacetone leads in a first stage to the formation of linear mono- and di-addition derivatives: $Ph_2Ge(H)(CH_2)_nCOCH_3$, $Ph_2Ge[(CH_2)_nCOCH_3]_2$ ($n = 2$ or 4). The monoaddition derivatives heated in the presence of AIBN lead through intramolecular addition of Ge—H to the carbonyl group to a heterocyclic compound with Ge—O—C bond (medium yield 70%) (151). Cf. addition reactions of organogermanium hydrides with carbonyl derivatives.

The radical addition reactions of diphenylgermane to acrylonitrile and allyl cyanide with formation of mono- and di-addition derivatives have been described recently (116).

Poly-additions of the phenylgermanes (Ph_2GeH_2 and $PhGeH_3$) to methyl acrylate, cyclohexene, vinyl methyl ether as well as to hexyne-1 have been studied (152).

Condensations of diphenylgermane and dibutylgermane on the α-ω-di-ynes, such as 1-5 hexadiyne, in spite of drastic reaction conditions (7 days at 120°C H_2PtCl_6 catalysis) lead to polymers with a low degree of polymerization, $+Ge(R)_2CH=CH-(CH_2)_2CH=CH+_n$ $(n \sim 35)$ (131).

(b) Alkylgermane additions

Addition reactions of alkylgermanes to unsaturated ethylenic and acetylenic substrates are also very numerous. Lesbre, Satgé and coworkers have studied them particularly beginning in 1958 (93, 95, 96, 105, 107, 108, 141–143, 148, 151). The results can be summarized in the following equations:

$$R_3GeH + CH_2=CHX \rightarrow R_3GeCH_2CH_2X + R_3Ge-\underset{\underset{CH_3}{|}}{CH}-X$$
$$(\text{traces})$$

$$X = -(CH_2)_5CH_3, -CN, -CH_2CN, -COOH, -COOR', -CH_2OH,$$
$$-CHO, -COCH_3, OC_4H_9, -CH_2NH_2, -CH_2SH, -GeR_3,$$
$$-CH_2GeR_3, -CH_2OGeR_3$$

$$R_3GeH + HC≡CY \rightarrow R_3GeCH_2=CHY$$

$$Y = H, -C_6H_5, -(CH_2)_3CH_3, -CH_2OH, -C(OH)(CH_3)_2, -CH_2Cl,$$
$$-Ge(C_4H_9)_3, -CH_2OGeR_3$$

Several polyaddition reactions of the di- and trihydrides R_2GeH_2 and $RGeH_3$ to acrylonitrile, methyl acrylate, vinyl methyl ketone, vinyl butyl ether have also been recorded. The action of acrylic acid and crotonaldehyde on dialkylgermanes, R_2GeH_2, leads only to the formation of the monoaddition derivative with a Ge—H bond (142):

$$R_2GeH_2 + CH_2=CH-X \rightarrow R_2Ge-\underset{\underset{H}{|}}{CH_2CH_2X}$$

These reactions represent a general method of synthesis of functionally substituted organogermanium derivatives, the organic function being fixed in the β position with respect to germanium. The acetylenic additions lead to organogermanium compounds containing a vinyl group substituted by various organic functions. The formation of 1:2 or 1:3 addition derivatives has also been reported in the reactions R_3GeH + acrylonitrile or R_3GeH + methyl acrylate. This fact can be attributed to a telomerization process (142).

These addition reactions of the alkylgermanes are not accompanied by secondary reactions of the organic functional groups. However, in the addition of triethylgermane to allyl (105, 121) and propargyl (105,

107, 108) chlorides, in the presence of chloroplatinic acid, a noticeable percentage of reduction of these halides to propene or propyne is observed. The reduction of the allyl halides (α = Cl, Br, I) by the trialkylgermanes is even exclusive in the absence of an addition catalyst (142). The catalytic dehydrocondensation reactions of the trialkylgermanes with certain unsaturated alcohols or acids as well as the additions of trialkylgermanes on the carbonyl group of unsaturated aldehydes and ketones will be described in a later paragraph.

These addition reactions are generally of the anti-Markownikoff type. The compounds obtained have a linear structure.

The use of radical reaction initiators such as benzoyl peroxide or ultraviolet radiation has been limited owing to the quick polymerization of derivatives such as acrylonitrile, acrylic acid and its esters. In the few cases which have been studied, however, (R_3GeH + allyl alcohol for example) these catalysts do not appear to be very active (142). In the case of vinyl derivatives with activated double bonds (acrylonitrile, acrylates, etc.) the presence of a catalyst is not absolutely indispensable and the additions are observed when the reagents are heated at reflux (142). However, the addition of chloroplatinic acid seems to improve the regularity and the yield of these addition reactions (139, 142).

The platinum catalysts (H_2PtCl_6 or Pt/asbestos) clearly favour the addition reactions of the trialkylgermanes to the allylic substituted derivatives CH_2=CH—CH_2X.

The high efficiency of chloroplatinic acid, an initiator of electrophilic reactions in comparable additions of silanes (167a), seems to point to an ionic addition mechanism which is favoured by the $\overset{\delta+}{Ge}$—$\overset{\delta-}{H}$ polarity in the trialkylgermanes. For polar additions a mechanism involving a four-centre transition state can be considered as one of the alternatives (cf. case of the alkylstannanes (175)):

$$R_3GeH + CH_2{=}CH{-}R' \rightarrow \begin{array}{c} R_3Ge\cdots H \\ \vdots \qquad \vdots \diagup^{H} \\ H_2C{=\!=\!=}C{\diagdown}_{R'} \end{array} \rightarrow R_3GeCH_2CH_2R'$$

In the case of a polar mechanism, the R_3Ge-electrophilic agent germanium group must attack the ethenoid carbon having the highest electron density, with formation of the linear derivative, owing to the electron-donating effect of alkyl substituents:

$$R_3\overset{\delta+}{Ge}{-}\overset{\delta-}{H} + \overset{\delta-}{C}H_2{=}\overset{\delta+}{C}H{\leftarrow}R \rightarrow R_3GeCH_2CH_2R$$

If a strongly electron-attracting substituent (—CN, —COOH, —COOR') is attached to the vinyl group, polarity is reversed and the formation of branched derivatives can be observed:

$$\overset{\delta+}{R_3Ge}\overset{\delta-}{-H} + \overset{\delta+}{CH_2}\overset{\delta-}{=CH} \rightarrowtail X \rightarrow R_3Ge-\underset{\underset{CH_3}{|}}{CH}-X$$

In fact, the formation of linear addition derivatives preponderates (142) or is exclusive (139).

In consideration of the low polarity of the germanium–hydrogen bond, the transition state is likely to depend more on the steric effects than on the polarization of the bonds. The hindering germanium group adds to the terminal carbon. However, addition by a radical mechanism can also lead to the same linear structure:

$$R_3GeH \rightarrow R_3Ge^{\cdot} + H^{\cdot}$$

$$R_3Ge^{\cdot} + CH_2=CH-R' \rightarrow R_3GeCH_2\dot{C}H-R'$$

$$R_3GeH + R_3GeCH_2\dot{C}HR' \rightarrow R_3GeCH_2CH_2R' + R_3Ge^{\cdot}$$

In fact, it seems that there is a superposition of two mechanisms and that the predominance of one over the other is closely linked with the experimental conditions, the polarity of the medium, the polarization of the unsaturated reagent, the catalysts, etc.

Dzhurinskaya, Mironov and Petrov also recorded the catalyzed addition reactions of triethylgermane by H_2PtCl_6 to allylic and propargylic alcohols, as well as to acrolein (35). There was a difference between $(C_2H_5)_3GeH$ and $(C_2H_5)_3SiH$ only in their reaction with allyl alcohol; while triethylsilane formed exclusively triethylallyloxysilane, $(C_2H_5)_3GeH$ added to the multiple bond:

$$CH_2=CH-CH_2OH + (C_2H_5)_3SiH \xrightarrow{\ H_2PtCl_6\ }$$
$$CH_2=CH-CH_2OSi(C_2H_5)_3 + H_2$$

Tributylgermane adds thermally to alkenylboronates in high yields. This reaction is only very partially inhibited by hydroquinone (18).

The addition reactions of trialkylgermanes to conjugated systems have been studied by several authors: triethylgermane reacts with 1,3-butadiene only in the presence of chloroplatinic acid. The addition is of the 1,4 type (105, 148). Chloroplatinic acid also favours the polar additions of the 1,4 type to vinyl methyl ketone (105):

$$R_3\overset{\delta+}{Ge}-\overset{\delta-}{H} + H_2C{=}CH-\underset{\underset{O}{\|}}{C}-CH_3 \xrightarrow{H_2PtCl_6} CH_3-CH{=}C\overset{\diagup CH_3}{\diagdown OGeR_3}$$

The addition to the double ethylenic bond is observed only in the absence of a catalyst (142). Fish and Kuivila (44, 45) undertook a comprehensive study in the same field. The chloroplatinic acid-catalyzed addition of trimethylgermane to allene, 1,3-butadiene, cyclopentadiene, 1,3-cyclohexadiene, 1,4-cyclohexadiene, 1,3-cyclooctadiene, 1,5-cyclooctadiene, 4-vinylcyclohexene and bicyclo(2,2,1)heptadiene-2,5 has been described. Structures and configurations of the adducts have been assigned on the basis of IR and NMR spectroscopy as well as chemical degradation. The addition of trimethylgermane to allene leads by the fixation on the germanium group to the terminal carbon or the central carbon to two isomers: trimethylallylgermane (60.5%) and trimethyl-prop-2-en-2-yl-germane (39.5%). The 1,4 addition to 1,3-butadiene leads to the *cis*- and *trans*-isomers of trimethylcrotylgermane. Besides the normal 1,4 (or 2,1) addition product, the unexpected formation of trimethyl-cyclopent-1-en-1-yl-germane, $Me_3Ge{-}\langle\!\langle\ \rangle$, is observed in the action of trimethyl-germane on cyclopentadiene.

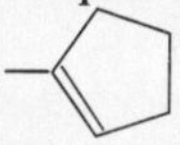

A rationale for the formation of this product can be found by assuming the formation of a bis-adduct of the trimethylgermane-platinum complex as shown in the following equation in which the complex is given a structure analogous to that used for the silane case by Ryan and Speier (141a):

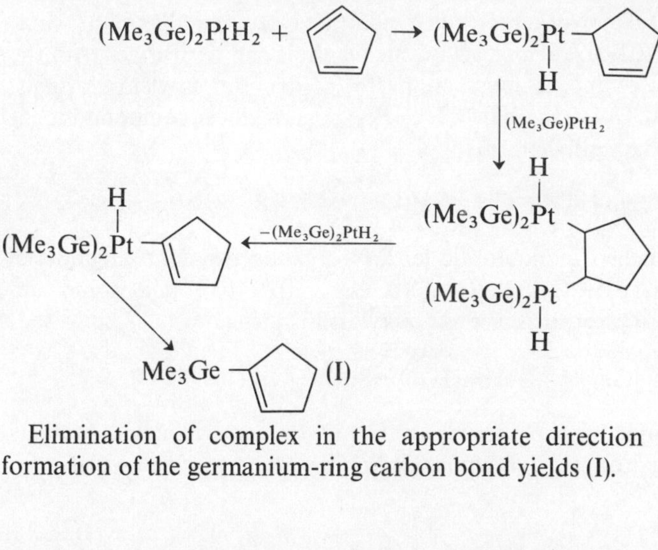

Elimination of complex in the appropriate direction followed by formation of the germanium-ring carbon bond yields (I).

The addition products with 1,4-cyclohexadiene and 1,3-cyclohexadiene are identical. This leads to the conclusion that 1-4-cyclohexadiene isomerizes to 1,3-cyclohexadiene which then reacts with the hydride to

form Me$_3$Ge —⬡ . The addition derivatives of trimethylgermane

on 1,3-cyclooctadiene or 1,5-cyclooctadiene are also identical. It was found that in the presence of catalytic amounts of chloroplatinic acid and trimethylgermane, 1,5-cyclooctadiene was isomerized to 1,3-cyclooctadiene. The addition of $(CH_3)_3GeH$ to norbornadiene yields 5-trimethylgermylnorbornene containing the "endo" and "exo" isomers in a ratio 5/1.

The 4-vinylcyclohexene reaction yielded one pure product. Trimethylgermane adds to the exocyclic double bond. The 4-vinylcyclohexene did not undergo isomerization (cf. addition Tables of alkylgermanium hydrides).

Cullen and Styan described in 1966 the reactions of the hydrides of Group IV to fluoroacetylenes (31) and perhalocyclobutenes (32). Hexafluoro-2-butyne and triethylgermane did not react in the dark at 20°C. However, on ultraviolet irradiation, the trans-adduct is formed in good percentage (92%) (31). Me$_3$GeH reacted with $\overline{FC\!=\!CFCF_2CF_2}$ to give the 1:1 adduct $(CH_3)_3Ge-\overline{CF-CHF-CF_2-CF_2}$.

The main products from the interaction of $(CH_3)_3GeH$ with $\overline{ClC\!=\!CClCF_2CF_2}$ are $(CH_3)_3GeCl$ and $(CH_3)_3Ge-\overline{C\!=\!CClCF_2CF_2}$; other minor products have been obtained (32) (see Tables). In the last two cases, the additions are observed after prolonged heating.

Shikhiev and coworkers studied additions of triethylgermane catalysed by H_2PtCl_6 to various acetylenic alcohols and derivatives with the aim of syntheses and chemical studies of germanium ethylenic formals, $ClCH_2CH_2OCH_2OCH_2CH\!=\!CHGeEt_3$ (1), germanium compounds containing secondary ethylenic alcohols,

$$Et_3GeCH\!=\!CH-CH(OH)-CHRR'\quad(164),$$

linear or branched monoatomic tertiary organogermanium alcohols of ethylene series $RR'C(OH)CH\!=\!CHGeEt_3$ (162, 163), and some unsaturated alkoxyderivatives of tetraalkyldisiloxanes,

$$(Et_3GeCH\!=\!CH-CH_2O-SiRR')_2O\quad(161, 162a).$$

Hydrogermylations of some acetylenic hydroxycycloalkane ethers (164a) and hydroxyalkylvinylphenylsulphide (90a) are also described.

Triethylgermane additions to several acetylenic, diacetylenic or ethylenic and acetylenic alcohols or glycols have been extensively studied and published by Gverdsiteli and coworkers. These studies concern particularly additions (catalysed by H_2PtCl_6) to the acetylenic tertiary alcohols

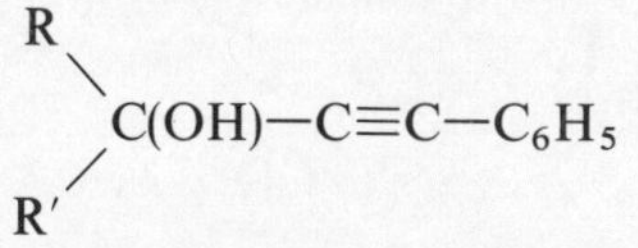

$$\begin{array}{c} R \\ \diagdown \\ \,\,C(OH)-C{\equiv}C-C_6H_5 \quad (65), \\ \diagup \\ R' \end{array}$$

the tetramethylbutynedioldiethylether (69), the asymmetrical acetylenic γ-glycol

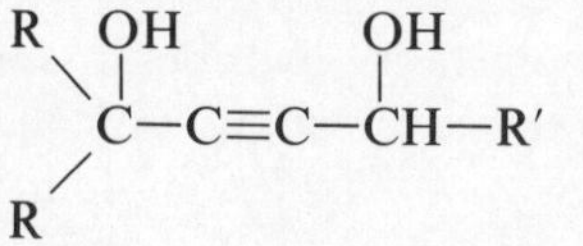

$$\begin{array}{c} R\quad OH \qquad\quad OH \\ \diagdown\,\,| \qquad\qquad\,\, | \\ \,\,C-C{\equiv}C-CH-R' \quad (67), \\ \diagup \\ R \end{array}$$

the vinylacetylenic carbinols $R-CHOH-C{\equiv}C-CH{=}CH_2$ (70, 71), the diacetylenic glycols

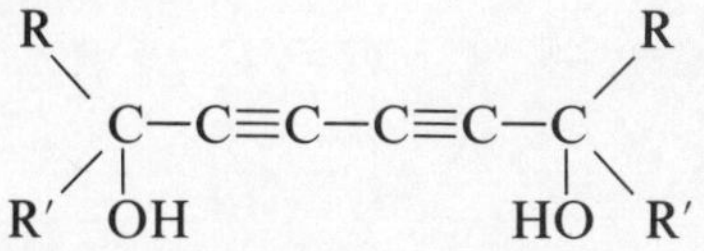

$$\begin{array}{c} R \qquad\qquad\qquad\quad R \\ \diagdown \qquad\qquad\qquad\, \diagup \\ \,\,C-C{\equiv}C-C{\equiv}C-C \quad (66,68), \\ \diagup\,\,| \qquad\qquad | \,\diagdown \\ R'\,\,\,OH \qquad\quad HO\,\,\,R' \end{array}$$

the γ-methoxyacetylenic glycols (69a), the phenylacetylenic carbinols (64a), and some unsaturated hydroxy-compounds (69b).The main results regarding these latter reactions have been transcribed in the Tables appended.

The addition of dibutylgermane to diethyldivinylgermane in the presence of benzoyl peroxide leads to 17% of cyclic derivative.

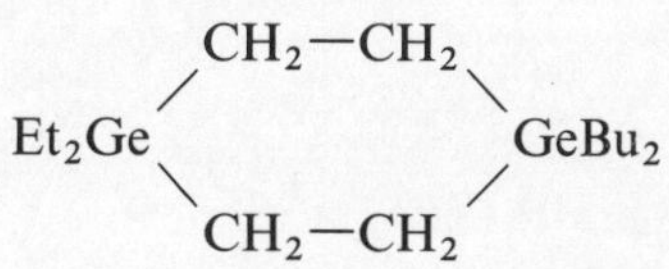

$$\begin{array}{c} CH_2-CH_2 \\ \diagup \qquad\qquad \diagdown \\ Et_2Ge \qquad\qquad\quad GeBu_2 \\ \diagdown \qquad\qquad \diagup \\ CH_2-CH_2 \end{array}$$

and a high percentage of polymers (112). Tri-n-butylgermane reacts with triethylethynylsilane in two directions, forming 1-triethylsilyl-2-tri-n-butylgermylethylene in accordance with the scheme (124):

$$(n\text{-}C_4H_9)_3GeH + (C_2H_5)_3SiC{\equiv}CH \rightarrow$$

$$(n\text{-}C_4H_9)_3GeCH{=}CHSi(C_2H_5)_3$$

and 1,2-bis(tri-*n*-butylgermyl)ethylene. The formation of the latter compound shows that in addition to the addition reaction, exchange of the ethynyl group bonded to the silicon for a hydrogen atom attached to the germanium also occurs and the resulting tri-*n*-butylethynylgermane reacts with tri-*n*-butylgermane with the formation of 1,2-bis(tri-*n*-butylgermyl)ethylene according to the scheme:

$$(n\text{-}C_4H_9)_3GeH + (C_2H_5)_3Si-C\equiv CH \rightarrow$$

$$(n\text{-}C_4H_9)_3GeC\equiv CH + (C_2H_5)_3SiH$$

$$(n\text{-}C_4H_9)_3GeH + (n\text{-}C_4H_9)_3GeC\equiv CH \rightarrow$$

$$(n\text{-}C_4H_9)_3GeCH=CHGe(C_4H_9\text{-}n)_3$$

The addition of organometallic hydrides to organometallic monoacetylenides proceeds in the *cis* position with the formation of *trans*-isomers, possibly through a π-complex or four-membered transition state according to (124):

$$R_3M-C\equiv CH + R_3MH \rightarrow$$

$$
\begin{array}{ccc}
R_3MC\equiv CH & \rightarrow \quad R_3MC=CH & \rightarrow \quad R_3M \qquad\qquad H \\
\downarrow & \vdots \qquad\vdots & \diagdown \qquad \diagup \\
H-MR_3 & H\cdots MR_3 & C=C \\
& & \diagup \qquad \diagdown \\
& & H \qquad\qquad MR_3
\end{array}
$$

Triethylgermane reacts with allyl- and vinylphosphonic dichlorides and with the corresponding esters at elevated temperatures in presence of catalytic amounts of chloroplatinic acid (34b):

$$(C_2H_5)_3GeH + CH_2=CH-CH_2P(O)X_2 \xrightarrow{H_2PtCl_6}$$

$$(C_2H_5)_3Ge(CH_2)_3P(O)X_2$$

$$X = Cl, OC_2H_5, OC_4H_9$$

$$(C_2H_5)_3GeH + CH_2=CHP(O)Cl_2 \xrightarrow{H_2PtCl_6}$$

$$(C_2H_5)_3GeCH_2CH_2P(O)Cl_2$$

4-3-5-2 With Carbonyl Derivatives. Lesbre and Satgé (97) described in 1962 the first condensation reactions of the alkylgermanium hydrides on the carbonyl group of the saturated aldehydes and ketones. At 150°C and in the presence of copper powder the trialkylgermanes, R_3GeH, add to heptanal, benzaldehyde, ethyl butyl ketone, methyl hexyl ketone and cyclohexanone with formation of trialkylalkoxygermanes:

$$R_3GeH + R'CHO \rightarrow R_3GeOCH_2R'$$

The additions proceed probably by an ionic mechanism and they give high product yields:

$$\overset{\delta+}{R_3Ge} - \overset{\delta-}{H} + \overset{\delta-}{O}=\overset{\delta+}{C}\diagup^{R'}_{\diagdown R''} \rightarrow R_3Ge-O-CH\diagup^{R'}_{\diagdown R''}$$

With the same catalyst, the trialkylgermanes add preferentially to the carbonyl group of ethylenic aldehydes and ketones such as undekenal and methylheptenone (144). Hexafluoroacetone adds to trimethyl-germane without catalyst to form the 1,1,1,3,3,3-hexafluoropropoxy derivative (30, 168):

$$(CH_3)_3GeH + CF_3COCF_3 \rightarrow (CH_3)_3Ge-OCH(CF_3)_2$$

To account for the direction of addition of the hexafluoroacetone to the Group IV hydrides, a reaction mechanism involving nucleophilic attack by the carbonyl oxygen on the central metal atom is postulated. Thus a five-coordinated intermediate of the type $(CH_3)_3H\bar{Ge}-O-\overset{+}{C}(CF_3)_2$ would be first formed, followed by Ge—H cleavage induced by the carbonyl carbon from the same intermediate.

On the other hand, the study of the triphenyl (and diphenylchloro-germane) condensation on the carbonyl group of saturated ketones and aldehydes (140) (cyclohexanone, methyl isobutyl ketone, benzaldehyde, isobutyraldehyde) allowed the demonstration in all cases of a radical addition mechanism. This is confirmed by the high activity of initiators such as UV radiation and AIBN and the complete inhibition of these reactions by "galvinoxyl". The catalysts of "ionic" reactions, such as $ZnCl_2$ and H_2PtCl_6, are without action in this type of reaction (140). The following reaction mechanism can be considered:

$$\geqslant Ge-H \xrightarrow[AIBN]{UV} \geqslant Ge^{\cdot} + H^{\cdot}$$

$$\geqslant Ge^{\cdot} + {>}C{=}O \longrightarrow \geqslant Ge-O-\dot{C}{<}$$

$$\geqslant Ge-O-\dot{C}{<} + \geqslant Ge-H \longrightarrow \geqslant Ge-O-CH{<} + \geqslant Ge^{\cdot}$$

With methyl-2-heptene-2-one-6, the steric hindrance around the double ethylenic bond directs the triphenylgermane addition toward the carbonyl group (151, 153):

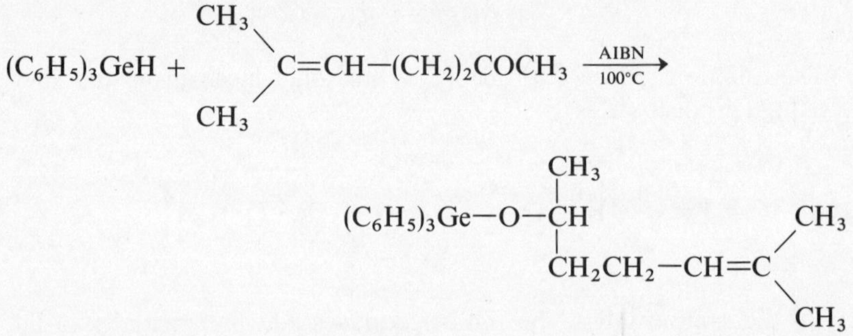

$$(C_6H_5)_3GeH + \underset{CH_3}{\overset{CH_3}{\diagdown}}C{=}CH{-}(CH_2)_2COCH_3 \xrightarrow[100°C]{AIBN}$$

$$(C_6H_5)_3Ge{-}O{-}\underset{\underset{CH_2CH_2-CH=C}{|}}{\overset{\overset{CH_3}{|}}{CH}}\underset{\diagdown CH_3}{\overset{\diagup CH_3}{}}$$

On the other hand, it is to be pointed out that the monoaddition derivatives of diphenylgermane to vinyl methyl ketone or allylacetone cyclize by a radical route, by an intramolecular addition of Ge—H to the carbonyl group and formation of a heterocyclic compound with Ge—O—C bond (medium yield 70%) (151). The formation of 20–30% of polymers having their origin in an intermolecular addition of Ge—H on the carbonyl of a neighbouring molecule is noted:

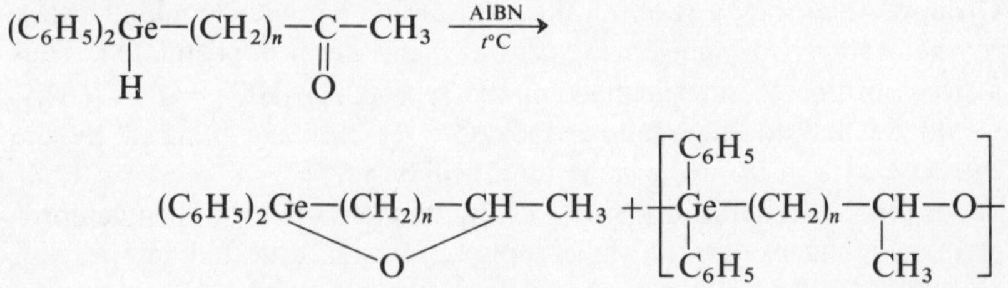

$$(C_6H_5)_2\underset{\underset{H}{|}}{Ge}{-}(CH_2)_n{-}\underset{\underset{O}{||}}{C}{-}CH_3 \xrightarrow[t°C]{AIBN}$$

$$(C_6H_5)_2Ge{-}(CH_2)_n{-}\underset{O}{\underbrace{\qquad}}CH{-}CH_3 + \left[\underset{\underset{C_6H_5}{|}}{\overset{\overset{C_6H_5}{|}}{Ge}}{-}(CH_2)_n{-}\underset{\underset{CH_3}{|}}{CH}{-}O\right]$$

$$(n = 2 \text{ or } 4)$$

The same type of reaction is observed with Et_2GeH_2 (106a).

Triethylgermane with $\overset{\delta+}{Ge}{-}\overset{\delta-}{H}$ polarity and triphenylgermane, mainly owing to steric hindrance, add on the carbonyl group of diphenylketene with formation of the corresponding enoxygermane (141, 151):

$$R_3GeH + Ph_2C{=}C{=}O \rightarrow R_3Ge{-}OCH{=}CPh_2 \qquad (R = Et \text{ or } Ph).$$

Trialkylgermanes add to ketene in 2–4 hours under UV light to form the $R_3Ge\underset{\underset{O}{||}}{C}{-}CH_3$ compounds (yields: R = Et, 60%; R = Pr, 70%; R = Bu, 84%) (83a).

4-3-5-3 With Unsaturated α-Oxides. Triethylgermane adds to α-oxides in the presence of chlorplatinic acid in the 1,4 position, like trialkylsilanes, forming triethylalkenoxygermanes (23):

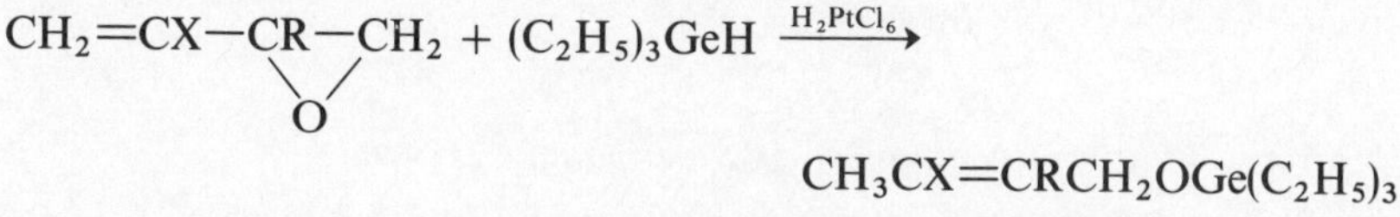

$$CH_2{=}CX{-}\underset{\displaystyle \diagdown \; O \; \diagup}{CR{-}CH_2} + (C_2H_5)_3GeH \xrightarrow{H_2PtCl_6}$$

$$CH_3CX{=}CRCH_2OGe(C_2H_5)_3$$

with $R = H$, $X = H$; $R = CH_3$, $X = H$; $R = H$, $X = Cl$.

Triphenylgermane adds to butadiene and isoprene oxides under the same conditions both to the oxide and to the ethylene bonds:

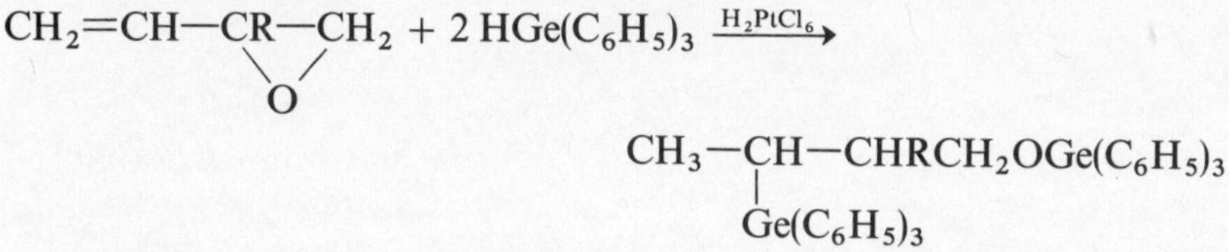

$$CH_2{=}CH{-}\underset{\displaystyle \diagdown \; O \; \diagup}{CR{-}CH_2} + 2\,HGe(C_6H_5)_3 \xrightarrow{H_2PtCl_6}$$

$$CH_3{-}\underset{\displaystyle Ge(C_6H_5)_3}{\underset{|}{CH}}{-}CHRCH_2OGe(C_6H_5)_3$$

Triphenyl-3-(triphenylgermyl)butoxygermane was also obtained from crotyl alcohol in the following way:

$$CH_3CH{=}CHCH_2OH \xrightarrow[Cu]{HGe(C_6H_5)_3}$$

$$CH_3CH{=}CHCH_2OGe(C_6H_5)_3 \xrightarrow[H_2PtCl_6]{HGe(C_6H_5)_3}$$

$$CH_3{-}\underset{\displaystyle Ge(C_6H_5)_3}{\underset{|}{CH}}{-}CH_2CH_2OGe(C_6H_5)_3$$

The addition of triethylgermane to butadiene, isoprene and chloroprene oxides in the presence of chloroplatinic acid takes place to the extent of 40–60%, and reduction of the oxides to alcohols and their isomerization to aldehydes takes place to the extent of 25–30%.

Compounds of the type $HM(Alkyl)_3$ ($M = Si, Ge, Sn$) reduce α-oxides of alka-1,3-dienes and can be arranged in the following sequence with respect to reducing power:

$$HSi(Alk)_3 < HGe(Alk)_3 < HSn(Alk)_3$$

Triethylgermane adds to piperylene monoxide principally in the 1-4 position in the presence of chloroplatinic acid (123a).

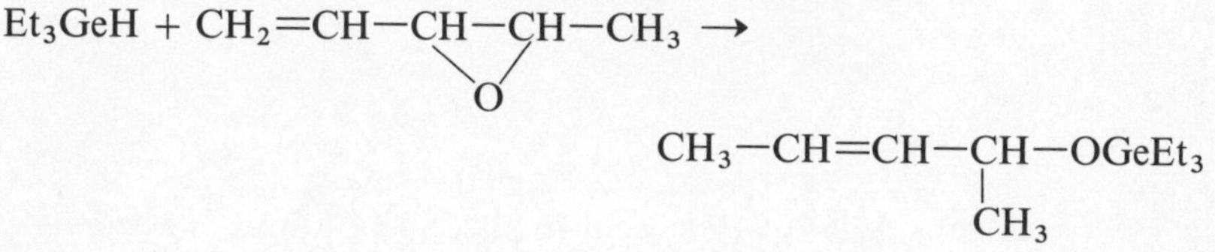

$$Et_3GeH + CH_2{=}CH{-}\underset{\displaystyle \diagdown \; O \; \diagup}{CH{-}CH}{-}CH_3 \rightarrow$$

$$CH_3{-}CH{=}CH{-}\underset{\displaystyle CH_3}{\underset{|}{CH}}{-}OGeEt_3$$

Besides the addition reaction, the reduction of the oxide to the alcohol (pent-3-ene-2-ol) was observed.

4-3-5-4   Addition Reactions of Alkylgermanium Hydrides

Table 4-5 (a–c) showing some addition reactions of alkylgermanium hydrides appears on the following pages.

Table 4-5

Addition Reactions of Alkylgermanium Hydrides

(a) To ethylenic compounds

UNSATURATED COMPOUND	GERMANE	CATALYST	REACTION TEMP. (°C)	TIME (h)	REACTION PRODUCTS	B.P. OR (M.P.) (°C/mm)	YIELD (%)	REFERENCES
$CH_2=CH-CH_2$	$(CH_3)_3GeH$	H_2PtCl_6	115	8.5	$(CH_3)_3GeCH_2CH=CH_2$ (60.5%) / $(CH_3)_3Ge-\overset{CH_3}{C}=CH_2$ (39.5%)	97–98/760	47.5	45
$CH_2=CHCN$	$(C_2H_5)_3GeH$	none	150	24	$(C_2H_5)_3GeCH_2CH_2CN$	225–27/760		93, 142
$CH_2=CHCN$	$(C_3H_7)_3GeH$	none	150	24	$(C_3H_7)_3GeCH_2CH_2CN$	152/11	24	93, 142
$CH_2=CH-CN$	$(C_4H_9)_3GeH$	none	to boil	5	$(C_4H_9)_3GeCH_2CH_2CN$	168–69/11	42	93, 142
$CH_2=CH-CN$	$(C_5H_{11})_3GeH$	none	150	10	$n\text{-}(C_5H_{11})_3GeCH_2CH_2CN$	193–94/11	40	93, 142
$CH_2=CH-CN$	$(C_3H_7)_2GeH_2$	none	150	20	$(C_3H_7)_2Ge(H)CH_2CH_2CN$	75/0.4	21	142
					$(C_3H_7)_2Ge(CH_2CH_2CN)_2$	140/0.4	31	
$CH_2=CH-CHO$	$(C_2H_5)_3GeH$	Pt/ asbestos	100	24	$(C_2H_5)_3GeCH_2CH_2CHO$	129–31/17	35	93
$CH_2=CH-CHO$	$(C_2H_5)_3GeH$	H_2PtCl_6	90	10	$(C_2H_5)_3GeCH_2CH_2CHO$	136–38/22	42	35
	$(n\text{-}C_4H_9)_3GeH$	Pt/ asbestos			$(n\text{-}C_4H_9)_3GeCH_2CH_2CHO$	125–28/1	60	93, 142
$CH_2=CHCH_2Cl$	$(C_2H_5)_3GeH$	H_2PtCl_6	to boil	2	$(C_2H_5)_3GeCH_2CH_2CH_2Cl$ (5%) / $(C_2H_5)_3GeCH=CH_2$ (26%) / $(C_2H_5)_3GeCl$ (55.5%)	—		121
$CH_2=CHCH_2OH$	$(C_2H_5)_3GeH$	H_2PtCl_6	90	4	$(C_2H_5)_3GeCH_2CH_2CH_2OH$	89–90/3	73	35
$CH_2=CHCH_2OH$	$(C_3H_7)_3GeH$	$(PhCOO)_2$	to boil	10	$(C_3H_7)_3GeCH_2CH_2CH_2OH$	175–80/20	14	93, 142
$CH_2=CHCH_2OH$	$(C_4H_9)_3GeH$	Pt/ asbestos	to boil	15	$(C_4H_9)_3GeCH_2CH_2CH_2OH$	153–55/3	57	93, 142
$CH_2=CHCH_2SH$	$(C_4H_9)_3GeH$	H_2PtCl_6	to boil	15	$(C_4H_9)_3GeCH_2CH_2CH_2SH$	150/15	30	142
$CH_2=CHCH_2NH_2$	$(C_4H_9)_3GeH$	H_2PtCl_6	80–130	14	$(C_4H_9)_3GeCH_2CH_2CH_2NH_2$	161/10	74	142
$CH_2=CH-COOH$	$(C_2H_5)_3GeH$	none	140–150	8	$(C_2H_5)_3GeCH_2CH_2COOH$	158–60/10	49	93, 142
$CH_2=CH-COOH$	$(C_3H_7)_2GeH_2$	none	140	18	$(C_3H_7)_2Ge(H)CH_2CH_2COOH$	140–42/0.5 (31)	30	142
$H_2C=CH-CH=CH_2$	$(CH_3)_3GeH$	H_2PtCl_6	140	10	$(CH_3)_3GeCH_2CH=CH-CH_3$ 60% cis-40% trans	125–27/760	65	45

Table 4-5(a)—continued

UNSATURATED COMPOUND	GERMANE	CATALYST	REACTION TEMP. (°C)	TIME (h)	REACTION PRODUCTS	B.P. OR (M.P.) (°C/mm)	YIELD (%)	REFERENCES
$H_2C=CH-CH=CH_2$	$(C_2H_5)_3GeH$	H_2PtCl_6	120	15	$(C_2H_5)_3GeCH_2CH=CHCH_3$ 52% cis-48% trans	90/20	36	105, 148
$H_2C=CHCOCH_3$	$(C_2H_5)_3GeH$	none	reflux	10	$(C_2H_5)_3GeCH_2CH_2COCH_3$	118/17	31	148
$H_2C=CH-COCH_3$	$(C_2H_5)_3GeH$	H_2PtCl_6	80	2	$(C_2H_5)_3Ge-OC(CH_3)=CH-CH-CH_3$ (85%) $(C_2H_5)_3GeCH_2CH_2COCH_3$ (15%)	100-30/20	75	105
$CH_2=CH-COCH_3$	$(C_4H_9)_2GeH_2$	none	to boil	10	$(C_4H_9)_2Ge(H)CH_2CH_2COCH_3$ (30%) $(C_4H_9)_2Ge(CH_2CH_2COCH_3)_2$ (27%)	121/12 155-56/0.4	—	142
$CH_2=C(CH_3)-CH_2OH$	$(C_4H_9)_3GeH$	H_2PtCl_6	120	10	$(C_4H_9)_3GeCH_2CH(CH_3)-CH_2OH$	171/13	47	143
$CH_3CH=CHCHO$	$(C_5H_{11})_2GeH_2$	Pt/asbestos	100	18	$(C_5H_{11})_2Ge(H)CH(CH_3)CH_2CHO$	139-40/10	44	142
$CH_2=CHCOOCH_3$	$(C_3H_7)_3GeH$	none	100	8	$(C_3H_7)_3GeCH_2CH_2COOCH_3$	150-54/18	34	93, 142
$CH_2=CHCOOCH_3$	$(C_4H_9)_3GeH$	none	100	8	$(C_4H_9)_3GeCH_2CH_2COOCH_3$	175-78/17	40	142
$CH_2=C(CH_3)CH_2Cl$	$(C_2H_5)_3GeH$	H_2PtCl_6	—	—	$(C_2H_5)_3GeCH_2CH(CH_3)CH_2Cl$ (41%) $(C_2H_5)_3GeCl$ (41%)	—	—	121
$CF=CF-CF_2CF_2$ (cyclic)	$(CH_3)_3GeH$	none	230	72	$(CH_3)_3GeCF-CFH-CF_2CF_2$ (40%)	118/752	86	32
$CCl=CCl-CF_2-CF_2$ (cyclic)	$(CH_3)_3GeH$	none	190	36	$(CH_3)_3Ge$[cyclobutane F_2 F_2]Cl (40%); $(CH_3)_3Ge$[cyclobutane F_2 H Cl] 4% cis-8% trans; $(CH_3)_3Ge$[cyclobutane F_2 Cl Cl H] (27%); $(CH_3)_3Ge$[cyclobutane F_2 H] (0.5%); $(CH_3)_3GeCl$ (20%)	78/50; 80-82/50; 86/50; —	20	32
$CH_2=CH-B$[$O-CH_2$ ring]	$(C_4H_9)_3GeH$	none	100	2	$(n-C_4H_9)_3GeCH_2CH_2CH_2B$[$O-CH_2$ ring]	129-31/0.5	95	18

Substrate	Ge–H reagent	Catalyst	Temp. (°C)	Time (h)	Product	b.p. (°C/mm)	Yield (%)	Ref.
 (cyclopentadiene ring)	$(CH_3)_3GeH$	H_2PtCl_6	175	5	$(CH_3)_3Ge$–(cyclopentene) (67.7%); $(CH_3)_3Ge$–(cyclopentene) (37.3%) 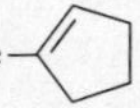	52–54/15	25	45
$H_2C=CHCOOC_2H_5$	$(CH_3)_3GeH$	H_2PtCl_6	to boil	2	$(CH_3)_3GeCH_2CH_2COOC_2H_5$	74–75/18	—	139
$H_2C=CH{-}COOC_2H_5$	$(C_2H_5)_3GeH$	H_2PtCl_6; Pt/asbestos	90–125	2	$(C_2H_5)_3GeCH_2CH_2COOC_2H_5$	118–18.5/19	—	139
$H_2C=CHCOOC_2H_5$	$(C_4H_9)_3GeH$	Pt/asbestos + Hydroquinone	100	10	$(C_4H_9)_3GeCH_2CH_2COOC_2H_5$	122–24/0.5	34	93, 142
$CH_2=CHCH_2OOCCH_3$	$(C_2H_5)_3GeH$	none	—	—	$(C_2H_5)_3GeCH_2CH_2CH_2OOCCH_3$	116.5/11	50	121
$CH_2=CH{-}CH_2B(OCH_3)_2$	$(C_2H_5)_3GeH$	H_2PtCl_6	to boil	17.5	$(C_2H_5)_3Ge(CH_2)_3B(OCH_3)_2$	118–20/12	77.8	120
 (fused diene ring)	$(CH_3)_3GeH$	H_2PtCl_6	140	10	$(CH_3)_3Ge$–(cyclohexene) 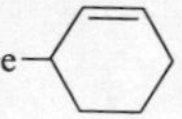	38–39/1.5	60	45
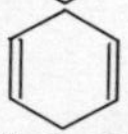 (fused diene ring)	$(CH_3)_3GeH$	H_2PtCl_6	140	10	$(CH_3)_3Ge$–(cyclohexene) 	34/1	49	45
$H_2C=CH{-}CH_2COOC_2H_5$	$(C_2H_5)_3GeH$	H_2PtCl_6	70–80	2	$(C_2H_5)_3Ge(CH_2)_3COOC_2H_5$	128/31/13	93	139
$C_4H_9OCH=CH_2$	$(C_4H_9)_3GeH$	none	130	8	$(C_4H_9)_3GeCH_2CH_2OC_4H_9$	187–88/18	72	142
$C_4H_9OCH=CH_2$	$n\text{-}C_7H_{15}GeH_3$	none	130–140	5	$n\text{-}C_7H_{15}Ge(CH_2CH_2OC_4H_9)_3$	180/0.4	70	142
(norbornadiene ring)	$(CH_3)_3GeH$	H_2PtCl_6	140	8	endo $Ge(CH_3)_3$ (75%); exo $Ge(CH_3)_3$ (15%); $Ge(CH_3)_3$ (10%)	83–84/1.7	80	45

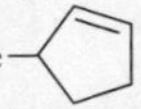

Table 4-5(a)—continued

UNSATURATED COMPOUND	GERMANE	CATALYST	REACTION TEMP. (°C)	TIME (h)	REACTION PRODUCTS	B.P. OR (M.P.) (°C/mm)	YIELD (%)	REFERENCES
(cyclooctadiene)	$(CH_3)_3GeH$	H_2PtCl_6	160	10	$(CH_3)_3Ge$—(cyclooctene)	78–78.5/0.75	30	45
(cyclooctatriene)	$(CH_3)_3GeH$	H_2PtCl_6	150	10	$(CH_3)_3Ge$—(cyclooctadiene)	47.5/0.2	42	45
(vinylcyclohexene)	$(CH_3)_3GeH$	H_2PtCl_6	150	8	$(CH_3)_3Ge$—CH_2CH_2—(cyclohexenyl)	93–94/0.3	47	45
$CH_3(CH_3)_3CH=CH_2$	$(C_2H_5)_3GeH$	Pt/ asbestos	to boil	—	$(C_2H_5)_3GeCH_2CH_2C_6H_{13}$	143/17	—	142
$(C_2H_5)_2Ge(CH=CH_2)_2$	$(C_4H_9)_2GeH_2$	Pt/ asbestos $(PhCOO)_2$	to boil	2	$(C_2H_5)_2Ge\,{<}^{CH_2CH_2-}_{CH_2CH_2-}{>}Ge(C_4H_9)_2$	120–21/0.7	17	112
$(C_2H_5)_3GeCH_2CH=CH_2$	$(C_2H_5)_3GeH$	Pt/ asbestos	to boil	—	$(C_2H_5)_3GeCH_2CH_2Ge(C_2H_5)_3$	275/760	78	111, 142
$(C_2H_5)_3GeCH_2CH=CH_2$	$(C_2H_5)_3GeH$	Pt/ asbestos	to boil	—	$(C_2H_5)_3GeCH_2CH_2CH_2Ge(C_2H_5)_3$	128–29/1.4	91	111, 115, 142
$(C_2H_5)_3GeOCH_2CH=CH_2$	$(C_2H_5)_3GeH$	H_2PtCl_6	150	15	$(C_2H_5)_3GeOCH_2CH_2CH_2Ge(C_2H_5)_3$	110/0.5	43	143
$CH_2=CH-P(O)Cl_2$	$(C_2H_5)_3GeH$	H_2PtCl_6	150–60	10	$(C_2H_5)_3GeCH_2CH_2P(O)Cl_2$	136/4.5	17.4	34b
$CH_2=CH-CH_2P(O)Cl_2$	$(C_2H_5)_3GeH$	H_2PtCl_6	160–65	10	$(C_2H_5)_3Ge(CH_2)_3P(O)Cl_2$	77/20	51.9	34b
$CH_2=CH-P(O)(OC_2H_5)_2$	$(C_2H_5)_3GeH$	H_2PtCl_6	150–60	10	$(C_2H_5)_3Ge(CH_2)_2P(O)(OC_2H_5)_2$	120/4	38	34b
$CH_2=CH-CH_2P(O)(OC_2H_5)_2$	$(C_2H_5)_3GeH$	H_2PtCl_6	160–65	10	$(C_2H_5)_3Ge(CH_2)_3P(O)(OC_2H_5)_2$	166/10	59.6	34b
$CH_2=CH-CH_2P(O)(OC_4H_9)_2$	$(C_2H_5)_3GeH$	H_2PtCl_6	200–20	10	$(C_2H_5)_3Ge(CH_2)_3P(O)(OC_4H_9)_2$	166/3	36.5	34b

(b) To acetylenic compounds

UNSATURATED COMPOUND	GERMANE	CATALYST	REACTION TEMP. (°C)	TIME (h)	REACTION PRODUCTS	B.P. OR (M.P.) (°C/mm)	YIELD (%)	REFERENCES
$CH\equiv CH$	$(C_2H_5)_3GeH$	H_2PtCl_6	100	4	$(C_2H_5)_3GeCH_2CH_2Ge(C_2H_5)_3$	150/0.2	75	95, 142
					$(C_2H_5)_3GeCH=CH_2$	—	12	
$HC\equiv CCH_2Cl$	$(C_2H_5)_3GeH$	H_2PtCl_6	exothermic	$\frac{1}{2}$	$(C_2H_5)_3GeCH=CHCH_2Cl$ 30% cis- 35% trans $\Bigg\}$ (35%) $(C_2H_5)_3Ge-C=CH_2$, $\;$ CH_2Cl	108–10/20	62	105, 107, 108
$HC\equiv CCH_2Cl$	$(C_2H_5)_3GeH$	H_2PtCl_6 solvent CH_3CN	80	1	$(C_2H_5)_3GeCl$	—	33	
					$(C_2H_5)_3GeCH=CHCH_2Cl$ 10% cis- 35% trans $\Bigg\}$ (55%) $(C_2H_5)_3Ge-CH_2Cl$, $=CH_2$	108–10/20	14	108
$HC\equiv CCH_2OH$	$(C_2H_5)_3GeH$	H_2PtCl_6	90	4	$(C_2H_5)_3GeCl$	—	56	35
					$(C_2H_5)_3GeCH=CHCH_2OH$	112–3/12	37	
$CH\equiv CCH_2OH$	$(C_4H_9)_3GeH$	H_2PtCl_6	100	6	$(C_4H_9)_3GeCH=CHCH_2OH$	105/0.7	80	95, 142
$F_3C-C\equiv C-CF_3$	$(C_2H_5)_3GeH$	UV	20	7 days	$(C_2H_5)_3Ge(CF_3)=C(H)CF_3$ 92% trans	82/26	—	31
$HC\equiv C-C(CH_3)_2OH$	$(C_4H_9)_3GeH$	H_2PtCl_6	exothermic	—	$(C_4H_9)_3GeCH=CHC(CH_3)_2OH$	120/1	90	95, 142
$CH\equiv C-C=CH_2$, $\;CH_3$	$(C_2H_5)_3GeH$	H_2PtCl_6	100 (autoclave)	15	$(C_2H_5)_3Ge-C=C-C=CH_2$ (14%), $\;CH_2CH_3$ $\;$ $(C_4H_9)_2Ge-CH=CH-C=CH_2$ (86%), $\;CH_3$ trans	102/18	50	106b
$HC\equiv C-CH_2-CH_2-C\equiv CH$	$(C_4H_9)_2GeH_2$	H_2PtCl_6	120	168	$+Ge(C_4H_9)_2CH=CH(CH_2)_2CH=CH+_n$ $n\sim 35$ Polymer	—	—	131
$CH_3(CH_2)_3C\equiv CH$	$(C_2H_5)_3GeH$	H_2PtCl_6	100	2	$(C_2H_5)_3GeCH=CHC_4H_9$	102/10	~ 100	95, 142
$CH_3(CH_2)_3C\equiv CH$	$(n\text{-}C_7H_{15})GeH_3$	H_2PtCl_6	200	10	$(n\text{-}C_7H_{15})Ge(CH=CHC_4H_9)_3$	174/0.4	70	95, 142
$C_6H_5C\equiv CH$	$(C_2H_5)_3GeH$	H_2PtCl_6	100	2	$(C_2H_5)_3GeCH=CH(C_6H_5)$	127/0.2	~ 100	95, 142

Starting compound	R_3GeH	Catalyst	Temp (°C)	n	Product	B.p./mm	Yield %	Ref
C$_3$H$_7$—CH(OH)—C≡CH	(C$_4$H$_9$)$_3$GeH	H$_2$PtCl$_6$	60–65	2	(C$_3$H$_7$)—CH(OH)—CH=CHGe(C$_4$H$_9$)$_3$	146–48/1	58.2	162
CH$_3$—C(C$_2$H$_5$)(OH)—C≡CH	(C$_4$H$_9$)$_3$GeH	H$_2$PtCl$_6$	60–65	2	CH$_3$—C(C$_2$H$_5$)(OH)—CH=CHGe(C$_4$H$_9$)$_3$	130–31/2	—	162
CH≡C—CHOH—CH$_2$CH$_2$CH$_3$	(C$_2$H$_5$)$_3$GeH	H$_2$PtCl$_6$	80	3	(C$_2$H$_5$)$_3$GeCH=CHCHOH—CH$_2$CH$_2$CH$_3$	88–89/0.1	—	164
CH≡CCHOH—CH(CH$_3$)	(C$_2$H$_5$)$_3$GeH	H$_2$PtCl$_6$	80	3	(C$_2$H$_5$)$_3$GeCH=CH—CHOH—CH—CH$_3$ (CH$_3$)	73–74/0.2	40	164
HOCH$_2$CH$_2$C≡C—CH=CH$_2$	(C$_2$H$_5$)$_3$GeH	H$_2$PtCl$_6$	exo-thermic	1	HOCH$_2$—CH$_2$—C—CH—CH=CH$_2$ (Ge(C$_2$H$_5$)$_3$)	129–31/4	—	71
					HOCH$_2$CH$_2$—CH—CH=CH—CH$_2$Ge(C$_2$H$_5$)$_3$ (Ge(C$_2$H$_5$)$_3$)	153–54/1	7	71
CH$_3$—CH(OH)—C≡C—CH=CH$_2$	(C$_2$H$_5$)$_3$GeH	H$_2$PtCl$_6$	exo-thermic	1	CH$_3$—CH—C=CH—CH=CH$_2$ (OH) (Ge(C$_2$H$_5$)$_3$)	99/2	54	71
					CH$_3$—CH—CH—CH=CH—CH$_2$Ge(C$_2$H$_5$)$_3$ (OH) (Ge(C$_2$H$_5$)$_3$)	147–50/2	5	71
ClCH$_2$CH$_2$OCH$_2$OCH$_2$C≡CH	(C$_2$H$_5$)$_3$GeH	H$_2$PtCl$_6$	70	4	ClCH$_2$CH$_2$OCH$_2$OCH$_2$CH=CHGe(C$_2$H$_5$)$_3$	127–29/2	55	1
CH$_3$ \ C—C≡C—CH=CH$_2$ / CH$_3$ (OH)	(C$_2$H$_5$)$_3$GeH	H$_2$PtCl$_6$	exo-thermic	1	CH$_3$ \ C—C=CH—CH=CH$_2$ (Ge(C$_2$H$_5$)$_3$) / CH$_3$ (OH)	112/6	70	70
C$_3$H$_7$—C(OH)C≡CH (CH$_3$)	(C$_2$H$_5$)$_3$GeH	H$_2$PtCl$_6$	120	2	C$_3$H$_7$—C(OH)CH=CH—Ge(C$_2$H$_5$)$_3$ (CH$_3$)	81–83/0.3	—	163
(n-C$_4$H$_9$)—C(OH)C≡CH (CH$_3$)	(C$_2$H$_5$)$_3$GeH	H$_2$PtCl$_6$	120	2	(n-C$_4$H$_9$)—C(OH)CH=CH—Ge(C$_2$H$_5$)$_3$ (CH$_3$)	88–89/0.2	—	163
(t-C$_4$H$_9$)—C(OH)C≡CH (CH$_3$)	(C$_2$H$_5$)$_3$GeH	H$_2$PtCl$_6$	120	2	(t-C$_4$H$_9$)—C(OH)CH=CHGe(C$_2$H$_5$)$_3$ (CH$_3$)	84–86/0.2	65.5	163

Table 4-5(b)—continued

UNSATURATED COMPOUND	GERMANE	CATALYST	REACTION TEMP. (°C)	TIME (h)	REACTION PRODUCTS	B.P. OR (M.P.) (°C/mm)	YIELD (%)	REFERENCES
(CH₃)₂C(OH)C≡C—C(OH)(CH₃)₂	(C₂H₅)₃GeH	H₂PtCl₆	140	5	(CH₃)₂C(OH)CH=C[Ge(C₂H₅)₃]C(OH)(CH₃)₂	133/0.6 (70)	40	95, 142
C₂H₅ \ C—C≡C—CH=CH₂ / OH CH₃	(C₂H₅)₃GeH	H₂PtCl₆	exo-thermic	1	C₂H₅ \ C—C=CH—CH=CH₂ / CH₃ OH Ge(C₂H₅)₃	98–100/3	79	70
C₃H₇—CH—C≡C—CH=CH₂ \ OH	(C₂H₅)₃GeH	H₂PtCl₆	exo-thermic	1	C₃H₇—CH—C=CH—CH=CH₂ \ OH Ge(C₂H₅)₃	123–24/4	56	71
					C₃H₇—CH—CH=CH—CH—CH₂Ge(C₂H₅)₃ \ OH Ge(C₂H₅)₃	188–89/4	11	
CH₂—CH₂ \ C—C≡C—CH=CH₂ / CH₂—CH₂ OH	(C₂H₅)₃GeH	H₂PtCl₆	exo-thermic	1	CH₂—CH₂ \ C—C=CH—CH=CH₂ / CH₂—CH₂ OH Ge(C₂H₅)₃	116–18/1	73	70
CH₂—CH₂ \ C—C≡C—CH=CH₂ / CH₂—CH₂ OH	(C₂H₅)₃GeH	H₂PtCl₆	exo-thermic	1	CH₂—CH₂ \ C—C=CH—CH=CH₂ / CH₂—CH₂ OH Ge(C₂H₅)₃	114–16/1	69	70
CH₃ OH CH₃ \	/ C—C≡C—C /	\ CH₃ OH CH₃	(C₂H₅)₃GeH	H₂PtCl₆	118	0.5	CH₃ OH Ge(C₂H₅)₃ \	/ C—C=CH—CH=CH₂ /	\ CH₃ OH CH₃	152/2	80	66
CH₃ OH \	C—C≡C—CH=CH₂ /	CH₃ Ge(C₂H₅)₃	(C₂H₅)₃GeH	H₂PtCl₆	120	1	CH₃ OH CH₃ \	/ C—C=CH—CH—C—GeEt₃ /	\ CH₃ GeEt₃ CH₃	163–65/2	13	66
CH—C≡C—CH=CH₂ \ OH	(C₂H₅)₃GeH	H₂PtCl₆	exo-thermic	1	CH—C=CH—CH=CH₂ \ OH Ge(Et)₃	163–64/3	41	71

Reactant	GeH reagent	Catalyst	Temp. (°C)	Time (hr)	Product	B.p./mm	Yield (%)	Ref.
CH_3, OH, CH_3 / $C-C{\equiv}C-C{\equiv}C-C$ / C_2H_5, C_2H_5 (diyne-diol)	$(C_2H_5)_3GeH$	H_2PtCl_6	120	1	$C_6H_5-CH-CH=CH-CH=CH-CH_2Ge(C_2H_5)_3$ (with OH, $Ge(C_2H_5)_3$)	198–200/2		9
					CH_3, HO, CH_3 / $C-C{\equiv}C-C=C-C$ / C_2H_5, C_2H_5, $Ge(C_2H_5)_3$	164/2	70	66
$(CH_3)_2C(OH)-C{\equiv}C(C_6H_5)$	$(C_2H_5)_3GeH$	H_2PtCl_6	exothermic	½	$(CH_3)_2C(OH)-C=CH(C_6H_5)$ / $Ge(C_2H_5)_3$	144–45/2	85	65
$(CH_3)(C_2H_5)C(OH)-C{\equiv}C(C_6H_5)$	$(C_2H_5)_3GeH$	H_2PtCl_6	exothermic	½	$(CH_3)(C_2H_5)C(OH)-C=CH-(C_6H_5)$ / $Ge(C_2H_5)_3$	152–53/2	80	65
cyclopentane–$COHC{\equiv}C-C_6H_5$	$(C_2H_5)_3GeH$	H_2PtCl_6	room temp.		cyclopentane $C-C=CH(C_6H_5)$ / OH $Ge(C_2H_5)_3$	164–66/2		64a
					cyclopentene $C-C=CH(C_6H_5)$ / $Ge(C_2H_5)_3$	159–60/2		64a
cyclohexane–$COHC{\equiv}CC_6H_5$	$(C_2H_5)_3GeH$	H_2PtCl_6	room temp.		cyclohexane $C-C=CHC_6H_5$ / OH $Ge(C_2H_5)_3$	173–75/2	80	64a
$(C_2H_5)_3SiC{\equiv}CH$	$(n\text{-}C_4H_9)_3GeH$	H_2PtCl_6	70–80	22	$(C_2H_5)_3SiCH=CHGe(C_4H_9)_3$	120–22/2	—	124
					$(C_4H_9)_3GeCH=CH-Ge(C_4H_9)_3$	150–53/2		
$(n\text{-}C_4H_9)_3GeC{\equiv}CH$	$(n\text{-}C_4H_9)_3GeH$	H_2PtCl_6	200	—	$(n\text{-}C_4H_9)_3GeCH=CH-Ge(C_4H_9\text{-}n)_3$	150/0.2	100	95, 142
$(C_2H_5)_3GeOCH_2C{\equiv}CH$	$(C_2H_5)_3GeH$	H_2PtCl_6	140	12	$(C_2H_5)_3GeOCH_2C{\equiv}CH=CHGe(C_2H_5)_3$	164–65/13	70	143
$[HC{\equiv}C-CR'R''OSi(CH_3)-O-R]_2$	$(C_2H_5)_3GeH$	H_2PtCl_6	reflux		$[(C_2H_5)_3GeHC=CH-CR'R''OSi(CH_3)-O-R]_2$, $R'=R''=H$	188–89/0.2	44.1	162a
					$R'=R''=H$, $R=Pr$	199/0.26	43.65	162a
					$R'=Me$, $R=Bu$, $R=Pr$	205–7/0.2	37.20	162a

(c) To the carbonyl group

UNSATURATED COMPOUND	GERMANE	CATALYST	REACTION TEMP. (°C)	TIME (h)	REACTION PRODUCTS	B.P. OR (M.P.) (°C/mm)	YIELD (%)	REFERENCES
CF_3COCF_3	$(CH_3)_3GeH$	none	20	15	$(CH_3)_3Ge-O-CH(CF_3)_2$	117/758	70	30, 168
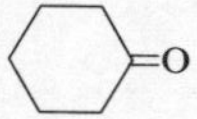 cyclohexanone	$(C_2H_5)_3GeH$	Cu	150	48	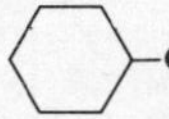 cyclohexyl–$OGe(C_2H_5)_3$	134/18	~100	97
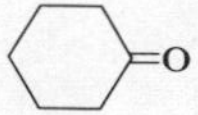 cyclohexanone	$(C_4H_9)_3GeH$	Cu	150	48	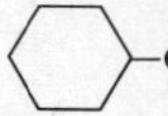 cyclohexyl–$OGe(C_4H_9)_3$	180/16	~100	97
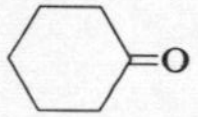 cyclohexanone	$(C_4H_9)_3GeH$	none	150	48	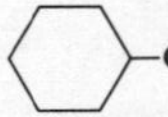 cyclohexyl–$OGe(C_4H_9)_3$	180/16	60	97
C_6H_5CHO	$(C_2H_5)_3GeH$	Cu	150	40	$C_6H_5CH_2OGe(C_2H_5)_3$	150–52/18	80	97
$C_4H_9COC_2H_5$	$(C_4H_9)_3GeH$	Cu	150	40	$(C_4H_9)_3GeOCH(C_2H_5)C_4H_9$	160/9	50	97
$CH_3(CH_2)_2CHO$	$(C_2H_5)_3GeH$	Cu	150	24	$(C_2H_5)_3GeO(CH_2)_6CH_3$	127/10	~100	97
$n\text{-}C_6H_{13}CO-CH_3$	$(C_4H_9)_3GeH$	Cu	150	40	$(C_4H_9)_3GeOCH(CH_3)C_6H_{13}$	180/12	50	97
$CH_2=CH-(CH_2)_8CHO$	$(C_2H_5)_3GeH$	Cu	150	24	$(C_2H_5)_3GeO(CH_2)_9CH=CH_2$	172–75/10	60	144
$\begin{array}{l}CH_3 \\ \quad\diagdown \\ \quad\quad C=CH-CH_2-CH_2CCH_3 \\ \quad\diagup \qquad\qquad\qquad\; \| \\ CH_3 \qquad\qquad\qquad\; O\end{array}$	$(C_4H_9)_3GeH$	Cu	160	48	$(C_4H_9)_3Ge-OCH(CH_2)_2CH=C(CH_3)_2$ (with CH_3)	177–80/10	70	144
$(C_6H_5)_2C=C=O$	$(C_2H_5)_3GeH$	none	exo-thermic	½	$(C_2H_5)_3GeO-CH=C(C_6H_5)_2$	139/0.09	80	141, 151

4-3-5-5   *Table*: Addition Reactions of Arylgermanium Hydrides

Table 4-6

Addition Reactions of Arylgermanium Hydrides

(a) To ethylenic compounds

UNSATURATED COMPOUND	GERMANE	CATALYST	REACTION TEMP. (°C)	TIME (h)	REACTION PRODUCTS	B.P. OR (M.P.) (°C/mm)	YIELD (%)	REFERENCES
$CH_2{=}CH{-}CN$	$(C_6H_5)_3GeH$	none	50-60	6	$(C_6H_5)_3GeCH_2CH_2CN$	(126-29)	83	73
$CH_2{=}CH{-}CN$	$(C_6H_5)_2GeH_2$	AIBN	95	60	$(C_6H_5)_2Ge(H)CH_2CH_2CN$	136/0.03	33	116
					$(C_6H_5)_2Ge(CH_2CH_2CN)_2$	203-5/0.01	21	73
$CH_2{=}CH{-}CONH_2$	$(C_6H_5)_3GeH$	none	50-60	21	$(C_6H_5)_3GeCH_2CH_2CONH_2$	(176-78)	—	73
$CH_2{=}CH{-}COCH_3$	$(C_6H_5)_3GeH$	none	60-70	18	$(C_6H_5)_3GeCH_2CH_2COCH_3$	(144-46)	53	73
$CH_2{=}CH{-}COCH_3$	$(C_6H_5)_2GeH_2$	none	80	18	$(C_6H_5)_2Ge(H)(CH_2)_2COCH_3$	$114/3.10^{-2}$	23	151
					$(C_6H_5)_2Ge(CH_2)_2CH{-}CH_3$ (epoxide, $-O-$)	$104/3.10^{-3}$	32	
$CH_2{=}CH{-}O{-}C_2H_5$	$C_6H_5GeH_3$	none	90	12	$(C_6H_5)Ge[(CH_2)_2{-}CO{-}CH_3]_2$	$180/2.10^{-2}$	6	
			130	12	Polymers		30	
			180	12	$(C_6H_5)Ge(H)(CH_2CH_2OC_2H_5)_2$	154/9	36	152
					$(C_6H_5)Ge(CH_2CH_2OC_2H_5)_3$	130/0.2	49	73
$CH_2{=}CHOCOCH_3$	$(C_6H_5)_3GeH$	none	50-60	6	$(C_6H_5)_3GeCH_2CH_2OOCCH_3$	(62-62.5)	54	73
$CH_2{=}CH{-}COOCH_3$	$(C_6H_5)_3GeH$	none	60-70	67	$(C_6H_5)_3GeCH_2CH_2COOCH_3$	(60.5-62.0)	50	73
$CH_2{=}CHCOOCH_3$	$(C_6H_5)_2GeH_2$	none	80	20	$(C_6H_5)_2Ge(H)CH_2CH_2COOCH_3$ (25%)	$160/3.10^{-2}$	75	152
					$(C_6H_5)_2Ge(CH_2CH_2COOCH_3)_2$ (50%)	$194/3.10^{-2}$		
$CH_2{=}CH{-}CH_2{-}CN$	$(C_6H_5)_2GeH_2$	AIBN	135	48	$(C_6H_5)_2Ge(H)CH_2CH_2CH_2CN$	163/0.05	35	116
					$(C_6H_5)_2Ge(CH_2CH_2CH_2{-}CN)_2$	226/0.04	50	
$CH_2{=}CH{-}CH_2{\cdot}B(OCH_3)_2$	$(C_6H_5)_3GeH$	none, solv.:ether	—	2	$(C_6H_5)_3Ge(CH_2)_3B(OCH_3)_2$	thick oil	100	120
$CH_2{=}CHCH_2CH_2CH_2OH$	$(C_6H_5)_3GeH$	none	50	4	$(C_6H_5)_3GeCH_2(CH_2)_4CH_2OH$	(58-61)	41	73
$CH_2{=}CH(CH_2)_3CH_3$	$(C_6H_5)_3GeH$	UV	30	1⅓	$(C_6H_5)_3Ge(C_6H_{13})$	(76)	95	140
cyclohexene	$(C_6H_5)_3GeH$	none	60-70	120	cyclohexyl$-Ge(C_6H_5)_3$	(144-47)	39	73
cyclohexene	$(C_6H_5)_3GeH$	$(PhCOO)_2$	75	48	cyclohexyl$-Ge(C_6H_5)_3$	(147-49.5)	—	46

Substrate	Ge—H compound	Catalyst	Temp.	Time	Product	b.p./(m.p.)	Yield %	Ref.
 (cyclohexene)	$C_6H_5GeH_3$	UV	—	48	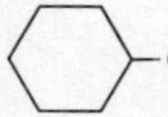 (cyclohexyl)—$Ge(C_6H_5)_3$	(147–49.5)	—	46
(cyclohexene)	$C_6H_5GeH_3$	none	90 / 150	12 / 24	$(C_6H_5)Ge(H)_2(C_6H_{11})$ (28%) / $(C_6H_5)Ge(H)(C_6H_{11})_2$ (48%)	68/0.1 / 132/0.2	76	152
$CH_2=CH(CH_2)_2C(=O)-CH_3$	$(C_6H_5)_3GeH$	AIBN	90	24	$(C_6H_5)_3Ge(CH_2)_4COCH_3$	(95)	92	151
$CH_2=CH(CH_2)_2COCH_3$	$(C_6H_5)_2GeH_2$	—	120	48	$(C_6H_5)_2Ge(H)(CH_2)_4COCH_3$ / $(C_6H_5)_2Ge[(CH_2)_4COCH_3]_2$	148/0.1 / 195/5.10^{-3}	72 / 13	151
$CH_3(CH_2)_5CH=CH_2$	$(C_6H_5)_3GeH$	none	110–15	120	$(C_6H_5)_3Ge(CH_2)_7CH_3$	(62–66)	29	73
$CH_3(CH_2)_5CH=CH_2$	$(C_6H_5)_3GeH$	$(PhCOO)_2$	75	24	$(C_6H_5)_3GeCH_2(CH_2)_6CH_3$	(71–72)	91	46
$CH_3(CH_2)_5CH=CH_2$	$(C_6H_5)_3GeH$	UV	—	48	$(C_6H_5)_3GeCH_2(CH_2)_6CH_3$	(69–70.5)	80	46
$C_6H_5-CH=CH_2$	$(C_6H_5)_3GeH$	none	120	overnight	$(C_6H_5)_3GeCH_2CH_2C_6H_5$	(145–46)	40	73
$C_{16}H_{33}CH=CH_2$	$(C_6H_5)_3GeH$	$(PhCOO)_2$	85	17½	$(C_6H_5)_3GeCH_2CH_2C_{16}H_{33}$	(75.5–76.5)	67	54
$(C_6H_5)_3GeCH_2CH=CH_2$	$(C_6H_5)_3GeH$	$(PhCOO)_2$	hexane reflux	18½	$(C_6H_5)_3GeCH_2(CH_2)_3Ge(C_6H_5)_3$	(134–36)	86.5	54
$(C_6H_5)_3SiCH_2CH=CH_2$	$(C_6H_5)_3GeH$	$(PhCOO)_2$	75	24	$(C_6H_5)_3Si(CH_2)_3Ge(C_6H_5)_3$	(134–35)	76	117

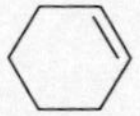

(b) *To acetylenic compounds*

UNSATURATED COMPOUND	GERMANE	CATALYST	REACTION TEMP. (°C)	TIME (h)	REACTION PRODUCTS	B.P. OR (M.P.) (°C/mm)	YIELD (%)	REFERENCES
$(CH_3)_2C(OH)C\equiv CH$	$(C_6H_5)_3GeH$	none	50-60	77	$(C_6H_5)_3GeCH=CHCOH(CH_3)_2$	(90.5-91.5)	49	73
$CH_3(CH_2)_3C\equiv CH$	$(C_6H_5)_2GeH_2$	none	70	24	$(C_6H_5)_2Ge[CH=CH(CH_2)_3CH_3]_2$	$138/8.10^{-3}$	84	152
$CH_3(CH_2)_3C\equiv CH$	$C_6H_5GeH_3$	none	70	24	$C_6H_5Ge[CH=CH(CH_2)_3CH_3]_3$	$126/10^{-3}$	74	152
$HC\equiv C-CH_2CH_2C\equiv CH$	$(C_6H_5)_2GeH_2$	H_2PtCl_6	120	168	$\dashv Ge(C_6H_5)_2CH=CH(CH_2)_2CH=CH\dashv_n$ $n = \sim 35$	—	—	131
$C_6H_5C\equiv CH$	$(C_6H_5)_3GeH$	none	135	96	$(C_6H_5)_3GeCH=CH(C_6H_5)$	146-49	12	73

(c) To the carbonyl compounds

UNSATURATED COMPOUND	GERMANE	CATALYST	REACTION TEMP. (°C)	TIME (h)	REACTION PRODUCTS	B.P. OR (M.P.) (°C/mm)	YIELD (%)	REFERENCES
CH_3\CH—CHO / CH_3/	$(C_6H_5)_3GeH$	AIBN	80	4	$(C_6H_5)_3GeOCH_2CH(CH_3)_2$	$146/5.10^{-3}$ (47–48)	93	140
CH_3\C=CH—C—CH_3 / CH_3/ ‖O	$(C_6H_5)_3GeH$	AIBN	180	48	$(C_6H_5)_3Ge$—O—$C(CH_3)$=CH—CH—$CH(CH_3)_2$ cis/trans: (40/60)	$148/5.10^{-3}$	60	151, 153
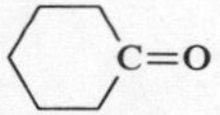 (cyclohexanone, ring =O)	$(C_6H_5)_3GeH$	AIBN	70	3	$(C_6H_5)_3GeOC_6H_{11}$	$160/3.10^{-3}$	92	140
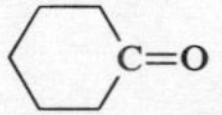 (cyclohexanone, ring =O)	$(C_6H_5)_2GeH_2$	AIBN	120	24	$(C_6H_5)_2Ge(OC_6H_{11})_2$	(57–58)	68	140
CH_3—C—CH_2—$CH(CH_3)_2$ ‖O	$(C_6H_5)_3GeH$	AIBN	120	10	$(C_6H_5)_3GeOCH(CH_3)CH_2CH(CH_3)_2$	$154/3.10^{-3}$	54	140
C_6H_5—CHO	$(C_6H_5)_3GeH$	AIBN	180	24	$(C_6H_5)_3GeOCH_2C_6H_5$	(78)	65	140
CH_3\C=CH—CH_2CH_2—C—CH_3 / CH_3/ ‖O	$(C_6H_5)_3GeH$	AIBN	100	10	$(C_6H_5)_3GeO$—$CH(CH_3)(CH_2)_2CH$=$C(CH_3)_2$	$164/10^{-3}$	82	151, 153
C_6H_5—C—C_6H_5 ‖O	$(C_6H_5)_3GeH$	AIBN	120	10	$(C_6H_5)_3GeOCH(C_6H_5)_2$	$215/3.10^{-2}$	58	152
$(C_6H_5)_2C$=C=O	$(C_6H_5)_3GeH$	none / AIBN	160 / 120	12 / 16	$(C_6H_5)_3Ge$—OCH=$C(C_6H_5)_2$ / $(C_6H_5)_2Ge(CH_2)_2CH$—CH_3 with O	$240\text{–}45/10^{-2}$ / $104/3.10^{-3}$	68 / 80	141, 151 / 151
$(C_6H_5)_2Ge(CH_2)_2COCH_3$ / H	—				$\left[\begin{smallmatrix} C_6H_5 & & CH_3 \\ Ge-(CH_2)_2 & & C-O \\ C_6H_5 & O & H \end{smallmatrix}\right]_n$	—	20	

Table 4-6(c)—continued

UNSATURATED COMPOUND	GERMANE	CATALYST	REACTION TEMP. (°C)	REACTION TIME (h)	REACTION PRODUCTS	B.P. OR (M.P.) (°C/mm)	YIELD (%)	REFERENCES
$(C_6H_5)_2Ge(H)(CH_2)_4COCH_3$	—	AIBN	100	16	$(C_6H_5)_2Ge(CH_2)_4CH{-}CH_3$ (Ge and CH bridged by O)	127/0.1	70	151
					$[{-}Ge(C_6H_5)_2{-}(CH_2)_4{-}CH(CH_3){-}O{-}]_n$	—	15	

4-3-6   Other reactions of organogermanium hydrides

4-3-6-1   With Diazo-derivatives, Carbenes and Halocarbenes (Insertion Reactions)

(a) Action of diazo derivatives on organogermanium hydrides

Diazomethane under UV irradiation reacts in ether with a series of trisubstituted germanes giving methyl derivatives in low yields (85):

Hydride	Product	Yield (%)
Et_3GeH	Et_3GeMe	9
$n-Pr_3GeH$	$n-Pr_3GeMe$	5
$n-Bu_3GeH$	$n-Bu_3GeMe$	2
Ph_3GeH	no reaction	

The decrease in the yields of methylenation product with the increased size of the substituent group can be interpreted as being due to steric shielding of the Ge—H group. However, trisubstituted germanes react more readily than the corresponding silanes. Triethylstannane is the most readily alkylated of the series of triethylhydrides:

$$Et_3SiMe\ 1\% > Et_3GeMe\ 9\% > Et_3SnMe\ 75\%$$

These reactions probably involve the insertion of methylene (85). Under the same conditions, diphenylgermane gives both diphenyl-methylgermane and diphenyldimethylgermane with 43 and 5% yields respectively (84, 150).

The action of the same quantity of diazomethane on diphenylgermane in the presence of copper powder leads to diphenylmethylgermane with a 25% yield; in this case, no trace of $(C_6H_5)_2Ge(CH_3)_2$ is detected (150). Phenylgermane, $PhGeH_3$, treated with a 50% diazomethane excess in an ethereal medium, leads after 6 hours of UV irradiation to 21% of unchanged $C_6H_5GeH_3$, 59% of $C_6H_5GeH_2(CH_3)$ and 14% of $C_6H_5GeH(CH_3)_2$ (150).

On the other hand, phenylmethylgermane could be isolated by the action of methylmagnesium iodide on the phenylhalogermanes Ph_2ClGeH, $PhClGeH_2$, $PhCl_2GeH$ (150).

The condensation of organogermanium hydrides with diazo derivatives such as ethyl diazoacetate, diazoacetone and diazoacetophenone in the presence of copper powder in ethereal or benzene solution allows the isolation in 30–40% yields of organic germanium derivatives having functional groups on the α-carbon atom (93, 142):

$$R_3GeH + N_2CHCOOC_2H_5 \rightarrow N_2 + R_3GeCH_2COOC_2H_5$$

$$R_3GeH + N_2CHCOCH_3 \rightarrow N_2 + R_3GeCH_2COCH_3$$

$$R_3GeH + N_2CHCOC_6H_5 \rightarrow N_2 + R_3GeCH_2COC_6H_5$$

This method which Lesbre (92) had earlier suggested for the analogous tin derivatives, R_3SnH, has been adopted again in 1964 by Rijkens and coworkers (139) for the synthesis of $Et_3MCH_2COOC_2H_5(M = Si, Ge)$.

The action of ethyl diazoacetate on the dialkylgermanes leads to the monoesters including a Ge—H bond (142):

$$R_2GeH_2 + N_2CHCOOC_2H_5 \rightarrow N_2 + R_2\underset{\underset{H}{|}}{Ge}-CH_2COOC_2H_5$$

Esters of the same type can be isolated in the action of the trialkyl-germanes and dialkylgermanes with esters of mercuri-bis-acetic acid (14) (102).

(b) Insertion reactions of carbene and halocarbenes on Ge—H bond

Seyferth and coworkers (158, 159) observed that reaction of phenyl-(trihalomethyl)mercury compounds, $C_6H_5HgCX_2Br$ (X = Cl and Br), with organosilicon and organogermanium hydrides in benzene at 80°C results in insertion of CCl_2 and CBr_2 into the Si—H and Ge—H bonds to give $Si-CX_2H$ and $Ge-CX_2H$ compounds:

$$(C_6H_5)_3GeH + C_6H_5HgCCl_2Br \rightarrow$$
$$(C_6H_5)_3GeCCl_2H \quad (88\%) + C_6H_5HgBr$$

$$(C_2H_5)_3GeH + C_6H_5HgCCl_2Br \rightarrow$$
$$(C_2H_5)_3GeCCl_2H \quad (83\%) + C_6H_5HgBr$$

Triethylgermane was found to be four times as reactive as triethyl-silane and triethylsilane was 0.8 times as reactive as cyclohexene. This procedure thus provided a useful general route to the preparation of dihalomethyl derivatives of silicon and germanium. (Dihalomethyl)-germanes were unknown prior to this study.

Two insertion mechanisms can be considered: passage through a transition state of type I

$$R_3M\text{--------}H \qquad\qquad R_3M{-}H \qquad (M = Si\ or\ Ge)$$
$$\underset{\underset{Cl}{\diagup}\ \underset{Cl}{\diagdown}}{C} \quad (I) \qquad\qquad \underset{\underset{Cl}{\diagup}\ \underset{Cl}{\diagdown}}{C} \quad (II)$$

or (II) nucleophilic attack by CCl_2 at silicon or germanium.

Either of these possibilities (I) or (II) would accommodate the experimentally observed greater reactivity of triethylgermane as compared with triethylsilane, but the transition state I is favoured in the case of CCl_2 insertion into the Si—H bond.

Phenyl(dihalomethyl)mercury compounds $PhHgCBr_2H$, $PhHgCClBrH$ have been found to transfer CHBr and CHCl, respectively, to organo-silicon hydrides forming (bromomethyl)- and (chloromethyl)silanes. (Bromomethyl)triethylgermane $(C_2H_5)_3GeCH_2Br$ also was prepared from triethylgermane by this procedure (157a).

The reactions of the halomethyl-mercury reagents $Hg(CH_2Br)_2$, ICH_2HgI/Ph_2Hg and $Hg(CH_2Br)_2/Ph_2Hg$ with a number of organo-silicon hydrides resulted in CH_2 insertion into the Si—H bonds to give methylsilanes. A similar reaction with triethylgermane gave triethyl-methylgermane. Although the mechanism of this new methylenation reaction is not known, a free-methylene process seems to be excluded and it is likely that a direct reaction between the halomethylmercury reagent and the hydride is involved (159a).

4-3-6-2 With Hydroxyl Compounds. In the presence of copper powder, alkylgermanium hydrides give dehydrocondensation reactions with certain hydroxyl derivatives such as water, alcohols, phenols, carboxylic acids and boric acid (98):

$$R_3GeH + R'OH \xrightarrow{Cu} R_3GeOR' + H_2$$

$$R_3GeH + R'COOH \xrightarrow{Cu} R_3GeO-\underset{\underset{O}{\|}}{C}-R' + H_2$$

$$3 R_3GeH + B(OH)_3 \xrightarrow{Cu} (R_3GeO)_3B + 3 H_2$$

These reactions can be extended to the glycols and diphenols:

$$2 R_3GeH + HO(CH_2)_nOH \xrightarrow{Cu}$$
$$R_3Ge-O(CH_2)_nO-GeR_3 + 2 H_2$$

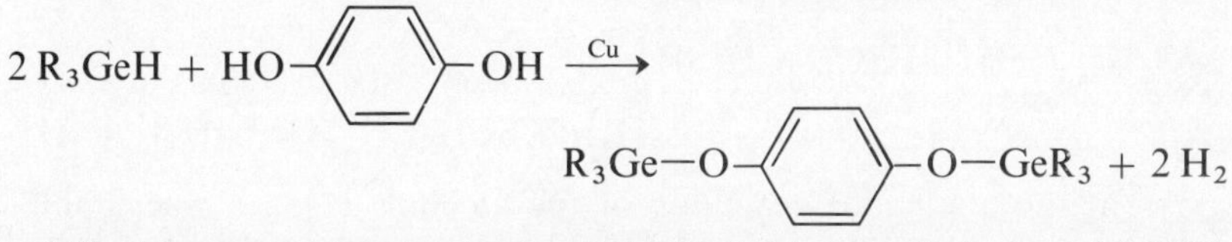

$$2 R_3GeH + HO-\langle\rangle-OH \xrightarrow{Cu}$$
$$R_3Ge-O-\langle\rangle-O-GeR_3 + 2 H_2$$

Dibutylgermane reacts as expected with 1,4-butanediol around 140°C. The cyclization reaction is observed with 80% yield (98):

$$(C_4H_9)_2GeH_2 + HO-(CH_2)_4-OH \rightarrow$$

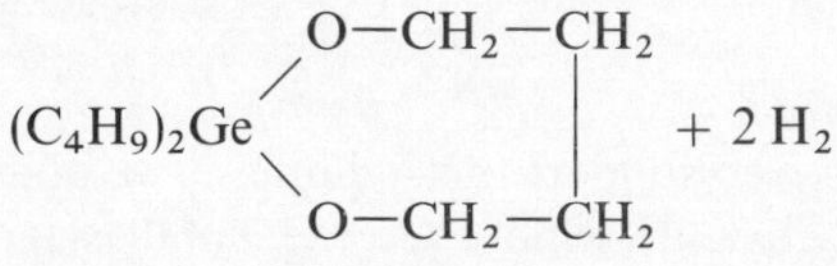

$$(C_4H_9)_2Ge\underset{\diagdown O-CH_2-CH_2}{\overset{\diagup O-CH_2-CH_2}{|}} + 2 H_2$$

In reactions of the same type, the use as catalyst of copper reduced by hydrogen has given better results (143). On reduced copper, dehydrocondensation reactions are observed already at ordinary temperatures with primary alcohols. With ethylenic alcohols such as the allylic alcohol, only the dehydrocondensation reaction is observed:

$$R_3GeH + CH_2{=}CH{-}CH_2OH \xrightarrow{\text{reduced Cu}}$$

$$R_3GeOCH_2CH{=}CH_2 + H_2$$

In this connection we wish to mention the addition of allyl alcohol with trialkylgermanes on platinum catalysts (Pt on asbestos and H_2PtCl_6) (35, 93, 142).

$$R_3GeH + CH_2{=}CH{-}CH_2OH \xrightarrow{\text{Pt}} R_3GeCH_2CH_2CH_2OH$$

Dehydrocondensation reactions are also noted between trialkylgermanes and methallyl, crotyl, furfuryl alcohols as well as with triethylgermyl-3-propene-2-ol-1 (143), $(C_2H_5)_3GeCH{=}CH{-}CH_2OH$.

Acetylenic alcohols dehydrocondense with trialkylgermanes with rather low yields (20–25%), and trialkylalkynoxygermanes are easier to obtain by a transalkoxylation reaction from trialkylmethoxygermanes, R_3GeOMe, and the corresponding acetylenic alcohols.

The dehydrocondensation reaction between dialkylgermanes and alcohols on copper (98) or Raney nickel (106a) lead to dialkylalkoxygermanes, $R_2(R'O)GeH$.

The thiols give condensation reactions of the same type on platinized asbestos (98) or, better, on reduced nickel (145):

$$R_3GeH + C_4H_9SH \rightarrow R_3GeSC_4H_9 + H_2$$

This type of reaction can be extended to the preparation of derivatives of the type $R_3GeS(CH_2)_nSGeR_3$ (145):

$$2\,(C_2H_5)_3GeH + HSCH_2CH_2SH \rightarrow$$

$$(C_2H_5)_3GeSCH_2CH_2SGe(C_2H_5)_3 + 2\,H_2$$

As in the analogous reactions of the alcohols (143), a nucleophilic substitution mechanism is supposed to take place, as the Ge—H bond of the trialkylgermanes is polarized in the $\overset{\delta+}{Ge}{-}\overset{\delta-}{H}$ direction (109):

$$R_3Ge\overset{\frown}{-}H$$
$$\uparrow \qquad \qquad \rightarrow R_3GeXR + H_2 \qquad (X = O, S)$$
$$RX^{\ominus}\text{------}\overset{\oplus}{H}$$

To support this hypothesis we point to the fact that triphenylgermane, Ph_3GeH, whose Ge—H bond is far less polarized, does not give any dehydrocondensation reaction of the same type under the same experimental conditions (152).

However, in a recent paper some Russian investigators recorded a catalytic dehydrocondensation reaction on copper reduced by hydrogen between the triphenylgermane and crotyl alcohol (23).

4-3-6-3 Catalytic Activity. α-Monoolefins are polymerized with a catalyst mixture containing R_3GeH and halides of titanium, vanadium, chromium, molybdenum in the liquid phase in an inert hydrocarbon solvent. The products have molecular weights above 1000. Other monomers polymerized were propylene with Me_3GeH and $TiCl_3$, 3-methyl-1-butene with Ph_3GeH and VCl_4, 4-methyl-1-pentene with tricyclohexylgermane and $MoCl_5$, and 4-4-dimethyl-1-pentene with trioctylgermane and chromyl chloride. The catalysts are claimed to be superior to Ziegler catalyst with respect to the crystallinity of the polymers produced (160).

Nametkin and coworkers (166) also used organosilicon, germanium and tin hydrides as reducing agents in the preparation of a two-component catalyst system containing titanium chloride. Maximum ethylene polymer yield was obtained with a mixture of triethyltin hydride and titanium chloride (which corresponds to the maximum content of divalent titanium).

Alkyl transfer from tetrabutylgermane to germanium tetrachloride was found to be catalyzed by germanium diiodide, compounds having a Ge—H function, reducing agents such as $LiAlH_4$ and Raney nickel (138). It was assumed that in these reactions a germanium dihalide acts as the common catalytically active species and germanium tetrachloride can be reduced to the bivalent state by reducing agents including organogermanium hydrides (142).

4-3-7 Asymmetric organogermanium hydrides

Brook and Peddle reported in 1963 (21) the first resolution of an asymmetric germanium compound, methyl-α-naphthylphenylgermane, into its enantiomers. The synthesis and resolution of methyl-α-naphthylphenylgermane was best accomplished by the following route:

$$\alpha\text{-NpPhMeGeBr} \xrightarrow{\text{NaOMe}} \alpha\text{-NpPhMeGeOMe} \xrightarrow{(-)\text{ Menthol}}$$

$$\alpha\text{-NpPhMeGeO}(-)\text{—}C_{10}H_{19}$$

one diastereoisomer $\xrightarrow{\text{LiAlH}_4}$ α-NpPhMeGeH $[\alpha]_D$ + 26.7

second diastereoisomer $\xrightarrow{\text{LiAlH}_4}$ α-NpPhMeGeH $[\alpha]_D$ − 25.5

At the same time, the resolution of ethyl-α-naphthyl phenylgermane has been reported by Bott, Eaborn and Varma (17).

Recrystallization of ethyl-α-naphthylphenylgermyl(−)menthoxide from pentane at −70°C gave a diastereoisomer which, on reduction with lithium aluminium hydride in ether, gave the (+)-hydride, $Et(\alpha\text{-}C_{10}H_7)PhGeH$. Crystallization of ethyl-1-naphthylphenylgermyl(−)-1-phenylethoxide at −15°C gave crystals from which the optically impure (−)-hydride was obtained by reduction. The (+)-hydride was converted into the (−)-chloride by treatment with chlorine in carbon tetrachloride and reduction of the chloride with lithium aluminium hydride gave the (−)-hydride. Only 7% of racemization occurred in this Walden cycle, the existence of which shows that substitution can proceed by at least two distinct mechanisms, one involving predominant inversion and the other predominant retention of configuration (39). Retentions or inversions of configuration of the same type had also been observed by Brook and Peddle (22, 133).

The absolute configurations of the (+) enantiomers of (+)methyl-α-naphthylphenylmethane, silane and germane and their specific rotations are given by Brook (20).

A linear plot of electronegativity of the central atom (C = 2.60, Si = 1.90) versus specific rotation predicts a value of 2.06 for the electronegativity of germanium, in excellent agreement with the value of 2.00 determined by Allred and Rochow from NMR data (2).

In the following reactions, the stereospecificities are fairly high: 90% in both Walden cycles (I) and (II) (37):

(I) $R_3GeH \xrightarrow{\text{n-BuLi}} R_3GeLi \xrightarrow{H_2O} R_3GeH$

$[\alpha]_D^{25}$ + 22.5   +17.8

(II) $R_3GeH \xrightarrow{\text{n-BuLi}} R_3GeLi \xrightarrow{CO_2} R_3GeCO_2H \xrightarrow{\text{n-BuLi}}$

$[\alpha]_D^{25}$ + 22.0   −9.8

$R_3GeLi \xrightarrow{H_2O} R_3GeH$

+17.4

The treatment of the hydride $(+)\text{-}R_3GeH$ with n-butyllithium in ether proceeds with retention of configuration (20, 22).

Correlation of the absolute configurations of R-(+)-methyl (α-naphthyl)phenylgermane and (+)-ethyl(α-naphthyl)phenylgermane has been accomplished through stereospecific metalation to give the germyllithium reagents followed by their reaction with either methyl or ethyl iodide to give optically-active ethylmethyl(α-naphthyl) phenylgermane (38):

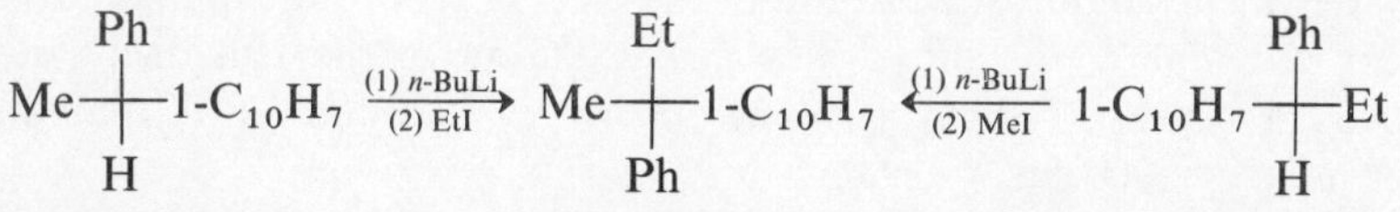

$$\underset{\text{S-}(-)\text{-II}}{\overset{\text{Ph}}{\underset{\text{H}}{\text{Me}\!-\!\!\!\!-\!\text{1-}C_{10}H_7}}} \xrightarrow[\text{(2) EtI}]{\text{(1) }n\text{-BuLi}} \underset{(+)\text{-(III)}}{\overset{\text{Et}}{\underset{\text{Ph}}{\text{Me}\!-\!\!\!\!-\!\text{1-}C_{10}H_7}}} \xleftarrow[\text{(2) MeI}]{\text{(1) }n\text{-BuLi}} \underset{R\text{-}(+)\text{-(I)}}{\overset{\text{Ph}}{\underset{\text{H}}{\text{1-}C_{10}H_7\!-\!\!\!\!-\!\text{Et}}}}$$

The stereochemical representations are based on the assumptions that metalation involves retention and the coupling with the alkyl iodides involves inversion of configuration (37).

Eaborn, Hill and Simpson (37a) recently have examined a range of reactions of optically-active ethyl-(1-naphthyl)-phenylgermanium compounds, R_3GeX, with nucleophilic reagents.

Loss of optical activity seems generally to be greater than in comparable reactions of methyl-(1-naphthyl)phenylsilicon compounds (164b). However, the Walden cycle (A) occurs with 90%, the cycle (B) with 68% and the cycle (C) with 72% overall stereospecificity.

(A) $(+)\text{-}R_3GeH \rightarrow (-)\text{-}R_3GeCl \rightarrow (-)\text{-}R_3GeH$

$[\alpha]_D^{25}$ $+22.5°C$ $-10.7°C$ $-17.9°C$

(B) $(+)\text{-}R_3GeH \rightarrow (-)\text{-}R_3GeCl \rightarrow (+)\text{-}R_3GeSPh$

$[\alpha]_D^{25}$ $+20.0°C$ $-9.2°C$ $+14.2°C$

$\rightarrow (+)\text{-}R_3GeH$

$+7.0°C$

(C) $(+)\text{-}R_3GeH \rightarrow (-)\text{-}R_3GeCl \rightarrow (-)\text{-}R_3GeNC_4H_4$

$[\alpha]_D^{25}$ $+21.5°C$ $-9.0°C$ $-1.3°C$

$\rightarrow (+)\text{-}R_3GeH$

$+9.5°C$

4-3-8   Spectral studies of organogermanium hydrides

4-3-8-1 IR and Raman Spectra. The infrared absorption spectra of the alkylgermanium hydrides R_3GeH, R_2GeH_2, $RGeH_3$ all present a strong absorption band in the zone of $2000-2100 \, cm^{-1}$ corresponding to the valence mode of vibration of the Ge—H group; the monohydrides near $2010 \, cm^{-1}$, the dihydrides near $2035 \, cm^{-1}$, the trihydrides near $2065 \, cm^{-1}$ (149). Trihexylgermane absorbs at $1980 \, cm^{-1}$ (47), dimethylgermane at $2060 \, cm^{-1}$ (193).

Ponomarenko and Egorov (135) reported that a simple empirical relation exists between the inductive effect of substituents bound to silicon and germanium and the oscillational frequencies of the bonds Si—H and Si—D, Ge—H and Ge—D. This relation is:

$$v = A\chi = A\Sigma\chi_R \quad (I)$$

where v is the frequency of valence oscillation of the bond Si—H or Si—D, Ge—H or Ge—D, A is a constant quantity equal to ~ 1011 for hydride and ~ 734 for deuterides, χ is the "effective" electronegativity of the silyl or germyl group, composed of the inductive constants (χ_r) of the three substituents bound with the silicon or the germanium.

This relation was refined by Ponomarenko, Zueva and Andreev for hydrides and deuterides of germanium (137) for the purpose of finding a measure of the inductive effect of the substituents in the number of organic compounds of germanium and silicon. To this end, 21 hydrides and deuterides of germanium were prepared and their combined scattering spectra were taken. The data obtained make it possible to express relation (I) for germanium deuterides in the following refined form:

$$v_{Ge-D} = 728\chi_{Ge} = 728\Sigma\chi_{GeR} \quad (II)$$

where χ_{Ge} is the "effective" electronegativity of the germyl group and χ_{GeR} gives the inductive constants of the substituents bound with the germanium. The value of A is equal to 1011 for both germanium hydrides and silicon hydrides.

Comparison of the inductive constants χ_{Ge-R} and χ_{Si-R} with the Taft polarity constants (σ^*) also makes it possible to find an interesting relation between the inductive constants of substituents in hydrides and deuterides of germanium and silicon and Taft's polarity constants.

The frequency of the Ge—H valence mode of vibration has been measured for about twenty alkylgermanes (R_nGeH_{4-n}, $n = 1, 2, 3$) and alkylhalogermanes $R_2(X)GeH$ by R. Mathis, Satgé and F. Mathis (110). This frequency is raised by $-I$ substituents on the Ge atom and lowered by $+I$ substituents. This substituent effect is additive and may be used for an estimation of inductive effect. The ν_{Ge-H} frequency in $R_1R_2R_3GeH$ can be approximately related to the sum of Taft σ^* coefficients for R_1, R_2 and R_3 by the equation:

$$\nu_{Ge-H}(cm^{-1}) = 2008 + 16.5\Sigma\sigma^*$$

The integrated absorption intensity (A_{Ge-H}) has been measured for the valence vibration band of Ge—H group in thirteen alkylgermanes or alkylhalogermanes dissolved in carbon tetrachloride or in heptane (109).

Electron-attracting ($-I$) substituents on the Ge atom enhance the absorption intensity, electron-repelling ($+I$) substituents lower it. For each solvent, an approximately rectilinear relation can be established between A_{Ge-H} in $R_1R_2R_3GeH$ and Taft's coefficients of atoms or group R_1, R_2, R_3.

$$A_{CCl_4} = -0.260\Sigma\sigma^* + 2.24$$

$$A_{heptane} = -0.227\Sigma\sigma^* + 1.94$$

The intensity of the Ge—H band diminishes with an increase in $\Sigma\sigma^*$; this is coherent with a $\overset{+}{Ge}-\overset{-}{H}$ dipole.

Many organogermanium hydrides have been examined by Cross and Glockling (28) and, in addition to the Ge—H stretching frequency near $2100\ cm^{-1}$, various deformation modes have been identified in simpler molecules (GeH_2 bend and wag near 860 and $780\ cm^{-1}$; GeH_3 rock near $600\ cm^{-1}$). The germanium–hydrogen stretch varies from $2175\ cm^{-1}$ in GeH_2F_2 to $1953\ cm^{-1}$ in $(Ph_3Ge)_3GeH$ (61). The Ge—H and Ge—D stretching and deformation frequencies for some hydrides and deuterides are given. The germanium–hydrogen deformation occurs near $700\ cm^{-1}$ in the monohydrides R_3GeH and 860–$870\ cm^{-1}$ in the di- and trihydrides R_2GeH_2 and $RGeH_3$ ($842\ cm^{-1}$ in CH_3GeH_3 (81)).

The germanium–deuterium stretch varies from $1473\ cm^{-1}$ in Ph_3GeD to $1433\ cm^{-1}$ in $i\text{-}Pr_3GeD$. Germanium–deuterium deformation modes occur in uncomplicated regions of the spectrum ($i\text{-}Pr_3GeD$: $508\ cm^{-1}$, Ph_3GeD: $526\ cm^{-1}$, $(PhCH_2)_3GeD$: $472\ cm^{-1}$, $Ph_2(Bu)GeD$: $525\ cm^{-1}$, $(PhCH_2)_2GeD_2$: 615–$495\ cm^{-1}$) and are easily recognized (28).

The infrared absorption spectra and the Raman spectra of the series $(CH_3)_nGeH_{4-n}$ ($n = 1, 2, 3$) have been studied by van de Vondel and van der Kelen (177). The attribution of the fundamental vibration frequencies has been carried out. This attribution has been facilitated by the measurement of the depolarization factor of several Raman frequencies. The theoretical calculations of the fundamental frequencies confirm this attribution.

Gas-phase and solid-film IR spectra of $CH_3CH_2GeH_3$ and $CH_3CH_2GeD_3$ are reported (103a) which, with the published Raman spectra (136), allow assignment of twenty-four fundamentals and tentative assignment of the remaining three.

4-3-8-2 NMR Spectra. The proton magnetic resonance spectra of a series of mono-, di- and trisubstituted hydrides of germanium have been recorded by Egorochkin and coworkers (41).

A connection of the chemical shifts of the protons directly joined to the germanium atom with the vibration frequencies of the Ge—H bond and the Taft constants of the substituents has been discovered.

The PMR spectra of the ethylgermanes $(C_2H_5)_3GeH$, $(C_2H_5)_2GeH_2$, $(C_2H_5)GeH_3$ have been studied by MacKay and Watt (103). In these compounds the Ge—H signal appears about 6ζ and shows the expected multiple structures from coupling with the methylene protons. There is little indication of interaction with the methyl protons, possibly a slight broadening of the peaks; any coupling is under 0.1 cps.

The positions of the Ge—H resonances in these $R_{4-n}GeH_n$ compounds move to low field with decreasing n, and when methyl is replaced by ethyl, as expected.

The coupling constants between the protons on germanium and the methylene protons $J(\underline{H}—Ge—C\underline{H})$ decreases with increasing number of alkyl groups as in the methylgermanes (174).

The coupling constants in the ethylgermanes are about 1 cps smaller than those in the methylgermanes.

COMPOUND	$\delta(Ge—H)$ (CPS)	$J(H_2C—Ge—H)$	CONCENTRA-TION (C_6H_6) (%)	REFERENCE
$(C_2H_5)_3GeH$	−235.5	2.6	10	103
	−230.2		50	—
$(C_2H_5)_2GeH_2$	−230.1	2.7	10	—
	−225.7		50	—
$(C_2H_5)GeH_3$	−213.0	3.0	10	—

With the ethyl- and butylgermanes R_nGeH_{4-n} R = Et, n-Bu ($n = 1, 2$ and 3), the results registered by Massol (105) on $\delta\underline{H}$—Ge are in good agreement with those of Egorochkin (41) and MacKay (103). The multiplicity Ge—H signals allow us to attribute to the spin coupling, between the $\underline{H}$—Ge proton and the protons of the methylene groups in α of the germanium $[J_{\underline{H}—Ge—CH_2}]$, values of 2.6 to 3 cps. The presence of halogen atoms linked to germanium in R_2XGeH and RX_2GeH gives a displacement of the $\underline{H}$—Ge signal towards the weaker fields. $\delta\underline{H}$—Ge increases with the electronegativity of the halogen on the one hand and the number of halogen atoms fixed on the germanium on the other hand (105, 106, 140).

$$\delta Ge—\underline{H} = 3.73 \text{ ppm (Et}_3\text{GeH), } 5.45 \text{ ppm (Et}_2\text{ClGeH),}$$

$$6.84 \text{ ppm (EtCl}_2\text{GeH), } 7.66 \text{ ppm (Cl}_3\text{GeH).}$$

However, there is no rectilinear relation between the chemical shift $\delta\underline{H}$—Ge and the sum of Taft coefficients in the series of alkylhalogermanes $R_nX_{3-n}GeH$ ($n = 0, 1, 2, 3$ and $X = $ Cl, Br, I). The influence of the halogens seems to be due not only to the inductive effect, but also to their magnetic anisotropy (105).

The NMR spectra of the methylgermanes $(CH_3)_3GeH$, $(CH_3)_2GeH_2$, CH_3GeH_3 have been registered and discussed by several authors (40, 41, 154, 174, 176). The values of the chemical displacements δCH_3, δGe—H as well as coupling constants $J(^{13}C—^1H)$ and $J(\underline{H}—C—Ge—\underline{H})$ were determined. A direct relation between the length of the bonds and the coupling constants has been established (154).

For methylgermanes, the constants of spin–spin interaction of the protons of the methyl groups and the proton directly connected with the germanium atom have been determined as 4 Hz on the average (41).

When the PMR spectra of several hydrides of silicon and germanium of similar structure and corresponding compounds of carbon were compared the following features were observed: in the methyl-substituted compounds with the general formulae $(CH_3)_3MH$, $(CH_3)_2MH_2$ and CH_3MH_3, where $M = $ C, Si and Ge, the screening of the protons of the methyl groups increases in the order C < Ge < Si; for the protons connected directly with the M atom the reverse order of screening is observed C > Ge > Si. The order of screening of the protons of the methyl groups and of Ge—H group is in agreement with the following relative electronegativities of the elements C > Ge > Si obtained from data on the PMR spectra of tetramethyl derivatives of the elements of Group IV B (2).

A relation, between the chemical shifts of the methyl protons and CH_2 protons of the ethyl groups, and the sums of the Taft constants of the three other substituents was detected for organogermanium compounds $CH_3GeR_1R_2R_3$ and $C_2H_5GeR_1R_2R_3$ including $(CH_3)_3GeH$, $(CH_3)_2GeH_2$ and $(CH_3)GeH_3$ (40).

The resonance contributions of polar structures to the bonding in methylgermanium hydrides and chlorides can be studied by PMR spectra and electrical dipole-moment measurements (176).

Proton magnetic resonance may be used to determine the relative acidities in liquid ammonia of germane GeH_4 and some of the methyl and ethyl derivatives of germane (16).

4-3-8-3   Mass Spectra. The mass spectra of numerous organic germanium derivatives comprising hydrides of the type R_3GeH, (Me_3GeH, Et_3GeH, Ph_3GeH, $(PhCH_2)_3GeH$) and deuterides such as $(PhCH_2)_3GeD$, Ph_2GeD_2 have been described by Glocking and Light (62).

4-3-8-4   Dipole-moment Measurements. The moment of the compound CH_3GeH_3 is 0.67 D according to the data of microwave spectra (11). Kartsev and coworkers (83) estimate the moments of some bonds in organogermanium compounds by comparing the dipole moments of various compounds of carbon, silicon and germanium and also on the basis of the comparatively small difference in the electronegativities of the last two elements and the similar values of their covalent radii. A comparison of this type leads to the following results for the moments of the bonds $\overset{+}{Ge}-\overset{-}{H} \sim 1.00$ D, $\overset{+}{Ge}-\overset{-}{C} \sim 0.65$ D, $\overset{+}{Ge}-\overset{-}{Cl} \sim 3.00$ D.

The dipole moments of methylgermanes are for CH_3GeH_3 $\mu = 0.64$ D, $(CH_3)_2GeH_2$ $\mu = 0.758$ D, $(CH_3)_3GeH$ $\mu = 0.6685$ D (176).

Stark-effect measurements gave a dipole moment of 0.616 D for dimethylgermane (171a).

Ulbricht and coworkers (173) using the value of 1.0 D for the bond moment of $Ge-H$ (83) and the dipole moment of methylgermanes $(CH_3)_nGeH_{4-n}$ ($n = 1, 2, 3$) (176) have calculated the group moments of $Ge-CH_3$ in these hydrides.

4-3-8-5   Other Physical-chemical Studies. The microwave spectrum of dimethylgermane is described by Thomas and Laurie (171a).

Normal vibrations and thermodynamic properties of CH_3GeH_3, $NCGeH_3$ and Ge_2H_6 and related deuterated derivatives are given by Galasso and coworkers (47a).

4-3-9   Organogermanium hydrides

Tables 4-7 (a–f) showing some physical properties of Organogermanium hydrides appear on the following pages.

Table 4-7

Organogermanium Hydrides

(a) R₃GeH

COMPOUND		B.P. OR (M.P.) (°C/mm)	n_D^{20}	d_4^{20}	REFERENCES
C₃H₁₀Ge	(CH₃)₃GeH	26/755.5 (−123.1)	1.3890	1.0128	45, 63, 136
C₃H₉DGe	(CH₃)₃GeD	26/758.5	1.3893	1.0207	136, 137
C₆H₁₆Ge	(C₂H₅)₃GeH	122/769	1.4330	1.0075	8, 35, 86, 137, 142
C₆H₁₅DGe	(C₂H₅)₃GeD	120/740	1.4333	1.0097	137
C₉H₂₂Ge	(n-C₃H₇)₃GeH	183/742	1.4441	0.9694	78, 82, 142
C₉H₂₂Ge	(i-C₃H₇)₃GeH	174/760	1.4505	0.9770	24, 111, 142
C₉H₂₁DGe	(i-C₃H₇)₃GeD	—	—	—	28
C₁₂H₂₈Ge	(n-C₄H₉)₃GeH	123/20	1.4508	0.9455	95, 142
C₁₅H₃₄Ge	(n-C₅H₁₁)₃GeH	150/11	1.4542	0.9310	93, 96, 142
C₁₅H₃₄Ge	(i-C₅H₁₁)₃GeH	140/17	1.4517	0.9238	93, 96, 142
C₁₈H₃₄Ge	(C₆H₁₁)₃GeH	(24–25)	—	—	79
C₁₈H₄₀Ge	(n-C₆H₁₃)₃GeH	169–70/9	1.4582	0.9228	47, 142
C₂₁H₂₂Ge	(C₆H₅CH₂)₃GeH	164/10⁻³ (80–82)	—	—	27, 28
C₂₁H₂₁DGe	(C₆H₅CH₂)₃GeD	170–76/10⁻³ (81)	—	—	27, 28
C₂₁H₄₆Ge	(n-C₇H₁₅)₃GeH	182/1.7	1.4600	0.9108	142
C₂₄H₅₂Ge	(n-C₈H₁₇)₃GeH	179–80/0.4	1.4610	0.9061	142
C₈H₂₀SiGeCl₂	CH₃Cl₂Si(CH₂)₃Ge(H)(C₂H₅)₂	125/25	1.1623	1.4682	113
C₈H₂₂SiGe	CH₃SiH₂(CH₂)₃Ge(H)(C₂H₅)₂	87/22	0.9793	1.4570	113

Table 4-7—continued

COMPOUND		B.P. OR (M.P.) (°C/mm)	n_D^{20}	d_4^{20}	REFERENCES
$CH_3Cl_2Si(CH_2)_3Ge(H)(C_4H_9)_2$	$C_{12}H_{28}SiGeCl_2$	160/18	1.0912	1.4699	113
$CH_3SiH_2(CH_2)_3Ge(H)(C_4H_9)_2$	$C_{12}H_{30}SiGe$	130/20	0.9453	1.4621	113
$(C_2H_5)_3Ge(CH_2)_3Ge(H)(C_4H_9)_2$	$C_{17}H_{40}Ge_2$	143/0.9	1.0627	1.4760	113
$(C_2H_5)_3Si(CH_2)_3Ge(H)(C_4H_9)_2$	$C_{17}H_{40}SiGe$	143/1.5	0.9406	1.4670	113
$(C_4H_9)_3ClSi(CH_2)_3Ge(H)(C_4H_9)_2$	$C_{19}H_{43}SiGeCl$	165/1.3	0.9814	1.4702	113
$(C_4H_9)_3SiH(CH_2)_3Ge(H)(C_4H_9)_2$	$C_{19}H_{44}SiGe$	148/1.3	0.9233	1.4660	113
$(C_6H_5)_2SiH(CH_2)_3Ge(H)(C_2H_5)_2$	$C_{20}H_{30}SiGe$	148/0.18	1.0752	1.5481	113
$(C_4H_9)_2CH_3Si(CH_2)_3Ge(H)(C_4H_9)_2$	$C_{20}H_{46}SiGe$	152/0.75	0.9246	1.4650	113
$(C_6H_5)_2CH_3Si(CH_2)_3Ge(H)(C_4H_9)_2$	$C_{24}H_{38}SiGe$	156/0.1	1.0261	1.5353	113
$(C_6H_5)_3GeH$	$C_{18}H_{16}Ge$	128–36/10^{-2} (41°–41°5) (47(α)) (27(β))	—	—	22a, 76, 80, 87
$(C_6H_5)_3GeD$	$C_{18}H_{15}DGe$	—	—	—	28
$(1\text{-}C_{10}H_7)_3GeH$	$C_{30}H_{22}Ge$	(249–50)	—	—	192
$(o\text{-}CH_3{-}C_6H_4)_3GeH$	$C_{21}H_{22}Ge$	120–50/10^{-3} (102–3)	—	—	28, 60
$(m\text{-}CH_3{-}C_6H_4)_3GeH$	$C_{21}H_{22}Ge$	160–70/10^{-3}	—	—	28, 60
$(p\text{-}CH_3{-}C_6H_4)_3GeH$	$C_{21}H_{22}Ge$	160/10^{-3} (81)	—	—	28, 60
$(mesityl)_3GeH$	$C_{27}H_{16}Ge$	—	—	—	28
$[(C_6H_5)_3Ge]_3GeH$	$C_{54}H_{46}Ge_4$	(192–94)	—	—	28, 61

(*b*) $R_2R'GeH$

COMPOUND		B.P. OR (M.P.) (°C/mm)	n_D^{20}	d_4^{20}	REFERENCES
$(CH_3)_2(C_2H_5)GeH$	$C_4H_{12}Ge$	62/755.5	1.4090	1.0158	136
$(CH_3)_2(C_2H_5)GeD$	$C_4H_{11}DGe$	60/737.4	1.4083	1.0262	136
$(i\text{-}C_3H_7)_2(CH_3)GeH$	$C_7H_{18}Ge$	—	—	—	156
$(C_6H_5)(CH_3)_2GeH$	$C_8H_{12}Ge$	80/50	1.5170	1.1583	150
$(C_2H_5)_2(i\text{-}C_5H_{11})GeH$	$C_9H_{22}Ge$	—	—	—	58, 59
$(n\text{-}C_4H_9)_2(H_2C{=}CHCH_2)GeH$	$C_{11}H_{24}Ge$	95–96/13	1.4608	0.9702	142
$(C_6H_5)_2(CH_3)GeH$	$C_{13}H_{14}Ge$	90/0.5	1.5830	1.2128	150
$(n\text{-}C_4H_9)_2(C_6H_5)GeH$	$C_{14}H_{24}Ge$	140/9	—	—	142
$(n\text{-}C_4H_9)_2(C_6H_5CH_2)GeH$	$C_{15}H_{26}Ge$	155/14	—	—	142
$(C_6H_5)_2C_4H_9GeH$	$C_{16}H_{20}Ge$	—	—	—	28
$(n\text{-}C_7H_{15})(C_6H_5CH_2)_2GeH$	$C_{21}H_{30}Ge$	180/1	1.514	1.079	142

(*c*) $RR'R''GeH$

COMPOUND		B.P. OR (M.P.) (°C/mm)	n_D^{20}	d_4^{20}	REFERENCES
$(+)(CH_3)(1{-}C_{10}H_7)(C_6H_5)GeH$	$C_{17}H_{16}Ge$	(74–75)	—	—	21, 22
$(-)(CH_3)(1{-}C_{10}H_7)(C_6H_5)GeH$	$C_{17}H_{16}Ge$	(74–75)	—	—	21
$(+)(C_2H_5)(1{-}C_{10}H_7)(C_6H_5)GeH$	$C_{18}H_{18}Ge$	(31–32.5)	—	—	39
$(-)(C_2H_5)(1{-}C_{10}H_7)(C_6H_5)GeH$	$C_{18}H_{18}Ge$	—	$n_D^{25} = 1.6226$	—	39
$(\pm)(C_2H_5)(1{-}C_{10}H_7)(C_6H_5)GeH$	$C_{18}H_{18}Ge$	140–41/0.01	$n_D^{25} = 1.6270$	—	39

(d) R_2GeH_2

COMPOUND		B.P. OR (M.P.) (°C/mm)	n_D^{20}	d_4^{20}	REFERENCES
$(CH_3)_2GeH_2$	C_2H_8Ge	3	—	—	63, 136, 137
$(CH_3)_2GeD_2$	$C_2H_6D_2Ge$	(−144.3)	—	—	136, 137
$(CD_3)_2GeH_2$	CH_2D_6Ge	6.5/745	—	—	63
$(C_2H_5)_2GeH_2$	$C_4H_{12}Ge$	74/760	1.4219	1.0390	136, 142, 149
$(C_2H_5)_2GeD_2$	$C_4H_{10}D_2Ge$	71.5/743.5	1.4200	1.0525	136, 137
$(n\text{-}C_3H_7)_2GeH_2$	$C_6H_{16}Ge$	126-27/760	1.4340	1.0030	142, 149
$(i\text{-}C_3H_7)_2GeH_2$	$C_6H_{16}Ge$	110-11/760	1.432	0.982	7
$(n\text{-}C_4H_9)_2GeH_2$	$C_8H_{20}Ge$	86/46	1.4428	0.9782	10, 142, 149
$(n\text{-}C_5H_{11})_2GeH_2$	$C_{10}H_{24}Ge$	92/12	1.4478	0.9595	142, 149
$(n\text{-}C_6H_{13})_2GeH_2$	$C_{12}H_{28}Ge$	113-14/8	1.4522	0.9484	142, 149
$(n\text{-}C_7H_{15})_2GeH_2$	$C_{14}H_{32}Ge$	148/10	1.4543	0.9348	142, 149
$(n\text{-}C_8H_{17})_2GeH_2$	$C_{16}H_{36}Ge$	164-65/9	1.4568	0.9274	142, 149
$(C_6H_5)_2GeH_2$	$C_{12}H_{12}Ge$	95/1	1.5935	—	22a, 75, 77
$(C_6H_5CH_2)_2GeH_2$	$C_{14}H_{16}Ge$	$80\text{-}85/10^{-3}$	—	—	27, 28
$(C_6H_5CH_2)_2GeD_2$	$C_{14}H_{14}D_2Ge$	—	—	—	27, 28

(e) $RR'GeH_2$

COMPOUND		B.P. OR (M.P.) (°C/mm)	n_D^{20}	d_4^{20}	REFERENCES
$(C_2H_5)(i\text{-}C_5H_{11})GeH_2$	$C_7H_{18}Ge$	—	—	—	58, 59
$(CH_3)(C_6H_5)GeH_2$	$C_7H_{10}Ge$	70/35	1.5292	1.2024	150
$(C_6H_5)(C_2H_5)GeH_2$	$C_8H_{12}Ge$	83/30	1.5250	1.1629	152

(f) RGeH₃

COMPOUND		B.P. OR (M.P.) (°C/mm)	n_D^{20}	d_4^{20}	REFERENCES
CH_6Ge	CH_3GeH_3	-34.1 (-154.5)	—	—	63, 136, 137, 171
CH_5D_3Ge	CH_3GeD_3	-23.5/752	—	—	136, 137
C_2H_6Ge	$CH_2{=}CHGeH_3$	-3.5 (extrapol.)	—	—	19
C_2H_8Ge	$C_2H_5GeH_3$	11.5/743.5	—	—	136, 137, 171
$C_2H_5D_3Ge$	$C_2H_5GeD_3$	11.3/748.5	—	—	136, 137
C_2H_8OGe	$CH_3OCH_2GeH_3$	44.4 (extrapol.)	—	—	50
C_3H_8Ge	$(CH_2{=}CHCH_2)GeH_3$	37/752.7	1.4315	1.0797	137
$C_3H_5D_3Ge$	$(CH_2{=}CHCH_2)GeD_3$	37/744	1.4320	1.1146	137
$C_3H_7GeF_3$	$CF_3CH_2CH_2GeH_3$	46/750.5	1.3530	1.3362	41, 137
$C_3H_{10}Ge$	$(n\text{-}C_3H_7)GeH_3$	41–42.5/757	1.4130	1.0391	78, 137
$C_3H_7D_3Ge$	$(n\text{-}C_3H_7)GeD_3$	41.5/754	1.4055	1.0508	137
$C_4H_{12}Ge$	$(n\text{-}C_4H_9)GeH_3$	74/760	1.4200	1.0220	10, 142, 149
$C_5H_{14}Ge$	$(n\text{-}C_5H_{11})GeH_3$	104–5/760	1.4302	1.0138	142, 149
$C_5H_{14}Ge$	$(i\text{-}C_5H_{11})GeH_3$	—	—	—	59
$C_6H_{16}Ge$	$(n\text{-}C_6H_{13})GeH_3$	128–29/760	1.4350	0.9972	142, 149
$C_7H_{18}Ge$	$(n\text{-}C_7H_{15})GeH_3$	85/74	1.4390	0.9819	142, 149
$C_8H_{20}Ge$	$(n\text{-}C_8H_{17})GeH_3$	80/31	1.4422	0.9717	142, 149
C_6H_8Ge	$C_6H_5GeH_3$	40/22	1.5353	1.2371	22a, 28, 150

References (4-2 and 4-3)

1. Ali-zade, I. G., M. I. Shikieva, M. A. Salimov, N. D. Abdullaev, and I. A. Shikhiev, *Dokl. Akad. Nauk SSSR*, **173**, 89 (1967).
2. Allred, A. L., and E. G. Rochow, *J. Inorg. Nucl. Chem.*, **5**, 269 (1958); **20**, 167 (1961).
3. Amberger, E., and H. Boeters, *Angew. Chem.*, **73**, 114 (1961).
4. Amberger, E., and J. Mühlhofer, *J. Organometal. Chem.*, **12**, 55 (1968).
5. Amberger, E., W. Stoeger, and R. Honigschmid-Grossich, *Angew. Chem. Internat. Ed.*, **5**, 522 (1966).
6. Anderson, H. H., *J. Am. Chem. Soc.*, **83**, 547 (1961).
7. Anderson, H. H., *J. Am. Chem. Soc.*, **78**, 1692 (1956).
8. Anderson, H. H., *J. Am. Chem. Soc.*, **79**, 326 (1957).
9. Anderson, H. H., *J. Am. Chem. Soc.*, **79**, 4913 (1957).
10. Anderson, H. H., *J. Am. Chem. Soc.*, **82**, 3016 (1960).
11. Barchukov, A. I., and A. M. Prokhorov, *Opt. i Spektroskopiya*, **4**, 799 (1958); *CA* **52**, 16875 (1958).
12. Baukov, Yu. I., I. Yu Belavin, and I. F. Lutsenko, *Zh. Obshch. Khim.*, **35**, 1092 (1965).
13. Baukov, Yu. I., and I. F. Lutsenko, *Zh. Obshch. Khim.*, **32**, 2747 (1962).
14. Baukov, Yu. I., and I. F. Lutsenko, *Zh. Obshch. Khim.*, **34**, 3453 (1964).
15. Berger, A., Fr. Pat. 1,429,930; *CA*, **65**, 18620 (1966).
15a. Berger, A., U.S. Pat., 3,401,183; *CA*, **70**, 337 (1969).
16. Birchall, T., and W. L. Jolly, *Inorg. Chem.*, **5**, 2177 (1966).
17. Bott, R. W., C. Eaborn, and I. D. Varma, *Chem. Ind.* (*London*), 614 (1963).
18. Braun, J., *Compt. Rend.*, **260**, 218 (1965).
19. Brinckman, F. E., and F. G. A. Stone, *J. Inorg. Nucl. Chem.*, **11**, 24 (1959).
20. Brook, A. G., *J. Am. Chem. Soc.*, **85**, 3051 (1963).
20a. Brooks, E. H., M. Elder, W. A. G. Graham, and D. Hall, *J. Am. Chem. Soc.*, **90**, 3587 (1968).
21. Brook, A. G., and G. S. D. Peddle, *J. Am. Chem. Soc.*, **85**, 1869 (1963).
22. Brook, A. G., and G. S. D. Peddle, *J. Am. Chem. Soc.*, **85**, 2338 (1963).
22a. Brooks, E. H., F. Glockling, and K. A. Hooton, *J. Chem. Soc.*, 4283 (1965).
23. Bryskovskaya, A. V., and V. M. Al'bitskaya, *Zh. Obshch. Khim.*, **37**, 1553 (1967).
23a. Carey, F. A., and H. S. Tremper, *Tetrahedron Letters*, 1645 (1969).
24. Carrick, A., and F. Glockling, *J. Chem. Soc.*, (A), 623 (1966).
25. Cradock, S., G. A. Gibbon, and C. H. van Dyke, *Inorg. Chem.*, **6**, 1751 (1967).
26. Creemers, H. M. J. C., and J. G. Noltes, *J. Organometal. Chem.*, **7**, 237 (1967).
27. Cross, R. J., and F. Glockling, *J. Chem. Soc.*, 4125 (1964).
28. Cross, R. J., and F. Glockling, *J. Organometal. Chem.*, **3**, 146 (1965).
29. Cross, R. J., and F. Glockling, *J. Chem. Soc.*, 5422 (1965).
30. Cullen, W. R., and G. E. Styan, *Inorg. Chem.*, **4**, 1437 (1965).
31. Cullen, W. R., and G. E. Styan, *J. Organometal. Chem.*, **6**, 117 (1966).
32. Cullen, W. R., and G. E. Styan, *J. Organometal. Chem.*, **6**, 633 (1966).
33. Curtis, M. D., *J. Am. Chem. Soc.*, **89**, 4241 (1967).
34. Dessy, R. E., W. Kitching, and T. Chivers, *J. Am. Chem. Soc.*, **88**, 453 (1966).
34a. Dutton, W. A., and M. Onyszchuk, *Inorg. Chem.*, **7**, 1735 (1968).
34b. Dzhurinskaya, N. G., S. A. Mikhailyants, and V. P. Evdakov, *Zh. Obshch. Khim.*, **38**, 1267 (1968).
35. Dzhurinskaya, N. G., V. F. Mironov, and A. D. Petrov, *Dokl. Akad. Nauk SSSR*, **138**, 1107 (1961).
36. Eaborn, C., W. A. Dutton, F. Glockling, and K. A. Hooton, *J. Organometal. Chem.*, **9**, 175 (1967).

37. Eaborn, C., R. E. E. Hill, and P. Simpson, *J. Organometal. Chem.*, **15**, P 1 (1968).
37a. Eaborn, C., R. E. E. Hill, and P. Simpson, *Chem. Comm.*, 1077 (1968).
38. Eaborn, C., R. E. E. Hill, P. Simpson, A. G. Brook, and J. MacRae, *J. Organometal. Chem.*, **15**, 241 (1968).
39. Eaborn, C., P. Simpson, and I. D. Varna, *J. Chem. Soc.*, (A), 1133 (1966).
40. Egorochkin, A. N., M. L. Khidekel', V. A. Ponomarenko, G. Ya Zueva, and G. A. Razuvaev, *Izv. Akad. Nauk SSSR Ser. Khim.*, 373 (1964).
41. Egorochkin, A. N., M. L. Khidekel', V. A. Ponomarenko, G. Ya Zueva, S. S. Svirezheva, and G. A. Razuvaev, *Izv. Akad. Nauk SSSR Ser. Khim.*, 1865 (1963).
42. English, W. D., *J. Am. Chem. Soc.*, **74**, 1927 (1952).
43. Finholt, A. E., A. C. Bond, K. E. Wilzbach, and H. I. Schlesinger, *J. Am. Chem. Soc.*, **69**, 2692 (1947).
44. Fish, R. H., Thesis, University of New Hampshire, U.S.A., (1965).
45. Fish, R. H., and M. G. Kuivila, *J. Org. Chem.*, **31**, 2445 (1966).
46. Fuchs, R., and H. Gilman, *J. Org. Chem.*, **22**, 1009 (1957).
47. Fuchs, R., and H. Gilman, *J. Org. Chem.*, **23**, 911 (1958).
47a. Galasso, V., A. Bigotto, and G. de Alti, *Z. Phys. Chem.*, **50**, 38 (1966).
48. Gee, W., R. A. Shaw, and B. C. Smith, *J. Chem. Soc.*, 2845 (1964).
49. Gibbon, G. A., Y. Rousseau, C. H. van Dyke, and G. J. Mains, *Inorg. Chem.*, **5**, 114 (1966).
50. Gibbon, G. A., J. T. Wang, and C. H. van Dyke, *Inorg. Chem.*, **6**, 1889 (1967).
51. Gilman, H., and J. Eisch, *J. Org. Chem.*, **20**, 763 (1965).
52. Gilman, H., and C. W. Gerow, *J. Am. Chem. Soc.*, **77**, 5509 (1955).
53. Gilman, H., and C. W. Gerow, *J. Am. Chem. Soc.*, **78**, 5435 (1956).
54. Gilman, H., and C. W. Gerow, *J. Am. Chem. Soc.*, **79**, 342 (1957).
55. Gilman, H., and S. P. Massie, *J. Am. Chem. Soc.*, **68**, 1128 (1946).
56. Gilman, H., and H. W. Melvin, *J. Am. Chem. Soc.*, **71**, 4050 (1949).
57. Gilman, H., and E. A. Zuech, *J. Org. Chem.*, **26**, 3035 (1961).
58. Glarum, S. N., Ph.D. Thesis, Brown University, May 1933.
59. Glarum, S. N., and C. A. Kraus, *J. Am. Chem. Soc.*, **72**, 5398 (1950).
60. Glockling, F., and K. A. Hooton, *J. Chem. Soc.*, 3509 (1962).
61. Glockling, F., and K. A. Hooton, *J. Chem. Soc.*, 1849 (1963).
62. Glockling, F., and J. R. C. Light, *J. Chem. Soc.*, (A), 717 (1968).
63. Griffiths, J. E., *Inorg. Chem.*, **2**, 375 (1963).
64. Griffiths, J. E., and M. Onyszchuk, *Can. J. Chem.*, **30**, 339 (1961).
64a. Gverdtsiteli, I. M., and L. V. Baramidze, *Zh. Obshch. Khim.*, **38**, 1598 (1968).
65. Gverdtsiteli, I. M., L. V. Baramidze, and M. V. Chelidze, *Zh. Obshch. Khim.*, **37**, 2654 (1967).
66. Gverdtsiteli, I. M., and M. A. Buachidze, *Dokl. Akad. Nauk SSSR*, **158**, 147 (1964).
67. Gverdtsiteli, I. M., and M. A. Buachidze, *Soobshch. Akad. Nauk Gruz.*, **37**, 59 (1965); *CA* **62**, 14716 (1965).
68. Gverdtsiteli, I. M., and M. A. Buachidze, *Soobshch. Akad. Nauk Gruz.*, **37**, 323 (1965); *CA*, **62**, 14719 (1965).
69. Gverdtsiteli, I. M., and M. A. Buachidze, *Soobshch. Akad. Nauk Gruz.*, **48**, 571 (1967); *CA*, **68**, 10175 (1968).
69a. Gverdtsiteli, I. M., and E. S. Gelashvili, *Soobshch. Akad. Nauk Gruz.*, **52**, 69 (1968); *CA*, **70**, 374 (1969).
69b. Gverdtsiteli, I. M., T. P. Guntsadze, and M. I. Gudavadze, *Soobshch. Akad. Nauk Gruz.*, **50**, 609 (1968); *CA*, **69**, 8153 (1968).
70. Gverdtsiteli, I. M., T. P. Guntsadze, and A. D. Petrov, *Dokl. Akad. Nauk SSSR*, **153**, 107 (1963).

71. Gverdtsiteli, I. M., T. P. Guntsadze, and A. D. Petrov, *Dokl. Akad. Nauk SSSR*, **157**, 607 (1964).
72. Harris, D. M., Ph.D. Thesis, University of Minnesota, 1950.
73. Henry, M. C., and M. F. Downey, *J. Org. Chem.* **26**, 2299 (1961).
74. Hugues, M. B., *Dissertation Abstr.*, **19**, 1921 (1958).
75. Johnson, O. H., and D. M. Harris, *J. Am. Chem. Soc.*, **72**, 5564 (1950).
76. Johnson, O. H., and D. M. Harris, *J. Am. Chem. Soc.*, **72**, 5566 (1950).
77. Johnson, O. H., and D. M. Harris, *Inorg. Syn.*, **5**, 74 (1957).
78. Johnson, O. H., and L. V. Jones, *J. Org. Chem.*, **17**, 1172 (1952).
79. Johnson, O. H., and W. H. Nebergall, *J. Am. Chem. Soc.*, **71**, 1720 (1949).
80. Johnson, O. H., W. H. Nebergall, and D. M. Harris, *Inorg. Syn.*, **5**, 76 (1957).
81. Jolly, W. L., *J. Am. Chem. Soc.*, **85**, 30 (1963).
82. Jones, L. V., *Dissertation Abstr.*, **13**, 308 (1953).
82a. Kaplan, L., *Chem. Comm.*, 106 (1969).
83. Kartsev, G. N., Ya. K. Syrkin, and V. F. Mironov, *Izv. Akad. Nauk SSSR Otd. Khim. Nauk*, 948 (1960).
83a. Kazankova, M. A., and I. F. Lutsenko, *Zh. Obshch. Khim.*, **39**, 926 (1969).
84. Kramer, K., and A. N. Wright, *Angew. Chem.*, **74**, 468 (1962).
85. Kramer, K., and A. N. Wright, *J. Chem. Soc.*, 3604 (1963).
86. Kraus, C. A., and E. A. Flood, *J. Am. Chem. Soc.*, **54**, 1635 (1932).
87. Kraus, C. A., and L. S. Foster, *J. Am. Chem. Soc.*, **49**, 457 (1927).
88. Kruglaya, O. A., N. S. Vyazankin, and G. A. Razuvaev, *Zh. Obshch. Khim.*, **35**, 394 (1965).
89. Kruglaya, O. A., N. S. Vyazankin, G. A. Razuvaev, and E. V. Mitrofanova, *Dokl. Akad. Nauk SSSR*, **173**, 834 (1967).
90. Kühlein, K., and W. P. Neumann, *Ann. Chem.*, **702**, 17 (1967).
90a. Kuliev, A. M., A. A. Dzhafarov, F. N. Mamedov, I. I. Namazov, and I. A. Aslanov, U.S.S.R. Pat. 229,513; *CA*, **71**, 338 (1969).
91. Labarre, J. F., M. Massol, and J. Satgé, *Bull. Soc. Chim. France*, 736 (1967).
92. Lesbre, M., and R. Buisson, *Bull. Soc. Chim. France*, 1204 (1957).
93. Lesbre, M., and J. Satgé, *Compt. Rend.*, **247**, 471 (1958).
94. Lesbre, M., and J. Satgé, *Bull. Soc. Chim. France*, 783 (1959).
95. Lesbre, M., and J. Satgé, *Compt. Rend.*, **250**, 2220 (1960).
96. Lesbre, M., and J. Satgé, *Compt. Rend.*, **252**, 1976 (1961).
97. Lesbre, M., and J. Satgé, *Compt. Rend.*, **254**, 1453 (1962).
98. Lesbre, M., and J. Satgé, *Compt. Rend.*, **254**, 4051 (1962).
99. Lesbre, M., J. Satgé, and M. Massol, *Compt. Rend.*, **256**, 1548 (1963).
100. Lesbre, M., J. Satgé, and M. Massol, *Compt. Rend.*, **257**, 2665 (1963).
101. Lesbre, M., J. Satgé, and M. Massol, *Compt. Rend.*, **258**, 2842 (1964).
102. Lutsenko, I. F., Yu. I. Baukov, and B. N. Khasapov, *Zh. Obshch. Khim.*, **33**, 2724 (1963).
103. MacKay, K. M., and R. Watt, *J. Organometal. Chem.*, **6**, 336 (1966).
103a. Mackay, K. M., and R. Watt, *Spectrochim. Acta*, **23A**, 2761 (1967).
104. Macklen, E. D., *J. Chem. Soc.*, 1989 (1959).
105. Massol, M., Thèse, Toulouse, France, 1967.
106. Massol, M., and J. Satgé, *Bull. Soc. Chim. France*, 2737 (1966).
106a. Massol, M., J. Satgé, and J. Barrau, *Compt. Rend.*, (C), **268**, 1710 (1969).
106b. Massol, M., J. Satgé, and Y. Cabadi, *Compt. Rend.*, (C), **268**, 1814 (1969).
107. Massol, M., J. Satgé, and M. Lesbre, *Compt. Rend.*, (C), **262**, 1806 (1966).
108. Massol, M., J. Satgé, and M. Lesbre, *J. Organometal. Chem.*, **17**, 25 (1969).
109. Mathis, R., M. Constant, J. Satgé, and F. Mathis, *Spectrochim. Acta*, **20**, 515 (1964).

110. Mathis, R., J. Satgé, and F. Mathis, *Spectrochim. Acta*, **18**, 1463 (1962).
111. Mazerolles, P., Dissertation, Toulouse, 1959.
112. Mazerolles, P., *Bull. Soc. Chim. France*, 1907 (1962).
113. Mazerolles, P., and J. Dubac, *Compt. Rend.*, (C), **265**, 403 (1967).
114. Mazerolles, P., J. Dubac, and M. Lesbre, *Tetrahedron Letters*, 255 (1967).
115. Mazerolles, P., and M. Lesbre, *Compt. Rend.*, **248**, 2018 (1959).
116. Mazerolles, P., M. Lesbre, and J. P. Lavergne, *Compt. Rend.*, (C), **266**, 639 (1968).
117. Meen, R. H., and H. Gilman, *J. Org. Chem.*, **22**, 684 (1957).
118. Mendelsohn, J. C., F. Metras, J. C. Lahournère, and J. Valade, *J. Organometal. Chem.*, **12**, 327 (1968).
119. Mendelsohn, J. C., F. Metras, and J. Valade, *Compt. Rend.*, **261**, 756 (1965).
120. Mikhailov, B. M., Yu. N. Bubnov, and V. G. Kiselev, *Izv. Akad. Nauk SSSR*, 68 (1965).
121. Mironov, V. F., N. G. Dzurinskaya, T. K. Gar, and A. D. Petrov, *Izv. Akad. Nauk SSSR Otd. Khim. Nauk*, 460 (1962).
122. Mironov, V. F., and A. L. Kravchenko, *Dokl. Akad. Nauk SSSR*, **158**, 656 (1964).
123. Mironov, V. F., and A. L. Kravchenko, *Zh. Obshch. Khim.*, **34**, 1356 (1964).
123a. Nekhorosheva, E. V., and V. M. Al'bitskaya, *Zh. Obshch. Khim.*, **38**, 1511 (1968).
124. Nesmeyanov, A. N., and A. E. Borisov, *Dokl. Akad. Nauk SSSR*, **174**, 96 (1967).
125. Nesmeyanov, A. N., A. E. Borisov, and N. V. Novikova, *Dokl. Akad. Nauk SSSR*, **172**, 1329 (1967).
126. Neumann, W. P., *Angew. Chem.*, **75**, 679 (1963).
127. Neumann, W. P., and K. Kühlein, *Ann. Chem.*, **683**, 1 (1965).
128. Neumann, W. P., and K. Kühlein, *Tetrahedron Letters*, 3419 (1966).
129. Neumann, W. P., B. Schneider, and R. Sommer, *Ann. Chem.*, **692**, 1 (1966).
130. Neumann, W. P., and R. Sommer, *Angew. Chem.*, **75**, 788 (1963).
131. Noltes, J. G., and G. J. M. van der Kerk, *Rec. Trav. Chim.*, **81**, 41 (1962).
132. Onyszchuk, M., *Angew. Chem.*, **75**, 577 (1963).
133. Peddle, G. J. D., *Dissertation Abstr.*, **25**, 1553 (1964).
134. Petukhov, G. G., S. S. Svirezheva, and O. N. Druzhkov., *Zh. Obshch. Khim.*, **36**, 914 (1966).
135. Ponomarenko, V. A., and Yu. Egorov., *Izv. Akad. Nauk SSSR Otd. Khim. Nauk*, 1133 (1960).
136. Ponomarenko, V. A., G. Ya Vzenkova, and Yu. P. Egorov, *Dokl. Akad. Nauk SSSR*, **122**, 405 (1958).
137. Ponomarenko, V. A., G. Ya Zueva, and N. S. Andreev, *Izv. Akad. Nauk SSSR Otd. Khim. Nauk*, 1758 (1961).
138. Rijkens, F., E. J. Bulten, W. Drenth, and G. J. M. van der Kerk, *Rec. Trav. Chim.*, **85**, 1223 (1966).
139. Rijkens, F., M. J. Janssen, W. Drenth, and G. J. M. van der Kerk, *J. Organometal. Chem.*, **2**, 347 (1964).
140. Rivière, P., and J. Satgé, *Bull. Soc. Chim. France*, 4039 (1967).
141. Rivière, P., and J. Satgé, *Compt. Rend.*, (C), **267**, 267 (1968).
141a. Ryan, J. W., and J. L. Speier, *J. Am. Chem. Soc.*, **86**, 895 (1964).
141b. Rustad, D. S., T. Birchall, and W. L. Jolly, *Inorg. Syn.*, **11**, 128 (1968).
142. Satgé, J., *Ann. Chim.* (*Paris*), **6**, 519 (1961).
143. Satgé, J., *Bull. Soc. Chim. France*, 630 (1964).
144. Satgé, J., and M. Lesbre, *Bull. Soc. Chim. France*, 703 (1962).
145. Satgé, J., and M. Lesbre, *Bull. Soc. Chim. France*, 2578 (1965).
146. Satgé, J., M. Lesbre, and M. Baudet, *Compt. Rend.*, **259**, 4733 (1964).

147. Satgé, J., and M. Massol, *Compt. Rend.*, **261**, 170 (1965).
148. Satgé, J., M. Massol, and M. Lesbre, *J. Organometal. Chem.*, **5**, 241 (1966).
149. Satgé, J., M. Mathis-Noel, and M. Lesbre, *Compt. Rend.*, **249**, 131 (1959).
150. Satgé, J., and P. Rivière, *Bull. Soc. Chim. France*, 1773 (1966).
151. Satgé, J., and P. Rivière, *J. Organometal. Chem.*, **16**, 71 (1969).
152. Satgé, J., and P. Rivière. Unpublished work.
153. Satgé, J., P. Rivière, and M. Lesbre, *Compt. Rend.*, (C), **265**, 494 (1967).
154. Schmidbaur, H., *Chem. Ber.*, **97**, 1639 (1964).
155. Schott, V. G., and C. Harzdorf, *Z. Anorg. Allgem. Chem.*, **307**, 105 (1960).
156. Semlyen, J. A., G. R. Walker, and C. S. G. Phillips, *J. Chem. Soc.*, 1197 (1965).
157. Seyferth, D., *J. Am. Chem. Soc.*, **79**, 2738 (1957).
157a. Seyferth, D., S. B. Andrews, and H. D. Simmons, *J. Organometal. Chem.*, **17**, 9 (1969).
158. Seyferth, D., and J. M. Burlitch, *J. Am. Chem. Soc.*, **85**, 2667 (1963).
159. Seyferth, D., J. M. Burlitch, H. Dertouzos, and H. D. Simmons Jr, *J. Organometal. Chem.*, **7**, 405 (1967).
159a. Seyferth, D., R. Damrauer, R. M. Turkel, and L. J. Todd, *J. Organometal. Chem.*, **17**, 367 (1969).
160. Shearer, N. H., and H. W. Coover, *CA*, **54**, 13732 (1960).
161. Shikhiev, I. A., G. F. Askerov, and Sh. V. Garaeva, *Zh. Obshch. Khim.*, **38**, 639 (1968).
162. Shikhiev, I. A., N. D. Abdullaev, and M. I. Aliev, *Zh. Obshch. Khim.*, **36**, 942 (1966).
162a. Shikhiev, I. A., G. F. Askerov, and Sh. V. Garaeva, *Zh. Obshch. Khim.*, **38**, 639 (1968).
163. Shikhiev, I. A., I. A. Aslanov, and N. T. Mekhmandarova, *Zh. Obshch. Khim.*, **36**, 1295 (1966).
164. Shikhiev, I. A., I. A. Aslanov, N. T. Mekhmandarova, and S. Sh. Verdieva, *CA*, **64**, 9760 (1966); *Azerb. Khim. Zh.* **42** (1965).
164a. Shikhiev, I. A., and M. M. Nasirova, *Azerb. Khim. Zh.*, 23 (1968); *CA*, **70**, 353 (1969).
164b. Sommer, L. H., *Stereochemistry Mechanism and Silicon*, McGraw-Hill, New York, 1965.
165. Sommer, R., P. Neumann, and B. Schneider, *Tetrahedron Letters*, 3875 (1964).
166. Sorokin, G. V., M. V. Pozdnyakova, N. I. Ter-Asaturova, V. N. Perchenko, and N. S. Nametkin, *Dokl. Akad. Nauk SSSR*, **174**, 376 (1967).
167. Spanier, E. J., and A. G. MacDiarmid, *Inorg. Chem.*, **2**, 215 (1963).
167a. Speier, J. L., J. A. Webster, and J. H. Barnes, *J. Am. Chem. Soc.*, **79**, 974 (1957).
168. Styan, G. E., *Dissertation Abstr.*, **27**, 1078 (1966).
169. Sujishi, S., and J. Keith, *J. Am. Chem. Soc.*, **80**, 4138, (1958).
170. Teal, G. K., Thesis, Brown University, May 1931.
171. Teal, G. K., and C. A. Kraus, *J. Am. Chem. Soc.*, **72**, 4706 (1950).
171a. Thomas, E. C., and W. Laurie, *J. Chem. Phys.*, **50**, 3512 (1969).
172. Tirsell, J. B., Thesis, *Dissertation Abstr.*, B, **27**, 3451 (1967).
173. Ulbricht, K., V. Vaisarova, V. Bazant, and V. Chvalovsky, *J. Organometal. Chem.*, **13**, 343 (1968).
174. Van der Kelen, G. P., L. Verdonck, and D. van de Vondel., *Bull. Soc. Chim. Belges*, **73**, 733 (1964).
175. Van der Kerk, G. J. M., and J. G. Noltes, *Ann. N. Y. Acad. Sci.*, **125**, 25 (1965).
176. Van de Vondel, D. F., *J. Organometal. Chem.*, **3**, 400 (1965).
177. Van de Vondel, D. F., and G. P. van der Kelen, *Bull. Soc. Chim. Belges*, **74**, 467 (1965).
178. Vyazankin, N. S., M. N. Bochkarev, and L. P. Sanina, *Zh. Obshch. Khim.*, **36**, 166 (1966).

179. Vyazankin, N. S., M. N. Bochkarev, and L. P. Sanina, *Zh. Obshch. Khim.*, **36**, 1154 (1966).
180. Vyazankin, N. S., M. N. Bochkarev, and L. P. Sanina, *Zh. Obshch. Khim.*, **36**, 1961 (1966).
181. Vyazankin, N. S., M. N. Bochkarev, and L. P. Sanina, *Zh. Obshch. Khim.*, **37**, 1037 (1967).
181a. Vyazankin, N. S., V. T. Bychkov, O. V. Linzina, and G. A. Razuvaev, *Zh. Obshch. Khim.*, **39**, 979 (1969).
182. Vyazankin, N. S., E. N. Gladyshev, and S. P. Korneva, *Zh. Obshch. Khim.*, **37**, 1736 (1967).
182a. Vyazankin, N. S., E. N. Gladyshev, S. P. Korneva, G. A. Razuvaev, and E. A. Arkhangel'skaya, *Zh. Obshch. Khim.*, **38**, 1803 (1968).
182b. Vyazankin, N. S., G. S. Kalinina, O. A. Kruglaya, and G. A. Razuvaev, *Zh. Obshch. Khim.*, **38**, 906 (1968).
183. Vyazankin, N. S., O. A. Kruglaya, G. A. Razuvaev, and G. S. Semchikova, *Dokl. Akad. Nauk SSSR*, **166**, 99 (1966).
184. Vyazankin, N. S., E. V. Mitrofanova, O. A. Kruglaya, and G. A. Razuvaev, *Zh. Obshch. Khim.*, **36**, 160 (1966).
185. Vyazankin, N. S., G. A. Razuvaev, and V. T. Bychkov, *Dokl. Akad. Nauk SSSR*, **158**, 382 (1964).
186. Vyazankin, N. S., G. A. Razuvaev, and V. T. Bychkov, *Izv. Akad. Nauk SSSR*, 1665 (1965).
187. Vyazankin, N. S., G. A. Razuvaev, and E. N. Gladyshev, *Dokl. Akad. Nauk SSSR*, **151**, 1326 (1963).
188. Vyazankin, N. S., G. A. Razuvaev, and E. N. Gladyshev, *Dokl. Akad. Nauk SSSR*, **155**, 830 (1964).
189. Vyazankin, N. S., G. A. Razuvaev, E. N. Gladyshev, and S. P. Korneva, *J. Organometal. Chem.*, **7**, 353 (1967).
190. Vyazankin, N. S., G. A. Razuvaev, S. P. Korneva, O. A. Kruglaya, and R. F. Galiulina, *Dokl. Akad. Nauk SSSR*, **158**, 884 (1964).
191. Vyazankin, N. S., G. A. Razuvaev, O. A. Kruglaya, and G. S. Semchikova, *J. Organometal. Chem.*, **6**, 474 (1966).
192. West, R., *J. Am. Chem. Soc.*, **74**, 4363 (1952).
193. West, R., *J. Am. Chem. Soc.*, **75**, 6080 (1953).
194. West, R., H. R. Hunt, and R. O. Whipple, *J. Am. Chem. Soc.*, **76**, 310 (1954).

5 Germanium–halide bond

5-1 ORGANOGERMANIUM HALIDES: PREPARATIONS AND PROPERTIES

Because of the many syntheses they make possible, the very reactive X_4Ge, X_3GeR, X_2GeR_2 and $XGeR_3$ halides take a most important part in organogermanium chemistry.

5-1-1 Preparations

5-1-1-1 By Formation of the Germanium–Halogen Bond

(a) Direct synthesis

Rochow (251, 253) noted that, in the presence of a copper catalyst, alkyl halides react with elemental germanium. This reaction produces a mixture of halogenated derivatives, their percentages within the mixture depending on the temperature and on the structure of the catalyst. In a study of direct methylchlorogermanes syntheses, Petrov (232), Ponomarenko (241) and later van de Vondel (317) reported that the formation of methyltrichlorogermane was favoured by higher temperature and an increased proportion of copper in the mixture.

Zueva and coworkers (348) carried out a series of experiments to define the optimum conditions for the direct synthesis of methylchlorogermanes and studied the effect of methyl chloride flow rate, reaction temperature and the influence of catalysts (Cu, Sb, As, $ZnCl_2$) in the contact mass. Their results showed that methyltrichlorogermane can be obtained in high yield (77%) under the following optimum conditions: $t = 480°C$; MeCl rate = 15 g/h; concentration of Cu, Sb, As and $ZnCl_2$ = 45, 0.004, 0.5 and 1%, respectively, for a reaction time of 5 h and a contact mass of 50 g (overall yield 90%). In this reaction, CH_3GeCl_3 and $(CH_3)_2GeCl_2$

are almost the sole reaction products: raising the temperature increases the overall yield of CH_3GeCl_3 and decreases the relative yield of $(CH_3)_2GeCl_2$; the activity of the contact mass decreases with time. The overall yield of methylchlorogermanes is strongly dependent on the reaction temperature (an increase of 9–10% in the yield was observed when raising the temperature by 20–30°C) but it is almost unaffected by $ZnCl_2$; but the presence of this last compound causes a 20–30% decrease of the yield of dimethyldichlorogermane and low ratios ($< 1\%$) of arsenic and antimony act in the same way. In the absence of these additional compounds $(CH_3)_2GeCl_2$ can be obtained in high yield (77%).

In further experiments (346) it was found that a 71–73% yield of dimethyldichlorogermane can be obtained (with 19–21% CH_3GeCl_3 and 6.2% $(CH_3)_3GeCl$) by action of methyl chloride at 400–450°C on a mixture composed of 50% Ge, 45% Cu and 0.5% Al. At 400°C propyl chloride gives 26% propyltrichlorogermane. The activity of this contact mass also changes with time: the trimethylchlorogermane ratio de-decreases while that of methyltrichlorogermane increases and the formation of dimethyldichlorogermane reaches a maximum ($\sim 80\%$) after 8 h.

Wieber and coworkers (329) studied the action of methyl chloride on germanium powder on glass wool at 450–550°C and obtained a greater amount of methyltrichlorogermane for a temperature of 505–510°C.

Zueva and coworkers (345) made numerous trials to find the best conditions for direct synthesis of diethyldichlorogermane. A 77% yield was reached by passing ethyl chloride (20 g/h) at 400°C over a mixture containing Ge (27.5 parts) fresh Cu (22.5 parts) and Nichrome column packing (48 parts) preheated for two hours at 380°C under nitrogen. The yield was improved ($\sim 10\%$) for a temperature of 400°C but it falls sharply under 350°C. When oxidized copper is used, ethyltrichlorogermane is formed preferentially.

In spite of the high reaction temperature (310–330°C), no isomerization seems to occur in the case of propyl chloride (256), but in the reaction of n-alkyl chlorides above n-C_3H_7Cl with Cu–Ge mixture, Vassiliou and Rochow (314) observed the formation of normal and isoalkylgermanium trichlorides. In similar experimental conditions, chlorocyclohexane gives only a poor yield of the corresponding derivative while phenyltri-chlorogermane can be prepared by the action of a $GeCl_4$–C_6H_5Cl mixture on Ge–Cu contact mass.

Comparative experiments with mixture of powders of copper and pure silicon and germanium in different ratios (7 to 45% Cu) (344) showed that germanium is more reactive towards methyl chloride than silicon: with germanium the reaction starts at 380–400°C while for silicon it begins only at 500–550°C. Moreover, a good rate and yield of methylchlorogermanes are obtained without any activator or promotor, whereas the synthesis of the methylchlorosilane analogues is satisfactory only in the presence of activating agents: $ZnCl_2$, As, Sb.

The reaction of methylene chloride with Ge/Cu at 350°C leads to a mixture of methyltrichlorogermane, bis(trichlorogermyl)methane and 1,1,3,3,5,5,-hexachloro-1,3,5-trigermacyclohexane (194):

$$CH_2Cl_2 \xrightarrow{Ge/Cu} CH_3GeCl_3 \, (27\%) + Cl_3GeCH_2GeCl_3 \, (23\%) +$$

$$\begin{array}{c} GeCl_2 \\ H_2C^{\diagup} \quad {}^{\diagdown}CH_2 \\ | \qquad | \\ Cl_2Ge_{\diagdown} \quad {}_{\diagup}GeCl_2 \\ CH_2 \end{array} \quad (19\%)$$

It should be noted that the last two derivatives are difficult to synthesize in other ways.

Direct synthesis of unsaturated germanium chlorides was first performed by Petrov and coworkers (232); good yields of allyltrichlorogermane and 2-methallyltrichlorogermane were obtained by passing allyl and 2-methallyl chlorides over a mixture of germanium powder and copper turnings at 340 ± 10°C.

Alkylbromogermanes also can be prepared by direct synthesis. In 1947 Rochow (251) obtained a mixture of methylbromogermanes by passing methyl bromide over germanium–copper mixture at 340–360°C. Ponomarenko (240), using the same process, observed that the relative yields of methyltribromogermane were greater the greater the amount of copper used:

$$Ge/Cu = 1 \qquad (CH_3)_2GeBr_2/CH_3GeBr_3 = 0.4$$
$$= 4 \qquad\qquad\qquad = 1.43$$

Allyl bromide gives in the same way allyltribromogermane at 300°C (188):

$$CH_2{=}CHCH_2Br + Ge/Cu \xrightarrow{300 \pm 10°C} CH_2{=}CHCH_2GeBr_3$$

With a germanium–copper–aluminium contact mass (346) methyl bromide gives at 450°C a mixture of $(CH_3)_2GeBr_2$ (73%) and CH_3GeBr_3 (27%). Higher alkyltribromogermanes ($C_3H_7GeBr_3$ and $C_4H_9GeBr_3$) were obtained by Zablotna (335) by the action of the corresponding alkyl bromides on germanium sponge at 300–340°C.

Zablotna, Akerman and Szuchnik (333, 334) investigated the action of alkyl iodides on the Ge–Cu mixture at different temperatures: their results prove that the catalytic action is not the same with alkyl iodides as with alkyl bromides and chlorides.

Under their experimental conditions (260°C < t < 350°C) methyl iodide leads mostly to the following derivatives: CH_3GeI_3, $(CH_3)_2GeI_2$

and $(CH_3)_3GeI$. Ethyl iodide leads especially to ethyltriiodogermane and to germanium tetraiodide:

$$Ge + 3\,C_2H_5I \rightarrow C_2H_5GeI_3 + 2\,C_2H_5^{\cdot}$$

$$Ge + 4\,C_2H_5I \rightarrow GeI_4 + 4\,C_2H_5^{\cdot}$$

Later (335), the same authors studied the action of propyl and butyl iodides on germanium sponge and noted an optimum temperature of 270–290°C in the preparation of alkyltriiodogermanes. $(CH_3)_3GeI$, $(CH_3)_2GeI_2$, CH_3GeI_3 and $C_2H_5GeI_3$ were also prepared by the same process (336).

(b) Condensation of a germanium (II) halide with halogenated derivatives

Germanium (II) iodide, which is the easiest of germanium (II) halides to produce and to isolate, will easily undergo oxidative addition with a great number of halogenated derivatives, thus leading to trihalogenated germanium (IV) derivatives. Flood (87, 88) carried out methyl and ethyl iodide additions to germanium (II) iodide at 120°C in sealed tubes:

$$GeI_2 + C_2H_5I \rightarrow I_3GeC_2H_5$$

This reaction is complete and applies to the other alkyl and aryl halides (57, 148). When a halide other than iodine is involved, a mixture of halogenated derivatives is produced:

$$GeI_2 + RX \rightarrow RGeI_3 + RGeI_2X + \cdots$$

The addition reactions make it possible to prepare organofunctional germanes: the reaction with $I(CH_2)_nI$ leads to a triiodinated compound with an iodomethyl substituent:

$$GeI_2 + ICH_2I \rightarrow I_3GeCH_2I$$

The same applies to dibrominated derivatives:

$$GeI_2 + Br(CH_2)_4Br \rightarrow X_3Ge(CH_2)_4Br \qquad (X = Br \text{ or } I)$$

For $n = 2$ the reaction is more complex. At 150°C in a sealed tube 1,2-diiodoethane produces the following reactions:

$$ICH_2CH_2I \rightarrow I_2 + CH_2{=}CH_2$$

$$GeI_2 + I_2 \rightarrow GeI_4$$

$$n\,GeI_2 + n\,CH_2{=}CH_2 \rightarrow (-CH_2-CH_2-\overset{\displaystyle I}{\underset{\displaystyle I}{Ge}}-)_n$$

Other organofunctional germanium triiodides have been obtained by the same process:

$$GeI_2 + ICH_2OCH_3 \rightarrow I_3GeCH_2OCH_3$$

$$GeI_2 + ICH_2COOC_2H_5 \rightarrow I_3GeCH_2COOC_2H_5$$

An interesting preparation of trichlorogermanium derivatives uses cesium trichlorogermanate as starting material (307). This compound, as with its tin (308) and lead (145) analogues, reacts with alkyl and aryl iodides in sealed tubes:

$$CsGeCl_3 + RI \rightarrow RGeCl_3 + CsI$$

This reaction has been reexamined and developed recently (243); these experiments, carried out in sealed tubes, in solid–liquid phase or in solution show that the yield of germanium trihalide is very dependent on the steric interference at the site of attachment of the iodine atom in the alkyl iodide:

$CH_2{=}CH{-}CH_2GeCl_3$, $n\text{-}C_4H_9GeCl_3$	60%
$sec\text{-}C_4H_9GeCl_3$	26%
$i\text{-}C_3H_7GeCl_3$	4%

(c) Cleavage reactions

Cleavage of Ge—C bonds

Halogens. Tetraalkylgermanes are chemically very stable and Ge—C bonds of these compounds are hardly sensitive to electrophilic reagents. Iodine does not seem to react directly with saturated tetraalkylgermanes. Tabern and coworkers (306) reported that at 50°C tetrapropylgermane dissolved iodine which was not observed to react; Anderson (7) reported the same observations after operating at reflux temperature; so did Dennis and Hance (66) with tetramethylgermane. However, Gilman and Leeper (102) have produced tributyliodogermane by direct action of iodine during eight hours on tetrabutylgermane at reflux temperature.

In the presence of a catalyst, iodine cleavage of Ge—C bonds is faster and also more regular. In the presence of aluminium iodide (5%) the reaction is nearly complete in just a few minutes (146). For the smallest R_4Ge compounds (tetramethyl- and tetraethylgermane) there is but one reaction:

(I) $R_4Ge + I_2 \rightarrow R_3GeI + RI$ $R = CH_3, C_2H_5$

In the case of heavier tetraalkylgermanes, the action of iodine is different and may involve reaction of two molecules per iodine molecule:

(II) $\qquad\qquad$ $2\,R_4Ge + I_2 \rightarrow 2\,R_3GeI + R{-}R$

Homologous compounds larger than tetra n-propylgermane are only involved in this second reaction. Tetrapropylgermane itself reacts both ways at the same time and distillation leads to a mixture of n-propyl iodide and alkanes. Different alkanes produced in reaction II are not pure, the catalyst then causing cracking and isomerization reactions. In the presence of a catalyst, and though it is harder, iodine may cleave two or even three Ge—C (alkyl) bonds. In this case also the halogen action varies as the number of substituted radicals increases. The reaction leading to the monoiodide:

$$(C_2H_5)_4Ge + I_2 \rightarrow (C_2H_5)_3GeI + C_2H_5I$$

gives ethyl iodide as the only organic product. Iodine, when made to react with the monoiodide, will do so in two different ways at the same time:

$$2\,(C_2H_5)_3GeI + I_2 \rightarrow 2\,(C_2H_5)_2GeI_2 + C_4H_{10}$$

$$(C_2H_5)_3GeI + I_2 \rightarrow (C_2H_5)_2GeI_2 + IC_2H_5$$

In the presence of greater quantities of aluminium iodide, iodine will react with the diiododiethylgermane but, in this case, ethyl iodide formation is not observed:

$$2\,(C_2H_5)_2GeI_2 \rightarrow 2\,C_2H_5GeI_3 + C_4H_{10}$$

Saturated mixed tetraalkylgermanes seem to be inert to iodine at 130°C. In the presence of aluminium iodide a reaction occurs and there is preferential cleavage of the less branched substituents (146, 163).

The weaker Ge—C (cyclobutane) and Ge—C (alkenyl) bonds are cleaved by iodine at 0°C in the absence of catalyst (169, 170):

$$R_2Ge\,\triangleleft\!\!\!\square\!\!\!\triangleright\!\!-R' + I_2 \rightarrow R_2GeCH_2CHCH_2I \qquad \begin{cases} R = Et,\ Bu \\ R' = H,\ CH_3 \end{cases}$$
$$\qquad\qquad\qquad\qquad\qquad \underset{I}{|} \quad \underset{R'}{|}$$

$$R_3GeCH_2CH{=}CH_2 + I_2 \rightarrow R_3GeI + CH_2{=}CH{-}CH_2I$$

Bromine, being more reactive than iodine, easily cleaves Ge—C (aryl) bonds, but its action on alkylgermanes is more difficult. This difference in reactivity is taken advantage of when preparing mixed derivatives:

$$Ar_4Ge \xrightarrow[\text{(2) RMgX}]{\text{(1) Br}_2} Ar_3GeR \xrightarrow[\text{(2) R'MgX}]{\text{(1) Br}_2} Ar_2GeRR' \cdots$$

Whereas the reaction of bromine with tetraphenylgermane (138) takes place rapidly within a few minutes, cleaving tetraethylgermane requires several days heating at 40°C (137):

$$(C_2H_5)_4Ge + Br_2 \rightarrow (C_2H_5)_3GeBr + C_2H_5Br$$

This reaction produces a mixture of bromides which cannot be separated by distillation. If the temperature is raised to increase the reaction rate, substitution reactions also appear:

$$(C_2H_5)_4Ge + Br_2 \rightarrow (C_2H_5)_3GeC_2H_4Br + HBr$$

It does not seem possible to continue direct bromination further than the monobromide step: C—H substitution reactions on the different alkyl substituents is about all that is achieved.

However, Flood and Horvitz (89) noted that bromine can cleave triethylchlorogermane (or triethylfluorogermane which is more susceptible) leading to alkyl bromide:

$$(C_2H_5)_3GeF + Br_2 \rightarrow (C_2H_5)_2GeBrF + C_2H_5Br$$

This process has been used in preparing dihalogenated derivatives: through hydrolysis, the bromofluoride is converted into the oxide, then an excess of the proper HX acid produces the halide. The presence of a catalyst may help considerably in bromine cleavage of tetraalkylgermanes: with steel powder Anderson (7) obtained a mixture of mono- and dibromide from the action of bromine on tetra-n-propylgermane at 45°C. Aluminium bromide seems more effective and permits the production of dibromides directly and in high yields (163, 204):

$$(C_2H_5)_4Ge \xrightarrow{Br_2, \ Al_2Br_6} (C_2H_5)_2GeBr_2 \quad (75\%)$$

Germacyclobutane and -pentene rings are cleaved by bromine even at $-80°C$. Allyl, methallyl- and 3-butenylgermanes are also cleaved without any catalyst at this temperature.

Hydrohalic acids and halogenated derivatives. Alkyl and cyclobutanic Ge—C bonds are easily cleaved by HX:

$$R_3GeCH_2-CH=CH_2 + HX \rightarrow R_3GeX + CH_3-CH=CH_2$$

$$R_2Ge\!\!\diamondsuit + HX \rightarrow R_2\underset{|}{\overset{}{Ge}}CH_2CH_2CH_3$$
$$X$$

Dialkylgermacyclopentanols give ω-olefinic-chain germanium halides (175):

$$R_2Ge \underset{\substack{| \\ OH}}{\overset{\substack{R' \\ |}}{\bigsqcup}}R'' + HX \rightarrow R_2Ge\underset{X}{\overset{|}{-}}CH_2\underset{R'}{\overset{|}{-}}CH\underset{R''}{\overset{|}{-}}C{=}CH_2 + H_2O$$

$$R', R'' = H, CH_3$$
$$X = F, Cl, Br, I$$

Mercuric chloride also cleaves these derivatives (170, 250).

$$R_3GeCH_2CH{=}CH_2 + HgCl_2 \rightarrow CH_2{=}CH{-}CH_2HgCl + R_3GeCl$$

Trimethylchlorogermane is formed together with methyl chloride and antimony trichloride by the action of antimony pentachloride on tetramethylgermane (259):

$$(CH_3)_4Ge + SbCl_5 \rightarrow CH_3Cl + SbCl_3 + (CH_3)_3GeCl$$

Tetramethylgermane is converted by gallium trichloride to trimethylchlorogermane (276):

$$(CH_3)_4Ge + GaCl_3 \rightarrow (CH_3)_3GeCl + CH_3GaCl_2$$

Phenyltrichlorogermane has been prepared in 75% yield from tetraphenylgermane and germanium tetrachloride redistribution at 350°C for 36 h (282):

$$(C_6H_5)_4Ge + 3 GeCl_4 \rightarrow 4 C_6H_5GeCl_3$$

In the presence of aluminium bromide, germanium tetrabromide takes part in redistribution reactions with tetraalkylgermanes (164):

$$(C_4H_9)_4Ge + GeBr_4 \xrightarrow{AlBr_3} 2 (C_4H_9)_2GeBr_2$$
$$3 (C_2H_5)_4Ge + GeBr_4 \xrightarrow{AlBr_3} 4 (C_2H_5)_3GeBr$$

Van der Kerk and coworkers (130, 248a) made a thorough study of the action of germanium tetrachloride on tetraalkyl- and tetraphenylgermanes in the presence of aluminium chloride. In particular, they observed that for temperatures higher than 120°C tetrabutylgermane is converted into products of intermediate degree of alkylation: approximately one butyl group is split off per mole of aluminium trichloride, apparently according to:

$$(C_4H_9)_4Ge + AlCl_3 \rightarrow (C_4H_9)_3GeCl + C_4H_9AlCl_2$$

But it is known that alkylaluminium halides are capable of alkylation reactions (181): consequently, it seems highly probable that the transfer of butyl groups proceeds via a butyl–aluminium chloride acting as an alkyl-transfer intermediate. Additional evidence is produced by the fact that similar reactions are reported when using trialkylaluminium or alkyl-aluminium chlorides instead of aluminium trichloride. Formation of butylaluminium derivatives during the reaction is balanced by a decrease in butyl radicals on the germanium centre: if the starting materials are used in exactly stoichiometric quantities, a mixture of butylgermanium halides with a lower average degree of alkylation is obtained. When this loss of butyl groups is compensated for by incorporating into the initial reaction mixture 10–20% of tetrabutylgermane in excess of the stoichiometric quantity, the formation of other butylgermanium chlorides than those sought can be prevented. Thus it is possible to prepare dibutyl-dichlorogermane and tributylchlorogermane in yields well above 80% by the reaction of appropriate quantities of tetrabutylgermane and germanium tetrachloride at 200°C in the presence of anhydrous aluminium trichloride. In these redistribution reactions the tendency to accept alkyl or aryl groups in the presence of aluminium trichloride decreases in the order:

$$R_2GeCl_2 \geqslant RGeCl_3 \gg GeCl_4$$

while the tendency to supply alkyl or aryl groups decreases in the same order:

$$R_4Ge > R_3GeCl > R_2GeCl_2$$

In both series the exchange velocity of alkyl or phenyl groups for chlorine atoms increases with degree of organic substitution at the germanium atom.

Rijkens explains the results that are reported by suggesting the presence of intermediates (complexes or transition states) of the following nature:

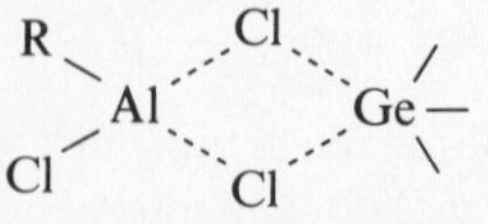

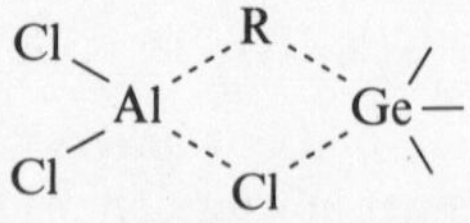

Intermediates of the type (II) are likely to occur in reactions in which an alkyl or an aryl group is transferred from an aluminium to a germanium atom and conversely. Replacement of electron-attracting chlorine atoms at the germanium atom by electron-donating alkyl groups makes the germanium atom less positive. The changes in the polarity of the germanium, having a much stronger influence on the more polar germanium–chlorine bond than on the germanium–carbon bond, weaken the bonds between germanium and the bridging chlorine atoms more than the germanium–carbon bonds in the intermediates (II). Thus these intermediates will become favoured relative to the intermediates I, when the number of carbon atoms increases.

It is interesting to note that in organosilicon chemistry a quite analogous order of reactivity has been observed in redistribution reactions catalyzed by aluminium trichloride and in alkylation by methylaluminium sesquichloride. On the other hand, alkylation of tin derivatives proceeds differently: $SnCl_4$ is alkylated faster than $RSnCl_3$ or R_2SnCl_2 whereas organolead compounds undergo redistribution reactions in the absence of a catalyst.

In further experiments (46, 156, 247), the Dutch group showed that the dealkylation of tetraalkylgermanes and hexaalkyldigermanes by germanium or tin tetrachloride is a convenient method for the preparation of germanium halides of the type R_3GeCl and R_3GeGeR_2Cl respectively:

$$R_3Ge(GeR_2)_nR + MCl_4 \xrightarrow{200°C} R_3Ge(GeR_2)_nCl + RMCl_3$$

$$(R = alkyl; n = 0,1; M = Ge, Sn)$$

Bulten and Noltes (47), when carrying out these reactions in polar solvents (CH_3COCl, CH_3NO_2), observed a considerable increase of the rate. This is in accordance with the polar mechanism, involving electrophilic attack by the metal atom of the tetrahalide on carbon, which had been previously proposed (46, 247).

Though nitromethane is sparingly soluble in organogermanes, the reactions are faster in this solvent than in acetyl chloride in which organogermanium compounds are soluble. The ease of dealkylation decreases in the series $C_2H_5 > C_3H_7 \approx C_4H_9 \approx C_5H_{11}$, but tin tetrachloride converts tetraamylgermane into triamylchlorogermane in less than two hours at 100°C with an excellent yield:

$$(C_5H_{11})_4Ge + SnCl_4 \xrightarrow{CH_3NO_2} C_5H_{11}SnCl_3 + (C_5H_{11})_3GeCl \quad (88\%)$$

Hexaethyldigermanes react faster than tetraalkylgermanes and, in contrast to the action of bromine (137), metal–metal bond cleavage does not occur in this case:

$$(C_2H_5)_3GeGe(C_2H_5)_3 + Br_2 \rightarrow 2(C_2H_5)_3GeBr$$

$$(C_2H_5)_3GeGe(C_2H_5)_3 + SnCl_4 \rightarrow$$

$$C_2H_5SnCl_3 + (C_2H_5)_3GeGe(C_2H_5)_2$$
$$|$$
$$Cl$$

The action of tin tetrachloride on mixed tetraalkylgermanes is very selective and allows the facile preparation of mixed substituted derivatives (47):

$$(CH_3)_3Ge(C_4H_9) \xrightarrow{SnCl_4} (CH_3)_2(C_4H_9)GeCl \xrightarrow{C_3H_7MgCl}$$

$$(CH_3)_2(C_3H_7)(C_4H_9)Ge \xrightarrow{SnCl_4} (CH_3)(C_3H_7)(C_4H_9)GeCl \xrightarrow{C_2H_5MgBr}$$

$$(CH_3)(C_2H_5)(C_3H_7)(C_4H_9)Ge \xrightarrow{SnCl_4} (C_2H_5)(C_3H_7)(C_4H_9)GeCl$$

In the above reactions methyltin trichloride is the only organotin compound formed. Identical cleavages are observed with mixed digermanes:

$$(CH_3)_3GeGe(C_2H_5)_3 \xrightarrow{SnCl_4} (CH_3)_2(Cl)GeGe(C_2H_5)_3 + CH_3SnCl_3$$

while with a silagermane an ethyl group linked to germanium is cleaved preferentially:

$$(CH_3)_3SiGe(C_2H_5)_3 \xrightarrow{SnCl_4} (CH_3)_3SiGe(Cl)(C_2H_5)_2 + C_2H_5SnCl_3$$

With a digermane having identical light groups on the two germanium atoms, a mixture of isomers is obtained:

$$(C_2H_5)_3GeGe(C_2H_5)_2(C_4H_9) \xrightarrow{SnCl_4} \begin{cases} (C_2H_5)_2\underset{\underset{Cl}{|}}{Ge}-Ge(C_2H_5)_2(C_4H_9) \\ \\ (C_2H_5)_3Ge-\underset{\underset{Cl}{|}}{Ge}(C_2H_5)(C_4H_9) \end{cases}$$

Mironov (204) has shown that alkyl halides cleave tetramethylgermane in the presence of catalytic amounts of aluminium halide:

$$(CH_3)_4Ge + iC_4H_7X \xrightarrow{AlX_3} (CH_3)_3GeX \quad (92\text{–}95\%) \quad X = Cl, Br$$

With trimethylphenylgermane, the phenyl group is removed:

$$(CH_3)_3Ge(C_6H_5) + C_4H_9X \xrightarrow{AlCl_3} (CH_3)_3GeX \quad X = Cl, I$$

An interesting preparation of mono- and dialkylgermanium chlorides involves the action of acetyl chloride with tetraalkylgermanes in the presence of aluminium chloride (260):

$$(CH_3)_4Ge + CH_3COCl \xrightarrow{AlCl_3} (CH_3)_3GeCl \quad (74\%)$$

$$(CH_3)_4Ge + 2 CH_3COCl \xrightarrow{AlCl_3} (CH_3)_2GeCl_2 \quad (70\%)$$

Cleavage of the Ge—O bond. $(R_3Ge)_2O$, $(R_2GeO)_n$ and $(RGeO)_2O$ germanium oxides which have basic character quickly and quantitatively react with hydrohalic acids. These reactions allow the exchange of one halide for another (5, 7, 137, 138):

$$R_3GeX \xrightarrow{\text{NaOH}} (R_3Ge)_2O \xrightarrow{\text{HX}'} R_3GeX'$$

$$R_2GeX_2 \xrightarrow{\text{NaOH}} (R_2GeO)_n \xrightarrow{\text{HX}'} R_2GeX'_2$$

$$RGeX_3 \xrightarrow{\text{NaOH}} (RGeO)_2O \xrightarrow{\text{HX}'} RGeX'_3$$

Cleavage of the Ge—H bond (see Chapter 4). The germanium–hydrogen bond is easily cleaved by halogens and a great number of inorganic and organic halides and these reactions allow the preparation of pure organogermanium halides.

With germanium polyhydrides it is possible to replace one or several hydrogen atoms by halogen atoms (15, 16, 161, 263):

$$R_nGeH_{4-n} + x\,R'X \rightarrow R_nGeH_{4-(n+x)}X_x + x\,R'H$$

Other cleavages (see the corresponding chapters). Ge—S, Ge—N and Ge—P linkages are easily cleaved by hydrogen halides, leading to the corresponding Ge—X derivatives. Ge–metal bonds are very sensitive to the action of halogens and the cleavage usually occurs under mild conditions without catalyst.

(d) Halogen–halogen exchange

Anderson (5, 13) classified germanium halides according to their reactivity towards silver halides:

$$\overset{\diagdown}{\underset{\diagup}{-}}Ge-I > \overset{\diagdown}{\underset{\diagup}{-}}Ge-Br > \overset{\diagdown}{\underset{\diagup}{-}}Ge-Cl > \overset{\diagdown}{\underset{\diagup}{-}}Ge-F$$

In this series any germanium halide, through a reaction with the proper silver halide, may be transformed into the halide following it, but cannot be transformed into a halide in front of it:

$$(C_2H_5)_3GeI + AgCl \rightarrow (C_2H_5)_3GeCl + AgI$$

This reaction has been extended to pseudo-halides (13). Gilman (91) observed a halogen–halogen exchange in the action of hexylmagnesium bromide on germanium tetrachloride:

$$3\,n\text{-}C_6H_{13}MgBr + GeCl_4 \rightarrow (n\text{-}C_6H_{13})_3GeBr + MgBr_2 + 2\,MgCl_2$$

5-1-1-2 By Transformation of an Organogermanium Halide

(a) Partial alkylation or arylation of germanium polyhalides

Grignard reagents are not suitable for preparing R_3GeX or R_2GeX_2 derivatives by partial alkylation of a germanium tetrahalide, for in this reaction there generally is formed a mixture of products. However, the action of a Grignard reagent in an ether solution on a large excess of germanium tetrachloride permits one to produce alkyltrichlorogermanes with a minimum of other products:

$$RMgX + GeCl_4 \rightarrow RGeCl_3 + MgClX$$

Organolithium compounds seem to react more progressively and enable one to produce partly substituted derivatives: thus trimethylchlorogermane has been prepared by the methylation of dimethyldichlorogermane with methyllithium (316):

$$(CH_3)_2GeCl_2 + CH_3Li \rightarrow (CH_3)_3GeCl + LiCl$$

Likewise the reaction of chloromethylmethyldichlorogermane with methyllithium takes place at $-30°C$ (327):

$$(CH_3)(CH_2Cl)GeCl_2 + LiCH_3 \rightarrow LiCl + (CH_3)_2(CH_2Cl)GeCl \quad (50\%)$$

The interaction between germanium tetrachloride and phenylbarenyllithium (2, 3 or 4 moles per mole of halide) resulted in the formation of only bis(phenylbarenyl)dichlorogermane (337):

$$2 C_6H_5-C \underset{B_{10}H_{10}}{\overline{\diagdown \diagup}} CLi + GeCl_4 \rightarrow \left[C_6H_5-C \underset{B_{10}H_{10}}{\overline{\diagdown \diagup}} C \right] GeCl_2 + 2 LiCl$$

Organomercuric compounds react with germanium tetrachloride leading to trichlorinated derivatives in very high yields (287):

$$(p\text{-}Z-C_6H_4)_2Hg + GeCl_4 \rightarrow p\text{-}Z-C_6H_4GeCl_3 + p\text{-}Z-C_6H_4HgCl$$

$$Z = H, Cl, F, CH_3, OCH_3$$

$$(CH_2{=}CH)_2Hg + GeCl_4 \rightarrow CH_2{=}CHHgCl + CH_2{=}CHGeCl_3$$

Alkylation of trichlorogermane with tetraorganotin compounds depends on the tin compound used. With tetramethyltin the reaction occurs very well in ether:

$$(CH_3)_4Sn + HGeCl_3 \rightarrow CH_3GeHCl_2 + CH_3SnCl$$
$$(80\%) \qquad (70\%)$$

but with tetraethyltin a 31 % yield of $C_2H_5GeCl_3$ is obtained in addition to $C_2H_5GeHCl_2$ (45%). Tetraethyllead gives somewhat better results

($C_2H_5GeHCl_2$ 60%; $C_2H_5GeCl_3$ 20%) and can be used to transform $GeCl_4$ into $C_2H_5GeCl_3$ in high yields. The reaction of tetraphenyltin with trichlorogermane gives phenyltrichlorogermane as the major product and only 5% yield of phenyldichlorogermane.

Mironov (190, 191) recently described a new process for synthesizing arylgermanium trihalides. It may be compared to a Würtz reaction: when heating a mixture of an aryl iodide and germanium tetrahalide with copper powder, arylgermanium trihalides are formed:

$$1\text{-}C_{10}H_7I + GeBr_4 \xrightarrow{Cu} 1\text{-}C_{10}H_7GeBr_3 \quad (64.5\%)$$

$$C_6H_5I + GeCl_4 \xrightarrow{Cu} C_6H_5GeCl_3 \quad (62.7\%)$$

$$C_6H_5I + GeBr_4 \xrightarrow{Cu} \begin{cases} C_6H_5GeBr_3 & (51\%) \\ (C_6H_5)_2GeBr_2 & (28\%) \end{cases}$$

(b) Addition reactions of hydrogenohalogermanes

Germanium halohydrides, and trichlorogermane in particular, easily add to a great number of unsaturated derivatives. These reactions are reported in detail in Chapter 4.

The $\overbrace{GeI_2 + IH}$ mixture reacts with $\diagup\hspace{-0.3em}\diagdown C{=}C{-}\underset{\underset{O}{\|}}{\overset{|}{C}}{-}$ conjugated struc-

tures leading to germanium triiodides with substituted functions within the chains in the β position:

$$\overbrace{GeI_2 + IH} + CH_2{=}CH{-}COCH_3 \rightarrow I_3GeCH_2CH_2COCH_3$$

$$\overbrace{GeI_2 + IH} + CH_2{=}CH{-}COOH \rightarrow I_3GeCH_2CH_2COOH$$

$$\overbrace{GeI_2 + IH} + CH_3CH{=}CH{-}CHO \rightarrow I_3Ge{-}\underset{\underset{CH_3}{|}}{CH}{-}CH_2CHO$$

$$\overbrace{GeI_2 + IH} + CH_2{=}CH{-}CH{=}CH{-}COOH \rightarrow$$

$$I_3GeCH \diagup[\overset{CH=CH_2}{}] \diagdown[\underset{CH_2-COOH}{}]$$

(c) Insertion in Ge—Cl bonds

Seyferth and Rochow (290) observed that diazomethane reacts with germanium and silicon halides: the reaction is carried out at low temperature ($-70°C$) in ether solution and in the presence of copper powder as catalyst.

With germanium tetrachloride the reaction proceeds readily and an excellent yield (94%) of methylenation product is obtained:

$$GeCl_4 + CH_2N_2 \xrightarrow{Cu} N_2 + Cl_3GeCH_2Cl$$

A small quantity of bis(chloromethyl)germanium dichloride is also produced as a by-product.

Similarly the methylenation of methylgermanium trichloride gives chloromethylmethylgermanium dichloride in good yields, while if two moles of CH_2N_2 per mole CH_3GeCl_3 is used, a mixture of mono- and dimethylenated products is obtained:

$$CH_3GeCl_3 + CH_2N_2 \xrightarrow{Cu} CH_3Ge(Cl_2)CH_2Cl \qquad (78\%)$$

$$CH_3GeCl_3 + 2\,CH_2N_2 \xrightarrow{Cu} \begin{cases} (CH_3)ClGe(CH_2Cl)_2 & (20.5\%) \\ (CH_3)Cl_2GeCH_2Cl & (33.5\%) \end{cases}$$

For further examples see reference 287.

5-1-2 Physical properties and characteristics

5-1-2-2 IR and Raman Spectra. In the Raman spectra of ethylgermanium chlorides the bands at 400 and 376 cm^{-1} were attributed to $\nu_{asym.}$ (Ge—Cl) and $\nu_{sym.}$ (Ge—Cl) respectively (53, 155). Similarly in the infrared spectrum of methyltrichlorogermane these bands appear at 430 and 403 cm^{-1} (18).

In their important study of infrared spectra of organogermanes, Cross and Glockling (61) give some germanium–halogen stretching frequencies:

	ν_{Ge-Cl}	ν_{Ge-Br}	ν_{Ge-I}
(i-C$_3$H$_7$)$_3$GeCl	369 s		
(C$_6$H$_5$)$_3$GeCl	379 s		
(C$_6$H$_5$CH$_2$)$_3$GeCl	362 m		
(C$_6$H$_5$)$_3$GeBr		313 s	
(C$_6$H$_5$CH$_2$)$_3$GeBr		251 s	
(C$_2$H$_5$)$_3$GeBr		269 m or 245 m	
(C$_6$H$_5$)$_2$GeBr$_2$		315 sh, m	
C$_6$H$_5$GeBr$_3$		327 s 228 w	
GeBr$_4$		330 s	
(C$_6$H$_5$)$_3$GeI			283 s
(CH$_3$GeI)$_3$			256 m
GeI$_4$			263 s

IR spectra of $(n\text{-}C_3H_7)_4Ge$, $(n\text{-}C_3H_7)_3GeCl$, $(n\text{-}C_3H_7)_2GeCl_2$ and $n\text{-}C_3H_7GeCl_3$ in the range 400–4000 cm^{-1} have been reported and the frequencies assigned (312); for $(n\text{-}C_3H_7)_4Ge$ and $(n\text{-}C_3H_7)_3GeCl$ the spectra at $-110°C$ are also given. The spectra show that several rotational isomers exist in the compounds: the Ge—C stretching vibration frequencies of the *l*-form appear between 550 and 600 cm^{-1} and those of the *trans* form between 640 and 670 cm^{-1}.

IR spectra of methyl-, isopropyl-, *n*-butyl, *sec*-butyl-, *tert*-butyl-, allyl-, phenyl- and benzyltrichlorogermanes have been recorded by Poskozim (243); the position of the asymmetric Ge—Cl stretch is invariant in these compounds and lies in the range 426–430 cm^{-1}.

5-1-2-2 NMR Spectra. NMR spectra of methylchlorogermanes (317) show an increasing chemical shift for methyl protons when the number of chlorine atoms linked to germanium increases:

	$\tau(C—H)$
CH_3GeCl_3	8.17
$(CH_3)_2GeCl_2$	8.70
$(CH_3)_3GeCl$	9.36

A similar variation is observed for the chemical shift of Ge—H in alkyl-halogermanes $R_nX_{3-n}GeH$ (X = Cl, Br, I) (161).

5-1-2-3 NQR Spectra. Measurements of nuclear quadrupole resonance on various organogermanium chlorides were made by Russian authors (20, 33, 34, 283, 318) and some results compared with those of carbon, silicon, tin and lead analogues (33). Voronkov (318) calculated for $Ge^{35}Cl_4$, $C_2H_5Ge^{35}Cl_3$, $(CH_3)_2Ge^{35}Cl_2$ and $F_3C(CH_2)_2Ge^{35}Cl_3$ the values of the mean frequencies by the equation: $v_m^{77} = 17.420 + 0.937\Sigma\sigma^*$ and observed good agreement with experimental data. In their recent tables, Biryukov and coworkers (35) give the frequencies relative to nineteen chlorogermanes.

5-1-2-4 UV Spectra. UV spectra of vinyl and allylgermanium trichlorides between 1700 and 2400 Å were recorded by Petukhov and coworkers (235) and compared with those of the corresponding silicon isologues. In order to examine the variation of the inductive effect of germanium atom on phenyl group, Marrot and coworkers (160) studied the UV spectra of phenylgermanium chlorides $(C_6H_5)_nGeCl_{4-n}$ in heptane between 2400 and 2800 Å. The spectra show an increasing electronegativity of the germanium with the number of chlorine atoms linked to it. The observed bathochromic shifts are attributed to mesomeric and inductive effects. A similar result was found for the isologous stannanes.

5-1-2-5   *Mass Spectra.* In their study of the mass spectra of organo-germanes (109), Glockling and Light observed that germanium tetra-chloride gave an exceptionally abundant molecular ion (16.9 %); for triphenyl germanium chloride the molecular ion was about five times as abundant as the $(C_6H_5)_3Ge^+$ ion whereas in the triphenylgermanium bromide, the molecular ion was relatively much weaker and was not detected in the iodide $(C_6H_5)_3GeI$. The molecular ion abundances are therefore in the same order as the Ge—X bond strengths.

The dominant process for all odd-electron ions is radical elimination by Ge—C cleavage and is frequently metastable-supported:

$$(CH_3)_3GeCl^{+\cdot} \rightarrow (CH_3)_2GeCl^+ + CH_3^{\cdot}$$

$$(C_2H_5)_3Ge^{35}Cl^{+\cdot} \rightarrow (C_2H_5)_2Ge^{35}Cl^+ + C_2H_5^{\cdot}$$

$$(C_2H_5)_3Ge^{79}Br^{+\cdot} \rightarrow (C_2H_5)_2Ge^{79}Br^+ + C_2H_5^{\cdot}$$

In contrast to tin compounds, the only triphenylgermanium halide showing a metastable peak for elimination of $X^{\cdot}$ from the molecule ion was the bromide:

$$(C_6H_5)_3Ge^{81}Br^{+\cdot} \rightarrow (C_6H_5)_3Ge^+ + {}^{81}Br^{\cdot}$$

Alkene elimination by Ge—C bond cleavage is also observed: with ethyl-germanium compounds for each ethyl group cleaved in this way a germanium–hydrogen bond is formed:

$$(C_2H_5)_2GeCl^+ \rightarrow C_2H_5Ge(H)Cl^+ + C_2H_4$$

The complete fragmentation schemes for $(CH_3)_3GeCl$, $(C_6H_5)_3GeCl$, $(C_6H_5)_3GeBr$ and $(C_6H_5)_3GeI$ are also given in this study.

5-1-2-6   *Dipole Moments.* The electrical dipole moment measurements for methylchlorogermanes in benzene solution were made by van de Vondel (317) and Kartsev (127, 129):

$$CH_3GeCl_3 \qquad \mu(D) = 2.70 \,(317) \quad 2.63 \,(129)$$
$$(CH_3)_2GeCl_2 \qquad\qquad\quad 3.14 \,(317) \quad 3.11 \,(129)$$
$$(CH_3)_3GeCl \qquad\qquad\quad 2.89 \,(317) \quad 2.78 \,(127)$$

Irisova (114) found 3.8 D for methylgermanium trifluoride. For ethyl-germanium chlorides, the following values are reported: 2.28 ± 0.09 D (226) and 2.87 D (129) for $C_2H_5GeCl_3$ and 3.19 D for $(C_2H_5)_2GeCl_2$ (129).

To examine the possible role of the $p_\pi - d_\pi$ dative character of Ge—Cl bond, Ulbricht and coworkers (313) have determined the dipole moments of propylchlorogermanes:

$$n\text{-}C_3H_7GeCl_3 \qquad \mu_{298} = 3.01$$

$$(n\text{-}C_3H_7)_2GeCl_2 \qquad\quad = 3.42$$

$$(n\text{-}C_3H_7)_3GeCl \qquad\quad = 2.82$$

With the aid of the moments of the Ge—CH_3 groups, the bond moments of the Ge—Cl bond were calculated for methylchlorogermanes:

$$CH_3GeCl_3: \qquad m(\overrightarrow{Ge\text{—}Cl}) = 3.03 \text{ D}$$

$$(CH_3)_2GeCl_2: \qquad\qquad = 3.07 \text{ D}$$

$$(CH_3)_3GeCl: \qquad\qquad = 3.17 \text{ D}$$

The comparison with the methylchlorosilane analogues:

$$CH_3SiCl_3: \qquad m(\overrightarrow{Si\text{—}Cl}) = 2.13 \text{ D}$$

$$(CH_3)_2SiCl_2: \qquad\qquad = 2.21 \text{ D}$$

$$(CH_3)_3SiCl: \qquad\qquad = 2.29 \text{ D}$$

shows the greater polarity of germanium bonds.

The dipole moments of $(CH_3)_3GeX$ halides (X = F, Cl, Br, I) were determined by Kartsev and coworkers (127) and the values of the moments compared with those for the corresponding $tert\text{-}C_5H_{11}$ halides and tri-methylsilicon analogues: from fluorine to iodine, μ changes in the same manner for R_3GeX and R_3CX but differently for R_3SiX. The values reported for the moments of Ge—F, Ge—Cl, Ge—Br and Ge—I are 2.80, 3.00, 3.15 and 3.10, respectively. The dipole moments of vinyl-tribromogermane (2.47 D) (128) and some chlorosubstituted germanium trichlorides were also measured: $ClCH_2GeCl_3$ (2.10 D) (128), $ClCH_2CH_2GeCl_3$ (2.41 D) (129), and $ClCH=CHGeCl_3$ (1.86 D) (128).

5-1-2-7 Racemization Mechanism of Asymmetric Chlorogermanes. Corriu and coworkers (50) extended their racemization mechanism of methyl-phenyl-α-naphthylchlorosilane to the germanium isologue. Methyl-phenyl-α-naphthylchlorogermane is optically stable in hydrocarbons as well as in chloroform and carbon tetrachloride. Racemization is not observed in ether, dioxane or anisole, but it is fast in tetrahydrofuran.

Racemization occurs, as with the corresponding chlorosilane, but more easily, with solvents with a nucleophilic atom and the order of racemization is the same:

$$MeOCH_2CH_2OMe > THF \gg Et_2O, \text{dioxan; also } C_6H_5COOC_2H_5$$

$$> CH_3COOC_2H_5$$

When lithium perchlorate is added to the solvent (ether or THF), the racemization rate increases owing to electrophilic catalysis by Li^+, allowing the Ge—Cl bond to be stretched. Alcohols and water in solution in benzene, dioxane and ether racemize chlorogermane without substitution. The mechanism of racemization probably involves the coordination of two molecules of solvent around the central atom and when it is carried out with alcohols it depends upon two factors: the nucleophilic power of the oxygen atom and the possible Ge—Cl bond stretching by hydrogen bonding.

5-1-2-8   Thermal Stability. Alkyl- and arylgermanium halides are usually very stable compounds. However, with substituted derivatives decomposition can occur in some cases: trifluoromethyltriiodogermane slowly decomposes at 180°C giving germanium tetraiodide, germanium tetrafluoride, fluoroolefins and cyclic fluorocarbons (57):

$$CF_3GeI_3 \rightarrow FGeI_3 + CF_2$$

$$4\,FGeI_3 \rightarrow 3\,GeI_4 + GeF_4$$

$$2\,CF_2 \rightarrow C_2F_4 \rightarrow (CF_2)_3 + (CF_2)_4$$

5-1-3   Chemical Properties

5-1-3-1   Hydrolysis of Germanium Halides. The complete hydrolysis of dimethylgermanium dichloride, investigated by Rochow and Allred (257), occurs according to the equations:

$$(CH_3)_2GeCl_2 \rightleftarrows (CH_3)_2Ge^{++} + 2\,Cl^-$$

$$(CH_3)_2Ge^{++} + H_2O \rightleftarrows (CH_3)_2Ge(OH)^+ + H^+$$

$$(CH_3)_2Ge(OH)^+ + H_2O \rightleftarrows (CH_3)_2Ge(OH)_2 + H^+$$

The comparison of these results with those obtained with dimethyltin dichloride shows that germanium has a surprisingly high electronegativity, dimethylgermanium dichloride behaving in solution like a strong acid while dimethyltin dichloride appears to be amphoteric.

The hydrolysis of trifluoromethylgermanium trihalides CF_3GeX_3 (X = I, Cl, F), was studied by Clark (57). At room temperature, aqueous sodium hydroxide gives rapid evolution of fluoroform (as with most perfluoroalkyl derivatives of Group IV and V elements) and silver oxide gives a similar reaction with the triiodo compound. With cold water they give stable clear aqueous solutions which possibly contain ionic species:

$$CF_3GeF_3 + 2\,H_2O \rightarrow CF_3GeF_3(OH)_2^{-2} + 2\,H^+$$

These ions probably undergo further hydrolysis producing fluoride ions:

$$CF_3GeF_3(OH)_2{}^{-2} + H_2O \rightarrow CF_3GeF_2(OH)_3{}^{-2} + H^+ + F^-$$

but this hydrolysis is reversible, addition of potassium fluoride solution to an aqueous solution of either trifluoromethyltrifluoro- or triiodo-germane leading in the two cases to the formation of potassium tri-fluoromethylpentafluorogermanate:

$$CF_3GeI_3 + 5\,F^- \rightarrow CF_3GeF_5{}^{-2} + 3\,I^-$$

Gingold and coworkers (103) reported that organogermanium halides form conducting solutions in dimethylformamide (as organosilicon halides) and undergo ionic metatheses. However, in further investigations with solutions of organometallic chlorides of Group IV B in the same solvent, Thomas and Rochow (311) showed that in rigorously purified DMF these chlorides did not dissociate into organogermanium cations and chloride ions and that conductances previously observed could be attributed to solvolysis of the solute by water present as impurity.

In contrast with triphenylmethyl chloride which dissolves in liquid hydrogen chloride forming a yellow, highly conducting solution (Ph_3C^+ ions), triphenylchlorogermane is sparingly soluble in the same solvent and no adduct could be obtained with boron trichloride (230).

Chipperfield and Prince (52) used fast-reaction techniques to follow the hydrolysis of triphenylchlorogermane and triphenylchlorosilane (and some of their derivatives) in acetone and in acetone–ether. These reactions have half-lives from 0.1 to 10 sec and at 25°C. In comparative experiments (acetone–ether 1/1) they observed that hydrolysis of the silyl derivative was complete whereas the germanium halide was only about 10% transformed, and that the germane was hydrolysed more slowly than the silane.

On the other hand, the electron-release to germanium or silicon by a p-methyl group decreases the hydrolysis rate (for the Ge compound $K_{PhH}/K_{PhMe} = 1.45$). They also found that the rate of hydrolysis of the germane in acetone was slower than in the less polar acetone–ether medium, and was greater by a factor of about 10^3 than the values previously reported for the same medium (123). This great difference can be explained by the different experimental conditions, the first study having been made near equilibrium.

5-1-3-2 Hydrolysis of ω-Chloroalkyltrichlorogermanes. Mironov (183) noted that in compounds of the type $Cl_3Ge(CH_2)_nCl$ only the three chlorine atoms linked to germanium are titrated by a 0.1N NaOH solution:

$$2\,Cl_3Ge(CH_2)_nCl + 6\,NaOH \rightarrow$$

$$6\,NaCl + [Cl(CH_2)_nGeO]_2O + 3\,H_2O \qquad (n \neq 2)$$

except in the case of 2-chloroethyltrichlorogermane where all four halogen atoms react; ethylene is evolved via β-elimination:

$$Cl_3GeCH_2CH_2Cl + 4\,NaOH \rightarrow CH_2{=}CH_2 + Ge(OH)_4 + 4\,NaCl$$

A similar difference is observed on treatment of these chlorides with Grignard reagents:

$$Cl_3Ge(CH_2)_nCl + 3\,RMgCl \rightarrow 3\,MgCl_2 + R_3Ge(CH_2)_nCl \qquad (n \neq 2)$$

$$Cl_3Ge(CH_2)_2Cl + 4\,RMgCl \rightarrow 4\,MgCl_2 + R_4Ge + CH_2{=}CH_2$$

Organogermanium halides react with hydrogen peroxide in the presence of tertiary amine or ammonia to form organogermanium peroxides (64b):

$$2\,R_3GeX + H_2O_2 + 2\,NR'_3 \rightarrow R_3GeOOGeR_3 + 2\,R'_3N, HX$$

5-1-3-3 Action of Alcohols, Thiols, Amines and Phosphines. A great number of germanium alkoxides were prepared with high yields by action of organogermanium halides with the corresponding alcohols:

$$R_nGeX_{4-n} + (4-n)R'OH \rightarrow R_nGe(OR')_{4-n} + (4-n)HX$$

Usually this reaction is carried out in the presence of pyridine or an aliphatic tertiary amine to remove the hydrogen halide formed. In the same way, Ge—S, Ge—N and Ge—P derivatives are obtained by the reactions of germanium halides with thiols, amines and phosphines. With appropriate bifunctional compounds, germanium heterocycles of various sizes are easily formed (see the corresponding Chapters).

5-1-3-4 Ge—X/Si—O Exchanges. Van Wazer and Moedritzer (212) studied the equilibrated system:

$$(CH_3)_2SiCl_2—(CH_3)_2GeCl_2—[(CH_3)_2SiO]—[(CH_3)_2GeO]—$$

which consists of a range of various chain, and some ring molecules resulting from scrambling of the bridging oxygen with the monofunctional chlorine atoms between the dimethylgermanium and dimethylsilicon moieties. They showed by analysis of PMR spectra of the methyl groups that there is a strong preference of the chlorine atoms for the dimethylgermanium and of the bridging group for the dimethylsilicon at equilibrium.

An identical result was observed for the equilibrated system $(CH_3)_2SiBr_2$—$(CH_3)_2GeBr_2$—$[(CH_3)_2SiO]$—$[(CH_3)_2GeO]$ (323): there is a very strong preference for the bromine atoms to link to the dimethyl germanium and for the bridging oxygens to link to the dimethylsilicon moiety.

5-1-3-5 Halogen–Halogen Exchange. Anderson (5, 13) observed that, when heating organogermanium halides with appropriate silver salts, "conversion" reactions could occur in the order:

$$I \rightarrow Br \rightarrow Cl \rightarrow F$$

Thus a silver halide should convert any compound to one on its right but not to one on its left:

$$R_3GeI + AgCl \rightarrow AgI + R_3GeCl$$

$$R_3GeCl + AgI \rightarrow no\ reaction$$

This reaction was applied to trifluoromethylgermanium derivatives by Clark and Willis (57):

$$CF_3GeI_3 + 3\,AgX \rightarrow CF_3GeX_3 + 3\,AgI \qquad X = F, Cl$$

and, in this case, the reaction is exothermic.

Trialkyliodogermane can be obtained by action of sodium iodide on trialkylchloro- or trialkylbromogermanes in acetone solution (91):

$$R_3GeX \xrightarrow{\text{acetone}} R_3GeI + NaX \qquad X = Cl, Br$$

The high rate of this reaction compared with that of *tert*-butyl halides in bimolecular nucleophilic displacement reaction is attributed to the lower steric hindrance at the large germanium atom, and to the polarization of the alkyl–germanium bonds, which tends to lower the electron density at the central atom, speeding SN^2 reactions (91). Antimony trifluoride transforms alkylgermanium chlorides, bromides or iodides into alkyl-germanium fluorides in high yields (10, 16, 240):

$$(n\text{-}C_4H_9)_2GeI_2 \xrightarrow{\text{SbF}_3} (n\text{-}C_4H_9)_2GeF_2 \qquad (99\%\ \text{yield})$$

and Mironov (204) prepared trimethylfluorogermane by reaction of trimethylbromogermane with aqueous hydrofluoric acid:

$$(CH_3)_3GeBr + HF \rightarrow (CH_3)_3GeF \qquad (40\%)$$

Germanium chlorides are converted to germanium bromides by the action of hydrobromic acid (188):

$$R_{4-n}GeCl_n + n\,HBr \rightarrow R_{4-n}GeBr + n\,HCl \qquad n = 1, 2, 3, 4$$

With vinyltrichloro- and allyltrichlorogermanes, even in the presence of benzoyl peroxide, there is no addition to the double bond:

$$Cl_3GeCH{=}CH_2 + 3\,HBr \rightarrow Br_3GeCH{=}CH_2 + 3\,HCl$$

$$Cl_3GeCH_2CH{=}CH_2 + 3\,HBr \rightarrow Br_3GeCH_2CH{=}CH_2 + 3\,HCl$$

It should be noted that with allyltrichlorosilane hydrobromic acid adds exclusively at the multiple bond (180)

$$Cl_3SiCH_2CH{=}CH_2 + HBr \rightarrow Cl_3SiCH_2CH_2CH_2Br$$

Only in the case of methallyltrichlorogermane do substitution and addition reactions occur simultaneously:

$$Cl_3GeCH_2{-}\underset{\underset{CH_3}{|}}{C}{=}CH_2 \xrightarrow{HBr} Br_3GeCH_2C(Br)(CH_3)_2$$

Halogen–halogen exchange is also observed in the action of Grignard reagents on germanium halides (10, 91, 240, 332):

$$R_3GeX + R'MgX' \rightarrow R_3GeX' + R'MgX$$

$$GeCl_4 + RMgBr \rightarrow [R_nGeCl_{4-n}] \xrightarrow{MgBrCl} R_nGeBr_{4-n} + MgCl_2$$

5-1-3-6 Reduction. Organogermanium halides R_nGeX_{4-n} are rapidly converted into the corresponding germanium hydrides by the action of lithium aluminium hydride, sodium borohydride or lithium hydride. The reaction, usually carried out in diethyl-, diisopropyl- or di-*n*-butyl-ether for $LiAlH_4$ and in dioxane or tetrahydrofuran with sodium boro-hydride, is generally quantitative. In the same way, the reduction by lithium deuteride (241) or lithium aluminium deuteride (69) leads to the corresponding Ge—D compounds:

$$R_2GeX_2 + 2\,LiD \rightarrow R_2GeD_2 + 2\,LiX$$

$$2\,\Big[GeI_2 + LiAlD_4 \rightarrow 2\,\Big[GeD_2 + AlI_3 + LiI$$

5-1-3-7 Metals. Organogermanium halides react easily with alkali metals giving organodigermanes by Würtz coupling. Kraus and Flood (137) obtained hexaethyldigermane by heating triethylgermanium bromide with sodium at elevated temperatures (270°C) without solvent:

$$2\,(C_2H_5)_3GeBr + 2\,Na \rightarrow 2\,NaBr + (C_2H_5)_3Ge{-}Ge(C_2H_5)_3$$

and Morgan and Drew (213) prepared hexaphenyldigermane by heating for three hours triphenylgermanium bromide with excess of sodium in dry xylene. Hexamethyldigermane is formed in the same manner by refluxing trimethylgermanium bromide with molten potassium (74 % yield) (44).

To estimate the relative reactivities of trimethylbromogermane and trimethylbromosilane in reaction with potassium, Mironov (204) carried out the reactions competitively. The highest percentage of the digermane in the resulting mixture shows the highest reactivity of trimethylbromo-germane:

$$(CH_3)_3GeBr + (CH_3)_3SiBr + K \rightarrow \begin{cases} (CH_3)_3Ge-Ge(CH_3)_3 & (70\%) \\ (CH_3)_3Ge-Si(CH_3)_3 & (20\%) \\ (CH_3)_3Si-Si(CH_3)_3 & (8\%) \end{cases}$$

A competitive reaction between triethylbromogermane and triethyl-bromosilane with sodium (296) leads to a similar result but the difference of reactivities seems to be smaller in this case:

$$(C_2H_5)_3GeBr + (C_2H_5)_3SiBr + Na \rightarrow \begin{cases} (C_2H_5)_3Ge-Ge(C_2H_5)_3 & (42\%) \\ (C_2H_5)_3Ge-Si(C_2H_5)_3 & (27\%) \\ (C_2H_5)_3Si-Si(C_2H_5)_3 & (26\%) \end{cases}$$

Lithium metal in tetrahydrofuran under argon atmosphere (296) reacts with triethylbromogermane giving hexaethyldigermane with 63 % yield. In the same solvent, but with an excess of lithium, arylgermanium halides give germyllithium compounds by cleavage of the digermane formed in the first step (94):

$$(C_6H_5)_3GeBr \xrightarrow[\text{THF}]{2\,Li} (C_6H_5)_3Ge-Ge(C_6H_5)_3 \xrightarrow[\text{THF}]{2\,Li} (C_6H_5)_3GeLi$$

A germanium compound with a Ge—Hg bond is formed by shaking under nitrogen for a week a mixture of trimethylgermanium bromide, 0.5 % sodium amalgam and cyclohexane (72):

$$(CH_3)_3GeBr + Na/Hg \rightarrow (CH_3)_3Ge-Hg-Ge(CH_3)_3 \quad (35\%)$$

Germanium–carbon bonds are easily formed by Würtz-coupling with sodium in the case of aryl compounds (73, 229):

$$GeCl_4 + 4\,C_6H_5Cl + 8\,Na \rightarrow 8\,NaCl + (C_6H_5)_4Ge$$

$$(C_2H_5)_3GeBr + XC_6H_4Cl + 2\,Na \rightarrow$$

$$NaCl + NaBr + (C_2H_5)_3GeC_6H_4X$$

$$(X = o\text{-}CH_3, m\text{-}CH_3, p\text{-}C_2H_5)$$

For the synthesis of aryltrihalogermanes by monoarylation of the tetra-halide, the use of copper is particularly suitable (191):

$$ArI + GeX_4 \xrightarrow{Cu} ArGeX_3 \quad (80\%)$$

The action of alkali metals with germanium polyhalides usually leads to cyclic or polymeric organogermanes depending on the metal used and the experimental conditions (136, 178, 280, 282). However, with lithium amalgam in diethyl ether in an inert atmosphere it is possible to remove only one halogen atom (178):

$$2\,C_6H_5GeBr_3 + 2\,Li \rightarrow 2\,LiBr + C_6H_5Br_2Ge\text{—}GeBr_2C_6H_5$$

$$2\,(C_6H_5)_2GeBr_2 + 2\,Li \rightarrow 2\,LiBr + (C_6H_5)_2BrGe\text{—}GeBr(C_6H_5)_2$$

The dehalogenation of 3-chloropropyldialkylchlorogermane by sodium in boiling toluene or xylene is an excellent method for the synthesis of dialkylgermacyclobutanes (169, 171):

$$R_2Ge(Cl)CH_2CH_2CH_2Cl + 2\,Na \rightarrow 2\,NaCl + R_2Ge\!\!\begin{array}{c}\diamond\end{array}$$

5-1-3-8 Action on Alkali Metal Derivatives. Germanium halides react easily with organolithium compounds at low or moderate temperature:

$$R_3GeX + R'Li \rightarrow R_3GeR' + LiX$$

and this reaction is frequently used for alkylation (80, 91, 101, 324, 327, 337), alkenylation (214, 222, 291, 294, 295), or alkynylation (32, 76, 112) of germanium compounds. The action of germanium polyhalides on dili-thium compounds permits the synthesis of cyclic or spirocyclic germanes (63, 100, 143). Germanium compounds with Ge—O, Ge—N and Ge—P bonds are commonly prepared from the corresponding lithium derivative:

$$ClCH_2Si(CH_3)_2OLi + (CH_3)_3GeCl \rightarrow$$
$$ClCH_2Si(CH_3)_2OGe(CH_3)_3 + LiCl \quad (328)$$

$$R_3GeCl + R'_2NLi \xrightarrow{heptane} R_3GeNR'_2 + LiCl \quad (265)$$

$$R_3GeCl + R'_2PLi \longrightarrow R_3GePR'_2 + LiCl \quad (43, 264)$$

Germanium–metal derivatives are also obtained with good yield by this method (45, 96, 97, 330):

$$3\,(CH_3)_3GeBr + Li_3Sb \rightarrow 3\,LiBr + [(CH_3)_3Ge]_3Sb \quad (85\% \text{ yield})$$

Sodium and potassium derivatives give similar reactions and are frequently used to prepare Ge—O, Ge—S, Ge—Se, Ge—N, Ge—As and Ge—metal compounds from germanium halides.

It should be noted that sometimes germanium and isologous silicon halides react differently with alkali salts. Thus, trimethylchlorogermane gives trimethylgermyl acetate with an aqueous solution of potassium acetate while trimethylchlorosilane gives hexamethyldisiloxane in the same conditions (204):

$$(CH_3)_3GeCl + CH_3COOK \xrightarrow{H_2O} CH_3GeOCOCH_3$$

$$(CH_3)_3SiCl + CH_3COOK \xrightarrow{H_2O} (CH_3)_3SiOSi(CH_3)_3$$

5-1-3-9 Silver Salts. In spite of their insolubility, a great number of silver salts react with germanium halides in inert solvents. Anderson extended the "conversion series" previously investigated for halides (5). He found that organogermanium esters are formed almost quantitatively from organogermanium iodide, sulphide, bromide, hydride, cyanide, chloride, isothiocyanate or isocyanate and the corresponding silver salts: with the oxide, the ester was obtained in approximately 60% and no reaction was observed with the fluoride (13). Therefore, the conversion series for organogermanium compounds is as follows:

$$I \rightarrow S \rightarrow Br \rightarrow CN \rightarrow (NCS \text{ and } Cl) \rightarrow$$

$$NCO \rightarrow (O \text{ and } OCOR) \rightarrow F$$

5-1-3-10 Grignard Reagents and Organozinc Derivatives. The great reactivity of these compounds causes usually a total alkylation or arylation of germanium halides: these reactions are detailed in Chapter 2.

5-1-3-11 Organomercuric Compounds. Germanium halides react easily with dialkyl- or diarylmercuric compounds at moderate temperatures without catalyst (39, 78, 262, 287):

$$GeCl_4 + (CH_2=CH)_2Hg \rightarrow Cl_3GeCH=CH_2 + ClHgCH=CH_2$$

$$GeI_2 + Ar_2Hg \rightarrow Ar_2GeI_2 + Ar_3GeI + Hg$$

Functionally substituted germanium derivatives had been prepared by this method: Baukov (27) studied the reaction of $Hg(CH_2CO_2CH_3)_2$ with triethyliodogermane and triethyliodosilane with the germanium halide; only the Ge—C derivative was formed:

$$(C_2H_5)_3GeI \xrightarrow[- IHgCH_2COOCH_3]{+ Hg(CH_2COOCH_3)_2} (C_2H_5)_3GeCH_2COOCH_3$$

while with the silicon iodide, depending on the solvents used, Si—O and Si—C products were obtained (28):

$$(C_2H_5)_3SiI \xrightarrow[- IHgCH_2COOCH_3]{+ Hg(CH_2COOCH_3)_2} \begin{cases} (C_2H_5)_3SiCH_2COOCH_3 \\ (C_2H_5)_3SiOC(OCH_3)=CH_2 \end{cases}$$

5-1-3-12 Organoaluminium Derivatives. Alkylation of germanium tetrachloride by trialkylaluminiums (338, 352) and trialkyltrihalodialuminiums (341) in stoichiometric amounts leads exclusively to the formation of tetraalkylgermanes. Mironov (181) observed that, with triethylaluminium and an excess of germanium tetrachloride in the presence of sodium chloride no alkylgermanium halide was formed but a mixture of tetraethyl germane, hexaethyldigermane, a germanium-containing polymer and unchanged germanium tetrachloride was obtained. He found that, in the absence of sodium chloride, germanium tetrachloride and triethylaluminium react exothermically giving a mixture distilling over the range 83–210°C. Prolonged heating of this mixture gives a complex containing, after redistribution, diethylgermanium dichloride and ethylaluminium dichloride (ratio 1:1), which forms with high yields diethyldichlorogermane by treatment with concentrated hydrochloric acid and triethylchlorogermane with heating in the presence of sodium chloride; while the action of triethylaluminium leads to a mixture of tetraethylgermane, hexaethyldigermane and polymers:

$$GeCl_4 + Al(C_2H_5)_3$$

$$\downarrow$$

$$[GeCl_4 \cdot Al(C_2H_5)_3]$$

$$\text{3 h} \downarrow \text{150°C}$$

$$(C_2H_5)_2GeCl_2 \xleftarrow{OH_2, HCl} [(C_2H_5)_2GeCl_2 \cdot Al(C_2H_5)Cl_2] \xrightarrow{NaCl} (C_2H_5)_3GeCl$$

$$(76\%) \qquad\qquad NaCl \downarrow Al(C_2H_5)_3 \qquad\qquad (85\%)$$

$$(C_2H_5)_4Ge + (C_2H_5)_3Ge—Ge(C_2H_5)_3 + \text{polymers}$$

Similar reactions with diethylchloroaluminium give corresponding complexes leading with high yields (78%) to diethyldichlorogermane by action of sodium chloride. In the same way tripropylchlorogermane (85%) triisobutylchlorogermane (67%) and diisobutyldichlorogermane (66%) were prepared by action of germanium tetrachloride with tripropyl-, triisobutyl-, and diisobutylchloroaluminium respectively.

To explain the formation of butane and polygermanes in side-reactions Mironov postulated the existence of Al—Ge bond derivatives:

$$\diagdown\!Ge\!-\!C_2H_5 + C_2H_5Al\diagup \rightarrow C_4H_{10} + \diagdown\!Ge\!-\!Al\diagup$$

$$\diagdown\!Ge\!-\!Al\diagup + \diagdown\!Ge\!-\!Cl \rightarrow \diagdown\!AlCl + \diagdown\!Ge\!-\!Ge\!\diagdown$$

$$C_2H_5\underset{|}{\overset{|}{Ge}}\!-\!\underset{|}{\overset{|}{Ge}}\!- + C_2H_5Al\diagup \rightarrow C_4H_{10} + \diagdown\!Al\!-\!\underset{|}{\overset{|}{Ge}}\!-\!\underset{|}{\overset{|}{Ge}}\!-$$

$$\diagdown\!Al\!-\!\underset{|}{\overset{|}{Ge}}\!-\!\underset{|}{\overset{|}{Ge}}\!- + ClGe\!\diagdown \rightarrow \diagdown\!AlCl + -\underset{|}{\overset{|}{Ge}}\!-\!\underset{|}{\overset{|}{Ge}}\!-\!\underset{|}{\overset{|}{Ge}}\!- \quad \text{etc.}$$

5-1-3-13 Organometallic Derivatives of Group IV Elements. The reaction of germanium halides with tetraarylgermanes (282) or tetraalkylgermanes in the presence of a catalyst (130, 164) is a convenient method for the synthesis of various germanium halides.

Germanium tetrachloride opens the germacyclobutane ring by cleavage of an intracyclic germanium–carbon bond (169):

$$GeCl_4 + R_2Ge\!\!\overset{\displaystyle \diagup\!\!\diagdown}{\diagdown\!\!\diagup}\!\! \rightarrow R_2Ge(CH_2)_3GeCl_3$$
$$\underset{Cl}{|}$$

The partial alkylation of germanium polyhalides with tin or lead tetraalkyls was performed by Mironov (202, 203). A complex disproportionation reaction takes place when tetramethyltin and trichlorogermane are mixed:

$$(CH_3)_4Sn + HGeCl_3 \rightarrow CH_4\ (30\%) + HCl_2GeCH_3\ (43\%)$$
$$+ CH_3GeCl_3\ (25\%) + (CH_3)_2SnCl_2\ (26\%)$$
$$+ (CH_3)_3SnCl\ (57\%)$$

If this reaction is carried out in ether (or with germachloroform etherate), only methyldichlorogermane and trimethyltin chloride are formed in yields of 80% and 70%, respectively:

$$(CH_3)_4Sn + HGeCl_3 \xrightarrow{\text{ether}} HGeCl_2CH_3 + (CH_3)_3SnCl$$

With tetraethyltin, trichlorogermane gives a mixture of ethyldichloro- and ethyltrichlorogermane:

$$(C_2H_5)_4Sn + HGeCl_3 \rightarrow HGeCl_2C_2H_5 + C_2H_5GeCl_3$$
$$\qquad\qquad (45\%) \qquad\qquad (31\%)$$
$$+ (C_2H_5)_3SnCl + H_2 + C_2H_6$$

The percentage of ethyldichlorogermane increases when tetraethyllead is used:

$$(C_2H_5)_4Pb + HGeCl_3 \rightarrow HGeCl_2(C_2H_5) + C_2H_5GeCl_3$$
$$ (60\%) (20\%)$$

A very good yield of ethyltrichlorogermane is obtained by alkylation of germanium tetrachloride with tetraethyllead (203):

$$(C_2H_5)_4Pb + GeCl_4 \rightarrow (C_2H_5)_3PbCl + C_2H_5GeCl_3 \quad (90\%)$$

Luijten (156) prepared butyltrichlorogermane by the action of tetra-butyltin on germanium tetrachloride:

$$(C_4H_9)_4Sn + GeCl_4 \rightarrow (C_4H_9)_3SnCl + C_4H_9GeCl_3$$

The partial arylation of trichlorogermane by tetraphenyltin gives the normal monoarylated derivative in only poor yield (203):

$$(C_6H_5)_4Sn + HGeCl_3 \rightarrow HGeCl_2C_6H_5 + C_6H_5GeCl_3$$
$$ (5\%) (45\%)$$

Functionally substituted organogermanium compounds have been prepared with 70–90% yields by reacting esters of (trialkylstannyl) acetic acid with halogermanes (158). With germanium tetrahalide, depending on the ratios of the reactants, one, two, three or four halogen atoms can be replaced by carbomethoxymethyl groups:

$$GeX_4 \xrightarrow[-R_3SnX]{+R_3SnCH_2COOCH_3} \begin{cases} 1:1 & X_3GeCH_2COOCH_3 \\ 1:2 & X_2Ge(CH_2COOCH_3)_2 \\ 1:3 & X\ Ge(CH_2COOCH_3)_3 \\ 1:4 & Ge(CH_2COOCH_3)_4 \end{cases}$$

Partially substituted germanium halides react in the same way:

$$Cl_nGe(CH_2COOCH_3)_{4-n} \xrightarrow[-R_3SnCl]{+R_3SnCH_2COOCH_3} Cl_{n-1}Ge(CH_2COOCH_3)_{5-n}$$

Alkoxyhalogermanes react easily:

$$(CH_3O)_3GeCl \xrightarrow[-R_3SnCl]{+R_3SnCOOCH_3} (CH_3O)_3GeCH_2COOCH_3$$

but with trialkylhalogermanes, the yield is dependent on the nature of the halogen:

$$(C_2H_5)_3GeX \xrightarrow[-R_3SnX]{+R_3SnCH_2COOR'} (C_2H_5)_3GeCH_2COOR' \quad R' = CH_3, C_4H_9$$

$$\text{Yield} \begin{cases} X = I & 80\% \\ X = Cl & 27\% \end{cases}$$

5-1-3-14 Insertion Reactions

(a) Diazomethane

Seyferth and Rochow (220) observed the first insertion reaction, of methylene group in a germanium–halogen bond, and noted that the higher the number of halogen atoms linked to the germanium, the easier was the reaction: thus germanium tetrachloride gives, at -60 to $-70°C$, chloromethyltrichlorogermane in excellent yield:

$$GeCl_4 + CH_2N_2 \xrightarrow{Cu} N_2 + Cl_3GeCH_2Cl \quad (94\%)$$

and a low yield of bis(chloromethyl)dichloro derivative as a by-product. Similarly, methyltrichlorogermane gives methyl (chloromethyl)-dichlorogermane in 78% yield:

$$CH_3GeCl_3 + CH_2N_2 \xrightarrow{Cu} N_2 + CH_3Ge(Cl)_2CH_2Cl$$

but dimethyldichlorogermane gives no reaction. With two molar quantities of diazomethane, methyltrichlorogermane gives a mixture of the mono and bis(chloromethyl) compounds:

$$CH_3GeCl_3 + 2CH_2N_2 \rightarrow \begin{cases} CH_3Ge(CH_2Cl)Cl_2 & (33.5\%) \\ CH_3Ge(CH_2Cl)_2Cl & (20.5\%) \end{cases}$$

Diazomethane gives also a methylenation reaction with phenylchlorogermanes (134). With phenyltrichlorogermane the bis(chloromethyl) derivative is obtained in 86% yield if an excess of diazomethane is used:

$$C_6H_5GeCl_3 + 2CH_2N_2 \xrightarrow{Cu} 2N_2 + C_6H_5Ge(CH_2Cl)_2Cl$$

while if the reactants are in equimolar amounts the monochloromethyl compound is formed in good yield:

$$C_6H_5GeCl_3 + CH_2N_2 \xrightarrow{Cu} N_2 + C_6H_5Ge(CH_2Cl)Cl_2 \quad (70\%)$$

With diphenyldichlorogermane the reaction gives only the monochloromethyl derivative in 26% yield:

$$(C_6H_5)_2GeCl_2 + CH_2N_2 \xrightarrow{Cu} N_2 + (C_6H_5)_2Ge(CH_2Cl)Cl$$

and triphenylchlorogermane does not react with diazomethane. Seyferth and Hetflejs (287) have studied the relative rates of conversion of substituted aryltrichlorogermanes to the aryl(chloromethyl)dichlorogermanes in competition experiments:

$$p\text{-}Z{-}C_6H_4GeCl_3 + CH_2N_2 \xrightarrow{-78°C} (p\text{-}Z\,C_6H_4)Cl_2GeCH_2Cl + N_2$$

$$Z = H, Cl, F, CH_3, OCH_3$$

The results show that electron-withdrawing substituents enhance the rate of the methylenation reaction, while those which supply electron density have a retarding effect:

Relative yields: $Z = OCH_3$: 1.0; CH_3: 1.29; H: 2.24; F: 5.37; Cl: 9.76

A rectilinear correlation of K_{rel} with Taft's σ substituent constants was found. These observations suggest that the mechanism of the methylenation of the aryltrichlorogermanes involves nucleophilic attack by diazomethane at germanium rather than a free radical process for the insertion of a —CH₂— group into the Ge—Cl bond:

$$\overset{\diagdown}{\underset{\diagup}{}}\text{Ge}-\text{Cl} + \overset{\ominus}{:}\text{CH}_2-\overset{\oplus}{\text{N}}\equiv\text{N} \xrightarrow{\text{slow}} \overset{\diagdown}{\underset{\diagup}{}}\text{Ge}-\underset{\underset{\oplus}{\text{CH}_2\text{N}\equiv\text{N}}}{\overset{|}{\text{Cl}}} \xrightarrow{\text{fast}} \overset{\diagdown}{\underset{\diagup}{}}\text{Ge}-\text{CH}_2\text{Cl} + \text{N}_2$$

or

$$\overset{\diagdown}{\underset{\diagup}{}}\text{Ge}-\text{Cl} + \overset{\ominus}{:}\text{CH}_2-\overset{\oplus}{\text{N}}\equiv\text{N} \rightarrow \overset{\diagdown}{\underset{\diagup}{}}\underset{\text{Cl}}{\overset{|}{\text{Ge}}} \leftarrow \overset{\ominus}{:}\text{CH}_2-\overset{\oplus}{\text{N}}\equiv\text{N} \rightarrow$$

$$\overset{\diagdown}{\underset{\diagup}{}}\text{Ge}-\text{CH}_2\text{Cl} + \text{N}_2$$

(b) Epoxides

Trimethylbromogermane, like trimethylbromosilane (95), reacts with ethylene oxide (204):

$$(CH_3)_3GeBr + \underset{\text{O}}{CH_2-CH_2} \rightarrow (CH_3)_3GeOCH_2CH_2Br \quad (82\%)$$

However, the analogous cleavage of tetrahydrofuran with trimethylbromogermane does not occur even in the presence of aluminium bromide (204), whereas under similar conditions $(CH_3)_3SiBr$ readily forms $(CH_3)_3SiOCH_2CH_2CH_2CH_2Br$ (342). Lavigne (142, 238, 239) studied the action of $GeCl_4$, $(C_3H_7)_3GeBr$, $(C_6H_5)_3GeCl$ and $(C_6H_5)_3GeBr$ with various epoxides and observed that the reaction rate constants with methoxy- and chloro-substituted propylene oxides are lower than that obtained with unsubstituted propylene oxide in spite of the fact that one of the substituents is electron-releasing and the other electron-withdrawing. With 3-chloro-1,2-epoxypropane, tripropyl bromogermane gives $(n\text{-}C_3H_7)_3GeOCH(CH_2Cl)CH_2Br$.

5-1-3-15   Inductive Effects of Ge—X *Groups.* The relative rates of chlorination (catalysed by $FeCl_3$ or iodine) of phenylmethylgermanium chlorides $C_6H_5Ge(CH_3)_{3-n}Cl_n$ have been studied, as well as the effect of their structure on the orientation in substitution of the phenyl group by Chvalovský and coworkers (54). Their results show that, when the chlorination is catalysed by iron, the trimethylgermyl group has a weak *o–p* directing effect: the remaining substituents have a *m* directing effect which increases with the number of halogen atoms bonded to the metal. When iodine is used as catalyst, all the substituted germyl groups have an *o–p* directing effect:

CATALYST	$FeCl_3$			I_2		
	o%	*m*%	*p*%	*o*%	*m*%	*p*%
$(CH_3)_3GeC_6H_5$	20	28	52	18	28	54
$(CH_3)_2Ge(C_6H_5)Cl$	15	50	35	17	29	54
$CH_3Ge(C_6H_5)Cl_2$	10	53	37	14	33	53
$Cl_3GeC_6H_5$	9	65	26	12	35	53

5-1-4   Organogermanium halides

Table 5–1 (a–l) showing some physical properties of organogermanium halides appears on the following pages.

Table 5-1
Organogermanium Halides

(a) *Organogermanium monofluorides* $-\overset{|}{\underset{|}{Ge}}-F$

COMPOUND	EMPIRICAL FORMULA	M.P. (°C)	B.P. (°C/mm)	n_D^{20}	d_4^{20}	REFERENCES
$FGe(CH_3)_3$	C_3H_9GeF	—	76/746, 77–78	1.3874, 1.3835	—, —	104, 105, 110, 204, 209, 240
$FGe(C_2H_5)_3$	$C_6H_{15}GeF$	—	149.5/751, 147–48/744	1.3863, 1.3855, 1.4221, 1.4206	1.2300, 1.2114, 1.5127	13, 17, 104, 105, 137, 146
$F{-}\underset{\displaystyle C_2H_5}{\overset{\displaystyle C_2H_5}{Ge}}{-}CH_2{-}\underset{\displaystyle CH_3}{CH}CH{=}CH_2$	$C_9H_{19}GeF$	—	80/18	1.4502	1.0951	175
$FGe(n\text{-}C_3H_7)_3$	$C_9H_{21}GeF$	−27.5	203	1.4340	1.074	5
$FGe(i\text{-}C_3H_7)_3$	$C_9H_{21}GeF$	−65	198	1.440	1.069	10
$FGe(n\text{-}C_4H_9)_3$	$C_{12}H_{27}GeF$	−12	245/760, 128–30/14	1.4419	1.038	7, 17
$FGe(C_6H_5)_3$	$C_{18}H_{15}GeF$	76.6	—	—	—	138
$FGe(C_6H_5)_2(n\text{-}C_6H_{13})$	$C_{18}H_{23}GeF$		124/0.2	1.5420	1.1733	249
$FGe(C_6H_{11})_3$	$C_{18}H_{33}GeF$	92	—	—	—	21
$FGe(C_6H_5CH_2)_3$	$C_{21}H_{21}GeF$	96	—	—	—	24

(b) Organogermanium difluorides   F—Ge—F (with F above and below)

COMPOUND	EMPIRICAL FORMULA	M.P. (°C)	B.P. (°C/mm)	n_D^{20}	d_4^{20}	REFERENCES
$F_2Ge(CH_3)_2$	$C_2H_6GeF_2$	—	112/750	1.3743	1.5726	240
$F_2Ge(n\text{-}C_3H_7)_2$	$C_6H_{14}GeF_2$	0.5	182.8	1.4128	1.248	8, 9
$F_2Ge(i\text{-}C_3H_7)_2$	$C_6H_{14}GeF_2$	−24	174	1.4146	1.222	10
$F_2Ge(n\text{-}C_4H_9)_2$	$C_8H_{18}GeF_2$	10	216	1.4222	1.183	16
$F_2Ge(C_6H_5)_2$	$C_{12}H_{10}GeF_2$	—	100/0.007	—	—	135, 179
$F_2Ge(C_6H_5)(n\text{-}C_6H_{13})$	$C_{12}H_{18}GeF_2$	—	86/0.2	1.4865	1.2568	249

(c) Organogermanium trifluorides   F—Ge—F (with F above and below)

COMPOUND	EMPIRICAL FORMULA	M.P. (°C)	B.P. (°C/mm)	n_D^{20}	d_4^{20}	REFERENCES
F_3GeCF_3	$CGeF_6$	3 (sealed tube)	−1.7	—	—	55, 57
F_3GeCH_3	CH_3GeF_3	38	96.5/751	—	—	240
$F_3GeC_2H_5$	$C_2H_5GeF_3$	15.5–6.5	112	—	—	87

(d) Organogermanium monochlorides —Ge—Cl

COMPOUND	EMPIRICAL FORMULA	M.P. (°C)	B.P. (°C/mm)	n_D^{20}	d_4^{20}	REFERENCES
ClGe(CH₂Cl)₂ CH₃	C₃H₇GeCl₃	—	95–97/30	1.5010 (25°)	—	290
(CH₃)₂Ge—CH₂Cl Cl	C₃H₈GeCl₂	—	148–49	—	—	67, 111, 204, 326, 327
ClGe(CH₃)₃	C₃H₉GeCl	—13 —14	115 102/760 98/736 98/736 97.8/750 96.5–97/730	1.4314 (29°) 1.4283 (29°) 1.4337 1.4337 1.4350	— 1.2382 (21.5°) 1.2493 1.2493 1.2435	82, 110, 204, 208, 240, 251, 252, 260, 272, 277, 317, 343
ClGe(C₂H₅)(CH₃)₂	C₄H₁₁GeCl	—	125/760	—	—	207, 241
(CH₃)₂(Cl)Ge—Ge(CH₃)₃	C₅H₁₅Ge₂Cl	—	98–100/746	1.4285	1.1763	47
ClGe(CH=CH₂)₃	C₆H₉GeCl	—	62–63/18	1.4911	—	284
ClGe(C₂H₅)₃	C₆H₁₅GeCl	—	175.9/760 —	1.4643 1.4650	1.175 —	10, 14, 17, 43, 73, 137, 162, 244, 248, 251, 263, 267, 279, 307, 319, 320, 321, 343

Structure	Molecular formula		bp (°C/mm)		n_D	d	Ref.
ClGe(CH$_3$)$_2$(C$_4$H$_9$)	C$_6$H$_{15}$GeCl	—	168	—	1.4490	—	47
ClGe(C$_2$H$_5$)$_2$(CH$_2$CH$_2$CH$_2$Cl)	C$_7$H$_{16}$GeCl$_2$	—	109/10	—	1.4862	1.2669	151
(C$_2$H$_5$)$_2$(Cl)Ge—Si(CH$_3$)$_3$	C$_7$H$_{19}$GeSiCl	—	126/65	—	1.4819	—	47
ClGe(C$_6$H$_5$)(CH$_2$Cl)$_2$	C$_8$H$_9$GeCl$_3$	—	93/0.1	—	1.5723	—	134
(CH$_3$)$_2$Ge—C$_6$H$_4$Cl-p | Cl	C$_8$H$_{10}$GeCl$_2$	—	116–18/12	—	1.5512	—	154
(C$_2$H$_5$)$_2$Ge—CH$_2$CH$_2$CH$_2$CN | Cl	C$_8$H$_{16}$GeClN	—	156/13	—	1.4823	1.2184	152
(C$_2$H$_5$)$_2$Ge—CH$_2$CH$_2$CH=CH$_2$ | Cl	C$_8$H$_{17}$GeCl	—	100/19	—	1.4742	1.1400	175
(C$_2$H$_5$)$_2$Ge—CH$_2$CH$_2$CH=CHCH$_3$ | Cl	C$_8$H$_{17}$GeCl	—	99–100/22 99.5/22	—	1.4818 1.4812	1.1524 1.1524	267 266
(C$_2$H$_5$)$_2$Ge—CH$_2$CH$_2$COCH$_3$ | Cl	C$_8$H$_{17}$GeClO	—	137/18	—	1.4816	1.2303	153, 267
(C$_2$H$_5$)$_2$Ge—CH$_2$CH$_2$OCOCH$_3$ | Cl	C$_8$H$_{17}$GeClO$_2$	—	145/20 132/15	—	1.4697 1.4675	1.2178 1.2437	153, 267
(C$_2$H$_5$)$_2$Ge—CH$_2$CH(CH$_3$)CH$_2$Cl | Cl	C$_8$H$_{18}$GeCl$_2$	—	119/10	—	1.4866	1.2406	68
(CH$_3$)(C$_3$H$_7$)(C$_4$H$_9$)GeCl	C$_8$H$_{19}$GeCl	—	104/26	—	1.4567	—	47
(C$_2$H$_5$)$_2$Ge—Ge(C$_2$H$_5$)$_2$ | Cl | Cl	C$_8$H$_{20}$Ge$_2$Cl$_2$	—	130–32/16	—	1.5197	—	46
(CH$_3$)$_2$Ge—Ge(C$_2$H$_5$)$_3$ | Cl	C$_8$H$_{21}$Ge$_2$Cl	—	108/14	—	1.5011	—	47

Table 5-1(d)—continued

COMPOUND	EMPIRICAL FORMULA	M.P. (°C)	B.P. (°C/mm)	n_D^{20}	d_4^{20}	REFERENCES
ClGe(CH₂COOCH₃)₃	C₉H₁₅GeClO₆	—	131–33/1	1.4860	1.4412	19, 158
(C₂H₅)₂Ge—CH₂CH=CCl—CH=CH₂ | Cl	C₉H₁₆GeCl₂	—	76/0.5	1.5257	1.2528	288
(C₂H₅)₂Ge—CH₂CH(CH₃)CH=CH₂ | Cl	C₉H₁₉GeCl		98/16	1.4739	1.1138	175
(C₃H₇)₂Ge—CH₂COOCH₃ | Cl	C₉H₁₉GeClO₂		82–83/1	1.4670	1.2003	26
(C₂H₅)₂Ge—CH₂CH₂COOC₂H₅ | Cl	C₉H₁₉GeClO₂		145/20	1.4697	1.2178	153, 267
(C₂H₅)₂Ge—CH₂CH(CH₃)OCOCH₃ | Cl	C₉H₁₉GeClO₂	—	130/18	1.4658	1.2101	153
ClGe(C₂H₅)(C₃H₇)(C⁴H₉)	C₉H₂₁GeCl	—	86–89/12	1.4611	—	47
ClGe(C₃H₇-n)₃	C₉H₂₁GeCl	–70	103–6/15 227/760 98–99/11 222/760	1.4641	1.100	5, 119, 124, 181, 248, 256
ClGe(C₃H₇-i)₃	C₉H₂₁GeCl	—	120–21/42	1.472	1.110	10, 177

Compound		b.p./mm	n_D	d	Ref.
$(C_2H_5)_2Ge$—$CH_2C(CH_3)$=$CClCH$=CH_2 $\quad\quad\quad\mid$ $\quad\quad\quad Cl$ $C_{10}H_{18}GeCl_2$	—	92/0.4	1.5242	1.2344	288
$(C_2H_5)_2Ge$—$CH_2CH(CH_3)$—$C(CH_3)$=CH_2 $\quad\quad\quad\mid$ $\quad\quad\quad Cl$ $(C_{10}H_{21}GeCl$	—	119/24	1.4774	1.1096	175
$(C_2H_5)_2Ge$—$CH_2C(CH_3)$=$C(CH_3)_2$ $\quad\quad\quad\mid$ $\quad\quad\quad Cl$ $C_{10}H_{21}GeCl$	—	111/10	1.4876	1.1149	266, 267
$ClGe[(CH_2)_3Cl]_2$ $\quad\mid$ $\quad C_4H_9$ $C_{10}H_{21}GeCl$	—	119/0.05	1.5002	1.2765	68
$(C_2H_5)_2(Cl)Ge(CH_2)_2Ge(Cl)(C_2H_5)_2$ $C_{10}H_{24}Ge_2Cl_2$	—	117/0.5 163/10	1.5018 1.5020	1.3196 1.3266	151, 161
$(C_2H_5)_3GeGe(Cl)(C_2H_5)_2$ $(C_2H_5)_2(Cl)GeCH_2C(CH_3)$=$C(Cl)C(CH_3)$=$CH_2$ $C_{10}H_{25}Ge_2Cl$	—	126–27/16	1.5092	—	46
$(C_4H_9)_2Ge$—CH_2CH_2CN $\quad\quad\quad\mid$ $\quad\quad\quad Cl$ $C_{11}H_{20}GeCl_2$ $C_{11}H_{22}GeClN$	— —	100/0.7 168.5/10	1.5112 1.4782	1.2098 1.1349	288 151, 267
$(C_4H_9)_2Ge$—CH_2COOCH_3 $\quad\quad\quad\mid$ $\quad\quad\quad Cl$ $C_{11}H_{23}GeClO_2$	—	93–95/1.5	1.4678	1.1557	26
$(n\text{-}C_4H_9)_2Ge$—$CH_2CH_2CH_2Cl$ $\quad\quad\quad\mid$ $\quad\quad\quad Cl$ $C_{11}H_{24}GeCl_2$	—	148/8	1.4820	1.1568	151
$ClGe(C_4H_9)_2$ $\quad\mid$ $\quad C_3H_7$ $C_{11}H_{25}GeCl$	—	124/13	1.4645	1.0593	171
$(C_4H_9)_2Ge$—$CH_2CH_2CH_2OH$ $\quad\quad\quad\mid$ $\quad\quad\quad Cl$ $C_{11}H_{25}GeClO$	—	116/0.5	1.4828	1.1377	151

Table 5-1(d)—continued

COMPOUND	EMPIRICAL FORMULA	M.P. (°C)	B.P. (°C/mm)	n_D^{20}	d_4^{20}	REFERENCES
$(C_2H_5)_2Ge$—CH_2CH_2—⟨cyclohexenyl⟩—Cl 	$C_{12}H_{23}GeCl$	—	162/18	1.4970	1.1463	266, 267
$ClGe[CH_2CH(CH_3)CH_2Cl]_2$—$C_4H_9$	$C_{12}H_{25}GeCl_3$	—	135/0.1	1.5008	1.2558	68
$(C_4H_9)_2Ge$—$CH_2CH(CH_3)CHO$, Cl	$C_{12}H_{25}GeClO$	—	151/13	1.4809	1.1440	153
$(C_4H_9)_2Ge$—$CH_2CH_2COOCH_3$, Cl	$C_{12}H_{25}GeClO_2$	—	164/12	1.4740	1.1454	153, 266, 267
$(C_4H_9)_2Ge$—$CH_2CH(CH_3)CH_2Cl$, Cl	$C_{12}H_{26}GeCl_2$	—	161/10	1.4820	1.1427	152
$ClGe(i\text{-}C_4H_9)_3$	$C_{12}H_{27}GeCl$	—	80–81/1.5 271/760	1.4659	1.0401	181
$ClGe(C_4H_9\text{-}n)_3$	$C_{12}H_{27}GeCl$	—	139–40/13	1.4652	1.054	7, 10, 17, 64a, 130, 150, 248, 263, 339, 340
$(C_2H_5)_4(C_4H_9)Ge_2Cl$	$C_{12}H_{29}Ge_2Cl$	—	114–6/0.5	1.5042	—	47
$(C_6H_5)_2Ge(Cl)CH_2Cl$	$C_{13}H_{12}GeCl_2$	—	100/0.004	1.5990	—	134
$(C_4H_9)_2Ge$—$(CH_2)_3OCOCH_3$, Cl	$C_{13}H_{27}GeClO_2$	—	186/18	1.4688	1.1256	153
$(C_4H_9)_2Ge$—$CH_2CH(CH_3)COOCH_3$, Cl	$C_{13}H_{27}GeClO_2$	—	173/18	1.4709	1.1323	153

Compound	Molecular formula	m.p.	b.p./mm	n_D	d	Refs.
$(C_4H_9)_2Ge-CH=CH-C_4H_9$ with Cl	$C_{14}H_{29}GeCl$	—	156.5/12	1.4735	1.0356	152
$(C_4H_9)_2Ge-CH_2C(CH_3)=C(CH_3)_2$ with Cl	$C_{14}H_{29}GeCl$	—	159/13	1.4837	1.0546	266, 267
$ClGe(C_6H_{11})(C_4H_9)_2$	$C_{14}H_{30}GeCl$	—	163.5/13	1.4891	1.0897	151
$(C_4H_9)_2Ge(Cl)(CH_2CH_2OC_4H_9)$	$C_{14}H_{31}GeClO$	—	164/10	1.4625	1.0489	151
$(C_2H_5)_3Ge(C_2H_5)_2Ge(C_2H_5)_2(Cl)$	$C_{14}H_{35}Ge_3Cl$	—	62–64/0.0005	1.5410	—	47
$ClGe(C_5H_9)_3$	$C_{15}H_{26}GeCl$	—	142–43/0.1	1.5299	1.1965	177
$ClGe(n-C_5H_{11})_3$	$C_{15}H_{33}GeCl$	—	110–14/0.3	1.4656	—	41, 248
$(C_4H_9)_2Ge(Cl)CH_2CH_2CH_2CH(OC_2H_5)_2$	$C_{15}H_{33}GeClO_2$	—	180/14	1.4640	1.0720	153
$(C_3H_7)_3GeGe(Cl)(C_3H_7)_2$	$C_{15}H_{35}Ge_2Cl$	—	110–12/0.4	1.5007	—	46
$(C_6H_5)_2Ge(Cl)CH_2CH_2COCH_3$	$C_{16}H_{17}GeClO$	—	140/0.003	1.5890	—	269, 271
$(C_4H_9)_2Ge(Cl)CH=CH-C_6H_5$	$C_{16}H_{25}GeCl$	—	140/0.4	1.5400	1.1293	151
$(C_4H_9)_2Ge(Cl)(CH_2)_7CH_3$	$C_{16}H_{35}GeCl$	—	134.5/0.6	1.4661	1.0078	151
$(C_4H_9)_2(Cl)GeGe(Cl)(C_4H_9)_2$	$C_{16}H_{36}Ge_2Cl_2$	—	133–38/0.16	1.5027	—	46
$(+)ClGe(C_6H_5)(CH_3)(\alpha-C_{10}H_7)$	$C_{17}H_{15}GeCl$	68–9	—	—	—	41
$ClGe(C_6F_5)_3$	$C_{18}GeClF_{15}$	103–4	—	—	—	80, 81
$ClGe(C_6H_5)_3$	$C_{18}H_{15}GeCl$	117–8 114–5	285/12	—	—	42, 117, 225, 248, 282, 310, 311
$(\pm)ClGe(C_2H_5)(C_6H_5)(1-C_{10}H_7)$	$C_{18}H_{17}GeCl$	88–90	—	—	—	74
$(-)ClGe(C_2H_5)(C_6H_5)(1-C_{10}H_7)$	$C_{18}H_{17}GeCl$	95–98	—	—	—	74
$(C_6H_5)_2(Cl)Ge(CH_2)_4COCH_3$	$C_{18}H_{21}GeClO$	—	115/0.005	1.5710	—	269
$ClGe(C_6H_4)_2(C_6H_{13})$	$C_{18}H_{23}GeCl$	—	148/0.1	1.5586	1.1842	249
$(C_6H_5)(Cl)Ge[(CH_2)_4COCH_3]_2$	$C_{18}H_{27}GeClO_2$	—	170/0.02	1.5255	—	269
$ClGe(C_6H_{11})_3$	$C_{18}H_{33}GeCl$	101–2 96	185/1	—	—	21, 73, 121, 177, 229, 258
$ClGe(n-C_6H_{13})_3$	$C_{18}H_{39}GeCl$	—	138–39/0.5	1.4661	0.989	91

Table 5-1(d)—continued

COMPOUND	EMPIRICAL FORMULA	M.P. (°C)	B.P. (°C/mm)	n_D^{20}	d_4^{20}	REFERENCES
$(C_6H_5)_2(Cl)GeCHCH_2CH_2COCH_3$ $\quad\quad\quad\|$ $\quad\quad CH(CH_3)_2$	$C_{20}H_{25}GeClO$	—	148/0.005	1.5720	—	269, 271
$(C_4H_9)_2Ge{-}Ge(C_4H_9)_3$ $\quad\quad\|$ $\quad\quad Cl$	$C_{20}H_{45}Ge_2Cl$	—	130–32/0.06	1.4932	—	46, 247
$ClGe(C_6H_5CH_2)_3$	$C_{21}H_{21}GeCl$	154–55	—	—	—	24
$ClGe(o{-}CH_3C_6H_4)_3$	$C_{21}H_{21}GeCl$	—	216–22/1	—	—	302, 304
$ClGe(m{-}CH_3C_6H_4)_3$	$C_{21}H_{21}GeCl$	79	221–24/2	—	—	302, 304
$ClGe(p{-}CH_3C_6H_4)_3$	$C_{21}H_{21}GeCl$	128–29	—	—	—	106, 303
$GeCl(n{-}C_7H_{15})_3$	$C_{21}H_{45}GeCl$	—	152/0.12	1.4671	—	41, 248

$$\text{(e) Organogermanium dichlorides } -\overset{\displaystyle Cl}{\underset{\displaystyle Cl}{Ge}}-$$

COMPOUND	EMPIRICAL FORMULA	M.P. (°C)	B.P. (°C/mm)	n_D^{20}	d_4^{20}	REFERENCES
Cl2Ge(CH2Cl)2	C2H4GeCl4	—	90/21	1.5176 (25°)	—	290
Cl2Ge(CH3)CH2Cl	C2H5GeCl3	—	71–74/40	1.4890 (25°)	1.642 (25°)	240, 290, 327
			155/760	1.4930	1.6694	
Cl2Ge(CH3)2	C2H6GeCl2	−22	156	—	—	82, 208, 209,
			122; 124	1.4600	1.492	220, 232, 240,
			127/727	1.4555	1.5053	251, 253, 257,
			119/735		1.4926	260, 272, 305,
						317, 348, 351
Cl2Ge(CH3)(C2H5)	C3H8GeCl2	—	149/750	1.4660	1.4381	181, 202, 327
Cl2Ge(C2H5)2	C4H10GeCl2	−37–39	172.8/763	1.4700	1.3738	113, 155, 181,
			168/760			248, 251, 254,
			175/758			255, 343
Cl2Ge(CH2COOCH3)2	C6H10GeCl2O4	—	95–96/0.5	1.4870	1.5430	19, 158
			106–10/2		1.5426	
Cl2Ge(n-C3H7)2	C6H14GeCl2	−45	209.5	1.4725	1.275	8, 119, 124, 256
Cl2Ge(i-C3H7)2	C6H14GeCl2	−52	203	1.4738	1.268	10, 17
Cl2Ge(CH2Cl)C6H4F-p	C7H6GeFCl3	—	—	1.5473	—	287
Cl2Ge(CH2Cl)C6H4Cl-p	C7H6GeCl4	—	—	1.5777	—	287
Cl2Ge(C6H5)CH2Cl	C7H7GeCl3	—	136/10	1.5658	—	134, 287
				1.5640 (25°)		

Table 5-1(e)—continued

COMPOUND	EMPIRICAL FORMULA	M.P. (°C)	B.P. (°C/mm)	n_D^{20}	d_4^{20}	REFERENCES
$Cl_2Ge(C_3H_7)(C_4H_9)$	$C_7H_{16}GeCl_2$	—	97/10	1.4716	1.2313	68
$Cl_2Ge(CH_2Cl)C_6H_4CH_3\text{-}p$	$C_8H_9GeCl_3$	—	—	1.5599	—	287
$Cl_2Ge(CH_2Cl)C_6H_4OCH_3\text{-}p$	$C_8H_9GeCl_3O$	—	—	1.5707	—	287
$Cl_2Ge(n\text{-}C_4H_9)_2$	$C_8H_{18}GeCl_2$	—	242, 107–8/8	1.4724	1.208	16, 130, 248, 263
$Cl_2Ge(i\text{-}C_4H_9)_2$	$C_8H_{18}GeCl_2$	—	62–3/1	1.4736	1.1894	181
$Cl_2Ge(C_6H_5)CH_2CH_2COCH_3$	$C_{10}H_{12}GeCl_2O$	—	108/0.01	1.5635	—	269, 271
$Cl_2Ge(C_4H_9)CH_2C{=}C(CH_3)_2$ $\quad CH_3$	$C_{10}H_{20}GeCl_2$	—	141/17	1.4953	1.1988	266, 267
$Cl_2Ge(C_6H_5)_2$	$C_{12}H_{10}GeCl_2$	9	223/12	1.5975	—	29, 111, 135, 179, 248, 254, 282
$Cl_2Ge(C_6H_5)(CH_2)_4COCH_3$	$C_{12}H_{16}GeCl_2O$	—	124/0.001	1.5400	—	269, 271
$Cl_2Ge(C_6H_5)C(CH_3)_2CH_2CH_2COCH_3$	$C_{12}H_{16}GeCl_2O$	67	—	—	—	269, 271
$Cl_2Ge(C_6H_5)C_6H_{13}$	$C_{12}H_{18}GeCl_2$	—	94/0.1	1.5280	1.2546	249
$Cl_2Ge(C_6H_{11})_2$	$C_{12}H_{22}GeCl_2$	—	—	—	—	281
$Cl_2Ge(C_6H_5)CHCH_2CH_2COCH_3$ $\quad CH(CH_3)_2$	$C_{14}H_{20}GeCl_2O$	—	122/0.01	1.5450	—	269, 271
$Cl_2Ge(C{-}C{-}C_6H_5)_2$ $B_{10}H_{10}$	$C_{16}H_{30}GeCl_2B_{20}$	205–6	—	—	—	337

(f) Organogermanium trichlorides

$$\text{—Ge—Cl} \quad \begin{matrix} \text{Cl} \\ | \\ | \\ \text{Cl} \end{matrix}$$

COMPOUND	EMPIRICAL FORMULA	M.P. (°C)	B.P. (°C/mm)	n_D^{20}	d_4^{20}	REFERENCES
Cl₃GeCF₃	CGeCl₃F₃	—	20/90	—	—	55, 57
Cl₃GeCCl₃	CGeCl₆	106–7	130/200	—	—	240
Cl₃GeCHCl₂	CHGeCl₅	—	168–69/760	1.5100	1.8166	71, 189
Cl₃GeCH₂Cl	CH₂GeCl₄	—	149/759	1.5003	1.8415	98, 189, 200, 240,
			60–65/35	1.4989 (25°)	1.833 (25°)	290
			60/31	1.5029	1.8438	
(Cl₃Ge)₂CH₂	CH₂Ge₂Cl₆	—	98–99/12	1.5300	2.0015	194, 307, 309
Cl₃GeCH₃	CH₃GeCl₃	—	111; 108;	1.4685	1.73 (24.5°)	82, 193, 208, 209,
			110/727	1.4685	1.7053	232, 243, 251,
						317, 329, 348
Cl₃GeC(Cl₂)CHCl₂	C₂HGeCl₇	—	123/25	1.5378	1.9243	186
Cl₃GeC(Cl)=CH₂	C₂H₂GeCl₄	—	151/753	1.5002	—	186, 206, 301
			152.5	1.4990	1.7396	
Cl₃GeCH=CHCl	C₂H₂GeCl₄	—	164–65/760	1.5139	1.7690	186
Cl₃GeCH(Cl)CH(Cl₂)	C₂H₂GeCl₆	—	92/9	1.5341	1.9000	186
Cl₃GeCH=CHGeCl₃	C₂H₂Ge₂Cl₆	71–72				193, 205, 215
Cl₃GeCH(Cl)CH(Cl)SiCl₃	C₂H₂GeCl₈Si	—	137/13	1.5345	1.8609	183
Cl₃GeCH=CH₂	C₂H₃GeCl₃	—	128.5/756	1.4815	1.6527	36, 39, 183, 189,
			127.5/745.5	1.4816	1.6520	193, 205, 232
Cl₃GeCH₂CF₂H	C₂H₃GeCl₃F₂	—	140/752	1.4490	1.7992	242

Table 5-1(f)—continued

COMPOUND	EMPIRICAL FORMULA	M.P. (°C)	B.P. (°C/mm)	n_D^{20}	d_4^{20}	REFERENCES
$Cl_3GeCH(Cl)CH_2Cl$	$C_2H_3GeCl_5$	—	88/12	1.5240	1.8390	186
$Cl_3GeCH_2CHCl_2$	$C_2H_3GeCl_5$	—	74/10	1.5176	1.8166	71
$Cl_3GeCH(BCl_2)CH_2BCl_2$	$C_2H_3GeCl_7B_2$	—	—	—	—	59
$Cl_3GeCH(Cl)CH_2GeCl_3$	$C_2H_3Ge_2Cl_7$		126/10	1.5461	1.9952	205
$Cl_3GeCH(Cl)CH_2SiCl_3$	$C_2H_3GeCl_7Si$		129/13	1.5268	1.8350	233
$Cl_3GeCH_2CH(Cl)SiCl_3$	$C_2H_3GeCl_7Si$		122/25	1.5200	1.7945	70
$Cl_3GeCH_2CH_2Br$	$C_2H_4GeCl_3Br$		73–74/8	1.5365	2.0211	195
$Cl_3GeCH(Cl)CH_3$	$C_2H_4GeCl_4$		167/746.5	1.4948	1.6975	126, 189, 232
$Cl_3GeCH_2CH_2Cl$	$C_2H_4GeCl_4$		188/746.5	1.5094	1.7587	126, 186, 189,
			75/15	1.5092	1.7637	232, 233
$Cl_3GeCH_2CH_2GeCl_3$	$C_2H_4Ge_2Cl_6$	56	130/12	—	—	193, 233
$Cl_3GeCH_2CH_2SiCl_3$	$C_2H_4GeCl_6Si$	35	120/15	—	—	233
$Cl_3GeC_2H_5$	$C_2H_5GeCl_3$	−33	140	1.4745	1.6091	9, 87, 155, 201,
			60/18	1.4719 (25°)	1.5953 (25°)	208, 226, 232,
			141/760	1.4750	1.6041	233, 240, 241,
						255, 307, 309
$Cl_3GeCH_2OCH_3$	$C_2H_5GeCl_3O$		61/15	1.4855	1.6154	216
$Cl_3GeCH(CH_3)CH_2GeCl_3$	$C_2H_6Ge_2Cl_6$		100/5	1.5400	1.8927	92
$Cl_3GeCCl{=}CHCF_3$ cis, trans	$C_3HGeCl_4F_3$		141–42	1.4430	1.7893	125
$Cl_3GeCH{=}CHCF_3$ cis, trans	$C_3H_2GeCl_3F_3$		124–26	1.4260	1.7188	125
$Cl_3GeCCl_2CH_2CF_3$	$C_3H_2GeCl_5F_3$		50/8	1.4600	1.8458	125
$Cl_3GeC{\equiv}CCH_3$	$C_3H_3GeCl_3$		54/20	1.4840	1.5953	29, 297
$Cl_3GeCHClCH_2CF_3$	$C_3H_3GeCl_4F_3$		46/8	1.4475	1.8047	125
$Cl_3GeCH_2CH_2CF_3$	$C_3H_4GeCl_3F_3$		143/758	1.4233	1.7105	125
			142	1.4240	1.7115	125, 242, 343

Formula	Empirical formula	m.p.	b.p./mm	n	d	References
$Cl_3GeCH_2CH_2CN$	$C_3H_4GeCl_3N$	37.7	135/22	1.5115	1.7514	233
$Cl_3GeCH_2CH_2COCl$	$C_3H_4GeCl_4O$	—	89–91/7	1.5455	1.8953	92, 182
$Cl_3GeC(CH_3)=CHGeCl_3$	$C_3H_4Ge_2Cl_6$	—	94–95/4.5	—	—	92
$Cl_3GeCH_2CH=CH_2$	$C_3H_5GeCl_3$	83–85	155.5/756; 153.8/743.5	1.4938; 1.4928	1.5480; 1.5274	182, 185, 187, 232, 242, 243
$Cl_3GeCH_2CH_2COOH$	$C_3H_5GeCl_3O_2$	—	70–71/6.5	1.4820	1.6760	182
$Cl_3GeCH_2COOCH_3$	$C_3H_5GeCl_3O_2$	—	68–70/8	1.4835	1.6765	19, 49, 158
$Cl_3GeCH(CH_2Cl)CH_2GeCl_3$	$C_3H_5Ge_2Cl_7$	—	140–43/7	1.5520	1.9394	185
$Cl_3GeCH_2CH_2CH(GeCl_3)_2$	$C_3H_5Ge_3Cl_9$	—	166.5–67/10	1.5660	2.0691	195
$Cl_3GeCH_2CHClCH_3$	$C_3H_6GeCl_4$	—	98/32	1.5007	1.6418	183, 185
$Cl_3GeCH(CH_3)CH_2Cl$	$C_3H_6GeCl_4$	—	68–70/6	1.5004	1.6442	183, 185
$Cl_3GeCH_2CH_2CH_2Cl$	$C_3H_6GeCl_4$	—	68–69/7; 78–80/8; 105/20	1.5036; 1.5065; 1.5070	1.6636; 1.6538	183, 184, 185, 232
$Cl_3GeCH_2CH_2CH_2GeCl_3$	$C_3H_6Ge_2Cl_6$	25.5	123/5	1.5314 (25°)	1.8705	71, 185
$Cl_3GeCH_2CH_2CH_2SiCl_3$	$C_3H_6GeCl_6Si$	—	107/4	1.4503	1.6538	185
$Cl_3Ge—C_3H_7\text{-}n$	$C_3H_7GeCl_3$	—	163.5/756; 167/763; 167/767	1.4779	1.5146	71, 119, 124, 242, 256
$Cl_3Ge—C_3H_7\text{-}i$	$C_3H_7GeCl_3$	—	48–50/20	1.4760	—	243, 256
$Cl_3GeCH_2CH_2(CH_3)SiCl_2$	$C_3H_7GeCl_5Si$	—	164.5/767	—	—	185
$Cl_3GeCH=CHCH=CH_2$	$C_4H_5GeCl_3$	—	121.5/19.5	1.5025	1.5883	195
$Cl_3GeCH_2CH=CHCH_2GeCl_3$	$C_4H_6Ge_2Cl_6$	56–58	53.5–54.5/7; 146–47/6.5	1.5250	1.5346	144
$Cl_3GeCH_2CH=CHCH_3$	$C_4H_7GeCl_3$	—	177/759; 83.5–84/32	1.5080; 1.5004	1.5127; 1.4953	71, 192, 195, 197
$Cl_3GeCH_2C(CH_3)=CH_2$	$C_4H_7GeCl_3$	—	66–67/17; 42/7; 172/760	1.5160; 1.4900	1.5594; 1.4787	185, 208, 232

Table 5-1(f)—continued

COMPOUND	EMPIRICAL FORMULA	M.P. (°C)	B.P. (°C/mm)	n_D^{20}	d_4^{20}	REFERENCES
$Cl_3GeCH_2CH_2COOCH_3$	$C_4H_7GeCl_3O_2$	—	127.5/12	1.4855	1.5923	71, 233
$Cl_3GeCH_2COOC_2H_5$	$C_4H_7GeCl_3O_2$	—	49–50/1.5	1.4750	1.5756	49, 158
$Cl_3GeCH(CH_3)OCOCH_3$	$C_4H_7GeCl_3O_2$	—	97/15	1.4868	1.5923	71
$Cl_3GeCH_2CH_2CHClCH_3$	$C_4H_8GeCl_4$	—	68/4	1.4990	1.5667	195
$Cl_3Ge(i\text{-}ClC_4H_8)$	$C_4H_8GeCl_4$	—	92/11	1.5000	1.5758	233
$Cl_3GeCH_2CH(CH_3)CH_2GeCl_3$	$C_4H_8Ge_2Cl_6$	—	156/20	1.5290	1.8010	185
$Cl_3Ge{-}C_4H_9\text{-}n$	$C_4H_9GeCl_3$	—	69/12; 184/760	1.4761; 1.4750	1.4496; 1.451	15, 71, 130, 149, 157, 243, 248, 268
$Cl_3Ge{-}C_4H_9\text{-}sec$	$C_4H_9GeCl_3$	—	184/752	—	1.4520	243
$Cl_3Ge{-}C_4H_9\text{-}i$	$C_4H_9GeCl_3$	95–98	92.5–94/68	—	—	92, 195, 243
$Cl_3GeC(CH_3)_3$	$C_4H_9GeCl_3$	66–67	—	—	—	
Cl_3Ge–(cyclopent-2-enyl)	$C_5H_7GeCl_3$	—	87/11; 63.5/5	1.5270; 1.5280	1.5327; 1.5621	71, 133, 144, 195, 199
$Cl_3Ge(CH_2)_3OCF_2CHFCl$	$C_5H_7GeCl_4O$	—	120–23/10	1.4523	1.6817	242
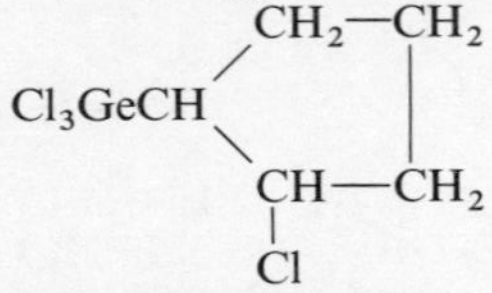 $Cl_3GeCH{<}$ (chlorocyclobutyl: $CH_2{-}CH_2$ / $CH{-}CH_2$, Cl)	$C_5H_8GeCl_4$	—	111–12/6.5	1.5295	1.6445	195
$Cl_3GeCH_2CH{=}C(CH_3)_2$	$C_5H_9GeCl_3$	—	—	—	—	173

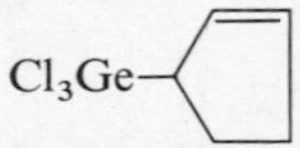

Molecular formula	Compound	M.p. (°C)	B.p. (°C/mm)	n_D	d	References
$C_5H_9GeCl_3$	$Cl_3GeCH_2CH{=}CHCH_2CH_3$	—	73.5/13	1.5035	1.4306	192
$C_5H_9GeCl_3$	$Cl_3GeCH(CH_3)CH{=}CHCH_3$	—	83–84.5/12	1.5070	1.5210	199
$C_5H_9GeCl_3$	$Cl_3GeC_5H_9$	29.8	107/9	1.4987	1.5752	198
$C_5H_9GeCl_3O_2$	$Cl_3GeC(CH_3)_2OCOCH_3$	—	127.5/12	1.4855	2.5392	233
$C_5H_9GeCl_3O_2$	$Cl_3GeCH_2CH_2CH_2OCOCH_3$	—	79–82/0.4	—	—	245, 246
$C_5H_{10}GeCl_4$	$Cl_3Ge(CH_2)_3CH(Cl)CH_3$	—	101–3/40	—	—	245, 246
$C_5H_{11}GeCl_3$	$Cl_3Ge{-}C_5H_{11}\text{-}n$	—	84–85/11	1.4933	1.3490	206
$C_5H_{11}GeCl_3Si$	$Cl_3GeCH{=}CHSi(CH_3)_3$	—	100–3/12	—	—	206
$C_5H_{11}Ge_2Cl_3$	$Cl_3GeCH{=}CHGe(CH_3)_3$	—	138–39/7	1.5570	1.8441	205, 206
$C_5H_{12}Ge_3Cl_6$	$(Cl_3Ge)_2CHCH_2Ge(CH_3)_3$	—	106–7/17	1.5342 (25°)	—	287
$C_6H_4GeFCl_3$	$Cl_3GeC_6H_4F\text{-}p$	—	116–7/10	1.5678 (25°)	1.6467	287, 298, 300
$C_6H_4GeCl_4$	$Cl_3GeC_6H_4Cl\text{-}p$	—	105–7/5	1.5738	—	—
$C_6H_5GeCl_3$	$Cl_3GeC_6H_5$	—	110–11/20; 115/19; 118–19/24	— ; 1.5532 (25°); 1.5540	1.5972	22, 23, 84, 140, 178, 190, 191, 225, 243, 248, 263, 280, 282, 287, 300, 307
$C_6H_{11}GeCl_3$	$Cl_3GeC_6H_{11}$	—	79/3	1.5130	1.4792	149, 185, 246, 268
$C_6H_{11}GeCl_3F_2$	$Cl_3GeC_6H_{11}F_2$	50–50.5	75–76/2	—	—	245, 246
$C_6H_{11}GeCl_3O$	$Cl_3GeC(CH_3)_2CH_2COCH_3$	46	109–11/2	—	—	198, 269, 271
$C_6H_{12}GeCl_4$	$Cl_3GeC_6H_{12}Cl$	—	89–91/0.2	—	—	245, 246
$C_6H_{13}GeCl_3$	$Cl_3GeC_6H_{13}\text{-}n$	—	97/14	1.4719 (25°)	—	83, 149, 245, 246, 263, 268
$C_7H_7GeCl_3$	$Cl_3GeC_6H_4CH_3\text{-}p$	—	129–31/24	1.5502 (25°)	—	22, 23, 287
$C_7H_7GeCl_3$	$Cl_3GeCH_2C_6H_5$	35–37; 33–36	115–16/12; 111–12/18	—	—	185, 216, 243
$C_7H_7GeCl_3O$	$Cl_3GeC_6H_4OCH_3\text{-}p$	—	98/10; 104–5/5	1.5610	—	287

Table 5-1(f)—continued

COMPOUND	EMPIRICAL FORMULA	M.P. (°C)	B.P. (°C/mm)	n_D^{20}	d_4^{20}	REFERENCES
$Cl_3GeC_7H_{15}$-n	$C_7H_{15}GeCl_3$	—	101–3/10 100/7.5	1.4742	1.3071	149, 195, 245, 246, 268
$Cl_3GeCH_2CH_2CHClCH_3$	$C_7H_{17}GeCl_4$	—	66–67/19.5	1.4505	1.1013	195
$Cl_3GeCH=CHC_6H_5$	$C_8H_7GeCl_3$	—	146–48/16	1.5833	1.5328	71
$Cl_3GeC_6H_4CH=CH_2$	$C_8H_7GeCl_3$	—	79/0.53	1.5855	1.5154	51
$Cl_3GeCH_2CH_2C_6H_5$	$C_8H_9GeCl_3$	—	132/10	1.5550	1.4816	71
Cl_3Ge 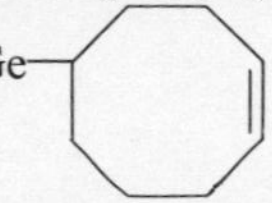 ⬡	$C_8H_{13}GeCl_3$	—	100–1/1.5	1.5310	1.4479	92
$Cl_3GeCH_2CH=CHCH_2CH_2CH=CHCH_2Ge(H)Cl_2$	$C_8H_{13}GeCl_5$	—	138–40/0.4	1.5488	—	217
$Cl_3GeC_8H_{17}$-n	$C_8H_{17}GeCl_3$	—	130/13 150/15	1.4720 —	1.2712	149, 233, 245, 246, 268
$Cl_3GeC_8H_{17}$-i	$C_8H_{17}GeCl_3$	—	78–82/0.6	—	—	245
$Cl_3GeC_8H_{17}$ (mixt)	$C_8H_{17}GeCl_3$	—	112–12.5/10	1.4838	—	218
$Cl_3GeCH(C_6H_5)CH_2$—CH_3	$C_9H_{11}GeCl_3$	—	125–26/8	1.5549	—	218
$Cl_3GeC_{10}H_{21}$-n	$C_{10}H_{21}GeCl_3$	—	87–90/0.1	—	—	245, 246
$Cl_3GeC(C_6H_5)_3$	$C_{19}H_{15}GeCl_3$	215–18	—	—	—	195

(g) *Organogermanium monobromides* $\overset{|}{\underset{|}{-\text{Ge}-}}\text{Br}$

COMPOUND	EMPIRICAL FORMULA	M.P. (°C)	B.P. (°C/mm)	n_D^{20}	d_4^{20}	REFERENCES
BrGe(CH$_3$)$_3$	C$_3$H$_9$GeBr	−25	— 113/735 113.7/760 115	1.4750 1.4672 1.4713 1.4660 1.4705 (18°)	— 1.5604 — 1.5486 1.544 (18°)	1, 58, 67, 84, 104, 204, 208, 240, 251, 252, 263, 299, 343
BrGe—CH=CHBr with CH$_3$ / CH$_3$	C$_4$H$_8$GeBr$_2$	123–24	—	—	—	316
BrGe(C$_2$H$_5$)(CH$_3$)$_2$	C$_4$H$_{11}$GeBr	—	143/775	1.4726	1.4952	241
BrGe(CH=CH$_2$)$_3$	C$_6$H$_9$GeBr	—	58/10	1.5057 (25°)	—	284
BrGe(C$_2$H$_5$)$_3$	C$_6$H$_{15}$GeBr	−33	190.9/760 84/24 68/10	1.489 1.4840 1.4836	1.1412 — 1.4009	14, 17, 43, 73, 104, 131, 137, 185, 187, 188, 244, 267, 279, 321, 322, 343
BrGe(C$_2$H$_5$)(C$_2$H$_5$)—CH$_2$CH$_2$CH$_2$Br	C$_7$H$_{16}$GeBr$_2$	—	111/1.5	1.5250	1.6640	169, 171

Table 5-1(g)—continued

COMPOUND	EMPIRICAL FORMULA	M.P. (°C)	B.P. (°C/mm)	n_D^{20}	d_4^{20}	REFERENCES
C₂H₅ | BrGe—CH₂CH=CH—CH₃ | C₂H₅	$C_8H_{17}GeBr$	—	112/22	1.5033	1.3525	266, 267
C₂H₅ | BrGe—CH₂CH₂CH=CH₂ | C₂H₅	$C_8H_{17}GeBr$	—	114/21	1.4951	1.3384	175
BrGe(C₄H₉)(C₂H₅)₂	$C_8H_{19}GeBr$	—	100/12	1.4828	1.3075	165
C₂H₅ | BrGe—CH₂CH(CH₃)CH=CH₂ | C₂H₅	$C_9H_{19}GeBr$	—	84/2 82.5/1.8 122/20	— 1.4936	— 1.3062	175
BrGe(C₃H₇)₃	$C_9H_{21}GeBr$	−47	242/760	1.4818 (25°)	—	5, 239
BrGe(C₃H₇-*i*)₃	$C_9H_{21}GeBr$	−45	112–13/12 234	1.4832 1.4852	1.282 1.231	10
BrGe(C₂H₅)₂ | C₆H₅	$C_{10}H_{15}GeBr$	—	141/18	1.5535	1.4354	68

Structure	Formula	mp	bp/mm	n_D	d	Ref.
$BrGe{-}CH_2CH(CH_3){-}C(CH_3){=}CH_2$ with C_2H_5, C_2H_5	$C_{10}H_{21}GeBr$	—	130/23	1.4963	1.2841	175
$BrGe{-}(CH_2)_2{-}GeBr$ with C_2H_5, C_2H_5, C_2H_5, C_2H_5	$C_{10}H_{24}Ge_2Br_2$	—	121/0.2	1.5315	1.5978	68
$BrGe(C_6H_5)(C_2H_5)(i{-}C_3H_7)$	$C_{11}H_{17}GeBr$	—	130–35/13 ; 74.5–76/0.15	— ; 1.5473 (25°)	—	74, 280
$BrGe(C_3H_7)(C_4H_9)_2$	$C_{11}H_{25}GeBr$	—	136–37/15	1.4816	1.2243	169, 171
$BrGe{-}CH_2CH_2CH_2{-}GeBr$ with C_2H_5, C_2H_5, C_2H_5, C_2H_5	$C_{11}H_{26}GeBr_2$	—	157/1.5	1.5252	1.5501	171a
$BrGeCH_2CH(CH_3)CH_2Br$ with C_2H_5, C_4H_9	$C_{12}H_{26}GeBr_2$	—	140–41/1	1.5108	1.4217	169
$BrGe(C_4H_9{-}n)_3$	$C_{12}H_{27}GeBr$	—	279/760 ; 143–44/10	— ; 1.4809	— ; 1.195	7, 64a, 132, 159
$BrGe(C_4H_9{-}i)_3$	$C_{12}H_{27}GeBr$	—	138–42/12	1.4808	1.194	248
$BrGe{-}CH_2CH_2CH_2CH_2{-}GeBr$ with C_2H_5, C_2H_5, C_2H_5, C_2H_5	$C_{12}H_{28}Ge_2Br_2$	—	158/0.7	1.5224	—	171a
$BrGe(C_2H_5)(C_6H_5)_2$	$C_{14}H_{15}GeBr$	67–70	127–29/0.2 (25°)	1.6025 (25°)	—	74
$BrGe(C_3H_7{-}i)(C_6H_5)_2$	$C_{15}H_{17}GeBr$	—	225–35/13	—	—	280

Table 5-1(g)—continued

COMPOUND	EMPIRICAL FORMULA	M.P. (°C)	B.P. (°C/mm)	n_D^{20}	d_4^{20}	REFERENCES
BrGe(C₂H₅)(C₆H₅)(C₆H₄CH₃-p)	$C_{15}H_{17}GeBr$	—	—	—	—	280
BrGe—(CH₂)₇CH₃	$C_{16}H_{35}GeBr$	—	123.5/0.3	1.4789	1.1258	151
BrGe(C₆H₅)(CH₃)(C₁₀H₇-α)	$C_{17}H_{15}GeBr$	58–60	—	—	—	40
BrGe(C₆F₅)₃	$C_{18}GeBrF_{15}$	105–7	—	—	—	81
BrGe(C₆H₅)₃	$C_{18}H_{15}GeBr$	138; 138.7	—	—	—	42, 43, 62, 65, 74, 90, 99, 118, 122, 132, 135, 136, 138, 213, 225, 245, 261, 302, 303, 304
BrGe(C₂H₅)(C₆H₅)(C₁₀H₇-1)	$C_{18}H_{17}GeBr$	76.5–78	156–58/0.02	1.6560 (25°)	—	74
BrGe(C₆H₁₁)₃	$C_{18}H_{33}GeBr$	109–10; 108	—	—	—	21, 121
BrGe(C₆H₁₃)₃	$C_{18}H_{39}GeBr$	—	143–45/0.5	1.4763 (27°)	1.117 (26°)	91
BrGe(C₆H₅)(C₆H₄CH₃-p)₂	$C_{20}H_{19}GeBr$	119	—	—	—	280
BrGe(C₆H₄CH₃-o)₃	$C_{21}H_{21}GeBr$	119–20	195/0.01; 205–10/1	—	—	106, 302, 303, 304
BrGe(C₆H₄CH₃-m)₃	$C_{21}H_{21}GeBr$	78–79	222–23/1	—	—	302, 303, 304
BrGe(C₆H₄CH₃-p)₃	$C_{21}H_{21}GeBr$	128–29; 121	—	—	—	106, 280, 302, 304

Compound	Formula				References
BrGe(C$_6$H$_5$CH$_2$)$_3$	C$_{21}$H$_{21}$GeBr	146–47 145	235–45/1	— —	24, 303
$\begin{array}{c} \text{C}_6\text{H}_5 \quad \text{C}_6\text{H}_5 \\ \mid \qquad \mid \\ \text{BrGe}\text{---}\text{GeBr} \\ \mid \qquad \mid \\ \text{C}_6\text{H}_5 \quad \text{C}_6\text{H}_5 \end{array}$	C$_{24}$H$_{20}$Ge$_2$Br	167–69	—	— —	178, 280
BrGe(C$_2$H$_5$)(C$_6$H$_5$C$_6$H$_4$)$_2$	C$_{26}$H$_{23}$GeBr	242–43.5	—	— —	280
BrGe(C$_{10}$H$_7$-α)$_3$	C$_{30}$H$_{21}$GeBr	—	—	— —	324
BrGe(C$_6$H$_5$—C$_6$H$_4$-p)$_3$	C$_{36}$H$_{27}$GeBr	242	—	— —	280

(h) Organogermanium dibromides

$$\text{Br}-\text{Ge}-\text{Br}$$

COMPOUND	EMPIRICAL FORMULA	M.P. (°C)	B.P. (°C/mm)	n_D^{20}	d_4^{20}	REFERENCES
Br$_2$Ge(CH$_3$)$_2$	C$_2$H$_6$GeBr$_2$	—	153/746 152.5/750 152.3/755	1.5268 1.5280 1.5265	2.1163 2.1145 2.1182	204, 208, 209, 240, 251, 299
Br$_2$Ge(C$_2$H$_5$)$_2$	C$_4$H$_{10}$GeBr$_2$	−33	202/760 82/12	1.5272 1.5359	1.8811 1.9854	85, 162, 164, 188, 343
Br$_2$Ge(CH$_2$COOCH$_3$)$_2$	C$_6$H$_{10}$GeBr$_2$O$_4$	—	114–16/2	1.5248	1.9220	48, 158
Br$_2$Ge(C$_3$H$_7$-n)$_2$	C$_6$H$_{14}$GeBr$_2$	−52	240.5	1.5173	1.689	8
Br$_2$Ge(C$_3$H$_7$-i)$_2$	C$_6$H$_{14}$GeBr$_2$	−22	234	1.519	1.670	10
Br$_2$Ge(CH$_3$)(C$_6$H$_5$)	C$_7$H$_8$GeBr$_2$	—	139–40/15	1.5962	—	40
Br$_2$Ge(C$_3$H$_7$)(C$_4$H$_9$)	C$_7$H$_{16}$GeBr$_2$	—	115/10	1.5125	1.6163	68
Br$_2$Ge[thienyl]$_2$	C$_8$H$_6$GeBr$_2$S$_2$	—	212–14/20	—	—	191
Br$_2$Ge(C$_2$H$_5$)(C$_6$H$_5$)	C$_8$H$_{10}$GeBr$_2$	—	70–80/0.02	1.5897 (25°)	—	74
Br$_2$Ge(C$_4$H$_9$-n)$_2$	C$_8$H$_{18}$GeBr$_2$	—	269 140/12 127–28/7	1.5109 1.5110	1.565	16, 150, 164, 165, 224, 263
Br$_2$Ge(C$_6$F$_5$)$_2$	C$_{12}$GeBr$_2$F$_{10}$	—	180–85/12	—	—	81
Br$_2$Ge(C$_6$H$_5$)$_2$	C$_{12}$H$_{10}$GeBr$_2$	298	208–10/24 120/0.007 140–45/0.1 —	1.6394 — — 1.6380	1.7779	117, 135, 147, 178, 179, 190, 191, 213, 261, 280

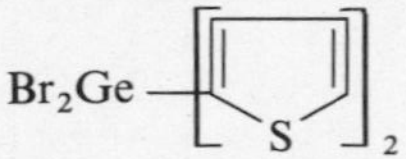

Br₂Ge——GeBr₂ | | C₆H₅ C₆H₅	$C_{12}H_{10}Ge_2Br_4$	113–17 115–19	—	—	—	178
Br₂Ge(C₆H₅)(n-C₆H₁₃)	$C_{12}H_{18}GeBr_2$	—	105/0.02	1.5600	—	249
Br₂Ge(C₆H₄CH₃-m)₂	$C_{14}H_{14}GeBr_2$	74.8–75.8	189–90/4	1.5665	—	303
Br₂Ge(C₆H₄CH₃-p)₂	$C_{14}H_{14}GeBr_2$	—	230–33/13	—	—	280

(i) Organogermanium tribromides —Ge—Br with Br and Br

COMPOUND	EMPIRICAL FORMULA	M.P. (°C)	B.P. (°C/mm)	n_D^{20}	d_4^{20}	REFERENCES
Br$_3$GeCH$_2$Br	CH$_2$GeBr$_4$	—	78.5–79/4	1.6327	2.970	200
Br$_3$GeCH$_3$	CH$_3$GeBr$_3$	—	168/750; 170	1.5770	2.6337	208, 209, 240, 251, 299
Br$_3$GeCH=CHGeBr$_3$	C$_2$H$_2$Ge$_2$Br$_6$	121.5–23	—	—	—	93, 196
Br$_3$GeCH=CH$_2$	C$_2$H$_3$GeBr$_3$	—	55/7	1.5883	2.5426	188
Br$_3$GeCH$_2$CH$_2$GeBr$_3$	C$_2$H$_4$Ge$_2$Br$_6$	122–24	—	—	—	93, 196
Br$_3$GeC$_2$H$_5$	C$_2$H$_5$GeBr$_3$	—	200/763	—	—	87
Br$_3$GeCH$_2$—CH=CH$_2$	C$_3$H$_5$GeBr$_3$	—	56/2; 72/7	1.5874; 1.5882	2.3274; 2.3410	77, 93, 185, 188, 196
Br$_3$GeCH$_2$COOCH$_3$	C$_3$H$_5$GeBr$_3$O$_2$	—	81–82/2	1.5613	2.3534	48, 158
Br$_3$GeCH$_2$CH$_2$CH$_2$Cl	C$_3$H$_6$GeBr$_3$Cl	—	109–11/6	1.5820	2.3383	93
Br$_3$GeC$_3$H$_{7}$-n	C$_3$H$_7$GeBr$_3$	—	50–55/5	1.5607	2.3354	119, 335
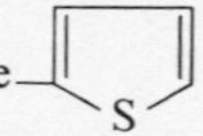 Br$_3$Ge—⟨S⟩	C$_4$H$_3$GeBr$_3$S	—	152–53/23	1.6495	2.4280	191
Br$_3$GeCH$_2$C(Br)(CH$_3$)$_2$	C$_4$H$_8$GeBr$_4$	—	123–28/8	1.5920	2.3822	188
Br$_3$GeC$_4$H$_{9}$-n	C$_4$H$_9$GeBr$_3$	—	237/760	1.5548	2.1480	15, 92, 188, 335
Br$_3$GeC(CH$_3$)$_3$	C$_4$H$_9$GeBr$_3$	133–34.5	90/9	1.5606	2.132	92

Compound		mp	bp	n_D	d	References
Br$_3$Ge–(cyclopentenyl) 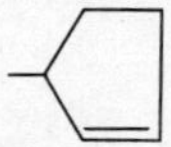	C$_5$H$_7$GeBr$_3$	—	104–5/4.5	1.6132	2.2666	92
Br$_3$GeC$_6$H$_4$I-*p*	C$_6$H$_4$GeBr$_3$I	73–74	—	—	—	191
Br$_3$GeC$_6$H$_4$F-*m*	C$_6$H$_4$GeBr$_3$F	—	151–52/23	1.6120	2.3250	191
Br$_3$Ge–C$_6$H$_4$–GeBr$_3$	C$_6$H$_4$Ge$_2$Br$_6$	187–88	—	—	—	191
Br$_3$GeC$_6$H$_5$	C$_6$H$_5$GeBr$_3$	—	120–22/13 160–61/24	1.6330	2.2641	22, 23, 43, 135, 178, 190, 191, 213, 280
Br$_3$Ge–(cyclohexyl) 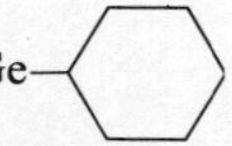	C$_6$H$_{11}$GeBr$_3$	—	113/5.5	1.5855	2.1094	93, 196
Br$_3$GeC$_6$H$_{13}$	C$_6$H$_{13}$GeBr$_3$	—	150/10	—	—	263
Br$_3$GeC$_6$H$_4$CH$_3$-*m*	C$_7$H$_7$GeBr$_3$	—	186–87/37	1.6215	2.1440	191
Br$_3$GeC$_6$H$_4$CH$_3$-*p*	C$_7$H$_7$GeBr$_3$	—	155–56/13	—	—	23
Br$_3$GeC$_{10}$H$_7$-1	C$_{10}$H$_7$GeBr$_3$	52.5–53.5	193–94/3.5	1.6988	2.1871	191

(j) Organogermanium monoiodides —Ge—I

COMPOUND	EMPIRICAL FORMULA	M.P. (°C)	B.P. (°C/mm)	n_D^{20}	d_4^{20}	REFERENCES
IGe(CH₃)₃	C₃H₉GeI	—	57–57.5/55	1.5156	2.6405 (23°)	204, 214, 289, 325, 333, 336, 343
			133.5/740	1.5190	1.7962	
			136	1.5159 (24°)	—	
IGe(CH=CH₂)₃	C₆H₉GeI	—	71–74/12	—	—	284
IGe(C₂H₅)₃	C₆H₁₅GeI	—	212.3/760	1.528	1.608	14, 17, 137, 146, 163
			89–91/13	1.5262	1.6185	163
C₂H₅ IGeCH₂CH₂CH=CH₂ C₂H₅	C₈H₁₇GeI	—	116/19	1.5288	1.5130	175
C₂H₅ IGeCH₂CH=CHCH₃ C₂H₅	C₈H₁₇GeI	—	126/22	1.5405	1.5501	267
C₂H₅ IGeCH₂CH(CH₃)CH=CH₂ C₂H₅	C₉H₁₉GeI	—	114/17	1.5202	1.4278	175

Structure	Molecular formula	m.p.	b.p./mm	n_D	d	References
$IGe(C_3H_7\text{-}n)_3$	$C_9H_{21}GeI$	−38	259/760; 122–23/10	1.5144; —	1.443; —	5
$IGe(C_3H_7\text{-}i)_3$	$C_9H_{21}GeI$	4	254; 137–38/13	1.524; 1.5098	1.446; 1.3543	10, 146
$\begin{array}{c}C_2H_5\\ \mid\\ IGeCH_2CH(CH_3)C(CH_3){=}CH_2\\ \mid\\ C_2H_5\end{array}$	$C_{10}H_{21}GeI$	—	150/17	1.5264	1.4456	175
$\begin{array}{c}C_4H_9\\ \mid\\ IGe(CH_2)_3I\\ \mid\\ C_4H_9\end{array}$	$C_{11}H_{24}GeI_2$	—	137/0.4	1.5603	1.7127	169
$IGe(C_3H_7)(C_4H_9)_2$	$C_{11}H_{25}GeI$	—	144/14	1.5082	1.3609	171
$IGe(C_4H_9\text{-}n)_3$	$C_{12}H_{27}GeI$	—	297/760; 163–64/15; 125–27/4	—; 1.508; 1.5098	—; 1.340; 1.3543	7, 102, 263
$IGe(C_4H_9\text{-}i)_3$	$C_{12}H_{27}GeI$	—	137–38/17	1.5023	1.2686	146, 163
$IGe(C_5H_{11}\text{-}n)_3$	$C_{15}H_{33}GeI$	—	184–85/11	1.5002	1.254	146, 150, 163
$IGe(C_5H_{11}\text{-}i)_3$	$C_{15}H_{33}GeI$	—	177–78/13	1.5002	1.2508	150, 263
$\begin{array}{c}C_4H_9\\ \mid\\ IGe(CH_2)_7CH_3\\ \mid\\ C_4H_9\end{array}$	$C_{16}H_{35}GeI$	—	143.5/0.5	—	—	151
$\begin{array}{c}C_4H_9\text{-}n\\ \mid\\ IGe\!-\!Ge(n\text{-}C_4H_9)_2I\\ \mid\\ C_4H_9\text{-}n\end{array}$	$C_{16}H_{36}Ge_2I_2$	—	—	—	—	116

Table 5-1(j)—continued

COMPOUND	EMPIRICAL FORMULA	M.P. (°C)	B.P. (°C/mm)	n_D^{20}	d_4^{20}	REFERENCES
IGe(C$_6$H$_4$Br-p)$_3$	C$_{18}$H$_{12}$GeIBr$_3$	170–71	—	—	—	78, 79
IGe(C$_6$H$_4$Cl-p)$_3$	C$_{18}$H$_{12}$GeICl$_3$	133–34.5	—	—	—	78, 79
IGe(C$_6$H$_5$)$_3$	C$_{18}$H$_{15}$GeI	156–57	—	—	—	43, 62, 78, 79, 117, 138
IGe(n-C$_6$H$_{13}$)(C$_6$H$_5$)$_2$	C$_{18}$H$_{23}$GeI	—	138/0.05	1.5998	1.4574	249
IGe(C$_6$H$_{11}$)$_3$	C$_{18}$H$_{33}$GeI	99–100	—	—	—	21, 121
IGe(n-C$_6$H$_{13}$)$_3$	C$_{18}$H$_{39}$GeI	—	154–55/0.5	1.4935 (26°)	1.188 (26°)	91
IGe(C$_6$H$_4$CH$_3$-m)$_3$	C$_{21}$H$_{21}$GeI	76–77.5	—	—	—	78, 79
IGe(C$_6$H$_5$CH$_2$)$_3$	C$_{21}$H$_{21}$GeI	141	—	—	—	24
IGe(C$_6$H$_4$OCH$_3$-p)$_3$	C$_{21}$H$_{21}$GeIO$_3$	114–15	—	—	—	78, 79
IGe(C$_6$H$_4$OC$_2$H$_5$-p)$_3$	C$_{24}$H$_{27}$GeIO$_3$	95.5–97	—	—	—	78, 79
IGe(C$_{10}$H$_7$-2)$_3$	C$_{30}$H$_{21}$GeI	174–76	—	—	—	78, 79

(k) *Organogermanium diiodides* I—Ge—I

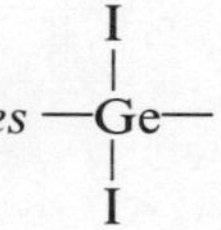

COMPOUND	EMPIRICAL FORMULA	M.P. (°C)	B.P. (°C/mm)	n_D^{20}	d_4^{20}	REFERENCES
$I_2Ge(CF_3)_2$	$C_2GeI_2F_6$	—	—	—	—	55, 57
$I_2Ge(CH_3)_2$	$C_2H_6GeI_2$	—	62–64/2; 194–99/759; 204.0	1.6385; 1.6381	2.5687; 2.8648 (23°)	209, 333, 336, 343
$I_2Ge(C_2H_5)_2$	$C_4H_{10}GeI_2$	–1–2	116–17/14; 252/759	1.6205	2.2771	113, 163
$I_2Ge(C_3H_7-n)_2$	$C_6H_{14}GeI_2$	–53.5	276.5	—	2.024	8
$I_2Ge(C_3H_7-i)_2$	$C_6H_{14}GeI_2$	–9	268	1.597	2.008	10
$I_2Ge(C_4H_9-n)_2$	$C_8H_{18}GeI_2$	—	304/760; 163/10	1.5770	1.863	16, 150, 263
$I_2Ge(C_5H_{11}-n)_2$	$C_{10}H_{22}GeI_2$	—	123/0.5	1.585	1.910	263
$I_2Ge(C_6H_4Br-p)_2$	$C_{12}H_8GeI_2Br_2$	170–71	—	—	—	78
$I_2Ge(C_6H_4Br-o)_2$	$C_{12}H_8GeI_2Br_2$	155.5–57.5	—	—	—	78, 79
$I_2Ge(C_6H_4Cl-o)_2$	$C_{12}H_8GeI_2Cl_2$	124–5.5	—	—	—	78, 79
$I_2Ge(C_6H_4Cl-m)_2$	$C_{12}H_8GeI_2Cl_2$	61–63	—	—	—	78, 79
$I_2Ge(C_6H_4Cl-p)_2$	$C_{12}H_8GeI_2Cl_2$	67.5–69	—	—	—	78, 79
$I_2Ge(C_6H_5)_2$	$C_{12}H_{10}GeI_2$	68–69; 70–71	—	—	—	78, 79, 117, 118
$I_2Ge(C_6H_5)(n-C_6H_{13})$	$C_{12}H_{18}GeI_2$	—	128/0.006	1.6260	1.8704	249
$I_2Ge(C_6H_4CH_3-o)_2$	$C_{14}H_{14}GeI_2$	85–86.5	—	—	—	78, 79
$I_2Ge(C_6H_4CH_3-p)_2$	$C_{14}H_{14}GeI_2$	85–86	—	—	—	78
$I_2Ge(C_6H_4OC_2H_5-o)_2$	$C_{16}H_{18}GeI_2O_2$	137–38	—	—	—	78, 79

(l) Organogermanium triiodides $\mathrm{I_3Ge{-}I}$

COMPOUND	EMPIRICAL FORMULA	M.P. (°C)	B.P. (°C/mm)	n_D^{20}	d_4^{20}	REFERENCES
I₃GeCF₃	CGeI₃F₃	8.4	40–42/0.001	1.6571	—	55, 56, 57
I₃GeCH₂I	CH₂GeI₄	—	147/0.6	>1.8	3.729	148
I₃GeCH₂GeI₃	CH₂Ge₂I₆	83	—	—	—	167
I₃GeCH₃	CH₃GeI₃	48	83–85/12	—	—	87, 88, 209, 333, 350
		47	237/752			
		44.0				
I₃GeC₂H₅	C₂H₅GeI₃	−1.5−−2.5	281/755	1.7486	2.021	87, 163, 334, 336
			142–3/14	1.70	2.9879	
			280			
I₃GeCH₂OCH₃	C₂H₅GeI₃O	—	115/1	1.750	3.027	148
I₃GeCH=CH₂	C₃H₅GeI₃	—	154/14	>1.7	2.908	167
I₃GeCH₂CH₂COOH	C₃H₅GeI₃O₂	112	—	—	—	174
I₃GeCH₂COOCH₃	C₃H₅GeI₃O₂	—	116–17/1	—	2.90	158
I₃GeCH₂CH₂I	C₃H₆GeI₄	—	178/1 (dec.)	>1.7	3.133	167
I₃GeC₃H₇-n	C₃H₇GeI₃	—	86/0.5	>1.7	2.8639	335, 336
I₃GeCH₂CH₂COCH₃	C₄H₇GeI₃O	—	136–37/0.01	>1.7	2.838	174
I₃GeCH(CH₃)CH₂CHO	C₄H₇GeI₃O	—	—	>1.7	2.807	174
I₃GeCH₂COOC₂H₅	C₄H₇GeI₃O₂	—	133/0.6	1.6945	2.761	148
I₃GeCH₂CH₂COOCH₃	C₄H₇GeI₃O₂	—	—	>1.7	2.716	174
I₃GeCH(CH₃)CH₂COOH	C₄H₇GeI₃O₂	78	—	—	—	174

Structure	Formula	mp (°C)	bp (°C/mm)	n_D	d	Ref.
I₃GeCH₂CH(CH₃)COOH	C₄H₇GeI₃O₂	63–64	—	—	—	174
I₃GeC₄H₉-n	C₄H₉GeI₃	—	310 (dec.)	—	—	15, 148, 335, 336
			119–21/1	>1.70	2.647	
			111/0.35		2.677	
I₃Ge(t-C₄H₉)₃	C₄H₉GeI₃	125	—	—	—	167
I₃Ge-(cyclopentyl)	C₅H₉GeI₃	—	140/0.8	>1.7	2.700	167
I₃GeCH₂CH(CH₃)COOCH₃	C₅H₉GeI₃O₂	55–56	—	1.678	2.520	174
I₃GeC₆H₅	C₆H₅GeI₃	—	—	—	—	23, 148
I₃GeCH(CH=CHCH₃)CH₂COOH	C₆H₉GeI₃O₂	—	—	—	—	174
I₃GeC(CH₃)₂CH₂COCH₃	C₆H₁₁GeI₃O	65	—	—	—	174
I₃GeC₆H₄CH₃-p	C₇H₇GeI₃	72	—	—	—	23
I₃GeCH₂Ge(C₂H₅)₃	C₇H₁₇Ge₂I₃	—	174/1.6	1.6616	2.3338	167

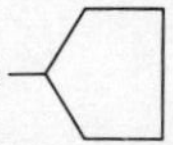

REFERENCES

1. Abel, E. N., D. A. Armitage, and D. B. Brady, *J. Org. Chem.*, **5**, 130 (1966).
2. Amberger, E., and H. Boeters, *Angew. Chem.*, **73**, 114 (1961).
3. Amberger, E., and R. W. Salazar, *Int. Symp. Organosilicon Chem. Sci. Commun. Suppl. Prague 1965*, 31–33.
4. Amberger, E., and R. W. Salazar, *J. Organometal. Chem.*, **8** (1), 111 (1967).
5. Anderson, H. H., *J. Am. Chem. Soc.*, **73**, 5440 (1951).
6. Anderson, H. H., *J. Am. Chem. Soc.*, **73**, 5798 (1951).
7. Anderson, H. H., *J. Am. Chem. Soc.*, **73**, 5800 (1951).
8. Anderson, H. H., *J. Am. Chem. Soc.*, **74**, 2370 (1952).
9. Anderson, H. H., *J. Am. Chem. Soc.*, **74**, 2371 (1952).
10. Anderson, H. H., *J. Am. Chem. Soc.*, **75**, 814 (1953).
11. Anderson, H. H., *J. Org. Chem.*, **20**, 536 (1955).
12. Anderson, H. H., *J. Org. Chem.*, **20**, 900 (1955).
13. Anderson, H. H., *J. Am. Chem. Soc.*, **78**, 1692 (1956).
14. Anderson, H. H., *J. Am. Chem. Soc.*, **79**, 326 (1957).
15. Anderson, H. H., *J. Am. Chem. Soc.*, **82**, 3016 (1960).
16. Anderson, H. H., *J. Am. Chem. Soc.*, **83**, 547 (1961).
17. Anderson, H. H., *Inorg. Chem.*, **3**, 910 (1964).
18. Aronson, J. R., and J. R. Durig, *Spectrochim. Acta*, **20**, 219 (1964).
19. Avdeeva, V. I., G. S. Burlachenko, Yu. I. Baukov, and I. F. Lutsenko, *Zh. Obshch. Khim.*, **36** (9), 1679 (1966).
20. Babouchkina, T. A., and G. K. Semin, *Zh. Strukt. Khim.*, **7**, 114 (1966).
21. Bauer, H., and K. Burschkies, *Chem. Ber.*, **65**, 956 (1932).
22. Bauer, H., and K. Burschkies, *Chem. Ber.*, **66**, 277 (1933).
23. Bauer, H., and K. Burschkies, *Chem. Ber.*, **66**, 1156 (1933).
24. Bauer, H., and K. Burschkies, *Chem. Ber.*, **67**, 1041 (1934).
25. Baukov, Yu. I., G. S. Burlachenko, I. Yu. Belavin, and I. F. Lutsenko, *Zh. Obshch. Khim.*, **36** (1) 153 (1966).
26. Baukov, Yu. I., I. Yu. Belavin, and I. F. Lutsenko, *Zh. Obshch. Khim.*, **35**, 1092 (1965).
27. Baukov, Yu. I., G. S. Burlachenko, and I. F. Lutsenko, *Zh. Obshch. Khim.*, **35** (7), 1173 (1965).
28. Baukov, Yu. I., G. S. Burlachenko, and I. F. Lutsenko, *Dokl. Akad. Nauk SSSR*, **157** (1), 119 (1964).
29. Behagel, O., and H. Seibert, *Chem. Ber.*, **66**, 922 (1933).
30. Benkeser, R. A., *J. Am. Chem. Soc.*, **78**, 682 (1956).
31. Benkeser, R. A., C. E. de Boer, R. E. Robinson, and D. M. Sauve, *J. Am. Chem. Soc.*, **78**, 682 (1956).
32. Birr, K. H., and D. Kräft, *Z. Anorg. Allgem. Chem.*, **311**, 235 (1961).
33. Biryukov, N. P., I. A. Safin, and M. G. Voronkov, *LSSR Zinatnu Akad. Vestis Khim. Ser.*, (2), 181 (1964).
34. Biryukov, N. P., M. G. Voronkov, and I. A. Safin, *Izvest. Akad. Nauk LSSR, Ser. Khim.*, 638 (1966).
35. Biryukov, N. P., M. G. Voronkov, and I. A. Safin, *Nuclear Quadrupole Resonance Frequencies Tables*, Leningrad (1968).
36. Bartocha, Bodo, U.S. Pat. 3,100,217 (Cl 260–448); (*CA*, **60**, 551 h).
37. Bott, R. W., C. Eaborn, K. C. Pande, and T. W. Swaddle, *J. Chem. Soc.*, 1217 (1962).
38. Bott, R. W., C. Eaborn, and T. W. Swaddle, *J. Organometal. Chem.*, **5**, 233 (1966).
39. Brinckman, F. E., and F. G. A. Stone, *J. Inorg. Nucl. Chem.*, **11**, 24 (1959).

40. Brook, A. G., and G. J. D. Peddle, *J. Am. Chem. Soc.*, **85**, 1869 (1963).
41. Brook, A. G., and G. J. D. Peddle, *J. Am. Chem. Soc.*, **85**, 2338 (1963).
42. Brooks, E. H., and F. Glockling, *J. Chem. Soc. A*, 1241 (1966).
43. Brooks, E. H., F. Glockling, and K. A. Hooton, *J. Chem. Soc.*, 4283 (1965).
44. Brown, M. P., and G. W. A. Fowles, *J. Chem. Soc.*, 2811 (1958).
45. Bulten, E. J., and J. G. Noltes, *Tetrahedron Letters*, (16), 1443 (1967).
46. Bulten, E. J., and J. G. Noltes, *Tetrahedron Letters*, (29), 3471 (1966).
47. Bulten, E. J., and J. G. Noltes, *J. Organometal. Chem.*, **15P**, 18 (1968).
48. Burlachenko, G. S., Dissertation, M.G.U. (1966).
49. Burlachenko, G. S., V. I. Avdeeva, Yu. I. Baukov, and I. F. Lutsenko, *Zh. Obshch. Khim.*, **35**, 1881 (1965).
50. Carre, F. H., R. J. P. Corriu, and R. B. Thomassin, *Chem. Commun.*, (10), 560 (1968).
51. Chernyshev, E. A. A. D. Petrov, and T. L. Krasnova, *Synthesis and Properties of Monomers*, Ed. "Nauk", 1964, 103.
52. Chipperfield, J. R., and R. H. Prince, *Proc. Chem. Soc.*, 385 (1960).
53. Chumaevskii, N. A., *Opt. Spectry (USSR)*, **13**, 68 (1962).
54. Chvalovsky, V., B. Lepeska, and J. Vĉelàk, *2nd Int. Symp. Organosilicon Compounds, Bordeaux, 1968.*
55. Clark, H. C. and C. J. Willis, *Proc. Chem. Soc.*, 282 (1960).
56. Clark, H. C., and C. J. Willis, *Amer. Chem. Soc. Abstr. Papers* (140), 24 M (1961).
57. Clark, H. C., and C. J. Willis, *J. Am. Chem. Soc.*, **84** (6), 898 (1962).
58. Clark, E. A., and A. Weber, *J. Chem. Phys.*, **45**, 1759 (1966).
59. Coyle, T. D., and J. J. Ritter, *J. Organometal. Chem.*, **12**, 269 (1968).
60. Cresswell, W. T., J. Leicester, and A. I. Vogel, *Chem. Ind. (London)*, 19 (1953).
61. Cross, R. J., and F. Glockling, *J. Organometal. Chem.*, **3**, 146 (1965).
62. Cross, R. J., and F. Glockling, *J. Chem. Soc.*, 5422 (1965).
63. Curtis, M. D., *J. Am. Chem. Soc.*, **89** (16), 4241 (1967).
64. Dailey, B. P., J. M. Mays, and C. H. Townes, *Phys. Rev.*, **76**, 136 (1949).
64a. Huy-Giao, Dao, *Compt. Rend.*, **260**, 6937 (1965).
64b. Davies, A. G., and C. D. Hall, *J. Chem. Soc.*, 3835 (1959).
65. Dennis, L. M., *Z. Anorg. Chem.*, **174**, 97 (1928).
66. Dennis, L. M., and F. E. Hance, *J. Phys. Chem.*, **30**, 1055 (1926).
67. Dennis, L. M., and W. I. Patnode, *J. Am. Chem. Soc.*, **52**, 2779 (1930).
68. Dubac, J., Dissertation, Toulouse, 1969.
69. Duffield, A. M., H. Budzikiewicz, and C. Djerassi, *J. Am. Chem. Soc.*, **87**, 2920 (1965).
70. Dzhurinskaya, N. G., Dissertation, Moscow, 1961.
71. Dzhurinskaya, N. G., V. F. Mironov, and A. D. Petrov, *Dokl. Akad. Nauk SSSR*, **138**, 1107 (1961).
72. Eaborn, C., W. A. Dutton, F. Glockling, and K. A. Hooton, *J. Organometal. Chem.*, **9**, 175 (1967).
73. Eaborn, C., and K. C. Pande, *J. Chem. Soc.*, 3200 (1960).
74. Eaborn, C., P. Simpson, and I. D. Varma, *J. Chem. Soc. A*, (8), 1133 (1966).
75. Eaborn, C., and D. R. M. Walton, *J. Organometal. Chem.*, **3**, 168 (1965).
76. Eaborn, C., and D. R. M. Walton, *J. Organometal. Chem.*, **4**, 217 (1965).
77. Egorov, Yu. P., L. A. Leites, I. D. Kravtsova, and V. F. Mironov, *Izv. Akad. Nauk SSSR*, (6), 1114 (1963).
78. Emel'yanova, L. I., and L. G. Makarova, *Izv. Akad. Nauk SSSR*, 2067 (1961).
79. Emel'yanova, L. I., V. N. Vinogradova, L. G. Makarova, and A. N. Nesmeyanov, *Izv. Akad. Nauk SSSR, Otd. Khim. Nauk*, 53 (1962).
80. Fenton, D. E., and A. G. Massey, *Chem. Ind. (London)*, 2100 (1964).

81. Fenton, D. E., A. G. Massey, and D. S. Urch, *J. Organometal. Chem.*, **6**, 352 (1966).
82. Finholt, A. E., *Nucl. Sci. Abstr.*, **6**, 617 (1957).
83. Fischer, A. K., R. C. West, and E. G. Rochow, *J. Am. Chem. Soc.*, **76**, 5878 (1954).
84. Fish, R. H., and H. G. Kuivila, *J. Org. Chem.*, **31**, 2445 (1966).
85. Flood, E. A., *J. Am. Chem. Soc.*, **54**, 1663 (1932).
86. Flood, E. A., *J. Am. Chem. Soc.*, **54**, 1667 (1932).
87. Flood, E. A., *J. Am. Chem. Soc.*, **55**, 4935 (1933).
88. Flood, E. A., K. L. Godfrey, and L. S. Foster, *Inorg. Syn.*, **3** (64), 1950.
89. Flood, E. A., and L. Horvitz, *J. Am. Chem. Soc.*, **55**, 2534 (1933).
90. Florinskii, F. S., *Zh. Obshch. Khim.*, **32**, 1443 (1962).
91. Fuchs, R., and H. Gilman, *J. Org. Chem.*, **23**, 911 (1958).
92. Gar, T. K., Dissertation, M, 1965.
93. Gar, T. K., and V. F. Mironov, *Izvest. Akad. Nauk SSSR*, (5), 855 (1965).
94. George, M. V., D. J. Peterson, and H. Gilman, *J. Am. Chem. Soc.*, **82**, 403 (1960).
95. Gilliam, W., R. Meals, and R. Sauer, *J. Am. Chem. Soc.*, **68**, 1161 (1946).
96. Gilman, H., and C. W. Gerow, *J. Am. Chem. Soc.*, **78**, 5823 (1956).
97. Gilman, H., and C. W. Gerow, *J. Org. Chem.*, **22**, 334 (1957).
98. Gilman, H., and C. W. Gerow, *J. Am. Chem. Soc.*, **79**, 342 (1957).
99. Gilman, H., and C. W. Gerow, *J. Org. Chem.*, **23**, 1582 (1958).
100. Gilman, H., and R. D. Gorsich, *J. Am. Chem. Soc.*, **80**, 1883 (1958).
101. Gilman, H., M. B. Hughes, and C. W. Gerow, *J. Org. Chem.*, **24**, 352 (1959).
102. Gilman, H., and R. W. Leeper, *J. Org. Chem.*, **16**, 466 (1951).
103. Gingold, K., E. G. Rochow, D. Seyferth, A. C. Smith, Jr., and R. West, *J. Am. Chem. Soc.*, **74**, 6306 (1952).
104. Gladshtein, B. M., I. P. Kulyulin, and L. Z. Soborovskii, *Zh. Obshch. Khim.*, **36** (3), 488 (1966).
105. Gladshtein, B. V., V. V. Rode, and L. Z. Soborovskii, *Zh. Obshch. Khim.*, **29**, 2155 (1959).
106. Glockling, F., and K. A. Hooton, *J. Chem. Soc.*, 3509 (1962).
107. Glockling, F., and K. A. Hooton, *Proc. Chem. Soc.*, 146 (1963).
108. Glockling, F., and J. R. C. Light, *J. Chem. Soc. A*, 623 (1967).
109. Glockling, F., and J. R. C. Light, *J. Chem. Soc. A*, 717 (1968).
110. Griffiths, J. F., and M. Onyszchuk, *Can. J. Chem.*, **39**, 339 (1961).
111. Gverditsiteli, N. M., T. P. Doksopulo, M. M. Menteshashvili, and I. I. Abkhazava, *Soobsh. Akad. Nauk Gruz.* **40**, 333 (1965).
112. Hartmann, H., H. Wagner, B. Karbstein, M. K. El A'ssar, and W. Reiss, *Naturwiss.*, **51**, 215 (1964).
113. Horvitz, L., and E. A. Flood, *J. Am. Chem. Soc.*, **55**, 5055 (1933).
114. Irisova, N. A., and M. E. Dianov, *Opt. i Spektroskopiya.*, **9**, 261 (1960).
115. Issleib, K., and B. Walther, *Angew. Chem. Int. Ed. Engl.*, **6** (1), 88 (1967).
116. Jacobs, G., *Compt. Rend.*, **238**, 1825 (1954).
117. Johnson, O. H., and D. M. Harris, *J. Am. Chem. Soc.*, **72**, 5564 (1950).
118. Johnson, O. H., and D. M. Harris, *J. Am. Chem. Soc.*, **72**, 5566 (1950).
119. Johnson, O. H., and L. V. Jones, *J. Org. Chem.*, **17**, 1172 (1952).
120. Johnson, O. H., and W. H. Nebergall, *J. Am. Chem. Soc.*, **70**, 1706 (1948).
121. Johnson, O. H., and W. H. Nebergall, *J. Am. Chem. Soc.*, **71**, 1720 (1949).
122. Johnson, O. H., W. H. Nebergall, and D. M. Harris, *Inorg. Syn.*, **5**, 76 (1957).
123. Johnson, O. H., and E. A. Schmall, *J. Am. Chem. Soc.*, **80**, 2931 (1958).
124. Jones, L. V., *Dissertation Abstr.*, **13**, 308 (1953).
125. Kadina, M. A., V. A. Ponomarenko, A. V. Kessenikh, and E. P. Prokof'ev, *Zh. Obshch. Khim.*, **37** (5), 1040 (1967).

126. Kadina, M. A., G. Ya. Zueva, and A. G. Kechina, *Izv. Akad. Nauk SSSR Ser. Khim.*, (12), 2215 (1966).
127. Kartsev, G. N., Ya. K. Syrkin, A. L. Kravchenko, and V. F. Mironov, *Zh. Strukt. Khim.*, **5**, 639 (1964).
128. Kartsev, G. N., Ya. K. Syrkin, A. L. Kravchenko, and V. F. Mironov, *Zh. Strukt. Khim.*, **5**, 942 (1964); *CA*, **61**, 8986.
129. Kartsev, G. N., Ya. K. Syrkin, and V. F. Mironov, *Izv. Akad. Nauk SSSR*, 948 (1960).
130. van der Kerk, G. J. M., F. Rijkens, and M. J. Janssen, *Rec. Trav. Chim.*, **81**, 764 (1962).
131. Kolesnikov, G. S., S. L. Davydova, and N. V. Klimentova, *Intern. Congress of Macromolecular Chemistry*, Ed. "Akad Nauk SSSR", 156 (1960); R. Zh. Khim., 5 R 69, (1961).
132. Kolesnikov, G. S., S. L. Davydova, and N. V. Klimentova, *Vysokomolekul. Soedin.*, **4**, 1098 (1962).
133. Korshak, V. V., A. M. Sladkov, and L. K. Luneva, *Izv. Akad. Nauk SSSR*, 728 (1962).
134. Kramer, K., and N. Wright, *Chem. Ber.*, **96**, 1877 (1963).
135. Kraus, C. A., and C. L. Brown, *J. Am. Chem. Soc.*, **52**, 3690 (1930).
136. Kraus, C. A., and C. L. Brown, *J. Am. Chem. Soc.*, **52**, 4031 (1930).
137. Kraus, C. A., and E. A. Flood, *J. Am. Chem. Soc.*, **54**, 1635 (1932).
138. Kraus, C. A., and L. S. Foster, *J. Am. Chem. Soc.*, **49**, 457 (1927).
139. Kraus, C. A., and H. S. Nutting, *J. Am. Chem. Soc.*, **54**, 1622 (1932).
140. Kraus, C. A., and C. B. Wooster, *J. Am. Chem. Soc.*, **52**, 372 (1930).
141. Kuehlein, K., and W. P. Neumann, *Ann. Chem.*, **702**, 17 (1967).
142. Lavigne, A. A., R. M. Pike, D. Monier, and C. T. Tabit, *Rec. Trav. Chim.*, **86**, 746 (1967).
143. Leavitt, F. C., T. A. Manuel, and F. Johnson, *J. Am. Chem. Soc.*, **81**, 3163 (1959).
144. Leites, L. A., T. K. Gar, and V. F. Mironov, *Dokl. Akad. Nauk SSSR*, **158**, 400 (1964).
145. Lesbre, M., *Compt. Rend.*, **204**, 1822 (1937).
146. Lesbre, M., and P. Mazerolles, *Compt. Rend.*, **246**, 1708 (1958).
147. Lesbre, M., P. Mazerolles, and G. Manuel, *Compt. Rend.*, **255**, 544 (1962).
148. Lesbre, M., P. Mazerolles, and G. Manuel, *Compt. Rend.*, **257**, 2303 (1963).
149. Lesbre, M., and J. Satgé, *Bull. Soc. Chim. France*, 783 (1959).
150. Lesbre, M., and J. Satgé, *Compt. Rend.*, **252**, 1976 (1961).
151. Lesbre, M., J. Satgé, and M. Massol, *Compt. Rend.*, **256**, 1548 (1963).
152. Lesbre, M., J. Satgé, and M. Massol, *Compt. Rend.* **257**, 2665 (1963).
153. Lesbre, M., J. Satgé, and M. Massol, *Compt. Rend.*, **258**, 2842 (1964).
154. Leusink, A. J., J. G. Noltes, H. A. Budding, and G. J. M. van der Kerk, *Rec. Trav. Chim.*, **83**, 844 (1964).
155. Lippincott, E. R., P. Mercier, and M. C. Tobin, *J. Phys. Chem.*, **57**, 939 (1953).
156. Luijten, J. G. A., and F. Rijkens, *Rec. Trav. Chim.*, **83**, 857 (1964).
157. Luijten, J. G. A., F. Rijkens, and G. J. M. van der Kerk, *Advan. Organometal. Chem.*, **3**, 397 (1965).
158. Lutsenko, I. F., Yu. I. Baukov, and G. S. Burlachenko, *J. Organometal. Chem.*, **6**, 496 (1966).
159. Lutsenko, I. F., Yu. I. Baukov, and B. N. Kasapov, *Zh. Obshch. Khim.*, **33**, 2724 (1963).
160. Marrot, J., J. C. Maire, and J. Cassan, *Compt. Rend.*, **260**, 3931 (1965).
161. Massol, M., Dissertation, Toulouse, 1967.
162. Massol, M., and J. Satgé, *Bull. Soc. Chim. France*, (9), 2737 (1966).
163. Mazerolles, P., Dissertation, Toulouse, 1959.
164. Mazerolles, P., *Bull. Soc. Chim. France*, 1911 (1961).
165. Mazerolles, P., *Bull. Soc. Chim. France*, 1907 (1962).
166. Mazerolles, P., *Bull. Soc. Chim. France*, 464 (1965).

167. Mazerolles, P., and coworkers, unpublished work.
168. Mazerolles, P., and J. Dubac, *Compt. Rend.*, **257**, 1103 (1963).
169. Mazerolles, P., J. Dubac, and M. Lesbre, *J. Organometal. Chem.*, **5**, 35 (1966).
170. Mazerolles, P., and M. Lesbre, *Compt. Rend.*, **248**, 2018 (1959).
171. Mazerolles, P., M. Lesbre, and J. Dubac, *Compt. Rend.*, **260**, 2255 (1965).
171a. Mazerolles, P., M. Lesbre, and Mme. M. Joanny, *J. Organometal. Chem.*, **16**, 227 (1969).
172. Mazerolles, P., M. Lesbre, and S. Marre, *Compt. Rend.*, **261**, 4134 (1965).
173. Mazerolles, P., and G. Manuel, *Bull. Soc. Chim. France*, 327 (1966).
174. Mazerolles, P., and G. Manuel, *Bull. Soc. Chim. France*, 2511 (1967).
175. Mazerolles, P., and G. Manuel, *Compt. Rend.*, **267**, 1158 (1968).
176. Mehrotra, R. C., and G. Chandra, *Proc. Natl. Acad. Sci. India Sect. A*, **35** (2), 148 (1966).
177. Mendelsohn, J. C., F. Métras, J. C. Lahournère, and J. Valade, *J. Organometal. Chem.*, **12**, 327 (1968).
178. Metlesics, W., and H. Zeiss, *J. Am. Chem. Soc.*, **82**, 3321 (1960).
179. Metlesics, W., and H. Zeiss, H., *J. Am. Chem. Soc.*, **82**, 3324 (1960).
180. Mironov, V. F., *Izv. Akad. Nauk SSSR*, 1862 (1959).
181. Mironov, V. F., L. M. Antipin, and E. S. Sobolev, *Zh. Obshch. Khim.*, **38** (2), 251 (1968).
182. Mironov, V. F., E. M. Berliner, and T. K. Gar, *Zh. Obshch. Khim.*, **37**, 962 (1967).
183. Mironov, V. F., and N. G. Dzhurinskaya, *Izv. Akad. Nauk SSSR*, 75 (1963).
184. Mironov, V. F., N. G. Dzhurinskaya, and T. K. Gar, *Synth. and Monomer Properties, Moscow*, 150 (1964).
185. Mironov, V. F., N. G. Dzhurinskaya, T. K. Gar, and A. D. Petrov, *Izv. Akad. Nauk SSSR Otd. Khim. Nauk*, 460 (1962).
186. Mironov, V. F., N. G. Dzhurinskaya, and A. D. Petrov, *Dokl. Akad. Nauk SSSR*, **131**, 98 (1960).
187. Mironov, V. F., N. G. Dzhurinskaya, and A. D. Petrov, *Izv. Akad. Nauk SSSR Otd. Khim. Nauk*, 2066 (1960).
188. Mironov, V. F., N. G. Dzhurinskaya, and A. D. Petrov, *Izv. Akad. Nauk SSSR Otd. Khim. Nauk*, 2095 (1961).
189. Mironov, V. F., Yu. P. Egorov, and A. D. Petrov, *Izv. Akad. Nauk SSSR Otd. Khim. Nauk*, 1400 (1959).
190. Mironov, V. F., and N. S. Fedotov, *Zh. Obshch. Khim.*, **34**, 4122 (1964).
191. Mironov, V. F., and N. S. Fedotov, *Zh. Obshch. Khim.*, **36** (3), 556 (1966).
192. Mironov, V. F., and T. K. Gar, *Dokl. Akad. Nauk SSSR*, **152**, (5), 1111 (1963).
193. Mironov, V. F., and T. K. Gar, *Izv. Akad. Nauk SSSR*, 1515 (1964).
194. Mironov, V. F., and T. K. Gar, *Izv. Akad. Nauk SSSR*, 1887 (1964).
195. Mironov, V. F., and T. K. Gar, *Izv. Akad. Nauk SSSR* 291 (1965).
196. Mironov, V. F., and T. K. Gar, *Izv. Akad. Nauk SSSR*, 755 (1965).
197. Mironov, V. F., and T. K. Gar, *Izv. Akad. Nauk SSSR Ser Khim.*, (3), 482 (1966).
198. Mironov, V. F., and T. K. Gar, *Zh. Obshch. Khim.*, **36**, 1719 (1966).
199. Mironov, V. F., T. K. Gar, and L. A. Leites, *Izv. Akad. Nauk SSSR Otd. Khim. Nauk*, 1387 (1962).
200. Mironov, V. F., and A. L. Kravchenko, *Izv. Akad. Nauk SSSR*, 1563 (1963).
201. Mironov, V. F., and A. L. Kravchenko, *Izv. Akad. Nauk SSSR*, 768 (1964).
202. Mironov, V. F., and A. L. Kravchenko, *Zh. Obshch. Khim.*, **34**, 1356 (1964).
203. Mironov, V. F., and A. L. Kravchenko, *Dokl. Akad. Nauk SSSR*, **158**, 656 (1964).
204. Mironov, V. F., and A. L. Kravchenko, *Izv. Akad. Nauk SSSR Ser. Khim.*, (6), 1026 (1965).

205. Mironov, V. F., A. L. Kravchenko, and A. L. Leites, *Izv. Akad. Nauk SSSR*, 1177 (1966).
206. Mironov. V. F., A. L. Kravchenko, and A. D. Petrov, *Dokl. Akad. Nauk SSSR*, **155**, 843 (1964).
207. Mironov, V. F., A. L. Kravchenko, and A. D. Petrov, *Izv. Akad. Nauk SSSR*, 1209 (1964).
208. Mironov, V. F., V. A. Ponomarenko, G. Vzenkova, I. E. Dolgii, and A. D. Petrov, *Chemistry and Practical Use of Organosilicon Compounds*, 1 L, Ed. CBTI, 1958, 189.
209. Moedritzer, K., *J. Organometal. Chem.*, **6** (3), 282 (1966).
210. Moedritzer, K., and J. R. van Wazer, *Inorg. Chem.*, **4**, (12), 1753 (1965).
211. Moedritzer, K., and J. R. van Wazer, *Inorg. Chem.*, **5** (4), 547 (1966).
212. Moedritzer, K., and J. R. van Wazer, *J. Polymer Sci.*, (A-1), 547 (1968).
213. Morgan, G. T., and H. D. Drew, *J. Chem. Soc.*, **127**, 1760 (1925).
214. Nametkin, N. S., S. G. Durgar'yan, and T. I. Tikhonova, *Dokl. Akad. Nauk SSSR*, **172** (3), 615 (1967).
215. Nefedov, O. M., and S. P. Kolesnikov, *Izv. Akad. Nauk SSSR*, 2068 (1963).
216. Nefedov, O. M., and S. P. Kolesnikov, *Izv. Akad. Nauk SSSR Ser. Khim.*, (2), 201–11, (1966).
217. Nefedov, O. M., S. P. Kolesnikov, A. S. Khachaturov, and A. D. Petrov, *Dokl. Akad. Nauk SSSR*, **154**, 1389 (1964).
218. Nefedov, O. M., S. P. Kolesnikov, and N. N. Nivotskaya, *Izv. Akad. Nauk SSSR Ser. Khim.*, (3), 579 (1965).
219. Nefedov, O. M., and M. N. Manakov, *Angew. Chem. Int. Ed. Engl.*, **5** (12), 1021 (1966).
220. Nefedov, O. M., M. N. Manakov, and A. D. Petrov, *Dokl. Akad. Nauk SSSR*, **147**, 1376 (1962).
221. Nesmeyanov, A. N., K. N. Anisimov, N. E. Kolobova, and F. S. Denisov, *Izv. Akad. Nauk SSSR Ser. Khim.*, (12), 2246 (1966).
222. Nesmeyanov, A. N., A. E. Borisov, and N. V. Novikova, *Dokl. Akad. Nauk SSSR*, **165** (2), 333 (1965).
223. Nesmeyanov, A. N., L. I. Emel'yanova, and L. G. Makarova, *Dokl. Akad. Nauk SSSR*, **122**, 403 (1958).
224. Noltes, J. G., and G. J. M. van der Kerk, *Rec. Trav. Chim.*, **81**, 41 (1962).
225. Orndorff, W. K., D. L. Tabern, and L. M. Dennis, *J. Am. Chem. Soc.*, **49**, 2512 (1927).
226. Osthoff, R. C., and E. G. Rochow, *J. Am. Chem. Soc.*, **74**, 845 (1952).
227. Patil, H. R. H., and W. A. G. Graham, *Inorg. Chem.*, **5** (8), 1401 (1966).
228. Patmore, D. J., and W. A. G. Graham, *Inorg. Chem.*, **5** (8), 1405 (1966).
229. Patmore, D. J., and W. A. G. Graham, *Inorg. Chem.*, **6** (5), 981 (1967).
230. Peach, M. E., and T. C. Waddington, *J. Chem. Soc.*, 1238 (1961).
231. Pearson, T. H., U.S. Pat. 3,283,019 (Cl 260–652) Nov. 1 (1966), Appl. Nov. 14 (1963); *CA*, **66**, 55008b (1966).
232. Petrov, A. D., V. F. Mironov, and I. E. Dolgii, *Izv. Akad. Nauk SSSR Otd. Khim. Nauk*, 1146 (1956).
233. Petrov, A. D., V. F. Mironov, and N. G. Dzhurinskaya, *Dokl. Akad. Nauk SSSR*, **128**, 302 (1959).
234. Petukhov, V. A., V. F. Mironov, and A. L. Kravchenko, *Izv. Akad. Nauk SSSR Ser. Khim.*, **1**, 156 (1966).
235. Petukhov, V. A., V. F. Mironov, and P. P. Shorygin, *Izv. Akad. Nauk SSSR*, (12), 2203 (1964).

236. Pike, R. M., and A. M. Dewidar, *Rec. Trav. Chim.*, **84**, 119 (1964).
237. Pike, R. M., and A. F. Fournier, *Rec. Trav. Chim.*, **81**, 475 (1962).
238. Pike, R. M., and A. A. Lavigne, *Rec. Trav. Chim.*, **82**, 49 (1963).
239. Pike, R. M., and A. A. Lavigne, *Rec. Trav. Chim.*, **83**, 883 (1964).
240. Ponomarenko, V. A., and G. Ya. Vzenkova, *Izv. Akad. Nauk SSSR Otd. Khim. Nauk*, 994 (1957).
241. Ponomarenko, V. A., G. Ya. Vzenkova, and Yu. P. Egorov, *Dokl. Akad. Nauk SSSR*, **122**, 405 (1958).
242. Ponomarenko, V. A., G. Ya. Zueva, and N. S. Andreev, *Izv. Akad. Nauk SSSR Otd. Khim. Nauk*, 1758 (1961).
243. Poskozin, P. S., *J. Organometal. Chem.*, **12**, 115 (1968).
244. Razuvaev, G. A., N. S. Vyazankin, O. D'yachkovskaya, I. G. Kiseleva, and Yu. I. Dergunov, *Zh. Obshch. Khim.*, **31**, 4056 (1961).
245. Reimschneider, R., K. Menge, and P. Klang, *Z. Naturforsch.*, **11**, 115 (1954).
246. Reimschneider, R., K. Menge, and P. Klang, *Z. Naturforsch.*, **11b**, 115 (1956).
247. Rijkens, F., E. J. Bulten, W. Drenth, and G. J. M. van der Kerk, *Rec. Trav. Chim.*, **85** (11), 1223 (1966).
248. Rijkens, F., and G. J. M. van der Kerk, *Investigation in the Field of Organogermanium Chemistry*, Germanium Research Committee, Utrecht, 1964.
248a. Rijkens, F., and G. J. M. van der Kerk, *Rec. Trav. Chim.*, **83**, 723 (1964).
249. Rivière, P., and J. Satgé, *Bull. Soc. Chim. France*, 4039 (1966).
250. Roberts, R. M. G., *J. Organometal. Chem.*, **12**, 97 (1968).
251. Rochow, E. G., *J. Am. Chem. Soc.*, **69**, 1729 (1947).
252. Rochow, E. G., *J. Am. Chem. Soc.*, **70**, 436 (1948).
253. Rochow, E. G., *J. Am. Chem. Soc.*, **70**, 1801 (1948).
254. Rochow, E. G., U.S. Pat. 2,451,871, Oct. 19 (1948); *CA*, **43**, 2631 (1949).
255. Rochow, E. G., *J. Am. Chem. Soc.*, **72**, 198 (1950).
256. Rochow, E. G., R. Ditchenko, and R. C. West, *J. Am. Chem. Soc.*, **73**, 5486 (1951).
257. Rochow, E. G., and A. L. Allred, *J. Am. Chem. Soc.*, **77**, 4489 (1955).
258. Rothermundt, M., and K. Burschkies, *Z. Immunitaetsforsch.*, **87**, 445 (1936).
259. Ruidisch, I., H. Schmidbaur, and H. Schumann, *Halogen. Chem.*, **2**, 233 (1967).
260. Sakurai, H., K. Tominaga, T. Watanabe, and M. Kumada, *Tetrahedron Letters*, (45), 5493 (1966).
261. Sarankina, S. A., and Z. M. Manulkin, *Zh. Obshch. Khim.*, **36**, 1299 (1966).
262. Sartori, P., and M. Weidenbruch, *Chem. Ber.*, **100** (6), 2049 (1967).
263. Satgé, J., *Ann. Chim. (Paris)*, **6**, 519 (1961).
264. Satgé, J., and C. Couret, *Compt. Rend.*, **264**, 2169–72 (1967).
265. Satgé, J., M. Lesbre, and M. Baudet, *Compt. Rend.*, **259**, 4733–36 (1964).
266. Satgé, J., and M. Massol, *Compt. Rend.*, **261**, 170 (1965).
267. Satgé, J., M. Massol, and M. Lesbre, *J. Organometal. Chem.*, **5**, 241 (1966).
268. Satgé, J., R. Mathis-Noël, and M. Lesbre, *Compt. Rend.*, **249**, 131 (1959).
269. Satgé, J., and P. Rivière, *J. Organometal. Chem.*, **16**, 71 (1969).
270. Satgé, J., and P. Rivière, *Bull. Soc. Chim. France*, (5), 1773 (1966).
271. Satgé, J., P. Rivière, and M. Lesbre, *Compt. Rend.*, **265**, 494 (1967).
272. Schmidt, M., and I. Ruidisch, *Z. Anorg. Allgem. Chem.*, **311**, 331 (1961).
273. Schmidt, M., and H. Ruf, *Angew. Chem.*, **73**, 64 (1961).
274. Schmidt, M., H. Schmidbaur, and I. Ruidisch, *Angew. Chem.*, **73**, 408 (1961).
275. Schmidbaur, H., and W. Findeiss, *Angew. Chem.*, **76**, 752 (1964).
276. Schmidbaur, H., and W. Findeiss, *Chem. Ber.*, **99** (7), 2187 (1966).
277. Schmidbaur, H., and M. Schmidt, *Chem. Ber.*, **94**, 1349 (1961).
278. Schmidbaur, H., and W. Tronich, *Chem. Ber.*, **100** (3), 1032 (1967).

279. Schott, G., and C. Harzdorf, *Z. anorg. allgem. Chem.*, **307**, 105 (1960).
280. Schwarz, R., and M. Lewinsohn, *Chem. Ber.*, **64**, 2352 (1931).
281. Schwarz, R., and W. Reinhardt, *Chem. Ber.*, **65**, 1743 (1932).
282. Schwarz, R., and M. Schmeisser, *Chem. Ber.*, **69**, 579 (1936).
283. Semin, G. K., *Radiospectroscopy of Solides Atomizdat*, 229 (1967).
284. Seyferth, D., *J. Am. Chem. Soc.*, **79**, 2378 (1957).
285. Seyferth, D., K. Brändle, and G. Raab, *Angew. Chem.*, **72**, 77 (1960).
286. Seyferth, D., R. J. Cross, and B. Prokai, *J. Organometal. Chem.*, **7**, P 20 (1967).
287. Seyferth, D., and J. Hetflejs, *J. Organometal. Chem.*, **11**, 253 (1968).
288. Seyferth, D., T. F. Jula, D. C. Mueller, P. Mazerolles, G. Manuel, and F. Thoumas, *J. Am. Chem. Soc.*, (in press).
289. Seyferth, D., and N. Kahlen, *J. Org. Chem.*, **25**, 809 (1960).
290. Seyferth, D., and E. G. Rochow, *J. Am. Chem. Soc.*, **77**, 907 (1955).
291. Seyferth, D., and L. G. Vaughan, *J. Organometal. Chem.*, **1**, 138 (1963).
292. Seyferth, D., and T. Wada, *Inorg. Chem.*, **1**, 78 (1962).
293. Seyferth, D., T. Wada, and G. Maciel, *Inorg. Chem.*, **1**, 232 (1962).
294. Seyferth, D., and M. A. Weiner, *J. Am. Chem. Soc.*, **83**, 3583 (1961).
295. Seyferth, D., and M. A. Weiner, *J. Org. Chem.*, **26**, 4797 (1961).
296. Shackelford, J. M., H. de Schmertzing, C. H. Heuter, and H. Podall, *J. Org. Chem.*, **28**, 1700 (1963).
297. Sharanina, L. G., V. S. Zavgorodnii, and A. A. Petrov, *Zh. Obshch. Khim.*, **36**, 1154 (1966).
298. Shikhiev, I. A., I. A. Aslanov, and B. G. Yusufov, *Zh. Obshch. Khim.*, **31**, 3647 (1961).
299. Shikhiev, I. A., M. F. Shostakovskii, N. B. Komarov, and I. A. Aslanov, *Zh. Obshch. Khim.*, **29**, 1549 (1959).
300. Shostakovskii, M. F., B. A. Sokolov, and L. T. Ermakova, *Zh. Obshch. Khim.*, **32** (5), 1714 (1962).
301. Simonnin, M. P., *J. Organometal. Chem.*, **5**, 155 (1966).
302. Simons, J. K., *J. Am. Chem. Soc.*, **55**, 3705 (1933).
303. Simons, J. K., *J. Am. Chem. Soc.*, **57**, 1299 (1935).
304. Simons, J. K., E. C. Wagner, and J. H. Muller, *J. Am. Chem. Soc.*, **55**, 3705 (1933).
305. Stavitskii, I. K., S. N. Borisov, V. A. Ponomarenko, N. G. Sviridova, and G. Ya. Zueva, *Vysokomolekul. Soedin.*, **1**, 1502 (1959).
306. Tabern, D. L., W. K. Orndorff, and L. M. Dennis, *J. Am. Chem. Soc.*, **47**, 2039 (1925).
307. Tchakirian, A., *Ann. Phys. (Paris)*, **12** (11), 415 (1939). ·
308. Tchakirian, A., M. Lesbre, and M. Lewinsohn, *Compt. Rend.*, **202**, 13 (1936).
309. Tchakirian, A., and M. Lewinsohn, *Compt. Rend.*, **201**, 835 (1935).
310. Thomas, A. B., and E. G. Rochow, *J. Inorg. Nucl. Chem.*, **4**, 205 (1957).
311. Thomas, A. B., and E. G. Rochow, *J. Am. Chem. Soc.*, **79**, 1843 (1957).
312. Ulbricht, K., and V. Chvalovsky, *J. Organometal. Chem.*, **12**, 105–113 (1968).
313. Ulbricht, K., V. Vaisarova, V. Bazant, and V. Chvalovsky, *J. Organometal. Chem.*, **13**, 343 (1968).
314. Vassiliou, E., and E. G. Rochow, U.S. Clearinghouse Fed. Sci. Tech. Inform. A.D. 661202; *CA*, **68**, 78387 e (1968).
315. Vilkov, L. V., and V. S. Mastryukov, *Zh. Strukt. Khim.*, **6** (6), 811 (1965).
316. Vol'pin, M. E., Yu. T. Struchkov, L. B. Vilkov, V. S. Mastryukov, V. G. Dulova, and D. N. Kursanov, *Izv. Akad. Nauk SSSR*, 2067 (1963).
317. van de Vondel, D. F., *J. Organometal. Chem.*, **3**, 400 (1965).
318. Voronkov, M. G., and N. P. Biryukov, *Teor. i Eksp. Khim. Akad. Nauk Ukr SSR*, **1** (1), 124 (1965).

319. Vyazankin, N. S., E. N. Gladyshev, and G. A. Razuvaev, *Dokl. Akad. Nauk SSSR,* **153**, 104 (1963).

320. Vyazankin, N. S., G. A. Razuvaev, V. T. Bychakov, and V. L. Zvezdin, *Izv. Akad. Nauk SSSR Ser. Khim.*, 562 (1966).

321. Vyazankin, N. S., G. A. Razuvaev, and O. S. Dvyachkovskaya, *Zh. Obshch. Khim.*, **33**, 613 (1963).

322. Vyazankin, N. S., G. A. Razuvaev, and E. N. Gladyshev, *Dokl. Akad. Nauk SSSR,* **151**, 1326 (1963).

323. van Wazer, J., and K. Moedritzer, *J. Am. Chem. Soc.*, **90** (1), 47 (1968).

324. West, R., *J. Am. Chem. Soc.*, **74**, 4363 (1952).

325. West, R., H. R. Hunt, and R. O. Wipple, *J. Am. Chem. Soc.*, **76**, 310 (1954).

326. Wieber, M., and C. D. Frohning, *Angew. Chem.*, **4** (12), 1096 (1965).

327. Wieber, M., and C. D. Frohning, *Z. Naturforsch.*, **21 B** (5), 492 (1966).

328. Wieber, M., and C. D. Frohning, *Angew. Chem.*, **5** (11), 966 (1966).

329. Wieber, M., C. D. Frohning, and M. Schmidt, *J. Organometal. Chem.*, **6** (4), 427 (1966).

330. Willemsens, L. C., and G. J. M. van der Kerk, *J. Organometal. Chem.*, **2**, 260 (1964).

331. Wittig, G., and L. Pohmer, *Chem. Ber.*, **89**, 1334 (1956).

332. Zablotna, R., *Bull. Acad. Pol. Sci.*, **12** (7), 475 (1964).

333. Zablotna, R., K. Akerman, and A. Szuchnik, *Bull. Acad. Pol. Sci.*, **12** (10), 695 (1964).

334. Zablotna, R., K. Akerman, and A. Szuchnik, *Bull. Acad. Pol. Sci.*, **13** (8), 527 (1965).

335. Zablotna, R., K. Akerman, and A. Szuchnik, *Bull. Acad. Pol. Sci. Ser. Sci. Chim.*, **14** (10), 731 (1966).

336. Zablotna, R., K. Akerman, and A. Szuchnik, Pol. Pat. 51,899, (*CA*, **67**, 108762 x).

337. Zakharkin, L. I., V. I. Bregadze, and O. Yu. Okhlobystin, *J. Organometal. Chem.*, **4** (3), 211 (1965).

338. Zhakharkin, L. I., and O. Yu. Okhlobystin, *Zh. Obshch. Khim.*, **31**, 3662 (1961).

339. Zakharkin, L. I., O. Yu. Okhlobystin, and B. N. Strunin, *Izv. Akad. Nauk SSSR*, 2002 (1962).

340. Zhakharkin, L. I., O. Yu. Okhlobystin, and B. N. Strunin, *Dokl. Akad. Nauk SSSR,* **144**, 1299 (1962).

341. Zhakharkin, L. I., O. Yu. Okhlobystin, and B. N. Strunin, *Zh. Prokl. Khim.*, **36**, 2034 (1963).

342. Zhakharkin, L. I., V. L. Stanko, and V. A. Brattsev, *Izv. Akad. Nauk SSSR*, 2079 (1961).

343. Zueva, G. Ya., Dissertation M, 1966.

344. Zueva, G. Ya., and N. V. Lukyankina, *2nd Int. Symp. of Organosilicon Compounds, Bordeaux, 1968.*

345. Zueva, G. Ya., N. V. Luk'yankina, A. G. Kechina, and V. A. Ponomarenko, *Izv. Akad. Nauk SSSR Ser. Khim.*, (10), 1843 (1966).

346. Zueva, G. Ya., N. V. Luk'yankina, and V. F. Ponomarenko, *Izv. Akad. Nauk SSSR Ser. Khim.*, (1), 192 (1967).

347. Zueva, G. Ya., I. F. Manucharova, I. P. Yakovlev, and V. A. Ponomarenko, *Izv. Akad. Nauk SSSR Neorg. Mater.*, **2**, 229 (1966).

348. Zueva, G. Ya., A. G. Pogorelov, V. I. Pisarenko, A. D. Snegova, and V. A. Ponomarenko, *Izv. Akad. Nauk SSSR Neorg. Mater.*, **2** (8), 1359 (1966).

349. Zueva, G. Ya., and V. A. Ponomarenko, *Izv. Akad. Nauk SSSR Neorg. Mater.*, **2**, 472 (1966).

350. *Inorg. Syn.*, **3**, p. 65.

351. Brit. Thomson-Houston Co. Ltd., Brit. Pat. 626,398; *CA* **44**, 2548a (1950).

352. Brit. Pat. 820,146 (1960).

6 Germanium–oxygen bond

6-1 ORGANOGERMANIUM OXIDES AND HYDROXIDES

6-1-1 Organogermanium hydroxides (organogermanols)

The hydrolysis of the organohalogermanes should in principle lead to the corresponding germanols:

$$R_3GeX \xrightarrow{\text{hydrolysis}} R_3GeOH$$

These germanols are, however, generally poorly stable and undergo dehydration to the corresponding oxides:

$$2\,R_3GeOH \rightarrow R_3Ge{-}O{-}GeR_3 + H_2O$$

$$R_2Ge(OH)_2 \rightarrow (R_2GeO) + H_2O$$

The only well-known and isolated germanols are triphenylgermanol, $(C_6H_5)_3GeOH$, tri(1-naphthyl)germanol, $(1\text{-}C_{10}H_7)_3GeOH$, tricyclohexylgermanol, $(C_6H_{11})_3GeOH$, tri(o-tolyl)germanol, $(o\text{-}CH_3C_6H_4)_3GeOH$, tris-(pentafluorophenyl)germanol, $(C_6F_5)_3GeOH$, and tri(i-propyl)-germanol, $(i\text{-}C_3H_7)_3GeOH$.

The triarylgermanols are generally more stable than the trialkylgermanols. Their relative stability can be explained by steric hindrance, which impedes their bimolecular dehydration. These germanols can be isolated after direct hydrolysis of the organohalogermanes or by indirect methods.

Brook and Gilman (39) prepared the triphenylgermanol by hydrolysis of triphenylbromogermane with aqueous alcoholic potassium hydroxide (yield: 84%). It has also been reported that the hydrolysis of the triphenyl-bromogermane leads to bis-(triphenylgermanium)oxide $[(C_6H_5)_3Ge]_2O$ (107, 158).

Triphenylgermanol is also obtained by the action of ammonium bromide on sodium triphenylgermanolate in liquid ammonia or by treating a benzene solution of the same germanolate with water (107). The sodium triphenylgermanolate was prepared by oxidizing triphenylgermyl-sodium in liquid ammonia. The same compound is obtained by carrying out the oxidation in the dry state or in benzene (107):

$$(C_6H_5)_3GeNa + \tfrac{1}{2}O_2 \rightarrow (C_6H_5)_3GeONa$$

Ph_3GeOH recrystallized from benzene melts at 134°C, but above this temperature it decomposes to water and $(Ph_3Ge)_2O$. It is, however, much more stable under ordinary conditions than the corresponding Ph_3GeNH_2 (109).

The formation of the triphenylgermanol has been reported in the hydrolysis by alcoholic potassium hydroxide of the germylphosphines $[(C_6H_5)_3Ge]_2PC_6H_5$ and $[(C_6H_5)_3Ge]_3P$ (227):

$$[(C_6H_5)_3Ge]_3P \xrightarrow{\text{KOH/O}_2} 3\,(C_6H_5)_3GeOH + H_3PO_4$$

as well as in the slow hydrolysis of triphenylgermane (7):

$$(C_6H_5)_3GeH + H_2O \rightarrow (C_6H_5)_3GeOH + H_2$$

The carbonation of triphenylgermyllithium gives Ph_3GeCO_2H which decarbonylates at its melting point (39):

$$Ph_3GeCO_2H \rightarrow CO + Ph_3GeOH$$

The triphenylgermanol thus produced is esterified by the remaining triphenylgermane carboxylic acid:

$$Ph_3GeOH + Ph_3GeCO_2H \rightarrow Ph_3GeCOOGePh_3$$

The saponification of the triphenylgermyltriphenylgermanecarboxylate allows the isolation of triphenylgermanol (39):

$$Ph_3GeCOOGePh_3 \xrightarrow{\text{KOH}} Ph_3GeOH + Ph_3GeCOOK$$

The tribenzyl derivative, $(C_6H_5CH_2)_3GeCO_2H$, lost CO at room temperature; the tribenzylgermanol produced is immediately esterified (55):

$$(PhCH_2)_3GeLi \xrightarrow{\text{CO}_2} (PhCH_2)_3GeCO_2H \rightarrow CO + (PhCH_2)_3GeOH$$
$$\xleftarrow{\quad (PhCH_2)_3GeCO_2H \quad}$$
$$(PhCH_2)_3GeCOOGe(CH_2Ph)_3$$

Tri-1-naphthylbromogermane, when titrated with dilute alcoholic sodium hydroxide, gave tri-1-naphthylgermanol which reacted quantitatively with the Karl Fisher reagent and was readily converted to the corresponding oxide (258).

Hydrolysis of tri-*m*-tolyl- and tri-*p*-tolylbromogermane by aqueous alkali or by alcoholic silver nitrate yields the corresponding tri-tolyl-germanium oxide, $(Ar_3Ge)_2O$. On the other hand, similar treatment of tri-*o*-tolylchlorogermane gives the hydroxide, $(o\text{-}CH_3\text{-}C_6H_4)_3GeOH$ (233).

In the hydrolysis of tris-(pentafluorophenyl)chlorogermane by cold distilled water, tris-(pentafluorophenyl)germanol, $(C_6F_5)_3GeOH$, is formed as an air-stable white solid. Above 130°C the germanol slowly loses water at atmospheric pressure to give bis-(tris-pentafluorophenyl)germanium oxide in virtually quantitative yield (78):

$$2\,(C_6F_5)_3GeOH \xrightarrow{130°C} (C_6F_5)_3Ge-O-Ge(C_6F_5)_3 + H_2O$$

The same oxide can be prepared in a single step by the reaction of tris-(pentafluorophenyl)chlorogermane with alcoholic silver nitrate, or moist pyridine, ammonia or triethylamine. Hydrolysis by distilled water of bis-(pentafluorophenyl)dibromogermane gave bis-(pentafluorophenyl)-germanium oxide $[(C_6F_5)_2GeO]_n$ (78).

Tricyclohexylgermanol (28, 101) has been prepared by reaction of tricyclohexylbromogermane with an alcoholic solution of silver nitrate. It also has been obtained by direct oxidation of tricyclohexylgermane in carbon tetrachloride or on short exposure to air (101):

$$(C_6H_{11})_3GeH \xrightarrow{O_2} (C_6H_{11})_3GeOH$$

Hydrolysis of triisopropylbromogermane by 6M aqueous sodium hydroxide furnished triisopropylgermanol, *i*-Pr_3GeOH, rather than the expected oxide, $(i\text{-}Pr_3Ge)_2O$, which results in reaction of the bromide with silver carbonate in dry hexane (15).

i-Pr_3GeOH as well as $(i\text{-}Pr_3Ge)_2O$, $(i\text{-}Pr_2GeO)_{3\text{ or }4}$ form in the hydrolysis of the reaction products between an excess of isopropylmagnesium chloride and germanium tetrachloride (19, 133). *i*-Pr_3GeOH lost water slowly below 200°C (15).

Dennis and Patnode report a possible formation of $(CH_3)_3GeOH$ in the hydrolysis of $(CH_3)_3GeBr$, but the germanol has not been isolated (69). Cryoscopic molecular weight determinations showed that the reaction $[(CH_3)_3Ge]_2O + H_2O \rightleftarrows 2\,(CH_3)_3GeOH$ is the reason for the water solubility of bis(trimethylgermanium) oxide. However, attempts to isolate Me_3GeOH from its aqueous solution by extraction with Et_2O always resulted in the isolation of $(Me_3Ge)_2O$ (219).

However, this lithium salt has been obtained:

$$Me_3GeOGeMe_3 + MeLi \rightarrow Me_4Ge + Me_3GeOLi$$

and the formation of trimethylgermanol has been reported in the hydrolysis of this salt (190):

$$(CH_3)_3GeOLi + H_2O \rightarrow (CH_3)_3GeOH + LiOH$$

Rochow and Allred (186) showed that dimethylgermanium dichloride dissociated and hydrolysed according to the equation:

$$(CH_3)_3GeCl_2 \rightleftarrows (CH_3)_2Ge^{++} + 2\,Cl^-$$

$$(CH_3)_2Ge^{++} + H_2O \rightleftarrows (CH_3)_2Ge(OH)^+ + H^+$$

$$(CH_3)_2Ge(OH)^+ + H_2O \rightleftarrows (CH_3)_2Ge(OH)_2 + H^+$$

Raman spectra have been obtained on aqueous solutions prepared from $[(CH_3)_2GeO]_4$; the solutions have been shown to contain tetrahedral $(CH_3)_2Ge(OH)_2$ and vibrational assignments have been made (242).

In the hydrolysis of esters of the type Et_3MCH_2COOEt (M = Si, Ge) Rijkens and coworkers (177, 179) report a partial hydrolysis of the M—C bond:

$$Et_3MCH_2COOEt \xrightarrow[OH^-]{H_2O} Et_3MOH + CH_3COOEt$$

but Et_3GeOH is not isolated.

6-1-1-1 Properties of Organogermanium Hydroxides. The tricyclohexylgermanol is reduced to tricyclohexylgermane by $LiAlH_4$:

$$4\,(C_6H_{11})_3GeOH \xrightarrow{\ LiAlH_4\ } 4\,(C_6H_{11})_3GeH + LiOH + Al(OH)_3$$

Tricyclohexylgermanol reacts with acetyl chloride to form tricyclohexylchlorogermane and with acetic anhydride to yield tricyclohexylacetoxygermane, $(C_6H_{11})_3GeOCOCH_3$ (101).

Tribenzylgermanol is esterified by phenylacetic acid (55). Passage of a stream of hydrogen sulphide into solution of triphenylgermanol in hot 95% ethanol furnished bis(triphenylgermanium)sulphide, $(Ph_3Ge)_2S$ (237). Triisopropylgermanol reacts readily with strong acids such as trifluoroacetic acid and sulphuric acid to yield the corresponding esters:

$$(i\text{-}C_3H_7)_3GeOCOCF_3 \quad \text{and} \quad [(i\text{-}C_3H_7)_3Ge]_2SO_4$$

However, triisopropylgermanol and weak organic acids such as formic or acetic acid do not yield pure esters (17).

Triisopropylgermanol reacts with isothiocyanic acid to give triisopropylgermanium isothiocyanate, $i\text{-}Pr_3Ge(NCS)$ (15). With $(Et_3Sn)_2O$ $i\text{-}Pr_3GeOH$ gives $(i\text{-}Pr_3Ge)_2O$ and Et_3SnOH (16).

Organoheterometalloxanes are prepared by dehydration of a mixture of hydroximetallic compounds. Ph_3GeOH heated under reflux in benzene, toluene or ligroin with Et_3SnOH or Ph_3PbOH gave $Ph_3GeOSnEt_3$ and

$Ph_3GeOPbPh_3$ ($\nu MOM'$: $Ge-O-Sn$ 810 cm^{-1}, $Ge-O-Pb$ 775 cm^{-1}) (67). Reactions of triorganogallium compounds with triorganogermanols gave organogallogermoxanes (26, 208):

$$2(C_6H_5)_3Ga + 2(C_6H_5)_3GeOH \rightarrow$$

$$2 C_6H_6 + [(C_6H_5)_2GaOGe(C_6H_5)_3]_2$$

A mixture of triphenylgermanol and anhydrous aluminium chloride produces bis-(triphenylgermanium)oxide in 95% yield and triphenyl-germanol with triphenylsilanol in presence of aluminium chloride produces tetraphenylbis-[(triphenylgermyl)oxy]disiloxane. Triphenylgermanol and diphenylsilanediol with the same catalyst give a cyclic germanosiloxane (68):

$$
\begin{array}{ccc}
& Ph & Ph \\
& | & | \\
Ph_3Ge-O-Si & -O-Si & -O-GePh_3 \\
& | & | \\
& O & O \\
& | & | \\
Ph-Si & -O-Si & -O-GePH_3 \\
& | & | \\
& Ph & Ph
\end{array}
$$

Triphenylhydroxycompounds, $(C_6H_5)_3MOH$ where M = C, Si and Ge, were condensed with N,N-diethylaminotrimethylsilane, $(CH_3)_3SiN-(C_2H_5)_2$. Diethylamine is formed in this reaction, as are the $(C_6H_5)_3MO-SiMe_3$ compounds.

The rate of condensation of the given triphenylhydroxide with amino-silane decreases in the order $(C_6H_5)_3SiOH > (C_6H_5)_3COH \approx (C_6H_5)_3GeOH$. This order can be correlated with the acid strength in this series (166).

The relative acidity and basicity of the triphenyl Group IV metal hydroxides have been examined by different authors. The acidity and basicity as hydrogen bond donors and acceptors have been measured for the compounds Ph_3MOH where M = C, Si, Ge, Sn, Pb. The trends in acidity and basicity in these compounds indicate that dative π-bonding from oxygen to M is strong in Ph_3SiOH, weaker in Ph_3GeOH and negligible in the other three compounds (260).

The relation between O–H stretching frequency and electronegativity in hydroxides of various elements has been established by West and Baney (259). The free OH frequencies for Ph_3MOH compounds (M = C, Si, Ge, Sn, Pb) are given. The fact that frequency enhancement is greater for triphenylsilanol than for triphenylgermanol and that no enhancement is found for the tin, lead and carbon compounds is consistent with studies of the acidity and basicity of these triphenylhydroxy compounds (260).

The acidity of the compounds R_3MOH was also examined by Matwiyoff and Drago (132) by investigating the change in OH stretching frequency upon hydrogen bonding to a Lewis base, diethyl ether and tetramethylene sulphoxide. The donor and acceptor properties of the compounds $(C_6H_5)_3MOH$ and $R_3MOC_6H_5$ (R = CH_3, C_2H_5 and M = C, Si, Ge, Sn and Pb) were investigated. Changes in the extent of π-bonding between the central element and oxygen are also proposed as the dominant effect giving rise to the observed trends. The hydroxy compounds of germanium are much poorer Lewis acids than those of the analogous silicon compounds. A considerable decrease in π-bonding is noted in the germanium compounds relative to silicon.

6-1-2   Organogermanium oxides

6-1-2-1   Bis-(triorganogermanium)   Oxides   (Hexaorganodigermoxanes).
The most usual approach for the preparation of bis-(triorganogermanium) oxides is the hydrolysis of the triorganogermanium monohalides:

$$R_3GeX \xrightarrow{\text{hydrolysis}} (R_3GeOH) \rightarrow (R_3Ge)_2O$$

The most frequently used hydrolysis agents are aqueous or alcoholic potassium hydroxide, aqueous sodium hydroxide, aqueous ammonia, silver carbonate, alcoholic silver nitrate, etc. The hydrolysis is followed by a digermoxane extraction with an organic solvent. With these different methods, the formation of $[(CH_3)_3Ge]_2O$ (69, 89, 219), $[(ClCH_2)(CH_3)_2$-$Ge]_2O$ (31, 263, 267), $(Et_3Ge)_2O$ (9, 106), $(i\text{-}Pr_3Ge)_2O$ (15), $(n\text{-}Pr_3Ge)_2O$ (10, 179), $(Bu_3Ge)_2O$ (12, 246), $[(n\text{-}C_5H_{11})_3Ge]_2O$, $[(i\text{-}C_5H_{11})_3Ge]_2O$, $[(n\text{-}C_6H_{13})_3Ge]_2O$ (179, 180), $[(C_6H_{11})_3Ge]_2O$ (84), $[(C_6H_5\text{—}CH_2)_3$-$Ge]_2O$ (29), $(Et_3Ge\text{—}GeEt_2)_2O$ (44), $(Me_3Ge\text{—}GeMe_2)_2O$ (43a) are cited in the literature. Oxides of type $[R_2(H)Ge]_2O$, $[R(H)_2Ge]_2O$ have been isolated by ammoniacal hydrolysis of alkyl- and phenylchlorogermanes, $R_2(Cl)GeH$ and $R(Cl)GeH_2$ (124, 126, 181, 204). However, Griffiths and Onyszchuk did not succeed in the preparation of $(CH_3GeH_2)_2O$ through the action of silver carbonate on $CH_3(Cl)GeH_2$ (89). Neither was bis-(n-butyldihydrogermanium) oxide, $Bu(H)_2GeOGe(H)_2Bu$, formed by hydrolysis of n-butylchlorogermane (22).

The oxidation of the Ge—H bonds of the alkylgermanium hydrides R_3GeH, R_2GeH_2, $RGeH_3$ by oxygen or slowly in the air leads quantitatively to the corresponding alkylgermanium oxides: $(R_3Ge)_2O$, R_2GeO, $(RGeO)_2O$ (194). Potassium permanganate in solution in acetone converts triethylgermane partially into oxide $(Et_3Ge)_2O$ (21).

The Ge—H bond of the trialkylgermanes is resistant to alkaline hydrolysis (224). However, water reacts quantitatively with triethylgermane in the presence of copper powder near 100°C (119):

$$2(C_2H_5)_3GeH + H_2O \xrightarrow{\ Cu\ } (C_2H_5)_3GeOGe(C_2H_5)_3 + H_2$$

The hydrolysis of many substituted germanes also leads to organo-germanium oxides. The hydrolysis of the germylamines $R_3GeNR'_2$, $(R_3Ge)_2NH$, $(R_3Ge)_3N$ (150, 178, 196, 203), Ph_3GeNH_2 (108, 109) and the germylazides $(CH_3)_3GeN_3$ (191) immediately yields the corresponding oxides $(R_3Ge)_2O$. The same oxides are formed quantitatively in the hydrolysis of the germylphosphines Ph_3GePPh_2, R_3GePPh_2 (40), $R_3GePR'_2$ (197).

The hydrolysis of bis-(trialkylgermyl)cadmium (253), tris-(trialkyl-germyl)bismuth (225) and tris-(trialkylgermyl)thallium (111) also leads to the trialkylgermanium oxides. These are formed also in the oxidation of the same derivatives as well as in the oxidation of bis-(trialkylgermyl)-mercury compounds (112, 173, 173a, 255):

$$2\,Et_3Ge-M-GeEt_3 + O_2 \;\rightarrow\; 2\,Et_3GeOGeEt_3 + 2\,M$$

$$(M = Hg\ or\ Cd)$$

$$2\,(Me_3Ge)_3Bi + 3\,H_2O \;\rightarrow\; 3\,(Me_3Ge)_2O + 2\,Bi + 3\,H_2$$

$$4\,(Et_3Ge)_3Tl + 3\,H_2O \;\rightarrow\; 4\,Tl + 6\,Et_3GeH + 3\,(Et_3Ge)_2O$$

Bis-(chloromethyldimethyl)germanium oxide results from the action of concentrated sulphuric acid on (chloromethyl)trimethylgermane followed by hydrolysis of the sulphonic ester (148):

$$(CH_3)_3GeCH_2Cl \xrightarrow[\ H_2O\ ]{\ H_2SO_4\ }$$

$$CH_4 + (ClCH_2)(CH_3)_2Ge-O-Ge(CH_3)_2(CH_2Cl)$$

The action of $n\text{-}Pr_3GeF$ on $(Et_3Sn)_2O$ gives $(n\text{-}Pr_3Ge)_2O$ and $EtSnF$ (16).

Bis-(triorganogermanium)oxides possess high thermal stability; they can be distilled at temperatures above 300°C without decomposition. They are little soluble or insoluble in water $[(Me_3Ge)_2O$ excepted] and soluble in the usual organic solvents.

Treatment of bis-(triorganogermanium) oxides with hydrohalic acids gives the corresponding organogermanium halides (10, 21, 84, 106, 134, 162):

$$(R_3Ge)_2O + 2\,HX \;\rightarrow\; 2\,R_3GeX + H_2O \qquad (X = F, Cl, Br, I).$$

The organohydrogermanium oxides, $[R_2(H)Ge]_2O$ and $[R(H)_2Ge]_2O$, on treatment with hydrohalic acids yield organohalogermanes, R_2XGeH and $RXGeH_2$, in high yields. The Ge—H bonds are preserved intact in these reactions (124, 126, 181, 204).

$(Et_3Ge)_2O$ reacts with formic, acetic and mercaptoacetic acid, as well as with acetic anhydride, giving the corresponding esters (9). Diphenyl-phosphonic acid reacts with $(Et_3Ge)_2O$ (40):

$$2\,Ph_2PO(OH) + (Et_3Ge)_2O \;\rightarrow\; 2\,Et_3GeOP(O)Ph_2 + H_2O$$

Boron trifluoride does not form a stable adduct with bis-(trimethyl-germanium) oxide. Instead, cleavage of the Ge—O—Ge linkage occurs, with quantitative formation of trimethylfluorogermane and boron trioxide (89):

$$(Me_3Ge)_2O \xrightarrow{\;BF_3\;} 2\,Me_3GeF + B_2O_3$$

Bis-(trialkylgermanium) oxides react readily with aluminium chloride (150):

$$R_3GeOGeR_3 \xrightarrow{\;AlCl_3\;} 2\,R_3GeCl$$

The same compounds are cleaved by organochlorogermanes:

$$R_3GeOGeR_3 + 2\,R'_3GeCl \;\rightarrow\; R'_3GeOGeR'_3 + 2\,R_3GeCl$$

and react also with trialkylchlorosilanes in the presence of ferric chloride:

$$(CH_3)_3GeOGe(CH_3)_3 + 2\,Et_3SiCl \;\rightarrow\; (Et_3Si)_2O + 2\,(CH_3)_3GeCl$$

It may be assumed that exchange proceeds through a four-membered transition state (150):

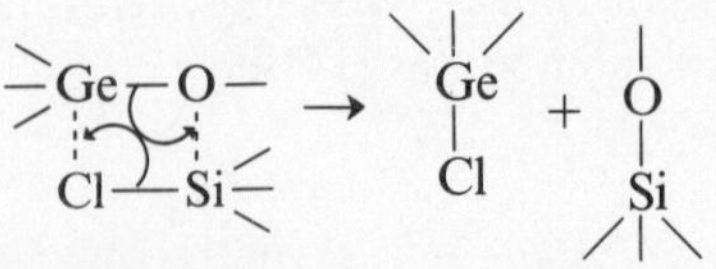

The alcoholysis of bis-(trialkylgermanium) oxides has also been reported (142, 149):

$$(Bu_3Ge)_2O + 2\,ROH \;\rightarrow\; 2\,Bu_3GeOR + H_2O$$

$$(Et_3Ge)_2O + 2\,HC\equiv C\!-\!CH_2OH \;\rightarrow\; 2\,Et_3GeOCH_2C\equiv CH + H_2O$$

(cf. organoalkoxygermanes, 6–2).

The quantitative thiolysis of the Ge—O—Ge bonds of $(Et_3Ge)_2O$ is observed with butanethiol (202):

$$(Et_3Ge)_2O + 2\,BuSH \;\rightarrow\; 2\,Et_3GeSBu + H_2O$$

The reactions between bis-(trimethylgermanium) oxide and *tert*-butane-thiol also give trimethyl(*tert*-butylthio)germane (2).

Bis-(chloromethyldimethylgermanium) oxide can be converted with potassium hydrogen sulphide into 2,2,5,5,tetramethyl-2,5,digerma-1,4-

dithiane (263) and with hydrogen sulphide into bis-(chloromethyldimethyl-germanium) sulphide (271):

$$[(ClCH_2)(CH_3)_2Ge]_2O + 2KSH \rightarrow \begin{array}{c} (CH_3)_2Ge \overset{S}{\diagdown} CH_2 \\ | \quad\quad | \\ H_2C \underset{S}{\diagdown} Ge(CH_3)_2 \end{array} + H_2O + 2KCl$$

$$[(ClCH_2)(CH_3)_2Ge]_2O + H_2S \rightarrow [(ClCH_2)(CH_3)_2Ge]_2S + H_2O$$

The aminolysis of the Ge—O—Ge bonds by pyrazole, imidazole and 1,2,4,triazole has been reported (178): (cf. Chapter 8, organogermanium-nitrogen bond):

$$(R_3Ge)_2O + 2H{-}N\diagdown^{\diagup} \rightarrow 2R_3Ge{-}N\diagdown^{\diagup} + H_2O$$

Bis-(triphenylgermanium) oxide, $[(C_6H_5)_3Ge]_2O$, has been reduced to triphenylgermane by lithium aluminium hydride (101). The bis-(triethyl-germanium) oxide reacts readily with lithium in ethylamine (106):

$$(Et_3Ge)_2O + 2Li \rightarrow Et_3GeLi + Et_3GeOLi$$

Metallic sodium reacts with bis-(triphenylgermanium) oxide in liquid ammonia with the formation of sodium triphenylgermanide and sodium triphenylgermanolate (109):

$$(Ph_3Ge)_2O + 2Na \rightarrow Ph_3GeNa + Ph_3GeONa$$

quantitative cleavage of $(Et_3Ge)_2O$ by potassium is observed in H.M.P.T. (hexamethylphosphoric triamide) (43a):

$$Et_3GeOGeEt_3 + 2K \rightarrow Et_3GeK + Et_3GeOK$$

Cleavage of bis-(trimethylgermanium) oxide with one molar equivalent of methyllithium in diethyl ether or in the ether–THF system produces lithium trimethylgermanolate and tetramethylgermane (188, 190):

$$Me_3GeOGeMe_3 + CH_3Li \rightarrow (CH_3)_3GeOLi + (CH_3)_4Ge$$

Cleavage of $(Me_3Ge)_2O$ and $(Et_3Ge)_2O$ by methylphosphonicdifluoride, $MePOF_2$, gave $MeP(O)F(OGeR_3)$ and R_3GeF (86).

The action of bromine on $(Et_2GeH)_2O$ leads to Et_2GeBr_2. This reaction probably takes place in two phases (126):

$$Et_2(H)GeOGe(H)Et_2 + 2Br_2 \rightarrow$$

$$Et_2(Br)GeOGe(Br)Et_2 + 2HBr \rightarrow 2Et_2GeBr_2 + H_2O$$

$(Et_3Ge)_2O$ reacts with AgF to form Et_3GeF in 60% yield (19).

6-1-2-2 Diorganogermanium Oxides. Hydrolysis of diorganodihaloger-manes produces the oxides of the general formula $(R_2GeO)_n$ existing in trimeric and tetrameric and polymeric cyclic form.

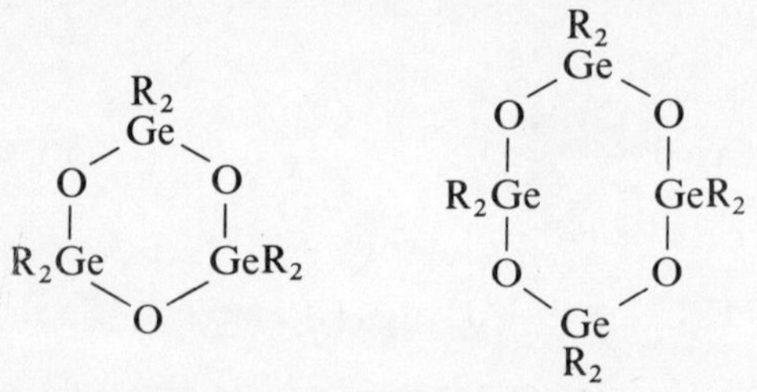

$$
\begin{array}{c}
R_2 \\
Ge \\
\diagup \quad \diagdown \\
O \qquad O \\
| \qquad\qquad | \\
R_2Ge \qquad\quad GeR_2 \\
\diagdown \quad \diagup \\
O
\end{array}
\qquad
\begin{array}{c}
R_2 \\
Ge \\
\diagup \quad \diagdown \\
O \qquad\qquad O \\
| \qquad\qquad\qquad | \\
R_2Ge \qquad\qquad GeR_2 \\
| \qquad\qquad\qquad | \\
O \qquad\qquad O \\
\diagdown \quad\quad \diagup \\
Ge \\
R_2
\end{array}
\qquad \text{and} \quad (R_2GeO)_n
$$

(Me_2GeO) (42, 190), (Et_2GeO) (9, 80, 134), $(Et(H)GeO)$ (124, 126), (n-$Pr_2GeO)$ (10, 13), (i-$Pr_2GeO)$ (15), (Bu_2GeO) (23, 134, 179, 246), $(Bu(H)$-$GeO)$ (124, 126), (Ph_2GeO) (99, 105, 146, 158), (Ar_2GeO) (75, 76), $(Ph(H)GeO)$ (181, 204) have thus been obtained. However, according to Morgan and Drew (158), diphenyldibromogermanium yielded on hydrolysis two well-defined complex substances derived by dehydration of the dihydroxide $(C_6H_5)_2Ge(OH)_2$. These products are:

$$HOGe(Ph)_2O-Ge(Ph)_2OGe(Ph)_2OGe(Ph)_2OH$$

and

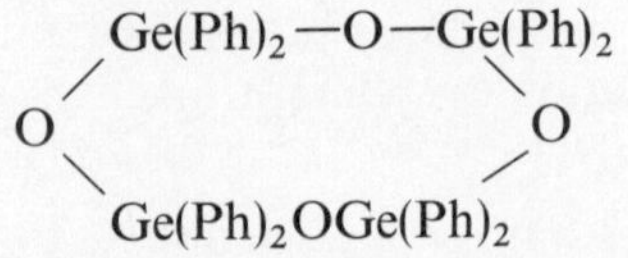

$$
\begin{array}{c}
Ge(Ph)_2-O-Ge(Ph)_2 \\
\diagup \qquad\qquad\qquad \diagdown \\
O \qquad\qquad\qquad\qquad O \\
\diagdown \qquad\qquad\qquad \diagup \\
Ge(Ph)_2OGe(Ph)_2
\end{array}
$$

The oxidation of dialkylgermanes R_2GeH_2 leads to the oxides R_2GeO (194). The action of gaseous oxygen on bis-(diethylhydrogermanium) oxide leads to the oxide $(Et_2GeO)_n$ (126):

$$(Et_2GeH)_2O \xrightarrow{O_2} (Et_2GeO)_n$$

The hydrolysis of the same dihydrides by 30% alcoholic potassium hydroxide yields the R_2GeO oxides (194). Dialkylgermanes react slowly with water in the presence of copper powder (119):

$$Bu_2GeH_2 + H_2O \xrightarrow{Cu} Bu_2GeO + 2H_2$$

Hydrolysis of dialkyldialkoxygermanes $R_2Ge(OR')_2$ also leads to (R_2GeO) oxides (129):

$$
Bu_2Ge(OR)_2 + H_2O \rightarrow Bu_2Ge\Big\langle{}^{OH}_{OR} + ROH
$$

$$
3\,Bu_2Ge\Big\langle{}^{OH}_{OR} \rightarrow (Bu_2GeO)_3 + 3\,ROH
$$

Dimethylgermanium oxide has been obtained in a crystalline polymeric form by hydrolysis of the sulphide Me_2GeS (185). Hydrolysis of diphenylgermanium imine leads to (Ph_2GeO) (105).

Diethylgermanium oxide exists in two polymeric forms. At room temperature the stable form is a white amorphous solid melting at about 175°C. The oxide also exists in an unstable liquid form which crystallises at 18°C (80).

Brown and Rochow have prepared and characterized trimeric, tetrameric and high-polymeric forms of dimethylgermanium oxide. At room temperature the high-polymeric form $(Me_2GeO)_n$ is the most stable and the trimeric form the least stable. At elevated temperature the vapours of dimethylgermanium oxide are trimeric. Infrared spectra in the 700 to 950 cm^{-1} region and cryoscopic as well as tensimetric measurements corroborate the existence of these forms of dimethylgermanium oxide (42).

The equilibria in carbon tetrachloride solutions between trimeric and tetrameric forms of dimethylgermanium oxide have been studied by proton NMR (152):

$$4\,(Me_2GeO)_3 \rightleftarrows 3\,(Me_2GeO)_4$$

The interconvertibility of three pure polymeric forms of the diphenylgermanium oxide is described by Metlesics and Zeiss (146). It can be summarized by the following scheme:

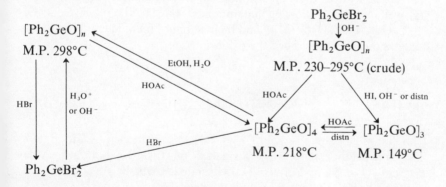

The treatment of the oxides R_2GeO by hydrohalic acids yields nearly quantitatively the dihalides R_2GeX_2 (13, 23, 80, 99, 105, 134, 135, 146, 246). HNCS reacts according to the same reaction (23). The alcoholysis of the Ge—O bonds of the R_2GeO oxides has been reported by Mehrotra and Mathur (142) (cf. synthesis of organoalkoxygermanes). The thiolysis of the Ge—O—Ge bonds of the cyclic trimer oxide Et_2GeO

by benzene–thiol has been observed (202):

$$\underset{\underset{O}{\overset{O}{\big|}}{\overset{\overset{\displaystyle Et_2}{\overset{\displaystyle Ge}{\diagup\quad\diagdown}}}{\underset{Et_2Ge\qquad\quad GeEt_2}{O\qquad\qquad O}}} + 6\,C_6H_5SH \;\rightarrow\; 3\,Et_2Ge(SC_6H_5)_2 + 3\,H_2O$$

The action of acetic anhydride on n-Pr_2GeO furnishes the diacetate (13). Tetrameric dimethylgermanium oxide and methyllithium in diethyl ether lead to quantitative yields of $(CH_3)_3GeOLi$ (188):

$$(Me_2GeO)_4 + 4\,LiCH_3 \;\rightarrow\; 4\,Me_3GeOLi$$

The redistribution reactions between $(Me_2GeO)_4$ and Me_2GeCl_2 have been studied by Moedritzer and van Wazer (153).

6-1-2-3 Germanoic Anhydrides $(RGeO)_2O$. The first anhydride of "germanoic acids", PhGeOOH, was mentioned by Morgan and Drew. Hydrolysis of phenyltribromogermane leads to "phenylgermanoic acid" whose composition varies between the limits C_6H_5GeOOH and $(C_6H_5GeO)_2O$ (158). Orndorff, Tabern and Dennis prepared phenyl-, benzyl-, p-tolyl- and dimethylaminophenylgermanoic anhydrides by ammoniacal hydrolysis of the corresponding trichlorides (162). Bauer and Burschkies described the phenyl- and naphthylgermanoic anhydrides (29) as well as the following compounds, $(p$-CH_3NH—$C_6H_4GeO)_2O$, $[p$-$(C_2H_5)_2N$—$C_6H_4GeO]_2O$ and their sulphur homologues. Burschkies (48) has moreover described the oxides:

$$[(CH_3)_2N\!-\!\!\overset{\overset{\displaystyle NO_2}{\big|}}{\underset{}{\bigcirc}}\!\!-\!GeO]_2O \quad\text{and}\quad [HOOC\!-\!\bigcirc\!-\!GeO]_2O$$

The first alkylgermanoic acid EtGeOOH was mentioned by Flood (80). The same author reports the formation of ethylgermanoic anhydride $(EtGeO)_2O$ in the hydrolysis of ethyltriiodogermane by silver oxide and in the hydrolysis of "ethylgermanium nitride" (81):

$$2\,C_2H_5GeN + 3\,H_2O \;\rightarrow\; (C_2H_5GeO)_2O + 2\,NH_3$$

$(EtGeO)_2O$ is a white solid which does not melt when heated to a temperature of 300°C. It is soluble in water and alcohol and insoluble in petroleum ether. $(n$-$C_3H_7GeO)_2O$ (100) and $(n$-$C_4H_9GeO)_2O$ (22, 179) are formed in the hydrolysis of the corresponding trichlorides $RGeCl_3$.

The action of the aryldiazonium fluoroborates on $GeCl_4$ in the presence of zinc powder leads to aryltrichlorogermanium compounds which have been isolated and analysed after treatment with a 20% solution of NaOH as the arylgermanoic anhydrides $(ArGeO)_2O$ (160).

The monoalkylgermanes such as the hexylgermane $n\text{-}C_6H_{13}GeH_3$, on treatment with an excess of 30% alcoholic potassium hydroxide, show at 80°C a slow and regular hydrogen release with formation of the corresponding oxide $(C_6H_{13}GeO)_2O$ (194). The oxidation of these hydrides $RGeH_3$ leads to the same oxides (194).

The germanoic anhydrides are white amorphous powders, soluble in some organic solvents and in sodium hydroxide solution. They decompose without melting at temperatures above 300°C. Their treatment with concentrated aqueous solutions of hydrohalic acids transform them quantitatively into trihalides.

$$(RGeO)_2O + 6\,HX \rightarrow 2\,RGeX_3 + 3\,H_2O$$

The compared reactivity of the oxides $(R_3Ge)_2O$, R_2GeO and $(RGeO)_2O$ with respect to formic, acetic acid and acetic anhydride show a progressive increase in acidity of these oxides. $(Et_3Ge)_2O$ is fairly basic. $(EtGeO)_2O$ is acidic (9).

Several organic germanium compounds, particularly some germanoic anhydrides, $(RGeO)_2O$, $R = C_6H_5$, $CH_3\text{—}C_6H_4\text{—}$, $(CH_3)_2N\text{—}C_6H_4\text{—}$, $Et_2N\text{—}C_6H_4\text{—}$, $Et_2N\text{—}C_6H_3\text{—}NO_2$, have been used for medical treatments in mice (187).

6-1-2-4 Heterogermoxanes (Ge—O—*metal compounds*). The first representatives of the alkylgermanosiloxanes $Me_3GeOSiMe_3$ and $(Me_3SiO)_2$-$GeMe_2$ were prepared by treating Me_3SiONa with Me_3GeCl and by reaction of Me_3SiOLi with Me_2GeCl_2. The infrared spectra of these derivatives have been described and a comparison is drawn with the analogous alkylsiloxanes and alkylgermoxanes (211). Lithium trimethylgermanolate reacts normally with trimethylchlorosilane to give $(CH_3)_3$-$SiOGe(CH_3)_3$ in 67% yield (230).

For the purpose of systematic investigations on the chemical and spectroscopic properties of alkylgermanosiloxanes, the compounds $Me_3SiOGeEt_3$, $Et_3SiOGeMe_3$, $(Me_3SiO)_3GeMe$ and $(Me_3SiO)_4Ge$ were synthesized from alkali trialkylsilanolates and alkylhalogermanes by Schmidbaur and Hussek (209).

Infrared and nuclear magnetic resonance spectra are reported, assigned and discussed. The ν_{as} for Si—O—Ge has been noted at 995 cm^{-1} in R_3Si—O—GeR_3, 1002 and 952 cm^{-1} in $(Me_3SiO)_2GeMe_2$, 1020 and 948 cm^{-1} in $(Me_3SiO)_3GeCH_3$, 1020 and 949 cm^{-1} in $[(CH_3)_3SiO]_4Ge$

(209). NMR spectra data of $Me_3SiOGeMe_3$ had already been given by Schmidbaur in 1963 (207).

$(CH_3)_3GeOSn(CH_3)_3$ can also be obtained quantitatively by the action of $(CH_3)_3GeOLi$ on trimethylchlorostannane (188). Lithium(chloromethyl)dimethylsilanolate serves as starting material for the synthesis of organofunctional heterosiloxanes (264):

$$(CH_3)_2(CH_2Cl)SiOLi + (CH_3)_3GeCl \rightarrow$$

$$(CH_3)_2(CH_2Cl)Si\!-\!O\!-\!Ge(CH_3)_3$$

$$(CH_3)_2(CH_2Cl)SiOLi + (CH_3)_2(CH_2Cl)GeCl \rightarrow$$

$$(CH_3)_2(CH_2Cl)Si\!-\!O\!-\!Ge(CH_2Cl)$$
$$H_3C \qquad CH_3$$

$$2\,(CH_3)_2(CH_2Cl)SiOLi + (CH_3)(CH_2Cl)GeCl_2 \rightarrow$$

$$\begin{array}{c}CH_3\\|\\(CH_2Cl)(CH_3)_2Si\!-\!O\!-\!Ge\!-\!OSi(CH_3)_2(CH_2Cl)\\|\\CH_2Cl\end{array}$$

Several organogermanosiloxanes are obtained by the interaction of triphenylgermanol, Ph_3GeOH, with triphenylsilanol and diphenylsilanediol in the presence of aluminium chloride (68) (cf. reactivity of organogermanium hydroxides, 6-1-1).

Treatment of dimethylgallium chloride, $(CH_3)_2GaCl$, with lithium trimethylgermanolate in diethyl ether afforded high yields of pentamethylgallogermoxane:

$$(Me_2GaCl)_2 + 2\,LiOGe(CH_3)_3 \rightarrow 2\,LiCl + (Me_2Ga\!-\!O\!-\!GeMe_3)_2$$

Infrared spectroscopic and X-ray analysis established that this compound possesses a planar inorganic skeleton:

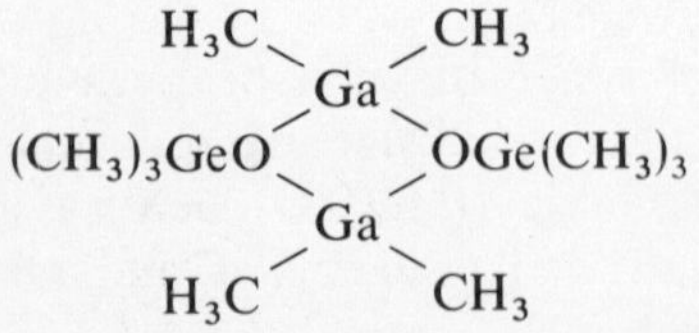

$$\begin{array}{c}H_3C \qquad\qquad CH_3\\ \diagdown\quad\diagup\\ Ga\\ \diagup\qquad\diagdown\\ (CH_3)_3GeO \qquad\qquad OGe(CH_3)_3\\ \diagdown\qquad\diagup\\ Ga\\ \diagup\qquad\diagdown\\ H_3C \qquad\qquad CH_3\end{array}$$

Another synthesis of $R_2Ga\!-\!O\!-\!GeR_3$ derivatives is the reaction of the compounds R_3GeOH with GaR_3 (26, 208):

$$2\,Ph_3GeOH + 2\,GaMe_3 \rightarrow 2\,CH_4 + (Ph_3GeOGaMe_2)_2$$

Pentamethylaluminogermoxane $(Me_2AlOGeMe_3)_2$ and pentamethyl-indogermoxane $(Me_2InOGeMe_3)_2$ were also synthesized from $[(Me)_2AlCl]_2$, $[(Me)_2InCl]_2$ and lithium trimethylgermanolate.

Physical and chemical properties as well as NMR and IR spectra of the compounds are described and their structure is established (26).

Me_3Si—O—$GeMe_3$ reacts with $AlCl_3$ to form Me_3GeCl and Me_3Si—O—$AlCl_3$ (212). The following reactions of hexamethylgermano-siloxanes have been mentioned by Schmidt and coworkers (222):

$(CH_3)_3Si$—O—$Ge(CH_3)_3$

$$+ POCl_3 \begin{array}{l} \xrightarrow{80\%} (CH_3)_3GeCl + (CH_3)_3SiOPOCl_2 \\ \xrightarrow{20\%} (CH_3)_3SiCl + (CH_3)_3GeOPOCl_2 \end{array}$$

$(CH_3)_3Si$—O—$Ge(CH_3)_3 + P_2O_3Cl_4 \rightarrow$

$$(CH_3)_3SiOPOCl_2 + (CH_3)_3GeOPOCl_2$$

Seyferth and Alleston have examined the reactions of phenyllithium with unsymmetrical hexa-organodisiloxanes and mixed Group IV oxides $(CH_3)_3Si$—O—$Ge(CH_3)_3$ and $(CH_3)_3Si$—O—$Sn(CH_3)_3$. These authors have found that nucleophilic attack by phenyllithium occurs exclusively at the more electrophilic Group IV atoms (230):

$(CH_3)_3Si$—O—$Ge(CH_3)_3 + C_6H_5Li \rightarrow$

$$(CH_3)_3GeC_6H_5 + (CH_3)_3SiOLi$$

$(CH_3)_3Si$—O—$Sn(CH_3)_3 + C_6H_5Li \rightarrow$

$$(CH_3)_3SnC_6H_5 + (CH_3)_3SiOLi$$

The action of organolithium compounds on $(CH_3)_3Ge$—O—$Sn(CH_3)_3$ takes place as expected on the tin atom (188).

Treatment of $Me_3P=CHSiMe_3$, organosilylalkylidene phosphorane, with Me_3Si—O—$GeMe_3$ gave $Me_3SiOSiMe_3$ and $Me_3P=CH$—$GeMe_3$ (214a).

6-1-3 Organogermanium oxides and hydroxides

Table 6-1(a–d) showing some physical properties of organogermanium oxides and hydroxides appears on the following pages.

Table 6-1
Organogermanium Oxides and Hydroxides

(a) R_3GeOH

COMPOUND	EMPIRICAL FORMULA	M.P. (°C)	B.P. (°c/mm)	n_D^{20}	d_4^{20}	REFERENCES
$(i\text{-}C_3H_7)_3GeOH$	$C_9H_{22}OGe$	—	$\begin{cases} 216 \\ 65\text{--}66/1 \end{cases}$	1.472	1.077	15–19
$(C_6F_5)_3$ GeOH	$C_{18}H_1F_{15}OGe$	—	115–17	—	—	78
$(C_6H_5)_3GeOH$	$C_{18}H_{16}OGe$	134.2	—	—	—	39, 107
$(C_6H_{11})_3GeOH$	$C_{18}H_{34}OGe$	175–76	—	—	—	28, 101
$(o\text{-}CH_3 \cdot C_6H_4)_3GeOH$	$C_{21}H_{22}OGe$	—	212–14/1	—	—	233
$(\alpha\text{-}C_{10}H_7)_3GeOH$	$C_{30}H_{22}OGe$	206–8	—	—	—	258

(b) (R₃Ge)₂O

COMPOUND	EMPIRICAL FORMULA	M.P. (°C)	B.P. (°C/mm)	n_D^{20}	d_4^{20}	REFERENCES
[(C₂H₅)GeH₂]₂O	C₄H₁₄OGe₂	—	145/760	1.4532	1.3489	126
[ClCH₂(CH₃)₂Ge]₂O	C₆H₁₆OCl₂Ge₂	—	74.5/4	1.4843	1.4676	31, 148
[(CH₃)₃Ge]₂O	C₆H₁₈OGe₂	−61.1	137.5/750	1.4302	1.2154	2, 86, 89, 190, 219
[(C₂H₅)₂GeH]₂O	C₈H₂₂OGe₂	—	97/21	1.4545	1.2115	126
[(C₄H₉)GeH₂]₂O	C₈H₂₂OGe₂	—	104/16	1.4590	1.2115	126
[(C₆H₅)GeH₂]₂O	C₁₂H₁₄OGe₂	—	90/0.01	1.5830	1.4340	181, 204
[(C₂H₅)₃Ge]₂O	C₁₂H₃₀OGe₂	−50	253.9/760	1.4614	1.147	9, 21, 86, 106
[n-C₄H₉)₂GeH]₂O	C₁₆H₃₈OGe₂	—	104/0.1	1.4604	1.0727	126
[n-C₃H₇)₃Ge]₂O	C₁₈H₄₂OGe₂	—	175/14	1.4648	1.068	10
[i-C₃H₇)₃Ge]₂O	C₁₈H₄₂OGe₂	—	119–20/1	1.4836	1.112	15
[(C₂H₅)₃Ge—Ge(C₂H₅)₂]₂O	C₂₀H₅₀OGe₄	—	162–65/0.3	1.5185	—	44
[n-C₄H₉)₃Ge]₂O	C₂₄H₅₄OGe₂	—	173–74/1	1.4652	1.027	12
[(C₆H₅)₂GeH]₂O	C₂₅H₂₂OGe₂	—	195/5.10⁻³	1.6130	1.3437	181, 204
[n-C₅H₁₁)₂Ge]₂O	C₃₀H₆₆OGe₂	—	156–58/0.06	1.4656	—	179, 180
[i-C₅H₁₁)₃Ge]₂O	C₃₀H₆₆OGe₂	—	160/0.2	1.4628	0.984	179, 180
[(C₆F₅)₃Ge]₂O	C₃₆OF₃₀Ge₂	270–71	—	—	—	78
[(C₆H₅)₃Ge]₂O	C₃₆H₃₀OGe₂	183–84	—	—	—	38, 85, 101, 107, 109, 158
[n- C₆H₁₃)₃Ge]₂O	C₃₆H₇₈OGe₂	—	210–11/0.4	1.4645 (25°)	0.963 (25°)	84, 179
[p-CH₃C₆H₄)₃Ge]₂O	C₄₂H₄₂OGe₂	147–48	—	—	—	233
[m-CH₃C₆H₄)₃Ge]₂O	C₄₂H₄₂OGe₂	125–25.2	—	—	—	76, 233
[(C₆H₅CH₂)₃Ge]₂O	C₄₂H₄₂OGe₂	134–35	—	—	—	29

(c) $(R_2GeO)_n$

COMPOUND	EMPIRICAL FORMULA	M.P. (°C)	B.P. (°C/mm)	n_D^{20}	d_4^{20}	REFERENCES
$[(CH_3)_2GeO]_4$	$C_8H_{24}O_4Ge_4$	92	86–96/1	—	—	{42, 153, 185, 186, 190}
$[(CH_3)_2GeO]_n$	$C_{12}H_{30}O_3Ge_3$	132–33	—	—	—	42
$[(C_2H_5)_2GeO]_3$	$C_{12}H_{30}O_3Ge_3$	18	—	—	—	80, 243
$[(C_2H_5)_2GeO]_4$	$C_{16}H_{40}O_4Ge_4$	29	128–29/3	1.4711	1.3582 (31°)	9
$[(C_4H_9)HGeO]_4$	$C_{16}H_{40}O_4Ge_4$	—	144/0.2	1.4810	1.3621	126
$[n\text{-}C_3H_7)_2GeO]_3$	$C_{18}H_{42}O_3Ge_3$	5.8	320	1.4730	1.240	13
$[i\text{-}C_3H_7)_2GeO]_3$	$C_{18}H_{42}O_3Ge_3$	44	321	—	—	15
$[(C_6H_5)HGeO]_4$	$C_{24}H_{24}O_4Ge_4$	—	—	1.615	—	181
$[n\text{-}(C_4H_9)_2GeO]_3$	$C_{24}H_{54}O_3Ge_3$	–17	180–82/1	1.4712	1.161	23
$[C_6F_5)_2GeO]_4$	$C_{48}F_{40}O_4Ge_4$	238–48	—	—	—	78
$[p\text{-}BrC_6H_4)_2OGe]_n$	$(C_{12}H_8Br_2OGe)_n$	282–88	—	—	—	76
$[m\text{-}ClC_6H_4)_2GeO]_4$	$C_{48}H_{32}Cl_{80}Ge_4$	158–59	—	—	—	76
$[p\text{-}ClC_6H_4)_2GeO]_4$	$C_{48}H_{24}Cl_8O_4Ge_4$	243–47	—	—	—	76
$[C_6H_5)_2GeO]_3$	$C_{36}H_{30}O_3Ge_3$	147–49	—	—	—	146
$[C_6H_5)_2GeO]_4$	$C_{48}H_{40}O_4Ge_4$	218	—	—	—	146, 158
$[C_6H_5)_2GeO]_n$	$(C_{12}H_{10}OGe)_n$	298	—	—	—	146, 150
$[p\text{-}CH_3C_6H_4)_2GeO]_2$	$C_{28}H_{28}O_2Ge_2$	221.5–22.5	—	—	—	75, 76
$[m\text{-}CH_3C_6H_4)_2GeO]_3$	$C_{42}H_{42}O_3Ge_3$	83–84	—	—	—	76
$[p\text{-}CH_3OC_6H_4)_2GeO]_n$	$(C_{14}H_{14}O_3Ge)_n$	281–301	—	—	—	76
$[p\text{-}C_2H_5OC_6H_4)_2GeO]_2$	$C_{32}H_{36}O_6Ge_2$	130–36	—	—	—	76
$[p\text{-}C_2H_5OC_6H_4)_2GeO]_4$	$C_{64}H_{72}O_{12}Ge_4$	177–79	—	—	—	76
$[\beta\text{-}C_{10}H_7)_2GeO]_n$	$(C_{20}H_{14}OGe)_n$	211–12.5	—	—	—	75, 76

(d) *Heterogermoxanes* (R₃Ge—O—metal compounds)

COMPOUND	EMPIRICAL FORMULA	M.P. (°C)	B.P. (°C/mm)	n_D^{20}	d_4^{20}	REFERENCES
(CH₂Cl)(CH₃)₂GeOSi(CH₃)₂(CH₂Cl)	$C_6H_{16}Cl_2OSiGe$	—	96/12	—	—	264
(CH₃)₃GeOSi(CH₃)₂(CH₂Cl)	$C_6H_{17}ClOSiGe$	—	62/12	—	—	264
(CH₃)₃GeOSi(CH₃)₃	$C_6H_{18}OSiGe$	−68	117/725	1.4038 (18°)	0.99 (20°)	209, 211
(CH₃)₃GeOSn(CH₃)₃	$C_6H_{18}SnOGe$	—	116–17	—	—	230
(CH₂Cl)(CH₃)₂Si O		—	51/12	—	—	188
(CH₂Cl)(CH₃)Ge O (CH₂Cl)(CH₃)₂Si	$C_8H_{21}Cl_5OSi_2Ge$	—	147/12	—	—	264
(CH₃)₂Ge[OSi(CH₃)₃]₂	$C_8H_{24}O_2Si_2Ge$	−61	165/725	—	—	209
			54.5/10.5	1.4064	—	211
(C₂H₅)₃GeOSi(CH₃)₃	$C_9H_{24}OSiGe$	—	33/1	—	—	209
(CH₃)₃GeOSi(C₂H₅)₃	$C_9H_{24}OSiGe$	—	34/1	—	—	209
(CH₃)₃Ge[OSi(CH₃)₃]₃	$C_{10}H_{30}O_3Si_3Ge$	—	77/10	—	—	209
[(CH₃)₃Ge—O—Al(CH₃)₂]₂	$C_{10}H_{30}Al_2O_2Ge_2$	64–66	50–55/1	—	—	26
[(CH₃)₃Ge—O—Ga(CH₃)₂]₂	$C_{10}H_{30}Ga_2O_2Ge_2$	42–44	84–85/1	—	—	26
[(CH₃)₃Ge—O—In(CH₃)₂]₂	$C_{10}H_{30}In_2O_2Ge_2$	14–16	98–100/1	—	—	26
[(CH₃)₃SiO]₄Ge	$C_{12}H_{36}O_4Si_4Ge$	−59	59/1	—	—	209
(C₆H₅)₃GeOSn(C₂H₅)₃	$C_{24}H_{30}SnOGe$	oil	198/725	—	—	67
[(CH₃)₃Ge—O—Ga(C₆H₅)₂]₂	$C_{30}H_{38}Ga_2O_2Ge_2$	186–88	167–69/0.04	—	—	26

Table 6-1 (d)—continued

COMPOUND	EMPIRICAL FORMULA	M.P. (°C)	B.P. (°C/mm)	n_D^{20}	d_4^{20}	REFERENCES
$(C_6H_5)_3GeOPb(C_6H_5)_3$	$C_{36}H_{30}PbOGe$	—	127.5–28.5	—	—	67
$[(C_6H_5)_3GeOGa(CH_3)_2]_2$	$C_{40}H_{42}Ga_2O_2Ge_2$	177–79	—	—	—	26
$[(C_6H_5)_3GeOGa(C_6H_5)_2]_2$	$C_{60}H_{50}Ga_2O_2Ge_2$	291–93	—	—	—	26
$(C_6H_5)_3GeOSi(C_6H_5)_2OSi(C_6H_5)_2OGe(C_6H_5)_3$	$C_{60}H_{50}O_3Si_2Ge_2$	90–97	—	—	—	68

6-2 ALKOXY- AND ORGANOALKOXYGERMANES

6-2-1 Syntheses

6-2-1-1 Action of Alkaline Alkoxides on Organohalogermanes. Tabern, Orndorff and Dennis were the first to synthesize tetraethoxygermane, $Ge(OC_2H_5)_4$, by the action of $GeCl_4$ on sodium ethoxide in an excess of absolute ethanol. Tetraethoxygermane is a colourless, mobile liquid which hydrolyses readily in the air with formation of germanium dioxide (241).

Tetraphenoxygermane, $(C_6H_5O)_4Ge$, is similarly synthesized by the action of $GeCl_4$ on sodium phenoxide in anhydrous benzene (229).

By the same method in anhydrous-alcohol medium, the derivatives $Ge(OR)_4$ with $R = CH_3$, C_2H_5, $n\text{-}C_3H_7$, $n\text{-}C_4H_9$, $n\text{-}C_5H_{11}$, $n\text{-}C_6H_{13}$, cyclohexyl- have been obtained (98).

West and coworkers synthesized the first alkylalkoxygermanes $CH_3Ge(OCH_3)_3$, $(CH_3)_2Ge(OCH_3)_2$, $(CH_3)_3GeOCH_3$ by the action of sodium methoxide on the methylchlorogermanes CH_3GeCl_3, $(CH_3)_2\text{-}GeCl_2$ and trimethyliodogermane (261). $EtGe(OMe)_3$, $PrGe(OMe)_3$, $Et_2Ge(OMe)_2$, $EtGe(OEt)_3$, $Me_2Ge(OEt)_2$, and $Me_2Ge(OPr)_2$ have been prepared by the same method (274). Several authors have afterwards synthesized in the same way $(CH_3O)_4Ge$ (228), $(t\text{-}BuO)_4Ge$ (238), $(C_2H_5)_3GeO\text{—}t\text{-}Bu$ (248), $(C_6H_5)_3GeOMe$ (164), $Et_3GeOCH_2CH{=}CH_2$ (195), $(CH_3)_3GeOCH_3$ (89), $Et_3GeO(CH_2)_6CH_3$, $Et_3GeOCH_2C_6H_5$, $Et_3GeOC_6H_{11}$ (118), H_3GeOPh (86a), $Me_3GeOC_6F_5$ (161a),

$\text{—OGe}(C_6H_5)_3$ (164),

(Et)(Ph)(1-naphthyl)Ge(−)menthoxide   (71),   (Et)(Ph)(1-naphthyl)-GeOCHMePh—(−) (71), and $Et_3GeGe(OMe)Et_2$ (43a).

The reaction at −80°C of methylbromogermane, CH_3GeH_2Br, with sodium methoxide produced methylmethoxygermane which decomposed slowly into methanol and white polymer of the type $(CH_3GeH)_x$ (89):

$$x\,CH_3GeH_2OCH_3 \;\rightarrow\; x\,CH_3OH + (CH_3GeH)_x$$

The methoxy- and methoxychlorogermanes, $R_2Ge(H)(OMe)$, $RGe(H)\text{-}(OMe)_2$ and $R(Cl)GeH(OMe)$ with $R = $ alkyl (or phenyl), synthesized by the direct action at low temperature of Na or Li methoxides on the chlorohydrides $R_2(H)GeCl$ and $R(H)GeCl_2$, have been characterized by "in situ" Ge—H addition to 1-alkenes. These alkoxyhydrides, generally poorly stable, decompose with formation of intermediates of germanium (II) which have been characterized by addition to conjugated dienes (128):

$\Sigma = \Sigma' = $ Et or Ph
$\Sigma = $ Et or Ph; $\quad \Sigma' = $ Cl
$\Sigma = $ Et or Ph; $\quad \Sigma' = $ OMe

6-2-1-2 Action of Organohalogermanes on Alcohols in the Presence of Bases. Pike and Fournier found that germanium tetrachloride did not react with ethanol even after refluxing the azeotrope at 65 to 77°C (168). Dimethyldichlorogermane did not react with isopropanol in refluxing hexane after one hour (123). On the other hand, Bradley, Kay and Wardlaw have been the first to prepare $(RO)_4Ge$ (R = Me, Et, n-Pr, n-Bu, i-Bu, s-Bu, n-C_5H_{11}) by direct action of germanium tetrachloride on alcohol in benzene in the presence of ammonia:

$$GeCl_4 + 4\,ROH + 4\,NH_3 \xrightarrow{C_6H_6} Ge(OR)_4 + 4\,NH_4Cl$$

In the preparation of tertiary alkoxides, the presence of pyridine followed by an ammonia treatment is necessary (36, 37). The dipyridine germanium tetrachloride adduct with boiling alcohols in the presence of further pyridine gives pyridine hydrochloride and germanium alkoxides $Ge(OR)_4$ (R = Et, n-Pr, n-Bu) (1). Alkoxychlorogermanes $(RO)_xGeCl_{4-x}$ (where x may be 1, 2, 3 and R = C_2H_5, i-C_3H_7 or t-C_4H_9) (168) were prepared by a modification of the method reported by Bradley and coworkers (37). Germanium tetraphenoxide can be conveniently prepared by passing ammonia in a mixture of germanium tetrachloride and phenol (excess) in benzene (137). Several diphenyl- and tributylalkoxides, $Ph_2Ge(OR)_2$, Bu_3GeOR (where R = Me, Et, n-Pr, n-Bu, t-Bu, s-Bu) (143), di-n-butyl-dialkoxygermanes, $Bu_2Ge(OR)_2$ (R = Me, Et, i-Pr, n-Bu, i-Bu, t-Bu, t-C_5H_{11}, sec-C_5H_{11}) (129), and butyltrialkoxygermanes, $BuGe(OR)_3$ (R = Me, Et, i-Pr) (145), have been synthesized by the same general method:

$$RGeCl_3 + 3 R'OH + 3 NH_3 \rightarrow RGe(OR')_3 + 3 NH_4Cl$$

However, with tertiary alcohols the reaction takes place in the presence of pyridine and ammonia. Zueva and coworkers also prepared a series of alkylalkoxygermanes including $EtGe(OPr)_3$, $Me_2Ge(OPr)_2$, $MeGe(OPh)_3$, $Me_2Ge(OPh)_2$, $MeGe(OCH_2CH=CH_2)_3$, $Me_2Ge(OCH_2-CH=CH_2)_2$, $Me_2Ge(o\text{-}t\text{-}Bu)_2$, $Me(Cl)Ge(o\text{-}t\text{-}Bu)_2$ and $[MeGe(o\text{-}t\text{-}Bu)_2]_2$-NH (274).

$(C_6H_5)_2GeCl(Ox)$ (HOx = 8-hydroxyquinoline) has been prepared by reaction of 8-hydroxyquinoline with $(C_6H_5)_2GeCl_2$ in alcoholic solution and addition of NH_3 (93). $Ph_3GeOC_6F_5$ is prepared by the action of Ph_3GeBr with C_6F_5OH in the presence of triethylamine (161a).

6-2-1-3 Transalkoxylation Reactions. Alkoxide exchange reactions can be carried out between germanium alkoxide and alcohols, followed by distillation of the lower-boiling alcohol. In this way tetraethoxygermane was converted to tetrabutoxygermane by alcohol interchange (36):

$$Ge(OEt)_4 + 4 ROH \rightarrow Ge(OR)_4 + 4 EtOH$$

$R = n\text{-}Bu; i\text{-}Bu; s\text{-}Bu.$

This reaction has been generalized for the preparation of $Ge(O-i\text{-}Pr)_4$ and $Ge(O-n\text{-}C_5H_{11})_4$ (37, 228). Diethoxydichlorogermane, on treatment with n-butanol and n-hexanol, gave di-n-butoxy- and di-n-hexoxydichlorogermane, respectively (168).

Mehrotra and Chandra prepared the series of ethoxyphenoxygermanes $(C_6H_5O)_xGe(OC_2H_5)_{4-x}$, $x = 1, 2, 3, 4$, by reacting stoichiometric ratios of phenol with tetraethoxygermane (137). (Diphenylmethoxy)triphenylgermane was prepared from Ph_3GeOMe and benzhydrol (164).

The transalkoxylation reactions of methoxy- and ethoxytrialkylgermanes and several ethylenic and acetylenic alcohols or glycols lead to the corresponding alkenoxy- and alkynoxyalkylgermanes in high yields (195). By the same methods, several dibutyldialkoxygermanes, $Bu_2Ge(OR)_2$ (130), and butyltrialkoxygermanes, $BuGe(OR)_3$ (145), have been synthesized.

Butyltrialkoxygermanes are also obtained from butylgermanoic anhydride (145):

$$(BuGeO)_2O + 6 BuOH \rightarrow 2 BuGe(OBu)_3 + 3 H_2O$$

In the reaction of the diethoxy- and di-isopropoxydibutylgermanes with higher alcohols the reactivity of these alcohols gradually decreases with the branching of the alkyl chain in the alcohol molecule. The reaction with n-butanol was completed within a short period of time; treatment of dibutyldiethoxygermane with *sec*-butanol was found to give mainly dibutylethoxy-*sec*-butoxygermane and further interchange could only be completed by the addition of an acid catalyst such as *p*-toluenesulphonic acid. With tertiary alcohols, the reaction did not appear to proceed at all without a catalyst (130).

Mehrotra and Mathur (142) have synthesized a number of dibutyl-, tributyl- and diphenylalkoxygermanes as well as their derivatives with glycols, acetylacetone, and substituted amine directly from their oxides. *p*-Toluenesulphonic acid was generally used as a catalyst in these reactions:

$$(Bu_3Ge)_2O + 2 ROH \rightarrow 2 Bu_3GeOR + H_2O$$

$$R_2GeO + 2 R'OH \rightarrow R_2Ge(OR')_2 + H_2O$$

$$(Bu_3Ge)_2O + R(OH)_2 \rightarrow (Bu_3GeO)_2R + H_2O$$

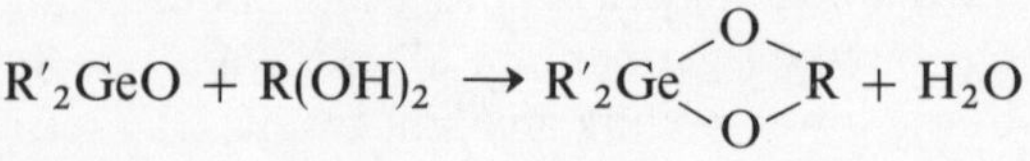

$$R'_2GeO + R(OH)_2 \rightarrow R'_2Ge{\overset{O}{\underset{O}{<>}}}R + H_2O$$

$$(Bu_3Ge)_2O + 2 Hacac \rightarrow 2 Bu_3Ge(acac) + H_2O$$

$$R'_2GeO + 2 Hacac \rightarrow R'_2Ge(acac)_2 + H_2O$$

$$(Bu_3Ge)_2O + (HO{-}CH_2CH_2)_2NH \rightarrow (Bu_3GeOCH_2CH_2)_2NH + H_2O$$

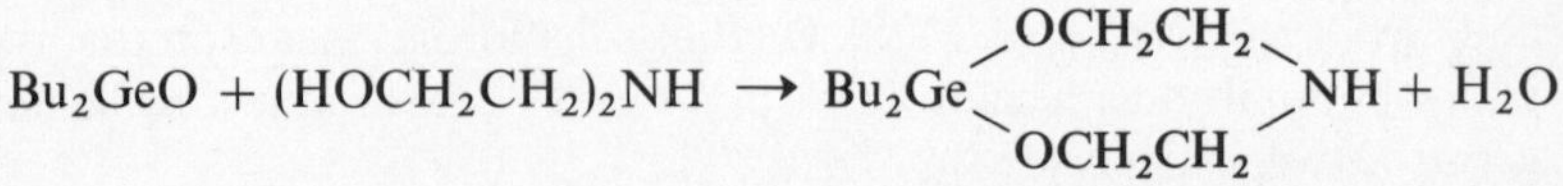

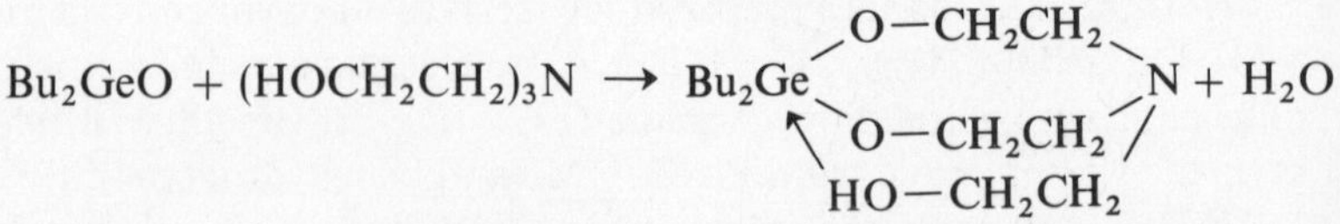

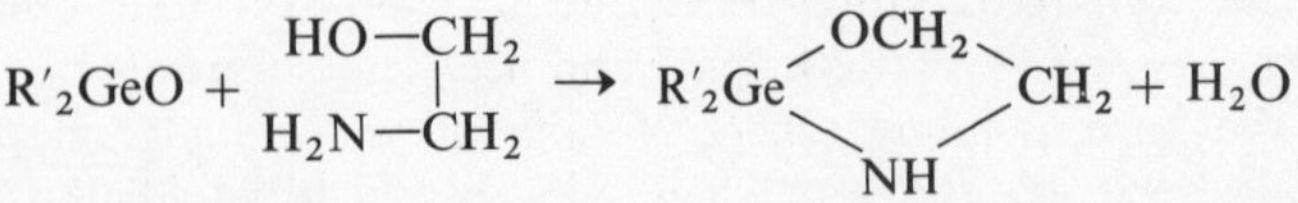

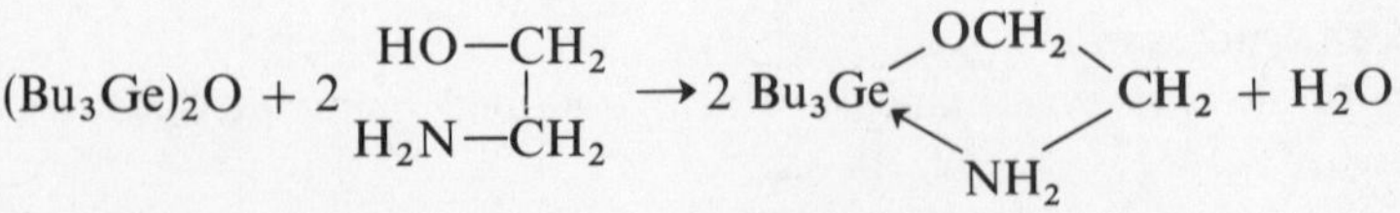

$$Bu_2GeO + (HOCH_2CH_2)_2NH \rightarrow Bu_2Ge{\overset{OCH_2CH_2}{\underset{OCH_2CH_2}{<>}}}NH + H_2O$$

$$Bu_2GeO + (HOCH_2CH_2)_3N \rightarrow Bu_2Ge{\overset{O-CH_2CH_2}{\underset{O-CH_2CH_2}{<>}}}N + H_2O$$
$$HO-CH_2CH_2$$

$$R'_2GeO + {\overset{HO-CH_2}{\underset{H_2N-CH_2}{|}}} \rightarrow R'_2Ge{\overset{OCH_2}{\underset{NH}{<>}}}CH_2 + H_2O$$

$$(Bu_3Ge)_2O + 2 {\overset{HO-CH_2}{\underset{H_2N-CH_2}{|}}} \rightarrow 2 Bu_3Ge{\overset{OCH_2}{\underset{NH_2}{<>}}}CH_2 + H_2O$$

1-organogermatranes were obtained by the same way (247a):

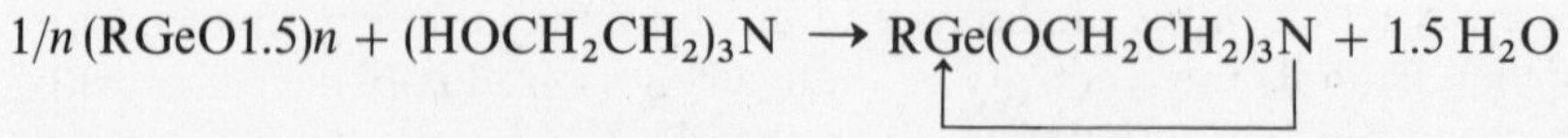

$$1/n (RGeO1.5)n + (HOCH_2CH_2)_3N \rightarrow RGe(OCH_2CH_2)_3N + 1.5 H_2O$$

$$R = CH_3, C_2H_5, C_6H_5, \alpha\text{-}C_{10}H_7$$

as well as the dimethoxydigermanes (43a):

$$\boxed{(Et_2GeGeEt_2O)_2} \xrightarrow[-H_2O]{MeOH} Et_2(MeO)GeGe(OMe)Et_2$$

The reactions of glycols with tri-n-butylalkoxygermanes lead to the bis-(tributylgermanium)glycolates which are colourless, viscous liquids (144):

$$2\,Bu_3GeOR + HO—X—OH \rightarrow Bu_3GeO—X—O—GeBu_3 + 2\,ROH$$

Reactions between tetraethoxygermane and various glycols in different stoichiometric ratios lead to alkylenedioxydiethoxy- and bis-alkylenedioxygermanes (139):

$$Ge(OEt)_4 + \begin{matrix} HO \\ \\ HO \end{matrix}\!\!\diagdown\!\!\diagup X \rightarrow Et_2Ge \begin{matrix} O \\ \\ O \end{matrix}\!\!\diagup\!\!\diagdown X + 2\,EtOH$$

$$Ge(OEt)_4 + 2 \begin{matrix} HO \\ \\ HO \end{matrix}\!\!\diagdown\!\!\diagup X \rightarrow X \begin{matrix} O \\ \\ O \end{matrix}\!\!\diagdown\!\!\diagup Ge \begin{matrix} O \\ \\ O \end{matrix}\!\!\diagdown\!\!\diagup X + 4\,EtOH$$

The study of these reactions has been pursued and completed by the latter authors, who described the formation of cyclic products in the reaction of dibutyldialkoxygermanes (130) and diphenyldialkoxygermanes (144) in the presence of p-toluenesulphonic acid as catalyst:

$$R_2Ge(OR')_2 + X(OH)_2 \rightarrow R_2Ge \begin{matrix} O \\ \\ O \end{matrix}\!\!\diagup\!\!\diagdown X + 2\,R'OH$$

Molecular weight determinations show that these compounds are monomeric. From the relatively stable dialkylmethoxyhydrogermanes, several transalkoxylation reactions with ethylenic and acetylenic alcohols lead to alkenoxy- and alkynoxyhydrogermanes:

$$R_2(H)Ge—O\!\!\left(\!C\!\right)_n\!\!—C{=}CH_2 \quad \text{and} \quad R_2(H)Ge—O\!\!\left(\!C\!\right)_n\!\!—C{\equiv}CH$$

The intramolecular addition of the Ge—H group to the terminal C=C or C≡C bonds leads to cyclic germanium ethers (127, 128):

$$R_2Ge \begin{matrix} \diagup \\ \diagdown \\ O \end{matrix}\!\!\rceil, \quad R_2Ge \begin{matrix} \diagup \\ \diagdown \\ O \end{matrix}\!\!\rceil, \quad R_2Ge \begin{matrix} \diagup \\ \diagdown \\ O \end{matrix}\!\!\rangle$$

Trimethylmethoxygermane is isolated in 77% yield in the reaction of trimethylcyclohexoxygermane with tributylmethoxytin (165).

6-2-1-4 Addition Reactions of Organogermanium Hydrides on the Carbonyl Group of Aldehydes and Ketones. Alkylgermanium hydrides add to the carbonyl group of saturated aldehydes and ketones near 150°C with formation of the corresponding alkoxygermanes (118):

$$R_3GeH + O=C\diagup_{\diagdown} \xrightarrow{Cu} R_3Ge-O-CH\diagup_{\diagdown}$$

In the presence of the same catalyst (Cu powder), a selective addition of the same hydrides to the carbonyl group of unsaturated aldehydes and ketones with formation of alkenoxygermanes is observed (201).

Triphenylgermane and diphenylchlorogermane add by a radical mechanism to the carbonyl group of saturated aldehydes and ketones (181). Triphenylgermane adds to the carbonyl group of methylheptenone and mesityl oxide (1,4-addition) to give enoxygermanes (205, 206). In the presence of chloroplatinic acid, triethylgermane adds in the 1,4 position to vinyl methyl ketone (124):

$$Et_3GeH + CH_2=CH-COCH_3 \xrightarrow{H_2PtCl_6} Et_3Ge-O-\underset{\underset{CH_3}{|}}{C}=CH-CH_3$$

Trimethylgermane adds to hexafluoroacetone to give (1,1,3,3,3-hexafluoro-propoxy)trimethylgermane (57, 239). Triethylgermane and triphenyl-germane add to the carbonyl group of diphenylketene with formation of vinyloxygermanes (182, 205):

$$R_3GeH + Ph_2C=C=O \rightarrow R_3GeOCH=CPh_2 \qquad (R = Et\ or\ Ph)$$

The monoaddition derivatives of the dihydrides R_2GeH_2, $RClGeH_2$, (R = Et or Ph) to several ethylenic ketones and aldehydes undergo cyclization (intramolecular addition of Ge—H on the carbonyl group) (127, 128, 205, 206):

$$R_2Ge\underset{H}{\overset{|}{+}}\overset{|}{C}\underset{n}{\overset{|}{+}}\underset{O}{\overset{||}{C}}- \xrightarrow[or\ \Delta]{AIBN} R_2Ge\underset{O}{\overset{}{\diagup}}\underset{n+1}{\overset{}{C}}-$$

(For further details cf. reactivity of organogermanium hydrides and organohalogermanium hydrides.)

6-2-1-5 Catalytic Dehydrocondensation Reactions between Organoger-manium Hydrides and Alcohols, Glycols, Phenols, etc. Alkoxygermanes can be isolated in high yield by catalytic dehydrocondensation on reduced copper powder of alkylgermanium hydrides with alcohols (118, 119, 195):

$$R_3GeH + R'OH \rightarrow R_3GeOR' + H_2$$

The dialkylgermanes, in the presence of copper (119) or on Raney nickel (127, 128) give dialkylalkoxygermanes:

$$R_2GeH_2 + R'OH \rightarrow R_2\underset{|}{Ge}-OR'$$
$$H$$

or dialkyldialkoxygermanes, $R_2Ge(OR')_2$ (119). Dibutylgermane and 1,4-butanediol give a cyclic derivative (yield: 80%):

$$Bu_2GeH_2 + HO(CH_2)_4OH \rightarrow Bu_2Ge \underset{\diagdown}{\overset{\diagup}{}} \begin{matrix} OCH_2-CH_2 \\ | \\ OCH_2-CH_2 \end{matrix}$$

Glycols and diphenols dehydrocondense with formation of derivatives of the type $R_3Ge-O-X-O-GeR_3$ (119) ($X = -(CH_2)_{\overline{2}}, -(CH_2)_{\overline{4}}, -C_6H_4-$).

In the presence of reduced copper, trialkylgermanes and several ethylenic alcohols give the dehydrocondensation reaction only, without any addition reaction, with formation of alkenoxygermanes (e.g. allyl, methallyl, crotyl, furfuryl alcohols and triethylgermyl-3-propene-2-ol-1) (195). A reaction of the same type has been reported between triphenylgermane and crotyl alcohol (43).

On the other hand, the formation of alkynoxygermanes from trialkylgermanes and acetylenic alcohols or glycols takes place in low yields (195).

The intramolecular catalytic dehydrocondensation reactions of germanium alcohols with a Ge—H bond gives germanium heterocycles with Ge—O bonds (127, 128):

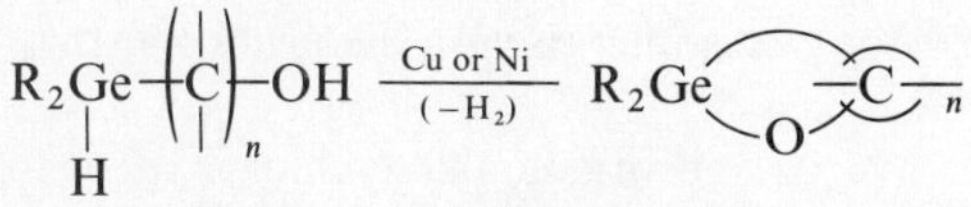

$$R_2\underset{|}{Ge} \left(\overset{||}{\underset{||}{C}} \right)_n OH \xrightarrow[(-H_2)]{Cu \text{ or } Ni} R_2Ge \underset{O}{\overset{}{\diagup}} \left(\overset{}{C} \right)_n$$
$$H$$

6-2-1-6 *Exchange and Redistribution Reactions.* Pike and Dewidar described the formation of tetraalkoxy- and chloroalkoxygermanes by treatment of germanium tetrachloride with trialkyl orthoformates in the presence of aluminium chloride (167):

$$\text{>}GeCl + HC(OR)_3 \rightarrow \text{>}Ge-OR + R-Cl + HCOOR$$

The reactions between the trialkylalkoxytins with germanium halides are a convenient way of replacing germanium-attached halogen by the alkoxy group (27, 29a):

$$GeCl_4 + 4\,Bu_3SnOR \rightarrow Ge(OR)_4 + 4\,Bu_3SnCl$$

$$(R = CH_3; C_2H_5)$$

$$Et_3GeI + Bu_3SnOCH_3 \rightarrow Et_3GeOCH_3 + Bu_3SnI$$

$$n\text{-}Pr_2GeHI + Bu_3SnOCH_3 \rightarrow n\text{-}Pr_2Ge(H)OCH_3 + Bu_3SnI$$

The yields of alkoxygermanes obtained by this method were 70–95 %. Germanium alkylacetates with alkoxygroups on the germanium atom were prepared by this method (27):

$$Cl_nGe(CH_2COOCH_3)_{4-n} \xrightarrow{R_3SnOCH_3}$$

$$(CH_3O)_nGe(CH_2COOCH_3)_{4-n} + nR_3SnCl$$

The reaction of trialkyl(1-cyclohexen-1-yloxy)tin with organochlorogermanes is a convenient method for the synthesis of O–germanium derivatives of 1-cyclohexen-1-ol (171):

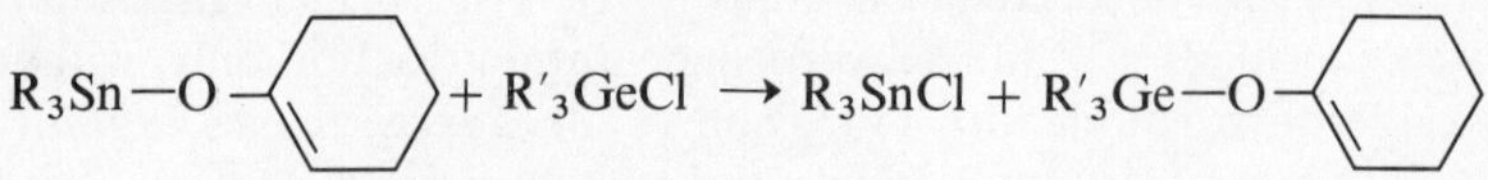

When tripropylstannylacetone is treated with trimethylbromogermane C– and O–germyl products are formed:

$$Me_3GeBr \xrightarrow[-Pr_3SnBr]{+Pr_3SnCH_2COCH_3}$$

$$Me_3GeCH_2COCH_3 + Me_3GeO\!-\!C(CH_3)\!=\!CH_2$$

The O-isomer content of the distillate was shown by NMR to be 10 %. The proportion of the two isomers is variable, according to the temperature (120).

Organoalkoxyhydro- and organoalkoxychlorohydrogermanes, $R_2Ge(H)OMe$, $RGe(H)(OMe)_2$, $R(Cl)Ge(H)OMe$, can be synthesized by exchange reactions (128):

$$R(H)GeCl_2 + R'_3GeOMe \rightarrow R(Cl)GeH(OMe) + R'_3GeCl$$

$$R = Et \text{ and } R' = Ph \quad \text{or} \quad R = Ph \text{ and } R' = Et$$

The action of triethylgermane at 180°C on Et_3SnOMe leads to the formation of Et_3GeOCH_3 in 48 % yield (249):

$$Et_3GeH + 2\,Et_3SnOMe \rightarrow Et_3GeOMe + CH_3OH + (Et_3Sn)_2$$

2-2′-Diphenylenedioxydimethylsilane and germanium tetrachloride give bis-(2,2′-diphenylenedioxy)germane (232):

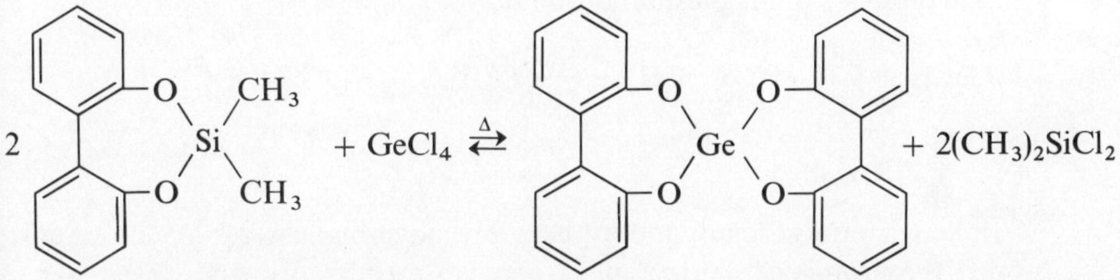

Monoalkoxytrichloroderivatives $ROGeCl_3$ can be conveniently prepared by refluxing appropriate tetraalkoxygermanes with excess germanium tetrachloride (138):

$$Ge(OR)_4 + 3\,GeCl_4 \rightarrow 4\,(RO)GeCl_3$$

Many papers deal with redistribution reactions between organogermanium compounds. The first reactions of this type with germanium–oxygen-bond derivatives have been studied by Moedritzer and van Wazer (153):

$$(CH_3)_2GeX_2 + (CH_3)_2Ge(OCH_3)_2 \rightleftarrows 2\,(CH_3)_2GeX(OCH_3)$$

$$(X = Cl, Br, I)$$

These authors have further studied the equilibria in the scrambling between dimethylgermanium of chlorine, bromine or iodine with either methoxy or methylthio groups by proton nuclear magnetic resonance. The scrambling of methoxy with the methylthio group is also reported (247). Equilibria of the systems $Ge(OMe)_4$—$GeCl_4$ and $Ge(OMe)_4$—$Ge(NMe_2)_4$ have also been studied by proton NMR spectra (46). All these reactions have been analysed in a summary by Moedritzer, "Redistribution Reactions of Organometallic Compounds of Silicon, Germanium, Tin and Lead" (151).

The equilibria in the systems CH_3GeCl_3—$CH_3Ge(OCH_3)_3$, $(CH_3)GeBr_3$—$(CH_3)Ge(OCH_3)_3$, $(CH_3)GeI_3$—$CH_3Ge(OCH_3)_3$ (154); $Ge(OCH_3)_4$—GeZ_4 ($Z = Cl$, Br, I or NCO) (33); $Ge(OCH_3)_4$—$Ge(SCH_3)_4$ (34); CH_3GeT_3—$(CH_3)_2GeZ_2$ ($T = OC_6H_5$, OCH_3, SCH_3; $Z = Cl$, Br, I) (156); Me_2SiCl_2, Me_2SiBr_2, Me_2GeI_2 and $Me_2Ge(OPh)_2$ (152a); Me_2GeCl_2, Me_2GeI_2, and $Me_2Ge(OPh)_2$ (156a); Me_2GeCl_2, Me_2GeBr_2, Me_2GeI_2 and $Me_2Ge(OPh)_2$ (156a) have been studied by the same methods.

6-2-1-7 Action of Halo- and Organohalogermanes on Epoxides. The action of halogermanes on cyclic ethers is a new approach to the synthesis of alkoxy-substituted germanium compounds. In a preliminary communication Pike and Lavigne reported the action of $GeCl_4$ on propylene oxide in the presence of magnesium bromide (169):

$$GeCl_4 + CH_3-CH-CH_2 \xrightarrow{\text{MgBr}_2} Cl_{(4-x)}Ge(OCH_2CHCl-CH_3)_x$$
$$\diagdown \diagup$$
$$O$$

$(x = 1, 2, 3 \text{ or } 4)$

Triphenylchlorogermane and triphenylbromogermane were also shown to open the propylene oxide and 1,2-epoxy-3-methoxypropane ring under similar conditions (169).

In the reaction of tri-n-propylbromogermane with propylene oxide the effects on the reaction rate of temperature, concentration of reacting species and the observed second-order kinetics indicate a bimolecular-type mechanism:

$$n\text{-Pr}_3GeBr + CH_2-CH-CH_3 \rightarrow n\text{-Pr}_3Ge-OCH_2CH(Br)CH_3$$
$$\diagdown \diagup$$
$$O$$

This reaction offers after hydrolysis a route to the synthesis of 2-halo-1-hydroxy substituted alkanes (170).

In the condensation of tri-n-propylbromogermane with ethylene oxide, propylene oxide, 3-chloro-1,2-epoxy propane and 3-methoxy-1,2-epoxy propane, the direction of ring opening is determined after hydrolysis and formation of the corresponding halohydrins. The structure of the compounds obtained in the condensation of n-Pr_3GeBr with 3-methoxy-1,2-epoxypropane, $(n\text{-}C_3H_7)_3GeOCH(CH_2OCH_3)CH_2Br$, and with 3-chloro-1,2-epoxy propane, $(n\text{-}C_3H_7)_3GeOCH(CH_2Cl)CH_2Br$, shows that the direction of the ring opening is opposite to that observed for propylene oxide and not that expected from the point of view of steric effects (114).

In the action of methyl- and ethylchlorogermanes on ethylene oxide, Zueva and Ponomarenko (275) observed that the alkylchlorogermanes react less actively than the corresponding alkylchlorosilanes and that the reactivity of alkylchlorogermanes increases with increasing numbers of chlorine atoms on the germanium:

$$(CH_3)_nGeCl_{4\text{-}n} + (4\text{-}n)H_2C-CH_2 \rightarrow (CH_3)_nGe(OCH_2CH_2Cl)_{4\text{-}n}$$
$$\diagdown \diagup$$
$$O$$

$AlCl_3$ and H_2PtCl_6 are used as catalysts in these condensation reactions (275). Mironov and Kravchenko have found that ethylene oxide reacts with $(CH_3)_3GeBr$ without catalyst (147). Cleavage of ethylene oxide and propylene oxide by diethylchlorogermane has also been reported (125):

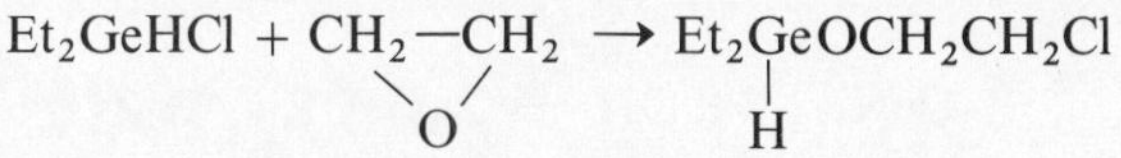

$$Et_2GeHCl + CH_2\!\!-\!\!CH_2 \underset{O}{\diagdown\!\diagup} \rightarrow Et_2Ge\,OCH_2CH_2Cl \atop H$$

Condensation products of several halogermanes with cyclohexene oxide have been recently identified and characterized by Lavigne and coworkers (115). The opening of the oxide linkage of cyclohexene oxide has been investigated through hydrolysis of the intermediate products (*cis-trans* cyclohexene halohydrin isomers) produced in these reactions (115).

6-2-1-8 Reactions of Triorganogermanes with Unsaturated α-Oxides. Triethylgermane adds to unsaturated α-oxides mainly in the 1,4 manner, forming triethylalkenoxygermanes (43):

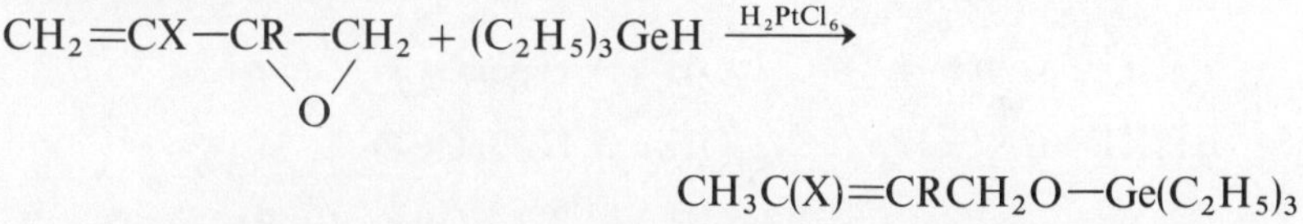

$$CH_2\!\!=\!\!CX\!\!-\!\!CR\!\!-\!\!CH_2 + (C_2H_5)_3GeH \xrightarrow{\;H_2PtCl_6\;}$$
$$\underset{O}{\diagdown\!\diagup}$$

$$CH_3C(X)\!\!=\!\!CRCH_2O\!\!-\!\!Ge(C_2H_5)_3$$

$$(R = H, X = H; R = CH_3, X = H; R = H, X = Cl)$$

Triphenylgermane adds to butadiene and isoprene oxides under the same conditions both to the oxide and to the C=C bonds (43):

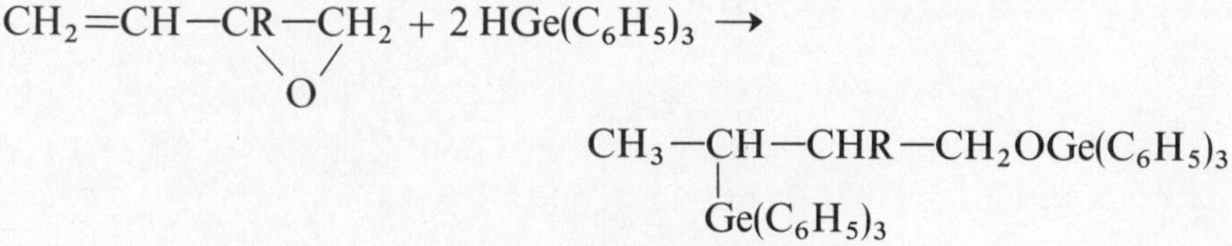

$$CH_2\!\!=\!\!CH\!\!-\!\!CR\!\!-\!\!CH_2 + 2\,HGe(C_6H_5)_3 \rightarrow$$
$$\underset{O}{\diagdown\!\diagup}$$

$$CH_3\!\!-\!\!CH\!\!-\!\!CHR\!\!-\!\!CH_2OGe(C_6H_5)_3 \atop Ge(C_6H_5)_3$$

$$(R = H \text{ and } R = CH_3).$$

The interaction of triethylgermane with piperylene oxide in the presence of chloroplatinic acid yields mainly 1,4 products,

$$CH_3\!\!-\!\!CH\!\!=\!\!CH\!\!-\!\!CH\!\!-\!\!O\!\!-\!\!GeEt_3 \quad (159a). \atop CH_3$$

6-2-1-9 Diverse Syntheses. Trialkylgermanes react with mercuri-bis-acetaldehyde to give vinyloxygermanes (30):

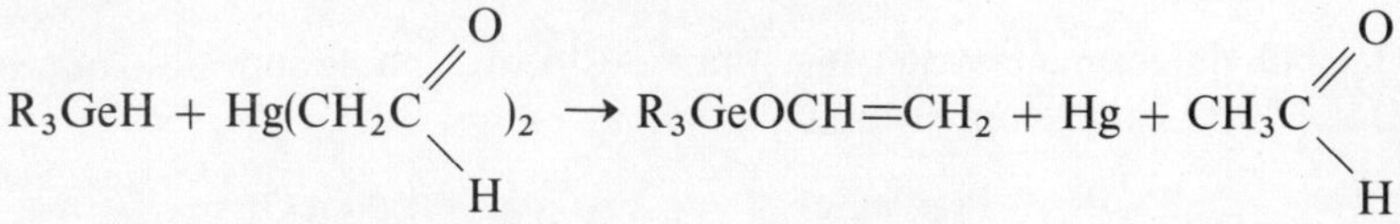

$$R_3GeH + Hg(CH_2C\overset{O}{\diagdown\!\!\diagdown}_{H})_2 \rightarrow R_3GeOCH\!\!=\!\!CH_2 + Hg + CH_3C\overset{O}{\diagdown\!\!\diagdown}_{H}$$

The alcoholysis of different germanium–heteroelement bonds leads to alkoxygermanes:

$$R_3GeNR' + R'OH \rightarrow R_3GeOR' + R'_2NH \quad (178, 203)$$

$$(R_3Ge)_2NH \text{ or } (R_3Ge)_3N + R'OH \rightarrow R_3GeOR' + NH_3 \quad (196)$$

$$(Et_3Ge)_2NH + HO-C_6H_4-OH \rightarrow$$

$$Et_3GeOC_6H_4OGeEt_3 + HN_3 \quad (196)$$

$$(Et_3Ge)_2NH + 2\,HC{\equiv}C-CH_2OH \rightarrow$$

$$2\,Et_3GeOCH_2C{\equiv}CH + NH_3 \quad (149)$$

$$(Et_2GeNH)_3 + 6\,CH_3OH \rightarrow$$

$$3\,Et_2Ge(OCH_3)_2 + 3\,NH_3 \quad (150)$$

$$H_3GeNCNGeH_3 + CH_3OH \rightarrow H_3GeOCH_3 \quad (54)$$

$$H_3GeNCNGeH_3 + CD_3OH \rightarrow H_3GeOCD_3 \quad (54)$$

$$R_3GePEt_2 + R'OH \rightarrow R_3GeOR' + Et_2PH \quad (197)$$

$$(Et_3Ge)_2Cd + C_3H_7OH \rightarrow$$

$$Et_3GeH + Et_3GeOC_3H_7 + Cd \quad (254)$$

$$Et_3GeSC_2H_5 + n\text{-}C_8H_{17}OH \rightarrow$$

$$Et_3GeOC_8H_{17} + C_2H_5SH \quad (202)$$

Triethyl-*tert*-butoxygermane was formed in the reaction of $(C_2H_5)_3Ge^{\cdot}$ radicals with di-*tert*-butyl peroxide (250):

$$(C_2H_5)_3Ge^{\cdot} + (CH_3)_3COOC(CH_3)_3 \rightarrow$$

$$(C_2H_5)_3GeOC(CH_3)_3 + (CH_3)_3CO^{\cdot}$$

The interaction of bis-(triethylgermyl)mercury with dicyclohexylperoxydicarbonate leads to an unstable compound which decarboxylates with formation of triethylcyclohexyloxygermane (256):

$$(C_2H_5)_3GeHgGe(C_2H_5)_3 + (C_6H_{11}OCOO)_2 \rightarrow$$

$$Hg + 2\,(C_2H_5)_3GeO-COOC_6H_{11}$$
$$\downarrow$$
$$(C_2H_5)_3GeOC_6H_{11} + CO_2$$

Hexaalkyldigermanes react instantaneously with potassium ethoxide in HMPT at room temperature (43a, 45):

$$EtOK + Et_3GeGeEt_3 \rightarrow Et_3GeK + Et_3GeOEt$$

Peddle and Ward (164) showed that α-germylcarbinols can rearrange in germylethers:

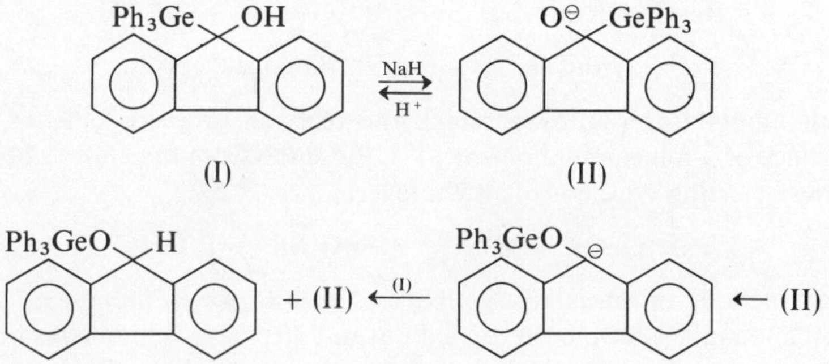

Thermal rearrangement of $(+)(Me)(Ph)(1\text{-naphthyl})GeCOOMe$ at 230–270°C gave 95% of $(-)(Me)(Ph)(1\text{-naphthyl})GeOMe$ and carbon monoxide (40a).

6-2-2   Properties

The Ge—O bond of the alkoxygermanes is sensitive to hydrolysis. Several papers report studies in this field. The hydrolysis of tetraethoxygermane in ambient air (241) or by means of water in alcoholic solution (113) leads to germanium dioxide. In the hydrolysis of $(i\text{-}C_3H_7O)_4Ge$, $(i\text{-}C_3H_7O)_3GeOGe(O—i\text{-}C_3H_7)_3$ and a polymer $[Ge_2O_3(i\text{-}C_3H_7)_2]_n$ can be isolated (228). Di-n- and sec-alkoxydibutylgermanes are readily and quantitatively hydrolysed to trimeric dibutylgermanium oxide by water at room temperature:

$$3\,Bu_2Ge(OR)_2 \xrightarrow{H_2O} (Bu_2GeO)_3 + 6\,ROH$$

di-n-Butyldi-$tert$-butoxygermane hydrolyses only in the presence of p-toluenesulphonic acid. Presumably nucleophilic attack by water on germanium is preceded by protonation of the oxygen atom (129). (Diphenylmethoxy)triphenylgermane is very sensitive to moisture and hydrolysed readily to bis(triphenylgermanium) oxide and benzhydrol (164). Upon hydrolysis the monomeric alkylalkoxygermanes polymerized to polyorganogermoxanes with loss of the alkoxy group (274). Methyl(β-chloroethoxy)germanes $(CH_3)_nGe(OCH_2CH_2Cl)_{4-n}$ $(n = 0, 1, 2, 3)$ are readily hydrolysed. $CH_3Ge(OCH_2CH_2Cl)_3$ gives an infusible polymer $[CH_3GeO_{1.5}]_n$ while $(CH_3)_2Ge(OCH_2CH_2Cl)_2$ forms polydimethylgermoxane $[(CH_3)_2GeO]_n$ or octamethyltetracyclogermoxane $[(CH_3)_2GeO]_4$ (275). Dimeric dibutylgermanium sulphide has been obtained by passing anhydrous hydrogen sulphide gas into a solution of di-*n*- and *sec*-alkoxydibutylgermanes in the parent alcohol (129):

$$Bu_2Ge(OR)_2 + 2\,H_2S \rightarrow Bu_2Ge(SH)_2 + 2\,ROH$$

$$2\,Bu_2Ge(SH)_2 \rightarrow (Bu_2GeS)_2 + 2\,H_2S$$

With dibutyldi-*tert*-butoxygermane the reaction proceeds only in the presence of *p*-toluenesulphonic acid (129). Thiolysis of the Ge—O bonds is an interesting synthesis of alkylthiogermanes (2, 202):

$$R_3GeOR' + R''SH \rightarrow R_3GeSR'' + R'OH$$

The reactions of tetraalkoxygermanes, $Ge(OR)_4$, with mono-, di- and triethanolamines have been carried out in different stoichiometric ratios in benzene media (140). With monoethanolamine the following reactions have been reported:

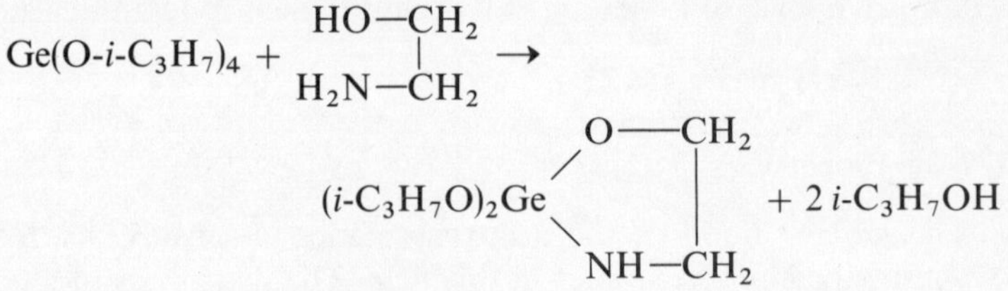

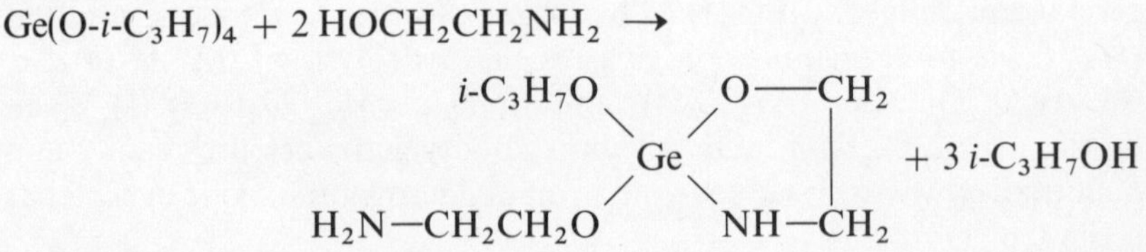

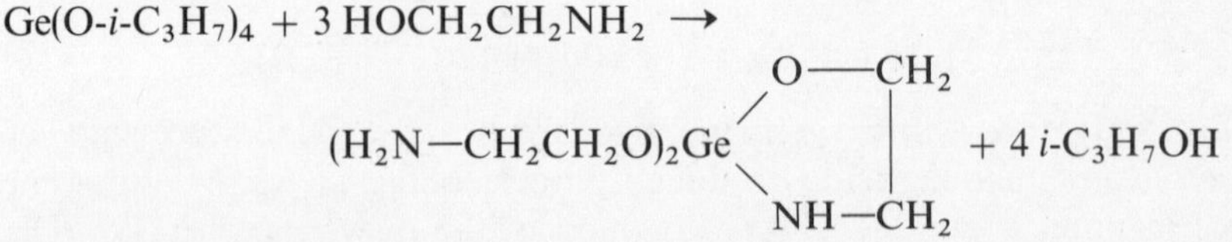

$$Ge(O\text{-}i\text{-}C_3H_7)_4 + \begin{array}{c} HO-CH_2 \\ | \\ H_2N-CH_2 \end{array} \rightarrow$$

$$(i\text{-}C_3H_7O)_2Ge \begin{array}{c} O-CH_2 \\ | \\ NH-CH_2 \end{array} + 2\,i\text{-}C_3H_7OH$$

$$Ge(O\text{-}i\text{-}C_3H_7)_4 + 2\,HOCH_2CH_2NH_2 \rightarrow$$

$$\begin{array}{c} i\text{-}C_3H_7O \\ \diagdown \\ Ge \\ \diagup \quad \diagdown \\ H_2N-CH_2CH_2O \end{array} \begin{array}{c} O-CH_2 \\ | \\ NH-CH_2 \end{array} + 3\,i\text{-}C_3H_7OH$$

$$Ge(O\text{-}i\text{-}C_3H_7)_4 + 3\,HOCH_2CH_2NH_2 \rightarrow$$

$$(H_2N-CH_2CH_2O)_2Ge \begin{array}{c} O-CH_2 \\ | \\ NH-CH_2 \end{array} + 4\,i\text{-}C_3H_7OH$$

With diethanolamine $Ge[(OCH_2CH_2)_2NH]_2$ is formed from $Ge(OC_2H_5)_4$; and one or two moles of triethanolamine with tetra-alkoxygermanes give the following reactions (140):

$$Ge(OR)_4 + (HOCH_2CH_2)_3N \rightarrow ROGe(OCH_2CH_2)_3N + 3\,ROH$$

$$Ge(OR)_4 + 2\,(HOCH_2CH_2)_3N \rightarrow$$

$$Ge[(O-CH_2CH_2)_2NCH_2CH_2OH]_2 + 4\,ROH$$

Reactions between tetraethoxygermanes and salicylic acid in different stoichiometric ratios have been studied by Mehrotra and Chandra (141).

The formation of different ethoxysalicylate derivatives of germanium isolated during these studies can be represented by the following equations:

$$\text{Ge(OEt)}_4 + \text{C}_6\text{H}_4\big\langle\begin{smallmatrix}\text{OH}\\\text{COOH}\end{smallmatrix} \xrightarrow{\text{cold}} (\text{EtO})_3\text{Ge}\overset{\text{H}}{\underset{\text{OOC}}{\overset{\text{O}}{\diamondsuit}}}\text{C}_6\text{H}_4 + \text{EtOH}$$

$$\text{Ge(OEt)}_4 + \text{C}_6\text{H}_4\big\langle\begin{smallmatrix}\text{OH}\\\text{COOH}\end{smallmatrix} \xrightarrow{\text{refluxed}} (\text{EtO})_2\text{Ge}\underset{\text{OOC}}{\overset{\text{O}}{\diamondsuit}}\text{C}_6\text{H}_4 + 2\,\text{EtOH}$$

$$\text{Ge(OEt)}_4 + 2\,\text{C}_6\text{H}_4\big\langle\begin{smallmatrix}\text{OH}\\\text{COOH}\end{smallmatrix} \;\; \text{C}_6\text{H}_4\big\langle\begin{smallmatrix}\\\text{COO}\end{smallmatrix} \longrightarrow \overset{\text{H}}{\text{O}}\;\overset{\text{OEt}}{\underset{|}{\text{Ge}}}\;\overset{\text{O}}{\underset{\text{OOC}}{}}\text{C}_6\text{H}_4 + 3\,\text{EtOH}$$

$$2\,\text{Ge(OEt)}_4 + 6\,\text{C}_6\text{H}_4\big\langle\begin{smallmatrix}\text{OH}\\\text{COOH}\end{smallmatrix} \longrightarrow$$

$$\text{C}_6\text{H}_4\big\langle\begin{smallmatrix}\text{H}\\\text{O}\\\text{COO}\end{smallmatrix}\;\overset{\text{OEt}}{\underset{|}{\text{Ge}}}\;\overset{\text{O}}{\underset{\text{OOC}}{}}\text{C}_6\text{H}_4 + (\text{EtO})\text{Ge}\left[\overset{\text{H}}{\underset{\text{OOC}}{\overset{\text{O}}{\diamondsuit}}}\text{C}_6\text{H}_4\right]_3$$

$$+\,\text{HOC}_6\text{H}_4\text{COOH} + 6\,\text{EtOH}$$

Triethoxygermane monosalicylate on being refluxed with benzene changes into diethoxygermane monosalicylate (141):

$$(\text{C}_2\text{H}_5\text{O})_3\text{Ge}\overset{\text{H}}{\underset{\text{OOC}}{\overset{\text{O}}{\diamondsuit}}}\text{C}_6\text{H}_4 \rightarrow (\text{C}_2\text{H}_5\text{O})_2\text{Ge}\underset{\text{OOC}}{\overset{\text{O}}{\diamondsuit}}\text{C}_6\text{H}_4 + \text{C}_2\text{H}_5\text{OH}$$

The reactions between organoalkoxygermanes and simple carboxylic acids (such as acetic and benzoic) or the hydroxycarboxylic acids (such as salicylic and mandelic acids) have also been studied by Mathur and

Mehrotra (131):

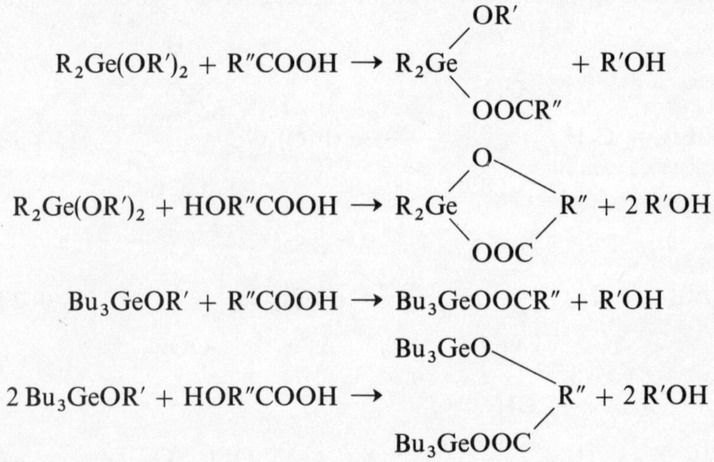

$$R_2Ge(OR')_2 + R''COOH \rightarrow R_2Ge\begin{array}{c} OR' \\ \diagup \\ \diagdown \\ OOCR'' \end{array} + R'OH$$

$$R_2Ge(OR')_2 + HOR''COOH \rightarrow R_2Ge\begin{array}{c} O \diagdown \\ \diagdown \\ OOC \diagup \end{array}R'' + 2\,R'OH$$

$$Bu_3GeOR' + R''COOH \rightarrow Bu_3GeOOCR'' + R'OH$$

$$2\,Bu_3GeOR' + HOR''COOH \rightarrow \begin{array}{c} Bu_3GeO \diagdown \\ \\ Bu_3GeOOC \diagup \end{array}R'' + 2\,R'OH$$

Dibutyldiethoxygermane, $Bu_2Ge(OEt)_2$, reacted with 2-mercaptoethanol, monothioglycerol and α- and β-mercaptopropionic acids in equimolecular proportion in benzene (240):

$$Bu_2Ge(OEt)_2 + HSCH_2CH_2OH \rightarrow$$
$$Bu_2Ge(SCH_2CH_2O) + 2\,EtOH$$

$$Bu_2Ge(OEt)_2 + HSCH_2CHOH-CH_2OH \rightarrow$$
$$Bu_2Ge(SCH_2CHOCH_2OH) + 2\,EtOH$$

$$Bu_2Ge(OEt)_2 + HSCH(CH_3)COOH \rightarrow$$
$$Bu_2Ge(SCH(CH_3)COO) + 2\,EtOH$$

$$Bu_2Ge(OEt)_2 + HSCH_2CH_2COOH \rightarrow$$
$$Bu_2Ge(SCH_2CH_2COO) + 2\,EtOH$$

The reaction of tetraisopropoxygermane with 2-mercaptoethanol in equimolecular ratio gave the bis derivative $Ge(SCH_2CH_2O)_2$ leaving the unchanged isopropoxygermane. The reactions between tetraisopropoxy-germane and α- and β-mercapto propionic acids gave generally cyclic products (240):

$$Ge(O\text{-}i\text{-}Pr)_4 + HSCH(CH_3)COOH \rightarrow$$
$$Ge(O\text{-}i\text{-}Pr)_2(SCH(CH_3)COO) + 2\,i\text{-}PrOH$$

$$Ge(O\text{-}i\text{-}Pr)_4 + 2\,HSCH(CH_3)COOH \rightarrow$$
$$Ge(O\text{-}i\text{-}Pr)(SCH(CH_3)COO)(HSCH(CH_3)COO) + 3\,i\text{-}PrOH$$

$$Ge(O\text{-}i\text{-}Pr)_4 + HSCH_2CH_2COOH \rightarrow$$

$$Ge(O\text{-}i\text{-}Pr)_2(SCH_2CH_2COO) + 2\,i\text{-}PrOH$$

$$Ge(O\text{-}i\text{-}Pr)_4 + 2\,HSCH_2CH_2COOH \rightarrow$$

$$Ge(SCH_2CH_2COO)_2 + 4\,i\text{-}PrOH$$

Many cleavage reactions of the Ge—O bond by different reagents such as hydrohalic acids, organic or mineral acids, acid chlorides or bromides, acid anhydrides, the Grignard reagents, the organolithiums, $LiAlH_4$, potassium etc. have been described in the literature:

$$Ge(OC_6H_5)_4 + 4\,HCl \rightarrow GeCl_4 + 4\,C_6H_5OH \quad (137)$$

$$R_3GeOR' + HBr \rightarrow R_3GeBr + R'OH \quad (123)$$

$$R_3GeOR' + 2\,HI \rightarrow R_3GeI + R'I + H_2O \quad (118)$$

$$R_3GeOR' + 2\,HCOOH \rightarrow$$

$$R_3GeOCOH + HCOOR' + H_2O \quad (118)$$

$$R_3GeOR' + 2\,C_6H_5SO_3H \rightarrow$$

$$R_3GeO_3SC_6H_5 + C_6H_5SO_3OR' \quad (118)$$

$$R_3GeOR' + C_6H_5COCl \rightarrow R_3GeCl + C_6H_5COOR' \quad (118)$$

$$Ge(OMe)_4 + CH_3COCl \rightarrow (MeO)_3GeCl \quad (27)$$

$$Ge(OR)_4 + CH_3COCl \rightarrow (RO)_3GeCl \quad (138)$$

$$Ge(OR)_4 + 2\,CH_3COCl \rightarrow (RO)_2GeCl_2 \quad (138)$$

$$Bu_2Ge(OR)_2 + CH_3COBr \rightarrow Bu_2Ge\underset{\diagdown OR}{\overset{\diagup Br}{}} + CH_3COOR \quad (129)$$

$$Bu_2Ge(OR)_2 + 2\,CH_3COBr \rightarrow Bu_2GeBr_2 + 2\,CH_3COOR \quad (129)$$

$$R_3GeOR' + (CH_3CO)_2O \rightarrow CH_3COOGeR_3 + CH_3COOR' \quad (118)$$

$$R_3GeOR' + CH_3MgI \rightarrow R_3GeCH_3 \quad (118)$$

$$(CH_3)_2Ge(OCH_3)_2 + CH_3MgI \text{ (stoichiom.)} \longrightarrow (CH_3)_3GeI \quad (261)$$

$$Ph_3GeOCHPh_2 \xrightarrow{BuLi} Ph_3GeBu \quad (164)$$

$$(C_2H_5)_3GeLi + (C_2H_5)_3GeOC_6H_{11} \longrightarrow$$

$$(C_2H_5)_3GeGe(C_2H_5)_3 + C_6H_{11}OLi \quad (249)$$

$$R_3GeOR' \xrightarrow{\text{LiAlH}_4} R_3GeH + R'OH \quad (118)$$

$$(-)R_3\overset{*}{Ge}OMe \xrightarrow[\text{LiAlH}_4]{} (-)R_3\overset{*}{Ge}H \quad (40a)$$

$$Et_3GeOPh + 2K \xrightarrow{\text{H·M·P·T}} Et_3GeK + PhOK \quad (160a)$$

An attempt to prepare the α-bromoether $Ph_3GeOCBrPh_2$ through the action of N-bromosuccinimide on (diphenylmethoxy)triphenylgermane instead gave benzophenone and triphenylbromogermane (164):

$$Ph_3GeOCHPh_2 \xrightarrow[\text{Bz}_2\text{O}_2]{\text{NBS}} (Ph_3GeOCBrPh_2) \rightarrow Ph_3GeBr + Ph_2CO$$

Treatment of 9-(triphenylgermoxy)fluorene with n-butyllithium followed by addition of benzyl chloride in acid medium leads to triphenylbutyl-germane bis(triphenylgermanium) oxide, 9-fluorenol and 9-benzyl-9-fluorenol. None of the Wittig-type rearrangement product 9-(triphenyl-germyl)-9-fluorenol was detected (164):

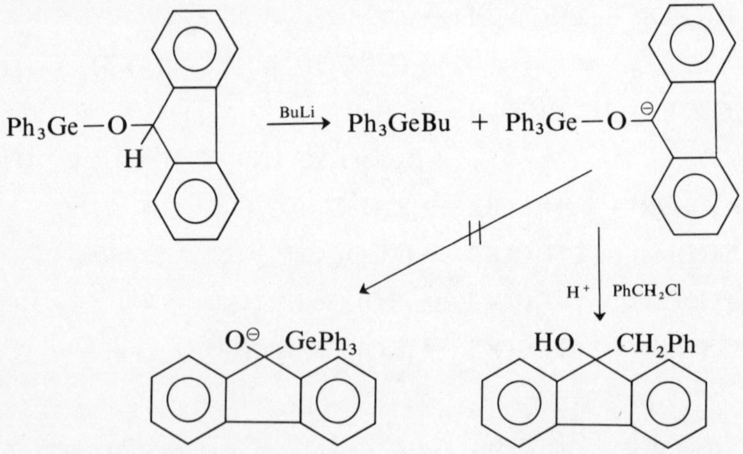

According to Koster-Pflugmacher and Termin the aminolithiums and sodium hexamethyldisilazane react with the Ge—O bonds of several alkoxy or phenoxygermanes (104):

$$(CH_3)_2Ge(OC_6H_5)_2 \xrightarrow{\text{LiN}(C_2H_5)_2}$$

$$(CH_3)_2Ge[N(C_2H_5)_2]_2$$

$$Ge(OC_2H_5)_4 + NaN[Si(CH_3)_3]_2 \text{ (excess)} \longrightarrow$$

$$(C_2H_5O)_3GeN[Si(CH_3)_3]_2$$

$$(CH_3)_2Ge(OC_6H_5)_2 + NaN[Si(CH_3)_3]_3 \longrightarrow$$

$$(CH_3)_2(C_6H_5O)GeN[Si(CH_3)_3]_2$$

The Ge—O bonds can be cleaved by certain hydrides with electrophilic activity, e.g.:

$$\text{PhCl}_2\text{Ge}-\text{O}-\overset{|}{\underset{|}{\text{C}}}\text{H} + \text{Ph(Cl)}_2\text{GeH} \rightarrow$$

$$\text{PhCl}_2\text{Ge}-\text{GeCl}_2\text{Ph} + \text{HO}-\overset{|}{\underset{|}{\text{C}}}\text{H} \quad (181)$$

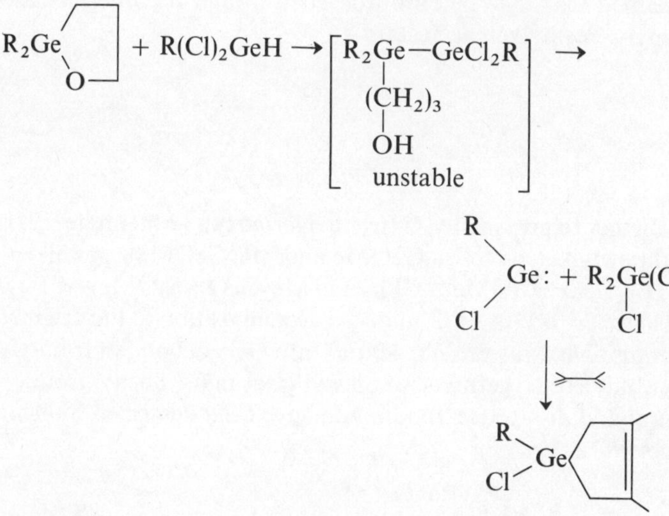

$$R_2Ge\underset{\text{O}}{\overset{}{\Big\langle}}\;\;\Big] + R(Cl)_2GeH \rightarrow \begin{bmatrix} R_2Ge-GeCl_2R \\ \overset{|}{(CH_2)_3} \\ \overset{|}{OH} \\ \text{unstable} \end{bmatrix} \rightarrow$$

$$\underset{Cl}{\overset{R}{\diagdown}}Ge: + R_2Ge(CH_2)_3OH$$

$$\Big\downarrow \;\;\nmid\mid$$

$$\underset{Cl}{\overset{R}{\diagdown}}Ge\Big\langle \quad (128)$$

Several methods for the determination of alkoxygermanes have been described in the literature. Ge—O—C and Ge—S—C linkages can be assayed directly by the perchloric acid-catalysed acetylation method of Fritz and Schenk (83). A second method for the direct assay of Ge—O—C linkages is by reaction with *in situ*-generated hydrogen bromide. HBr reacts with the alkoxygermane to form a bromogermane and the corresponding parent alcohol. The first excess of hydrogen bromide is detected by a blue BZL indicator end point. This method allows a quantitative determination of alkoxygermane in the presence of alkoxysilanes and alcohols (123).

Recently, Klimova and coworkers (102) used a modification of the Zeisel method for the micro-determination of alkoxygermanes. A mixture of potassium iodide and orthophosphoric acid was proposed for the decomposition of alkoxy compounds.

Some insertion reactions on the Ge—O bond of the alkoxygermanes have been reported in the literature.

The treatment of $ROGeCl_3$ with ketene gave Cl_3GeCH_2COOR (R = Me, or Et). These compounds are also prepared from $GeCl_4$ and esters of trialkylstannylacetic acid (47, 121). Methyl-3-(triethylgermyloxy)-3-buten-oate was isolated in 70% yield from the reaction mixture obtained by saturating trimethylmethoxygermane with ketene:

$$Et_3GeOCH_3 \xrightarrow{2\ CH_2=C=O} \underset{\underset{OGeEt_3}{|}}{CH_2=C-CH_2COOCH_3}$$

It can be assumed that the reaction proceeds through the intermediate o-triethylgermyl-o-methylketene acetal:

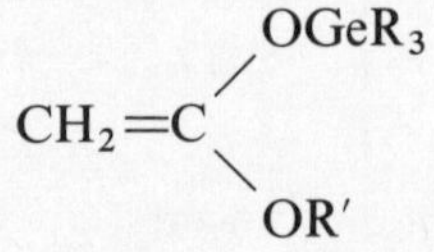

$$CH_2=C\overset{\displaystyle OGeR_3}{\underset{\displaystyle OR'}{\big\langle}}$$

which reacts further to give methyl-3-(triethylgermoxy)-3-butenoate (121). Addition of diphenylketene to Et_3GeOMe and Me_3GeOMe gives alken-oxygermanes $Et_3GeO-C(OMe)=CPh_2$ and $Me_3GeO-C(OMe)=CPh_2$. Both, heated *in vacuo* to 110–50°C, undergo decomposition to the original starting materials (29b). Insertion reactions into Ge—O bond of tributyl-methoxy or tributylethoxygermane of phenyl isocyanate, phenyl isothio-cyanate, chloral and di-p-tolylcarbodiimide have been described by Ishii and coworkers (96):

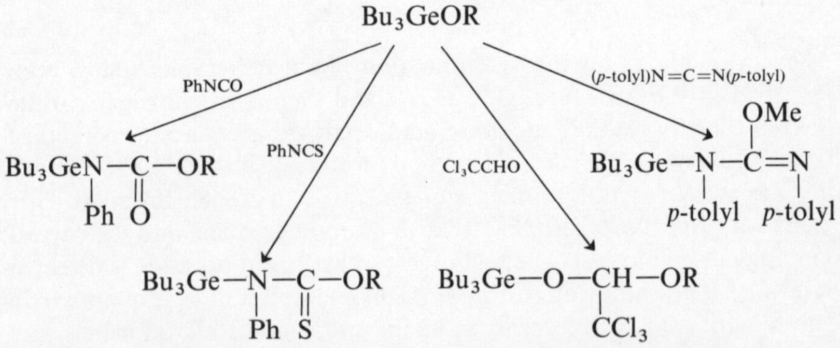

$$Bu_3GeOR$$

PhNCO

(p-tolyl)N=C=N(p-tolyl)

PhNCS

Cl₃CCHO

$$\underset{\underset{Ph\ \ O}{|\ \ \ |}}{Bu_3GeN-C-OR}$$

$$\underset{\underset{p\text{-tolyl}\ \ p\text{-tolyl}}{|\qquad\ \ |}}{Bu_3Ge-N-C=N}\overset{OMe}{}$$

$$\underset{\underset{Ph\ \ S}{|\ \ \ |}}{Bu_3Ge-N-C-OR}$$

$$\underset{\underset{CCl_3}{|}}{Bu_3Ge-O-CH-OR}$$

No reaction could be observed between tributylmethoxygermane and carbon disulphide (96). Insertion reactions of aldehydes and ketones of various structure have been observed by Satgé and Dousse with formation of the corresponding germanium acetals and ketals (200):

$$R_3GeOR' + R''CHO \rightarrow \underset{\underset{R''}{|}}{R_3Ge-O-CH-OR'}$$

$$R_3GeOR' + R''_2CO \rightarrow R_3Ge-O-C'(R'')_2-OR'$$

These reactions are in some cases catalysed by H_2PtCl_6 and $ZnCl_2$. These acetals are generally not very stable. The decomposition scheme is the following:

$$2\,R_3Ge{-}O{-}\underset{\underset{R''}{|}}{CH}{-}OR' \;\rightarrow\; (R_3Ge)_2O + R''CHO + R''CH\overset{OR'}{\underset{OR'}{\diagdown}}$$

Hexafluoroacetone and the propoxygermane $(CH_3)_3GeOCH(CF_3)_2$ form a $1:1$ complex. The structure of this complex is under discussion and several hypotheses have been advanced (57):

$$(CH_3)_3\overset{-}{Ge}{-}\underset{\underset{O{-}C(CF_3)_2}{|}}{\overset{+}{O}}{-}CH(CF_3)_2$$

$$(CH_3)_3\;(CH_3)_3Ge{-}\underset{\underset{(CF_3)_2}{\overset{\equiv}{\underset{+}{O}}}}{\overset{CH(CF_3)_2}{\overset{\equiv}{O}}}\;\underset{(CH_3)_3}{\overset{}{\underset{\equiv}{C}}}{-}O{-}\overset{-}{Ge}{-}\left(\underset{(CF_3)_2\;(CH_3)_3}{\overset{CH(CF_3)_2}{O{-}\underset{+}{C}{-}O{-}\overset{-}{Ge}}}\right)_n$$

$$(CH_3)_3Ge{-}OC(CF_3)_2{-}OCH(CF_3)_2$$

Trimethylmethoxygermane and boron trifluoride form a $1:1$ addition compound which sublimes without decomposition. Dative $d\pi{-}p\pi$ bonding, if present in the Ge—O linkage of trimethylmethoxygermane, does not appear to affect appreciably the electron-donor activity of the oxygen toward boron trifluoride (89). The reactions of $PhOGeH_3$ with boron trifluoride and boron trichloride are described by Glidewell and co-workers (86a). With BCl_3 cleavage of the Ge—O bond occurs; with BF_3, acid-catalysed disproportionation with formation of monogermane and $GeH_x(OPh)_{4-x}$ ($x = 0$ or 1) is noted.

6-2-3   Physical-chemical studies of germanium–oxygen compounds (alkoxy- and organoalkoxygermanes, organogermanium oxides)

6-2-3-1   Infrared and Raman Spectra. Many IR studies have been made in the series of organoalkoxygermanes and organogermanium oxides. The frequencies attributable to the germanium–oxygen stretch vary greatly with the structure of the molecule. Thus, tetraalkoxygermanes, $Ge(OR)_4$, show two bands at $1040\ cm^{-1}$ and $680\ cm^{-1}$ (98), while the trimeric and tetrameric forms of diphenylgermanium oxide show only one band at $850\ cm^{-1}$ (146). In contrast, dimethylgermanium oxide shows variations in the band position between 738–868 cm^{-1}, depending on the degree of polymerization (41). A similar complexity is indicated for the di-iso-propylgermanium oxide (i-Pr$_2$GeO)$_n$ (two bands: 844 and 789 cm^{-1}) (56). (Et$_3$Ge)$_2$O shows one band (855 cm^{-1}) (51) as do (Ph$_3$Ge)$_2$O at 858 cm^{-1} (56) and [(PhCH$_2$)$_3$Ge]$_2$O at 926 cm^{-1} (56), whereas bis-(trimethyl-germanium) oxide, [(CH$_3$)$_3$Ge]$_2$O, displays four bands between 755 and 870 cm^{-1} (41). IR data of some organogermanium oxides published by MacKay and Watt (122) are in good agreement with those of Cross and Glockling (56).

The IR spectra of two bis-(trialkylgermanium) oxides, [(CH$_3$)$_3$Ge]$_2$O and [n-Bu$_3$Ge]$_2$O, and ten trialkylalkoxygermanes have been reported recently by Valade and coworkers (116); v_a Ge—O—Ge (794–841 cm^{-1}) and v_S Ge—O—Ge (400–467 cm^{-1}) as well as the v_a Ge—O—C (972–1101 cm^{-1}) and v_S Ge—O—C (582–653 cm^{-1}) have been assigned. The Raman spectra of some derivatives have also been recorded.

IR studies in the alkoxyalkyl(aryl)germanes series were reported recently, by Mathur and coworkers (131a).

IR and Raman investigations indicate a highly bent skeleton for H$_3$Ge—O—GeH$_3$ as well as H$_3$GeSGeH$_3$ in contrast to the near-linear skeleton of disilylether H$_3$Si—O—SiH$_3$ (88). H$_3$GeOCD$_3$ and H$_3$GeSCH$_3$ have also nonlinear molecule skeletons. The Ge—O—Ge bond angle in (H$_3$Ge)$_2$O has been calculated as $139° \pm 6°$ using the valence force-field approach (53). These results are interpreted in terms of the smaller tendency of germanium to participate in (p $\rightarrow$ d) π-bonding.

The observed shifts in the C–D stretching frequency in mixtures of base and Cl$_3$CD have been used by Abel and coworkers (3) to compare the donor properties of a variety of organometallic bases. These include ethers such as R$_3$GeOR and (R$_3$Ge)$_2$O, sulphides and amines in which one or more (CH$_3$)$_3$M-groups (M = Si, Ge, Sn, Pb) are attached to the donating atom. In the ethers [(CH$_3$)$_3$M]$_2$O the drop in basicity from the carbon to the silicon compounds is largely due to the effective π-bonding of two silicon atoms which can interact with both lone pairs of oxygen. For the germanium and tin analogues, the basicities are both much higher than the carbon compounds, again presumably due to a large inductive contribution and low or non-existent π-bonding (3).

Ulbricht and coworkers (244) determined by IR spectroscopy the relative basicities of the oxygen atoms of methylethoxygermanes, propyl-ethoxygermanes and some organosilicon analogues as proton acceptors during the formation of a hydrogen bond with phenol, pyrrole, ethanol and phenylacetylene. The high relative basicities of oxygen in different alkylethoxygermane series as compared with derivatives of silicon and carbon and the dependence of basicity on the degree of substitution of the individual series can be accounted for by differences in the electro-negativity of the central atoms, by the inductive effect of the substituents, and by the (p–d) π-interaction of the Si–O bond.

6-2-3-2   UV Spectra. The ultraviolet spectra of $(R_3X)_2O$ and $(R_3X)_2S$ where R is an alkylgroup and X = C, Si, Ge or Sn have been determined in pure hexane over the wavelength range 195–350 nm. The tin compounds absorb at about 204 nm but analogous C, Si or Ge compounds are transparent in this region (59).

6-2-3-3   NMR Spectra. A relation between the chemical shifts of the methyl protons and of the CH_2 protons of the ethyl groups and the sums of the Taft constants of the three other substituents was detected for organogermanium compounds $CH_3GeR'R''R'''$ and $C_2H_5GeR'R''R'''$ including $(CH_3)_2Ge(OC_2H_5)_2$, $(CH_3)_2Ge(OCH_3)_2$, $CH_3Ge(OC_2H_5)_3$, $CH_3Ge(OCH_3)_3$, $(C_2H_5)_3GeOCH_3$, $C_2H_5Ge(OCH_3)_3$, and C_2H_5Ge-$(OC_2H_5)_3$ (72). Nuclear magnetic resonance spectra of some methyl-germanium halides, oxides, sulphides and selenides are described by Schmidbaur and coworkers (209). The authors report that their results can no longer be interpreted in terms of hyperconjugation or π-bonding, since these concepts are not likely to play an important role in the bonding of germanium.

The PMR spectra of the compounds $(C_2H_5)_{4-n}GeX_n$ are reported for X = Cl, Br, I, H and O. This series comprises the oxides $(Et_3Ge)_2O$, $(Et_2GeO)_n$ and $(EtGeO)_2O$. The effects of substitution on chemical shifts of the CH_3 and CH_2 protons is discussed and it is concluded that the principal factor is the magnetic-anisotropy effect of the substituents X. Anisotropy effects at the germanium atom, inductive effects of X and intramolecular dispersion forces may make smaller contributions (122).

In a study by PMR of the structure of diheteroorganic compounds $(Et_3Si)_2M$, $(Et_3Ge)_2M$ and $(Et_3Sn)_2M$ with M = O, S, Se, Te, the authors report that an evaluation of effects of the interaction $p\pi$–$d\pi$ on the chemical shifts of CH_2 and CH_3 protons of the ethyl group does not seem possible in the germanium and tin compounds (74).

The proton chemical shifts and ^{13}C—^1H coupling constants of compounds belonging to the series $Me_{4-n}X(OMe)_n$ where $n = 1, 4$ and X = C, Si or Ge have been measured by Cumper and coworkers (58); the results indicate the presence of $p\pi$–$d\pi$ bonding only between silicon and oxygen atoms. In contrast to the conclusions of these authors, Egorochkin and coworkers (73) found in an NMR study of the derivatives $(CH_3)_{4-n}MR_n$ (M = C, Si, Ge, Sn) that the chemical shifts of the methyl protons are influenced not only by the inductive effects and the effect of magnetic anisotropy of the substituents, but also, in the case

R = $-OCH_3$, $-OC_2H_5$, $-CH=CH_2$, by the effect of $d\pi-p\pi$ conjugation.

NMR and IR spectra of $Me_3GeOC_6F_5$ and $Ph_3GeOC_6F_5$ also have been reported (161a).

6-2-3-4   Dipole Moment Measurements. The electric dipole moments of compounds with bond between O or S atoms and C, Si, Ge or Sn atoms have been measured by Cumper and coworkers (60). The Me_3XO- and Me_3XS-group moments are deduced and evidence obtained for $p\pi-d\pi$ interaction only in the Si—O bond.

The same authors have measured the electric moments of the compounds $Me_{4-n}Ge(OMe)_n$ and $Me_{4-n}Ge(SMe)_n$. The results indicate that the methyl groups rotate freely about the Ge—O bond, but in the case of Ge—S the dipole moments calculated for fixed molecular conformations are in better agreement with the experimental results (61).

The dipole moments of methylethoxygermanes and propylethoxygermanes have been determined by Ulbricht and coworkers (245). The experimental values are used for calculating the moments of the Ge—O bond and its magnitude is compared with that of the Si—O bond.

6-2-3-5   Mass Spectroscopy. Mass spectra of miscellaneous organogermanes including $(Me_3Ge)_2O$, $(Et_3Ge)_2O$ and $(Ph_3Ge)_2O$ are described by Glockling and Light (87). Mass spectrometric analysis of some chlorogermanium alkoxides are described by Sara and coworkers (191a).

6-2-3-6   Other Physical-Chemical Studies. The molecular parachors of $GeCl_4$, $GeBr_4$, $Ge(Et)_4$ and $Ge(OEt)_4$ have been determined and the atomic parachor for germanium calculated (231). The densities and surface tensions of the tetraalkoxygermanes $(RO)_4Ge$ were measured between 20°C and 40°C and molecular volumes and parachors were calculated by Bradley and coworkers (37).

Vapour pressures and viscosities have been measured over a range of temperature for $(RO)_4Ge$ where R = Me, Et, *n*-Pr, *-i*-Pr, *n*-Bu, *i*-Bu, *s*-Bu and *t*-Bu. Energies and entropies of vaporization have been calculated together with energies and entropies of activation for viscous flow. The behaviour of alkoxygermanes in the liquid state is discussed (35).

Glidewell and coworkers (86b) found, by electron diffraction study, the following values for Ge—X bond length and Ge—X—Ge angle (X = O, S) in $(H_3Ge)_2O$ and $(H_3Ge)_2S$:

$(GeH_3)_2O$	1.774 Å	125.6 degrees
$(GeH_3)_2S$	2.205 Å	99.1 degrees

6-2-4   Alkoxy- and organoalkoxygermanes

Table 6-2(*a–d*) showing some physical properties of alkoxy- and organoalkoxygermanes appears on the following pages.

Table 6-2
Alkoxy- and Organoalkoxygermanes
(a) $Ge(OR')_4$

COMPOUND	EMPIRICAL FORMULA	M.P. (°C)	B.P. (°C/mm)	n_D^{20}	d_4^{20}	REFERENCES
$(CH_3O)_4Ge$	$C_4H_{12}O_4Ge$	−18	145–46/760	1.4015	1.3254	27, 98, 228; 37
$(C_2H_5O)_4Ge$	$C_8H_{20}O_4Ge$	−72	184/741; 71–72/9	1.4073; 1.4049	1.1395; 1.1337	1, 27, 36, 37, 98, 241, 244
$C_6H_5OGe(OC_2H_5)_3$	$C_{12}H_{20}O_4Ge$	—	204/0.5	—	—	137
$[Cl(CH_2)_3O]_4Ge$	$C_{12}H_{24}O_4Cl_4Ge$	—	186–88/6	1.4815	1.5115	275
$(n\text{-}C_3H_7O)_4Ge$	$C_{12}H_{28}O_4Ge$	—	108–9.5/9	1.4172 (25°)	1.0580 (25°)	1, 37, 98, 167
$(i\text{-}C_3H_7O)_4Ge$	$C_{12}H_{28}O_4Ge$	—	108–11/30	1.4141 (25°)	1.0245 (25°)	37, 167, 228
$(n\text{-}C_4H_9O)_4Ge$	$C_{16}H_{36}O_4Ge$	—	143/8	1.4255 (25°)	1.0173 (25°)	1, 36, 37, 98, 167
$(i\text{-}C_4H_9O)_4Ge$	$C_{16}H_{36}O_4Ge$	—	264.5/760	—	1.0054 (25°)	36, 37
$(sec\text{-}C_4H_9O)_4Ge$	$C_{16}H_{36}O_4Ge$	—	136–39/54	1.4291 (25°)	1.0164 (25°)	36, 37, 167
$(tert\text{-}C_4H_9O)_4Ge$	$C_{16}H_{36}O_4Ge$	44	224/760	—	1.0574 (25°)	36, 37, 238
$(C_6H_5O)_3Ge(OC_2H_5)$	$C_{20}H_{20}O_4Ge$	—	222/0.3	—	—	137
$(n\text{-}C_5H_{11}O)_4Ge$	$C_{20}H_{44}O_4Ge$	—	120/0.03	1.4335 (25°)	0.9930 (25°)	37, 98
$(C_2H_5C(CH_3)_2{-}O)_4Ge$	$C_{20}H_{44}O_4Ge$	—	—	—	1.0425 (25°)	36
$(C_6H_5O)_4Ge$	$C_{24}H_{20}O_4Ge$	—	220/0.3	—	—	137, 229
$(C_6H_{11}O)_4Ge$	$C_{24}H_{44}O_4Ge$	—	166–67/0.01	—	—	98
$(n\text{-}C_6H_{13}O)_4Ge$	$C_{24}H_{52}O_4Ge$	—	150–53/0.04	1.4372 (25°)	0.9682 (25°)	98

(b) RGe(OR')$_3$

COMPOUND	EMPIRICAL FORMULA	M.P. (°C)	B.P. (°C/mm)	n_D^{20}	d_4^{20}	REFERENCES
(CH$_3$O)$_3$GeCl	C$_3$H$_9$O$_3$ClGe	—	135–36/760	1.4150	1.4340	27
CH$_3$Ge(OCH$_3$)$_3$	C$_4$H$_{12}$O$_3$Ge	—	136.5–38/760	1.4053	1.264	261
C$_2$H$_5$Ge(OCH$_3$)$_3$	C$_5$H$_{14}$OGe	—	154/761	1.4178	1.2446	274
(CH$_3$O)$_3$GeCH$_2$COOCH$_3$	C$_6$H$_{14}$O$_5$Ge	—	67–68/1	1.4400	1.3596	121, 127
(C$_2$H$_5$O)$_3$GeCl	C$_6$H$_{15}$ClO$_3$Ge	—	181.5–183	—	—	168
C$_3$H$_7$Ge(OCH$_3$)$_3$	C$_6$H$_{16}$O$_3$Ge	—	169/736.8	1.4185	1.2024	274
CH$_3$Ge(OCH$_2$CH$_2$Cl)$_3$	C$_7$H$_{15}$Cl$_3$O$_3$Ge	—	134–38/2	1.4795	1.4557	275
CH$_3$Ge(OC$_2$H$_5$)$_3$	C$_7$H$_{18}$O$_3$Ge	—	165/736.7	1.4128	1.1280	274
			43–44/3	1.4102 (25°)	1.1317 (25°)	244
C$_4$H$_9$Ge(OCH$_3$)$_3$	C$_7$H$_{18}$O$_3$Ge	—	76–78/9	1.4245	—	145
C$_2$H$_5$Ge(OC$_2$H$_5$)$_3$	C$_8$H$_{20}$O$_3$Ge	—	180/761	1.4178	1.1105	274
(n-C$_3$H$_7$O)$_3$GeCl	C$_9$H$_{21}$ClO$_3$Ge	—	75–76/6.5	1.4258 (25°)	—	167
(i-C$_3$H$_7$O)$_3$GeCl	C$_9$H$_{21}$ClO$_3$Ge	—	81–83/10	1.4135 (27°)	—	138, 168
n-C$_3$H$_7$Ge(OC$_2$H$_5$)$_3$	C$_9$H$_{22}$O$_3$Ge	—	55–56/3	1.4239 (25°)	1.1225 (25°)	244
CH$_3$Ge(OCH$_2$—CH=CH$_2$)$_3$	C$_{10}$H$_{18}$O$_3$Ge	—	91–94/7	1.4540	1.1427	274
C$_4$H$_9$Ge(OC$_2$H$_5$)$_3$	C$_{10}$H$_{24}$O$_3$Ge	—	90–92/8	1.4330	—	145
C$_2$H$_5$Ge(OC$_3$H$_7$)$_3$	C$_{11}$H$_{26}$O$_3$Ge	—	118.5–20/25	1.4258	1.0563	274
(n-C$_4$H$_9$O)$_3$GeCl	C$_{12}$H$_{27}$ClO$_3$Ge	—	110–15/2.1	—	—	138
(t-C$_4$H$_9$O)$_3$GeCl	C$_{12}$H$_{27}$ClO$_3$Ge	—	72.5–74/0.8	1.4227 (25°)	—	168
C$_4$H$_9$Ge(O-i-C$_3$H$_7$)$_3$	C$_{13}$H$_{30}$O$_3$Ge	—	94–95/7.5	1.4151	—	145
C$_4$H$_9$Ge(O—n-C$_3$H$_7$)$_3$	C$_{13}$H$_{30}$O$_3$Ge	—	95–97/2.5	1.4250	—	145
C$_4$H$_9$Ge(O—n-C$_4$H$_9$)$_3$	C$_{16}$H$_{36}$O$_3$Ge	—	111–14/0.5	1.4310	—	145
C$_4$H$_9$Ge(O-sec-C$_4$H$_9$)$_3$	C$_{16}$H$_{36}$O$_3$Ge	—	96–98/1	1.4260	—	145
C$_4$H$_9$Ge(O-tert-C$_4$H$_9$)$_3$	C$_{16}$H$_{36}$O$_3$Ge	—	99–100/4.5	1.4280	—	145
CH$_3$Ge(O—C$_6$H$_5$)$_3$	C$_{19}$H$_{18}$O$_3$Ge	—	217/3	1.3010	1.5835	274

(c) $R_2Ge(OR')_2$

COMPOUND	EMPIRICAL FORMULA	M.P. (°C)	B.P. (°C/mm)	n_D^{20}	d_4^{20}	REFERENCES
$(C_2H_5O)_2GeCl_2$	$C_4H_{10}Cl_2O_2Ge$	—	157–59/760	—	—	138, 168
$(CH_3)_2Ge(OCH_3)_2$	$C_4H_{12}O_2Ge$	—	116/755	1.4130	1.2207	153, 261, 274
$(CH_3)_2Ge(OCH_2CH_2Cl)_2$	$C_6H_{14}Cl_2O_2Ge$	—	150–53/28	1.4695	1.3842	275
$(n\text{-}C_3H_7O)_2GeCl_2$	$C_6H_{14}Cl_2O_2Ge$	—	115–18/90	1.4327 (25°)	—	167
$(i\text{-}C_3H_7O)_2GeCl_2$	$C_6H_{14}Cl_2O_2Ge$	—	59/61/8.2	1.4228 (27°)	—	138, 168
$(C_2H_5)_2Ge(OCH_3)_2$	$C_6H_{16}O_2Ge$	—	159/755	1.4369	1.0692	202, 274
						150
$(CH_3)_2Ge(OC_2H_5)_2$	$C_6H_{16}O_2Ge$	—	139/745.5	1.4128	1.1129	274
			130/760	1.4119 (25°)	1.1142 (25°)	244
$(CH_3)_2Ge(OCH_2CH{=}CH_2)_2$	$C_8H_{16}O_2Ge$		92/30	1.4470	1.1270	274
$(CH_3O)_2Ge(CH_2COOCH_3)_2$	$C_8H_{16}O_6Ge$		101–2/2	1.4630	1.3683	27
$(n\text{-}C_4H_9O)_2GeCl_2$	$C_8H_{18}Cl_2O_2Ge$		102–5/6	—	—	138, 168
$(t\text{-}C_4H_9O)_2GeCl_2$	$C_8H_{18}Cl_2O_2Ge$		59–60/0.5	1.4313 (25°)	—	168
$(CH_3)_2Ge(OC_3H_7\text{-}n)_2$	$C_8H_{20}O_2Ge$		174/755	1.4212	1.0658	274
$(CH_3)_2Ge(OC_3H_7\text{-}i)_2$	$C_8H_{20}O_2Ge$		92–93/92	1.4097 (25°)		123
$(CH_3)(Cl)Ge(O{-}C_4H_9\text{-}t)_2$	$C_9H_{21}ClO_2Ge$		80–2/15	1.1277	1.4320	274
$(CH_3OOCCH_2)_2Ge(OCH_3)_2$	$C_{10}H_{16}O_6Ge$		101–2/2	1.4630	1.3683	27
$(CH_3)_2Ge(OC_4H_9\text{-}tert)_2$	$C_{10}H_{24}O_2Ge$		70–72/13	1.4220	1.0181	274
$(n\text{-}C_3H_7)_2Ge(OC_2H_5)_2$	$C_{10}H_{24}O_2Ge$		61/3	1.4315 (25°)	1.0419 (25°)	244
$(C_4H_9)_2Ge(OCH_3)_2$	$C_{10}H_{24}O_2Ge$		113–16/13	1.4365	—	129, 142
$(n\text{-}C_6H_{13}O)_2GeCl_2$	$C_{12}H_{26}Cl_2O_2Ge$		120–25/0.5	1.4414 (28°)	—	168
$(C_4H_9)_2Ge(OC_2H_5)_2$	$C_{12}H_{28}O_2Ge$		106/5.9	1.4357	—	129, 142
$(CH_3)_2Ge(OC_6H_5)_2$	$C_{14}H_{16}O_2Ge$		176–77/14	1.5633	1.2754	123, 274
$(CH_3O)_2Ge(C_6H_5)_2$	$C_{14}H_{16}O_2Ge$		148/2.2	1.5564 (25°)	—	123, 143
$(C_4H_9)_2Ge(OC_3H_7\text{-}n)_2$	$C_{14}H_{32}O_2Ge$		127/4.8	1.4357	—	129, 142

Table 6-2 (e)—continued

COMPOUND	EMPIRICAL FORMULA	M.P. (°C)	B.P. (°C mm)	n_D^{20}	n_4^{20}	REFERENCES
$(C_4H_9)_2Ge(OC_3H_7-i)_2$	$C_{14}H_{32}O_2Ge$	—	103–8/5	1.4350	—	129, 142
$(C_4H_9)_2Ge(OC_2H_5)(OC_4H_9\text{-}tert)$	$C_{14}H_{32}O_2Ge$	—	123–25/5.5	1.4370	—	142
$(C_6H_5)_2Ge(OC_2H_5)_2$	$C_{16}H_{20}O_2Ge$	—	156–58/3	1.5400	—	142
$(C_4H_9)_2Ge(OC_4H_9\text{-}n)_2$	$C_{16}H_{36}O_2Ge$	—	106–8/0.5	1.4388	—	129, 130, 142
$(C_4H_9)_2Ge(OC_4H_9\text{-}i)_2$	$C_{16}H_{36}O_2Ge$	—	114–16/2	1.4390	—	129
$(C_4H_9)_2Ge(OC_4H_9\text{-}s)_2$	$C_{16}H_{36}O_2Ge$	—	110/2.4	1.4373	—	129, 130, 142
$(C_4H_9)_2Ge(OC_4H_9\text{-}t)_2$	$C_{16}H_{36}O_2Ge$	—	105–8/2.5	1.4355	—	129, 142
$(C_6H_5)_2Ge(OC_3H_7)_2$	$C_{18}H_{24}O_2Ge$	—	129–30/0.7	1.5247	—	143
$(C_6H_5)_2Ge(O\!-\!i\text{-}C_3H_7)_2$	$C_{18}H_{24}O_2Ge$	—	122–25/0.6	1.5230	—	145
$(C_4H_9)_2Ge(OCHC_3H_7)_2$ | CH_3	$C_{18}H_{40}O_2Ge$	—	142–45/4	1.4400	—	129
$(t\text{-}C_5H_{11}O)_2Ge(C_4H_9)_2$	$C_{18}H_{40}O_2Ge$	—	141/4.7	1.4452	—	129, 130
$(C_6H_5)_2Ge(O\!-\!n\text{-}C_4H_9)_2$	$C_{20}H_{28}O_2Ge$	—	154–55/0.6	1.5240	—	143
$(C_6H_5)_2Ge(OC_4H_9\text{-}s)_2$	$C_{20}H_{28}O_2Ge$	—	124–25/0.2	1.5130	—	143
$(C_6H_5)_2Ge(OC_4H_9\text{-}t)_2$	$C_{20}H_{28}O_2Ge$	—	125/0.4	1.5240	—	142, 143
$(C_4H_9)_2Ge(OC_6H_{11})_2$	$C_{20}H_{40}O_2Ge$	—	149/0.5	1.4773	1.0468	119
$(C_6H_5)_2Ge(OC_6H_{11})_2$	$C_{24}H_{32}O_2Ge$	57–58	—	—	—	181

(d) R₃GeOR'

COMPOUND	EMPIRICAL FORMULA	M.P. (°C)	B.P. (°C/mm)	n_D^{20}	d_4^{20}	REFERENCES
$C_2H_5OGeCl_3$	$C_2H_5OCl_3Ge$	—	40/36	—	—	168
n-$C_3H_7OGeCl_3$	$C_3H_7OCl_3Ge$	—	97	1.4480 (25°)	—	167
i-$C_3H_7OGeCl_3$	$C_3H_7OCl_3Ge$	—	130–32/725	—	—	138, 168
$(CH_3)Cl_2GeOCH_2CH_2Cl$	$C_3H_7OCl_3Ge$	—	81/22	1.4750	1.5897	275
n-$C_4H_9OGeCl_3$	$C_4H_9OCl_3Ge$	—	70–75/20	—	—	138
t-$C_4H_9OGeCl_3$	$C_4H_9OCl_3Ge$	—	37–39/4.5	—	—	168
$(CH_3)_2ClGeOCH_2CH_2Cl$	$C_4H_{10}OClGe$	—	103–5/4	1.4670	1.4276	275
$(CH_3)_3GeOCH_3$	$C_4H_{12}OGe$	−102.2	87–88/753	1.401 (25°)	1.075 (25°)	89, 261
$(CH_3)_3GeOCH_2CH_2Cl$	$C_5H_{13}OClGe$	—	153/763	1.4421	1.2304	275
$(CH_3)_3GeOCH_2CH_2Br$	$C_5H_{13}OBrGe$	—	64/13	1.4670	1.4823	147
$(CH_3)_3GeOC_2H_5$	$C_5H_{14}OGe$	—	100–1	1.4067	1.06	2, 244
$(C_2H_5)_2Ge(H)OCH_3$	$C_5H_{14}OGe$	—	123.5/760	1.4310	1.1070	127
$(CH_3)_3GeOCH(CF_3)_2$	$C_6H_{10}F_6Ge$	—	117/758	—	—	57
$(C_6H_5)(Cl)_2GeOCH_3$	$C_7H_8Cl_2OGe$	—	122/20	1.5310	1.4764	181
$(C_2H_5)_2Ge(H)OCH_2$—$CH{=}CH_2$	$C_7H_{16}OGe$	—	56/12	1.4513	1.0900	127
$(C_2H_5)_3GeOCH_3$	$C_7H_{18}OGe$	—	163/755	1.4362	1.0682	25, 195
$(C_3H_7)_2Ge(H)OCH_3$	$C_7H_{18}OGe$	—	79–81/45	1.4355	1.0514	29a
$(CH_3)_3GeOC(CF_3)_2$—SC_2H_5	$C_8H_{14}SOF_6Ge$	—	88/15	1.4088 (22°)	1.38	6
$(C_2H_5)_2Ge(H)OCH_2CH{=}CH$—$CH_3$	$C_8H_{18}OGe$	—	71/10	1.4521	1.0501	127
$(C_2H_5)_2Ge(H)OCH(CH_3)$—$CH{=}CH_2$	$C_8H_{18}OGe$	—	69/15	1.4427	1.0630	127
$(C_2H_5)_2Ge(H)OCH_2$—$C(CH_3){=}CH_2$	$C_8H_{18}OGe$	—	65/14	1.4496	1.0573	127
$(C_2H_5)_3GeOC_2H_5$	$C_8H_{20}OGe$	—	170/750	1.4360	1.0395	195
$(CH_3)_3GeOC_6F_5$	$C_9H_9F_5OGe$	—	41/0.6	1.4402 (25°)	—	161a
$(C_2H_5)_3GeOCH_2C{\equiv}CH$	$C_9H_{18}OGe$	—	92/17	1.4600	1.0953	195

Table 6-2 (d)—continued

COMPOUND	EMPIRICAL FORMULA	M.P. (°C)	B.P. (°C/mm)	n_D^{20}	d_4^{20}	REFERENCES
(C₂H₅)₃GeOCH₂CH=CH₂	C₉H₂₀OGe	—	86/20	1.4458	1.0411	195
(C₂H₅)₂Ge(H)OCH(CH₃)C(CH₃)=CH₂	C₉H₂₀OGe	—	67/11	1.4455	1.0244	127
(C₂H₅)₃GeOCH₂CH₂CH₃	C₉H₂₂OGe	—	190/760	1.4388	1.0250	119
(C₂H₅)₃GeOCH(CH₃)₂	C₉H₂₂OGe	—	75/20	1.4339	1.0152	195
(C₆H₅)(Cl)₂GeOCH₂CH(CH₃)₂	C₁₀H₁₄OCl₂Ge	—	68/0.2	1.5100	1.3179	181
(CH₃OCOCH₂)₃GeOCH₃	C₁₀H₁₈OGe	—	131–33/1	1.4750	1.3615	27
(C₂H₅)₃GeOCH₂CH=C(Cl)CH₃	C₁₀H₂₁OClGe	—	120–23/15	1.4655	1.1272	43
(C₂H₅)₃GeOCH₂CH=CH—CH₃	C₁₀H₂₂OGe	—	103/16	1.4522	1.0367	195
(C₂H₅)₃GeO(CH₂)₂CH₃	C₁₀H₂₄OGe	—	209/755	1.4406	1.0155	119
(C₂H₅)₃GeOC(CH₃)₃	C₁₀H₂₄OGe	—	75.5/14	1.4380	1.0064	195
(C₂H₅)₃GeOCH₂—C(cyclic: CH=CH / CH—O)	C₁₁H₂₀O₂Ge	—	118/12	1.4769	1.1486	195
(C₂H₅)₃Ge—O—C(=CH₂)—CH₂COOCH₃	C₁₁H₂₂O₃Ge	—	81–83/1	1.4615	1.1272	121
(C₃H₇)₃GeOCH=CH₂	C₁₁H₂₄OGe	—	88–91/7	1.4565	1.0177	30
(C₂H₅)₃GeOCH₂C(CH₃)=CH—CH₃	C₁₁H₂₄OGe	—	110–14/15	1.4558	1.0349	43
(C₂H₅)₂Ge—CH₂—CH(CH₃)CH=CH₂ (OC₂H₅)	C₁₁H₂₄OGe	—	90/10	1.4544	1.0235	136
(n-C₃H₇)₃GeOC₂H₅	C₁₁H₂₆OGe	—	88/10	1.4405 (25°)	0.9916 (25°)	244

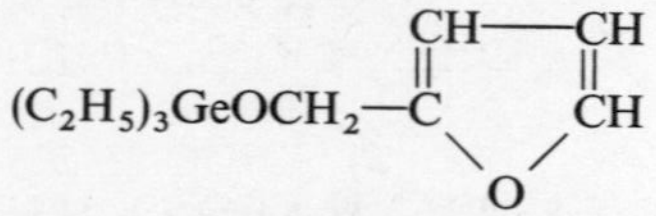

Compound	Empirical formula	m.p.	b.p./mm	n_D	d	Ref.
$(C_6H_5)(Cl)_2GeOC_6H_{11}$	$C_{12}H_{16}OCl_2Ge$	—	108/0.1	1.5350	1.3503	181
$(C_2H_5)_3GeOC_6H_5$	$C_{12}H_{20}OGe$	—	140/18	1.5102	1.1329	119
$(C_2H_5)_3GeO$⟨cyclohexenyl⟩ 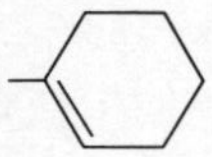	$C_{12}H_{24}OGe$	—	93–94/2.5	1.4820	1.1084	171
$(C_2H_5)_3GeO$⟨cyclohexyl, Cl⟩ 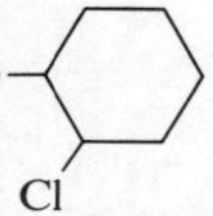	$C_{12}H_{25}OClGe$	—	121–25/2.5	1.4765 (25°)	—	115
$(C_2H_5)_3GeO$⟨cyclohexyl, Br⟩ 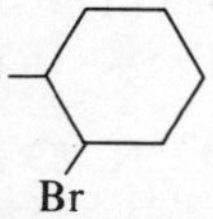	$C_{12}H_{25}OBrGe$	—	134–36/2.5	1.4890 (25°)	—	115
$(C_2H_5)_3GeOC_6H_{11}$	$C_{12}H_{26}OGe$	—	134/18	1.4695	1.0612	118
$(C_2H_5)_3Ge-O-\underset{\parallel}{C}-P(C_2H_5)_2$ $\;CH_2$	$C_{12}H_{27}POGe$	—	89/0.35	—	—	199
$(n\text{-}C_3H_7)_3GeOCH_2-CHBr-CH_3$	$C_{12}H_{27}BrOGe$	—	134–38/2.5	1.4778 (25°)	1.1103	170
$(C_2H_5)_3GeOCH_2C_6H_5$	$C_{13}H_{22}OGe$	—	150–52/18	1.5030	0.9882	118
$(C_4H_9)_3GeOCH_3$	$C_{13}H_{30}OGe$	—	136/16	1.4502	0.9852	142, 143, 195
$(C_2H_5)_3GeO(CH_2)_6CH_3$	$C_{13}H_{30}OGe$	—	127/10	1.4460	0.9879	118
$(C_4H_9)_3GeOCH=CH_2$	$C_{14}H_{30}OGe$	—	86.5–90/1	1.4580		30
$(C_2H_5)_3GeOCH=CHCH(CH_3)-P(C_2H_5)_2$	$C_{14}H_{31}POGe$	—	93–95/0.15	—	—	198
$(C_4H_9)_3GeOC_2H_5$	$C_{14}H_{32}OGe$	—	129/10	1.4481	0.9718	123, 142, 143, 195
$(C_2H_5)_3Ge-O-CH-P(C_2H_5)_2$ $\;(CH_2)_2CH_3$	$C_{14}H_{33}POGe$	—	85/0.15	—	—	198
$(C_2H_5)_3GeO-(CH_2)_2-OGe(C_2H_5)_3$	$C_{14}H_{34}O_2Ge_2$	—	162/11	1.4640	1.1490	119

Table 6-2 (d)—continued

COMPOUND	EMPIRICAL FORMULA	M.P. (°C)	B.P. (°C/mm)	n_D^{20}	d_4^{20}	REFERENCES
$(C_4H_9)_3GeOCH_2C{\equiv}CH$	$C_{15}H_{30}OGe$	—	151/25	1.4617	1.0033	195
$(n\text{-}C_3H_7)_3GeO$—[cyclohexyl, Br] 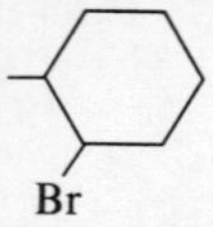	$C_{15}H_{31}OBrGe$	—	150–55/2.8	1.4880 (25°)	—	115
$(n\text{-}C_3H_7)_3GeO$—[cyclohexyl, Cl] 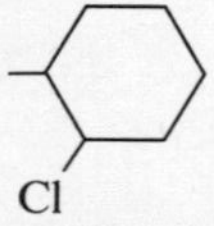	$C_{15}H_{31}OClGe$	—	135–38/2.2	1.4742 (27°)	—	115
$(C_2H_5)_3Ge{-}O{-}CH_2{-}C{\equiv}C{-}Ge(C_2H_5)_3$	$C_{15}H_{32}OGe_2$	—	173/15	1.4789	1.1351	195
$(C_4H_9)_2Ge(H)O(CH_2)_6CH_3$	$C_{15}H_{34}OGe$	—	166/18	1.4489	0.9614	119
$(C_4H_9)_3GeOC_3H_7\text{-}i$	$C_{15}H_{34}OGe$	—	113–14/3.8	1.4400	—	142, 143
$(C_2H_5)_3GeOCH_2CH{=}CHGe(C_2H_5)_3$	$C_{15}H_{34}OGe_2$	—	164–65/13	1.4777	1.1204	195
$(C_2H_5)_3GeO(CH_2)_3Ge(C_2H_5)_3$	$C_{15}H_{36}OGe_2$	—	108/0.45	1.4694	1.1057	195
$(C_6H_5)_2(Cl)GeOC_4H_9$	$C_{16}H_{19}OClGe$	—	$106/4.10^{-2}$	1.5560	1.2432	181
$(C_4H_9)_3GeOCH_2C(CH_3){=}CH_2$	$C_{16}H_{34}OGe$	—	151/12	1.4569	0.9722	195
$[(C_2H_5)_3Ge{-}O{-}CH_2{-}C{\equiv}]_2$	$C_{16}H_{34}O_2Ge_2$	—	128/0.3	1.4795	1.1621	195
$(C_4H_9)_3GeOC_4H_9\text{-}t$	$C_{16}H_{36}OGe$	—	116/4	1.4420	—	142, 143
$(C_2H_5)_3GeOCH_2CH{=}CHCH_2OGe(C_2H_5)_3$	$C_{16}H_{36}O_2Ge_2$	—	119/0.2	1.4738	1.1366	195
$(C_2H_5)_3GeO(CH_2)_4OGe(C_2H_5)_3$	$C_{16}H_{38}O_2Ge_2$	—	151/1	1.4650	1.1191	119
$(C_2H_5)_3Ge{-}O{-}CH{-}P(C_2H_5)_2$ $\quad\mid$ $\quad C_6H_5$	$C_{17}H_{31}POGe$	—	117/0.09	—	—	198

Structure	Empirical formula	m.p.	b.p./mm	n_D	d	Ref.
$(C_4H_9)_3Ge-O-C(CF_3)_2-N(CH_3)_2$	$C_{17}H_{33}NOF_6Ge$	—	110/0.01	1.4220	—	5
$(C_6H_5)_2Ge(Cl)OC_6H_{11}$	$C_{18}H_{21}OClGe$	—	$136/3.10^{-3}$	1.5720	—	181
$(C_2H_5)_3GeO-C_6H_4-OGe(C_2H_5)_3$	$C_{18}H_{34}O_2Ge_2$	—	157/0.35	1.5103	1.1931	119
$(i\text{-}C_4H_9)_3GeO$ 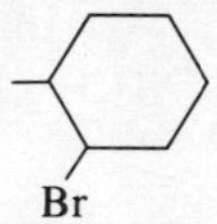 [cyclohexyl–Br]	$C_{18}H_{37}OBrGe$	—	105–10/1	1.4854 (26°)	—	115
$(C_4H_9)_3GeOC_6H_{11}$	$C_{18}H_{38}OGe$	—	180/16	1.4680	0.9963	118
$[(C_2H_5)_3Ge-O-CH(CH_3)-C\equiv]_2$	$C_{18}H_{38}O_2Ge_2$	—	117–18/0.15	1.4710	1.1151	195
$(C_6H_5)_3GeOCH_3$	$C_{19}H_{18}OGe$	66–67	—	—	—	38
$(C_2H_5)_3Ge-OCH=CH-CH-C_6H_5$, $P(C_2H_5)_2$	$C_{19}H_{33}POGe$	—	113/0.15	—	—	198
$(C_4H_9)_3GeO-CH(CH_2)_3CH_3$, C_2H_5	$C_{19}H_{42}OGe$	—	160/9	1.4532	0.9504	118
$(C_6H_5)_3GeOC_2H_5$	$C_{20}H_{20}OGe$	83–85	—	—	—	91
$(C_2H_5)_3Ge-O-CH=C(C_6H_5)_2$	$C_{20}H_{26}OGe$	—	139/0.09	1.5740	1.1411	182, 205
$(n\text{-}C_4H_9)_3Ge-O-CH_2CH_2-C\equiv C-(CH_2)_3CH_3$	$C_{20}H_{40}OGe$	—	150/0.4	1.4655	0.9684	195
$(C_2H_5)_3Ge-O-CH(CH_2)_5CH_3$, CH_3	$C_{20}H_{44}OGe$	—	180/12	1.4540	0.9452	118
$(C_6H_5)_3GeOCH_2CHBrCH_2OCH_3$	$C_{22}H_{23}O_2BrGe$	91–94	—	—	—	169
$(C_6H_5)_3Ge-O-CH_2-CH(CH_3)_2$	$C_{22}H_{24}OGe$	47	$146/5.10^{-3}$	—	—	181
$(C_6H_5)_3GeOC_6F_5$	$C_{24}H_{15}F_5OGe$	99–100	—	—	—	161a
$(C_6H_5)_3GeO$ 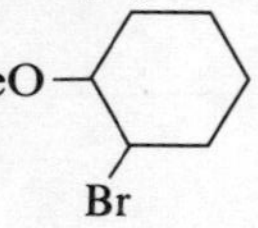 [cyclohexyl–Br]	$C_{24}H_{25}OBrGe$	179–80	—	—	—	115

Table 6-2 (d)—continued

COMPOUND	EMPIRICAL FORMULA	M.P. (°C)	B.P. (°C mm)	n_D^{20}	d_4^{20}	REFERENCES
$(C_6H_5)_3GeOC_6H_{11}$	$C_{24}H_{26}OGe$	—	$160/3.10^{-3}$	1.5945	—	181
$(C_6H_5)_3GeOC(CH_3)=CHCH(CH_3)_2$	$C_{24}H_{26}OGe$	—	$148/5.10^{-3}$	1.5832	—	205, 206
$(C_6H_5)_3GeOCH(CH_3)CH_2CH(CH_3)_2$	$C_{24}H_{28}OGe$	—	$154/3.10^{-2}$	1.5700	1.1462	181
$(C_2H_5)_3Ge-O-C=P(C_2H_5)_2$ $C(C_6H_5)_2$	$C_{24}H_{35}POGe$	—	$154/4.10^{-2}$	1.5793	—	199
$(C_6H_5)_3Ge-O-CH_2-C_6H_5$	$C_{25}H_{22}OGe$	78	—	—	—	181
$(C_6H_5)_3Ge-O-CH-CH_2CH_2-CH=C(CH_3)_2$ CH_3	$C_{26}H_{30}OGe$	—	$164/10^{-3}$	1.5780	—	205, 206
$[(C_4H_9)_3GeO-CH_2-]_2$	$C_{26}H_{58}O_2Ge_2$	—	180/0.4	1.4625	—	142, 144
$(C_4H_9)_3GeO-(CH_2)_3OGe(C_4H_9)_3$	$C_{27}H_{60}O_2Ge_2$	—	180/0.3	1.4630	—	144
$(C_4H_9)_3GeO-CH-CH_2-OGe(C_4H_9)_3$ CH_3	$C_{27}H_{60}O_2Ge_2$	—	167-69/0.2	1.4580	—	142, 144
$(C_4H_9)_3GeO-CH-(CH_2)_2-OGe(C_4H_9)_3$ CH_3	$C_{28}H_{62}O_2Ge_2$	—	175/0.2	1.4635	—	144
$(C_4H_9)_3GeO(CH_2)_4OGe(C_4H_9)_3$	$C_{28}H_{62}O_2Ge_2$	—	187/0.25	1.4650	—	144
$[(C_4H_9)_3Ge-O-CH(CH_3)-]_2$	$C_{28}H_{62}O_2Ge_2$	—	165/0.2	1.4625	—	144
$(C_4H_9)_3GeO-(CH_2)_2-O-(CH_2)_2-OGe(C_4H_9)_3$	$C_{28}H_{62}O_3Ge_2$	—	185/0.1	1.4650	—	144
$[(C_4H_9)_3GeOCH_2-CH_2]_2NH$	$C_{28}H_{63}NO_2Ge_2$	—	152-55/0.6	1.4600	—	142

Compound	Formula	m.p.	b.p./mm	n_D		Ref.
$(C_4H_9)_3GeO-(CH_2)_5-OGe(C_4H_9)_3$	$C_{29}H_{64}O_2Ge_2$	—	186/0.2	1.4660	—	144
$[(C_4H_9)_3GeO-CH_2-CH_2-CH_2-CH_2-]_2$	$C_{30}H_{66}O_2Ge_2$	—	184/0.5	1.4605	—	144
$[(C_4H_9)_3GeO-C(CH_3)_2-]_2$	$C_{30}H_{66}O_2Ge_2$	—	141/0.2	1.4625	—	144
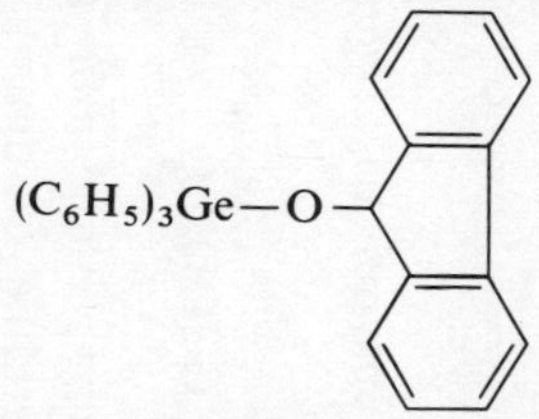	$C_{31}H_{24}OGe$	124-26	—	—	—	164
$(C_6H_5)_3GeOCH(C_6H_5)_2$	$C_{31}H_{26}OGe$	—	220-40/1	—	—	164
$(C_6H_5)_3GeOCH=C(C_6H_5)_2$	$C_{32}H_{26}OGe$	—	240-45/10^{-2}	1.6410	—	182
$(C_6H_5)_3GeO(CH_2)_2CH(CH_3)CH(CH_3)Ge(C_6H_5)_3$	$C_{40}H_{38}OGe_2$	183-5	—	—	—	43
$(C_6H_5)_3GeOCH_2CH(CH_3)CH(CH_3)Ge(C_6H_5)_3$	$C_{41}H_{40}OGe_2$	176	—	—	—	43

$(C_6H_5)_3Ge-O-$ (fluorenyl)

6-3   ORGANOGERMANIUM PEROXIDES

Dannley and Farrant have recently reported the synthesis of triphenyl-germylhydroperoxide by reaction of 98% hydrogen peroxide with tri-phenylbromogermane and dry ammonia gas in ether, or better with triphenylgermanol or triphenylgermanium oxide (65% yield) (62). Quite stable, the compound does not decompose on melting (135–136.5°C). Thermal decomposition in dichlorobenzene begins at 150°C and gives the following products:

$$Ph_3GeOOH \rightarrow O_2 + Ph_3GeOH + (Ph_2GeO)_x + PhOH$$

This reaction has a first order kinetic dependence on hydroperoxide concentration. Base-catalysed decompositions of some hydroperoxides were reported, in the presence of pyridine or triethylamine. Davies and Hall suggested the following mechanism (65):

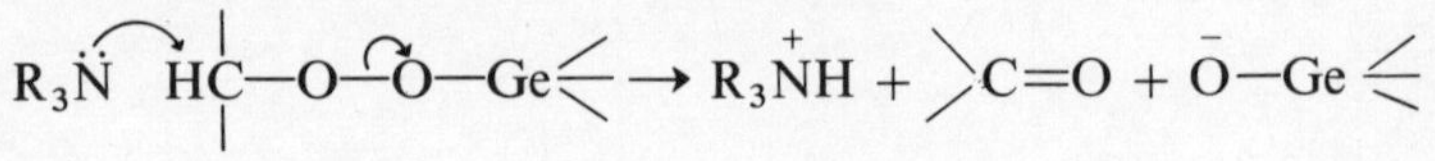

$$R_3\overset{..}{N} \quad HC-O-O-Ge \Big\langle \rightarrow R_3\overset{+}{N}H + \Big\rangle C=O + \overset{-}{O}-Ge \Big\langle$$

6-3-1   Symmetrical organogermanium peroxides

Symmetrical organogermanium peroxides (bis-triallyl, bis-tripropyl, bis-triphenylgermanium peroxides) were prepared in good yields by the nucleophilic substitution reaction (176):

$$2\,R_3GeX + H_2O_2 \xrightarrow{\ NH_3\ } R_3GeOOGeR_3 + 2\,HX$$

$$(X = Cl, Br, NH_2, OR)$$

These compounds are relatively stable under anhydrous conditions: bis(triphenylgermanium) peroxide can be kept at room temperature for several weeks. They hydrolyse readily, giving hydrogen peroxide and trialkylgermanium hydroxides:

$$R_3GeOOGeR_3 + 2\,H_2O \rightarrow 2\,R_3GeOH + H_2O_2$$

6-3-2   Unsymmetrical organogermanium peroxides

Davies and Hall were the first to prepare unsymmetrical germanium peroxides by the following reaction (64):

$$R_{4-n}GeCl_n + n\,R'OOH \rightarrow n\,HCl + R_{4-n}Ge(OOR')_n \qquad n = 1\text{--}4$$

The reactions were carried out in cyclohexane or pentane in the presence of ammonia or triethylamine, or better pyridine or dimethylaniline

(Rieche and Dahlmann) (174, 175). Cyclic organogermanium peroxides are prepared from the corresponding halides with hydroperoxides in the presence of tertiary amines:

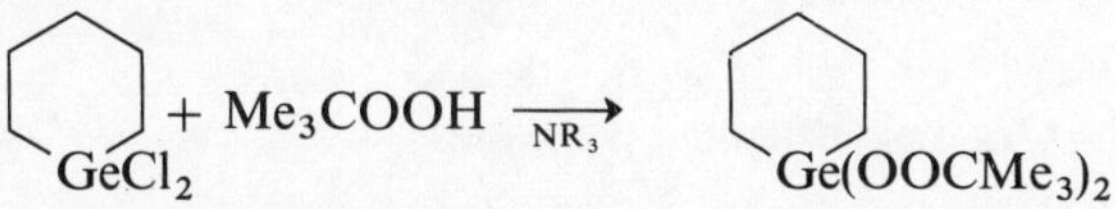

$$\text{GeCl}_2 + \text{Me}_3\text{COOH} \xrightarrow{NR_3} \text{Ge(OOCMe}_3)_2$$

Similarly trialkyl or triarylgermanium bromides react with the anhydrous sodium salt of an alkylhydroperoxide:

$$R_3GeBr + NaOOR' \rightarrow R_3GeOOR' + NaBr$$

Triphenylgermanium peroxides Ph_3GeOOR ($R = CMe_3$, CMe_2Ph) and bis-(triphenylgermanium) peroxide may be obtained by converting triphenylgermanium bromide to the amine, then reacting it with an alkylhydroperoxide:

$$Ph_3GeBr \xrightarrow{NH_3} Ph_3GeNH_2$$

$$Ph_3GeNH_2 + ROOH \longrightarrow NH_3 + Ph_3GeOOR$$

The reaction proceeds rapidly even at room temperature. Organogermanium alkoxides react also with organic hydroperoxides; good yields are obtained with trialkylgermanium alkoxides (235):

$$R_3GeOR'' + R'OOH \rightarrow R_3GeOOR' + R''OH$$

Organogermanium peroxides react readily with water and with anhydrous hydrogen chloride:

$$R_3GeOOR' + H_2O \rightarrow R_3GeOH + R'OOH$$

$$R_3GeOOR' + HCl \rightarrow R_3GeCl + R'OOH$$

As with organosilicon peroxides, evidence of 1,2 rearrangements has been found. Thus, the reaction between germanium tetrachloride and 1-methyl-1-phenylethyl hydroperoxide yields, in the course of a few days, non peroxidic compounds. Davies and Hall suggest a nucleophilic migration of the phenyl group from carbon to oxygen (65):

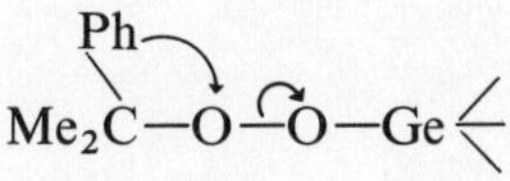

$$\overset{Ph}{\underset{Me_2C-O-O-Ge}{\diagdown}}$$

The thermal decomposition of germanium peroxides begins below 70°C (O—O homolysis). These compounds seem to be less liable to detonate than the silicon peroxides. Some of them are found to catalyse the polymerization of vinyl polymers (175, 176).

These compounds are listed in the following table.

6-3-3   Organogermanium peroxides

Table 6-3(*a–b*) showing some physical properties of organogermanium peroxides appears on the following pages.

Table 6-3
Organogermanium Peroxides

(a) R₃GeOOGeR₃

COMPOUND	EMPIRICAL FORMULA	M.P. (°C)	B.P. (°C/mm)	n_D^{20}	d_4^{20}	REFERENCES
(C₂H₅)₃GeOOGe(C₂H₅)₃	C₁₂H₃₀O₂Ge₂	—	56–57/0.05	1.4603	—	176
(C₃H₇)₃GeOOGe(C₃H₇)₃	C₁₈H₄₂O₂Ge₂	—	65/0.1–0.05	1.4608 (26°)	—	65, 64
(C₆H₅)₃GeOOGe(C₆H₅)₃	C₃₆H₃₀O₂Ge₂	—	126–28	—	—	176

(b) $R_{4-n}Ge(OOR')_n$

COMPOUND	EMPIRICAL FORMULA	M.P. (°C)	B.P. (°C/mm)	n_D^{20}	d_4^{20}	REFERENCES
$(C_2H_5)_3GeOOC(CH_3)_3$	$C_{10}H_{24}O_2Ge$	—	78–79/14–15	1.4368	—	176
$OOGe(CH_3)_3$ 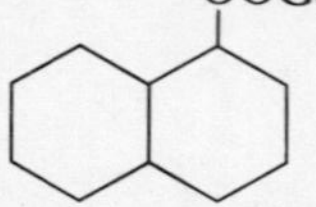 (decahydronaphthyl)	$C_{13}H_{26}O_2Ge$	—	65/0.5	—	—	65
$(CH_2)_5{-}Ge(OOC_4H_9\text{-}tert)_2$	$C_{13}H_{28}O_4Ge$	—	60/0.001	1.4553	—	65
			65/0.2	1.4553	—	64
$(C_3H_7)_3GeOOC(CH_3)_3$	$C_{13}H_{30}O_2Ge$	—	35/0.001	1.4383 (26°)	—	176
$OOGe(C_3H_7)_3$ 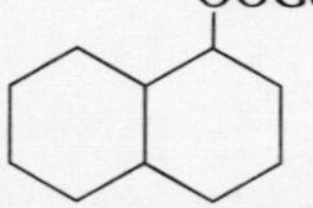 (decahydronaphthyl)	$C_{19}H_{38}O_2Ge$	—	70/0.01	1.4779 (26°)	—	64, 65
$(C_6H_5)_3GeOOC(CH_3)_3$	$C_{22}H_{24}O_2Ge$	—	55–57	—	—	174, 176
$(C_3H_7)_2Ge(OOC_{10}H_{17})_2$	$C_{26}H_{48}O_4Ge$	—	>95/0.01	—	—	64, 65
$(C_6H_5)_3GeOOC_9H_9O$	$C_{27}H_{24}O_3Ge$	—	159–61	—	—	176
$C_9H_9O=$isochroman)						
$(C_6H_5)_3GeOOC(CH_3)_2C_6H_5$	$C_{27}H_{26}O_2Ge$	—	106–8	—	—	176
$(C_6H_5)_3GeOOC_{10}H_{11}$	$C_{28}H_{26}O_2Ge$	—	112–14	—	—	176
$(C_6H_5)_3GeOOC(C_6H_5)_3$	$C_{37}H_{30}O_2Ge$	—	193–95	—	—	176
$Ge(OOC_{10}H_{17})_4$	$C_{40}H_{68}O_8Ge$	—	84–85	—	—	64, 65

6-4   ORGANOGERMANIUM ESTERS OF ORGANIC ACIDS

Many organic germanium esters have been described in the literature and there are several methods of synthesis to obtain them. One of the most frequently used is the action of acids and of their metallic salts (Ag, Pb, Na, K, Hg, Tl etc.) on organogermanium halides. The action of the acids and of their derivatives on the organogermanium alkoxides, oxides and hydrides is largely used. Cleavage of several germanium–heteroelement linkages by acids as well as transesterification reactions are also reported.

6-4-1   Syntheses

6-4-1-1   Action of Metallic Salts of Carboxylic Acids on Organogermanium Halides. Very many germanium esters can be prepared by the action of silver salts on organogermanium halides in organic solvents (CCl_4, benzene, ether, etc.). In this way are prepared amongst others trimethyl-tri-*i*-butyl-, tripentyl-, tri-*i*-pentyl-, and trihexyl-germanium acetates (179), triethylgermanium α-chloropropionate, *n*-valerate (18), triethylgermanium benzoate (20), ethylgermanium triacetate, tripropionate, tri-*n*-butyrate and tri-*n*-valerate (14), tri-*i*-propylgermanium acetate, propionate, *n*-butyrate, *n*-valerate, monochloroacetate, β-chloropropionate, α-chloropropionate and benzoate (17), di-*i*-propylgermanium diacetate, dipropionate, di-*n*-butyrate and dibenzoate (19), tributylgermanium acetate (12), the trifluoro-acetates $Ge(OCOCF_3)_4$, $CH_3Ge(OCOCF_3)_3$, $(CH_3)_2Ge(OCOCF_3)_3$ and $(CF_3)Ge(OCOCF_3)_3$ (92), triphenylgermanium benzoate, salicylate, mandelate, methacrylate and tri-*p*-tolylgermanium benzoate (237) and $Et_3GeGe(OOCCH_3)Et_2$ (43a).

Ethyltriiodogermane reacts with lead formate in benzene to give the triformate $EtGe(OCOH)_3$ (14); di-*n*-propyldiiodogermane and lead formate did not react.

Dimethylgermanium formate, acetate and succinate are prepared by reaction of sodium salts of the corresponding acids with an aqueous solution of $(CH_3)_2GeCl_2$ (186).

$(CH_3CH_2COO)_4Ge$, $[CH_3(CH_2)_2COO]_4Ge$, $(CH_3CH{=}CH{-}COO)_4$-Ge, $(CH_2{=}\underset{\underset{CH_3}{|}}{C}{-}COO)_4Ge$, $[Ph{-}O{-}\underset{\underset{CH_3}{|}}{CH}{-}COO]_4Ge$, $(PhCH_2CH_2{-}$ $COO)_4Ge$, $[(CH_3)_2CHCOO]_4Ge$, $[CH_3(CH_2)_{16}COO]_4Ge$, $[(CH_3)_3C{-}COO]_4Ge$, and $(HSCH_2CH_2COO)_4Ge$ have been prepared in 63.4 to 94.2% yields from tetrachlorogermane and the sodium salts of the organic acids (156b).

An aqueous solution of potassium acetate reacts with $(CH_3)_3GeCl$ to give $(CH_3)_3GeOCOCH_3$ (87% yield) (147).

The trifluoroacetates of germanium are obtained in 25% yield on reaction of germanium tetrachloride with mercury (II) trifluoroacetate in anhydrous trifluoroacetic acid (192, 193).

Thallous acetate and germanium tetrachloride react in acetic anhydride furnishing germanium tetraacetate $Ge(OCOCH_3)_4$ (217).

6-4-1-2 Action of Carboxylic Acids on Organogermanium Halides.
Germanium tetrachloride does not react with acetic acid, but the organogermanium esters form through direct action of the organogermanium halides on the carboxylic acids in the presence of base (Et_3N) (269).

6-4-1-3 Action of Acids and their Derivatives on Organogermanium Hydroxides, Oxides or Alkoxides. Triisopropylgermanium trifluoroacetate, dichloroacetate, trichloroacetate, bromoacetate, iodoacetate are isolated after the action of the corresponding acids on triisopropylgermanol (17).

The action of acetic anhydride on tricyclohexylgermanol leads to tricyclohexylgermanium acetate (101).

Another largely used method is the action of carboxylic acids or their anhydrides on the organogermanium oxides. The basic bis-(trialkylgermanium) oxides react readily, the dialkylgermanium oxides less quickly, and the acid germanoic anhydrides $(R_3GeO)_2O$ do not react:

$$(R_3Ge)_2O + 2\,R'COOH \rightarrow 2\,R_3GeOCOR' + H_2O$$

$$R_2GeO + 2\,R'COOH \rightarrow R_2Ge(OCOR')_2 + H_2O$$

The reactions of various organic acids on $(Et_3Ge)_2O$ (9, 18) and $(Et_2GeO)_4$ (9), $(n\text{-}Pr_3Ge)_2O$ (11), $(n\text{-}Bu_3Ge)_2O$ (12) have been described as well as the action of the acetic anhydride on $(Me_2GeO)_4$ (14), $(n\text{-}Pr_2GeO)$ (13), $(n\text{-}Bu_2GeO)$, $[(n\text{-}C_6H_{13})_2GeO]$, and $(Ph_3Ge)_2O$ (179):

$$(R_2GeO) + (CH_3CO)_2O \rightarrow R_2Ge(OCOCH_3)_2$$

The reactions of tetraethoxygermane, $Ge(OEt)_4$, with salicylic acid lead to various ethoxysalicylates of germanium (141). The reactions between organoalkoxygermanes and simple carboxylic acids and hydroxylcarboxylic acid have been studied (131):

$$R_3GeOR + R'CO_2H \rightarrow R_2GeOCOR' + ROH$$

The reaction of the dialkoxygermanes can lead to alkoxyesters:

$$R_2Ge(OR')_2 + RCO_2H \rightarrow R_2Ge\begin{matrix} \diagup OR' \\ \diagdown OOCR \end{matrix}$$

Alkylgermanium mercaptopropionates are isolated in the action of mercaptopropionic acid on $Bu_2Ge(OEt)_2$ and $Ge(O-i-Pr)_4$ (240) (cf. properties of organoalkoxygermanes).

6-4-1-4   Action of Acids and their Derivatives on Organogermanium Hydrides. Triethylgermane reacts with CF_3COOH, C_2F_5COOH and $n-C_3F_7COOH$ to yield the corresponding esters. The weak acid CH_3COOH does not react at an appreciable rate (21). However, acetic acid and acrylic acid react with trialkylgermanes in the presence of copper powder (119):

$$R_3GeH + R'COOH \rightarrow R_3GeOCOR' + H_2$$

$$(R' = CH_3;\ CH_2{=}CH{-})$$

Et_3GeH reacts with $Hg(OCOCH_3)_2$ to give $Et_3Ge(OCOCH_3)$ with 60% yield (21). Silver acetate reacts with diisopropylgermane to give diacetate $(i-C_3H_7)_2Ge(OCOCH_3)_2$ (16% yield) (19).

6-4-1-5   Reactions of Various Organogermanes with Carboxylic Acids and their Derivatives. Allyl and propargyltriorganogermanes are readily cleaved by organic acids such as CCl_3COOH, $CHCl_2COOH$ and $HCOOH$ (133):

$$R_3GeCH_2C{\equiv}CH + R'CO_2H \rightarrow R_3GeOCOR' + CH_3C{\equiv}CH$$

Cleavage of 1-trimethylgermyl-2 butene by CF_3COOH leads to trimethyl-germanium trifluoroacetate and 1-butene (mechanism SE2)(79):

$$Me_3GeCH_2CH{=}CH{-}CH_3 + CF_3COOH \rightarrow$$

$$Me_3GeOCOCF_3 + CH_3CH_2CH{=}CH_2$$

Reaction of tetraphenylgermanium and trifluoroacetic acid in anhydrous benzene gives at 80°C triphenylgermanium trifluoroacetate quantitatively (192, 193).

Germacyclobutanes react easily with mono- and dichloroacetic acids (135a):

$$R_2Ge{<}\!\!\!\square \xrightarrow{R'CO_2H} R_2(n-Pr)GeOCOR' \qquad (R' = CH_2Cl, CHCl_2)$$

Cleavage of diethylgermacyclopentanol by acetic and dichloroacetic acid leads to the ethylenic esters according the reaction (136):

$$(C_2H_5)_2Ge{<}\!\!\!\underset{OH}{\overset{R}{\boxed{}}}\!\!\!{\cdot}R' + HX \rightarrow (C_2H_5)_2Ge{-}CH_2{-}\underset{R}{CH}{-}\underset{R'}{C}{=}CH_2 + H_2O$$
$$ X$$

$$(X = CH_3COO^- \text{ or } CHCl_2COO^-)$$

$Ph_3Ge—SnPh_3$ reacts with excess anhydrous acetic acid to form quantitatively the solid hexaacetate $(AcO)_3Ge—Sn'(OAc)_3$, stable up to 360°C (262). The action of acids and acid anhydrides on organogermylamines or phosphines leads quantitatively to the corresponding germanium esters (196, 197, 199, 203):

$$R_3GeNMe_2 + R'COOH \rightarrow R_3GeOCOR' + Me_2NH$$

$$R_3GePR'_2 + R'COOH \rightarrow R_3GeOCOR' + R'_2PH$$

$$R_3GeNMe_2 + (CH_3CO)_2O \rightarrow R_3GeOCOCH_3 + CH_3CONMe_2$$

$$R_3GePEt_2 + (CH_3CO)_2O \rightarrow R_3GeOCOCH_3 + CH_3COPEt_2$$

A systematic investigation of the ester group in the organogermanium "conversion series" for reactions using silver salts together with other reactions has been made by Anderson (19). The main reactions for germanium esters are given below.

	REGENT	GERMANIUM PRODUCTS	YIELD (%)
$(i\text{-}Pr)_2Ge(NCO)_2$	$AgOCO\text{-}n\text{-}C_3H_7$	$(i\text{-}Pr)_2Ge(OCO\text{-}n\text{-}C_3H_7)_2$	78
$(n\text{-}Pr)_3GeCN$	$AgOCOC_2H_5$	$(n\text{-}Pr)_3GeOCOC_2H_5$	80
$(i\text{-}Pr)_2Ge(NCS)_2$	$AgOCOCH_3$	$(i\text{-}Pr)_2Ge(OCOCH_3)_2$	80
$[(i\text{-}Pr)_2GeS]_2$	$AgOCOCH_3$	$(i\text{-}Pr)_2Ge(OCOCH_3)_2$	

Compounds of the $(Et_3M)_3M'$ type (where M = Si, Ge, Sn; M' = Bi, Sb) react vigorously with benzoyl peroxide with cleavage of the M—M'—M bonds and formation of the corresponding benzoates (251, 257):

$$2(Et_3M)_3M' + 3(C_6H_5COO)_2 \rightarrow 2M' + 6(C_2H_5)_3MOCOC_6H_5$$

The interaction of benzoyl peroxide with bis-(triethylgermyl)mercury (255), bis-(triethylgermyl)cadmium (48a, 253) and tris-(triethylgermyl)thallium, (111, 252) proceeds in the same manner. In these last reactions, however, on account of the interaction of the metal with the peroxide, cadmium benzoate and thallium benzoate are formed instead of metallic cadmium and thallium.

Glacial acetic acid reacts with $(Et_3Ge)_2Cd$ (254), $(Et_3Ge)_2Hg$ (110) and $(Et_3Ge)_3Bi$ (257) with formation of metal and equimolecular amounts of triethylgermane and triethylgermylacetate:

$$CH_3COOH + Et_3GeMGeEt_3 \rightarrow M + Et_3GeH + Et_3GeOCOCH_3$$

$$(M = Cd, Hg)$$

$$2(Et_3Ge)_3Bi + 3 CH_3COOH \rightarrow 2 Bi + 3 Et_3GeOCOCH_3 + 3 Et_3GeH$$

6-4-1-6 Transesterification Reactions. In transesterification reactions the less volatile acids displace the more volatile acids from the ester, apparently without regard for dissociation constants:

$$R_3GeOCOR' + R''COOH \rightarrow R_3GeOCOR'' + R'COOH$$

Many reactions of this type have been described (11–14, 18–20, 172).

6-4-1-7 Various Syntheses. The action of CO_2 on the trialkylgermyl-amines R_3GeNMe_2 or R_3GeNEt_2 leads to trialkylgermanium dimethyl or diethyl-carbamate, $R_3Ge-O-\overset{\|}{\underset{O}{C}}-NR'_2$ (183, 203).

The reaction of trimethylgermyldiethylamine with β-propiolactone gives the trimethylgermyl ester of N,N-diethyl-β-alanine (97). The tri-phenylgermyltriphenylgermane carboxylate forms in the thermal decomposition of triphenylgermane carboxylic acid followed by esterification of triphenylgermanol (38).

6-4-2 Properties

The germanium esters of organic acids possess high thermal stability. However, certain decomposition reactions have been reported. Organo-germanium formates are not all completely stable and ethylgermanium triformate loses water and carbon monoxide and has no definite boiling point (14). The tri-*i*-propylgermanium haloacetates and halopropionates decomposed in approximately the same relative degree of ease as the corresponding free acids do. The trifluoroacetate was highly stable. The bromoacetate and especially the iodoacetate were relatively unstable. The β-chloropropionate was much less stable than the α-chloropropionate. The thermal decomposition of the β-chloropropionate yields propenoic acid and the corresponding organogermanium chloride (17):

$$(i\text{-}C_3H_7)_3GeOCOCH_2CH_2Cl \rightarrow (i\text{-}C_3H_7)_3GeCl + CH_2=CH-COOH$$

$Ge(OCOCF_3)_4$ decomposes slowly at room temperature with removal of trifluoroacetic anhydride and formation of solid polymeric germoxanes

(192, 193). Triphenylgermyltriphenylgermane carboxylate

$$Ph_3Ge-\underset{\underset{O}{\|}}{C}-O-Ge-Ph_3$$

rearranges at elevated temperatures:

$$Ph_3Ge-\underset{\underset{O}{\|}}{C}-O-GePh_3 \rightarrow CO + Ph_3Ge-O-GePh_3$$

Despite the fact that bonds to germanium are generally somewhat weaker than the corresponding bond to silicon, triphenylgermane carboxylic acid and its derivatives are somewhat more stable toward rearrangement than are its silicon analogues (38).

The germanium esters of carboxylic acids are rather readily hydrolysed with formation of the corresponding acids and organogermanium oxides (9, 217). The hydrolysis of triorganogermanium acetates in aqueous dioxane is much faster than that of the analogous silicon compounds (171a).

The rates of hydrolysis of compounds $(n\text{-}Pr)_3MOCOCH_{(3-n)}Me_n$ (M = Si and Ge) (n = 0 to 3) have been measured in aqueous dioxane. The entropy of activation is shown to be the main factor determining the rate differences; pK_a values of the substituted acetic acids correlate well with $\log(K_1)$. A transition state similar to a solvated proton and a solvated carboxylate ion is suggested (172).

The hexaacetate $(AcO)_3Ge-Sn(OAc)_3$ with liquid hydrogen chloride at low temperatures give the solid chloride $Cl_3GeSnCl_3$. By means of $LiAlH_4$ the acetate and chloride have been converted to $H_3Ge-SnH_3$ (262).

Exchange reactions between $EtGe(OCOR)_3$ and Ph_2SiCl_2 are reported (14):

$$2\,EtGe(OCOCH_3)_3 + 3\,Ph_2SiCl_2 \rightarrow$$

$$2\,EtGeCl_3 + 3\,Ph_2Si(OCOCH_3)_2$$

$$2\,EtGe(OCOC_3H_7)_3 + 3\,Ph_2SiCl_2 \rightarrow$$

$$2\,EtGeCl_3 + 3\,Ph_2Si(OCOC_3H_7)_2$$

Triethylgermanium acetate reacts with silver fluoride with formation of Et_3GeF (80 % yield) (19).

Tetrapropionoxygermane reacts with the Grignard reagent, ethylmagnesium bromide, to produce 3-pentanone, 3-ethyl-3-pentanol, tetraethylgermane and bis-(triethylgermanium) oxide. Tetrapropionoxygermane reacts with phenylmagnesium bromide to form propiophenone, ethyldiphenylcarbinol, triphenylgermanol and bis-(triphenylgermanium) oxide (156b).

Triphenylgermanium methacrylate monomer is readily amenable to both block and solution polymerization. The polymer obtained from block polymerization is a hard, clear thermoplastic mass that is soluble in dimethylformamide. The polymer obtained in toluene solution, using azobisisobutyronitrile as the initiator, is a colourless powder with M.P. 167–180°C (82).

The infrared spectra of $H_3GeOCOCH_3$ and $(CH_3)_3GeOCOCH_3$ have been measured by Srivastava and Onyszchuk (236). Carbonyl absorption bands shift to lower frequencies and carbon–oxygen (skeletal) absorptions to higher frequencies as M changes from C to Si to Ge in the series of acetates $(CH_3)_3MOCOCH_3$. These results suggest that the Ge—O bond is more polar than the Si—O bond, and the order of electron release in acetate derivatives is $(CH_3)_3Ge$, $(CH_3)_3Si$, $(CH_3)_3C$ which is consistent with the electronegativity order C > Si > Ge (236).

Trifluoroacetates of germanium are characterized by high-frequency carbonyl absorptions (C=O) at $1770\,cm^{-1}$ (192, 193).

Application of the empirical equation:

$$\text{B.P. of } R_nGe(OCOR')_{4-n} = n(0.250)\,(\text{B.P. of } R_4Ge)$$

$$+ (4-n)(k)\,(\text{B.P. of } R'COOH)$$

furnishes adequate results (root mean square error, 6.3°C, standard deviation 6.6°C) in calculations of normal boiling points of 46 organogermanium carboxylates; all temperatures are in K while k is 0.3425 for $RGe(OCOR')_3$ and $R_2Ge(OCOR')_2$, 0.3767 or 0.3527 for $R_3GeOCOR'$ (25).

6-4-3 Organogermanium esters of carboxylic acids

Table 6-4 showing some physical properties of organogermanium esters of carboxylic acids appears on the following pages.

Table 6-4

Organogermanium Esters of Carboxylic Acids

COMPOUND	EMPIRICAL FORMULA	M.P. (°C)	B.P. (°C/mm)	n_D^{20}	d_4^{20}	REFERENCES
$(C_2H_5)Ge(OCOH)_3$	$C_5H_8O_6Ge$	—	118/9	1.452	1.617	14
$(CH_3)_3GeOCOCF_3$	$C_5H_9O_2F_3Ge$	—	71/760	1.3820	—	79
$(CH_3)_3GeOCOCH_3$	$C_5H_{12}O_2Ge$	—	127–28/760	1.4211	1.1941	147, 179, 236
$(CH_3)_2Ge(OCOCF_3)_2$	$C_6H_6O_4F_6Ge$	—	167/755	—	—	92
$(CH_3)_2Ge(OCOCH_3)_2$	$C_6H_{12}O_4Ge$	—	188/760	—	—	14
$(C_2H_5)_2Ge(OCOH)_2$	$C_6H_{12}O_4Ge$	—	210	1.4454	1.3665	9
$CH_3Ge(OCOCF_3)_3$	$C_7H_3O_6F_9Ge$	—	167/760	—	—	92
$CF_3Ge(OCOCF_3)_3$	$C_7O_6F_{12}Ge$	—	120/758	—	—	92
$(C_2H_5)_3GeOOCH$	$C_7H_{16}O_2Ge$	—	185.7/768	1.4436	1.1672	9
$(CF_3COO)_4Ge$	$C_8F_{12}O_8Ge$	—	25/10^{-2}	—	—	192, 193
			140–41/760	—	—	92
$(CH_3COO)_4Ge$	$C_8H_{12}O_8Ge$	156	—	—	—	217
$C_2H_5Ge(OCOCH_3)_3$	$C_8H_{14}O_6Ge$	49	249/760	1.444	1.393	14
$(C_2H_5)_3GeOCOCCl_3$	$C_8H_{15}O_2Cl_3Ge$	—	125–26/3	1.4790	1.368	18
$(C_2H_5)_3GeOCOCF_3$	$C_8H_{15}O_2F_3Ge$	—	183/760	1.402	1.272	21
$(C_2H_5)_3GeOCOCHCl_2$	$C_8H_{16}O_2Cl_2Ge$	—	248/760	1.4672	1.304	18
$(C_2H_5)_2Ge(OCOCH_3)_2$	$C_8H_{16}O_4Ge$	—	217	1.4404	1.260	9
$(C_2H_5)_3GeOCOCH_2Br$	$C_8H_{17}O_2BrGe$	—	244/760	1.4842	1.423	18
$(C_2H_5)_3GeOCOCH_2Cl$	$C_8H_{17}O_2ClGe$	—	234/760	1.4672	1.243	18
$(C_2H_5)_3GeOCOCH_2I$	$C_8H_{17}O_2IGe$	—	254/760	1.5112	1.593	18
$(C_2H_5)_3GeOCOCH_3$	$C_8H_{18}O_2Ge$	—	190–91	1.4413	1.1299	9, 21, 257
$(C_2H_5)_3GeOOCC_2F_5$	$C_9H_{15}O_2F_5Ge$	—	189.6/760	1.3844	1.324	21
$(C_2H_5)_3GeOCOCH_2CN$	$C_9H_{17}NO_2Ge$	—	98–99/1	1.464	1.194	18
$(C_2H_5)_3GeOCO—CH{=}CH_2$	$C_9H_{18}O_2Ge$	—	85/10	1.4480	1.1141	119

Compound	Formula		bp/mm	n	d	Ref.
$(C_2H_5)_3GeOCOCHBrCH_3$	$C_9H_{19}O_2BrGe$	—	74–75/1	1.4725	1.370	18
$(C_2H_5)_3GeOCOCH_2Br$	$C_9H_{19}O_2BrGe$	—	85–86/1	1.4832	1.376	18
$(C_2H_5)_3GeOCOCHClCH_3$	$C_9H_{19}O_2ClGe$	—	235/760	1.4619	1.201	18
$(C_2H_5)_3GeOCOCH_2CH_2Cl$	$C_9H_{19}O_2ClGe$	—	236/760	1.4672	1.210	18
$(C_2H_5)_3GeOCOCH_2CH_2I$	$C_9H_{19}O_2IGe$	—	94–96/1	1.5002	1.544	18
$(C_2H_5)_3GeOOCC_2H_5$	$C_9H_{20}O_2Ge$	—	205/760	1.4481	1.109	18
$(C_2H_5)_3GeOCOC_3F_7\text{-}n$	$C_{10}H_{15}O_2F_7Ge$	—	201/760	1.378	1.383	21
$(n\text{-}C_3H_7)_2Ge(OCOCH_2Cl)_2$	$C_{10}H_{18}O_4Cl_2Ge$	—	143–45/2	1.4793	1.374	13
$(i\text{-}C_3H_7)_2Ge(OCOCH_2Cl)_2$	$C_{10}H_{18}O_4Cl_2Ge$	—	298/760	1.480	1.372	19
$(n\text{-}C_3H_7)_2Ge(OCOCH_3)_2$	$C_{10}H_{20}O_4Ge$	35.6	244.6	—	—	13
$(i\text{-}C_3H_7)_2Ge(OCOCH_3)_2$	$C_{10}H_{20}O_4Ge$	—	236	1.445	1.193	19
$(C_2H_5)_3GeOCOC_3H_7\text{-}n$	$C_{10}H_{22}O_2Ge$	—	221/760	1.4486	1.084	18
$(C_3H_7)_3GeOOCH$	$C_{10}H_{22}O_2Ge$	—	233/760	1.4505	1.094	11
$(C_2H_5)_2N(CH_2)_2COOGe(CH_3)_3$	$C_{10}H_{23}NO_2Ge$	—	86.9–87.2/3	—	—	96, 97
$(C_2H_5)_2GeOCOCHCl_2$ $\quad CH_2{-}CH(CH_3)CH{=}CH_2$	$C_{11}H_{20}O_2Ge$	—	168/23	1.4793	1.2460	136
$C_2H_5Ge(OCOC_2H_5)_3$	$C_{11}H_{20}O_6Ge$	—	256/760	1.4434	1.271	14
$(i\text{-}C_3H_7)_3GeOCOCCl_3$	$C_{11}H_{21}O_2Cl_3Ge$	—	107–9/1	1.4850	1.291	17
$(n\text{-}C_3H_7)_3GeOCOCF_3$	$C_{11}H_{21}O_2F_3Ge$	—	226/760	1.4103	1.189	11
$(i\text{-}C_3H_7)_3GeOCOCF_3$	$C_{11}H_{21}O_2F_3Ge$	—	220/760	1.4200	1.178	17
$(C_2H_5)_2Ge{-}OCOCH_3$ $\quad CH_2{-}CH(CH_3)CH{=}CH_2$	$C_{11}H_{22}O_2Ge$	—	104/16	1.4617	1.0885	136
$(n\text{-}C_3H_7)_3GeOCOCHCl_2$	$C_{11}H_{22}O_2Cl_2Ge$	—	150/10	1.4708	1.226	11
$(i\text{-}C_3H_7)_3GeOCOCHCl_2$	$C_{11}H_{22}O_2Cl_2Ge$	—	92–93/1	1.4772	1.236	17
$(n\text{-}C_3H_7)_3GeOCOCH_2Br$	$C_{11}H_{23}O_2BrGe$	—	149/10	1.4778	1.325	11
$(i\text{-}C_3H_7)_3GeOCOCH_2Br$	$C_{11}H_{23}O_2BrGe$	—	98–99/1	1.4872	1.331	17
$(n\text{-}C_3H_7)_3GeOCOCH_2Cl$	$C_{11}H_{23}O_2ClGe$	—	143/10	1.4640	1.166	11

Table 6-4—continued

COMPOUND	EMPIRICAL FORMULA	M.P. (°C)	B.P. (°C/mm)	n_D^{20}	d_4^{20}	REFERENCES
$(i\text{-}C_3H_7)_3GeOCOCH_2Cl$	$C_{11}H_{23}O_2ClGe$	—	101–3/1	1.4722	1.182	17
$(i\text{-}C_3H_7)_3GeOCOCH_2I$	$C_{11}H_{23}O_2IGe$	—	107–8/1	1.5090	1.465	17
$(n\text{-}C_3H_7)_3GeOCOCH_3$	$C_{11}H_{24}O_2Ge$	—	236/760	1.4464	1.071	11, 172
			115/14	—		
$(i\text{-}C_3H_7)_3GeOCOCH_3$	$C_{11}H_{24}O_2Ge$	—	229/760	1.4532	1.070	17
$(C_2H_5)_3GeOCOC_4H_9$	$C_{11}H_{24}O_2Ge$	—	230/760	1.4467	1.062	18
$(C_2H_5)_3GeOOCN(C_2H_5)_2$	$C_{11}H_{25}NO_2Ge$	—	115/5	1.4529	1.0906	203
$(i\text{-}C_3H_7)_2Ge(OCOCH_2CH_2Cl)_2$	$C_{12}H_{22}O_4Cl_2Ge$	—	140–42/1	1.474	1.294	19
$(C_4H_9)_2Ge(OCOCH_3)_2$	$C_{12}H_{24}O_4Ge$	—	144–46/18	1.4452	1.444	179
			127/5	1.4475	—	131
$(i\text{-}C_3H_7)_2Ge(OCOC_2H_5)_2$	$C_{12}H_{24}O_4Ge$	—	94–95/1	1.447	1.146	19
$(i\text{-}C_3H_7)_3GeOCOCH_2CH_2Cl$	$C_{12}H_{25}O_2ClGe$	—	97–99/1	1.4717	1.154	17
$(i\text{-}C_3H_7)_3GeOCOCHClCH_3$	$C_{12}H_{25}O_2ClGe$	—	95–97/1	1.4672	1.144	17
$(n\text{-}C_3H_7)_3GeOCOC_2H_5$	$C_{12}H_{26}O_2Ge$	—	246/760	1.4473	1.054	11, 172
$(i\text{-}C_3H_7)_3GeOCOC_2H_5$	$C_{12}H_{26}O_2Ge$	—	242/760	1.4538	1.059	17
$(C_4H_9)_2Ge{<}^{OC_2H_5}_{OOCCH_3}$	$C_{12}H_{26}O_3Ge$	—	110/4	1.4400	—	131
$(C_2H_5)_3GeOCOC_6H_5$	$C_{13}H_{20}O_2Ge$	—	105–7/1	1.513	1.172	20, 253, 255
$o\text{-}NH_2C_6H_4COOGe(C_2H_5)_3$	$C_{13}H_{21}NO_2Ge$	—	174–76/1	1.544	1.215	20
$(n\text{-}C_3H_7)_3GeOCOCH(CH_3)_2$	$C_{13}H_{28}OGe$	—	109/10	—	—	172
$(n\text{-}C_4H_9)_3GeOCOH$	$C_{13}H_{28}O_2Ge$	—	149–51/16	1.4538	1.051	12
$(i\text{-}C_3H_7)_3GeOCO\text{-}n\text{-}C_3H_7$	$C_{13}H_{28}O_2Ge$	—	180–82/98	1.5480	1.045	17

Compound	Molecular formula		b.p. (°C/mm)	n_D	d	Ref.
$C_2H_5Ge(OOC\text{-}n\text{-}C_3H_7)_3$	$C_{14}H_{26}O_6Ge$	—	136–37/2	1.4432	1.186	14
$(n\text{-}C_4H_9)_3GeOOCCF_3$	$C_{14}H_{27}O_2F_3Ge$	—	147–49/19	1.419	1.144	12
$(i\text{-}C_3H_7)_2Ge(OOCC_3H_7\text{-}n)_2$	$C_{14}H_{28}O_4Ge$	—	106–8/1	1.452	1.112	19
$(n\text{-}C_4H_9)_3GeOOCCH_3$	$C_{14}H_{30}O_2Ge$	—	147–48/14	1.4514	1.027	12
$(i\text{-}C_4H_9)_3GeOOCCH_3$	$C_{14}H_{30}O_2Ge$	—	80/0.2	1.4635	—	131
$(i\text{-}C_3H_7)_3GeOOCC_4H_9\text{-}n$	$C_{14}H_{30}O_2Ge$	—	135–38/12	1.4508	1.026	179
$(n\text{-}C_3H_7)_3GeOCOC(CH_3)_3$	$C_{14}H_{30}O_2Ge$	—	82–83/1	1.4560	1.034	17
$(C_2H_5)_3GeSCH_2COOGe(C_2H_5)_3$	$C_{14}H_{32}O_2SGe_2$	—	120/10	—	1.2224	172
$(C_6H_5)_2Ge(OOCCH_3)_2$	$C_{16}H_{16}O_4Ge$	—	326	1.4993	—	9
$(C_6H_5)_2Ge(OOCCH_3)_2$	$C_{16}H_{16}O_4Ge$	—	150/0.2	—	—	131
$(C_6H_5)_2Ge\big(OC_2H_5\big)\big(OOCCH_3\big)$	$C_{16}H_{18}O_3Ge$	—	140/0.4	1.5557	—	131
$(i\text{-}C_3H_7)_3GeOCOC_6H_5$	$C_{16}H_{26}O_2Ge$	—	130–32/1	1.5138	1.132	17
$(C_6H_{13})_2Ge(OCOCH_3)_2$	$C_{16}H_{32}O_4Ge$	—	110–15/0.14	1.4496	—	179
$(C_2H_5)_3GeOCO\text{—}CH_2$ $(C_2H_5)_3GeOCO\text{—}CH_2$	$C_{16}H_{34}O_4Ge_2$	—	130–31/0.25	1.4679	—	196
$(C_4H_9)_2Ge\big(OC_2H_5\big)\big(OOCC_6H_5\big)$	$C_{17}H_{28}O_3Ge$	—	140/0.2	1.4955	—	131
$C_2H_5Ge(OOCC_4H_9\text{-}n)_3$	$C_{17}H_{32}O_6Ge$	—	305/760	1.4456	1.136	14, 25
$(n\text{-}C_5H_{11})_3GeOCOCH_3$	$C_{17}H_{36}O_2Ge$	—	97–100/0.05	1.4532	1.004	179
$(i\text{-}C_5H_{11})_3GeOCOCH_3$	$C_{17}H_{36}O_2Ge$	—	157–58/14	1.4512	0.996	179
$(C_4H_9)_3GeOOCC_6H_5$	$C_{19}H_{32}O_2Ge$	—	168/1.5	1.4930	—	131
$(n\text{-}C_4H_9)_3GeOOCC_6H_{13}$	$C_{19}H_{40}O_2Ge$	—	147–48/1	1.4539	0.988	12

Table 6-4—continued

COMPOUND	EMPIRICAL FORMULA	M.P. (°C)	B.P. (°C/mm)	n_D^{20}	d_4^{20}	REFERENCES
$(C_6H_5)_3GeOCOCF_3$	$C_{20}H_{15}O_2F_3Ge$	120	$120/10^{-3}$	—	—	192, 193
$(C_6H_5)_3GeOCOCH_3$	$C_{20}H_{18}O_2Ge$	83	—	—	—	179, 237
$(i\text{-}C_3H_7)_2Ge(OOCC_6H_5)_2$	$C_{20}H_{24}O_4Ge$	54	186–88/1	—	—	19
$(C_6H_{11})_3GeOCOCH_3$	$C_{20}H_{36}O_2Ge$	82–83	—	—	—	101
$(C_6H_{13})_3GeOCOCH_3$	$C_{20}H_{42}O_2Ge$	—	138/0.075	1.4554	0.980	179
$(C_6H_5)_2Ge$ (OC$_2$H$_5$) (OOCC$_6$H$_5$)	$C_{21}H_{20}O_3Ge$	—	180/0.2	1.5600	—	131
$(C_6H_5)_3GeOOCC(CH_3)=CH_2$	$C_{22}H_{20}O_2Ge$	86–88	—	—	—	82
$(C_4H_9)_2Ge(OOCC_6H_5)_2$	$C_{22}H_{28}O_4Ge$	—	198/0.4	—	—	131
$(C_7H_{15})_3GeOCOCH_3$	$C_{23}H_{48}O_2Ge$	—	146–50/0.03	1.4576	0.965	179
$(C_6H_5)_3GeOOCC_6H_5$	$C_{25}H_{20}O_2Ge$	112–13	—	—	—	237
$(C_6H_5)_3GeOOC(OH)C_6H_4$	$C_{25}H_{20}O_3Ge$	146–47	—	—	—	237
$(C_6H_5)_3GeOOCCH(OH)C_6H_5$	$C_{26}H_{22}O_3Ge$	111–12	—	—	—	237
$(C_6H_5)_2Ge(OOCC_6H_5)_2$	$C_{26}H_{20}O_4Ge$	160	—	—	—	131
$(p\text{-}C_6H_4CH_3)_3GeOOCC_6H_5$	$C_{28}H_{26}O_2Ge$	—	—	—	—	237
$(C_4H_9)_3GeO$—C$_6$H$_4$—$(C_4H_9)_3GeOOC$	$C_{29}H_{58}O_3Ge$	—	160/0.5	1.5065	—	131
$(C_6H_5CH_2)_3GeOOCCH_2C_6H_5$	$C_{29}H_{28}O_2Ge$	146–48	—	—	—	55

Compound	Formula	m.p./b.p.	b.p./mm	n_D	Ref.
$(C_4H_9)_3GeOOC\!-\!CH\!-\!OGe(C_4H_9)_3$ $\quad\quad\quad\quad\quad\quad\; \vert$ $\quad\quad\quad\quad\quad\;\; C_6H_5$	$C_{32}H_{60}O_3Ge_2$	—	203/0.4	1.4875	131
$(C_6H_5)_3GeCOOGe(C_6H_5)_3$	$C_{37}H_{30}O_2Ge_2$	163–64	—	—	38, 39
$(C_6H_5CH_2)_3GeOOCGe(CH_2C_6H_5)_3$	$C_{43}H_{42}O_2Ge_2$	77–78	—	—	55

6-5   ORGANOGERMANIUM ESTERS OF INORGANIC ACIDS

The general methods of preparation of these esters are the action of organogermanium halides on the silver salts of the acids, the action of the acids or their derivatives (anhydrides or chlorides) on the oxides $(R_3Ge)_2O$ and R_2GeO, or on the organogermanium hydrides. The transesterification reactions from organic esters and some various reactions such as the cleavage of various linkages (Ge—C, Ge—S, etc.) by the acids or their derivatives have also been described.

6-5-1   Syntheses

6-5-1-1   Action of Metallic Salts of Inorganic Acids on Organogermanium Halides. Me_3GeCl in inert organic solvents treated with $AgNO_3$, Ag_3PO_4, $AgClO_4$, Ag_3AsO_4, Ag_2SeO_4, Ag_3VO_4, Ag_2CrO_4, and $AgReO_4$ gave Me_3GeONO_2, $(Me_3GeO)_3PO$ (220), $Me_3GeOClO_3$ (189), $[(Me)_3Ge-O]_3AsO$, $[Me_3GeO]_2SeO_2$ (221, 223), $(Me_3GeO)_3VO$, $(Me_3GeO)_2CrO_2$, and Me_3GeReO_4 (218), respectively. Triphenylgermanium sulphate, selenite, selenate and tri-*p*-tolylgermanium sulphate are also obtained through the action of Ph_3GeBr or (*p*-tolyl)$_3GeBr$ on the corresponding silver salts (237). Dimethylgermanium chromate was prepared by the action of Me_2GeCl_2 on an aqueous solution of K_2CrO_4 (186).

6-5-1-2   Action of Acids and Their Derivatives on Organogermanium Oxides. Sulphuric acid reacts with $(Et_3Ge)_2O$ and (Et_2GeO) to give $(Et_3Ge)_2SO_4$ and $(Et_2GeSO_4)_2$ respectively (8). Dimeric dimethylgermanium sulphate is probably an eight-membered ring:

$$
\begin{array}{ccccc}
 & O \diagdown \quad \diagup O & & & \\
\text{Et} & \quad S & & \text{Et} & \\
\diagdown & O{-}S{-}O & & \diagup & \\
& \text{Ge} & & \text{Ge} & \\
\diagup & & & \diagdown & \\
\text{Et} & O{-}S{-}O & & \text{Et} & \\
& O \diagup \quad \diagdown O & & &
\end{array}
$$

This compound is a crystalline white solid; $(Et_3Ge)_2SO_4$ is a colourless liquid readily miscible in organic solvents and concentrated sulphuric acid (8). *i*-Pr_3GeOH also reacts with H_2SO_4 (17):

$$2\,i\text{-}Pr_3GeOH + H_2SO_4 \;\rightarrow\; (i\text{-}Pr_3Ge)_2SO_4 + 2\,H_2O$$

The action of various acid anhydrides on $(Me_3Ge)_2O$ leads to the esters of the corresponding acids already isolated by the action of Me_3GeCl on the silver salts:

$$3\,(Me_3Ge)_2O + As_2O_5 \;\rightarrow\; 2\,(Me_3GeO)_3AsO \quad (221)$$

$$(Me_3Ge)_2O + SeO_3 \;\rightarrow\; (Me_3GeO)_2SeO_2 \quad (221)$$

$(Me_3GeO)_3VO$, $(Me_3GeO)_2CrO_2$, Me_3GeReO_4, $(Me_3GeO)_2SeO_2$, $(Me_3GeO)_3PO$, $Me_3GeOClO_3$ and $Me_3Ge\,OSO_2CH_3$ were also prepared by the action of V_2O_5, CrO_3, Re_2O_7 (218), SeO_3 (223), P_2O_5 (220), Cl_2O_7 (189) and $(MeSO_2)_2O$ (214) on $(Me_3Ge)_2O$. CrO_3 cleaves $Me_3SiOGeMe_3$ with the formation of explosive $Me_3SiOCrO_2OGeMe_3$ (213). $(Me_3Ge)_2O$ and $POCl_3$ give Me_3GeCl and $Me_3GeOPOCl_2$. $(Me_3Ge)_2O$ and $P_2O_3Cl_4$ give a high yield of $Me_3GeOPOCl_2$. $Me_3SiOGeMe_3$ and $POCl_3$ yield 80% of Me_3GeCl and $Me_3SiOPOCl_2$ and 20% Me_3SiCl and $Me_3Ge\text{-}OPOCl_2$, whereas $Me_3SiOGeMe_3$ and $P_2O_4Cl_4$ give equimolecular amounts of $Me_3GeOPOCl_2$ and $Me_3SiOPOCl_2$ (222).

6-5-1-3 Action of Acids and Their Derivatives on Organogermanium Hydrides. 100% H_2SO_4 reacts with Et_3GeH to give $(Et_3Ge)_2SO_4$ (70% yield). Reaction between Et_3GeH and $HgSO_4$ gives also $(Et_3Ge)_2SO_4$, metallic mercury and hydrogen (21). Triethylgermanium benzene sulphonate forms after 6 hours refluxing of triethylgermane with benzene sulphonic acid in toluene solution (194):

$$Et_3GeH + C_6H_5SO_3H \;\rightarrow\; Et_3GeSO_3C_6H_5 + H_2$$

Boric acid reacts with triethylgermane in the presence of copper powder to lead to triethylgermanium borate (119):

$$3\,Et_3GeH + B(OH)_3 \;\rightarrow\; (Et_3GeO)_3B + 3\,H_2$$

6-5-1-4 Transesterification Reactions. Many transesterification reactions have been described between the organogermanium esters of organic acids and various mineral acids. Organogermanium sulphates such as (n-$Pr_3Ge)_2SO_4$ (11), (i-$Pr_3Ge)_2SO_4$ (17), $(Me_2GeSO_4)_2$ (14), (n-$Pr_2GeSO_4)_2$ (13) and (i-$Pr_2GeSO_4)_2$ (19) are prepared by the general reactions:

$$2\,R_3GeOCOR' + H_2SO_4 \;\rightarrow\; (R_3Ge)_2SO_4 + 2\,HOCOR'$$

$$R_2Ge(OCOR')_2 + H_2SO_4 \;\rightarrow\; (R_2GeSO_4) + 2\,R'COOH$$

Five alkylgermanium alkane sulphonates have been prepared by transesterification from triethylgermanium acetate, diisopropylgermanium acetate, tri-n-butylgermanium trifluoroacetate and methane and ethanesulphonic acid (24).

6-5-1-5   Various Syntheses. The reaction between trimethyl-methylthio-germane and dimethyl sulphate gives in high yield (trimethylgermyl) methyl sulphate (91):

$$(CH_3)_3GeSCH_3 + 2\,(CH_3O)_2SO_2 \rightarrow$$

$$(CH_3)_3GeOSO_2OCH_3 + [(CH_3)_3S]^+ + [(CH_3)_3SO_4]^-$$

Concentrated sulphuric acid attacks diethylgermacyclopentane with opening of the cycle (135):

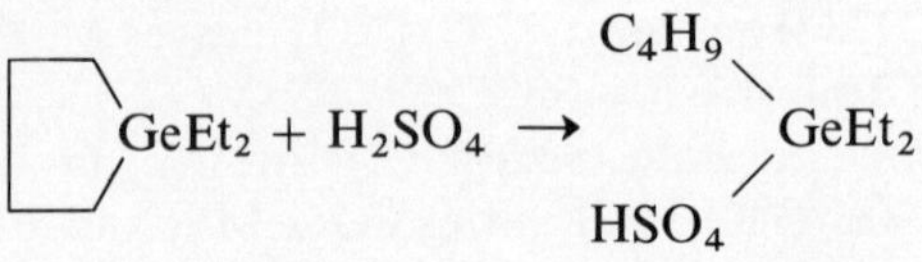

$$\boxed{}\!\!\diagup GeEt_2 + H_2SO_4 \rightarrow \quad \begin{array}{c} C_4H_9 \\[-2pt] \diagdown \\[-4pt] \diagup GeEt_2 \\[-2pt] HSO_4 \end{array}$$

Several insertion reactions of SO_2 or SO_3 lead to the corresponding sulphinates, sulphonates or sulphates:

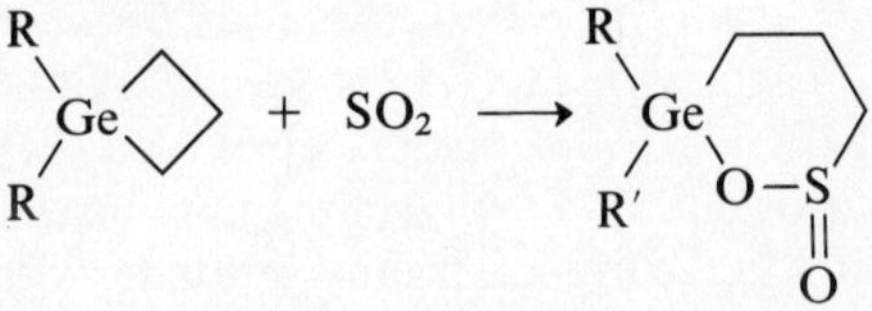

$$\begin{array}{c} R \\ \diagdown \\ Ge \\ \diagup \\ R \end{array}\!\!\diamondsuit + SO_2 \longrightarrow \begin{array}{c} R \\ \diagdown \\ Ge \\ \diagup \\ R' \quad O-S \\ \quad\quad \parallel \\ \quad\quad O \end{array} \qquad (70)$$

$$p\text{-}Et_3Ge-C_6H_4-GeEt_3 + SO_3 \rightarrow$$

$$p\text{-}Et_3Ge-C_6H_4-SO_2-OGeEt_3 \quad (32)$$

$$Me_4Ge + SO_3 \rightarrow Me_3GeOSO_2Me \quad (214)$$

$$2\,Me_4Ge + 3\,SO_3 \rightarrow$$

$$MeSO_2OSO_2Me + (Me_3GeO)_2SO_2 \quad (214)$$

$$Me_3SiOGeMe_3 + SO_3 \rightarrow Me_3SiOSO_2-OGeMe_3 \quad (213)$$
$$t° \downarrow$$
$$(Me_3SiO)_2SO_2 + (Me_3GeO)_2SO_2$$

Oxidation of R_3GePPh_2 by dry oxygen gives the monomeric phosphorus (V) esters $R_3GeOP(O)Ph_2$ which are also obtained from diphenylphosphinic acid and bis-(triethylgermanium) oxide (40):

$$Et_3GePPh_2 \xrightarrow{O_2} Et_3Ge-O-P(O)Ph_2 \leftarrow (Et_3Ge)_2O + 2\,Ph_2P(O)OH$$

In the same way the oxidation of the Ge—As bond derivatives leads to the derivatives of the type $Ph_3Ge-O-As(C_6H_5)_2$ (226).
$$\downarrow$$
$$O$$

6-5-2   Properties

The organogermanium esters of inorganic acids hydrolyse readily with formation of the corresponding acid and organogermanium oxides or tri-organogermanols:

$$Et_3GeOP(O)Ph_2 \xrightarrow{H_2O} (Et_3Ge)_2O + Ph_2P(O)OH \quad (40)$$

$$(Me_3GeO)_2SeO_2 \xrightarrow{H_2O} 2\,Me_3GeOH + H_2SeO_4 \quad (223)$$

$$Me_3GeONO_2 \xrightarrow{H_2O} Me_3GeOH + HNO_3 \quad (220)$$

$$(Me_3GeO)_3PO \xrightarrow{H_2O} Me_3GeOH + 3\,H_3PO_4 \quad (220)$$

$$Me_3GeOClO_3 \xrightarrow{H_2O} Me_3GeOH + HClO_4 \quad (189)$$

$$Me_3SiOSO_2OGeMe_3 \xrightarrow{H_2O} Me_3GeOH + Me_3SiOH + H_2SO_4 \quad (213)$$

$$Me_3SiOCrO_2GeMe_3 \xrightarrow{H_2O} Me_3GeOH + Me_3SiOH + H_2CrO_4 \quad (209)$$

Several cleavage or exchange reactions have been studied from esters of inorganic acids:

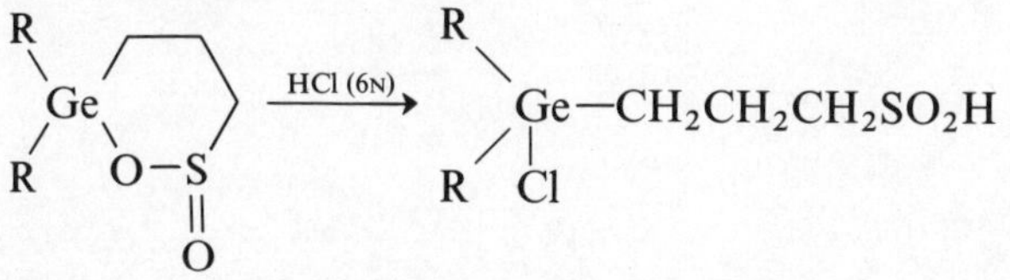

Bis-(trimethylgermyl) sulphate $(Me_3GeO)_2SO_2$ was converted with Me_3SiOLi to Li_2SO_4 and $Me_3SiOGeMe_3$ (213).

Exchange reactions between $R_3GeO—SO_2R'$ and $HgCl_2$, n-Bu_2GeBr_2, Et_3SnF, Ph_2SnCl_2, $AgCl$, $HgBr_2$, KI, KCN, Ph_3SiNCO, $KSCN$, $ZnCl_2$, SbF_3, $PbCl_2$, Na_2SO_3, $NaCl$, $KOCOCH_3$ show replacement of an —SO_2R group attached to germanium by fluoride, chloride, bromine, iodide, cyanide, isocyanate, isothiocyanate, acetate, and oxide (24).

6-5-3   Organogermanium esters of inorganic acids

Table 6-5 showing some physical properties of organogermanium esters of inorganic acids appears on the following pages.

Table 6-5
Organogermanium Esters of Inorganic Acids

COMPOUND	EMPIRICAL FORMULA	M.P. (°C)	B.P. (°C/mm)	n_D^{20}	d_4^{20}	REFERENCES
(CH$_3$)$_3$GeONO$_2$	C$_3$H$_9$NO$_3$Ge	5	62/11	1.4436 (18°)	—	220
(CH$_3$)$_3$GeReO$_4$	C$_3$H$_9$O$_4$ReGe	108	115-25/1.5	—	—	218
(CH$_3$)$_3$GeOClO$_3$	C$_3$H$_9$O$_4$ClGe	5-6	91-92/2	—	—	189
(CH$_3$)$_3$GeOSO$_2$CH$_3$	C$_4$H$_{12}$O$_3$SGe	—	56-57/0.1	—	—	214
(CH$_3$)$_3$GeOSO$_2$OCH$_3$	C$_4$H$_{12}$O$_4$SGe	—	55-57/0.2	—	—	91
[(CH$_3$)$_3$GeO]$_2$SO$_2$	C$_6$H$_{18}$O$_4$SGe$_2$	138-39	—	—	—	213
[(CH$_3$)$_3$GeO]$_2$CrO$_2$	C$_6$H$_{18}$O$_4$CrGe$_2$	61	97-98.5/1	—	—	218
(CH$_3$)$_3$GeOSO$_2$OSi(CH$_3$)$_3$	C$_6$H$_{18}$O$_4$SSiGe	92-98	—	—	—	213
(CH$_3$)$_3$SiOCrO$_2$OGe(CH$_3$)$_3$	C$_6$H$_{18}$O$_4$CrSiGe	-3-+2	—	—	—	213
[(CH$_3$)$_3$GeO]$_2$SeO$_2$	C$_6$H$_{18}$O$_4$SeGe$_2$	147-48	135/1	—	—	221, 222, 223
(C$_2$H$_5$)$_3$GeOSO$_2$CH$_3$	C$_7$H$_{18}$OSGe	—	116-17/1	1.4650	1.286	24
(C$_2$H$_5$)$_3$GeOSO$_2$C$_2$H$_5$	C$_8$H$_{20}$O$_3$SGe	—	125-26/1	1.4651	1.253	24
(i-C$_3$H$_7$)$_2$Ge(SO$_3$CH$_3$)$_2$	C$_8$H$_{20}$O$_6$S$_2$Ge$_2$	—	180-82/1	1.4770	1.426	24
[C$_2$H$_5$)$_2$GeOSO$_2$O]$_2$	C$_8$H$_{20}$O$_8$S$_2$Ge$_2$	115.5-16.5	—	—	—	8
[(CH$_3$)$_3$GeO]AsO	C$_9$H$_{27}$O$_4$AsGe$_3$	46-47	130-31/1.5	—	—	221
[(CH$_3$)$_3$GeO]$_3$VO	C$_9$H$_{27}$O$_4$VGe$_3$	-18	106-12/1.5	—	—	218
[(CH$_3$)$_3$GeO]$_3$PO	C$_9$H$_{27}$O$_4$PGe$_3$	37	102-3/1	—	—	220
(C$_2$H$_5$)$_3$GeSO$_3$C$_6$H$_5$	C$_{12}$H$_{20}$O$_3$SGe	38	160/1.3	—	—	194
[n-C$_3$H$_7$)$_2$GeSO$_4$]$_2$	C$_{12}$H$_{28}$O$_8$S$_2$Ge$_2$	129	—	—	—	13
[i-C$_3$H$_7$)$_2$GeSO$_4$]$_2$	C$_{12}$H$_{28}$O$_8$S$_2$Ge$_2$	115	—	—	—	19
(C$_2$H$_5$)$_3$GeOSO$_2$OGe(C$_2$H$_5$)$_3$	C$_{12}$H$_{30}$O$_4$SGe$_2$	-4	165/3	1.475 (22°)	1.333 (22°)	8, 21
(n-C$_4$H$_9$)$_3$GeOSO$_2$CH$_3$	C$_{13}$H$_{30}$O$_3$SGe	—	150-52/1	1.4641	1.131	24
(n-C$_4$H$_9$)$_3$GeOSO$_2$C$_2$H$_5$	C$_{14}$H$_{32}$O$_3$SGe	—	149-51/1	1.4654	1.117	24
(C$_2$H$_5$)$_3$GeOP(O)(C$_6$H$_5$)$_2$	C$_{18}$H$_{25}$O$_2$PGe	—	160-62/10^{-3}	—	—	40
p-(C$_2$H$_5$)$_3$GeC$_6$H$_4$SO$_2$OGe(C$_2$H$_5$)$_3$	C$_{18}$H$_{34}$O$_3$SGe$_2$	41-43	192-94/0.02	—	—	32

Compound	Formula	m.p.	b.p.	n	d	Ref.
[(n-C₃H₇)₃Ge]₂SO₄	$C_{18}H_{42}O_4SGe_2$	—	180–82/1	—	1.186 (40°)	11
[(i-C₃H₇)₃Ge]₂SO₄	$C_{18}H_{42}O_4SGe_2$	—	380	1.482	1.217	17, 24
B[O—Ge(C₂H₅)₃]₃	$C_{18}H_{45}BO_3Ge_3$	—	182–84/1	1.467	1.1723	119
[(C₆H₅)₃Ge]₂SO₄	$C_{36}H_{30}O_4SGe_2$	225–26	128/0.25	—	—	237
[(C₆H₅)₃Ge]₂SeO₃	$C_{36}H_{30}O_3SeGe_2$	159	—	—	—	237
[(C₆H₅)₃Ge]₂SeO₄	$C_{36}H_{30}O_4SeGe_2$	172	—	—	—	237
[(p-C₆H₄CH₃)₃Ge]₂SO₄	$C_{42}H_{42}O_4SGe_2$	147–49	—	—	—	237

6-6   MISCELLANEOUS COMPOUNDS WITH Ge—O BONDS

Germanium tetrachloride and germanium tetrabromide react with acetylacetone with evolution of HCl to lead to the germanium bis-acetylacetonedihalide derivatives (157):

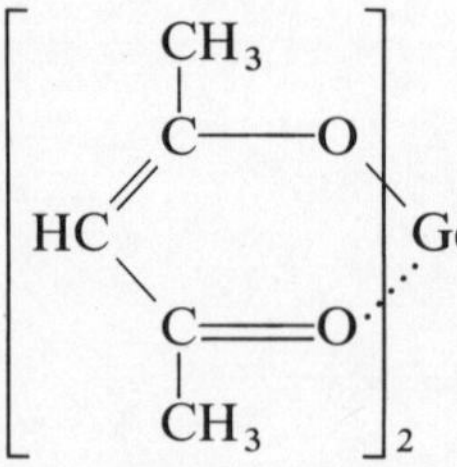

$$
\left[
\begin{array}{c}
CH_3 \\
| \\
C \!\!-\!\! O \\
\diagup \!\! \diagdown \\
HC \quad\quad GeX_2 \\
\diagdown \\
C \!=\! O \\
| \\
CH_3
\end{array}
\right]_2
\equiv (acac)_2 GeX_2
$$

The coordination number of germanium in these compounds is 6.

Germanium tetrachloride or germanium tetrabromide and copper acetylacetone react in chloroform to form intermediate unstable adducts which decompose yielding germanium tris-acetylacetone derivatives of the type:

$$[acac]_3 Ge[CuCl_2]$$

$$[acac]_3 Ge[CuBr_2]$$

$$[acac]_3 Ge[CuBr_3]$$

$$[acac]_3 Ge[Cu_2Br_3]$$

The analogous octahedral complex $[CH(CC_2H_5O)(CCH_3O)]_2GeCl_2$ is formed with β-propionylacetone (157).

Tris-acetylacetonate derivatives were also obtained by Heinrich and Muller (159):

$$(acac)_2GeCl_2 + (CH_3CO)_2CH_2 + FeCl_3 \rightarrow [(acac)_3Ge]^+(FeCl_4)^-$$

Chelated germanium compounds were prepared from $GeCl_4$ and β-diketones such as 1-phenyl-1,3-butanedione, 1,3-diphenyl-1,3-propanedione and 4,4,4-trifluoro-1-(2-thienyl)-1,3-butanedione. These complexes and also the 2,4-pentanedione complex were investigated by IR spectroscopy and by dipole moment measurements. The high values of the dipole moments point to the formation of an ionic bond as the result of the displacement of a chloride ion from the sphere of the complex. The complexes are assigned the structure of a trigonal pyramid (ionic form), which is possibly in equilibrium with a cis-octahedral form (163). Nuclear magnetic resonance spectra of $(acac)_2GeCl_2$ complexes confirm the *cis*-configuration (234).

$GeCl_4$ reacts in tetrachloroethane with bis-(diphenylphosphinyl)imine $HN(Ph_2PO)_2$ to lead to tris-(imidophosphinato)cationic complexes of germanium (216):

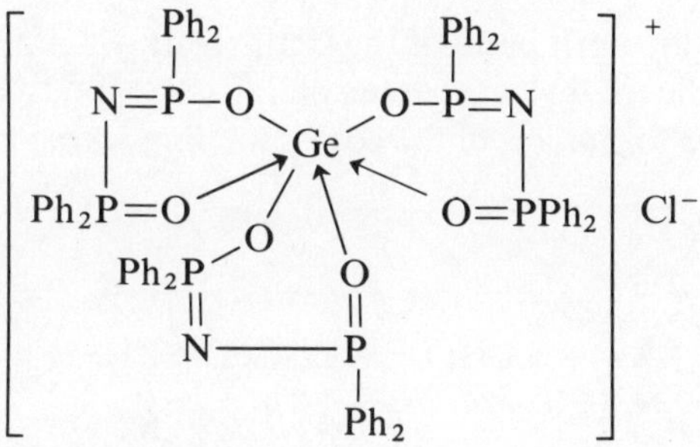

Bis-(cyclohexyl-dioxy)germane reacts with sodium methylate to form five-coordinate adduct (159):

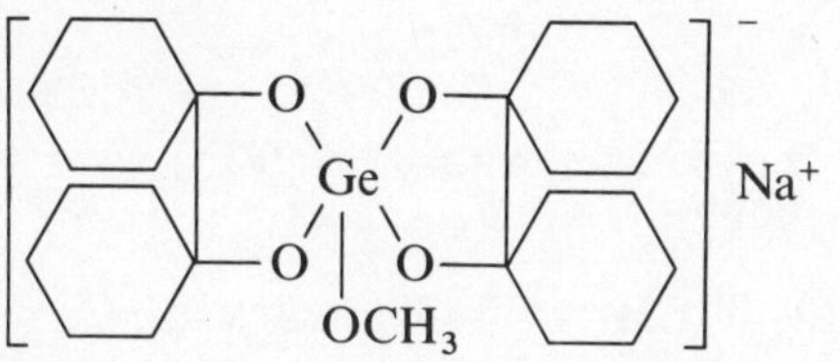

Catechol and germanium tetrachloride in the presence of pyridine yield an octahedral complex:

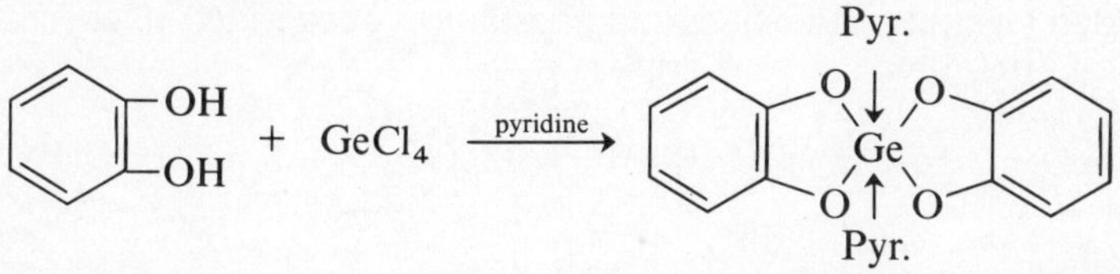

Heating the dipyridine complex in dimethylformamide precipitates a new complex (276):

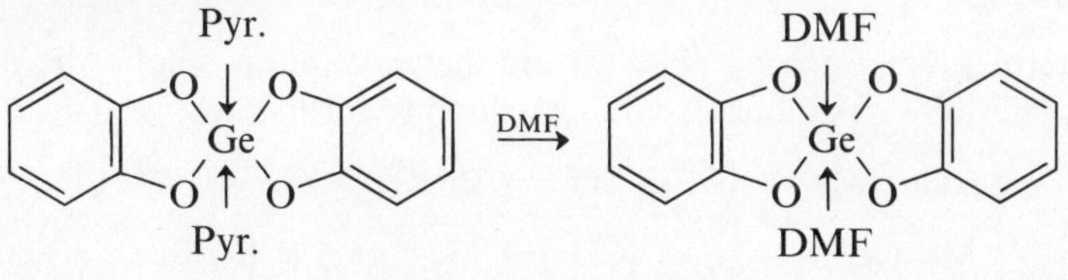

The action of trimethylchlorogermane on the carbonyl group of diarylketones in the presence of magnesium leads to organogermanium derivatives of the corresponding pinacols (49):

$$2\,Ar\!-\!\underset{\underset{O}{\|}}{C}\!-\!Ar' + 2\,Me_3GeCl + Mg \xrightarrow{\text{THF}} Ar\!-\!\underset{\underset{OGeMe_3}{|}}{\overset{\overset{Ar'}{|}}{C}}\!\!-\!\!\!-\!\!\!-\!\underset{\underset{OGeMe_3}{|}}{\overset{\overset{Ar'}{|}}{C}}\!-\!Ar$$

The germanium-nitrogen bond of the trialkylgermylamines (184) and the Ge—P bond of the trialkylgermylphosphines (198) add to the C=O bond of aldehydes with formation of the corresponding alkoxygermanes:

$$Et_3GeNR'_2 + RCHO \rightarrow Et_3Ge\!-\!O\!-\!\underset{\underset{R}{|}}{C}H\!-\!NR'_2$$

$$Et_3GePR'_2 + RCHO \rightarrow Et_3GeO\!-\!\underset{\underset{R}{|}}{C}H\!-\!PR'_2$$

1,4-Addition is observed with α-ethylenic aldehydes (198):

$$Et_3GePEt_2 + RCH\!=\!CH\!-\!CHO \rightarrow Et_3Ge\!-\!O\!-\!CH\!=\!CH\!-\!\underset{\underset{PEt_2}{|}}{C}HR$$

Ketene and diphenylketene with Et_3GePEt_2 give the corresponding enoxygermanes (199):

$$Et_3GePEt_2 + CH_2\!=\!C\!=\!O \rightarrow Et_3Ge\!-\!O\!-\!\underset{\underset{CH_2}{\|}}{C}\!-\!PEt_2$$

The phosphorus-containing enoxygermanes hydrolyse to the corresponding trialkylgermanium oxides and phosphorus-substituted aldehydes and ketones (198, 199):

$$Et_3GeO\!-\!CH\!=\!CH\!-\!\underset{\underset{PEt_2}{|}}{C}H\!-\!R \xrightarrow{H_2O} (Et_3Ge)_2O + R\!-\!\underset{\underset{PEt_2}{|}}{C}H\!-\!CH_2CHO$$

$$Et_3Ge\!-\!O\!-\!\underset{\underset{CH_2}{\|}}{C}\!-\!PEt_2 \xrightarrow{H_2O} (Et_3Ge)_2O + Et_2P\!-\!\underset{\underset{O}{\|}}{C}\!-\!CH_3$$

The germanium–nitrogen bond (5) and the germanium–sulphur bond (6) add to hexafluoroacetone in similar fashion to form alkoxygermanes:

$$Bu_3GeNMe_2 + (CF_3)_2CO \rightarrow Bu_3GeOC(CF_3)_2NMe_2$$

$$Me_3GeSEt + (CF_3)_2CO \rightarrow Me_3GeOC(CF_3)_2SEt$$

REFERENCES

1. Abel, E. W., *J. Chem. Soc.*, 3746 (1958).
2. Abel, E. W., D. A. Armitage, and D. B. Brady, *J. Organometal. Chem.*, **5**, 130 (1966).
3. Abel, E. W., D. A. Armitage, and D. B. Brady, *Trans. Faraday Soc.*, **62**, 3459 (1966).
4. Abel, E. W., D. A. Armitage, and S. P. Tyfield, *J. Chem. Soc. A*, 554 (1967).
5. Abel, E. W., and J. P. Crow, *J. Chem. Soc. A*, 1361 (1968).
6. Abel, E. W., D. J. Walker, and J. N. Wingfield, *J. Chem. Soc. A*, 1814 (1968).
7. Amberger, E., W. Stoeger, and R. Hönigschmid-Grossich, *Angew. Chem.*, **78**, 459 (1966); internat. edn **5**, 522 (1966).
8. Anderson, H. H., *J. Am. Chem. Soc.*, **72**, 194 (1950).
9. Anderson, H. H., *J. Am. Chem. Soc.*, **72**, 2089 (1950).
10. Anderson, H. H., *J. Am. Chem. Soc.*, **73**, 5440 (1951).
11. Anderson, H. H., *J. Am. Chem. Soc.*, **73**, 5798 (1951).
12. Anderson, H. H., *J. Am. Chem. Soc.*, **73**, 5800 (1951).
13. Anderson, H. H., *J. Am. Chem. Soc.*, **74**, 2370 (1952).
14. Anderson, H. H., *J. Am. Chem. Soc.*, **74**, 2371 (1952).
15. Anderson, H. H., *J. Am. Chem. Soc.*, **75**, 814 (1953).
16. Anderson, H. H., *J. Org. Chem.*, **19**, 1766 (1954).
17. Anderson, H. H., *J. Org. Chem.*, **20**, 536 (1955).
18. Anderson, H. H., *J. Org. Chem.*, **20**, 900 (1955).
19. Anderson, H. H., *J. Am. Chem. Soc.*, **78**, 1692 (1956).
20. Anderson, H. H., *J. Org. Chem.*, **21**, 869 (1956).
21. Anderson, H. H., *J. Am. Chem. Soc.*, **79**, 326 (1957).
22. Anderson, H. H., *J. Am. Chem. Soc.*, **82**, 3016 (1960).
23. Anderson, H. H., *J. Am. Chem. Soc.*, **83**, 547 (1961).
24. Anderson, H. H., *Inorg. Chem.*, **3**, 910 (1964).
25. Anderson, H. H., *J. Chem. Eng. Data*, **12**, 371 (1967).
26. Armer, B., and H. Schmidbaur, *Chem. Ber.*, **100**, 1521 (1967).
27. Avdeeva, V. I., G. S. Burlachenko, Yu. I. Baukov, and I. F. Lutsenko, *Zh. Obshch. Khim.*, **36**, 1679 (1966).
28. Bauer, H., and K. Burschkies, *Chem. Ber.*, **65**, 956 (1932).
29. Bauer, H., and K. Burschkies, *Chem. Ber.*, **67**, 1041 (1934).
29a. Baukov, Yu. I., G. S. Burlachenko, I. Yu. Belavin, and I. F. Lutsenko, *Zh. Obshch. Khim.*, **38**, 1899 (1968).
29b. Baukov, Yu. I., G. S. Burlachenko, A. S. Kostyuk, and I. F. Lutsenko, *Zh. Obshch. Khim.*, **39**, 467 (1969).
30. Baukov, Yu. I., and I. F. Lutsenko, *Zh. Obshch. Khim.*, **32**, 3838 (1962).
31. Becke-Goehring, M., and D. Schläfer, *Z. Naturforsch.*, **21b**, 492 (1966).
32. Bott, R. W., C. Eaborn, and T. Hashimoto, *J. Chem. Soc.*, 3906 (1963).
33. Bottei, R. S., and L. J. Kuzma, *J. Inorg. Nucl. Chem.*, **30**, 415 (1968).
34. Bottei, R. S., and L. J. Kuzma, *J. Inorg. Nucl. Chem.*, **30**, 2345 (1968).
35. Bradley, D. C., L. J. Kay, J. D. Swanwick, and W. Wardlaw, *J. Chem. Soc.*, 3656 (1958).
36. Bradley, D. C., L. J. Kay, and W. Wardlaw, *Chem. Ind. (London)*, 746 (1953).
37. Bradley, D. C., L. J. Kay, and W. Wardlaw, *J. Chem. Soc.*, 4916 (1956).
38. Brook, A. G., *J. Am. Chem. Soc.*, **77**, 4827 (1955).
39. Brook, A. G., and H. Gilman, *J. Am. Chem. Soc.*, **76**, 77 (1954).
40. Brooks, E. H., F. Glockling, and K. A. Hooton, *J. Chem. Soc.*, 4283 (1965).
40a. Brook, A. G., and G. J. D. Peddle, *J. Am. Chem. Soc.*, **85**, 2338 (1963).
41. Brown, M. P., R. Okawara, and E. G. Rochow, *Spectrochim. Acta*, **16**, 595 (1960).

42. Brown, M. P., and E. G. Rochow, *J. Am. Chem. Soc.*, **82**, 4166 (1960).
43. Bryskovskaya, A. V., and V. M. Al'Bitskaya, *Zh. Obshch. Khim.*, **37**, 1553 (1967).
43a. Bulten, E. J., Thesis, Utrecht, 1969.
44. Bulten, E. J., and J. G. Noltes, *Tetrahedron Letters*, 3471 (1966).
45. Bulten, E. J., and J. G. Noltes, *J. Organometal. Chem.*, **11**, P 19 (1968).
46. Burch, G. M., and J. R. van Wazer, *J. Chem. Soc. A*, 586 (1966).
47. Burlachenko, G. S., V. I. Avdeeva, Yu. I. Baukov, and I. F. Lutsenko, *Zh. Obshch. Khim.*, **35**, 1881 (1965); *CA*, *Phys. Chem. Sect.*, **64**, 1607 (1966).
48. Burschkies, K., *Chem. Ber.*, **69**, 1143 (1936).
48a. Bychkov, V. T., O. V. Linzina, N. S. Vyazankin, and G. A. Razuvaev, *Izv. Akad. Nauk SSSR, Ser. Khim.*, 2141 (1968).
49. Calas, R., N. Duffaut, C. Biran, P. Bourgeois, F. Pisciotti, and J. Dunogues, *Compt. Rend. (C)*, **267**, 322 (1968).
50. Chipperfield, J. R., and R. H. Prince, *J. Chem. Soc.*, 3567 (1963).
51. Chumaevskii, N. A., *Opt. Spectry. (USSR)*, **13**, 68 (1962).
52. Clark, E. R., *J. Inorg. Nucl. Chem.*, **25**, 353 (1963).
53. Cradock, S., *J. Chem. Soc. A*, 1426 (1968).
54. Cradock, S., and E. A. V. Ebsworth, *J. Chem. Soc. A*, 1423 (1968).
55. Cross, R. J., and F. Glockling, *J. Chem. Soc.*, 4125 (1964).
56. Cross, R. J., and F. Glockling, *J. Organometal. Chem.*, **3**, 146 (1965).
57. Cullen, W. R., and G. E. Styan, *Inorg. Chem.*, **4**, 1437 (1965).
58. Cumper, C. W. N., A. Melnikoff, E. F. Mooney, and A. I. Vogel, *J. Chem. Soc. B*, 874 (1966).
59. Cumper, C. W. N., A. Melnikoff, and A. I. Vogel, *J. Chem. Soc. A*, 242 (1966).
60. Cumper, C. W. N., A. Melnikoff, and A. I. Vogel, *J. Chem. Soc. A*, 246 (1966).
61. Cumper, C. W. N., A. Melnikoff, and A. I. Vogel, *J. Chem. Soc. A*, 323 (1966).
62. Dannley, R. L., and G. Farrant, *J. Am. Chem. Soc.*, **88**, 627 (1966).
63. Davidson, W. E., K. Hills, and M. C. Henry, *J. Organometal. Chem.*, **3**, 285 (1965).
64. Davies, A. G., and C. D. Hall, *Chem. Ind. (London)*, 1695 (1958).
65. Davies, A. G., and C. D. Hall, *J. Chem. Soc.*, 3835 (1959).
66. Davies, A. G., and P. G. Harrison, *J. Organometal. Chem.*, **10**, P 31 (1967).
67. Davies, A. G., P. G. Harrison, and T. A. G. Silk, *Chem. Ind. (London)*, 949 (1968).
68. Delman, A. D., A. A. Stein, B. B. Simms, and R. J. Katzenstein, *J. Polymer Sci.*, **4**, 2307 (1966).
69. Dennis, L. M., and W. I. Patnode, *J. Am. Chem. Soc.*, **52**, 2779 (1930).
70. Dubac, J., and P. Mazerolles, *Compt. Rend. C*, **267**, 411 (1968).
71. Eaborn, C., P. Simpson, and I. D. Varma, *J. Chem. Soc. A*, 1133 (1966).
72. Egorochkin, A. N., M. L. Khidekel', V. A. Ponomarenko, G. Ya Zueva, and G. A. Razuvaev, *Izv. Akad. Nauk SSSR Ser. Khim.*, 373 (1964).
73. Egorochkin, A. N., M. L. Khidekel', and G. A. Razuvaev, *Izv. Akad. Nauk SSSR Ser. Khim.*, 437 (1966).
74. Egorochkin, A. N., N. S. Vyazankin, G. A. Razuvaev, O. A. Kruglaya, M. N. Bochkarev, *Dokl. Akad. Nauk SSSR*, **170**, 333 (1966).
75. Emelyanova, L. I., and L. G. Makarova, *Izv. Akad. Nauk SSSR*, 2067 (1961).
76. Emelyanova, L. I., V. N. Vinoghradova, L. G. Makarova, and A. N. Nesmeyanov, *Izv. Akad. Nauk SSSR Otd. Khim. Nauk*, 53 (1962).
77. Fenton, D. E., and A. G. Massey, *Chem. Ind. (London)*, **51**, 2100 (1964).
78. Fenton, D. E., A. G. Massey, and D. S. Urch, *J. Organometal. Chem.*, **6**, 352 (1966).
79. Fish, R. H., and H. G. Kuivilia, *J. Org. Chem.*, **31**, 2445 (1966).
80. Flood, E. A., *J. Am. Chem. Soc.*, **54**, 1663 (1932).
81. Flood, E. A., *J. Am. Chem. Soc.*, **55**, 4935 (1933).

82. Florinskii, F. S., *Zh. Obshch. Khim.*, **32**, 1443 (1962).
83. Fritz, J. S., and G. H. Schenk, *Anal. Chem.*, **31**, 1808 (1959).
84. Fuchs, R., and H. Gilman, *J. Org. Chem.*, **23**, 911 (1958).
85. Gilman, H., and C. W. Gerow, *J. Am. Chem. Soc.*, **78**, 5435 (1956).
86. Gladshtein, B. H., I. P. Kulyulin, and L. Z. Soborovskii, *Zh. Obshch. Khim.*, **36**, 488 (1966).
86a. Glidewell, C., and D. W. H. Rankin, *J. Chem. Soc. A*, 753 (1969).
86b. Glidewell, C., D. W. H. Rankin, A. G. Robiette, and G. M. Sheldrick, *Inorg. Nucl. Chem. Letters*, **5**, 417 (1969).
87. Glockling, F., and J. R. C. Light, *J. Chem. Soc. A*, 717 (1968).
88. Goldfarb, T. D., and S. Sujishi, *J. Am. Chem. Soc.*, **86**, 1679 (1964).
89. Griffiths, J. E., and H. Onyszchuk, *Can. J. Chem.*, **39**, 339 (1961).
90. Harwood, J. H., *Chem. Process Eng.*, **48**, 60 (1967).
91. Hooton, K. A., and A. L. Allred, *Inorg. Chem.*, **4**, 671 (1965).
92. Hota, N. K., and C. J. Willis, *Can. J. Chem.*, **46**, 3921 (1968).
93. Huber, F., *Angew. Chem. Int. Ed.*, **4**, 1089 (1965).
94. Huber, F., and R. Kaiser, *Z. Naturforsch.*, **20b**, 1011 (1965).
95. Huber, F., and R. Kaiser, *Z. Naturforsch.*, **20b**, 1437 (1965).
96. Ishii, Y., K. Itoh, A. Nakamura, and S. Sakaï, *Chem. Comm.*, 224 (1967).
97. Itoh, K., S. Sakaï, and Y. Ishii, *Tetrahedron Letters*, 4941 (1966).
98. Johnson, O. H., and H. E. Fritz, *J. Am. Chem. Soc.*, **75**, 718 (1953).
99. Johnson, O. H., and D. M. Harris, *J. Am. Chem. Soc.*, **72**, 5564 (1950).
100. Johnson, O. H., and L. V. Jones, *J. Org. Chem.*, **17**, 1172 (1952).
101. Johnson, O. H., and W. H. Nebergall, *J. Am. Chem. Soc.*, **71**, 1720 (1949).
102. Klimova, V. A., K. S. Zabrodina, and N. L. Shitikova, *Izv. Akad. Nauk SSSR Ser. Khim.*, 178 (1965).
103. Kolesnikov, G. S., S. L. Davydova, and N. V. Klimentova, *Syntheses and Properties of Monomers*, Ed. Sciences 1964; *Ref. Zh. khim.*, **15**, J.363 (1963).
104. Koster-Pflugmacher, A., and E. Termin, *Naturwiss.*, **51**, 554 (1964).
105. Kraus, C. A., and C. L. Brown, *J. Am. Chem. Soc.*, **52**, 3690 (1930).
106. Kraus, C. A., and E. A. Flood, *J. Am. Chem. Soc.*, **54**, 1635 (1932).
107. Kraus, C. A., and L. S. Foster, *J. Am. Chem. Soc.*, **49**, 457 (1927).
108. Kraus, C. A., and H. S. Nutting, *J. Am. Chem. Soc.*, **54**, 1622 (1932).
109. Kraus, C. A., and C. B. Wooster, *J. Am. Chem. Soc.*, **52**, 372 (1930).
110. Kruglaya, O. A., N. S. Vyazankin, and G. A. Razuvaev, *Zh. Obshch. Khim.*, **35**, 394 (1965).
111. Kruglaya, O. A., N. S. Vyazankin, G. A. Razuvaev, and E. V. Mitrofanova, *Dokl. Akad. Nauk SSSR*, **173**, 834 (1967).
112. Kühlein, K., W. P. Neumann, and H. P. Becker, *Angew. Chem.*, **79**, 870 (1967); internat. edn **6**, 876 (1967).
113. Laubengayer, A. W., and P. L. Brandt, *J. Am. Chem. Soc.*, **54**, 549 (1932).
114. Lavigne, A. A., R. M. Pike, D. Monier, and C. T. Tabit, *Rec. Trav. Chim.*, **86**, 746 (1967).
115. Lavigne, A. A., J. Tancrède, R. M. Pike, and C. T. Tabit, *J. Organometal. Chem.*, **15**, 57 (1968).
116. Lebedeff, M., A. Marchand, and J. Valade, *Compt Rend. (C)*, **267**, 813 (1968).
117. Lesbre, M., and R. W. Russell, *Bull. Soc. Chim. (France)*, 566 (1959).
118. Lesbre, M., and J. Satgé, *Compt. Rend.*, **254**, 1453 (1962).
119. Lesbre, M., and J. Satgé, *Compt. Rend.*, **254**, 4051 (1962).
120. Lutsenko, I. F., Yu. I. Baukov, I. Yu. Belavin, and A. N. Tvorogov, *J. Organometal. Chem.*, **14**, 229 (1968).

121. Lutsenko, I. F., Yu. I. Baukov, and G. S. Burlachenko, *J. Organometal. Chem.*, **6**, 496 (1966).
122. MacKay, K. M., and R. Watt, *J. Organometal. Chem.*, **6**, 336 (1966).
123. Magnuson, J. A., and E. W. Knaub, *Anal. Chem.*, **37**, 1607 (1965).
124. Massol, M., Thesis. Toulouse (France), 1967.
125. Massol, M., and J. Satgé, *Bull. Soc. Chim. France*, 1714 (1964).
126. Massol, M., and J. Satgé, *Bull. Soc. Chim. France*, 2737 (1966).
127. Massol, M., J. Satgé, and J. Barrau, *Compt. Rend. C*, **268**, 1710 (1969).
128. Massol, M., J. Satgé, P. Rivière, and J. Barrau, Abstract of 4th Symp. of Organo-metallic Chemistry, Bristol (1969). *Progress in Organometallic Chemistry*, Ed. Bruce, M. I., and F. G. A. Stone, H₃ and *J. Organometal. Chem.*, **22**, 599 (1970).
129. Mathur, S., G. Chandra, A. K. Rai, and R. C. Mehrotra, *J. Organometal. Chem.*, **4**, 294 (1965).
130. Mathur, S., G. Chandra, A. K. Rai, and R. C. Mehrotra, *J. Organometal. Chem.*, **4**, 371 (1965).
131. Mathur, S., and R. C. Mehrotra, *J. Organometal. Chem.*, **7**, 227 (1967).
131a. Mathur, S., R. Ouaki, V. K. Mathur, R. C. Mehrotra, and J. C. Maire, *Indian J. Chem.*, **7**, 284 (1969).
132. Matwiyoff, N. A., and R. S. Drago, *J. Organometal. Chem.*, **3**, 393 (1965).
133. Mazerolles, P., Thesis, Toulouse, 1959.
134. Mazerolles, P., *Bull. Soc. Chim. France*, 1911 (1961).
135. Mazerolles, P., *Bull. Soc. Chim. France*, 1907 (1962).
135a. Mazerolles, P., J. Dubac, and M. Lesbre, *J. Organometal. Chem.*, **5**, 35 (1966).
136. Mazerolles, P., and G. Manuel, *Compt. Rend. (C)*, **267**, 1158 (1968).
137. Mehrotra, R. C., and G. Chandra, *J. Indian Chem. Soc.*, **39**, 235 (1962).
138. Mehrotra, R. C., and G. Chandra, *Rec. Trav. Chim.*, **82**, 683 (1963).
139. Mehrotra, R. C., and G. Chandra, *J. Chem. Soc.*, 2804 (1963).
140. Mehrotra, R. C., and G. Chandra, *J. Ind. Chem.*, **3**, 497 (1965).
141. Mehrotra, R. C., and G. Chandra, *Proc. Natl. Acad. Sci. India Sect. A*, **35**, 148 (1966).
142. Mehrotra, R. C., and S. Mathur, *J. Organometal. Chem.*, **6**, 11 (1966).
143. Mehrotra, R. C., and S. Mathur, *J. Indian Chem. Soc.*, **43**, 489 (1966).
144. Mehrotra, R. C., and S. Mathur, *J. Organometal. Chem.*, **6**, 425 (1966).
145. Mehrotra, R. C., and S. Mathur, *J. Organometal. Chem.*, **7**, 233 (1967).
146. Metlesics, W., and H. Zeiss, *J. Am. Chem. Soc.*, **82**, 3324 (1960).
147. Mironov, V. F., and A. L. Kravchenko, *Izv. Akad. Nauk SSSR Ser. Khim.*, 1026 (1965).
148. Mironov, V. F., A. L. Kravchenko, and A. D. Petrov, *Izv. Akad. Nauk SSSR, Ser. Khim.*, 1209 (1964).
149. Mironov, V. F., E. S. Sobolev, and L. M. Antipin, *Zh. Obshch. Khim.*, **37**, 1707 (1967).
150. Mironov, V. F., E. S. Sobolev, and L. M. Antipin, *Zh. Obshch. Khim.*, **37**, 2573 (1967).
151. Moedritzer, K., *Organometal. Chem. Rev.*, **1**, 179 (1966).
152. Moedritzer, K., *J. Organometal. Chem.*, **5**, 254 (1966).
152a. Moedritzer, K., L. C. D. Groenweghe, and J. R. van Wazer, *J. Phys. Chem.*, **72**, 4380 (1968).
153. Moedritzer, K., and J. R. van Wazer, *Inorg. Chem.*, **4**, 1753 (1965).
154. Moedritzer, K., and J. R. van Wazer, *J. Inorg. Nucl. Chem.*, **29**, 1571 (1967).
155. Moedritzer, K., and J. R. van Wazer, *J. Inorg. Nucl. Chem.*, **29**, 1577 (1967).
156. Moedritzer, K., and J. R. van Wazer, *J. Organometal. Chem.*, **13**, 145 (1968).
156a. Moedritzer, K., and J. R. van Wazer, *J. Chem. Soc. A*, 1124 (1969).
156b. Moore, M., and F. C. Lanning, *Trans. Kansas Acad. Sci.*, **70**, 426 (1967).
157. Morgan, G. T., and H. D. K. Drew, *J. Chem. Soc.*, **125**, 1261 (1924).

158. Morgan, G. T., and H. D. K. Drew, *J. Chem. Soc.*, **127**, 1760 (1925).

159. Muller, R., and L. Heinrich, *Chem. Ber.*, **95**, 2276 (1962).

159a. Nekhorosheva, E. V., and V. M. Al'bitskaya, *Zh. Obshch. Khim.*, **38**, 1511 (1968).

160. Nesmeyanov, A. N., L. I. Emel'ianova, and L. G. Makarova, *Dokl. Akad. Nauk SSSR*, **122**, 403 (1958).

160a. Normant, H. and Th. Cuvigny, *Bull. soc. chim. France*, 3344 (1966).

161. Noltes, J. G., H. A. Budding, and G. J. M. van der Kerk, *Rec. Trav. Chim.*, **79**, 408 (1960).

161a. Oliver, A. J., and W. A. G. Graham, *J. Organometal. Chem.*, **19**, 17 (1969).

162. Orndorff, W. R., D. L. Tabern, and L. M. Dennis, *J. Am. Chem. Soc.*, **49**, 2512 (1927).

163. Osipov, O. A., V. L. Shelepina, and O. E. Shelepin, *Zh. Obshch. Khim.*, **36**, 264 (1966).

164. Peddle, G. J. D., and J. E. H. Ward, *J. Organometal. Chem.*, **14**, 131 (1968).

165. Pereyre, M., B. Bellegarde, and J. Valade, *Compt. Rend. C*, **265**, 939 (1967).

166. Pike, R. M., *Rec. Trav. Chim.*, **80**, 885 (1961).

167. Pike, R. M., and A. M. Dewidar, *Rec. Trav. Chim.*, **84**, 119 (1964).

168. Pike, R. M., and A. F. Fournier, *Rec. Trav. Chim.*, **81**, 475 (1962).

169. Pike, R. M., and A. A. Lavigne, *Rec. Trav. Chim.*, **82**, 49 (1963).

170. Pike, R. M., and A. A. Lavigne, *Rec. Trav. Chim.*, **83**, 883 (1964).

171. Ponomarev, S. V., Yu. I. Baukov, O. V. Dudukina, I. V. Petrosyan, and L. I. Petrovskaya, *Zh. Obshch. Khim.*, **37**, 2204 (1967).

171a. Prince, R. H., and R. E. Timms, *Inorg. Chim. Acta*, **1**, 129 (1967).

172. Prince, R. H., and R. E. Timms, *Inorg. Chim. Acta*, **2/3**, 257 (1968).

173. Razuvaev, G. A., Yu. A. Alexandrov, V. N. Glushakova, and G. N. Figurova, *J. Organometal. Chem.*, **14**, 339 (1968).

173a. Razuvaev, G. A., Yu. A. Aleksandrov, V. N. Glushakova, and G. N. Figurova, *Dokl. Akad. Nauk SSSR*, **180**, 1119 (1968).

174. Rieche, A., and J. Dahlmann, *Angew. Chem.*, **71**, 194 (1959).

175. Rieche, A., and J. Dahlmann, *Monatsber. Deut. Akad. Wiss. Berlin*, **1**, 491 (1959); *CA* **55**, 18640 (1961).

176. Rieche, A., and J. Dahlmann, *Ann. Chem.*, **675**, 19 (1964).

177. Rijkens, F., M. J. Janssen, W. Drenth, and G. J. M. van der Kerk, *J. Organometal. Chem.*, **2**, 347 (1964).

178. Rijkens, F., M. J. Janssen, and G. J. M. van der Kerk, *Rec. Trav. Chim.*, **84**, 1597 (1965).

179. Rijkens, F., and G. J. M. van der Kerk, *Investigations in the Field of Organogermanium Chemistry*, Germanium Research Committee, Utrecht, 1964.

180. Rijkens, F., and G. J. M. van der Kerk, *Rec. Trav. Chim.*, **83**, 723 (1964).

181. Rivière, P., and J. Satgé, *Bull. Soc. Chim. France*, 4039 (1967).

182. Rivière, P., and J. Satgé, *Compt. Rend. C*, **267**, 267 (1968).

183. Rivière-Baudet, M., and J. Satgé, *Bull. Soc. Chim. France*, 1356 (1969).

184. Rivière-Baudet, M., and J. Satgé, unpublished work.

185. Rochow, E. G., *J. Am. Chem. Soc.*, **70**, 1801 (1948).

186. Rochow, E. G., and A. L. Allred, *J. Am. Chem. Soc.*, **77**, 4489 (1955).

187. Rothermundt, M., and K. Burschkies, *Z. Immunitaetsforsch.*, **87**, 445 (1936).

188. Ruidisch, I., and M. Schmidt, *Angew. Chem.*, **75**, 575 (1963).

189. Ruidisch, I., and M. Schmidt, *Z. Naturforsch.*, **18b**, 508 (1963).

190. Ruidisch, I., and M. Schmidt, *Chem. Ber.*, **96**, 821 (1963).

191. Ruidisch, I., and M. Schmidt, *J. Organometal. Chem.*, **1**, 493 (1964).

191a. Sara, A. N., O. H. J. Christie, and K. Taugbol, *Chem. Ind.* (*London*), 725 (1969).

192. Sartori, P., and M. Weidenbruch, *Angew. Chem.*, **77**, 1138 (1965).

193. Sartori, P., and M. Weidenbruch, *Chem. Ber.*, **100**, 2049 (1967).

194. Satgé, J., *Ann. Chim.* (*Paris*), **6**, 519 (1961).
195. Satgé, J., *Bull. Soc. Chim. France*, 630 (1964).
196. Satgé, J., and M. Baudet, *Compt. Rend. C*, **263**, 435 (1966).
197. Satgé, J., and C. Couret, *Compt. Rend. C*, **264**, 2169 (1967).
198. Satgé, J., and C. Couret, *Compt. Rend. C*, **267**, 173 (1968).
199. Satgé, J., and C. Couret, *Bull. Soc. Chim. France*, 333 (1969).
200. Satgé, J., and G. Dousse, unpublished work.
201. Satgé, J., and M. Lesbre, *Bull. Soc. Chim. France*, 703 (1962).
202. Satgé, J., and M. Lesbre, *Bull. Soc. Chim. France*, 2578 (1965).
203. Satgé, J., M. Lesbre, and M. Baudet, *Compt. Rend.*, **259**, 4733 (1964).
204. Satgé, J., and P. Rivière, *Bull. Soc. Chim. France*, 1773 (1966).
205. Satgé, J., and P. Rivière, *J. Organometal. Chem.*, **16**, 71 (1969).
206. Satgé, J., P. Rivière, and M. Lesbre, *Compt. Rend. C*, **265**, 494 (1967).
207. Schmidbaur, H., *J. Am. Chem. Soc.*, **83**, 2336 (1963).
208. Schmidbaur, H., and B. Armer, *Angew. Chem.*, **78**, 305 (1966).
209. Schmidbaur, H., and H. Hussek, *J. Organometal. Chem.*, **1**, 235 (1964).
210. Schmidbaur, H., and I. Ruidisch, *Inorg. Chem.*, **3**, 599 (1964).
211. Schmidbaur, H., and M. Schmidt, *Chem. Ber.*, **94**, 1138 (1961).
212. Schmidbaur, H., and M. Schmidt, *Chem. Ber.*, **94**, 1349 (1961).
213. Schmidbaur, H., and M. Schmidt, *Chem. Ber.*, **94**, 2137 (1961).
214. Schmidbaur, H., L. Sechser, and M. Schmidt, *J. Organometal. Chem.*, **15**, 77 (1968).
214a. Schmidbaur, H., and W. Tronich, *Chem. Ber.*, **101**, 3545 (1968).
215. Schmidbaur, H., and W. Wolfsberger, *Chem. Ber.*, **101**, 1664 (1968).
216. Schmidpeter, A., and K. Stoll, *Angew. Chem.*, **80**, 558 (1968).
217. Schmidt, M., C. Blohm, and G. Jander, *Angew. Chem.*, **A59**, 233 (1947).
218. Schmidt, M., and I. Ruidisch, *Angew. Chem.*, **73**, 408 (1961).
219. Schmidt, M., and I. Ruidisch, *Z. Anorg. Allgem. Chem.*, **311**, 331 (1961).
220. Schmidt, M., and I. Ruidisch, *Chem. Ber.*, **95**, 1434 (1962).
221. Schmidt, M., I. Ruidisch, and H. Schmidbaur, *Chem. Ber.*, **94**, 2451 (1961).
222. Schmidt, M., H. Schmidbaur, and I. Ruidisch, *Angew. Chem.*, **73**, 408 (1961).
223. Schmidt, M., H. Schmidbaur, I. Ruidisch, and P. Bornmann, *Angew. Chem.*, **73**, 408 (1961).
224. Schott, G., and C. Harzdorf, *Z. Anorg. Allgem. Chem.*, **307**, 105 (1960).
225. Schumann-Ruidisch, I., and H. Blass, *Z. Naturforsch.*, **B22**, 1081 (1967).
226. Schumann, H., and M. Schmidt, *Inorg. Nucl. Chem. Letters*, **1**, 1 (1965).
227. Schumann, H., P. Schwabe, and M. Schmidt, *Inorg. Nucl. Chem. Letters*, **2**, 309 (1966).
228. Schwarz, R., and K. G. Knauff, *Z. Anorg. Allgem. Chem.*, **275**, 193 (1954).
229. Schwarz, R., and W. Reinhardt, *Chem. Ber.*, **65**, 1743 (1932).
230. Seyferth, D., and D. L. Alleston, *Inorg. Chem.*, **2**, 418 (1963).
231. Sidgwick, N. V., and A. W. Laubengayer, *J. Am. Chem. Soc.*, **54**, 948 (1932).
232. Silcox, C. M., and J. J. Zuckerman, *J. Am. Chem. Soc.*, **88**, 168 (1966).
233. Simons, J. K., E. C. Wagner, and J. H. Muller, *J. Am. Chem. Soc.*, **55**, 3705 (1933).
234. Smith, J. A. S., and E. J. Wilkins, *J. Chem. Soc. A*, 1749 (1966).
235. Sosnovsky, G., and J. H. Brown, *Chem. Rev.*, **66**, 529 (1966).
236. Srivastava, T. N., and M. Onyszchuk, *Can. J. Chem.*, **41**, 1244 (1963).
237. Srivastava, T. N., and S. K. Tandon, *Z. Anorg. Allgem. Chem.*, **353**, 87 (1967).
238. Stipanits, P., and F. Hecht, *Monatsh. Chem.*, **88**, 892 (1957).
239. Styan, G. E., Thesis, *Dissertation Abstr. B*, **27**, 1078 (1966).
240. Sukhani, D., V. D. Gupta, and R. C. Mehrotra, *Australian J. Chem.*, **21**, 1175 (1968).
241. Tabern, D. L., W. R. Orndorff, and L. M. Dennis, *J. Am. Chem. Soc.*, **47**, 2039 (1925).
242. Tobias, R. S., and S. Hutcheson, *J. Organometal. Chem.*, **6**, 535 (1966).

243. Trautman, C. E., and H. A. Ambrose, US Pat. 2,416,360 (1947); *CA*, **42**, 2760g (1948).
244. Ulbricht, K., M. Jakoubkova, and V. Chvalovsky, *Collection Czech. Chem. Commun.*, **33**, 1693 (1968).
245. Ulbricht, K., V. Vaisarova, V. Bazant, and V. Chvalovsky, *J. Organometal. Chem.*, **13**, 343 (1968).
246. van der Kerk, G. J. M., F. Rijkens, and M. J. Janssen, *Rec. Trav. Chim.*, **81**, 764 (1962).
247. van Wazer, J. R., K. Moedritzer, and L. C. D. Groenweghe, *J. Organometal. Chem.*, **5**, 420 (1966).
247a. Voronkov, M. G., G. Zelcans, V. F. Mironov, J. Bleidelis, and A. Kemme, *Khim. Geterotsikl. Soedin.*, 227 (1968).
248. Vyazankin, N. S., T. N. Brevnova, and G. A. Razuvaev, *Zh. Obshch. Khim.*, **37**, 2334 (1967).
249. Vyazankin, N. S., E. N. Gladyshev, and S. P. Korneva, *Zh. Obshch. Khim.*, **37**, 1736 (1967).
250. Vyazankin, N. S., E. N. Gladyshev, and G. A. Razuvaev, *Dokl. Akad. Nauk SSSR*, **153**, 104 (1963).
251. Vyazankin, N. S., O. A. Kruglaya, G. A. Razuvaev, and G. S. Semchikova, *Dokl. Akad. Nauk SSSR*, **166**, 99 (1966).
252. Vyazankin, N. S., E. V. Mitrofanova, O. A. Kruglaya, and G. A. Razuvaev, *Zh. Obshch. Khim.*, **35**, 160 (1966).
253. Vyazankin, N. S., G. A. Razuvaev, and V. T. Bychkov, *Dokl. Akad. Nauk SSSR*, **158**, 382 (1964).
254. Vyazankin, N. S., G. A. Razuvaev, and V. T. Bychkov, *Izv. Akad. Nauk SSSR*, 1665 (1965).
255. Vyazankin, N. S., G. A. Razuvaev, and E. N. Gladyshev, *Dokl. Akad. Nauk SSSR*, **151**, 1326 (1963).
256. Vyazankin, N. S., G. A. Razuvaev, E. N. Gladyshev, and T. G. Gurikova, *Dokl. Akad. Nauk SSSR*, **155**, 1108 (1964).
257. Vyazankin, G. S., G. A. Razuvaev, O. A. Kruglaya, and G. S. Semchikova, *J. Organometal. Chem.*, **6**, 474 (1966).
258. West, R., *J. Am. Chem. Soc.*, **74**, 4363 (1952).
259. West, R., and R. H. Baney, *J. Phys. Chem.*, **64**, 822 (1960).
260. West, R., R. H. Baney, and D. L. Powell, *J. Am. Chem. Soc.*, **82**, 6269 (1960).
261. West, R., H. R. Hunt, and R. O. Whipple, *J. Am. Chem. Soc.*, **76**, 310 (1954).
262. Wiberg, E., E. Amberger, and H. Cambensi, *Z. Anorg. Allgem. Chem.*, **351**, 164 (1967).
263. Wieber, M., and C. D. Frohning, *Angew. Chem. Int. Ed.*, **4**, 1096 (1965).
264. Wieber, M., and C. D. Frohning, *Angew. Chem.*, **78**, 1022 (1966).
265. Wieber, M., and C. D. Frohning, *J. Organometal. Chem.*, **8**, 459 (1967).
266. Wieber, M., and M. Schmidt, *J. Organometal. Chem.*, **1**, 93 (1963).
267. Wieber, M., and M. Schmidt, *Angew. Chem.*, **75**, 1116 (1963).
268. Wieber, M., and M. Schmidt, *Z. Naturforsch.*, **18b**, 847 (1963).
269. Wieber, M., and M. Schmidt, *Z. Naturforsch.*, **18b**, 848 (1963).
270. Wieber, M., and M. Schmidt, *Z. Naturforsch.*, **18b**, 849 (1963).
271. Wieber, M., and G. Schwarzmann, *Monatsh. Chem.*, **99**, 255 (1968).
272. Yoder, C. M. S., Thesis, *Dissertation Abstr.*, **28**, 90 (1967).
273. Yoder, C. M. S., and J. J. Zuckermann, *Inorg. Chem.*, **6**, 163 (1967).
274. Zueva, G. Ya, I. F. Manucharova, I. P. Yakovlev, and V. A. Ponomarenko, *Izv. Akad. Nauk SSSR Neorg. Mater.*, **2**, 229 (1966).
275. Zueva, G. Ya, and V. A. Ponomarenko, *Izv. Akad. Nauk SSSR Neorg. Mater.*, **2**, 472 (1966).

7 Germanium–sulphur, germanium–selenium, germanium–tellurium bond

7-1 ORGANOGERMANIUM SULPHIDES AND ORGANOGERMANIUM THIOLS

7-1-1 Triorganogermanium thiols, R_3GeSH

Triphenylgermanethiol $(C_6H_5)_3GeSH$ was synthesized for the first time by Henry and Davidson (39):

$$(C_6H_5)_3GeBr + H_2S + C_5H_5N \rightarrow (C_6H_5)_3GeSH + [C_5H_5NH]Br$$

This thiol is a white, crystalline compound stable in dry air. In the presence of moist air a slight evolution of hydrogen sulphide is noted. Oxidation of triphenylgermanethiol with iodine produced bis-triphenyl-germanium disulphide, $Ph_3GeSSGePh_3$.

Reactions between the thiol and RX compounds give in the presence of pyridine the expected products, Ph_3GeSR:

$$(C_6H_5)_3GeSH + RX \xrightarrow{C_5H_5N} (C_6H_5)_2GeSR + HX$$

Reaction of triphenylgermanethiol with dithiocyanogen did not yield the expected sulphur thiocyanogen compound but rather the simpler germanium–thiocyanogen compound and elemental sulphur:

$$Ph_3GeSH + (SCN)_2 \rightarrow [Ph_3GeSSCN] + HSCN$$
$$\downarrow \text{decomp.}$$
$$(Ph_3GeSCN) + S$$

Successful attempts to synthesize diphenylgermanedithiol from $(C_6H_5)_2GeBr_2$ have not been realized. There is some evidence that it may exist in solution. However, it is analogous to carbon gem-thiols and not expected to exhibit a high degree of stability (39).

492

Triethylgermanethiol is formed by heating an equimolecular mixture of triethylgermane and sulphur at 140° for four hours (yield 49.5%) (84, 85):

$$Et_3GeH + \tfrac{1}{8}S_8 \rightarrow Et_3GeSH$$

Et_3GeSH was converted to $(Et_3Ge)_2S$ (yield 57%) at 130°C (28 h).

The general mechanism of formation of R_3MXH type derivatives and their transformation into an $(R_3M)_2X$ compound seems to be the following (56):

$$R_3MH + X_n \rightarrow R_3MX_nH \xrightarrow{(n-1)R_3MH} nR_3MXH$$

$$R_3MXH + R_3MH \rightarrow (R_3M)_2X + H_2$$

$$2\,R_3MXH \rightarrow (R_3M)_2X + H_2X$$

$$M = Si,\ Ge,\ Sn \qquad X = S,\ Se,\ Te$$

The reactivity of these derivatives varies in the following order:

$$S > Se > Te \qquad Sn > Ge > Si$$

Triethylgermanethiol reacts with lithium in THF medium:

$$2\,Et_3GeSH + 2\,Li \rightarrow H_2 + 2\,Et_3GeSLi$$

and with diethylmercury at elevated temperature (88):

$$2\,Et_3GeSH + Et_2Hg \rightarrow 2\,C_2H_6 + HgS + (Et_3Ge)_2S$$

The hydrogen bonded to sulphur can be quantitatively determined by the Chugaev–Tserivitinov method (88). Et_3GeSH shows absorption bands at $2555\ cm^{-1}$ (ν_{S-H}) and $408\ cm^{-1}$ (ν_{Ge-S}) (32, 85). The PMR spectra of compounds R_3GeXH (X = S, Se) are described by Egorochkin and coworkers (31, 33).

A study was made by the PMR method of hydrogen bond formation between the compounds Et_3MXH (M = Si or Ge and X = S or Se) and also C_4H_9SeH and the proton acceptor solvents benzene, dioxane, nitromethane, pyridine, dimethylformamide and triethylamine. It was shown that the strength of the hydrogen bond in these solvents diminishes in the order:

$$Et_3SiSH > Et_3GeSH > Et_3SiSeH > Et_3GeSeH$$

The high electronegativity of the triethylsilyl group as compared with triethylgermyl in compounds Et_3MXH is possibly explained by the $d\pi–p\pi$ effect which is more marked for the silicon compounds (31). The greater tendency of sulphur towards $d\pi–p\pi$ interaction as compared with that of selenium leads to a more significant decrease in electron density at protons bonded to sulphur. This together with the somewhat greater electronegativity of sulphur as compared with that of selenium is responsible for the location of the signal of hydrogen Ge—S—$\underline{H}$ at a considerably weaker field than the Ge—Se—$\underline{H}$ hydrogen resonance (33).

7-1-2 R_3GeS–metal compounds

The sodium salt of triphenylgermanethiol may be prepared by addition of a benzene solution of triphenylbromogermane to a suspension of sodium sulphide in ethanol (38):

$$Ph_3GeBr + Na_2S \rightarrow Ph_3GeSNa + NaBr$$

Refluxing in anhydrous ethanol produces bis-(triphenylgermanium) sulphide and sodium sulphide. The reaction is reversible since bis-(triphenylgermanium) sulphide when refluxed with excess sodium sulphide produces the starting material:

$$2(C_6H_5)_3GeSNa \xrightarrow[\text{Na}_2\text{S}]{\text{C}_2\text{H}_5\text{OH}} (C_6H_5)_3GeSGe(C_6H_5)_3 + Na_2S$$

The same method was applied in the preparation of disodium diphenylgermanedithiol (39):

$$Ph_2GeBr_2 + 2\,Na_2S \rightarrow Ph_2Ge(SNa)_2 + 2\,NaBr$$

This compound is in fact the trihydrate $Ph_2Ge(SNa)_2 \cdot 3\,H_2O$. The dehydrated product is not as stable as the hydrate and seems to decompose (39). $(CH_3)_3GeSLi$ is obtained in the reaction of methyllithium with the dimethylgermanium sulphide trimer:

$$[(CH_3)_2GeS]_3 + 3\,LiCH_3 \rightarrow 3\,(CH_3)_3GeSLi$$

Triphenylgermyllithium reacts with sulphur in THF, forming Ph_3GeSLi. Nucleophilic attack on the sulphur by the Ph_3Ge^- ion may be postulated (77). Triethylgermanethiol reacts with lithium in THF to form Et_3GeSLi (88).

R_3GeS–metal compounds are sensitive to solvolysis and oxidation. An increase in temperature causes their rearrangement:

$$2\,(CH_3)_3GeSLi \xrightarrow{>85°C} [(CH_3)_3Ge]_2S + Li_2S$$

or their decomposition (62). These compounds have not been isolated. They are, however, very reactive and their isolation is not necessary. In solution, they react readily with organic halides and triorganometal chlorides of germanium, tin and lead to lead to the organogermanium thio-compounds, R_3GeSR (38, 40, 77, 88), or the symmetrical and unsymmetrical sulphides:

$$R_3GeSLi + R'_3MCl \rightarrow R_3GeSMR'_3$$

$$(M = Ge, Si, Sn, Pb) \quad (62, 77, 88)$$

$$2\,Ph_3GeSNa + Ph_2GeBr_2 \rightarrow Ph_2Ge(SGePh_3)_2 \quad (38)$$

$$Ph_2Ge(SNa)_2 + 2\,Ph_3GeBr \rightarrow Ph_2Ge(SGePh_3)_2 \quad (39)$$

Ph_3GeSNa, when treated with CS_2 and methyl iodide, did not produce the expected organogermane-substituted carbonic acid trithioester. Instead, bis-(triphenygermanium) sulphide and disodium trithiocarbonate were formed (38):

$$2\,(C_6H_5)_3GeSNa + CS_2 \begin{cases} \rightarrow [2\,(C_6H_5)_3GeSC(S)SNa] \rightarrow \\ \qquad\qquad [(C_6H_5)_3Ge]_2S + Na_2CS_3 \\ \\ \rightarrow [(C_6H_5)_3Ge]_2S + Na_2S \xrightarrow{\;CS_2\;} \end{cases}$$

Ph_3GeSNa reacts with sulphonyl chlorides or hydrogen peroxide to form the symmetrical disulphide (38, 39):

$$Ph_3GeSNa + RSO_2Cl \rightarrow [Ph_3GeSSO_2R] \xrightarrow{RSO_2Cl} [Ph_3GeS]_2$$

$$(R = CH_3-, \ p\text{-}CH_3-C_6H_4-)$$

$$Ph_3GeSNa \xrightarrow{H_2O_2} Ph_3GeSSGePh_3$$

On the other hand, tests carried out with a view to preparing the same disulphide by halogenation of Ph_3GeSNa failed. Instead, cleavage of the germanium–sulphur bond was observed (39).

Reaction of sodium triphenylgermanethiol with sulphur dichloride, SCl_2, produces bis-(triphenylgermanium) trisulphide, $(Ph_3Ge)_2S_3$ (38).

Reaction of the trihydrate of $Ph_2Ge(SNa)_2$ with methyl iodide produces impure diphenyldi(methylthio)germane, $Ph_2Ge(SMe)_2$. Benzoyl chloride (and nitrobenzoyl chloride) react with the same trihydrate to form diphenylgermanium sulphide (Ph_2GeS) and benzoic acid (or p-nitrobenzoic acid). It is probable that the halide reacted with the water of the hydrate to form hydrochloric acid, which then interacted with the sodium salt to form diphenylgermanium sulphide $[(C_6H_5)_2GeS]_n$.

7-1-3 Organogermanium sulphides

7-1-3-1 Bis-(triorganogermanium) Sulphides, $(R_3Ge)_2S$. Several methods of synthesis allow the preparation of bis-(triorganogermanium) sulphides.

The action of H_2S on organogermanium halides in the presence of a base may be mentioned (4, 105):

$$2 R_3GeCl + H_2S \rightarrow (R_3Ge)_2S + 2 HCl$$

With chloromethyldimethylchlorogermane, $(ClCH_2)(CH_3)_2GeCl$, in a second step the two carbon-bound chlorine atoms of these compounds may be substituted by sulphur, yielding germanium-containing heterocycles:

$$[(ClCH_2)(CH_3)_2Ge]_2S + H_2S + 2(C_6H_5)_3N \rightarrow$$

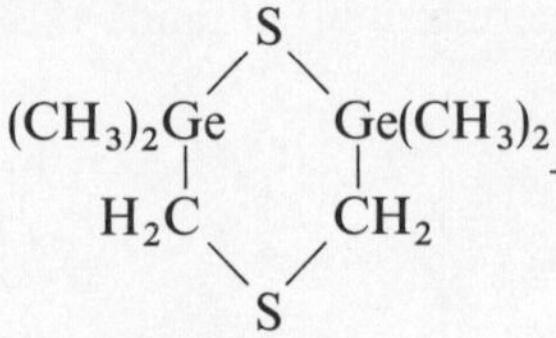

$$
\begin{array}{c}
S \\
(CH_3)_2Ge \quad\quad Ge(CH_3)_2 \\
| \quad\quad\quad\quad | \\
H_2C \quad\quad\quad CH_2 \\
S
\end{array}
+ 2[(C_2H_5)_3N]\cdot HCl
$$

The same hereocycle is formed in the reaction of 1,3-bis-(chloromethyl)-tetramethylgermasiloxane, $(ClCH_2)(CH_3)_2-Si-O-Ge(CH_3)_2(CH_2Cl)$, with H_2S in the presence of triethylamine (105a). The bis-trialkyl- and triarylgermanium sulphides have been obtained by Burschkies (23) via action of an aqueous or alcoholic Na_2S solution on the organobromogermanes:

$$2 R_3GeBr + Na_2S \rightarrow R_3GeSGeR_3 + 2 NaBr$$

Under the same experimental conditions, tricyclohexylbromogermane leads to tricyclohexylgermanium disulphide, $(C_6H_{11})_3GeSSGe(C_6H_{11})_3$ (23). The bis-(trialkylgermanium) sulphides have been isolated in good yield via the action of trialkyliodogermanes on silver sulphide (27, 65):

$$2 R_3GeI + Ag_2S \rightarrow R_3GeSGeR_3 + 2 AgI$$

The five-membered ring $Et_2Ge\!-\!-\!GeEt_2$ reacts with sulphur and selenium

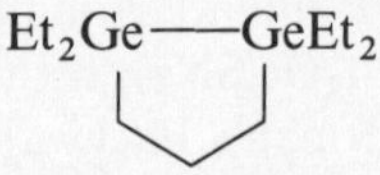

giving the corresponding six-membered ring heterocycles (45a):

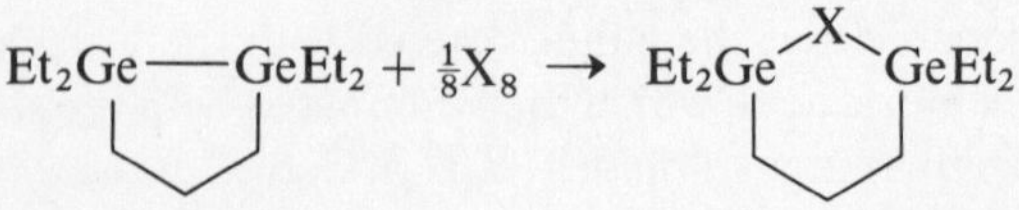

$$Et_2Ge\!-\!-\!GeEt_2 + \tfrac{1}{8}X_8 \rightarrow Et_2Ge\overset{X}{\diagup}\diagdown GeEt_2 \quad X = S, Se$$

Bis-(triphenylgermanium) sulphide was obtained by passing a stream of hydrogen sulphide into a solution of triphenylgermanol in hot 95% ethanol (79).

$(Et_3Ge)_2S$ forms in 20% yield when Et_3GeH is heated with elemental sulphur (4.5 hours at 170°C) (83).

The action of triphenylbromogermane on lithium triphenyltin sulphide leads to triphenyltin triphenylgermanium sulphide (75):

$$Ph_3SnSLi + BrGePh_3 \rightarrow Ph_3Sn-S-GePh_3 + LiBr$$

The action of the derivatives R_3GeSLi on the halides R_3MX (M = Si, Ge, Sn) leads to the symmetrical or unsymmetrical sulphides (62, 77, 88):

$$R_3GeSLi + R'_3MCl \rightarrow R_3GeSMR'_3 \qquad (M = Ge, Si, Sn, Pb)$$

The bis-(triorganogermanium) sulphides have high thermal stability. However, $(CH_3)_3GeSSi(CH_3)_3$ and $(CH_3)_3GeSSn(CH_3)_3$ undergo a thermal disproportionation leading to the symmetric sulphides $[(CH_3)_3$-$Ge]_2S$, $[(CH_3)_3Si]_2S$ (62) and $[(CH_3)_3Sn]_2S$ (73a).

The halogenation of the compounds $(Et_3M)_2M'$, where M = Si, Ge, Sn; M' = S, Se, Te, by bromine in benzene medium studied by Vyazankin and coworkers (89), leads to the formation of Et_3MBr and the free chalcogen (89).

The reactions of benzoyl peroxide with the same compounds $(Et_3M)_2M'$ take place at room temperature with the separation of sulphur, selenium or tellurium and the formation of the corresponding benzoate Et_3M-$OCOC_6H_5$. Acetyl benzoyl peroxide and dicyclohexyl percarbonate react similarly (87). The reaction of organogermanium sulphides with mercurated organic derivatives is a new method for the synthesis of functionally substituted organic derivatives of germanium (54):

$$(R_3Ge)_2S + Hg(CH_2COR')_2 \rightarrow 2 R_3GeCH_2COR' + HgS$$

$$(R = Alk., Ar.; R' = H, Alk., OAlk.)$$

Quantitative thiolysis of the Ge—S bonds of the bis-(trialkylgermanium) sulphides is observed near 180–190°C (65):

$$(Et_3Ge)_2S + 2 C_6H_5SH \rightarrow 2 Et_3GeSC_6H_5 + SH_2$$

7-1-3-2 Diorganogermanium Sulphides, (R_2GeS). Dimethyldichloroger-mane dissolved in water and treated with hydrogen sulphide gives di-methylgermanium sulphide $[(CH_3)_2GeS]$ (94.5% yield). It crystallizes in flat plates and hydrolyses very slowly in moist air to liberate hydrogen sulphide. Hydrolysis also is slow in boiling water but more rapid in dilute acids or dilute solutions of hydrogen peroxide and leads to dimethyl-germanium oxide $[(CH_3)_2GeO]_4$ (58, 106, 107). $[Me_2GeS]$ is trimeric in benzene solution according to cryoscopic measurements (22). Inter-action between $(i\text{-}C_3H_7)_2GeI_2$ and silver sulphide yields $[(i\text{-}C_3H_7)_2GeS]_2$ (13). Dimeric dibutylgermanium sulphide has been obtained by pass-ing anhydrous hydrogen sulphide gas into a solution of di-n- and

-*sec*-alkoxydibutylgermanes in the parent alcohol:

$$Bu_2Ge(OR)_2 + 2\,H_2S \rightarrow Bu_2Ge(SH)_2 + 2\,ROH$$

$$2\,Bu_2Ge(SH)_2 \rightarrow (Bu_2GeS)_2 + 2\,H_2S$$

In the case of di-*tert*-butoxydibutylgermane the reaction proceeds only in the presence of *p*-toluenesulphonic acid (43). The reaction of the tetra-organogermanes ($R = C_4H_9$ or C_6H_5) with elementary sulphur has been studied by Schmidt and Schumann (70). In the action of sulphur on Bu_4Ge, the formation of $[Bu_2GeS]$ is observed:

$$3\,(C_4H_9)_4Ge + 6\,S \xrightarrow{250°C} \begin{array}{c} Bu_2 \\ | \\ Ge \\ S \diagup \quad \diagdown S \\ | \qquad | \\ Bu_2Ge \diagdown \quad \diagup GeBu_2 \\ S \end{array} + 3\,(C_4H_9)_2S$$

The transitory formation of Bu_3GeSBu and $Bu_2Ge(SBu)_2$ has also been considered, but the latter derivative degrades thermally:

$$Bu_2Ge(SBu)_2 \rightarrow \tfrac{1}{3}(Bu_2GeS)_3 + Bu_2S$$

The action of sulphur on tetraphenylgermane at 270°C, after an incomplete reaction, results in the formation of germanium metal and Ph_2S after thermal decomposition of the previously formed $(Ph_2GeS)_3$ (70).

Benzoyl chloride (and *p*-nitrobenzoyl chloride) react with the trihydrate of $Ph_2Ge(SNa)_2$ to form diphenylgermanium sulphide and benzoic acid (or *p*-nitrobenzoic acid). The sulphide exists in both dimeric and trimeric forms (39):

$$\begin{array}{c} Ph \diagdown \quad S \quad \diagup Ph \\ Ge \qquad Ge \\ Ph \diagup \quad S \quad \diagdown Ph \end{array} \qquad \begin{array}{c} S \\ Ph_2Ge \diagup \quad \diagdown GePh_2 \\ | \qquad | \\ S \diagdown \quad \diagup S \\ Ge \\ Ph_2 \end{array}$$

Thus far, there is no evidence for the existence of an open chain structure of the type:

$$\begin{array}{ccccccc} Ph & & Ph & & Ph & & Ph \\ | & & | & & | & & | \\ -Ge & -S- & Ge & -S- & Ge & -S- & Ge- \\ | & & | & & | & & | \\ Ph & & Ph & & Ph & & Ph \end{array}$$

Dimethylgermanium sulphide dissolved in ethanol and treated with zinc amalgam and 12N HCl gives dimethylgermane and hydrogen sulphide

(94). In the "conversion series" reactions studied by Anderson (13), the following results were observed:

$$[i\text{-}Pr_2GeS]_2 + AgOCOCH_3 \rightarrow i\text{-}Pr_2Ge(OCOCH_3)_2$$

$$[(i\text{-}Pr_2GeS]_2 + AgCN \rightarrow i\text{-}Pr_2Ge(CN)_2$$

$$[i\text{-}Pr_2GeS]_2 + AgBr \rightarrow i\text{-}Pr_2GeBr_2$$

In exchange reactions between $(CH_3)_2GeX_2$ (X = Cl, Br, I) and $[(CH_3)_2GeS]$ it is shown that the resulting equilibrium compositions are mixtures of chain and ring molecules, the trimeric compound $[(CH_3)_2GeS]_3$ being present in large amount. Equilibrium constants are given for the distribution of structure-building units between the various sized chains and the trimeric ring (49).

The equilibrated reaction mixture between hexamethylcyclotrisilthiane and hexamethylcyclotrigermanium trisulphide shows a permutation of Me_2Si and Me_2Ge moieties between these and other ring structures (50):

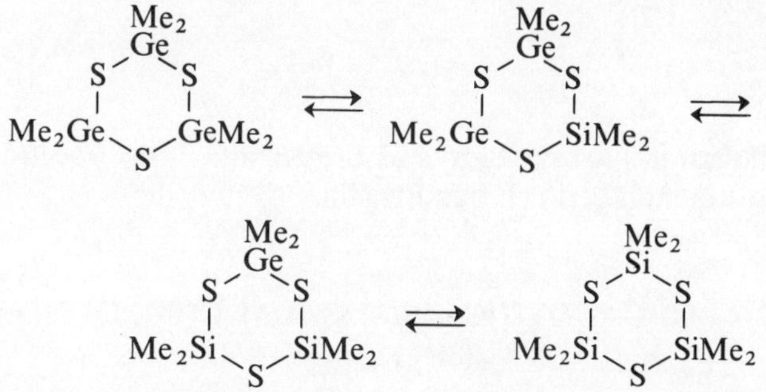

7-1-3-3 Organogermanium Sulphides, $(RGeS)_2S$.

The first organogermanium sulphides of the type $(RGeS)_2S$ were obtained by Bauer and Burschkies by the action of H_2S on an acetic-acid solution of germanoic anhydrides, $(RGeO)_2O$:

$$R = C_6H_5-, \ p\text{-}CH_3-C_6H_4-, \ \alpha\text{-}C_{10}H_7-, \ p\text{-}(CH_3)_2N-C_6H_4-,$$

$$p\text{-}(C_2H_5)_2N-C_6H_4-$$

These compounds are white powders which are not very water-soluble and do not melt. They are soluble in alkaline bases and alkaline sulphides. In moist air, they decompose gradually with release of H_2S (19). Two of these derivatives (R = C_6H_5 and $(CH_3)_2N-C_6H_4-$) as well as $[(PhCH_2)_3Ge]_2S$ have been used to treat infections in mice. Their toxicity is lower than that of similar tin and lead compounds (60).

Tetramethylgermanium hexasulphide has been described by Moedritzer who prepared it via the following reaction:

$$4\,CH_3GeBr_3 + 6\,H_2S + 12\,N(C_2H_5)_3 \rightarrow$$

$$(CH_3GeS_{1.5})_4 + 12\,[N(C_2H_5)_3H]Br$$

This sulphide is a stable, crystalline, high-melting material which because of its molecular weight and mass-spectroscopic cracking pattern must consist of four $(CH_3GeS_{1.5})$ moieties probably arranged in a structure of type I or II.

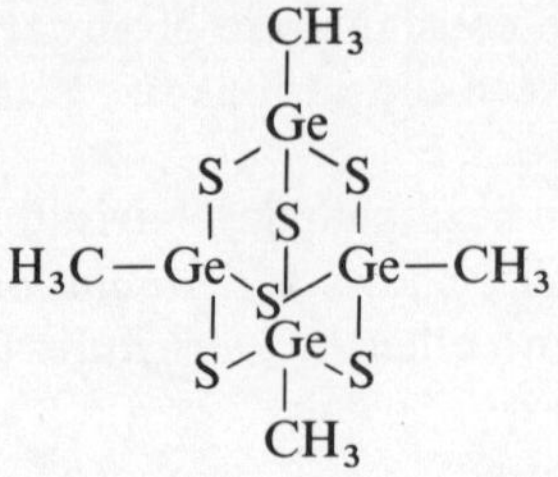

(I) 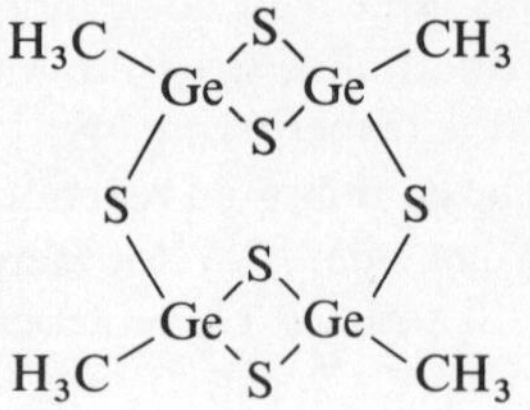   (II)

Crystallographic X-ray study and the IR and PMR spectra do not permit a distinction between I and II (48).

7-2   ORGANOTHIOGERMANIUM COMPOUNDS
$R_nGe(SR')_{4-n}$ $(n = 0-3)$

7-2-1   Synthesis

Many methods of synthesis for the derivatives of the type $R_nGe(SR')_{4-n}$ $(n = 0-3)$ have been described in the literature. We have just seen that compounds of the type R_3GeS–metal (I) were valuable agents of synthesis for the derivatives of the type R_3GeSR'. The other methods of synthesis used are action on organogermanium halides by the metal thiolates or thiols in the presence of bases, the cleavage reactions of the germanium–oxygen bonds of the oxides, alkoxides, esters etc. caused by the thiols, the reactions of the organogermanium hydrides with thiols or organic disulphides, the exchange or redistribution reactions, and lastly various reactions involving thiolysis of the Ge—S, Ge—N, Ge—P etc. bonds or sulphur insertion into Ge—C bonds.

7-2-1-1   Reaction of Derivatives of Type R_3GeSLi (or Na) with Organic Halogen Derivatives. The reactions of this type have already been recorded in the study of the reactivity of R_3Ge-S--metal derivatives. Noteworthy are the reactions of Ph_3GeSNa with derivatives of the type RX:

$$R = -CH_3, \text{-}n\text{-}C_4H_9, -CH_2-C_6H_5, -COC_6H_5,$$

$$-COC_6H_4-NO_2\text{-}p, -CH_2SCH_3 \text{ and } -GePh_3 \text{ (38)}$$

according to the general reaction:

$$Ph_3GeSNa + RX \rightarrow Ph_3GeSR$$

To be mentioned also are the reactions of Ph_3GeSLi with $PhCOCl$ and the Ph_3MCl derivatives (M = Si, Ge, Sn) (77), Et_3GeSLi with C_4H_9Br, $C_6H_5CH_2Cl$, Et_3MX (M = Ge, Sn) (88).

7-2-1-2 Action of Organogermanium Halides with Metal Thiolates. In 1932, Schwarz and Reinhardt prepared germanium tetrathiophenoxide, $Ge(SC_6H_5)_4$, via the reaction of sodium (thio)phenoxide with germanium tetrachloride, as well as by the Grignard reaction between thiophenol and $GeCl_4$ (78). Backer and Stienstra prepared a large series of derivatives of the type $Ge(SR)_4$ by the action of thiols on $GeCl_4$ in absolute alcohol in the presence of sodium (17, 18). Under the same experimental conditions, the 2,2'-dimercaptodiethyl ether, $O(CH_2CH_2SH)_2$, leads to the spirocyclic ether (17):

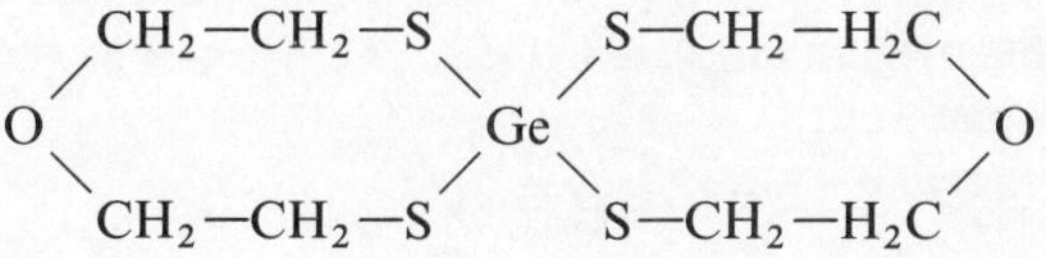

$$O \underset{CH_2-CH_2-S}{\overset{CH_2-CH_2-S}{\Big\langle}} \quad Ge \quad \underset{S-CH_2-H_2C}{\overset{S-CH_2-H_2C}{\Big\rangle}} O$$

and the di-thiol ethane of the corresponding spirocyclic thioether (15):

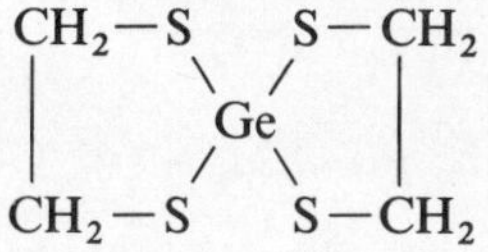

$$\underset{CH_2-S}{\overset{CH_2-S}{\Big|}} \quad Ge \quad \underset{S-CH_2}{\overset{S-CH_2}{\Big|}}$$

Germyl methyl sulphide, H_3GeSCH_3, (91, 93), germyl phenyl sulphide (36) and trimethyl(ethylthio)germane, Me_3GeSEt, (3) have been prepared by the reaction of H_3GeCl, H_3GeBr and Et_3GeBr with the sodium (or K) methyl, ethyl or phenyl thiolates. $(CH_3)_3GeSC_6F_5$ and $(C_6H_5)_3GeSC_6F_5$ are prepared by reaction of C_6F_5SLi with R_3GeX (54a). $Ph_3GeSSnPh_3$ is prepared by the action of Ph_3GeBr on Ph_3SnSLi (74). Lead thiolates react in benzene medium with alkylchloro- and bromogermanes with average yields of 65 % (49, 65):

$$2 R_3GeCl + Pb(SR')_2 \rightarrow 2 R_3GeSR' + PbCl_2$$

$$R_2GeCl_2 + Pb(SR')_2 \rightarrow R_2Ge(SR')_2 + PbCl_2$$

Diethylchlorogermane gives a product containing both a Ge—H and a Ge—S bond (65):

$$2 Et_2GeHCl + Pb(SC_2H_5)_2 \rightarrow 2 Et_2HGeSC_2H_5 + PbCl_2$$

7-2-1-3   Action of Thiols on Organogermanium Halides in the Presence of Bases. No reaction occurs when organogermanium halides are treated with thiols in boiling benzene, but good yields of the thio derivatives are obtained when triphenylchlorogermane or triphenylbromogermane is allowed to react at room temperature with thiols in the presence of a tertiary base (triethylamine or pyridine) (30):

$$Ph_3GeX + RSH + base \rightarrow (C_6H_5)_3GeSR + base\text{-}HX$$

(R = —$(CH_2)_{10}CH_3$, —CH_2CH_2OH, —$CH_2CH=CH_2$, —CH_2COOH,

—$COCH_3$, —CH_2COOCH_3, —CH_2CONH_2, —C_6H_5,

—C_6F_5, p-ClC_6H_4—, 2-$C_{10}H_7$—, 17β-mercaptotestosterone)

Diphenyldiorganothiogermanes, $R_2Ge(SR)_2$, are prepared readily from diphenyldihalogermanes by this reaction using two moles of thiol (30). Dithiobis-(triphenylgermanes), Ph_3GeS—R—S—$GePh_3$, are also obtained in the same way:

(R = —CH_2—CH_2—, —$CH_2C_6H_4$—O—$C_6H_4CH_2$—,

—COC_6H_4CO—)

The pyridine–germanium tetrachloride adduct with methanethiol, ethanethiol and butane-1-thiol and further pyridine give tetra-alkylthiogermanes, $(MeS)_4Ge$ (40), $(EtS)_4Ge$, $(BuS)_4Ge$ (1):

$$GeCl_4(C_5H_5N)_2 + 4 RSH + 2 C_5H_5N \rightarrow Ge(SR)_4 + 4 [C_5H_5NH]Cl$$

The phenylmethylthiogermanes, Ph_3GeSMe, $Ph_2Ge(SMe)_2$, $PhGe(SCH_3)_3$, and trimethylmethylthiogermane, $(CH_3)_3GeSMe$, have been prepared from the corresponding bromides and thiol in the presence of pyridine (40).

A number of di- and tributylorganothiogermanes, $Bu_2Ge(SR)_2$ and $Bu_3Ge(SR)$, have been prepared by passing ammonia into or by adding triethylamine to a mixture of the butylchlorogermane and the thiol in benzene (47):

$$Bu_{4-n}GeCl_n + nRSH + nNH_3 \text{ (or } nEt_3) \longrightarrow$$

$$Bu_{4-n}Ge(SR)_n + nNH_4Cl \text{ (or } NEt_3, HCl) \qquad (n = 1 \text{ and } 2)$$

$$(R = Et, Pr, i\text{-Pr}, Bu, i\text{-Bu}, t\text{-Bu}, n\text{-}C_{12}H_{25}, C_6H_5CH_2\text{—}, C_6H_5)$$

Germanium tetrathiolates, $Ge(SR)_4$, have been synthesized by the same ammonia method in good yields ($R = Et, n\text{-Pr}, i\text{-Pr}, n\text{-Bu}, i\text{-Bu}, n\text{-}C_{12}H_{25},$ $C_6H_5CH_2\text{—}, C_6H_5$). In the case of *tert*-butanethiol, the final product obtained as a distillable solid corresponded to $NH_2Ge(S\text{—}t\text{-}C_4H_9)_3$ (46).

p-p'-Oxybis-(α-toluenethiol) and 1,5-pentanedithiol reacted with diphenyl dihalogermanes to give gums which were soluble in most organic solvents:

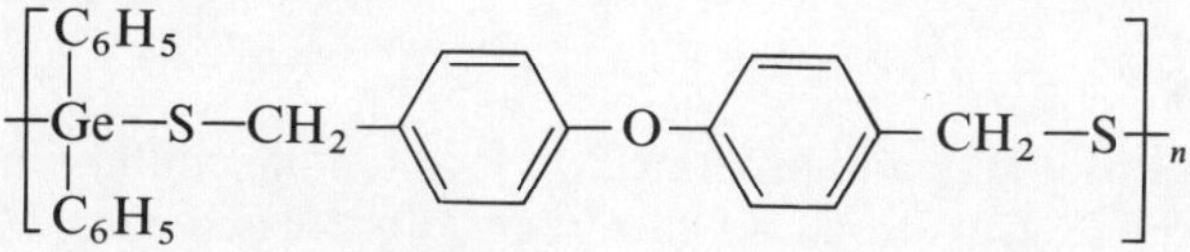

$$n \approx 50$$

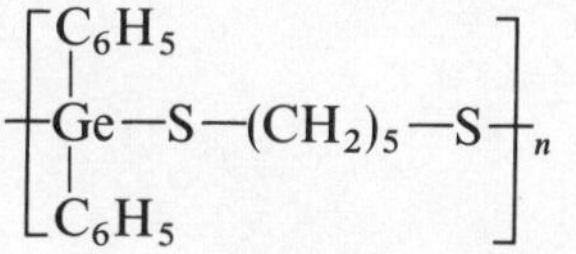

$$n \approx 60$$

Germanium tetrachloride reacts with these thiols to give insoluble white solids which did not melt below 300°C (30).

The synthesis of many germanium heterocycles from dithiols or difunctional systems and di- or tetra-halogermanes is briefly recorded below.

The reaction of ethanedithiol, thioglycolic acid and 2-hydroxyethanethiol with diphenyldihalogermanes in the presence of base gave heterocyclic compounds (30):

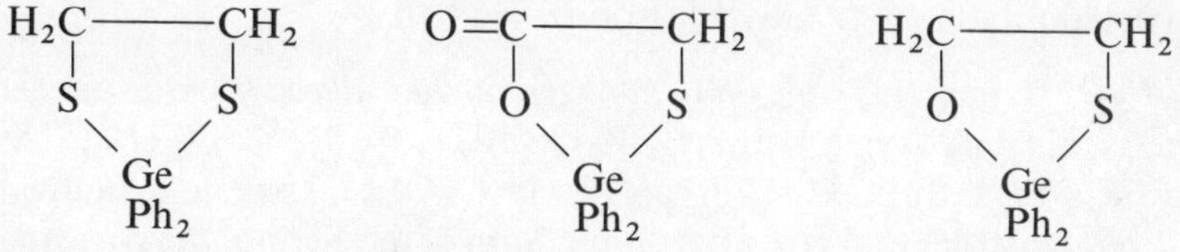

The Ge-dimethyl analogues of these heterocyclic compounds have also been prepared by Wieber and Schmidt (97–99, 102) from dimethyldichlorogermane.

Dimethylgermanium ring compounds from thiosalicylic acid and o-aminothiophenol (99, 100, 102) are obtained in the same way:

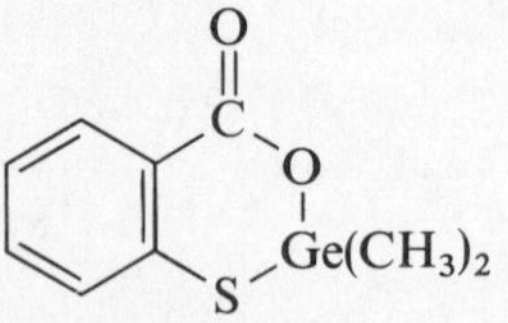

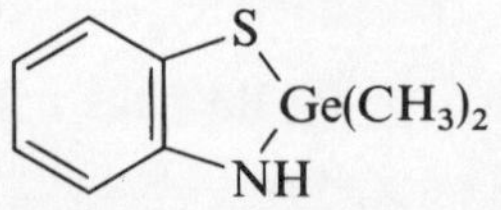

Dimethyldichlorogermane and (chloromethyl)dimethylchlorogermane react with 1,3-propanedithiol and 3,4-dimercaptotoluene to give six-membered ring heterocycles (103) or bicyclic heterocycles (96, 101, 104):

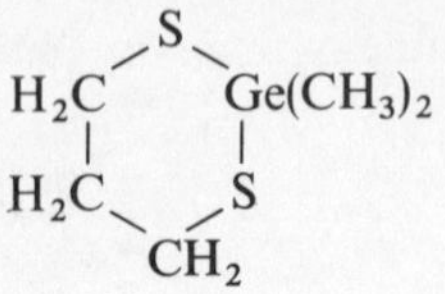

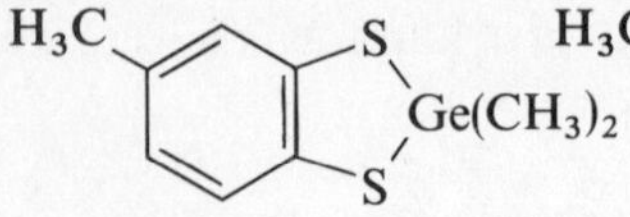

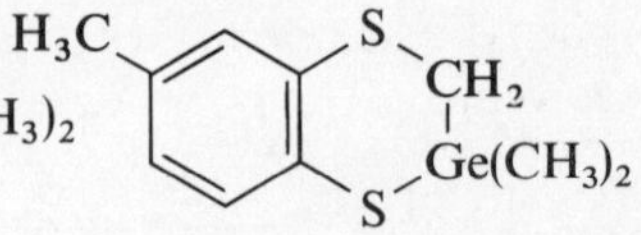

From the reaction of germanium tetrachloride and ethanedithiol in the presence of pyridine, Davidson and coworkers (30) isolated compounds with spiran-type structure:

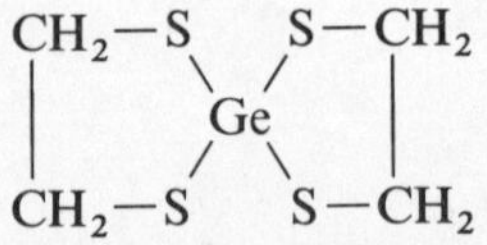

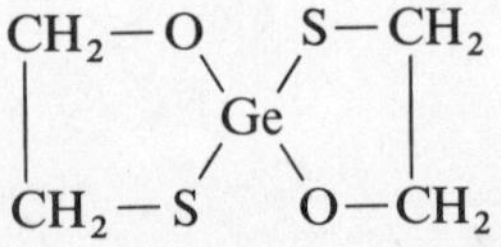

A new synthesis of the compounds $(CH_3S)_nM(CH_3)_{4-n}$ (M = Ge, Sn, $n = 1, 2, 3$) was described by van der Berghe and coworkers (81). The methylgermanium halide is dissolved in a small volume of water and methylmercaptan is passed at atmospheric pressure into the solution, while this is slowly being neutralized with sodium hydroxide solution.

The reactions between the thiol Ph_3GeSH and RX compounds give the expected products Ph_3GeSR in good yield (39).

7-2-1-4   Action of Thiols on Organogermanium Alkoxides, Oxides, Esters. The alkylalkoxygermanes, R_3GeOCH_3, $R_2Ge(OCH_3)_2$ (65), $Bu_2Ge(OEt)_2$, $Bu_2Ge(O\text{-}n\text{-}Pr)_2$ (47), Me_3GeOEt (3), when heated with a not very volatile thiol give after elimination of the alcohol the corresponding alkylthiogermanes:

$$R_3GeOR' + R''SH \rightarrow R_3GeSR'' + R'OH$$

$$R_2Ge(OR')_2 + 2\,R''SH \rightarrow R_2Ge(SR'')_2 + 2\,R'OH$$

This displacement reaction requires generally simple heating (3, 65); however, with the dibutyldialkoxygermanes (47) the presence of p-toluenesulphonic acid as catalyst seems necessary.

Tetraisopropoxygermane has been found to interchange its isopropoxy groups with thiols in the presence of p-toluene sulphonic acid (46):

$$Ge(O\text{-}i\text{-}Pr)_4 + 4\,RSH \rightarrow Ge(SR)_4 + 4\,i\text{-}PrOH$$

$$(R = n\text{-}C_4H_9, n\text{-}C_{12}H_{25}, -CH_2C_6H_5)$$

An attempt was also made to isolate mixed derivatives containing Ge—S and Ge—O linkages by the reaction of $Ge(O\text{-}i\text{-}Pr)_4$ with butanethiol in $1:1$ molar ratio. However, purification by distillation under reduced pressure always led to disproportionation (46). Dibutyldiethoxygermane, $Bu_2Ge(OEt)_2$, with 2-mercaptoethanol, mono-thioglycerol and α- and β-mercaptopropionic acids in equimolecular proportions in benzene solution, give monomeric heterocycles (80):

$$Bu_2Ge\begin{array}{c} O-CH_2 \\ \diagdown\quad | \\ S-CH_3 \end{array} \qquad Bu_2Ge\begin{array}{c} O-CH-CH_2OH \\ \diagdown\qquad | \\ S-CH_2 \end{array} \qquad Bu_2Ge\begin{array}{c} S-CH-CH_3 \\ \diagdown\qquad | \\ O-C=O \end{array}$$

$$Bu_2Ge\begin{array}{c} S-CH_2 \\ \diagdown\qquad \diagup CH_2 \\ O-C \\ \diagdown\!\!\backslash O \end{array}$$

The reaction between tetraisopropoxygermane, $Ge(O\text{-}i\text{-}Pr)_4$, and 2-mercaptoethanol α- and β-mercaptopropionic acids also were studied (80) (cf. properties of alkoxygermanes).

Several exchange reactions of the same type have been described by Anderson using triethylgermanium acetate. The thiols less volatile than acetic acid furnish triethylorganothiogermanes through removal of acetic acid by fractional distillation (12):

$$Et_3GeOCOCH_3 + RSH \rightarrow Et_3GeSR + CH_3COOH$$

$(R = C_6H_5-,\quad o\text{-}CH_3-C_6H_4-,\quad m\text{-}CH_3-C_6H_4-,\quad C_6H_5CH_2-,$
$o\text{-}NH_2-C_6H_4-,\quad \beta\text{-}C_{10}H_7,\quad C_4H_3OCH_2-,\quad n-C_6H_{13}-, n\text{-}C_7H_{15}-,$
$n\text{-}C_{12}H_{25}-)$

With m-toluenethiol and o-aminothiophenol this reaction yielded small amounts of thioacetic acid besides the normal products (12):

$$(C_2H_5)_3GeOCOCH_3 + RSH \rightarrow (C_2H_5)_3GeOR + CH_3COSH$$

Alkylthiogermanes can be prepared by thiolysis of the Ge—O linkages of the alkylgermanium oxides $(R_3Ge)_2O$ or R_2GeO (3, 12, 65):

$$(R_3Ge)_2O + 2 R'SH \rightarrow 2 R_3GeSR' + H_2O$$

$$R_2GeO + 2 R'SH \rightarrow R_2Ge(SR')_2 + H_2O$$

However, in the reaction of butanethiol with Bu_2GeO or $(Bu_3Ge)_2O$ the presence of p-toluenesulphonic acid as catalyst seems to be necessary (47), and the action of a volatile thiol such as EtSH on $(Et_3Ge)_2O$ gives practically no result (12).

All of these experiments would appear to indicate the favoured formation of the germanium–sulphur over germanium–oxygen bonds.

7-2-1-5   Exchange Reactions. Abel reported an exchange reaction between trimethylbromogermane and (ethylthio)trimethylsilane (3):

$$Me_3GeBr + Me_3SiSEt \rightarrow Me_3GeSEt + Me_3SiBr$$

Exchange of the methylgermanium moieties of the monofunctional thiomethyl-group with either chlorine, bromine, iodine, isocyanate and methoxy groups has been studied by proton nuclear magnetic resonance. The following systems have been studied:

$$CH_3GeT_3 \; v. \; CH_3SiZ_3 \quad (T = OCH_3, SCH_3, N(CH_3)_2;$$

$$Z = Cl, Br, I) \quad (51)$$

$$(CH_3)_2GeX_2 \; v. \; (CH_3)_2Ge(SCH_3)_2 \quad (X = Cl, Br \; or \; I) \quad (49, 82)$$

$$(CH_3)_2Ge(SCH_3)_2 \; v. \; (CH_3)_2Ge(OCH_3)_2 \quad (82)$$

$$CH_3GeX_3 \; v. \; CH_3Ge(SCH_3)_3 \quad (X = Cl, Br, I) \quad (52)$$

$$Ge(SCH_3)_4 \; v. \; GeZ_4 \quad (Z = Cl, Br, I \; or \; NCO) \quad (20)$$

$$Ge(SCH_3)_4 \; v. \; Ge(OCH_3)_4 \quad (20)$$

$$(CH_3)_3GeBr_2 \; v. \; (CH_3)Ge(SCH_3)_3 \quad (53)$$

In halogen v. methylthio exchange, halogen attachment to the dimethylgermanium moiety is favoured. However, these exchange reactions cannot be used as methods of preparation for mixed methylthiogermanes.

7-2-1-6   Reactions of Organogermanium Hydrides with Thiols. Triethylbutylthiogermane is obtained by the catalytic dehydrocondensation reaction, in the presence of a platinum catalyst, between triethylgermane and butanethiol (41):

$$Et_3GeH + C_4H_9SH \xrightarrow{\text{Pt}} Et_3GeSC_4H_9 + H_2 \quad (75\%)$$

The rate of this reaction is accelerated in the presence of reduced nickel (65). As in the analogous reactions with alcohols (cf. synthesis of alkoxy-germanes) a mechanism of nucleophilic substitution is considered:

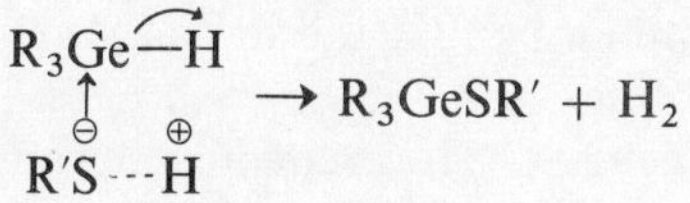

$$R_3Ge\overset{\frown}{}H$$
$$\underset{R'S\cdots H}{\overset{\uparrow}{\underset{\ominus\quad\oplus}{}}} \rightarrow R_3GeSR' + H_2$$

The same dehydrocondensation reaction is extended to the preparation of derivatives of the type $R_3GeS(CH_2)_nSGeR_3$ (65):

$$2\,R_3GeH + HS(CH_2)_nSH \rightarrow R_3GeS(CH_2)_nSGeR_3 + 2\,H_2$$

7-2-1-7 Miscellaneous Syntheses. Triethylgermane reacts without catalyst on dimethyldisulphide with formation of methanethiol and triethyl-methylthiogermane (65):

$$Et_3GeH + CH_3S\!-\!SCH_3 \rightarrow Et_3GeSCH_3 + CH_3SH$$

Primary, secondary and tertiary thiols react quantitatively with the germanium–nitrogen bond of the trialkylgermylamines (3, 63, 65, 66):

$$R_3GeNR'_2 + R''SH \rightarrow R_3GeSR'' + R'_2NH$$

$$(R_3Ge)_2NH + 2\,R'SH \rightarrow 2\,R_3GeSR' + NH_3$$

The same cleavage reactions are observed with the triorganogermylphosphines (64):

$$Et_3GePEt_2 + R'SH \rightarrow Et_3GeSR' + Et_2PH$$

Sulphur and selenium insert into the germacyclobutane ring to give the germatetrahydrothiophenes and germatetrahydroselenophenes (44, 45):

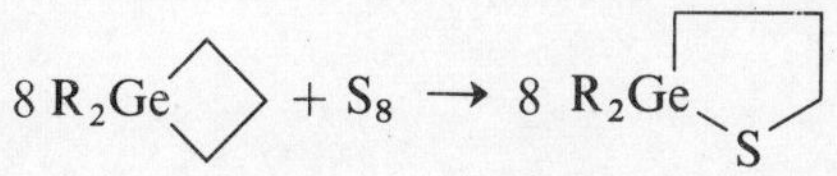

$$8\,R_2Ge\!\!\diamondsuit + S_8 \rightarrow 8\,R_2Ge\underset{S}{\bigcup} \quad (R = Me,\,n\text{-Bu})$$

The action of sodium sulphide on the dihalogenated derivatives $R_2Ge(Cl)CH_2CH_2CH_2Cl$, $R_2Ge(Br)CH_2CH(CH_3)CH_2Br$ leads to the same cyclic sulphides (45):

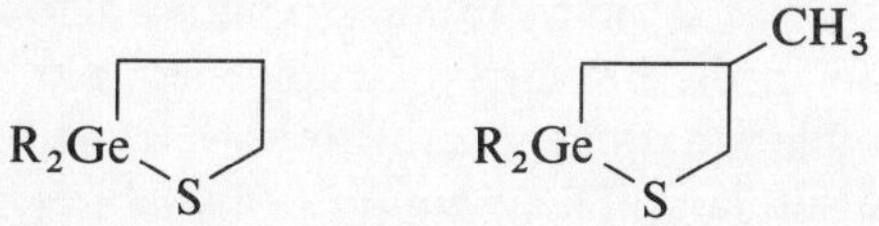

7-2-2   Properties

Thiogermanium derivatives of type R_3GeSR', $R_2Ge(SR')_2$ are thermally stable. Many of them can be distilled without change (30).

The germanium–sulphur bond of the derivatives R_3GeSR' is much more stable towards hydrolysis than the Si—S bond of the analogous compounds of silicon (65). $Ge(SC_4H_9)_4$ remains unchanged when its benzene or ethanol solutions are shaken with water. This behaviour of mercaptogermanes is in sharp contrast to that of alkoxygermanes which are highly susceptible to moisture.

The thermal stability of the metal–sulphur bond in the trialkyl(aryl)-organothioderivatives, R_3MSR', decreases in the series Si, Ge, Sn and Pb while the hydrolytic stability increases. These experiments appear to indicate the favoured formation of germanium–sulphur over germanium–oxygen linkages under these conditions. This can possibly be ascribed to the π-bonding tendency of the filled d orbitals of germanium (46). Among other reasons, such a change in the reactivity may be due to a possible reversed order of metal–oxygen and metal–sulphur bond strength in going from silicon to tin.

Alcoholysis of the Ge—S linkage of triethylethylthiogermane by primary alcohols is observed near 170°C (65):

$$Et_3GeSC_2H_5 + n\text{-}C_8H_{17}OH \rightarrow Et_3GeOC_8H_{17} + C_2H_5SH$$

The reaction of excess anhydrous ethanol with $Ge(SC_3H_7)_4$ is achieved after 60 hours in the presence of a catalyst (p-toluenesulphonic acid). It was found to take place more readily than the corresponding reaction of tin:

$$Ge(S\text{-}n\text{-}C_3H_7)_4 + 4\,EtOH \rightarrow Ge(OEt)_4 + 4\,n\text{-}C_3H_7SH$$

It appears that the ease of conversion of metal–sulphur to metal–oxygen bonds follows the order Si > Ge > Sn in such derivatives. Trimethyl- and triethylorganothiogermanes, R_3GeSR', react slowly with higher boiling mercaptans (3, 12, 65):

$$Et_3GeSC_4H_9 + C_6H_5CH_2SH \xrightarrow{170°C} Et_3GeSCH_2C_6H_5 + C_4H_9SH$$

The difference of boiling points between the original thiol and the thiol formed must be sufficiently great.

The thiolysis reaction between $Bu_2Ge(S\text{-}n\text{-}C_3H_7)_2$ and $Ge(S\text{-}n\text{-}C_3H_7)_4$ with higher thiols is slower than the corresponding alcoholysis and could only be completed in the presence of p-toluenesulphonic acid (46, 47). Primary and secondary amines react with both trimethylethylthiosilane and hexamethyldisilthiane (2). In contrast, aniline does not react with triethylmethylthiogermane (65). Air oxidation is reported not to affect alkylthiogermanes but triphenylmethylthiogermane is oxidized at 0°C by either hydrogen peroxide or nitrogen dioxide to give bis-(triphenylgermanium) oxide; no germylsulphoxide or sulphone is obtained. The same thio derivative undergoes a rapid reaction with aqueous silver nitrate solution to yield bis-(triphenylgermanium) oxide and silver (I) methanethiolate (40):

$$2\,(C_6H_5)_3GeSCH_3 + 2\,AgNO_3 + H_2O \rightarrow$$
$$[(C_6H_5)_3Ge]_2O + 2\,AgSCH_3 + 2\,HNO_3$$

The Ge—S bond is cleaved by various reagents such as methyl iodide (40), $LiAlH_4$ (65), n-butyllithium (40), ethylmagnesium bromide (65), benzoic acid, Ph_2SiCl_2 and dimethyl sulphate (40):

$$(CH_3)_3GeSCH_3 + 2\,CH_3I \longrightarrow$$
$$(CH_3)_3GeI + [(CH_3)_3S]^+I^-$$

$$R_3GeSR' \xrightarrow{\ LiAlH_4\ } R_3GeH + R'SH$$

$$(C_6H_5)_3GeSCH_3 + n\text{-}C_4H_9Li \longrightarrow$$
$$(C_6H_5)_3GeC_4H_9 + CH_3SLi$$

$$(C_2H_5)_3GeSCH_3 + EtMgBr \longrightarrow$$
$$(C_2H_5)_4Ge + CH_3SMgBr$$

$$(C_2H_5)_3GeSCH_2OC_4H_3 + C_6H_5COOH \longrightarrow$$
$$(C_2H_5)_3GeOCOC_6H_5 + C_4H_3OCH_2SH$$

$$(C_2H_5)_3Ge\text{-}\beta SC_{10}H_7 \xrightarrow{\ Ph_2SiCl_2\ } (C_2H_5)_3GeCl$$

$$(CH_3)_3GeSCH_3 + 2\,(CH_3O)_2SO_2 \longrightarrow$$
$$(CH_3)_3GeOSO_2OCH_3 + [(CH_3)_3S]^+[CH_3SO_4]^-$$

$Et_3GeSCH_2COOGeEt_3$ is cleaved by $HgCl_2$ with formation of Et_3GeCl and $Hg(SCH_2COO)$ (11) and by H_2SO_4 to give triethylgermanium sulphate $(Et_3Ge)_2SO_4$ and β-mercaptoacetic acid (10).

Insertion of hexafluoroacetone into a germanium–sulphur bond of Me_3GeSEt gives alkoxygermanes (7):

$$Me_3GeSEt + (CF_3)C{=}O \rightarrow Me_3Ge{-}O{-}C(CF_3)_2SEt_2$$

H_3GeSCH_3 is thermally stable at room temperature but decomposes in the presence of B_2H_6 or BF_3 to form GeH_4 and unidentified solid material

(93). Studies of interaction of B_2H_6 with $(H_3Ge)_2S$ and H_3GeSCH_3, for investigation of the Lewis basicity of the sulphur atom in the Ge—S—Ge and Ge—S—C linkages, indicate that both $(GeH_3)_2S$ and H_3GeSCH_3 are weaker bases than $(CH_3)_2S$. However, the relative basicities of the two germanium compounds could not be distinguished (93).

Germyl phenyl sulphide, H_3GeSPh, reacts with boron trifluoride to give monogermane and a pale yellow solid which might be $GeH_x[S(C_6H_5)]_{4-x}$ ($x = c.$ 2). Germyl phenyl sulphide gives a solid adduct with boron trichloride but this decomposes to give germyl chloride, monogermane, dichlorogermane and hydrogen chloride (36).

Determination of alkoxy- and mercaptogermanes by acetylation and with hydrogen bromide has been reported by Magnuson and Knaub (42), (cf. properties of alkoxygermanes).

Application of the empirical equation: (B.P. of $R_3GeSR' = 0.750$ (B.P. of R_4Ge) + 0.527 (B.P. of $R'SH$)) furnishes satisfactory results (root mean square error 6.3°C, standard deviation 6.6°C) in calculations of normal boiling points of nine triethylorganothiogermanes (14).

7-2-3 Physical-chemical studies in the organogermanium–sulphur compound series: organothiogermanes and organogermanium sulphides

7-2-3-1   IR and Raman Spectra. Infra-red spectra of organogermanium sulphides $(H_3Ge)_2S$ (25, 37, 92), $(Et_3Ge)_2S$, $Et_3GeSSiEt_3$, $Et_3GeSSnEt_3$ and their analogues with selenium and tellurium (32), $Ph_3GeSMPh_3$ where M = Ge, Sn, Pb (73), $(Me_2GeS)_3$ (21, 71), $(Bu_2GeS)_3$ (71), $(Ph_2GeS)_3$ (70, 71) of organothiogermanes H_3GeSCH_3 (25, 92, 93), H_3GeSPh (36), $Ph_2Ge(SMe)_2$, $PhGe(SMe)_3$, $(CH_3S)_4Ge$ (40) $(CH_3)_3GeSC_6F_5$, $(C_6H_5)_3$-$GeSC_6F_5$ (54a) are reported in the literature.

The bands at 416 cm^{-1} in $(Ph_2GeS)_3$ (70) and 417 cm^{-1} in $(Ph_3Ge)_2S$ (73) have been attributed to the asymmetric Ge—S—Ge stretching frequencies. The following data are also recorded: ν_sGe—S—Ge at 385 cm^{-1} in $(Ph_3Ge)_2S$; ν_sGe—S at 404 cm^{-1} in $Ph_3GeSSnPh_3$, 400 cm^{-1} in $Ph_3GeSPbPh_3$ (73), and at 410 cm^{-1} in H_3GeSCH_3 (25). The absorptions at 410 cm^{-1} for $Ph_2Ge(SMe)_2$, 410 and 420 cm^{-1} for $PhGe(SMe)_3$, 409, 421, 439 cm^{-1} for $(CH_3S)_4Ge(40)$, 410 cm^{-1} for $Et_3GeSCH_2C_6H_5$ (88) can also be attributed to the Ge—S—C system. The vibrational spectra (IR and Raman) of digermyl sulphide $H_3GeSGeH_3$ (37) and methylthiogermane H_3GeSCH_3 (25) indicate a highly-bent skeleton for these two compounds. The bond angle calculated (valence force field calculations) for $(H_3Ge)_2S$ is 100°. This result is interpreted in terms of the weak tendency of germanium to participate in (p → d)π bonding.

The infrared spectral data of $(Me_2GeS)_3$, $(Bu_2GeS)_3$, $(Ph_2GeS)_3$ and analogues compounds of tin and lead indicate that strong (p → d)π

double-bond character for the Ge—S, Sn—S and Pb—S linkages would not be observed (71).

Wang and van Dyke have investigated the hydrogen bonding properties of Me_2S, $MeSGeH_3$ and $(H_3Ge)_2S$ in order to determine the relative basicities of simple organic sulphides and germanium sulphides. These results indicate that simple sulphur derivatives of germane are weaker bases than Me_2S. It appears that π-bonding is present in the germanium–sulphur linkage and that it is important enough to lower the Lewis basicities of the compounds $MeSGeH_3$ and $(H_3Ge)_2S$ below the basicities of analogous organic sulphides (92).

In the studies of relative basicities of organometallic ethers, amines and sulphides carried out by Abel and coworkers (4), the observed shifts in the C—D stretching frequency in mixtures of organometallic bases and deuteriochloroform indicate that, in contrast to the ethers, the organogermanium sulphides are only of a base strength similar to that of the carbon analogues, which appears to indicate that some π-bonding between germanium and sulphur may take place neutralizing any inductively-increased basicity (4).

The general rules for the infrared spectra of 1,3-bis-metallorganic compounds of the types Et_3GeXH, $Et_3GeXGeEt_3$ and $Et_3GeXMEt_3$ where M = Si, Ge, Sn, and X = S, Se, Te, were discussed by Egorochkin and coworkers (32). It was shown that in bimetalorganic compounds the stretching vibration frequencies of Ge—C bonds depend on the electronegativity of the X atom.

$v_{as}[Ge(C_6H_5)_3]$ and $v_s[Ge(C_6H_5)_3]$ in hexaphenylsulphides $Ph_3GeSGePh_3$, $Ph_3GeSSnPh_3$, $Ph_3GeSPbPh_3$ are given by Schumann and Schmidt (72).

7-2-3-2 UV Spectra. The UV spectra of $(R_3Ge)_2S$ compounds (R = alkyl) and $Ge(SMe)_4$ show that these compounds are transparent in the 195–350 nm region (27).

7-2-3-3 NMR Spectra. A study of the structure of di-heteroorganic compounds $(Et_3M)_2S$ (M = Si, Ge, Sn) (X = O, S, Se, Te) by proton magnetic resonance indicates the following order of decreasing tendency of elements of Groups IV and VI toward $d\pi$–$p\pi$ interaction: Si > Ge > Sn O > S > Se > Te (33).

NMR spectroscopic investigations on methylgermanium halides, oxides, sulphides and selenides have shown that proton signals of these compounds are shifted to lower fields with increasing atomic radii of the substituents and $J(H^1—C^{13})$ coupling constants increase in this series in this direction (F → I, O → Se). The anisotropy or dispersion effect of the ligands seems to play the most important part in these correlations (69).

A small down-field shift is observed in the value of δ for the methyl protons in the series Ph_3GeSCH_3, $Ph_2Ge(SCH_3)_2$, $PhGe(SCH_3)_3$ and $Ge(SCH_3)_4$. It should be noted that the nature of the central atom has little effect upon the methyl–sulphur resonance. δ for $(CH_3)_3SiS\underline{CH}_3$ is -115.9 cps and for $(CH_3)_3GeSCH_3$ it is -118.0 cps (40).

The PMR spectra of the compounds $(CH_3S)_nM(CH_3)_{4-n}$ (M = Ge, Sn) are discussed by van der Berghe and coworkers (81). The chemical shift data as well as the coupling constants $J(\underline{Sn}-C-\underline{H})$ and $J(\underline{Sn}-S-C-\underline{H})$ reveal a certain degree of metal–sulphur π-bonding increasing with increasing value of n.

The proton chemical shifts and $^{13}C-^1H$ coupling constants of compounds $Me_{4-n}X(OMe)_n$ and $Me_{4-n}X(SMe)_n$ where $n = 1$–4 and X = C, Si, Ge have been measured by Cumper and coworkers (26). With the exception of the $Me_{4-n}Si(OMe)_n$ compounds, an increase in the number of methoxy- or methylthio-groups progressively decreases the shielding of the X—Me protons as would be expected on increasing the number of electronegative oxygen or sulphur atoms attached to the central atom X. An increase in the number of methoxy or methylthio-groups in the molecules results in a de-shielding of the protons of these groups.

Proton magnetic resonance shielding for the series of compounds $(CH_3)_3MSM(CH_3)_3$ and $(CH_3)MSCH_3$ (M = C, Si, Ge, Sn and Pb) are recorded and discussed by Abel and Brady (6). The τ values for the M–methyl protons in both series of sulphur compounds follow quantitatively the order Si > Sn > Ge > Pb > C previously established for the tetramethyl derivatives of the Group IV B elements (8). The chemical shifts of the S–methyl protons in the $(CH_3)_3MSCH_3$ series vary over a much smaller range and except for the slight anomaly in the case of carbon, the τ values decrease steadily down the group. Owing to the changes taking place in the electronegativity, possible π-bonding, bond strengths, diamagnetic anisotropy etc., it would be difficult to ascribe such a steady decrease to any one reason (2, 6).

Proton nuclear magnetic resonance of $(H_3Ge)_2S$ and H_3GeSCH_3 (93) and long-range H—H spin–spin coupling in methyl germyl sulphide, H_3GeSCH_3, (91) have been reported by Wang and van Dyke. The spectrum of H_3GeSCH_3 consisted of two resonances at $\tau 7.9(\pm 0.01)$ $(CH_3$ protons) and 5.52 ± 0.01 $(GeH_3$ protons) each of which was $1:3:3:1$ quartet. The measured coupling constant $J(H-H')$ is 0.60 ± 0.05 Hz; no $^{73}Ge-H$ coupling was observed.

The basicity of a number of series of organometallic bases has also been studied by their hydrogen bonding with chloroform and the shift of the chloroform proton resonance (2, 5). This nuclear resonance shift method for the measurements of the base strengths of the isostructural organometallic series including $(Me_3Ge)_2S$ and Me_3GeSEt largely confirms the information obtained by infrared spectra (4).

7-2-3-4   Dipole Moment Measurements. The electric dipole moments of compounds of the series $Me_{4-n}M(SMe)_n$ where $n = 1$–4 and $M = Si$, Ge, Sn have been measured by Cumper and coworkers (29). The Me_3GeS— group moment (1.67 D) is deduced from the dipole moment of bis-(trimethylgermanium) sulphide $(Me_3Ge)_2S$ (28).

7-2-3-5   Mass Spectroscopy. Mass spectra of H_3GeSCH_3 are described by Wang and van Dyke (93).

7-2-3-6   Other Physical-Chemical Studies. Measurements of the magnetic susceptibility of various germanium derivatives comprising $(C_6H_5S)_4Ge$ have been made by Pascal and coworkers (55).

The molecular susceptibility K, the magnetic contribution ΣKA of the linked atoms to germanium and the value of the magnetic coefficient KGe are given.

7-3   MISCELLANEOUS ORGANOGERMANIUM COMPOUNDS WITH Ge–S BONDS

The reaction of $GeCl_4$ with toluene-3,4-dithiol in appropriate solvents leads to a racemic mixture of the enantiomers of the compound:

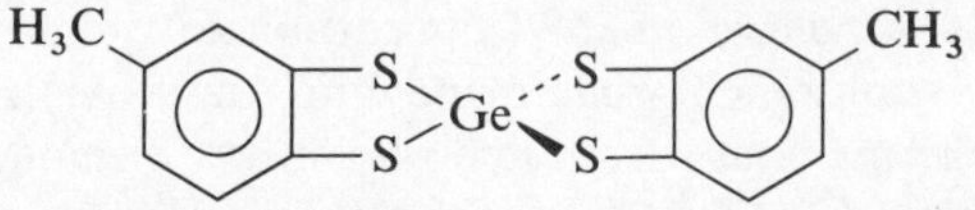

Partial resolution of the enantiomers was accomplished on a column of activated, optically active α-quartz. The IR, UV and NMR spectra of this compound are discussed (34, 35).

1,3-Bis-(chloromethyl)tetramethyldigermoxane can be converted with potassium hydrogen sulphide into 2,2,5,5-tetramethyl-2,5-digerma-1,4-dithiane (95):

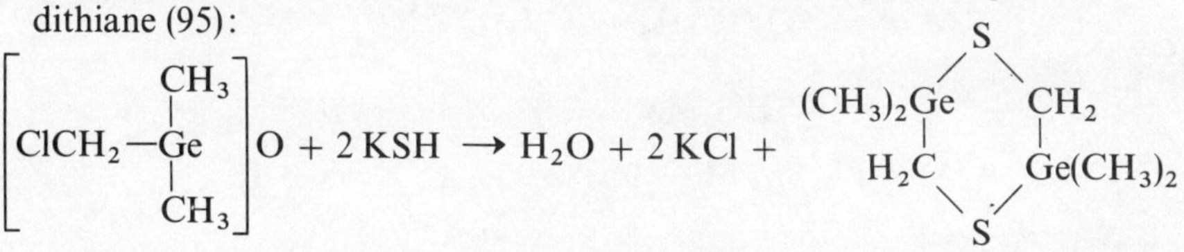

In the action of CS_2 on the trialkylgermylamines, the formation of trialkylgermanium dithiocarbamates is noted (57, 66):

$$Et_3GeNMe_2 + CS_2 \rightarrow Et_3Ge-S-\underset{\underset{S}{\|}}{C}-NMe_2$$

In the addition of PhNCS to $(Et_3Ge)_2NH$ as well as in the action of MeNCS on $(Et_3Ge)_2NMe$ the formation of addition derivatives with Ge—S bonds which are not very stable is noted (57):

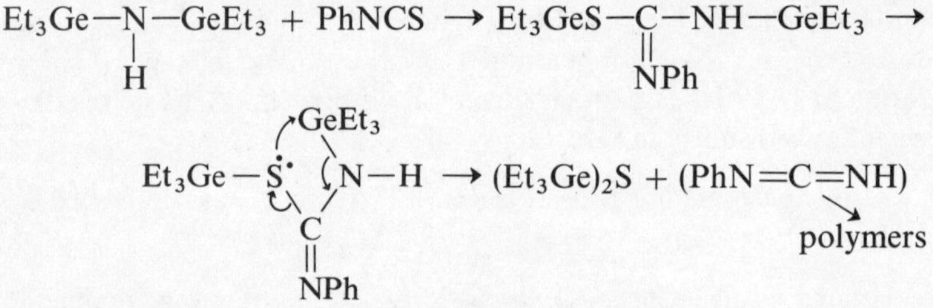

$$Et_3Ge\underset{\underset{H}{|}}{-N}-GeEt_3 + PhNCS \rightarrow Et_3GeS\underset{\underset{NPh}{||}}{-C}-NH-GeEt_3 \rightarrow$$

$$Et_3Ge-\overset{\overset{\displaystyle GeEt_3}{\diagup}}{\underset{\underset{NPh}{||}}{\underset{C}{S}}}\diagdown\underset{\diagup}{N}-H \rightarrow (Et_3Ge)_2S + (PhN{=}C{=}NH)$$

$$\rightarrow \text{polymers}$$

$$(Et_3Ge)_2NMe + MeNCS \rightarrow Et_3Ge-S\underset{\underset{NMe}{||}}{-C}\overline{\qquad}\underset{\underset{Me}{|}}{N}-GeEt_3$$

$$\downarrow$$

$$(Et_3Ge)_2S + (MeN{=}C{=}NMe)_n$$

The insertion of CS_2 into the Ge—P linkage of the trialkylgermyl-phosphines has been reported (64):

$$Et_3GePEt_2 + CS_2 \rightarrow Et_3GeS\underset{\underset{S}{||}}{-C}-PEt_2$$

The reactions of potassium ethyl thiocarbonate and potassium ethyl-xanthate with triphenylchlorogermane and diphenyldichlorogermane lead to organogermanium-substituted derivatives of mono- and dithio-carbonates $(C_6H_5)_3Ge-S-C(O)-OC_2H_5$, $(C_6H_5)_3Ge-S-C(S)OC_2-H_5$ and $(C_6H_5)_2Ge[S-C(S)OC_2H_5]_2$.

The IR spectra of the new compounds are discussed (70a). Studies of displacement reactions between mono- or dithiocarbonic acid derivatives of IV B elements are described by Schmidt and Jaggard (69a):

$$(C_6H_5)_3M-S-C(O)OC_2H_5 + (C_6H_5)_3M'X$$

$$\uparrow\downarrow$$

$$(C_6H_5)_3MX + C_6H_5M'-S-C(O)OC_2H_5$$

$$(M = Pb, Sn; \ M' = Pb, Sn, Ge, Si; \ X = Cl, Br)$$

$$(C_6H_5)_3M-S-C(S)OC_2H_5 + (C_6H_5)_3M'X$$

$$\uparrow\downarrow$$

$$(C_6H_5)_3MX + (C_6H_5)_3M'-S-C(S)OC_2H_5$$

$$(M = Pb, Sn, Ge; \ M' = Pb, Sn, Ge, Si; \ X = Cl, Br)$$

The reactions between organic thio-acids and germanoic acid in aqueous solution has been investigated by Clark (24).

Germanic acid reacts with thioglycollic acid, 1-mercaptopropionic acid and thiomalic acid to form 1:2 complexes:

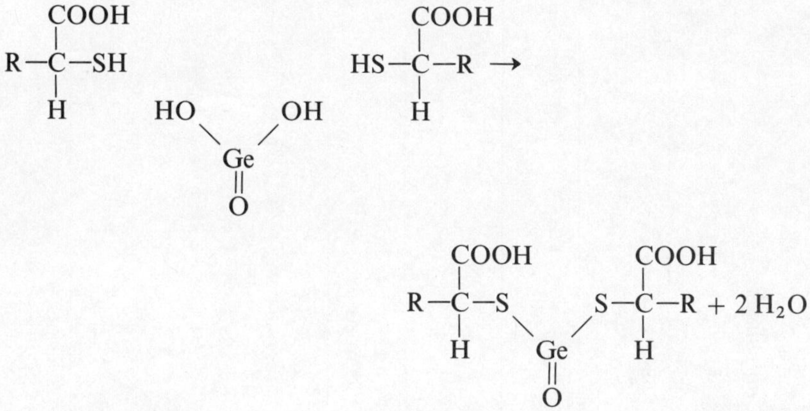

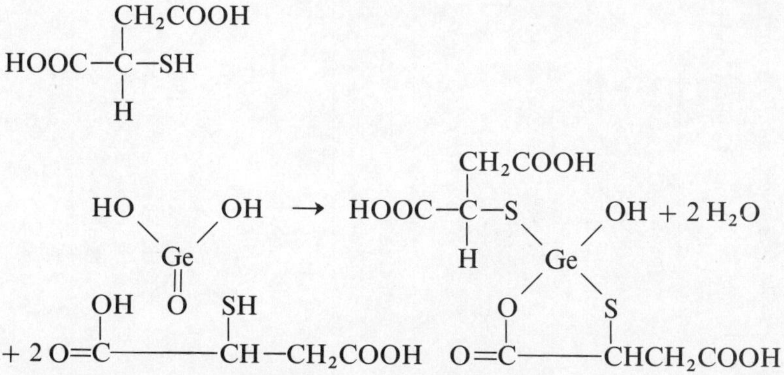

Dimethylgermanium dithiocyanate, $(CH_3)_2Ge(SCN)_2$, was prepared in aqueous solution by the action of thiocyanate ions on dimethyldichlorogermane (59). Triphenylgermanium thiocyanate, Ph_3GeSCN, is obtained from the reaction of triphenylbromogermane with lead thiocyanate (39).

7-4 ORGANOGERMANIUM–SULPHUR COMPOUNDS

Table 7-1(a–k) showing some physical properties of organogermanium-sulphur compounds appears on the following pages.

Table 7-1

Organogermanium–Sulphur Compounds

(a) R_3GeSH

COMPOUND	EMPIRICAL FORMULA	M.P. (°C)	B.P. (°C/mm)	n_D^{20}	d_4^{20}	REFERENCES
$(C_2H_5)_3GeSH$	$C_6H_{16}SGe$	—	86/35	1.4852	1.1270	84, 85
$(C_6H_5)_3GeSH$	$C_{18}H_{16}SGe$	110–14	—	—	—	39

(b) $(R_3Ge)_2S$

COMPOUND	EMPIRICAL FORMULA	M.P. (°C)	B.P. (°C/mm)	n_D^{20}	d_4^{20}	REFERENCES
$[(CH_2Cl)(CH_3)_2Ge]_2S$	$C_6H_{16}Cl_2SGe_2$	—	139–41/12	—	—	105
$[(CH_3)_3Ge]_2S$	$C_6H_{18}SGe_2$	−22	68/12 71/7.5	1.4980 (21°)	1.278	4, 27, 62
$[(C_2H_5)_3Ge]_2S$	$C_{12}H_{30}SGe_2$	—	95–98/1.5	1.5111	1.186	23, 65, 83, 85
$[(C_4H_9)_3Ge]_2S$	$C_{24}H_{54}SGe_2$	—	190–91/0.3	1.4940	1.0448	65
$[(C_6H_5)_3Ge]_2S$	$C_{36}H_{30}SGe_2$	138	—	—	—	23, 38, 77
$[(p\text{-}CH_3C_6H_4)_3Ge]_2S$	$C_{42}H_{42}SGe_2$	156–57	—	—	—	23
$[(C_6H_5CH_2)_3Ge]_2S$	$C_{42}H_{42}SGe_2$	124	—	—	—	23
$[(C_6H_5C_6H_4)_3Ge]_2S$	$C_{72}H_{54}SGe_2$	238	—	—	—	23

(c) $(R_3Ge)_2S_2$

COMPOUND	EMPIRICAL FORMULA	M.P. (°C)	B.P. (°C/mm)	n_D^{20}	d_4^{20}	REFERENCES
$[(C_6H_5)_3Ge]_2S_2$	$C_{36}H_{30}S_2Ge_2$	171–72	—	—	—	38, 39
$[(C_6H_{11})_3Ge]_2S_2$	$C_{36}H_{66}S_2Ge_2$	87–88	—	—	—	23

(d) $(R_2GeS)_n$

COMPOUND	EMPIRICAL FORMULA	M.P. (°C)	B.P. (°C/mm)	n_D^{20}	d_4^{20}	REFERENCES
$[(CH_3)_2GeS]_3$	$C_6H_{18}S_3Ge_3$	55.5	302 110/1	—	—	58, 62, 69, 106
$[(i\text{-}C_3H_7)_2GeS]_2$	$C_{12}H_{28}S_2Ge_2$	—	117–21/1 312	1.5510	1.3270	13
$[(C_4H_9)_2GeS]_2$	$C_{16}H_{36}S_2Ge_2$	—	176–80/0.5	—	—	43
$[(C_6H_5)_2GeS]_2$	$C_{24}H_{20}S_2Ge_2$	203	—	—	—	39
$[(n\text{-}C_4H_9)_2GeS]_3$	$C_{24}H_{54}S_3Ge_2$	—	222–25/1	—	—	70
$[(C_6H_5)_2GeS]_3$	$C_{36}H_{30}S_3Ge_3$	170	—	—	—	39

(e) (RGeS)$_2$S

COMPOUND	EMPIRICAL FORMULA	M.P. (°C)	B.P. (°C/mm)	n_D^{20}	d_4^{20}	REFERENCES
(CH$_3$GeS$_{1.5}$)$_4$	C$_4$H$_{12}$S$_6$Ge$_4$	345–46	—	—	—	48
[(C$_6$H$_5$)GeS]$_2$S	C$_{12}$H$_{10}$S$_3$Ge$_2$	—	—	—	—	19, 60
(p-CH$_3$C$_6$H$_4$GeS)$_2$S	C$_{14}$H$_{14}$S$_3$Ge$_2$	—	—	—	—	19
(p-(CH$_3$)$_2$NC$_6$H$_4$GeS)$_2$S	C$_{16}$H$_{20}$N$_2$S$_3$Ge$_2$	—	—	—	—	19
(α-C$_{10}$H$_7$GeS)$_2$S	C$_{20}$H$_{14}$S$_3$Ge$_2$	—	—	—	—	19
[p-(C$_2$H$_5$)$_2$NC$_6$H$_4$GeS]$_2$S	C$_{20}$H$_{28}$N$_2$S$_3$Ge$_2$	—	—	—	—	19

(f) R$_3$GeSMR$_3$

COMPOUND	EMPIRICAL FORMULA	M.P. (°C)	B.P. (°C/mm)	n_D^{20}	d_4^{20}	REFERENCES
(CH$_3$)$_3$GeSSi(CH$_3$)$_3$	C$_6$H$_{18}$SiSGe	−27	63/10	—	—	62
(CH$_3$)$_3$GeSSn(CH$_3$)$_3$	C$_6$H$_{18}$SnSGe	−8	89–90/12	—	—	62
(C$_2$H$_5$)$_3$GeSSn(C$_2$H$_5$)$_3$	C$_{12}$H$_{30}$SnSGe	—	100–3/1	1.5279	1.2990	84, 86
(C$_6$H$_5$)$_3$GeSSn(C$_6$H$_5$)$_3$	C$_{36}$H$_{30}$SnSGe	136	—	—	—	73, 74, 75, 77
(C$_6$H$_5$)$_3$GeSPb(C$_6$H$_5$)$_3$	C$_{36}$H$_{30}$PbSGe	128–29	—	—	—	77

(g) Ge(SR')$_4$

COMPOUND	EMPIRICAL FORMULA	M.P. (°C)	B.P. (°C/mm)	n_D^{20}	d_4^{20}	REFERENCES
(CH$_3$S)$_4$Ge	C$_4$H$_{12}$S$_4$Ge	−3	139/4 85/0.4	1.6379 (25°)	1.4346	15, 17, 18, 27, 40
(C$_2$H$_5$S)$_4$Ge	C$_8$H$_{20}$S$_4$Ge	—	164.5–65/5 151/1	1.5886	1.2574 (25°)	1, 17, 18, 46
(n-C$_3$H$_7$S)$_4$Ge	C$_{12}$H$_{28}$S$_4$Ge	—	191.5/4.5 148–50/0.4	1.5612 (25°)	1.1662 (25°)	17, 18, 46
(i-C$_3$H$_7$S)$_4$Ge	C$_{12}$H$_{28}$S$_4$Ge	15	163/4 128/0.4	1.5535 (25°)	1.1478 (25°)	17, 18, 46
(n-C$_4$H$_9$S)$_4$Ge	C$_{16}$H$_{36}$S$_4$Ge	—	222.5/4.5 200/0.5	1.5450	1.1072 (25°)	1, 17, 18, 46
(i-C$_4$H$_9$S)$_4$Ge	C$_{16}$H$_{36}$S$_4$Ge	—	199.5/4.5 162–65/0.5	1.5381 (25°)	1.0984 (25°)	17, 18, 46
(sec-C$_4$H$_9$S)$_4$Ge	C$_{16}$H$_{36}$S$_4$Ge	—	200.5/4	1.5497 (25°)	1.1119 (25°)	17, 18
(tert-C$_4$H$_9$S)$_4$Ge	C$_{16}$H$_{36}$S$_4$Ge	172–73	141/0.8	—	—	16, 17, 18, 46
(n-C$_5$H$_{11}$S)$_4$Ge	C$_{20}$H$_{44}$S$_4$Ge	—	240.5/3.5	1.5336 (25°)	1.0697 (25°)	18
(p-BrC$_6$H$_4$S)$_4$Ge	C$_{24}$H$_{16}$Br$_4$S$_4$Ge	196–96.5	—	—	—	17, 18
(C$_6$H$_5$S)$_4$Ge	C$_{24}$H$_{20}$S$_4$Ge	101.5	255/0.3	—	—	17, 18, 46, 78
(C$_6$H$_{11}$S)$_4$Ge	C$_{24}$H$_{44}$S$_4$Ge	88 (monoclinic) 84 (tetragonal)	—	—	1.259 (15°) 1.270 (15°)	18
(p-CH$_3$C$_6$H$_4$S)$_4$Ge	C$_{28}$H$_{28}$S$_4$Ge	110–11	—	—	—	17, 18
(p-tert-C$_4$H$_9$—C$_6$H$_4$S)$_4$Ge	C$_{40}$H$_{52}$S$_4$Ge	155–56	—	—	—	18
(n-C$_{12}$H$_{25}$S)$_4$Ge	C$_{48}$H$_{100}$S$_4$Ge		decomp.	—	—	46
(n-C$_{16}$H$_{33}$S)$_4$Ge	C$_{64}$H$_{132}$S$_4$Ge	50–51	—	—	—	18

(h) RGe(SR')₃

COMPOUND	EMPIRICAL FORMULA	M.P. (°C)	B.P. (°C/mm)	n_D^{20}	d_4^{20}	REFERENCES
$CH_3Ge(SCH_3)_3$	$C_4H_{12}S_3Ge$	—	130/15	1.5908	1.39	81
$(C_6H_5)Ge(SCH_3)_3$	$C_9H_{14}S_3Ge$	—	110/0.2	—	—	40

(i) R₂Ge(SR')₂

COMPOUND	EMPIRICAL FORMULA	M.P. (°C)	B.P. (°C/mm)	n_D^{20}	d_4^{20}	REFERENCES
$(CH_3)_2Ge(SCH_3)_2$	$C_4H_{12}S_2Ge$	—	42/0.1 85/15	1.5470	1.30	49, 81
$(CH_3)_2Ge(SC_4H_9)_2$	$C_{10}H_{24}S_2Ge$	—	93/0.05	1.5083 (25°)	—	42
$(C_4H_9)_2Ge(SC_2H_5)_2$	$C_{12}H_{28}S_2Ge$	—	158/10	1.5132	1.0782	47, 65
$(C_6H_5)_2Ge(SCH_3)_2$	$C_{14}H_{16}S_2Ge$	—	132–35/0.2	—	—	40
$(C_4H_9)_2Ge(SC_3H_7\text{-}n)_2$	$C_{14}H_{32}S_2Ge$	—	115/0.2	—	—	47
$(C_4H_9)_2Ge(SC_3H_7\text{-}i)_2$	$C_{14}H_{32}S_2Ge$	—	114–15/0.5	—	—	47
$(C_6H_5)_2Ge(SCH_2CONH_2)_2$	$C_{16}H_{18}N_2O_2S_2Ge$	140	—	—	—	30
$(C_2H_5)_2Ge(SC_6H_5)_2$	$C_{16}H_{20}S_2Ge$	—	153–54/0.15	1.6242	1.2470	65
$(C_4H_9)_2Ge(SC_4H_9\text{-}n)_2$	$C_{16}H_{36}S_2Ge$	—	140–43/0.5	—	—	47
$(C_4H_9)_2Ge(SC_4H_9\text{-}i)_2$	$C_{16}H_{36}S_2Ge$	—	141/0.7	—	—	47
$(C_4H_9)_2Ge(SC_4H_9\text{-}t)_2$	$C_{16}H_{36}S_2Ge$	—	145/1.5	—	—	47
$(C_6H_5)_2Ge(SCH_2CH{=}CH_2)_2$	$C_{18}H_{20}S_2Ge$	—	—	1.6229 (28.5°)	—	30
$(C_4H_9)_2Ge(SC_6H_5)_2$	$C_{20}H_{28}S_2Ge$	—	204–6/1.2	—	—	47

COMPOUND	EMPIRICAL FORMULA	M.P. (°C)	B.P. (°C/mm)	n_D^{20}	d_4^{20}	REFERENCES
$(C_4H_9)_2Ge(SCH_2C_6H_5)_2$	$C_{22}H_{32}S_2Ge$	—	233–35/1.7	—	—	47
$(C_6H_5)_2Ge(SC_6H_4—Cl\text{-}p)_2$	$C_{24}H_{18}Cl_2S_2Ge$	—	—	1.6751 (22°)	—	30
$(C_6H_5)_2Ge(SC_6H_4—NO_2\text{-}p)_2$	$C_{24}H_{18}N_2O_4S_2Ge$	114–15	—	—	—	30
$(C_6H_5)_2Ge(SC_6H_5)_2$	$C_{24}H_{20}S_2Ge$	—	—	1.6776(22°)	—	30
$(C_6H_5)_2Ge(SC_6H_4—NH_2\text{-}p)_2$	$C_{24}H_{22}N_2S_2Ge$	115	—	—	—	30
$(C_6H_5)_2Ge(SCOC_6H_5)_2$	$C_{26}H_{20}O_2S_2Ge$	150–51	—	—	—	30
$(C_6H_5)_2Ge(SCH_2C_6H_5)_2$	$C_{26}H_{24}S_2Ge$	50–51	—	—	—	30
$(C_6H_5)_2Ge(S\text{-}2\text{-}C_{10}H_7)_2$	$C_{32}H_{24}S_2Ge$	90–91	—	—	—	30
$(C_4H_9)_2Ge(SC_{12}H_{25}\text{-}n)_2$	$C_{32}H_{64}S_2Ge$	—	238/0.6	—	—	47
$(C_6H_5)_2Ge[S(CH_2)_{10}CH_3]_2$	$C_{34}H_{56}S_2Ge$	—	—	—	—	30
$(C_6H_5)_2Ge[SGe(C_6H_5)_3]_2$	$C_{48}H_{40}S_2Ge_3$	162	—	1.5343 (28.5°)	—	38

(j) R_3GeSR'

COMPOUND	EMPIRICAL FORMULA	M.P. (°C)	B.P. (°C/mm)	n_D^{20}	d_4^{20}	REFERENCES
$(CH_3)_3GeSCH_3$	$C_4H_{12}SGe$	—	135/734; 129–31	1.4792	1.18	40, 81
$(CH_3)_3GeSC_2H_5$	$C_5H_{14}SGe$	—	148	1.4779	1.10	3
$(C_2H_5)_2(H)GeSC_2H_5$	$C_6H_{16}SGe$	—	83/25	1.4905	1.1214	65
$(C_2H_5)_3GeSCH_3$	$C_7H_{18}SGe$	—	84/15	1.4920	1.1133	65
$(CH_3)_3GeSC_4H_9\text{-}tert$	$C_7H_{18}SGe$	—	25–26/0.1	1.4729 (22°)	1.08	3
$(CH_3)_3GeSC_4H_9\text{-}n$	$C_7H_{18}SGe$	—	62/8	1.4736 (22°)	1.08	3
$(C_2H_5)_3GeSC_2H_5$	$C_8H_{20}SGe$	—	93/10	1.4900	1.0868	65
$(CH_3)_3GeSC_6F_5$	$C_9H_9F_5SGe$	—	52/0.5	1.4920 (25°)	—	54a
$(CH_3)_3GeSC_6H_5$	$C_9H_{14}SGe$	—	37/0.001	1.5564	1.20	3
$(C_2H_5)_3GeSCH(CH_3)_2$	$C_9H_{22}SGe$	—	116/17	1.4868	1.0590	65
$(C_2H_5)_3GeSC_4H_9\text{-}n$	$C_{10}H_{24}SGe$	—	120/13	1.4868	1.0520	41, 65

Table 7-1 (j)—continued

COMPOUND	EMPIRICAL FORMULA	M.P. (°C)	B.P. (°C/mm)	n_D^{20}	d_4^{20}	REFERENCES
$(C_2H_5)_3GeSCH_2$—(furyl ring: CH=CH, CH, C–O)	$C_{11}H_{20}OSGe$	—	276/760 130–32/1	1.522 —	1.177 —	12
$(C_2H_5)_3GeSC_6H_5$	$C_{12}H_{20}SGe$	—	112–13/1 286/760	1.553	1.153	12
$(C_2H_5)_3GeSC_6H_4NH_2\text{-}o$	$C_{12}H_{21}NSGe$	—	163–64/1 326/760	1.583 —	1.197 —	12
$(C_2H_5)_3GeSC_6H_{13}\text{-}n$	$C_{12}H_{28}SGe$	—	108–9/1 277/760	1.488 —	1.029 —	12
$(C_2H_5)_3GeSCH_2C_6H_5$	$C_{13}H_{22}SGe$	—	130–31/1 305/760	1.549 —	1.139 —	12
$(C_2H_5)_3GeSC_6H_4CH_3\text{-}o$	$C_{13}H_{22}SGe$	—	123–24/1 298/760	1.553 —	1.141 —	12
$(C_2H_5)_3GeSC_6H_4CH_3\text{-}m$	$C_{13}H_{22}SGe$	—	143–45/1 300/760	1.550 —	1.131 —	12
$(C_2H_5)_3GeSC_7H_{15}\text{-}n$	$C_{13}H_{30}SGe$	—	117–18/2 288/760	1.489 —	1.019 —	12
$(C_4H_9)_3GeSC_2H_5$	$C_{14}H_{32}SGe$	—	164/17	1.4830	1.0082	47, 65
$(C_2H_5)_3GeSC_8H_{17}\text{-}n$	$C_{14}H_{32}SGe$	—	161/7	1.4839	1.0050	65
$(C_2H_5)_3GeSCH_2COOGe(C_2H_5)_3$	$C_{14}H_{32}O_2SGe_2$	—	326 158.5–59.5/4	1.4993	1.2224	9
$(C_4H_9)_3GeSC_3H_7\text{-}n$	$C_{15}H_{34}SGe$	—	115–16/0.5	—	—	47
$(C_4H_9)_3GeSC_3H_7\text{-}i$	$C_{15}H_{34}SGe$	—	120/1	—	—	47
$(C_2H_5)_3GeS\text{-}\beta\text{-}C_{10}H_7$	$C_{16}H_{22}SGe$	—	195–97/1 367/760	1.613 —	1.184 —	12

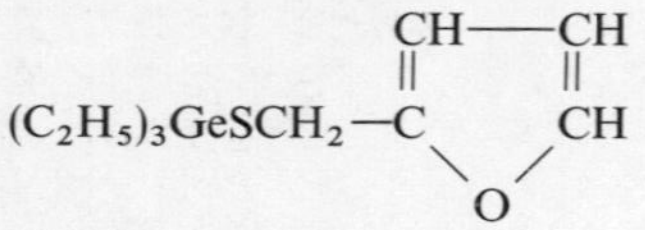

Compound	Molecular formula	m.p.	b.p.	n_D	d	References
$(C_4H_9)_3GeSC_4H_9\text{-}n$	$C_{16}H_{36}SGe$	—	169/9	1.4824	0.9855	47, 65
$(C_4H_9)_3GeSC_4H_9\text{-}i$	$C_{16}H_{36}SGe$	—	136–38/2.0	—	—	47
$(C_4H_9)_3GeSC(CH_3)_3$	$C_{16}H_{36}SGe$	—	169/18	1.4835 (25°)	0.9828 (25°)	47, 65
$(C_4H_9)_3GeSC_6H_5$	$C_{18}H_{32}SGe$	—	134/0.12	1.5280	1.0555	47, 65
$(C_2H_5)_3GeSC_{12}H_{25}\text{-}n$	$C_{18}H_{40}SGe$	—	184–86/1; 357/760	1.481	0.975	12
$(C_6H_5)_3GeSCH_3$	$C_{19}H_{18}SGe$	87–88	—	—	—	38, 40
$(C_4H_9)_3GeSCH_2C_6H_5$	$C_{19}H_{34}SGe$	—	171–74/1.2	—	—	47
$(C_6H_5)_3GeSCOCH_3$	$C_{20}H_{18}OSGe$	105	—	—	—	30
$(C_6H_5)_3GeSCH_2COOH$	$C_{20}H_{18}O_2SGe$	149–50	—	—	—	30
$(C_6H_5)_3GeSCH_2CONH_2$	$C_{20}H_{19}NOSGe$	151	—	—	—	30
$(C_6H_5)_3GeSCH_2SCH_3$	$C_{20}H_{20}S_2Ge$	63	—	—	—	38
$(C_6H_5)_3GeSCH_2CH_2OH$	$C_{20}H_{20}OSGe$	oil	—	—	—	30
$(C_4H_9)_3GeSC_8H_{17}\text{-}n$	$C_{20}H_{44}SGe$	—	143/0.2	1.4810	0.9648	65
$(C_6H_5)_3GeSCH_2COOCH_3$	$C_{21}H_{20}O_2SGe$	44–45	—	—	—	30
$(C_6H_5)_3GeSCH_2CH{=}CH_2$	$C_{21}H_{20}SGe$	62	—	—	—	30
$(C_6H_5)_3GeSC_4H_9\text{-}n$	$C_{22}H_{24}SGe$	—	147–50/0.05	1.6135	—	38
$(C_6H_5)_3GeSC_6F_5$	$C_{24}H_{15}F_5SGe$	96–97	—	—	—	30
$(C_6H_5)_3GeC_6H_4Cl\text{-}p$	$C_{24}H_{19}ClSGe$	101–2	—	—	—	30
$(C_6H_5)_3GeSC_6H_5$	$C_{24}H_{20}SGe$	96	—	—	—	30, 40
$(C_4H_9)_3GeSC_{12}H_{25}\text{-}n$	$C_{24}H_{52}SGe$	—	192–93/0.4	—	—	47
$(C_6H_5)_3GeSC(=O)C_6H_4NO_2\text{-}p$	$C_{25}H_{19}NO_3SGe$	151	—	—	—	38
$(C_6H_5)_3GeSC(=O)C_6H_5$	$C_{25}H_{20}OSGe$	145.5; 142–43	—	—	—	38, 39, 77
$(C_6H_5)_3GeSCH_2C_6H_5$	$C_{25}H_{22}SGe$	98.5	—	—	—	38, 39
$(C_6H_5)_3GeS\text{-}2\text{-}C_{10}H_7$	$C_{28}H_{22}SGe$	88–89	—	—	—	30
$(C_6H_5)_3GeS(CH_2)_{10}CH_3$	$C_{29}H_{38}SGe$	56	—	—	—	30
$(C_6H_5)_3GeS\text{-}17\text{-}\beta$ (mercaptotestosterone)	$C_{37}H_{42}OSGe$	188	—	—	—	30

(k) *Miscellaneous compounds with Ge—S bonds in non-cyclic systems*

COMPOUND	EMPIRICAL FORMULA	M.P. (°C)	B.P. (°C/mm)	n_D^{20}	d_4^{20}	REFERENCES
$(CH_3)_2Ge(SCN)_2$	$C_4H_6N_2S_2Ge$	45.5–47	266–68	—	—	59
$(C_2H_5)_3GeS-\overset{\parallel}{C}-N(CH_3)_2$ ($=S$)	$C_9H_{21}NS_2Ge$	—	179–80/16	1.5700	1.2156	57, 66
$(C_2H_5)_3GeS-\overset{\parallel}{C}-N(C_2H_5)_2$ ($=S$)	$C_{11}H_{25}NS_2Ge$	—	122/0.4	1.5572	1.1618	66
$(C_2H_5)_3GeS-\overset{\parallel}{C}-P(C_2H_5)_2$ ($=S$)	$C_{11}H_{25}PS_2Ge$	—	98/0.1	1.5493	1.1694	64
$(C_2H_5)_3Ge-S(CH_2)_2S-Ge(C_2H_5)_3$	$C_{14}H_{34}S_2Ge_2$	—	165/17	1.5149	1.1899	65
$(C_6H_5)_3GeSNa$	$C_{18}H_{15}NaSGe$	185–95	—	—	—	38, 39
$(C_6H_5)_2Ge[SC(S)OC_2H_5]_2$	$C_{18}H_{20}O_2S_4Ge$	90	—	—	—	70a
$(C_6H_5)_3GeSCN$	$C_{19}H_{15}NSGe$	105	—	—	—	39
$(C_6H_5)_3GeSC(S)OC_2H_5$	$C_{21}H_{20}OS_2Ge$	71	—	—	—	70a
$(C_6H_5)_3GeSCH_2CH_2SGe(C_6H_5)_3$	$C_{38}H_{34}S_2Ge_2$	152	—	—	—	30
$(C_6H_5)_3GeSCC_6H_4CSGe(C_6H_5)_3$ ($=O \quad =O$)	$C_{44}H_{34}O_2S_2Ge$	255–57	—	—	—	30
$[(C_6H_5)_3GeSCH_2C_6H_4]_2O$	$C_{50}H_{42}OS_2Ge$	—	syrup	—	—	30

7-5   ORGANOGERMANIUM–SELENIUM AND -TELLURIUM COMPOUNDS

Backer and Hurenkamp (16) prepared the first selenium derivatives of germanium by the following two reactions:

$$GeCl_4 + 4\,RSeNa \rightarrow Ge(SeR)_4 + 4\,NaCl$$

$$GeCl_4 + 4\,BrMgSeR \rightarrow Ge(SeR)_4 + 4\,MgBrCl$$

The first method affords better yields. The following compounds have been isolated: $Ge(SeR)_4$ with R = t-Bu, C_6H_5—, —$C_6H_4CH_3$-p, —C_6H_4CH-$(CH_3)_2$-p, —C_6H_4—t-Bu-p, —C_6H_4Cl-p, α-$C_{10}H_7$, —C_6H_{11}. Their crystallographic properties are analogous to those of the corresponding sulphur compounds (16).

Dimethyldichlorogermane reacts with sodium selenide in benzene with formation of cyclo-tris-dimethylgermanium selenide $[(CH_3)_2GeSe]_3$ (61, 67, 68):

$$3\,(CH_3)_2GeCl_2 + 3\,Na_2Se \rightarrow 6\,NaCl +$$

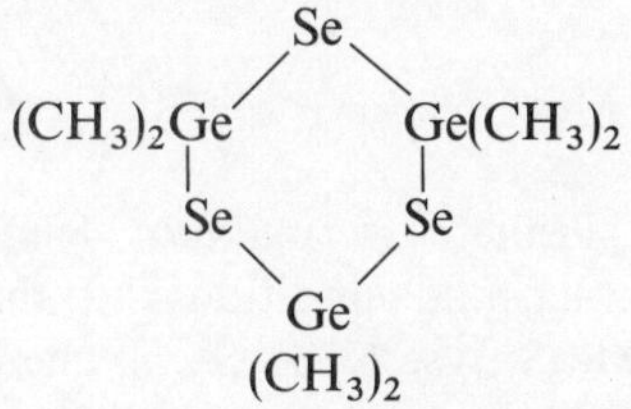

Besides the cyclic trimer, high polymers of dimethylgermanium selenide of variable ring size $[(CH_3)_2GeSe]_n$ ($n > 3$) and a very few open chain Cl-terminated compounds are formed:

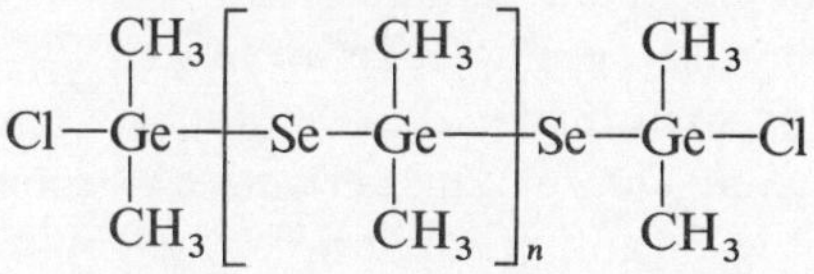

After distillation of the trimer, the mixture of the two other derivatives remains in the form of oil to wax.

$[(CH_3)_2GeSe]_3$ is scarcely attacked by cold water; on heating, hydrolysis takes place with formation of H_2Se.

Lithium trimethylgermanium selenide is formed by the reaction of polymeric dimethylgermanium selenide with methyllithium in quantitative yields (61):

$$[(CH_3)_2GeSe]_3 + 3\,CH_3Li \rightarrow 3\,(CH_3)_3GeSeLi$$

$$[(CH_3)_2GeSe]_n + n\,CH_3Li \rightarrow n\,(CH_3)_3GeSeLi$$

$$Cl-\underset{\underset{CH_3}{|}}{\overset{\overset{CH_3}{|}}{Ge}}\left[-Se-\underset{\underset{CH_3}{|}}{\overset{\overset{CH_3}{|}}{Ge}}\right]_n-Se-\underset{\underset{CH_3}{|}}{\overset{\overset{CH_3}{|}}{Ge}}-Cl + (n+3)CH_3Li$$

$$\downarrow$$

$$(n+1)(CH_3)_3GeSeLi + (CH_3)_4Ge + 2\,LiCl$$

Above 65°C $(CH_3)_3GeSeLi$ is decomposed thermally into lithium selenide and bis-(trimethylgermanium) selenide:

$$2\,(CH_3)_3GeSeLi \rightarrow (CH_3)_3GeSeGe(CH_3)_3 + Li_2Se$$

The latter compound may also be synthesized by the reaction of trimethylchlorogermane with lithium trimethylgermanium selenide.

Trimethylchlorosilane and trimethylchlorostannane react in a similar way:

$$Me_3GeSeLi + Me_3XCl \rightarrow LiCl + Me_3GeSeXMe_3 \qquad (X = Si, Sn)$$

Trimethylgermanium trimethyltin selenide disproportionates even under mild conditions into the symmetrical compounds $(Me_3Ge)_2Se$ and $(Me_3Sn)_2Se$ (73a). Some chemical and physical properties including IR and HNMR spectra of the latter compounds have been reported (61).

Triphenylgermyllithium reacts with selenium and tellurium, forming $Ph_3GeSeLi$ and $Ph_3GeTeLi$ respectively. These lithium compounds react easily with triphenylbromogermane and triphenyltin and triphenyllead chloride to give the expected products: $(Ph_3Ge)_2Se$, $Ph_3GeSeSnPh_3$, $Ph_3GeSePbPh_3$, $(Ph_3Ge)_2Te$, $Ph_3GeTeSnPh_3$ and $Ph_3GeTePbPh_3$ (77).

$Ph_3SnSeGePh_3$ is also formed in the reaction between triphenylbromogermane with $Ph_3SnSeLi$ (74, 76). This derivative is decomposed by diluted acids with release of H_2Se and precipitation of selenium (76).

In this series the sulphur compounds are colourless, the selenium compounds have a yellow tinge and the tellurium compounds are yellow.

Unlike the Ge—S—Ge group, Ge—Se—Ge is very sensitive to hydrolysis. These compounds are soluble in anhydrous solvents such as benzene, dioxane, THF and chloroform.

Infrared spectra of organometal sulphur, selenium and tellurium compounds are described by Schumann and Schmidt (72, 73). ν_{AS} GePh$_3$ ν_S GePh$_3$ in the hexaphenyl derivatives $(Ph_3Ge)_2Se$, $(Ph_3Ge)_2Te$, $Ph_3GeSeSnPh_3$. $Ph_3GeSePbPh_3$, $Ph_3GeTeSnPh_3$, and $Ph_3GeTePbPh_3$ are given (72). Skeletal vibrations of the sulphur compounds occur between 250 and 400 cm^{-1} (Cs Br region); their assignments are given (73).

Corresponding bands are to be expected at still longer wavelength for the Se and Te compounds (in the region of $250\,cm^{-1}$).

Many investigations concerning the organogermanium derivatives of selenium and tellurium have also been carried out by Razuvaev, Vyazankin and coworkers of the University of Gorkii (USSR).

Triethylgermane and selenium at 140°C for four hours give Et_3GeSeH in 63% yield (84, 85):

$$R_3GeH + Se \rightarrow R_3GeSeH$$

This reaction has been generalized to various trialkylgermanes ($R = i\text{-}C_3H_7$, $-C_6H_{11}$)(86). This reaction may be accompanied by the following reactions leading to formation of symmetrical compounds (86):

$$2\,R_3GeSeH \rightarrow H_2Se + R_3GeSeGeR_3$$

$$R_3GeSeH + R_3GeH \rightarrow R_3GeSeGeR_3 + H_2$$

Et_3GeSeH was then converted into $(Et_3Ge)_2Se$ at 130°C (28 h) (yield 37%). Et_3GeSeH shows an absorption band ν_{Se-H} at $2303\,cm^{-1}$ (85), and its PMR spectra are given (31) (cf. organogermanium thiols). Triethylgermyl-hydroselenide reacts with triethylstannane with formation of triethyl-germyl-triethylstannyl selenide (85):

$$Et_3GeSeH + Et_3SnH \rightarrow H_2 + Et_3GeSeSnEt_3$$

Triethylgermylhydroselenide reacts with lithium in THF medium:

$$2\,Et_3GeSeH + 2\,Li \rightarrow H_2 + 2\,Et_3GeSeLi$$

$Et_3GeSeLi$ reacts with halogen derivatives (88):

$$Et_3GeSeLi + RCl(RBr) \rightarrow LiCl(LiBr) + Et_3GeSeR$$

$Et_3GeSeLi$ reacts with 1,2 dibromoethane with formation of ethylene, bis-(triethylgermyl)selenide, LiBr and selenium. $Et_3GeSeCH_2CH_2SeGeEt_3$ is not detected in the reaction products.

The synthesis of $Et_3GeSeBu$ has been recently described by Vyazankin and coworkers (90) in the reaction between butylselenol and bis-(triethyl-germyl)mercury or ethyl(triethylgermyl) mercury $EtHg-GeEt_3$ (90):

$$Hg(GeEt_3)_2 + BuSeH \rightarrow BuSeGeEt_3 + Et_3GeH + Hg$$

In Et_3GeSH or Et_3GeSeH, the hydrogen bonded to the sulphur or selenium can be quantitatively determined by the Chugaev–Tserevitinov method:

$$Et_3GeSeH + CH_3MgI \rightarrow Et_3GeSeMgI + CH_4$$

$Et_3GeSeMgI$ is identified in this reaction by Et_3GeBr which leads to $(Et_3Ge)_2Se$. Et_3GeSeH and bromine in C_6H_6 give Se and Et_3GeBr (yield 75–85%) (89).

Triethylgermylhydroselenide, Et_3GeSeH, reacts with diethylmercury much more readily than triethylgermane thiol (88):

$$2 Et_3GeSeH + Et_2Hg \rightarrow 2 C_2H_6 + HgSe + (Et_3Ge)_2Se$$

Under UV light Et_3GeSeH reacts with $PhCH{=}CH_2$ and $CH_2{=}CH{-}COOEt$ to form 1:1 adducts (82a):

$$Et_3GeSeH + CH_2{=}CHY \rightarrow Et_3GeSeCH_2CH_2Y$$

The compounds $(Et_3Ge)_2X$ in which $X = $ Se or Te may be synthesized by heating elementary Se and Te with triethylgermane and by reaction of diethylselenide with two molecular proportions of Et_3GeH (83, 85):

$$2 Et_3GeH + Se \rightarrow Et_3GeSeGeEt_3 + H_2 \qquad (18.2\% \text{ yield})$$

$$2 Et_3GeH + Et_2Se \rightarrow Et_3GeSeGeEt_3 + 2 C_2H_6 \qquad (45.5\% \text{ yield})$$

$$2 Et_3GeH + Te \rightarrow Et_3GeTeGeEt_3 + H_2 \qquad (60.4\% \text{ yield})$$

In the reaction between triethylgermane and tellurium, the unsymmetrical product Et_3GeTeH is not isolated (86).

Bis-(triethylgermyl) telluride is formed when triethylgermane is heated with diethyl telluride (7 h at 140°C). The ethyl groups are replaced successively (84, 86):

$$(C_2H_5)_2Te \xrightarrow{Et_3GeH} Et_3GeTeC_2H_5 \xrightarrow{Et_3GeH} (Et_3Ge)_2Te$$

The same derivative is formed in the action of Et_3GeH on $(Et_3Si)_2Te$ (56, 87a):

$$(Et_3Si)_2Te + 2 Et_3GeH \rightarrow (Et_3Ge)_2Te + 2 Et_3SiH$$

By the reaction between $Et_3GeTeC_2H_5$ and Et_3SnH, $Et_3GeTeSnEt_3$ is formed in 62% yield (84, 86). $(Et_3Ge)_2Te$ reacts with Et_3SnH at 170°C to give Et_3GeH and $(Et_3Sn)_2Te$ (19a). From these derivatives several exchange reactions are observed (56, 87a):

$$(Et_3Ge)_2Te + Se \rightarrow (Et_3Ge)_2Se + Te$$

$$(Et_3Ge)_2Te + S \rightarrow (Et_3Ge)_2S + Te$$

$$(Et_3Ge)_2Se + S \rightarrow (Et_3Ge)_2S + Se$$

$$(Et_3Ge)_2Te + 2 Et_3SnH \rightarrow (Et_3Sn)_2Te + 2 Et_3GeH$$

Selenium as well as sulphur inserts into the Ge—C bond of germacyclo-butanes (45):

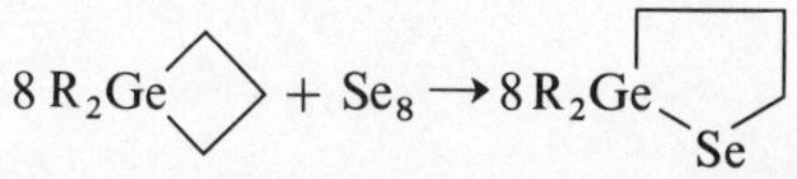

$$8\,R_2Ge\!\!\left\langle\rule{0pt}{10pt}\right\rangle + Se_8 \rightarrow 8\,R_2Ge\underset{Se}{\underline{\qquad}}$$

It will be useful to record the reactions that have already been mentioned with compounds of the type $[R_3M]_2M'$ (M = Si, Ge, Sn; M' = S, Se, Te) (cf. bis-(triorganogermanium) sulphide); the cleavage by bromine leads to the formation of Et_3MBr and free Se or Te (89). $(Et_3Ge)_2Se$ and dry HCl give Et_3GeCl and H_2Se (19a). The reactions of the same derivatives $[R_3M]_2M'$ with benzoyl peroxide, acetyl benzoyl peroxide lead to the corresponding esters with separation of selenium or tellurium. The reaction of dicyclohexyl percarbonate with $(Et_3Ge)_2Te$ is as follows:

$$(C_6H_{11}OCOO)_2 + (Et_3Ge)_2Te \;\rightarrow\; Te + 2\,CO_2 + 2\,Et_3GeOC_6H_{11}$$

owing to the decomposition of the initially formed $Et_3GeOCOOC_6H_{11}$ (87). The action of benzoyl peroxide on $Et_3GeTeEt$ is more complex and only benzoyloxytriethylgermane is isolated; no free tellurium appears (87). Digermylselenide $(H_3Ge)_2Se$ and digermyltelluride $(H_3Ge)_2Te$ have been prepared by exchange with the analogous silyl compounds. The selenide has also been made from digermylcarbodi-imide and hydrogen selenide. The compounds were characterized spectroscopically and by their reaction with hydrogen iodide. The vibrational spectra of the compounds indicate that, as expected, the heavy-atom skeletons are bent at the Group VI atoms (25a).

7-5-1 Organogermanium–selenium and –tellurium compounds

Table 7-2 showing some physical properties of organogermanium-selenium and -tellurium compounds appears on the following pages.

Table 7-2
Organogermanium–Selenium and –Tellurium Compounds

COMPOUND	EMPIRICAL FORMULA	M.P. (°C)	B.P. (°C/mm)	n_D^{20}	d_4^{20}	REFERENCES
$(C_2H_5)_3GeSeH$	$C_6H_{16}SeGe$	—	95–97/35	1.5058	1.3700	84, 85
$[(CH_3)_3Ge]_2Se$	$C_6H_{18}SeGe_2$	–12	50/1	—	—	61
$[(CH_3)_2GeSe]_3$	$C_6H_{18}Se_3Ge_3$	53	—	—	—	67, 68
$(i\text{-}C_3H_7)_3GeSeH$	$C_9H_{22}SeGe$	—	111–13/15	1.5129	—	86
$(C_2H_5)_3GeSeC_4H_9\text{-}n$	$C_{10}H_{24}SeGe$	—	105–6/6	1.5035	—	19a
$(C_2H_5)_3GeSeCH_2CH_2COOC_2H_5$	$C_{11}H_{24}O_2SeGe$	—	123–5/5	1.5003	—	82a
$[(C_2H_5)_3Ge]_2Se$	$C_{12}H_{30}SeGe_2$	—	129–30/3; 124–25/1.5	1.5287	1.328	83, 85
$(C_2H_5)_3GeSeSn(C_2H_5)_3$	$C_{12}H_{30}SnSeGe$	—	111/0.5	1.5470	—	84, 85
$(C_2H_5)_3GeSeCH_2C_6H_5$	$C_{13}H_{22}SeGe$	—	118–20/1	1.5652	—	88
$(C_2H_5)_3GeSeCH_2CH_2C_6H_5$	$C_{14}H_{24}SeGe$	—	153–5/5	1.5577	—	82a
$(CH_3)_3SiSeGe(CH_3)_3$	$C_6H_{18}SiSeGe$	–17, –19	79/12	—	—	61
$Ge(Se\text{-}tert\text{-}C_4H_9)_4$	$C_{16}H_{36}Se_4Ge$	192–93	—	—	—	16
$(cyclo\text{-}C_6H_{11})_3GeSeH$	$C_{18}H_{34}SeGe$	80–83	—	—	—	86
$Ge(SeC_6H_4Cl\text{-}p)_4$	$C_{24}H_{16}Se_4Cl_4Ge$	178–79	—	—	—	16
$Ge(SeC_6H_5)_4$	$C_{24}H_{20}Se_4Ge$	119–19.5	—	—	—	16
$Ge(SeC_6H_{11})_4$	$C_{24}H_{44}Se_4Ge$	79	—	—	—	16
$Ge(SeC_6H_4CH_3\text{-}p)_4$	$C_{28}H_{28}Se_4Ge$	105.5–6	—	—	—	16
$[(C_6H_5)_3Ge]_2Se$	$C_{36}H_{30}SeGe_2$	150	—	—	—	77
$(C_6H_5)_3GeSeSn(C_6H_5)_3$	$C_{36}H_{30}SnSeGe$	133; 144–45	—	—	—	74, 76, 77
$(C_6H_5)_3GeSePb(C_6H_5)_3$	$C_{36}H_{30}PbSeGe$	119	—	—	—	77
$Ge(SeC_6H_4\text{-}iso\text{-}C_3H_7\text{-}p)_4$	$C_{36}H_{44}Se_4Ge$	113	—	—	—	16
$Ge(SeC_{10}H_7\text{-}\alpha)_4$	$C_{40}H_{28}Se_4Ge$	130–31	—	—	—	16

Compound	Molecular formula	m.p. [°C]	b.p. [°C/torr]	n_D		Ref.
Ge(SeC₆H₄-tert-C₄H₉-p)₄	$C_{40}H_{56}Se_4Ge$	141	—	—	—	16
(C₂H₅)₃GeTeC₂H₅	$C_8H_{20}TeGe$	—	61/1	1.5458	—	84, 86
(C₂H₅)₃GeTeSn(C₂H₅)₃	$C_{12}H_{30}TeSnGe$	—	126–28/1	1.5723	—	84, 86
[(C₂H₅)₃Ge]₂Te	$C_{12}H_{30}TeGe_2$	—	112–15/1	1.5601	—	83, 86
		—	113–16/1	1.5610	—	
[(C₆H₅)₃Ge]₂Te	$C_{36}H_{30}TeGe_2$	120	—	—	—	77
(C₆H₅)₃GeTeSn(C₆H₅)₃	$C_{36}H_{30}SnTeGe$	142–46	—	—	—	77
(C₆H₅)₃GeTePb(C₆H₅)₃	$C_{36}H_{30}PbTeGe$	115–17	—	—	—	77
[(cyclo-C₆H₁₁)₃Ge]₂Te	$C_{36}H_{66}TeGe_2$	128–29	—	—	—	86

REFERENCES

1. Abel, E. W., *J. Chem. Soc.*, 3746 (1958).
2. Abel, E. W., and D. A. Armitage, *Advan. Organometal. Chem.*, **5**, 1 (1967).
3. Abel, E. W., D. A. Armitage, and D. B. Brady, *J. Organometal. Chem.*, **5**, 130 (1966).
4. Abel, E. W., D. A. Armitage, and D. B. Brady, *Trans. Faraday Soc.*, **62**, 3459 (1966).
5. Abel, E. W., D. A. Armitage, and S. P. Tyfield, *J. Chem. Soc. A*, 554 (1967).
6. Abel, E. W., and D. B. Brady, *J. Organometal. Chem.*, **11**, 145 (1968).
7. Abel, E. W., D. J. Walker, and J. N. Wingfield, *J. Chem. Soc. A*, 1814 (1968).
8. Allred, A. L., and E. G. Rochow, *J. Inorg. Nucl. Chem.*, **5**, 269 (1968).
9. Anderson, H. H., *J. Am. Chem. Soc.*, **72**, 2089 (1950).
10. Anderson, H. H., *J. Am. Chem. Soc.*, **73**, 5798 (1951).
11. Anderson, H. H., *J. Am. Chem. Soc.*, **75**, 1576 (1953).
12. Anderson, H. H., *J. Org. Chem.*, **21**, 869 (1956).
13. Anderson, H. H., *J. Am. Chem. Soc.*, **78**, 1692 (1956).
14. Anderson, H. H., *J. Chem. Eng. Data*, **12**, 371 (1967).
15. Backer, H. J., and W. Drenth, *Rec. Trav. Chim.*, **70**, 559 (1951).
16. Backer, H. J., and J. B. G. Hurenkamp, *Rec. Trav. Chim.*, **61**, 802 (1942).
17. Backer, H. J., and F. Stienstra, *Rec. Trav. Chim.*, **52**, 1033 (1933).
18. Backer, H. J., and F. Stienstra, *Rec. Trav. Chim.*, **54**, 607 (1935).
19. Bauer, H., and K. Burschkies, *Chem. Ber.*, **65**, 956 (1932).
19a. Bochkarev, M. N., L. P. Sanina, and N. S. Vyazankin, *Zh. Obshch. Khim.*, **39**, 135 (1969).
20. Bottei, R. S., and L. J. Kuzma, *J. Inorg. Nucl. Chem.*, **30**, 2345 (1968).
21. Brown, M. P., R. Okawara, and E. G. Rochow, *Spectrochim. Acta*, **16**, 595 (1960).
22. Brown, M. P., and E. G. Rochow, *J. Am. Chem. Soc.*, **82**, 4166 (1960).
23. Burschkies, K., *Chem. Ber.*, **69**, 1143 (1936).
24. Clark, E. R., *J. Inorg. Nucl. Chem.*, **25**, 353 (1963).
25. Cradock, S., *J. Chem. Soc. A*, 1426 (1968).
25a. Cradock, S., and E. V. A. Ebsworth, *J. Chem. Soc. A*, 1628 (1969).
26. Cumper, C. W. N., A. Melnikoff, E. F. Mooney, and A. I. Vogel, *J. Chem. Soc. B*, 874 (1966).
27. Cumper, C. W. N., A. Melnikoff, and A. I. Vogel, *J. Chem. Soc. A*, 242 (1966).
28. Cumper, C. W. N., A. Melnikoff, and A. I. Vogel, *J. Chem. Soc. A*, 246 (1966).
29. Cumper, C. W. N., A. Melnikoff, and A. I. Vogel, *J. Chem. Soc. A*, 323 (1966).
30. Davidson, W. E., K. Hills, and M. C. Henry, *J. Organometal. Chem.*, **3**, 285 (1965).
31. Egorochkin, A. N., N. S. Vyazankin, M. N. Bochkarev, V. T. Bychkov, and A. I. Burov, *Zh. Obshch. Khim.*, **38**, 396 (1968).
32. Egorochkin, A. N., N. S. Vyazankin, M. N. Bochkarev, and S. Ya. Khorshev, *Zh. Obshch. Khim.*, **37**, 1165 (1967).
33. Egorochkin, A. N., N. S. Vyazankin, G. A. Razuvaev, O. A. Kruglaya, and M. N. Bochkarev, *Dokl. Akad. Nauk SSSR*, **170**, 333 (1966).
34. Fink, F. H., *Dissertation Abstr. B*, **27**, 1413 (1966).
35. Fink, F. H., J. A. Turner, and D. A. Payne, *J. Am. Chem. Soc.*, **88**, 1571 (1966).
36. Glidewell, C., and D. W. H. Rankin, *J. Chem. Soc. A*, 753 (1969).
37. Goldfarb, T. D., and S. Sujishi, *J. Am. Chem. Soc.*, **86**, 1679 (1964).
38. Henry, M. C., and W. E. Davidson, *J. Org. Chem.*, **27**, 2252 (1962).
39. Henry, M. C., and W. E. Davidson, *Can. J. Chem.*, **41**, 1276 (1963).
40. Hooton, K. A., and A. L. Allred, *Inorg. Chem.*, **4**, 671 (1965).
41. Lesbre, M., and J. Satgé, *Compt. Rend.*, **254**, 4051 (1962).

42. Magnuson, J. A., and E. W. Knaub, *Anal. Chem.*, **37**, 1607 (1965).
43. Mathur, S., G. Chandra, A. K. Rai, and R. C. Mehrotra, *J. Organometal. Chem.*, **4**, 294 (1965).
44. Mazerolles, P., J. Dubac, and M. Lesbre, *J. Organometal. Chem.*, **5**, 35 (1966).
45. Mazerolles, P., J. Dubac, and M. Lesbre, *J. Organometal. Chem.*, **12**, 143 (1968).
45a. Mazerolles, P., M. Lesbre, and M. Joanny, *J. Organometal. Chem.*, **16**, 227 (1969).
46. Mehrotra, R. C., V. D. Gupta, and D. Sukhani, *J. Inorg. Nucl. Chem.*, **29**, 83 (1967).
47. Mehrotra, R. C., V. D. Gupta, and D. Sukhani, *J. Organometal. Chem.*, **9**, 263 (1967).
48. Moedritzer, K., *Inorg. Chem.*, **6**, 1248 (1967).
49. Moedritzer, K., and J. R. van Wazer, *J. Am. Chem. Soc.*, **87**, 2360 (1965).
50. Moedritzer, K., and J. R. van Wazer, *Inorg. Chim. Acta*, **1**, 152 (1967).
51. Moedritzer, K., and J. R. van Wazer, *Inorg. Chim. Acta*, **1**, 407 (1967).
52. Moedritzer, K., and J. R. van Wazer, *J. Inorg. Nucl. Chem.*, **29**, 1571 (1967).
53. Moedritzer, K., and J. R. van Wazer, *J. Organometal. Chem.*, **13**, 145 (1968).
54. Nguen, D. K., V. S. Fainberg, Yu. I. Baukov, and I. F. Lutsenko, *Zh. Obshch. Khim.*, **38**, 191 (1968).
54a. Oliver, A. J., and W. A. G. Graham, *J. Organometal. Chem.*, **19**, 17 (1969).
55. Pascal, P., A. Pacault, and A. Tchakirian, *Compt. Rend.*, **226**, 849 (1948).
56. Razuvaev, G. A., *Conférence à l'Assemblée Annuelle de la Société Chimique de France*, Toulouse, May, 1969.
57. Rivière-Baudet, M., and J. Satgé, *Bull. Soc. Chim. France*, 1356 (1969).
58. Rochow, E. G., *J. Am. Chem. Soc.*, **70**, 1801 (1948).
59. Rochow, E. G., and A. L. Allred, *J. Am. Chem. Soc.*, **77**, 4489 (1955).
60. Rothermündt, M., and K. Burschkies, *Z. Immunitaetsforsch.*, **87**, 445 (1936).
61. Ruidisch, I., and M. Schmidt, *J. Organometal. Chem.*, **1**, 160 (1963).
62. Ruidisch, I., and M. Schmidt, *Chem. Ber.*, **96**, 1424 (1963).
63. Satgé, J., and M. Baudet, *Compt. Rend. C*, **263**, 435 (1966).
64. Satgé, J., and C. Couret, *Compt. Rend. C*, **264**, 2169 (1967).
65. Satgé, J., and M. Lesbre, *Bull. Soc. Chim. France*, 2578 (1965).
66. Satgé, J., M. Lesbre, and M. Baudet, *Compt. Rend.*, **259**, 4733 (1964).
67. Schmidt, M., and H. Ruf, *Angew. Chem.*, **73**, 64 (1961).
68. Schmidt, M., and H. Ruf., *J. Inorg. Nucl. Chem.*, **25**, 557 (1963).
69. Schmidbaur, H., and I. Ruidisch, *Inorg. Chem.*, **3**, 599 (1964).
69a. Schmidt, M., and J. F. Jaggard, *J. Organometal. Chem.*, **17**, 283 (1969).
70. Schmidt, M., and H. Schumann, *Z. Anorg. Allgem. Chem.*, **325**, 130 (1963).
70a. Schmidt, M., H. Schumann, F. Gliniecki, and J. F. Jaggard, *J. Organometal. Chem.*, **17**, 277 (1969).
71. Schumann, H., *Z. Anorg. Allgem. Chem.*, **354**, 192 (1967).
72. Schumann, H., and M. Schmidt, *J. Organometal. Chem.*, **3**, 485 (1965).
73. Schumann, H., and M. Schmidt, *Angew. Chem.*, **77**, 1049 (1965); *Int. Ed.*, **4**, 1007 (1965).
73a. Schumann, H., and I. Schumann-Ruidisch, *J. Organometal. Chem.*, **18**, 355 (1969).
74. Schumann, H., K. F. Thom, and M. Schmidt, *Angew. Chem.*, **75**, 138 (1963); *Int. Ed.*, **2**, 99 (1963).
75. Schumann, H., K. F. Thom, and M. Schmidt, *J. Organometal. Chem.*, **1**, 167 (1963).
76. Schumann, H., K. F. Thom, and M. Schmidt, *J. Organometal. Chem.*, **2**, 361 (1964).
77. Schumann, H., K. F. Thom, and M. Schmidt, *J. Organometal. Chem.*, **4**, 22 (1965).
78. Schwarz, R., and W. Reinhardt, *Chem. Ber.*, **65**, 1743 (1932).
79. Srivastava, T. N., and S. K. Tandon, *Z. Anorg. Allgem. Chem.*, **353**, 87 (1967).
80. Sukhani, D., V. D. Gupta, and R. C. Mehrotra, *Australian J. Chem.*, **21**, 1175 (1968).

81. Van der Berghe, E. V., D. F. van de Vondel, and G. P. van der Kelen, *Inorg. Chim. Acta*, **1**, 97 (1967).

82. Van Wazer, J. R., K. Moedritzer, and L. C. D. Groenweghe, *J. Organometal. Chem.*, **5**, 420 (1966).

82a. Vyazankin, N. S., M. N. Bochkarev, and L. P. Maiorova, *Zh. Obshch. Khim.*, **39**, 468 (1969).

83. Vyazankin, N. S., M. N. Bochkarev, and L. P. Sanina, *Zh. Obshch. Khim.*, **36**, 166 (1966).

84. Vyazankin, N. S., M. N. Bochkarev, and L. P. Sanina, *Zh. Obshch. Khim.*, **36**, 1154 (1966).

85. Vyazankin, N. S., M. N. Bochkarev, and L. P. Sanina, *Zh. Obshch. Khim.*, **36**, 1961 (1966).

86. Vyazankin, N. S., M. N. Bochkarev, and L. P. Sanina, *Zh. Obshch. Khim.*, **37**, 1037 (1967).

87. Vyazankin, N. S., M. N. Bochkarev, and L. P. Sanina, *Zh. Obshch. Khim.*, **37**, 1545 (1967).

87a. Vyazankin, N. S., M. N. Bochkarev, and L. P. Sanina, *Zh. Obshch. Khim.*, **38**, 414 (1968).

88. Vyazankin, N. S., M. N. Bochkarev, L. P. Sanina, A. N. Egorochkin, and S. Ya. Khorshev, *Zh. Obshch. Khim.*, **37**, 2576 (1967).

89. Vyazankin, N. S., L. P. Sanina, G. S. Kalinina, and M. N. Bochkarev, *Zh. Obshch. Khim.*, **38**, 1800 (1968).

90. Vyazankin, N. S., I. A. Vostokov, and V. T. Bychkov, *Zh. Obshch. Khim.*, **38**, 2485 (1968).

91. Wang, J. T., and C. H. van Dyke, *Chem. Commun.*, 612 (1967).

92. Wang, J. T., and C. H. van Dyke, *Chem. Commun.*, 928 (1967).

93. Wang, J. T., and C. H. van Dyke, *Inorg. Chem.*, **7**, 1319 (1968).

94. West, R., *J. Am. Chem. Soc.*, **75**, 6080 (1953).

95. Wieber, M., and C. D. Frohning, *Angew. Chem. Int. Ed.*, **4**, 1096 (1965).

96. Wieber, M., and C. D. Frohning, *J. Organometal. Chem.*, **8**, 459 (1967).

97. Wieber, M., and M. Schmidt, *Z. Naturforsch.*, **18b**, 846 (1963).

98. Wieber, M., and M. Schmidt, *Z. Naturforsch.*, **18b**, 847 (1963).

99. Wieber, M., and M. Schmidt, *Z. Naturforsch.*, **18b**, 848 (1963).

100. Wieber, M., and M. Schmidt, *Z. Naturforsch.*, **18b**, 849 (1963).

101. Wieber, M., and M. Schmidt, *J. Organometal. Chem.*, **2**, 129 (1964).

102. Wieber, M., and M. Schmidt, *Angew. Chem. Int. Ed.*, **3**, 153 (1964).

103. Wieber, M., and M. Schmidt, *J. Organometal. Chem.*, **1**, 336 (1964).

104. Wieber, M., and M. Schmidt, *Angew. Chem.*, **76**, 573 (1964); *Int. Ed.*, **3**, 657 (1964).

105. Wieber, M., and G. Schwarzmann, *Monatsh. Chem.*, **99**, 255 (1968).

105a. Wieber, M., and G. Schwarzmann, *Monatsh. Chem.*, **100**, 68 (1969).

106. Brit. Pat. 654,571 (1951); *CA* **46**, 4561 (1952).

107. US Pat. 2,506,386; *CA*, **44**, 7344 (1950).

8 Germanium–nitrogen and germanium–phosphorus bond

8-1 ORGANOGERMANIUM–NITROGEN COMPOUNDS

As early as 1927, C. A. Kraus and coworkers were the first to describe germanium–nitrogen derivatives. However, this area remained unexplored for a long time and, until recently (1960) only a score of organogermanyl-amines had been described in the literature, along with some organo-germanium isocyanides, isocyanates and isothiocyanates. But the growing interest in the Si—N and Sn—N bonds recently gave rise to a rapid development of the study of germanium–nitrogen bond derivatives, and now about 180 publications in that field may be cited.

8-1-1 Formation reactions of the germanium–nitrogen bond

8-1-1-1 Ammonolysis of Halo- and Organohalogermanes. In an important investigation about organic germanium derivatives, Kraus and coworkers described, in 1927, the first organogermanium derivatives with a german-ium–nitrogen bond (69).

Triphenylbromogermane $(C_6H_5)_3GeBr$ is very soluble in liquid am-monia and highly solvolysed in it. After the solvent has been evaporated, tris-(triphenylgermyl)amine $[(C_6H_5)_3Ge]_3N$ is obtained (82). Highsmith and Sisler have repeated the Kraus and Wooster experiments and indicate that the reaction of triphenylbromogermane and ammonia yields bis-(triphenylgermyl)amine instead of tris-(triphenylgermyl)amine (61).

Kraus and Wooster isolated the triphenylgermylamine, $(C_6H_5)_3GeNH_2$, by action of ammonia gas on triphenylbromogermane in organic solvents, ether, benzene or petroleum ether (84). This amine, which remains one of the sole known derivatives containing the Ge—NH_2 group, is extremely

535

sensitive to hydrolysis:

$$2\,(C_6H_5)_3GeNH_2 + H_2O \rightarrow [(C_6H_5)_3Ge]_2O + NH_3$$

It also is possible to obtain triphenylgermylamine by the action of potassium amide on triphenylbromogermane in liquid ammonia:

$$(C_6H_5)_3GeBr + KNH_2 \rightarrow (C_6H_5)_3GeNH_2 + KBr$$

An excess of potassium amide leads to the formation of the potassium salt of the amine:

$$(C_6H_5)_3GeNH_2 + KNH_2 \rightarrow (C_6H_5)_3GeNHK + NH_3$$

Triphenylgermylamine has a great tendency to lose ammonia with formation of bis-(triphenylgermyl)amine:

$$2\,(C_6H_5)_3GeNH_2 \rightarrow [(C_6H_5)_3Ge]_2NH + NH_3$$

A complete conversion of $(C_6H_5)_3GeNH_2$ to $[(C_6H_5)_3Ge]_3N$ is obtained by heating the primary amine to above 200°C at atmospheric pressure (84).

The action of triethylchlorogermane $(C_2H_5)_3GeCl$, on lithium amide $LiNH_2$ in a THF medium gives only $(Et_3Ge)_2NH$ (113). The very unstable initially formed primary amine, Et_3GeNH_2, is likely to become quantitatively transformed into bis-(triethylgermyl)amine, $(Et_3Ge)_2NH$, with ammonia evolution.

These easy deamination reactions show that germylamines are clearly different from their carbon analogues.

Triphenylgermylamine also is formed by the action of triphenylgermylsodium, Ph_3GeNa, in liquid ammonia, on some aryl halides (159) or on either methylene chloride or bromide (83, 167):

$$Ph_3GeNa + C_{10}H_7Br + NH_3 \rightarrow NaBr + C_{10}H_8 + Ph_3GeNH_2$$

$$2\,Ph_3GeNa + CH_2Cl_2 + NH_3 \rightarrow 2\,NaCl + Ph_3GeMe + Ph_3GeNH_2$$

The methylene bromide reaction with sodium germanyl is complex. Methylgermane is probably formed, along with H_3GeNH_2 (167):

$$2\,NaGeH_3 + CH_2Br_2 + NH_3 \rightarrow H_3GeNH_2 + H_3GeCH_3 + 2\,NaBr$$

H_3GeNH_2 and $H_2Ge(NH_2)_2$ have been found in the acid hydrolysis of $CaGe/Ca_3N_2$ (109).

Action of liquid ammonia on triethylbromogermane only gives the "$Et_3GeBr\cdot NH_3$" adduct. In the presence of sodium, a stronger hydrogen halide acceptor than NH_3, ammonolysis leads to a $(Et_3Ge)_2NH$ (81).

Using the same synthesis, Rijkens, van der Kerk and coworkers isolated bis-(tributylgermyl)amine, $(Bu_3Ge)_2NH$, from tributylchlorogermane (88, 105, 106). Bis-(trimethylgermyl)amine is isolated by action of gaseous ammonia on trimethylchlorogermane in anhydrous ether solution (111). Satgé and Baudet in a study of this same reaction observed, at $-60°C$, the formation of a substantial amount of tris-(trimethylgermyl)amine (113):

$$Me_3GeCl + NH_3 \xrightarrow[-60°C]{ether} (Me_2Ge)_2NH \ (50\%) + (Me_3Ge)_3N \ (32\%)$$

The simultaneous formation of the two derivatives is also observed in Me_3GeNMe_2 ammonolysis (113):

$$Me_3GeNMe_2 + NH_3 \xrightarrow[0°C]{ether} (Me_3Ge)_2NH \ (32\%) + (Me_3Ge)_3N \ (28\%)$$

It is interesting to note that Et_3GeCl and Et_3GeNMe_2 ammonolysis and Et_3GeCl action with Li_3N or $(Et_3Ge)_2NLi$ do not lead to the tertiary amine $(Et_3Ge)_3N$, probably because of steric hindrances (113). This assumption seems quite plausible since Massol and Satgé have shown in the ammonolysis of the ethylchlorogermanes Et_2GeHCl and $EtGeH_2Cl$, that ethyl substitution by hydrogen favoured the formation of the trigermanyl tertiary amine. Diethylchlorogermane leads to a mixture of $(Et_2GeH)_2NH$ and $(Et_2GeH)_3N$ and ethylchlorogermane leads to $(EtGeH_2)_3N$ only (90).

Tris-(trimethylgermyl)amine formed in Me_3GeCl or Me_3GeNMe_2 ammonolysis could be isolated by the action of trimethylchlorogermane on lithium nitride, Li_3N, in THF and also by the action of trimethylchlorogermane on the lithium derivative of bis-(triethylgermyl)amine (113):

$$[(CH_3)_3Ge]_2NLi + (CH_3)_3GeCl \rightarrow [(CH_3)_3Ge]_3N + LiCl$$

This germanium tertiary amine has been identified as the derivative already isolated by Ruidisch and Schmidt in CH_3Li action on $(Me_2GeCl)_3N$, a product of the partial Me_2GeCl_2 ammonolysis (111).

A total ammonolysis of diphenyldichlorogermane has been observed in liquid ammonia (79):

$$(C_6H_5)_2GeCl_2 + 3 NH_3 \rightarrow [(C_6H_5)_2GeNH] + 2 NH_4Cl$$

This "imine", a very viscous colourless liquid soluble in organic solvents, certainly does not exist as a monomer. The authors do not give the molecular weight but it may be supposed that this compound is a cyclic trimer or tetramer. It is very easily hydrolysed, giving the corresponding oxide $[(C_6H_5)_2GeO]$, and ammonia is released.

Similarly ammonolysis in the presence of sodium and diethyldibromo-germane, Et_2GeBr_2, and dibutyldichlorogermane, Bu_2GeCl_2, leads to $(Et_2GeNH)_{3 \, or \, 4}$ (43) and $(Bu_2GeNH)_3$ (105, 106), cyclic trimers having the general form:

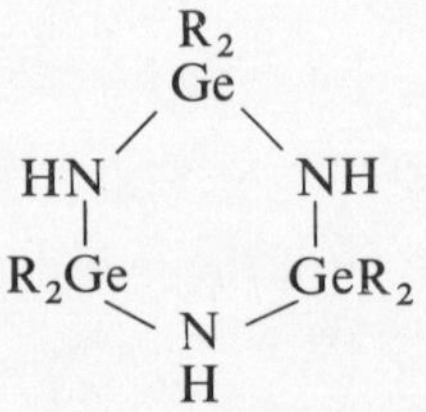

$$\begin{array}{c} R_2 \\ Ge \\ {}_{HN} \diagup \quad \diagdown {}_{NH} \\ | \qquad \qquad | \\ R_2Ge \qquad GeR_2 \\ \diagdown \quad \diagup \\ N \\ H \end{array}$$

They are colourless liquids which are very quickly hydrolysed in air. Ammonolysis of ethyltribromo- or triiodogermane immediately gives a white ammonia-insoluble precipitate (44):

$$C_2H_5GeI_3 + 4\,NH_3 \rightarrow C_2H_5GeN + 3\,NH_4I$$

The polymeric "nitride" on hydrolysis gives ethylgermanoic anhydride:

$$2\,C_2H_5GeN + 3\,H_2O \rightarrow (C_2H_5GeO)_2O + 2\,NH_3$$

Onyszchuk (99) in a recent study on ammonolysis of GeH_3F, GeH_3Cl, GeH_2Cl_2, Me_3GeBr, Ph_2GeCl_2 and Et_2GeCl_2 at $-78°C$ reports the formation of adducts with one or two base molecules in a first reaction stage. Then, at a temperature above $-78°C$ the substitutions on germanium take place that lead, with H_3GeF and Me_3GeBr, to quaternary ammonium salts $[H_3GeNH_3]^+F^-$ or $[Me_3GeNH_3]^+Br^-$. The system $H_3GeCl/N(CH_3)_3$ leads to $(GeH_2)x$ and $HN(CH_3)_3^+Cl^-$.

If diorganodichlorogermanes lead to dimers or trimers, $(R_2GeNH)_n$, dichlorogermane H_2GeCl_2 forms with NH_3 at $-78°C$ $(H_2GeNH)_n$ but this derivative immediately decomposes to Ge and NH_3 (99).

Chloromethyldimethylchlorogermane reacts with NH_3 in mild conditions by substitution of Ge—Cl to give bis-[(chloromethyl)dimethyl-germyl]amine (183):

$$2\,(ClCH_2)(CH_3)_2GeCl + 3\,NH_3 \rightarrow$$

$$[(ClCH_2)(CH_3)_2Ge]_2NH + 2\,NH_4Cl$$

In a second step, the two carbon-bound chlorine atoms of these compounds may be substituted by sulphur yielding germanium-containing heterocycles:

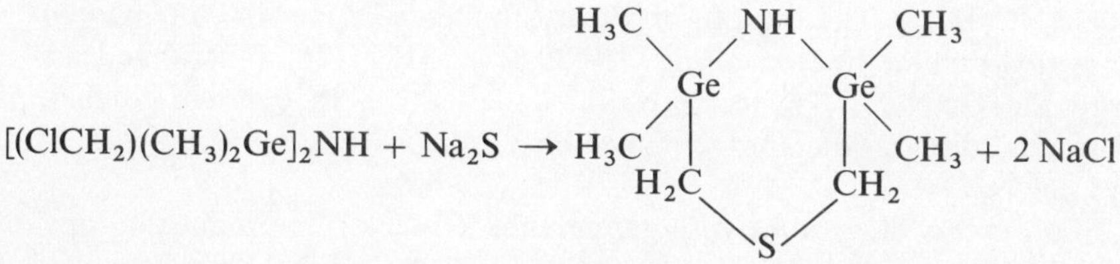

$$[(ClCH_2)(CH_3)_2Ge]_2NH + Na_2S \rightarrow$$

$$+ 2\,NaCl$$

$$[(ClCH_2)(CH_3)_2Ge]_2NH + 2\,H_2S + (C_2H_5)_3N \rightarrow$$

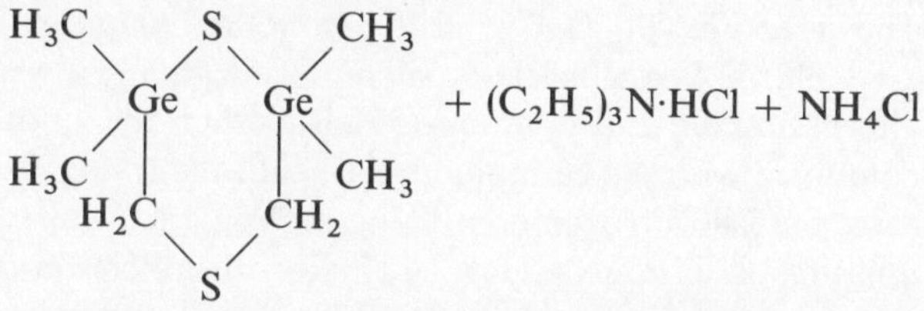

$$+ (C_2H_5)_3N \cdot HCl + NH_4Cl$$

8-1-1-2   Aminolysis of Halo- and Organohalogermanes. Germanium tetra-halides as well as organohalogermanes react with amines. In a first step adducts are formed. Most of these, however, are stable only at temperatures well below 0°C. Adducts of halogermanes with ammonia, hydrazines and amines which are stable at room temperature are listed on page 37 in the work of Rijkens and van der Kerk (106) and in Gielen and Sprecher's article (52).

Many adducts of halogermanes with ammonia, primary and secondary amines are converted, with elimination of quaternary ammonium salts, into organogermylamines:

$$\equiv GeX + HNR_2 \rightarrow \equiv GeX \cdot HNR_2 \xrightarrow{\ HNR_2\ } \equiv GeNR_2 + R_2NH \cdot HX$$

By interaction between $GeCl_4$ and aniline or between $GeCl_4$ and cyclohexylamine, Davidson prepared the compounds $ClGe(NHC_6H_5)_3$, $Cl_2Ge(NHR)_2$ (R = phenyl or cyclohexyl) and $Ge(NHC_6H_{11})_4$ (28). But generally the derivatives containing two or more $-NH_2$ or $-NHR$ groups bonded with a germanium atom are unstable. This has been illustrated by the formation of the $[R_2GeNH]$ and $[RGeN]$ type derivatives in ammonolysis of organogermanium di- and trihalides, as well as by the formation of $Ge(NH)_2$ in $GeCl_4$ (154, 175) and GeI_4 (70) ammonolysis. The $(RHN)_4Ge$ type derivatives lead to derivatives of the $Ge(NR)_2$ type (176).

Thomas and Southwood, in particular, report, in the action of $GeCl_4$ on aniline, the formation of germanium diphenyldiimide dihydrochloride,

$Ge(NC_6H_5, HCl)_2$. The action of ethylamine on $GeCl_4$ would result in an unstable adduct and, finally, the germanium diethyldiimide $Ge(NC_2H_5)_2$. The action of diethylamine leads to the germanium diethyl-diimide hydrochloride $Ge(NC_2H_5)_2$, HCl (176). Piperidine leads to the $Ge(NC_5H_{10})_4$ tetrasubstituted derivative. In contrast to these results, diphenyldichlorogermane, Ph_2GeCl_2, is not ammonolysed by $C_2H_5NH_2$ (80).

An excess of pyridine with germanium tetrachloride produces a dipyridine germanium tetrachloride adduct $GeCl_4 \cdot 2\,C_5H_5N$ in excellent yield (98%). Tetrakis-diethylaminogermane $Ge[N(C_2H_5)_2]_4$ was formed by the interaction of the complex with further pyridine and diethylamine (1). The reaction of $GeCl_4$ with monomethylamine in ether leads to $(Cl_2GeNMe)_3$, a trimeric cyclic molecule having a six-atom backbone of alternating Ge and N atoms (36). But when the amount of methylamine combined with the germanium tetrachloride is progressively decreased to zero, the last nitrogen-containing molecule to vanish in the equilibrated mixtures is $Cl_3GeN(CH_3)GeCl_3$. For all molar ratios of amine to germanium tetrachloride from 0 to 3, these two compounds are the predominant species at equilibrium (37).

The reactions of $GeBr_4$, $EtGeCl_3$, Bu_2GeCl_2 with such secondary amines as dimethylamine and diethylamine in benzene, hexane or cyclohexane, respectively, lead to the $Ge(NR_2)_4$, $RGe(NR_2)_3$ and $R_2Ge(NR_2)_2$ types (9, 13).

Germanium tetrachloride reacts with a number of N,N'-disubstituted ethylenediamines to give spiro-imidazolidines with germanium as the central spiro atom. The analogous silicon compounds also were prepared (186, 187).

The N,N'-dialkylcompounds were prepared by the reaction of the corresponding diamine with the tetrachloride in benzene at 40–60°C:

$$6\,RNH{-}CH_2CH_2{-}NHR + GeCl_4 \xrightarrow{\text{benzene}} \begin{array}{c} R{-}N{\diagdown}{\diagup}N{-}R \\ Ge \\ R{-}N{\diagup}{\diagdown}N{-}R \end{array}$$

$$+\ 4\,RNHCH_2CH_2NHR \cdot HCl$$

Aryl derivatives were synthesized in the presence of triethylamine using either benzene or dioxane as a solvent. The spiro structure has been confirmed by IR and NMR spectroscopy. NMR spectra of the alkyl derivatives show a single sharp peak for the methylene bridge protons of the ring. This has been interpreted as indicating either the planarity of the ring systems or the rapid inversion of tetrahedral nitrogen atoms. The IR spectra of this five-membered germanium ring show a sharp absorption in the $1330 \pm 10 \text{ cm}^{-1}$ region associated with the imidazolidine ring, and a single absorption in the $930–870 \text{ cm}^{-1}$ region that may be assigned to the germanium–nitrogen asymmetrical stretching vibration (186, 190).

However an attempt to prepare imidazolidine by amination of dimethyldichlorogermane with N,N′-dimethylenediamine in the presence of triethylamine produced N,N′-bis-[(dimethylchlorogermyl)-N,N′-dimethylethylenediamine (186, 189):

$$2 \, (CH_3)_2 GeCl_2 + 3 \, CH_3 - \underset{\substack{| \\ H}}{N} \quad \underset{\substack{| \\ H}}{N} - CH_3 \; \rightarrow$$

$$CH_3 - \underset{\substack{| \\ (CH_3)_2GeCl}}{N} \quad \underset{\substack{| \\ ClGe(CH_3)_2}}{N} - CH_3 + 2 \, CH_3 - \underset{\substack{| \\ H}}{N} \quad \underset{\substack{| \\ H}}{N} - CH_3 \cdot HCl$$

The same material was obtained in the reaction between N,N′-dimethylethylenediamine and dimethyl(diethylamino)chlorogermane (186, 189):

$$2 \, (CH_3)_2 \underset{\substack{| \\ Cl}}{Ge} - N(C_2H_5)_2 + CH_3 - \underset{\substack{| \\ H}}{N} \quad \underset{\substack{| \\ H}}{N} - CH_3 \; \rightarrow$$

$$CH_3 - \underset{\substack{| \\ (CH_3)_2GeCl}}{N} \quad \underset{\substack{| \\ ClGe(CH_3)_2}}{N} - CH_3 + 2 \, HN(C_2H_5)_2$$

where transamination was shown to proceed in preference to direct amination on the same germanium atom.

Recently, Yoder and Zuckermann (189) found that the direct amination of dimethyldichlorogermane with diethylamine replaces only one chlorine atom on germanium to give dimethylchlorodiethylaminogermane as the sole product:

$$(CH_3)_2GeCl_2 + 2 \, HN(C_2H_5)_2 \underset{0°C}{\overset{\text{benzene}}{\rightleftharpoons}}$$

$$(CH_3)_2Ge[N(C_2H_5)_2]Cl + (C_2H_5)_2NH_2Cl$$

An analogous reaction with dimethylamine at $-60°C$ resulted in a mixture of dimethylchlorodimethylaminogermane and the disubstituted bis-(dimethylamino)dimethylgermane (189).

On the other hand a complete aminolysis of dimethyldichlorogermane by o-phenylenediamine, o-aminophenol and o-aminothiophenol in the presence of triethylamine is observed (182):

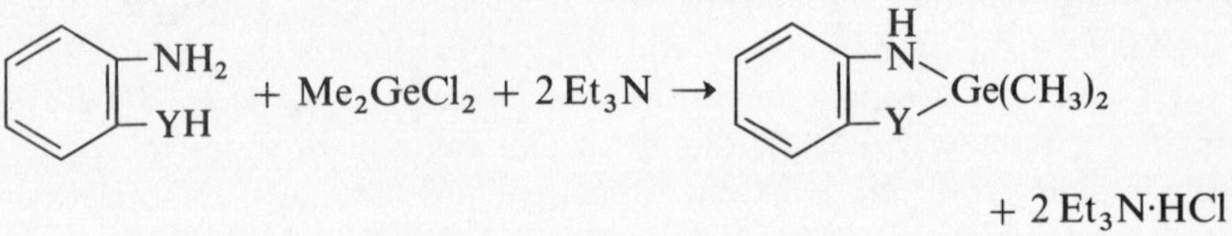

$$+ 2 Et_3N \cdot HCl$$

(with Y = NH, O or S)

Aminolysis of the same halide by methylamine in ether gives the cyclic trimer of the dimethylgermylmethylamine in 80% yield (137, 143):

$$3 (CH_3)_2GeCl_2 + 9 CH_3NH_2 \rightarrow$$

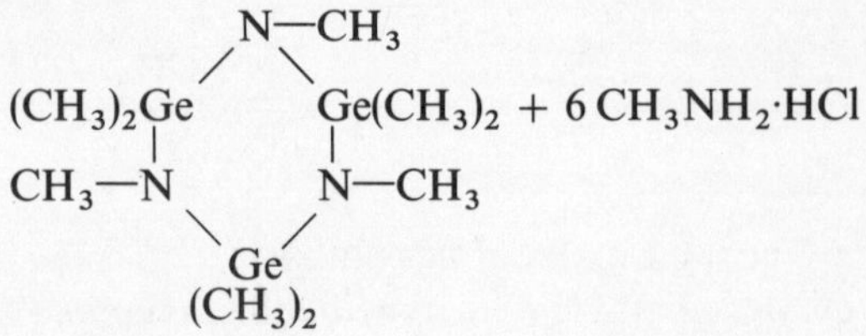

$$+ 6 CH_3NH_2 \cdot HCl$$

This compound is quantitatively cleaved by methyllithium in ether to give the lithium trimethylgermylmethylamine:

The action of tributylchlorogermane on cyanamide in an ether-alcohol solution in the presence of C_2H_5ONa does not lead to the disubstituted derivative $[(C_4H_9)_3Ge]_2NCN$, but to its isomer bis-(tributylgermyl)carbodiimide $(C_4H_9)_3Ge—N=C=N—Ge(C_4H_9)_3$ (105, 106).

8-1-1-3 Synthesis of Ge—N Bond Derivatives by Transamination Reactions. The first transamination reactions in the organogermanium series were described in 1964 (106, 121). We have already reported the synthesis of $(Et_3Ge)_2NH$, $(Me_3Ge)_2NH$ and $(Me_3Ge)_3N$ by the action of ammonia gas on Et_3GeNMe_2 and Me_3GeNMe_2 in an ether solution (113).

Organogermylamines R_3GeNHR' (R' = hexyl or phenyl) are formed in good yields by a mild heating of trialkyldimethylaminogermanes with the corresponding primary amines, with a dimethylamine release (121):

$$R_3GeNMe_2 + R'NH_2 \rightarrow R_3GeNHR' + Me_2NH$$

Rijkens and van der Kerk in particular investigated transamination reactions of bis-(trialkylgermyl)amines, $(R_3Ge)_2NH$ (106). In such a case, a higher temperature (160–170°C) and the presence of an $(NH_4)_2SO_4$ catalyst are required:

$$(Bu_3Ge)_2NH + 2\,C_8H_{17}NH_2 \rightarrow 2\,Bu_3GeNHC_8H_{17} + NH_3$$

$$(Bu_3Ge)_2NH + 2\,H{-}N{\Big\langle} \rightarrow 2\,Bu_3Ge{-}N{\Big\langle} + 2\,NH_3$$

$$(H{-}N{\Big\langle} = \text{pyrrolidine and 2,4-dimethyl pyrrole})$$

Bis-(trimethylgermyl)amine seems more reactive than its analogous derivatives with a heavier radical, R (e.g. R = Bu) since the same transamination reaction with octylamine is observed between 50 and 100°C in the absence of a catalyst (113):

$$(Me_3Ge)_2NH + 2\,n\text{-}C_8H_{17}NH_2 \rightarrow 2\,Me_3GeNHC_8H_{17} + NH_3$$

Tris-(trimethylgermyl)amine gives the same transamination reaction but at a higher temperature (120–180°C):

$$(Me_3Ge)_3N + 3\,C_8H_{17}NH_2 \rightarrow 3\,Me_3GeNHC_8H_{17} + NH_3$$

Moreover, in these reactions, the formation of $(Me_3Ge)_2NC_8H_{17}$ from the condensation of the previously formed N-trimethylgermyl-octylamine has been observed (113):

$$2\,Me_3GeNHC_8H_{17} \rightarrow (Me_3Ge)_2NC_8H_{17} + C_8H_{17}NH_2$$

These transamination reactions can be extended to the synthesis of N-germylamides (121):

$$Et_3GeNMe_2 + H_2N{-}\underset{\underset{O}{\|}}{C}{-}R' \rightarrow$$

$$Et_3Ge{-}\underset{\underset{H}{|}}{N}{-}\underset{\underset{O}{\|}}{C}{-}R' + Me_2NH \qquad (R' = H \text{ or } CH_3)$$

Even a double transamination reaction is observed with acetamide (114):

$$Et_3Ge{-}\underset{\underset{H}{|}}{N}{-}\underset{\underset{O}{\|}}{C}{-}CH_3 + Et_3GeNMe_2 \rightarrow$$

$$(Et_3Ge)_2N{-}\underset{\underset{O}{\|}}{C}{-}CH_3 + Me_2NH$$

The phenylhydrazine N-trialkylgermanium derivatives are easily obtained by interaction, at a moderate temperature, between R_3GeNMe_2 or $(R_3Ge)_2NH$ with phenylhydrazine (115):

$$Et_3GeNMe_2 + H_2N-NH-C_6H_5 \rightarrow$$

$$Et_3GeNH-NH-C_6H_5 + Me_2NH$$

$$(Et_3Ge)_2NH + 2 H_2N-NH-C_6H_5 \rightarrow$$

$$2 Et_3GeNH-NH-C_6H_5 + NH_3$$

These derivatives are colourless, viscous liquids, sensitive to hydrolysis. They become covered with a deep blue layer by oxidation at the contact interface with air, probably owing to the production of diazenes: $EtGeN=N-C_6H_5$.

The diazenes are obtained in another way: by a method similar to the one described by Wannagat and Krüger (180) for the analogous silicon derivatives; the dilithium compound of N-triethylgermyl-N'phenylhydrazine is obtained by the action of methyllithium at 0°C:

$$Et_3GeNH-NH-C_6H_5 + 2 CH_3Li \xrightarrow{0°C} Et_3GeN-N-C_6H_5$$
$$\underset{Li}{|} \quad \underset{Li}{|}$$

and by treatment of the dilithium derivative with a bromine solution in ether at $-70°C$ (115):

$$Et_3GeN-N-C_6H_5 + Br_2 \rightarrow Et_3GeN=N-C_6H_5 + 2 LiBr$$
$$\underset{Li}{|} \quad \underset{Li}{|}$$

Triethylgermylphenyldiazene is a dark blue liquid distillable under reduced pressure, but very sensitive to hydrolysis. It has not yet been possible to isolate it as a pure compound. Its main impurity is $Et_3GeNHC_6H_5$, which is probably formed by the action of bromine on the monolithium derivative, $Et_3Ge-NH-N-C_6H_5$, with release of
$$\underset{Li}{|}$$
nitrogen.

The UV absorption spectrum of N-triethylgermyl-N'-phenyldiazene is very similar to that of silicon analogues (85) and shows an absorption maximum at 225 nm with a shoulder at 275 nm, and another maximum at 573 nm which seems to be very characteristic of the $Ge-N=N-$ group. The IR spectrum shows, at 1455 and 920 cm^{-1}, the typical bands of the unsymmetrical $-N=N-$ substituted group (114, 115).

Recently Yoder and Zukermann synthesized germanium imidazolidines by transamination between dimethylbis-(diethylamino)germane and N—N′dimethylethylenediamine (186, 188):

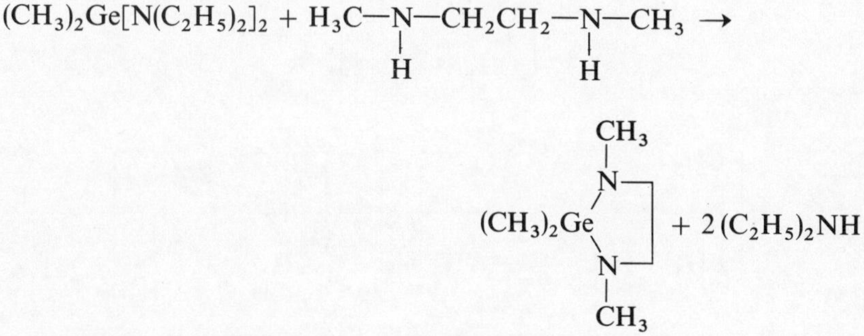

$$(CH_3)_2Ge[N(C_2H_5)_2]_2 + H_3C—\underset{\underset{H}{|}}{N}—CH_2CH_2—\underset{\underset{H}{|}}{N}—CH_3 \rightarrow$$

$$(CH_3)_2Ge\left\langle \begin{array}{c} \overset{\displaystyle CH_3}{|} \\ N— \\ \\ N— \\ | \\ CH_3 \end{array} \right. + 2\,(C_2H_5)_2NH$$

Germanium imidazolidines are stable as the monomers. However, in the presence of impurities, particularly Bronsted acids such as ammonium sulphate, these compounds undergo polymerization:

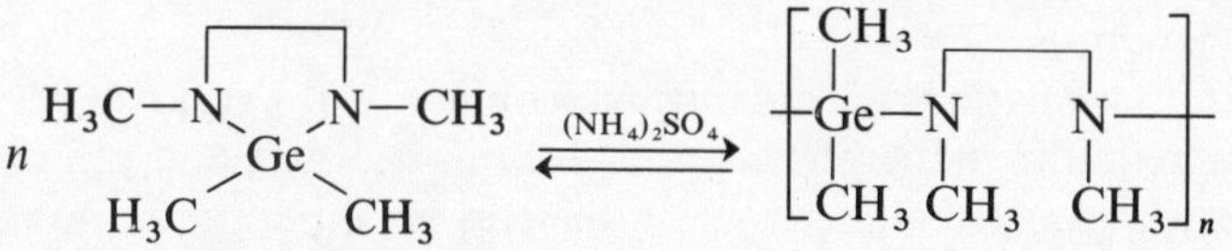

$$n\; \begin{array}{c} H_3C—N \\ \quad\quad\quad N—CH_3 \\ \quad Ge \\ H_3C \quad\quad CH_3 \end{array} \xrightarrow{\;(NH_4)_2SO_4\;} \left[\begin{array}{c} \overset{\displaystyle CH_3}{|} \\ +Ge—N \quad\quad N— \\ | \quad\; | \quad\quad | \\ CH_3\; CH_3 \quad CH_3 \end{array} \right]_n$$

Polymerization is reversible, the monomer being regenerated on attempted distillation. Transamination of dimethylbis-(diethylamino)-germane with piperazine can be summarized by the following formulation (189):

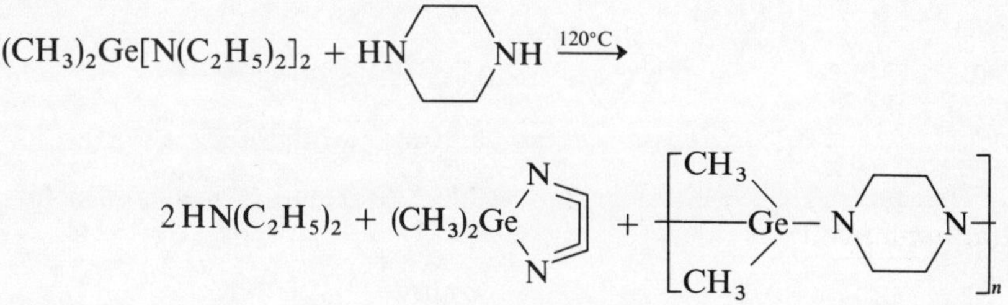

$$(CH_3)_2Ge[N(C_2H_5)_2]_2 + HN\overset{\frown}{\underset{\smile}{}}NH \xrightarrow{120°C}$$

$$2\,HN(C_2H_5)_2 + (CH_3)_2Ge\left\langle \begin{array}{c} N \\ \\ \\ N \end{array} \right] + \left[\begin{array}{c} CH_3 \\ \diagdown \\ —Ge—N\overset{\frown}{\underset{\smile}{}}N— \\ \diagup \\ CH_3 \end{array} \right]_n$$

The reaction of dimethylbis-(diethylamino)germane with ethylene-diamine at 120°C released diethylamine quantitatively and produced a paste with a mixture of various polymers (189).

The NMR and IR spectra and the elementary analyses enable one to establish that the different polymers formed have the following structures:

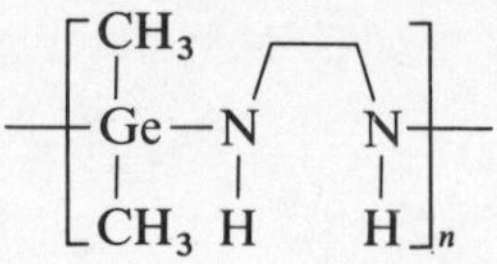

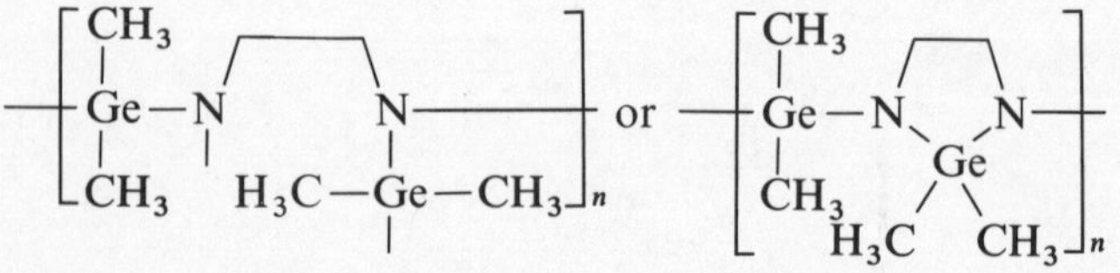

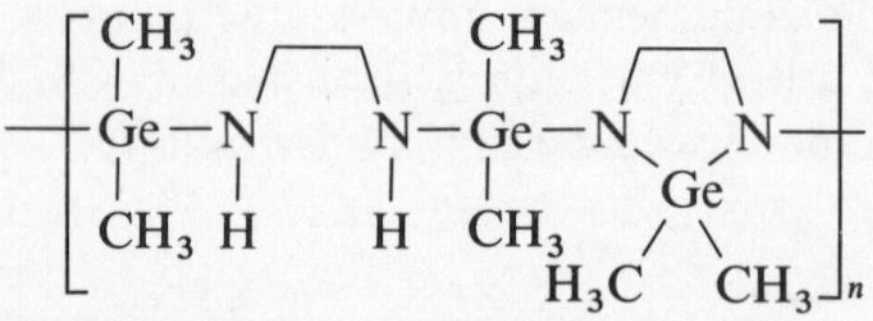

$$
\left[\begin{array}{c} CH_3 \\ | \\ Ge-N \qquad N \\ | \qquad | \qquad | \\ CH_3 \ H \qquad H \end{array} \right]_n
$$

$$
\left[\begin{array}{c} CH_3 \\ | \\ Ge-N \qquad N \\ | \qquad | \\ CH_3 \quad H_3C-Ge-CH_3 \\ | \end{array} \right]_n \quad \text{or} \quad \left[\begin{array}{c} CH_3 \\ | \\ Ge-N \qquad N \\ | \qquad \diagdown Ge \diagup \\ CH_3 \quad H_3C \qquad CH_3 \end{array} \right]_n
$$

$$
\left[\begin{array}{c} CH_3 \qquad\qquad CH_3 \\ | \qquad\qquad\quad | \\ Ge-N \qquad N-Ge-N \qquad N \\ | \qquad | \qquad | \qquad | \quad \diagdown Ge \diagup \\ CH_3 \ H \qquad H \ CH_3 \ H_3C \qquad CH_3 \end{array} \right]_n
$$

The transamination of organosilyldiamines with various ethylene-diamines requires moderately strong heating (150°C) and ammonium sulphate as a catalyst, without which the conversion does not proceed at a reasonable rate. With bis-(diethylamino)diorganogermanes, transamination proceeds at about 100°C in the absence of a catalyst. Tin-diamines undergo transamination at room temperature even exothermically. Thus, the ease of transamination of $(CH_3)_2M[N(C_2H_5)_2]_2$ by ethylene diamines varies in the order:

$$M:Si \qquad < \ Ge \qquad < \ Sn$$

$$\approx 150°C \qquad \approx 100°C \qquad \approx 30°C$$

$$\text{catalyst} \qquad \text{no catalyst} \quad \text{exothermic}$$

The authors (189) consider a possible SN^2 reaction mechanism for transamination:

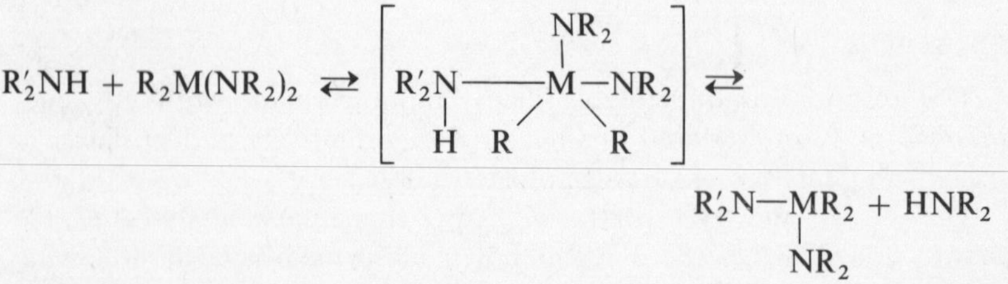

$$
R'_2NH + R_2M(NR_2)_2 \rightleftarrows \left[\begin{array}{c} NR_2 \\ | \\ R'_2N\text{———}M-NR_2 \\ | \quad \diagup \qquad \diagdown \\ H \quad R \qquad R \end{array} \right] \rightleftarrows
$$

$$
R'_2N-MR_2 + HNR_2 \\
\ \ \ \ | \\
\ \ NR_2
$$

8-1-1-4 Reactions of Halo- and Organohalogermanes with Metal Derivatives of Amines. We have already reported the synthesis of Ph_3GeNH_2 (84) and $(Et_3Ge)_2NH$ (113) from triorganochlorogermanes and potassium amide or lithium amide.

This type of synthesis of Ge—N bond derivatives has become generalized for metal derivatives of amines $\diagdown N—M$ (M = Li, Na, K):

$$R_xGeX_{4-x} + (4 - x)M—N \diagup \!\!\!\!\diagdown \; \rightarrow \; R_xGe(—N \diagdown)_{4-x} + (4 - x)MX$$

Tetra N-pyrrolylgermane, $Ge(NC_4H_4)_4$, was among the first derivatives isolated by this method from $GeCl_4$ and C_4H_4NK (153). By the same method Rijkens and coworkers prepared in toluene N-tributylgermylpyrrole, $Bu_3GeNC_4H_4$, N-triphenylgermylpyrrole, $Ph_3GeNC_4H_4$, dibutyldi-N-pyrrylgermane, $Bu_2Ge(NC_4H_4)_2$, as well as N-tributylgermylsuccinimide and phthalimide, $Bu_3GeNC_4H_4O_2$ and $Bu_3GeNC_8H_4O_2$.

The lithium or sodium salts of the amine are more generally used (105, 106). N-Triphenylgermyldiphenylamine, Ph_3GeNPh_2, has been isolated in about 25% yields by interaction of triphenylchlorogermane with Ph_2NLi in ether solution or with Ph_2NNa in THF (14).

From the lithium derivatives of dimethyl- and diethylamine and from the corresponding organochlorogermanes Me_3GeNMe_2 (86), Et_3GeNMe_2, Bu_3GeNMe_2, Et_3GeNEt_2 (27, 121), Bu_3GeNEt_2 (105, 106, 121), Me_3GeNEt_2 (2), $Me_2Ge(NEt_2)_2$ (188), $Ph_2Ge(NMe_2)_2$ (189) have been isolated.

The lithium trimethylgermylmethylamide is obtained by cleaving dimethylgermylmethylamine trimer with CH_3Li (137, 143):

$$[(CH_3)_2GeNCH_3]_3 + 3\,LiCH_3 \; \rightarrow \; 3\,(CH_3)_3GeN \diagup \!\!\!\!\diagdown \begin{smallmatrix} Li \\[4pt] CH_3 \end{smallmatrix}$$

It is highly reactive and, with trimethylchlorosilane, -germane, -stannane and -plumbane affords very good yields of the ammonia derivatives, trimethylsilyl-, trimethylgermyl-, trimethylstannyl- and trimethylplumbyl-trimethylgermylmethylamine (137, 143):

$$(CH_3)_3Ge—N \diagup \!\!\!\!\diagdown \begin{smallmatrix} Li \\[4pt] CH_3 \end{smallmatrix} + (CH_3)_3MCl \; \rightarrow \; (CH_3)_3Ge—\overset{\underset{\displaystyle CH_3}{|}}{N}—M(CH_3)_3$$

$$(M = Si, Ge, Sn, Pb)$$

In dry organic solvents, trimethylchlorogermane reacts with hexamethyldisilazanesodium to give good yields of hexamethyltrimethylgermyldisilazane (127, 128):

$$[(CH_3)_3Si]_2NNa + (CH_3)_3GeCl \rightarrow [(CH_3)_3Si]_2N-Ge(CH_3)_3 + NaCl$$

Dimethyldichlorogermane also reacts with $[(CH_3)_3Si]_2NNa$ giving the corresponding bis-(disilazyl)germane (127):

$$(CH_3)_2GeCl_2 + 2\,[(CH_3)_3Si]_2NNa \rightarrow$$

$$\underset{\underset{CH_3}{|}}{\overset{\overset{CH_3}{|}}{[(CH_3)_3Si]_2N-Ge-N[Si(CH_3)_3]_2}} + 2\,NaCl$$

Et$_3$SiNHLi and dimethyldichlorogermane at room temperature yield bis-(triethylsilylamino)dimethylgermane:

$$2\,(C_2H_5)_3SiNHLi + (CH_3)_2GeCl_2 \rightarrow$$

$$\underset{\underset{CH_3}{|}}{\overset{\overset{CH_3}{|}}{(C_2H_5)_3SiNH-Ge-NHSi(C_2H_5)_3}} + 2\,LiCl$$

A metalation with BuLi gives the dilithium derivative which reacts with dimethyldichlorostannane to give 1,3-bis-triethylsilyl-2,2,4,4-tetramethyl-cyclo-2-germa-4-stanna-1-3-diazane (125).

$$\underset{\underset{Li}{|}\;\;\underset{CH_3}{|}\;\;\underset{Li}{|}}{\overset{\overset{CH_3}{|}}{(C_2H_5)_3SiN-Ge-NSi(C_2H_5)_3}} + (CH_3)_2SnCl_2 \rightarrow$$

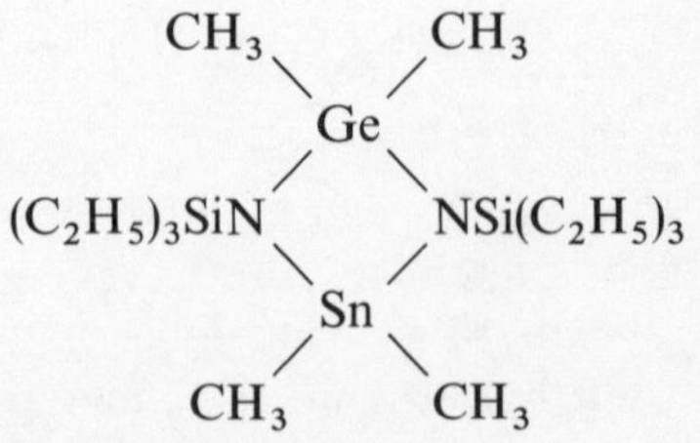

$(C_2H_5)_3SiNHLi$ and trimethylchlorogermane give (triethylsilyl) (triethylgermyl)amine:

$$(C_2H_5)_3SiNHLi + (CH_3)_3GeCl \rightarrow \underset{\underset{H}{|}}{(C_2H_5)_3Si-N-Ge(CH_3)_3}$$

A new metalation with a BuLi followed by treating trimethylchlorostannane in Et_2O at room temperature yielded (triethylsilyl)(trimethylgermyl)(trimethylstannyl)amine (125):

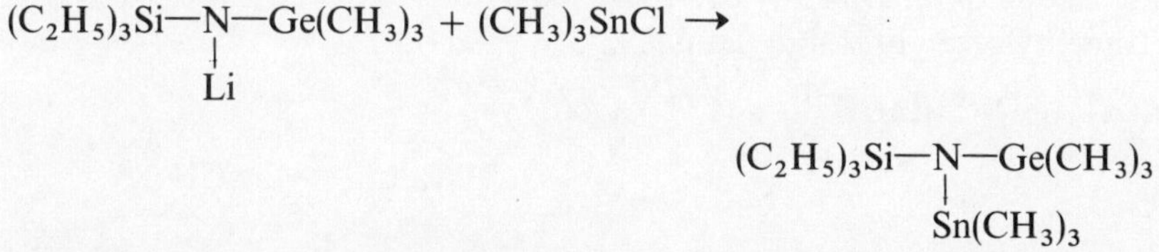

$$(C_2H_5)_3Si\text{—}\underset{\underset{Li}{|}}{N}\text{—}Ge(CH_3)_3 + (CH_3)_3SnCl \rightarrow$$

$$(C_2H_5)_3Si\text{—}\underset{\underset{Sn(CH_3)_3}{|}}{N}\text{—}Ge(CH_3)_3$$

On metalation with butyllithium, (trimethylsilylmethylamino)dimethyl silylamine (I) was converted to lithium-(trimethylsilylmethylamino)-dimethylsilylamide (II).

Treating this salt with trimethylchlorogermane in ether gives 50% yield of (trimethylsilylmethylamino)-dimethylsilyl-(trimethylgermyl)-amine (III) (124):

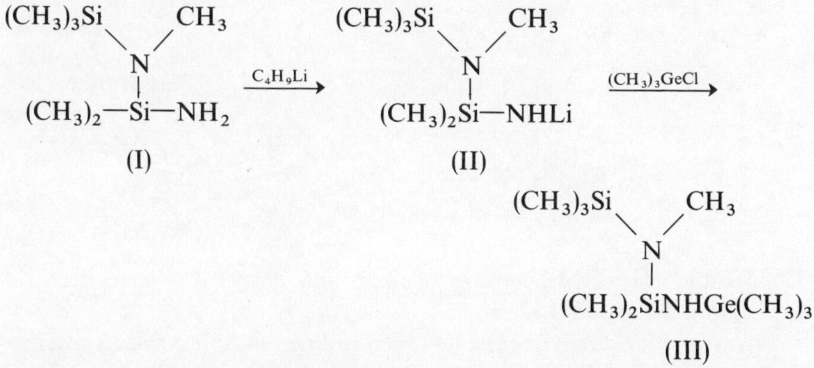

(CH₃)₃Si, CH₃ — N — (CH₃)₂—Si—NH₂ (I) →(C₄H₉Li) (CH₃)₃Si, CH₃ — N — (CH₃)₂Si—NHLi (II) →((CH₃)₃GeCl)

(CH₃)₃Si, CH₃ — N — (CH₃)₂SiNHGe(CH₃)₃ (III)

Further metalation of (III) with methyllithium gives lithium-(trimethylsilylmethylamino)dimethylsily(trimethylgermyl)amide IV which with trimethyltin chloride affords (trimethylsilylmethylamino)dimethylsilyl-(trimethylgermyl)-(trimethylstannyl)amine (V) (124):

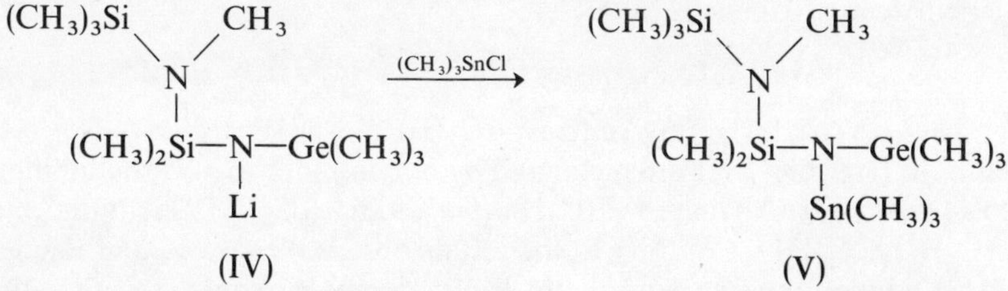

(CH₃)₃Si, CH₃ — N — (CH₃)₂Si—N—Ge(CH₃)₃ | Li (IV) →((CH₃)₃SnCl) (CH₃)₃Si, CH₃ — N — (CH₃)₂Si—N—Ge(CH₃)₃ | Sn(CH₃)₃ (V)

The synthesis and the properties of the metal-organic derivatives of N-trimethylsilyl-*tert*-butylamine, $(CH_3)_3SiNH\text{-}t\text{-}Bu$, have been described by Schumann-Ruidisch and coworkers (144). The lithium derivative of this silylamine, $(CH_3)_3SiN(Li)t$-Bu, leads by reaction with trimethyl-chlorosilane to bis-(N-trimethylsilyl)-*tert*-butylamine$[(CH_3)_3Si]_2N$—$C(CH_3)_3$ in a 50% yield. On the other hand, under the same conditions, the action of the trimethylchlorogermane leads to the N-trimethylsilyl-N-trimethylgermyl-*tert*-butylamine in 5% yield only.

$(CH_3)_3SiN(Li)C(CH_3)_3 + (CH_3)_3GeCl \rightarrow$

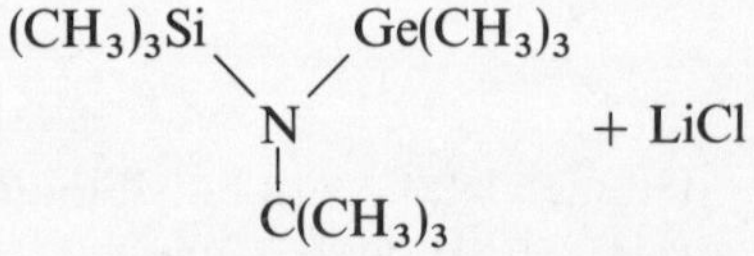

$$(CH_3)_3Si \diagdown \quad \diagup Ge(CH_3)_3$$
$$N \qquad\qquad + LiCl$$
$$| $$
$$C(CH_3)_3$$

The yield could be increased up to 25% via the N-trimethylsilyl-N-dimethylchlorogermyl-*tert*-butylamine methylated by methyllithium:

$(CH_3)_3SiN(Li)C(CH_3)_3 + (CH_3)_2GeCl_2 \rightarrow$

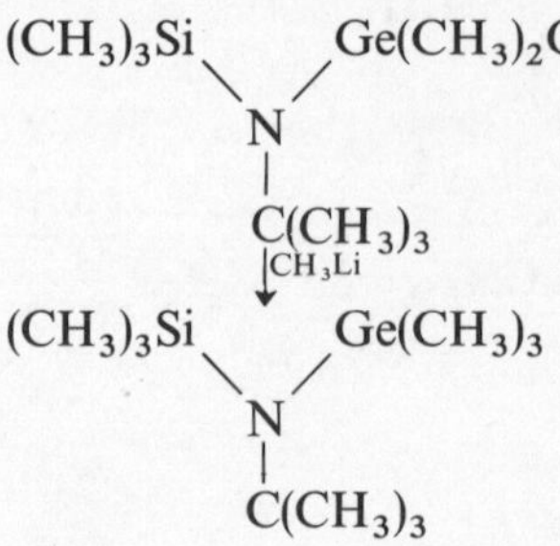

$$(CH_3)_3Si \diagdown \quad \diagup Ge(CH_3)_2Cl$$
$$N$$
$$|$$
$$C(CH_3)_3$$
$$\xrightarrow{[CH_3Li]}$$
$$(CH_3)_3Si \diagdown \quad \diagup Ge(CH_3)_3$$
$$N$$
$$|$$
$$C(CH_3)_3$$

The IR and ^{1}H—NMR spectra of these new derivatives have been recorded and discussed (144).

Tris-(trimethylgermyl)amine has been isolated in 55% yield by interaction, in a THF medium, of trimethylchlorogermane with lithium nitride (113):

$$3\ Me_3GeCl + Li_3N \xrightarrow{\ THF\ } (Me_3Ge)_3N + 3\ LiCl$$

as well as by the action of Me_3GeCl on the lithium derivative of bis-(triethylgermyl)amine (113):

$$(Me_3Ge)_2NLi + Me_3GeCl \xrightarrow{\ ether\ } (Me_3Ge)_3N + LiCl$$

The bis-(trialkylgermyl)amines $[(C_2H_5)_3Ge]_2NH$, $[(C_3H_7)_3Ge]_2NH$ and $[(C_4H_9)_3Ge]_2NH$ are metalated by butyllithium to give the lithium-bis-(trialkylgermyl)amides $[(C_2H_5)_3Ge]_2NLi$, $[(C_3H_7)_3Ge]_2NLi$ and $[(C_4H_9)_3Ge]_2NLi$. Contrary to the results observed by Satgé and Baudet (113), these amides react with trialkylbromogermanes to give the tris-(trialkylgermyl)amines $[(C_2H_5)_3Ge]_3N$, $[(C_3H_7)_3Ge]_3N$ and $[(C_4H_9)_3Ge]_3N$ (75).

The reaction of diphenyldichlorogermane with the azobenzene dilithium adduct gives octaphenyl-1,2,4,5-tetraaza-3,6-digermacyclohexane (51):

$$2\,C_6H_5N(Li)\!-\!N(Li)C_6H_5 + 2\,Ph_2GeCl_2 \rightarrow$$

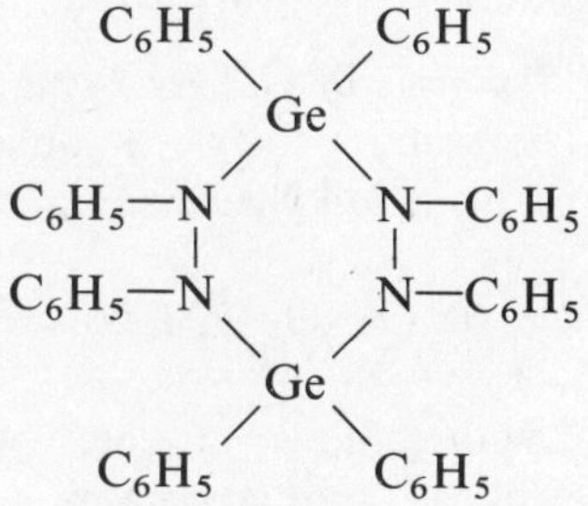

N-dilithiopentafluoroaniline reacts with diphenyldichlorogermane to give a cyclodigermazane which is the first four-membered germanium–nitrogen ring reported (60):

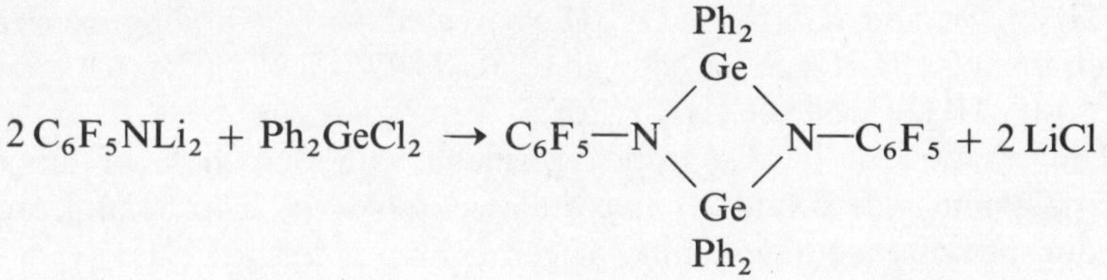

$$2\,C_6F_5NLi_2 + Ph_2GeCl_2 \rightarrow$$

Reaction of 1,3-bis-(methylamino)pentamethyldisilazane (after metalation or in the presence of triethylamine) with dichloroderivatives of germanium yields cyclosilazanes with germanium as a member of the ring (179):

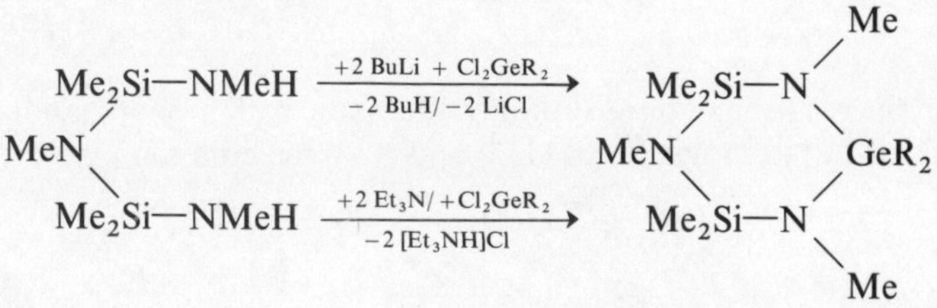

$$(R = Cl,\ n\text{-Bu})$$

Lastly, the aminomagnesium derivatives of secondary and primary amines are equally used with excellent results in the synthesis of organoaminogermane (113, 121):

$$R_3GeCl + R'_2NMgBr \xrightarrow{THF} R_3GeNR'_2 + MgBrCl$$

$$R_3GeCl + R'NHMgBr \xrightarrow{THF} R_3GeNHR' + MgBrCl$$

$$2\,R_3GeCl + C_6H_5N(MgBr)_2 \xrightarrow{THF} (R_3Ge)_2NC_6H_5 + 2\,MgBrCl$$

8-1-1-5 Synthesis of Ge—N *Bond Derivatives from* Ge—O *Bond Compounds.* The reaction between some amines having a marked "acidic" nature and bis-(trialkylgermanium) oxides:

$$(R_3Ge)_2O + 2\,HN\!\!\diagup^{\diagup} \rightarrow 2\,R_3GeN\!\!\diagup^{\diagup} + H_2O$$

at 150°C, with removal of water by slow addition of dry toluene, has enabled Rijkens and coworkers to isolate the *N*-trialkylgermanium compounds of pyrazole, imidazole and 1,2,4-triazole (105, 106).

Dimethyldiphenoxygermane $(CH_3)_2Ge(OC_6H_5)_2$ when added to an ether solution of $(C_2H_5)_2NLi$ in excess leads, after boiling for a few hours, to dimethylbis-(diethylamino)germanium $(CH_3)_2Ge[N(C_2H_5)_2]_2$. $(C_2H_5O)_4Ge$ and $(CH_3)_2Ge(OC_6H_5)_2$ treated with a boiling benzene solution of $[(CH_3)_3Si]_2NNa$ give $(C_2H_5O)_3GeN[Si(CH_3)_3]_2$ and $(CH_3)_2(C_6H_5O)GeN[Si(CH_3)_3]_2$ (76).

The reaction of tetraisopropoxygermane with one mole of monoethanolamine was found to liberate two moles of isopropanol, and diisopropoxymonoethanolamine of germanium is formed (91):

$$(i\text{-}C_3H_7O)_4Ge + \begin{matrix} HO{-}CH_2 \\ | \\ H_2N{-}CH_2 \end{matrix} \rightarrow$$

$$2\,i\text{-}C_3H_7OH + [i\text{-}C_3H_7O]_2Ge\!\!\diagup^{O{-}CH_2}_{\diagdown NH{-}CH_2}\!\!\Big|$$

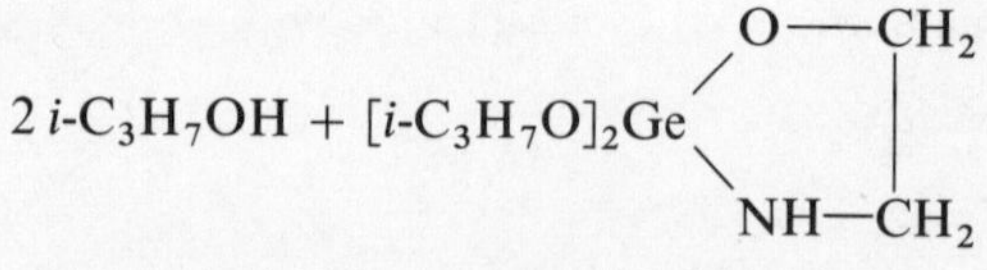

When the reaction of tetraisopropoxygermane with monoethanolamine was carried out in molar ratios (1:2) and (1:3) the compounds:

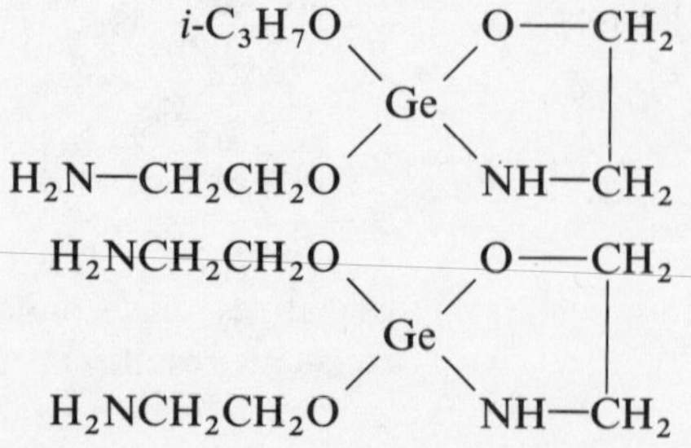

$$\begin{matrix} i\text{-}C_3H_7O & & O{-}CH_2 \\ \diagdown & & \diagup & | \\ & Ge & \\ \diagup & & \diagdown & \\ H_2N{-}CH_2CH_2O & & NH{-}CH_2 \end{matrix}$$

$$\begin{matrix} H_2NCH_2CH_2O & & O{-}CH_2 \\ \diagdown & & \diagup & | \\ & Ge & \\ \diagup & & \diagdown & \\ H_2NCH_2CH_2O & & NH{-}CH_2 \end{matrix}$$

were obtained in quantitative yield (91). These derivatives are white solids, slightly soluble or almost insoluble in benzene.

8-1-1-6 Other Syntheses of the Ge—N Bond. When azobenzene and triphenylgermyllithium in $1:1$ molar ratio were allowed to react in tetrahydrofuran, an 85% yield of NN′-diphenyl-N-triphenylgermylhydrazine was obtained (50):

$$C_6H_5N{=}NC_6H_5 + (C_6H_5)_3GeLi \rightarrow C_6H_5{-}\underset{\underset{Ge(C_6H_5)_3}{|}}{\overset{\overset{Li}{|}}{N}}{-}N{-}C_6H_5 \xrightarrow{H_2O}$$

$$C_6H_5{-}\underset{\underset{Ge(C_6H_5)_3}{|}}{\overset{\overset{H}{|}}{N}}{-}N{-}C_6H_5$$

The reaction of triphenylgermyllithium with azoxybenzene also gives N,N′-diphenyl-N-triphenylgermylhydrazine. In this reaction it has been observed that azoxybenzene is first reduced to azobenzene which then reacts with germyllithium compounds to give the corresponding addition products (50).

Redistribution reactions between $Ge(NMe_2)_4$ and $GeCl_4$ and between $Ge(OMe)_4$ and $Ge(NMe_2)_4$ have been investigated by Burch and van Wazer (19); $ClGe(NMe_2)_3$, $Cl_2Ge(NMe_2)_2$ and $Cl_3Ge(NMe_2)$ are formed in the first reaction, but large deviations from statistically random interchange of the substituent were observed in this system. In the second reaction, the formation of $(Me_3O)_3GeNMe_2$, $(MeO)_2Ge(NMe_2)_2$ and $(MeO)Ge(NMe_2)_3$ is observed, in a variable amount, according to the relative proportions in the initial reactants. These exchange reactions are followed by NMR.

A new synthesis reaction of the Ge—N bond by dehydrocondensation, over a cobalt catalyst, of trialkylgermanes with primary amines, has been described (115):

$$Et_3GeH + R'NH_2 \xrightarrow[160°C]{Co} Et_3GeNHR + H_2$$

N-organosilylketimines and a few germanium and tin analogues were prepared by reaction of the lithium derivatives of ketimines, $\diagdown C{=}NLi$, with chlorosilanes, -germanes and -stannanes (87):

$$Ph_2C{=}NLi + R_3GeX \rightarrow Ph_2C{=}N{-}GeR_3 \qquad (R = Me, Ph)$$
$$(X = Cl, Br)$$

Reaction of N-phosphinyllithium amide with Me_3MCl affords di-(*tert*-butyl)phosphinyltrimethylmetal amides:

$$[(CH_3)_3C]_2P-NH-Li \xrightarrow[LiCl]{(CH_3)_3MCl} [(CH_3)_3C]_2P-NH-M(CH_3)_3$$

$$(M = Si, Ge, Sn, Pb)$$

Further metalation of these compounds and treatment with $Me_3M'Cl$ ($M' = Si, Ge, Sn$) gives N—P—organometalphosphinimine (129, 130):

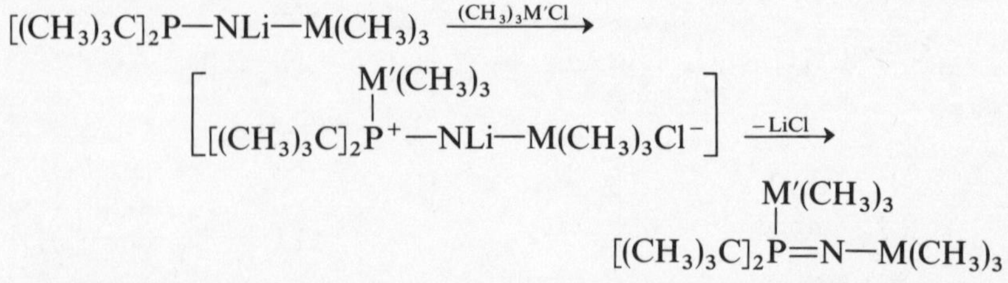

$$[(CH_3)_3C]_2P-NLi-M(CH_3)_3 \xrightarrow{(CH_3)_3M'Cl}$$

$$\left[\begin{array}{c} M'(CH_3)_3 \\ | \\ [(CH_3)_3C]_2P^+-NLi-M(CH_3)_3Cl^- \end{array} \right] \xrightarrow{-LiCl}$$

$$\begin{array}{c} M'(CH_3)_3 \\ | \\ [(CH_3)_3C]_2P=N-M(CH_3)_3 \end{array}$$

Oxidation of aminophosphines by trimethylsilylazide yields N-silylated aminophosphinimines (for example: $(Me_3C)_2P(=NSiMe_3)(NHSiMe_3)$) which can be alcoholized, metalated (by *n*-BuLi) and transformed into different substituted N- and N,N'-organometallic aminophosphinimines (131):

$$\left. \begin{array}{l} (Me_3C)_2P(=NMMe_3)(NHR) \\ (Me_3C)_2P(=N-MMe_3)(NRM'Me_3) \\ (Me_3C)_2P(=N-MMe_3)(NMe_2) \end{array} \right\} \begin{array}{l} (R = H \text{ or } CH_3) \\ (M = M' = Si, Ge, Sn) \end{array}$$

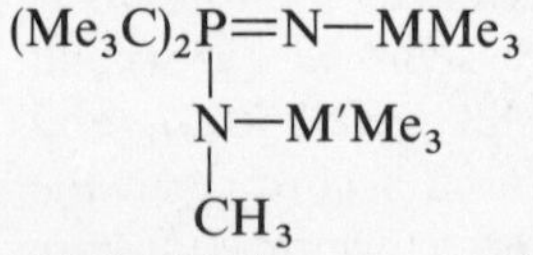

$$\begin{array}{c} (Me_3C)_2P=N-MMe_3 \\ | \\ N-M'Me_3 \\ | \\ CH_3 \end{array}$$ forms stable N—N'-organometallic aminophos-

phinimine isomers and can be used as ^1H NMR reference signal to identify Me_3M- ligands which are bonded to imine or amine nitrogen atoms (131).

Scherer and Hornig (126) recently described the synthesis and NMR spectral properties of silicon, germanium, tin and lead benzamidines. In the addition of lithium-N-trimethylsilylalkylamide on benzonitrile the N'-lithium-N-trimethylsilyl-N-alkyl-benzamidines are isolated in good yields:

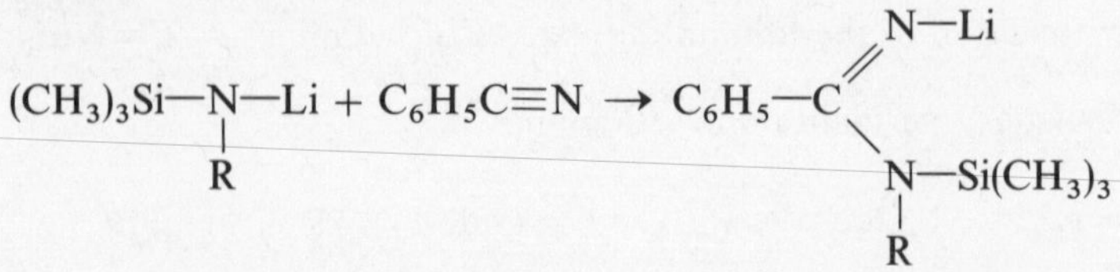

$$(CH_3)_3Si-N-Li + C_6H_5C\equiv N \rightarrow C_6H_5-C \begin{array}{c} \nearrow N-Li \\ \\ \searrow N-Si(CH_3)_3 \end{array}$$
$$\begin{array}{cc} | & \qquad\qquad\qquad\qquad | \\ R & \qquad\qquad\qquad\qquad R \end{array}$$

$$(R = CH_3, C_2H_5)$$

These benzamidines react with the organohalides R_3MCl (M = Si, Ge, Sn, Pb) according to the following scheme:

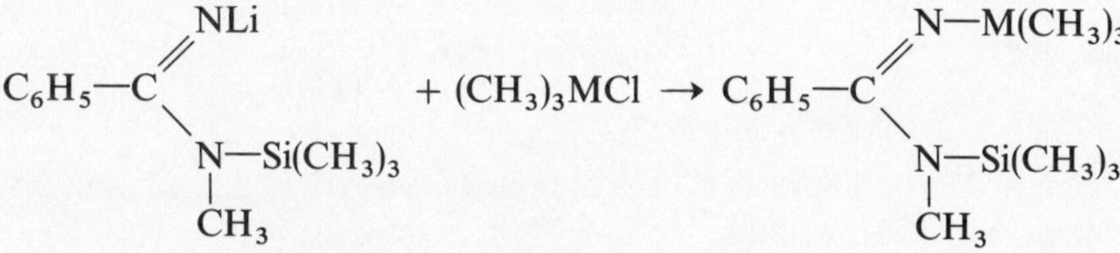

$$C_6H_5-C\begin{array}{c} \diagup\diagup NLi \\ \diagdown N-Si(CH_3)_3 \\ | \\ CH_3 \end{array} + (CH_3)_3MCl \rightarrow C_6H_5-C\begin{array}{c} \diagup\diagup N-M(CH_3)_3 \\ \diagdown N-Si(CH_3)_3 \\ | \\ CH_3 \end{array} \quad (I)$$

NMR study of these derivatives allows it to be established that the M = Si compound exists in two isomeric forms:

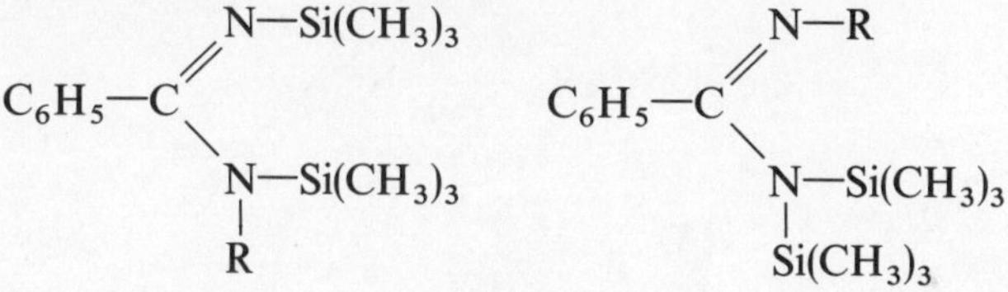

$$C_6H_5-C\begin{array}{c} \diagup\diagup N-Si(CH_3)_3 \\ \diagdown N-Si(CH_3)_3 \\ | \\ R \end{array} \qquad C_6H_5-C\begin{array}{c} \diagup\diagup N-R \\ \diagdown N-Si(CH_3)_3 \\ | \\ Si(CH_3)_3 \end{array}$$

The derivatives with M = Ge, Sn, Pb exist only in a single form of the type:

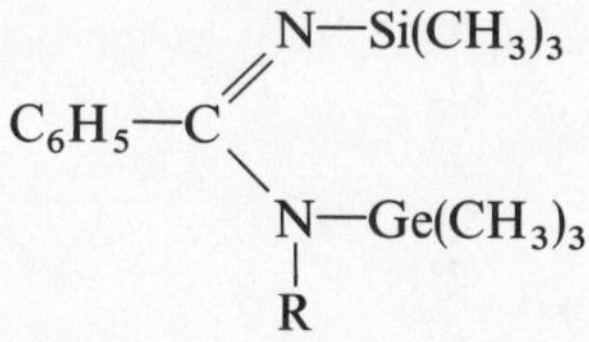

$$C_6H_5-C\begin{array}{c} \diagup\diagup N-Si(CH_3)_3 \\ \diagdown N-Ge(CH_3)_3 \\ | \\ R \end{array}$$

It seems in the latter cases that the reaction (I) takes place according to the above scheme but that an irreversible exchange reaction occurs afterwards, probably via a penta-coordinated germanium:

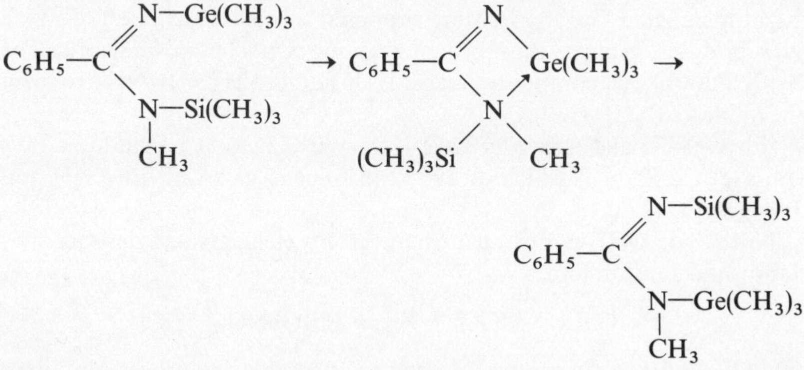

$$C_6H_5-C\begin{array}{c} \diagup\diagup N-Ge(CH_3)_3 \\ \diagdown N-Si(CH_3)_3 \\ | \\ CH_3 \end{array} \rightarrow C_6H_5-C\begin{array}{c} N \\ \diagdown \\ Ge(CH_3)_3 \\ \diagup \\ N \\ \diagdown \\ (CH_3)_3Si \qquad CH_3 \end{array} \rightarrow$$

$$C_6H_5-C\begin{array}{c} \diagup\diagup N-Si(CH_3)_3 \\ \diagdown N-Ge(CH_3)_3 \\ | \\ CH_3 \end{array}$$

The N'-lithium-N,N-dimethyl benzamidine forms as an adduct from $(CH_3)_2NLi$ and benzonitrile. $(CH_3)_3GeCl$ reacts on this derivative and leads to the corresponding benzamidine (126):

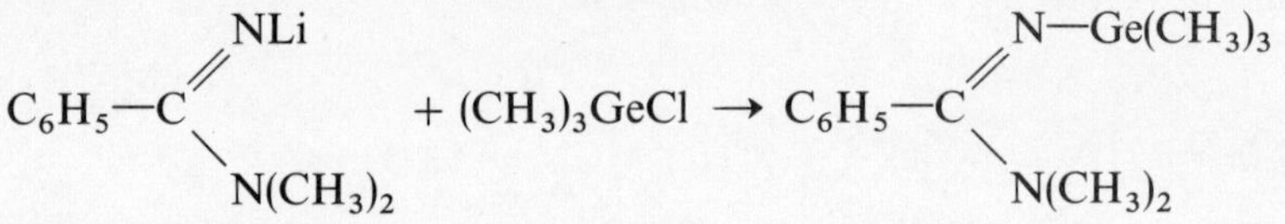

$$C_6H_5-C\begin{matrix}NLi\\\\N(CH_3)_2\end{matrix} + (CH_3)_3GeCl \rightarrow C_6H_5-C\begin{matrix}N-Ge(CH_3)_3\\\\N(CH_3)_2\end{matrix}$$

Digermylcarbodiimide has been obtained from germyl fluoride and bis-(trimethylsilyl)carbodiimide.

The vibrational spectra imply that (I) above is a carbodiimide (A) rather than a cyanamide (B) (24):

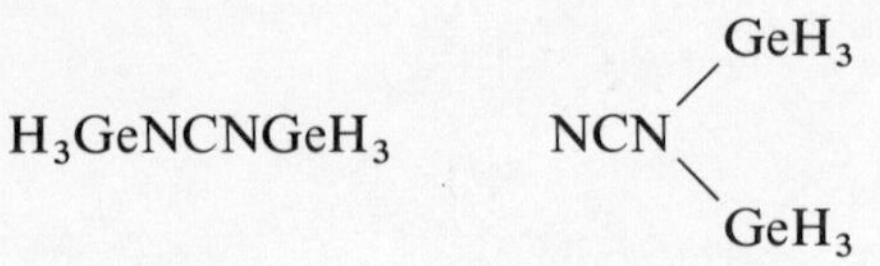

$$H_3GeNCNGeH_3 \qquad NCN\begin{matrix}GeH_3\\\\GeH_3\end{matrix}$$

(A) (B)

(I)

The ammonolysis of tetraphenylgermane $Ge(C_6H_5)_4$ in liquid NH_3 in the presence of KNH_2 yields the potassium salt of an imidogermazane having probably $K_3(Ge_4N_3NH)(NH)_4$ as the smallest possible structure unit:

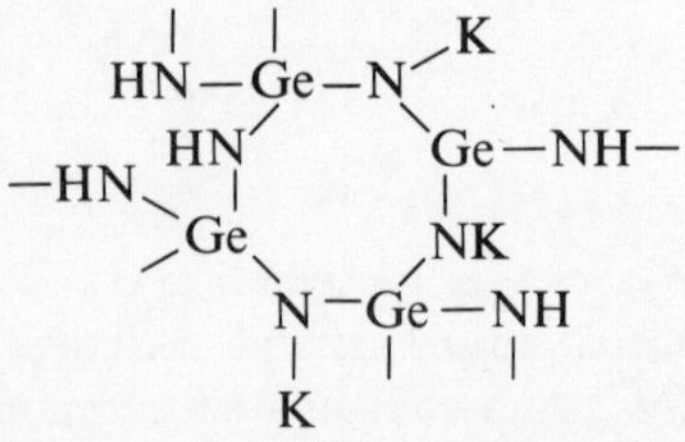

Eight-membered Ge_4N_4 rings are supposed to exist which are linked via NH bridging groups, thus bringing about a polymeric network. The X-ray amorphous compound reacts with liquid NH_4NO_3/NH_3 to form the imide $Ge(NH)_2$ (139).

$(Ph_3Ge)_2Cd$ and excess of Et_2NH at 100°C in toluene solution gave $(Ph_3Ge)_2Cd \cdot Et_2NH$, which at 170°C in toluene gave Cd, Ph_3GeH and Ph_3GeNEt_2 (178).

Nitroso-derivatives of fourth main group elements are described by Jappy and Preston (68).

$$Ph_3GeLi + NOCl \xrightarrow[<-30°]{THF} Ph_3GeNO + LiCl$$

IR, UV and visible spectra of these compounds have been discussed (68).

Preparation and properties of a new organogermanium dicyanamide $(C_6H_5)_3GeN(CN)_2$ are reported by Köhler and Beck (78). The structure of this compound has been discussed using results of infrared-spectroscopic measurements.

8-1-2 Properties of germanium–nitrogen bond derivatives

The extreme sensitivity of the germanium–nitrogen bond to hydrolysis had been previously reported by several authors (69). It seems established that hydrolysis as well as the action of protonic reagents proceeds by an electrophilic attack on the nitrogen atom:

$$R_3GeNR'_2 + A\!-\!H \rightarrow R_3GeA + R'_2NH$$

The nucleophilic action of substrate A^- on the germanium atom has a much smaller influence on the reaction process.

According to Rijkens and coworkers (88, 105, 106) the following mechanism may be expected:

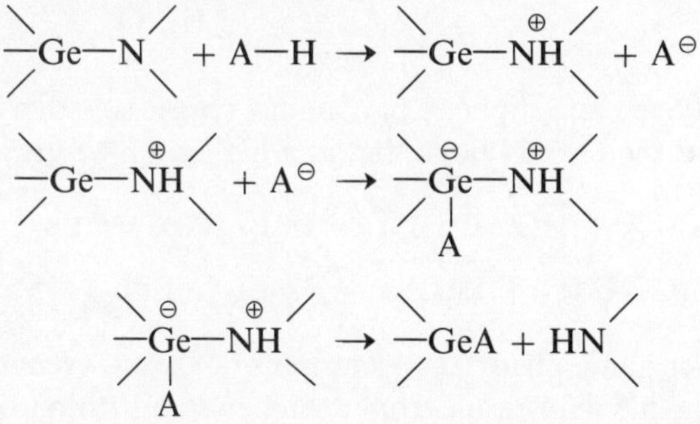

But the reaction is also able to proceed in one step with the formation of the following transition state:

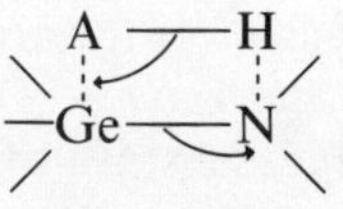

The evidence for the electrophilic attack by H^+ was established by the same authors (88, 105, 106), who showed that the order of reactivity of phenol, ethyl alcohol, tert-butanol, phenylacetylene and octylamine on various organogermylamines corresponds to their acidity and does not follow the order of the increasing nucleophilic character. It appears that the germanium–nitrogen bond of organogermylamines, $R_3GeNR'_2$, is very reactive to reagents with some acidic character like phenols, alcohols, thiols, and, of course, to organic acids (105, 120, 121):

$$R_3GeNR_2 + R'OH \rightarrow R_3GeOR' + R_2NH$$

$$R_3GeNR_2 + R'SH \rightarrow R_3GeSR' + R_2NH$$

$$R_2GeNR_2 + R'COOH \rightarrow R_3GeO\!-\!\underset{\underset{O}{\|}}{C}\!-\!R' + R_2NH$$

The reaction with thiols seems to be among the best synthetic methods for organothiogermanium compounds with a Ge—S bond. Primary, secondary and tertiary thiols all react with equal ease with the Ge—N bond (120, 121).

The transamination–elimination reactions with primary amines and amides belonging to the same mechanism are also very easy to effect (121):

$$R_3GeNMe_2 + R'NH_2 \rightarrow R_3GeNHR' + Me_2NH$$

$$Et_3GeNMe_2 + CH_3CONH_2 \rightarrow Et_3Ge\!-\!\underset{\underset{H}{|}}{N}\!-\!\underset{\underset{O}{\|}}{C}\!-\!CH_3 + Me_2NH$$

Phenylacetylene and diphenylphosphine, reagents with a low protonic activity, cleave the Ge—N bond, but at a higher temperature (121):

$$Bu_3GeNMe_2 + HC\equiv C\!-\!Ph \xrightarrow{150°C} Bu_3GeC\equiv CPh$$

$$Et_3GeNMe_2 + Ph_2PH \xrightarrow{170°C} Et_3GePPh_2 + Me_2NH$$

On the other hand, phenylacetylene does not react, even at 200°C, with $(Bu_3Ge)_2NH$: this may result from a higher stabilization of the Ge—N bonds by interaction $p\pi$–$d\pi$ between the free electron pair of nitrogen and the empty 4d orbitals of germanium (105). This interaction, particularly clear in the silylamines with the participation of silicon 3d orbitals, accounts for the decreasing sensitivity towards hydrolysis of the R_3SiNR_2, $R_3SiNHSiR_3$ and $(R_3Si)_3N$ derivatives. The $(H_3Si)_3N$ and $(Me_3Si)_3N$ molecules remain particularly insensitive towards water (35).

$(Me_3Ge)_3N$ hydrolysis in air seems slower than that of $(Me_3Ge)NH$ (111) but Satgé and Baudet have observed that the trimethylgermylamines of the three types Me_3GeNMe_2, $(Me_3Ge)_2NH$, $(Me_3Ge)_3N$ remain very sensitive towards active protonic reagents like water or alcohols.

Diphenylphosphine also cleaves $(Me_3Ge)_2NH$ and $(Me_3Ge)_3N$ at about 200°C:

$$(Me_3Ge)_2NH \text{ or } (Me_3Ge)_3N + Ph_2PH \rightarrow Me_3GePPh_2 + NH_3$$

A rather clear difference in reactivity is only observed in the action of octylamine (113):

REAGENT	REACTION TEMPERATURE (°C)	REACTION PRODUCTS	YIELD (%)
Me_3GeNMe_2 $n\text{-}C_8H_{17}NH_2$	25–60	$(Me_3Ge)_2NC_8H_{17}$	82
$(Me_3Ge)_2NH$ "	50–100	"	60
$(Me_3Ge)_3N$ "	120–180	"	54

The stabilization of Ge—N bonds by $p\pi$–$d\pi$ interactions between the free electron pair of nitrogen and the germanium 4 d orbitals seems to increase in the germylamines from $R_3GeNR'_2$ to $(R_3Ge)_3N$, but, however, it remains rather low and, at all times, much less significant than in analogous silylamines (113).

One must note, however, the surprising stability towards hydrolysis of trimethylgermyl-bis-(trimethylsilyl)amine, $[(CH_3)_3Si]_2NGe(CH_3)_3$, reported by Scherer and Schmidt (127, 128) and the incomprehensible stability towards boiling aqueous 3N sodium hydroxide recently reported for $(CH_3)_2Ge[N(C_2H_5)_2]_2$ and $RR'_2GeN[Si(CH_3)_3]_2$ (R and R' = OC_2H_5 and R = OC_6H_5, R' = CH_3) (76).

Digermylcarbodiimide reacts readily with protonic acids, ROH(R = H, CH_3, CD_3), to give $(H_3Ge)_2O$ (R = H) or $H_3GeOR(R = CH_3, CD_3)$ (24). This appears to be a reaction of some importance in the preparation of germyl compounds. These compounds, which were readily obtained in this way, have been prepared only with difficulty and in low yields by other routes. Digermylcarbodiimide reacts also with silyl bromide to give germyl bromide and disilylcarbodiimide (24).

Tributylgermyldiethylamine does not react, even above 200°C, with tributylgermane (105) and this seems to confirm the $R_3\overset{\delta+}{Ge}$—$\overset{\delta-}{H}$ polarity of the trialkylgermane Ge—H bond.

The electrophilic attack on nitrogen is highly dependent on the availability of the nitrogen free electron-pair. This availability can be reduced by participation of the lone pair in π-bonding with a neighbouring atom or group. This last assumption has been confirmed by the very low reactivity of pyrrole and indole compounds where the electron pair takes part in the π-system of the ring. N-triphenylgermylpyrrole, indeed, can be dissolved in acetone and precipitated by addition of water without decomposition. N-tributylgermylpyrrole is not appreciably hydrolysed upon heating with an equivalent amount of water in a Carius tube at 180°C (105, 106). Similarly, N-tributylgermyl-substituted succinimide and phthalimide are stable towards ethyl alcohol and do not react or only react very slowly with phenol. The comparatively low reactivity of the succinimide and phthalimide derivatives can also be understood in terms of a decreased availability of the free electron pair at the nitrogen atom owing to resonance between the nitrogen atom and the neighbouring carbonyl groups (105, 106).

The high level of reactivity of the pyrazole, imidazole and triazole derivatives in which the electron pair of the nitrogen atom linked with germanium is even more involved in the π-system of the ring is surprising. These ring-systems, however, have a second nitrogen atom: the free electron-pair of this nitrogen atom is not involved in the aromatic π-system and, consequently, is very prone to electrophilic attack. The reaction sequence for these molecules is probably (105, 106):

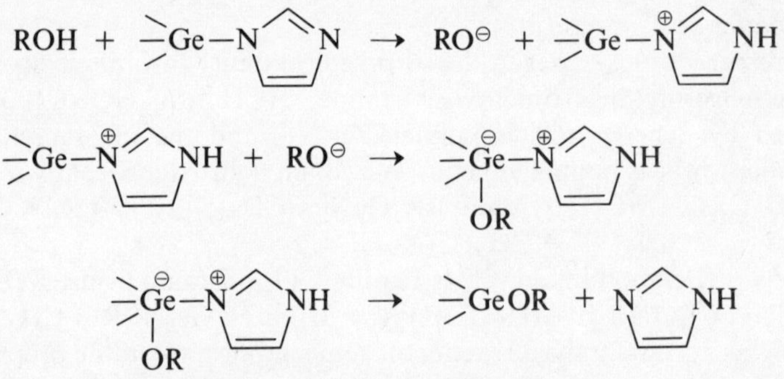

Alternatively, when reactions occur in one step a transition state as for example:

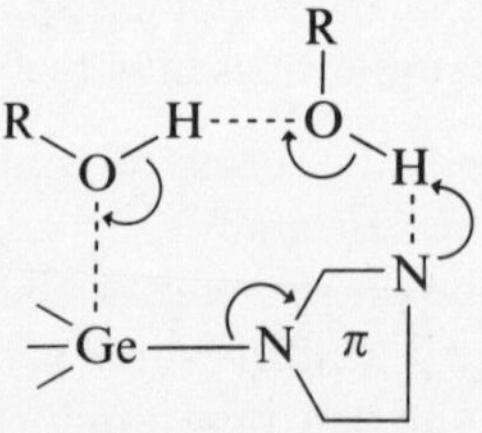

may be involved.

N,N'-Diphenyltriphenylgermylhydrazine,

$$(C_6H_5)_3GeN(C_6H_5)\!-\!NH(C_6H_5),$$

has been isolated in good yields after hydrolysis of

$$(C_6H_5)_3GeN(C_6H_5)NLi(C_6H_5)$$

with water or with a saturated aqueous solution of ammonium chloride (50). In this compound, the basicity of the nitrogen atom linked with germanium is lowered by resonance with the phenyl group.

Mironov and coworkers have recently published some cleavage reactions of bis-(trialkylgermyl) amines (96, 97):

$$(R_3Ge)_2NH + 2 HC \equiv C—C_6H_5 \xrightarrow{150–180°C}$$

$$2 R_3GeC \equiv C—C_6H_5 + NH_3 \quad (R = CH_3, C_2H_5)$$

$$(Et_3Ge)_2NH + 2 HC \equiv C—SiMe_3 \rightarrow 2 Et_3GeC \equiv C—SiMe_3 + NH_3$$

$$(Et_3Ge)_2NH + 2 HC \equiv CCH_2OH \rightarrow 2 Et_3GeOCH_2C \equiv CH + NH_3$$

$$(Et_3Ge)_2NH + HC \equiv C—CH_2Cl \rightarrow Et_3GeCl + polymers$$

$$2 Me_3GeNHSiMe_3 + 2 HC \equiv C—C_6H_5 \rightarrow$$

$$2 Me_3GeC \equiv C—C_6H_5 + [Me_3Si]_2NH + NH_3$$

$$R_3GeNHGeR_3 \xrightarrow{AlCl_3} 2 R_3GeCl$$

$$(R_3Ge)_2NH + 2 R_3'GeCl \leftrightarrows R_3'GeNHGeR_3' + 2 R_3GeCl$$

$$(Me_3Ge)_2NH + 2 Et_3SiCl \leftrightarrows Et_3SiNHSiEt_3 + 2 (CH_3)_3GeCl$$

$$(Et_3Ge)_2NH + AlEt_3 \rightarrow Et_2AlN[GeEt_3]_2 + C_2H_6$$

$$(Et_2GeNH)_3 + 6 CH_3OH \rightarrow 3 Et_2Ge(OCH_3)_2 + 3 NH_3$$

Triphenyl(diethylamino)germane reacts with sodium acetylide and triphenylchlorosilane to give (triphenylsilyl)(triphenylgermyl)acetylene:

$$Ph_3GeNEt_2 + NaC \equiv CH \rightarrow Ph_3GeC \equiv CNa + Et_2NH$$

$$Ph_3GeC \equiv CNa + Ph_3SiCl \rightarrow Ph_3GeC \equiv CSiPh_3 + NaCl$$

Alternatively, the reaction could proceed by a second course in which the organometalalkyne is the primary product, which then reacts with the organometalamino compounds (42):

$$Ph_3SiCl + NaC \equiv CH \rightarrow Ph_3SiC \equiv CH + NaCl$$

$$Ph_3GeNEt_2 + HC \equiv C—SiPh_3 \rightarrow Ph_3GeC \equiv C—SiPh_3 + Et_2NH$$

Whereas trialkyltindiethylamines and triphenyltinhydride react exothermally, the reaction of tributylgermyldiethylamine and triphenyltin hydride proceeds only slowly in refluxing butyronitrile (26):

$$(C_4H_9)_3GeN(C_2H_5)_2 + (C_6H_5)_3SnH \rightarrow$$

$$(C_4H_9)_3Ge—Sn(C_6H_5)_3 + HN(C_2H_5)_2$$

The diminished reactivity of organogermanium–nitrogen compounds is consistent with the occurrence of significant $d\pi$–$p\pi$ interaction in the Ge—N as contrasted with the Sn—N bond.

The synthesis of pentabutylgermyltin monohydride has been achieved by reaction between tributylgermyldimethylamine and dibutyltin dihydride:

$$(C_4H_9)_3GeN(CH_3)_2 + (C_4H_9)_2SnH_2 \rightarrow$$

$$(C_4H_9)_3GeSn(C_4H_9)_2H + (CH_3)_2NH$$

The Ge—N bond of dialkylaminogermanes $Ge(NMe_2)_4$ and $EtGe(NMe_2)_3$ is also broken by benzoyl chloride or hydrogen iodide, $GeCl_4$ and $EtGeI_3$ being formed (9).

The reaction between silicon and germanium imidazolidines and phenyldichlorophosphine oxide produces a new phosphorus (V)—imidazolidine, the 1,3-dimethyl-2-phenyldiazaphosphole oxide with the release of dimethyldichlorosilane or germane (186, 188):

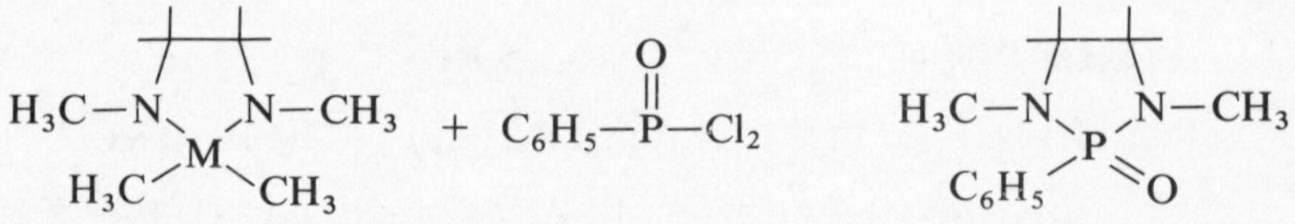

(M = Si, Ge) $+ (CH_3)_2MCl_2$

Chloramine cleaves the germanium–nitrogen bond in trialkylgermylamines:

$$3\,R_3GeN(CH_3)_2 + 3\,NH_2Cl \rightarrow$$

$$3\,R_3GeCl + 3\,(CH_3)_2NH + NH_3 + N_2$$

$$3\,(R_3Ge)_2NH + 6\,NH_2Cl \rightarrow 6\,R_3GeCl + 5\,NH_3 + 2\,N_2$$

The nonreactivity of bis-(triphenylgermyl)amine compared to the reactivity of the corresponding bis-(triethylgermyl)amine with chloramine could be the result of an inductive effect of the phenyl groups. Dimethylchloroamine does not react with the trialkylgermylamines but cleaves readily the tin–nitrogen bond of dimethyl(tri-n-butylstannyl)amine (62).

Cleavage reactions of Ge—N bonds in organogermylamines by halides of Group III–IV are reviewed by Scherer (123).

The action of acetic anhydride on Bu_3GeNMe_2 leads to tributylgermaniumacetate and dimethylacetamide (121):

$$Bu_3GeNMe_2 + (CH_3CO)_2O \rightarrow Bu_3Ge\!-\!O\!-\!\underset{\underset{O}{\|}}{C}\!-\!CH_3 + CH_3CONMe_2$$

Satgé, Lesbre and Baudet (121) observed an important reactivity level in the Ge—N bond of trialkylgermyldialkylamines, $R_3GeNR'_2$, towards some multiple-bond derivatives like CO_2, CS_2, PhNCO, PhNCS. The formation of insertion derivatives by the opening of a double bond in the reagent and addition to the Ge—N bond has been observed. In the first two instances, the formation of trialkylgermanium dialkylcarbamates and dialkyldithiocarbamates (yield 90%) has been observed:

$$Et_3Ge\!-\!NEt_2 + O\!=\!C\!=\!O \rightarrow Et_3Ge\!-\!O\!-\!\underset{\underset{O}{\|}}{C}\!-\!NEt_2$$

$$Et_3Ge\!-\!NMe_2 + S\!=\!C\!=\!S \rightarrow Et_3Ge\!-\!S\!-\!\underset{\underset{S}{\|}}{C}\!-\!NMe_2$$

On the addition of phenylisocyanates and phenylisothiocyanates, urea and thiourea derivatives have been isolated:

$$Et_3Ge\!-\!NEt_2 + PhN\!=\!C\!=\!O \rightarrow Et_3Ge\!-\!\underset{\underset{Ph}{|}}{N}\!-\!\underset{\underset{O}{\|}}{C}\!-\!NEt_2$$

$$Et_3Ge\!-\!NR_2 + PhN\!=\!C\!=\!S \rightarrow Et_3Ge\!-\!\underset{\underset{Ph}{|}}{N}\!-\!\underset{\underset{S}{\|}}{C}\!-\!NR_2$$

$$(R = Me, Et)$$

Their hydrolysis leads, through the breaking of the Ge—N bond, to triethylgermanium oxide and to the corresponding urea and thiourea derivatives:

$$2Et_3Ge\!-\!\underset{\underset{Ph}{|}}{N}\!-\!\underset{\underset{O}{\|}}{C}\!-\!NEt_2 + H_2O \rightarrow (Et_3Ge)_2O + 2\,H\!-\!\underset{\underset{Ph}{|}}{N}\!-\!\underset{\underset{O}{\|}}{C}\!-\!NEt_2$$

$$2\,Et_3Ge\!-\!\underset{\underset{Ph}{|}}{N}\!-\!\underset{\underset{S}{\|}}{C}\!-\!NMe_2 + H_2O \rightarrow (Et_3Ge)_2O + 2\,HN\!-\!\underset{\underset{Ph}{|}}{}\underset{\underset{S}{\|}}{C}\!-\!NMe_2$$

All these aminogermylation reactions show strong similarities with the recently described aminosilylation (17) and aminostannylation (48, 71) reactions.

The study of the reactions of CO_2, CS_2, PhNCO, PhNCS with trialkylgermylamines has been extended to the germylamines R_3GeNHR', $(R_3Ge)_2NH$ and $(R_3Ge)_2NR'$ (107); the mono- and di-insertion derivatives have been characterized. Their structure has been established by means of

NMR and IR spectroscopy, as well as chemically by identification of their hydrolysis and thermal decomposition products. The following results have been observed:

$$Et_3Ge-\underset{\underset{H}{|}}{N}-GeEt_2 + 2\,CO_2 \rightarrow Et_3Ge-O-\underset{\underset{O}{||}}{C}-\underset{\underset{H}{|}}{N}-\underset{\underset{O}{||}}{C}-O-GeEt_3$$

$$\downarrow 100-150°C$$

$$CO_2 + Et_3GeNCO + \tfrac{1}{2}(Et_3Ge)_2O + \tfrac{1}{2}H_2O$$

$$Et_3Ge-\underset{\underset{H}{|}}{N}-GeEt_3 + CS_2 \rightarrow (Et_3Ge-S-\underset{\underset{S}{||}}{C}-\underset{\underset{H}{|}}{N}-GeEt_3) \overset{40°C}{\rightarrow}$$

[unstable]

$$Et_3GeNCS + \tfrac{1}{2}(Et_3Ge)_2S + \tfrac{1}{2}H_2S$$

The action of one molecule of phenylisocyanate on Et_3GeNMe_2, $Et_3GeNHPh$, $(Et_3Ge)_2NH$ and $(Et_3Ge)_2NEt$ leads via the insertion into a Ge—N bond to stable monoaddition derivatives (107):

$$Et_3Ge-\underset{\underset{Ph}{|}}{N}-\underset{\underset{O}{||}}{C}-NMe_2 \qquad\qquad Et_3Ge-\underset{\underset{Ph}{|}}{N}-\underset{\underset{O}{||}}{C}-\underset{\underset{H}{|}}{N}-Ph$$

(I)   (II)

$$Et_3Ge-\underset{\underset{Ph}{|}}{N}-\underset{\underset{O}{||}}{C}-\underset{\underset{H}{|}}{N}-GeEt_3 \qquad\qquad Et_3Ge-\underset{\underset{Ph}{|}}{N}-\underset{\underset{O}{||}}{C}-\underset{\underset{Et}{|}}{N}-GeEt_3$$

(III)   (IV)

On the other hand, the addition of a second phenyl isocyanate molecule leads to unstable diaddition derivatives:

$$I + PhNCO \rightarrow Et_3Ge-\underset{\underset{Ph}{|}}{N}-\underset{\underset{O}{||}}{C}-\underset{\underset{Ph}{|}}{N}-\underset{\underset{O}{||}}{C}-NMe_2 \overset{t°}{\rightleftarrows}$$

$$Et_3Ge-\underset{\underset{Ph}{|}}{N}-\underset{\underset{O}{||}}{C}-NMe_2 + PhNCO$$

(I)

$$III + PhNCO \rightarrow Et_3Ge-\underset{\underset{Ph}{|}}{N}-\underset{\underset{O}{||}}{C}-\underset{\underset{H}{|}}{N}-\underset{\underset{O}{||}}{C}-\underset{\underset{Ph}{|}}{N}-GeEt_3 \overset{t°}{\rightarrow}$$

$$Et_3Ge-\underset{\underset{Ph}{|}}{N}-\underset{\underset{O}{||}}{C}-\underset{\underset{H}{|}}{N}-Ph + Et_3GeNCO$$

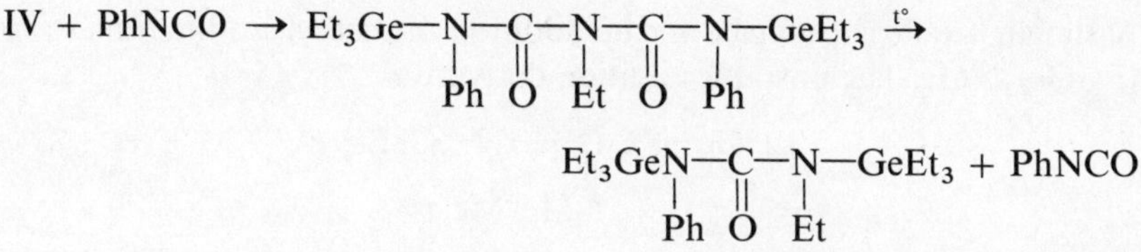

$$\text{IV} + \text{PhNCO} \rightarrow \underset{\substack{\;\;|\;\;\;||\;\;\;|\;\;\;||\;\;\;|\\ \text{Ph} \;\; \text{O} \;\; \text{Et} \;\; \text{O} \;\; \text{Ph}}}{\text{Et}_3\text{Ge}-\text{N}-\text{C}-\text{N}-\text{C}-\text{N}-\text{GeEt}_3} \overset{t^\circ}{\rightarrow}$$

$$\underset{\substack{\;\;|\;\;\;||\;\;\;|\\ \text{Ph} \;\; \text{O} \;\; \text{Et}}}{\text{Et}_3\text{GeN}-\text{C}-\text{N}-\text{GeEt}_3} + \text{PhNCO}$$

$$\text{(IV)}$$

In the action of phenyl isothiocyanate on $\text{Et}_3\text{GeNMe}_2$, $(\text{Et}_3\text{Ge})_2\text{NMe}$, and $(\text{Et}_3\text{Ge})_2\text{NH}$ only the monoaddition could be demonstrated:

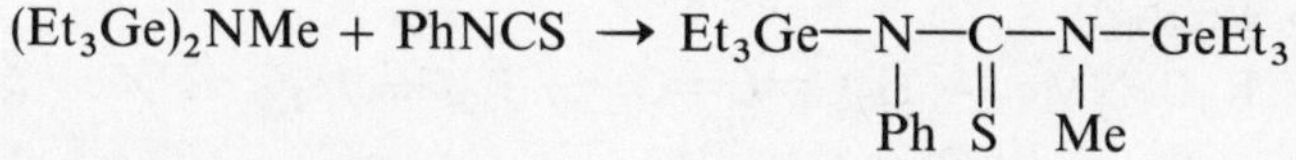

$$(\text{Et}_3\text{Ge})_2\text{NMe} + \text{PhNCS} \rightarrow \underset{\substack{\;\;|\;\;\;||\;\;\;|\\ \text{Ph} \;\; \text{S} \;\; \text{Me}}}{\text{Et}_3\text{Ge}-\text{N}-\text{C}-\text{N}-\text{GeEt}_3}$$

At high temperature (10 days at 150°C) a stoichiometric mixture of $(\text{Et}_3\text{Ge})_2\text{NMe}$ and PhNCS leads exclusively to bis-(triethylgermyl)-sulphide $(\text{Et}_3\text{Ge})_2\text{S}$, and yellow resinous polymers. The IR spectrum of the mixture measured after 1 hour at 150°C shows the characteristic bands at 408 cm^{-1} (Ge—S—Ge) and 2129 cm^{-1} (—N=C=N—) which can be attributed to N-phenyl-N′-methylcarbodiimide. According to these results, the authors (107) suggest the following reaction scheme:

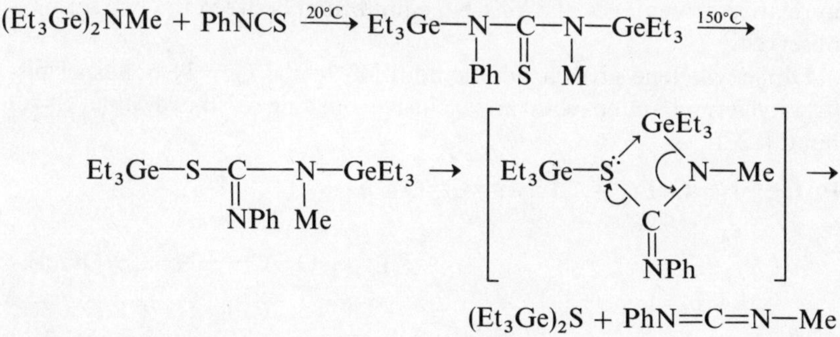

$$(\text{Et}_3\text{Ge})_2\text{NMe} + \text{PhNCS} \xrightarrow{20^\circ\text{C}} \underset{\substack{\;\;|\;\;\;||\;\;\;|\\ \text{Ph} \;\; \text{S} \;\; \text{M}}}{\text{Et}_3\text{Ge}-\text{N}-\text{C}-\text{N}-\text{GeEt}_3} \xrightarrow{150^\circ\text{C}}$$

$$\underset{\substack{\;\;||\;\;\;|\\ \text{NPh} \;\; \text{Me}}}{\text{Et}_3\text{Ge}-\text{S}-\text{C}-\text{N}-\text{GeEt}_3} \rightarrow \left[\begin{array}{c} \text{GeEt}_3 \\ \nwarrow \\ \text{Et}_3\text{Ge}-\text{S}\;\;\;\;\text{N}-\text{Me} \\ \searrow\;\;\swarrow \\ \text{C} \\ || \\ \text{NPh} \end{array} \right] \rightarrow$$

$$(\text{Et}_3\text{Ge})_2\text{S} + \text{PhN}=\text{C}=\text{N}-\text{Me}$$

The addition of phenyl isothiocyanate to bis-(triethylgermyl)amine, $(\text{Et}_3\text{Ge})_2\text{NH}$, takes place almost exclusively with opening of the double C=S bond of the isothiocyanate:

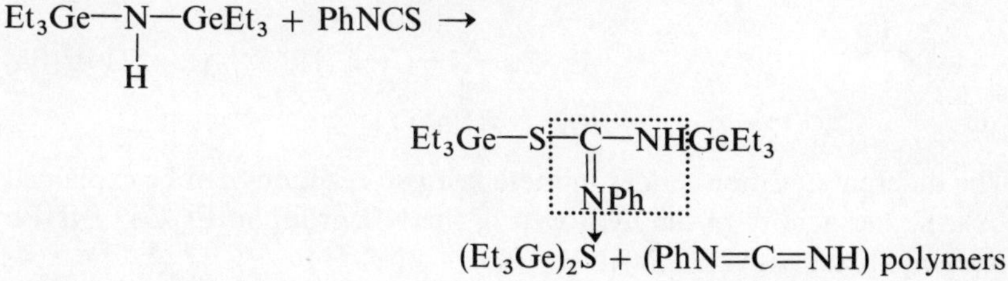

$$\underset{\substack{|\\ \text{H}}}{\text{Et}_3\text{Ge}-\text{N}-\text{GeEt}_3} + \text{PhNCS} \rightarrow$$

$$\underset{\substack{||\\ \text{NPh}}}{\text{Et}_3\text{Ge}-\text{S}-\text{C}-\text{NH}-\text{GeEt}_3}$$

$$(\text{Et}_3\text{Ge})_2\text{S} + (\text{PhN}=\text{C}=\text{NH}) \text{ polymers}$$

A similar result is obtained in the addition of methyl isothiocyanate to $(Et_3Ge)_2NMe$. The unstable addition derivative

$$Et_3Ge-S-\underset{\underset{NMe}{\parallel}}{C}-\underset{\underset{Me}{\mid}}{N}-GeEt_3$$

shows a strong absorption at $1610\ cm^{-1}$ which can be attributed to the $C=N-$ group (107).

Another aminogermylation reaction has been observed with ketene which easily adds to the Ge—N bond of N-trialkylgermyldialkylamines (77, 115, 122):

$$R_3GeNMe_2 + H_2C=C=O \rightarrow R_3GeCH_2\underset{\underset{O}{\parallel}}{C}-NMe_2$$

On the other hand, diphenylketene gives with Et_3GeNMe_2 two types of addition (122) by opening of the double carbon–carbon bond and by opening of the double $C=O$ bond, with formation of:

$$Et_3GeC(Ph)_2-\underset{\underset{O}{\parallel}}{C}-NMe_2 \quad \text{and} \quad Et_3Ge-O-\underset{\underset{CPh_2}{\parallel}}{C}-NMe_2$$

in relative percentages of 57/43. No equilibrium between the two forms is observed.

Diphenylketene gives a double addition to the Ge—N bonds of bis-(triethylgermyl)amine with an exclusive opening of the double $C=O$ bond (122):

$$Et_3Ge-\underset{\underset{H}{\mid}}{N}-GeEt_3 + 2\,Ph_2C=C=O \rightarrow$$

$$Et_3GeO-\underset{\underset{CPh_2}{\parallel}}{C}-\underset{\underset{H}{\mid}}{N}-\underset{\underset{CPh_2}{\parallel}}{C}-OGeEt_3$$

On the other hand, addition of ketene and diphenylketene to the N—H bond of the N-triethylgermyl-phenylamine is observed:

$$Et_3Ge-\underset{\underset{Ph}{\mid}}{N}-H + R_2C=C=O \rightarrow$$

$$Et_3Ge-\underset{\underset{Ph}{\mid}}{N}-\underset{\underset{O}{\parallel}}{C}-CHR_2 \quad (R = H \text{ or } Ph)$$

The different addition modes in these last two reactions can be explained by a higher acidity of the hydrogen of the NH group in $Et_3Ge-NHPh$ in comparison with $(Et_3Ge)_2NH$ (122).

Fission of germanium–nitrogen bonds with hexafluoroacetone produced the expected alkoxygermane (3):

$$Bu_3GeNMe_2 + (CF_3)_2CO \rightarrow Bu_3GeOC(CF_3)_2NMe_2$$

Thiobenzoyl isocyanate forms 1/1 adducts with trimethylgermyl-dialkylamine. In this adduct the trimethylmetal group can migrate between an oxygen and a nitrogen atom (64):

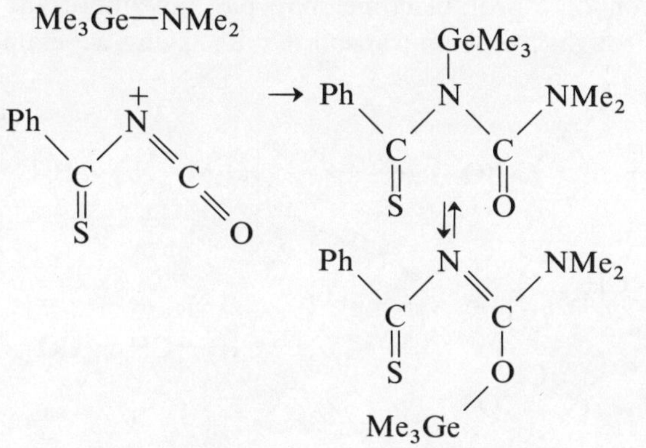

Et_3GeNMe_2 and $CH_2 = \underset{\underset{CH_3}{|}}{C}OOCCF_3$ react at 10°C to yield 82%

$Et_3GeO\underset{\underset{CH_3}{|}}{C} = CH_2$ and CF_3CONMe_2 (89). Bu_3GeNEt_2 with chloral or di-p-tolylcarbodiimide give a 1.1 adduct (63).

A carbon–carbon insertion is observed with acrylonitrile (114, 116) and acetylene dicarboxylic ester (21, 49):

$$Et_3GeNMe_2 + CH_2{=}CH{-}CN \rightarrow Et_3Ge\underset{\underset{CN}{|}}{C}H{-}CH_2{-}NMe_2$$

$$Me_3GeNMe_2 + EtO_2C{-}C{\equiv}C{-}CO_2Et \rightarrow$$

$$Me_3Ge{-}\underset{\underset{CO_2Et}{|}}{C}{=\!=\!=}\underset{\underset{CO_2Et}{|}}{C}{-}NMe_2$$

Yoder and Zuckerman have reported insertion reactions of carbon dioxide and phenylisocyanate on Ge—N bonds of N,N'-dialkylgermanium imidazolidine (191).

As with trimethylsilyldiethylamine, the reaction of trimethylgermyl-diethylamine with β-propiolactone gives the trimethylgermylester of

N,N-diethyl-β-alanine (65):

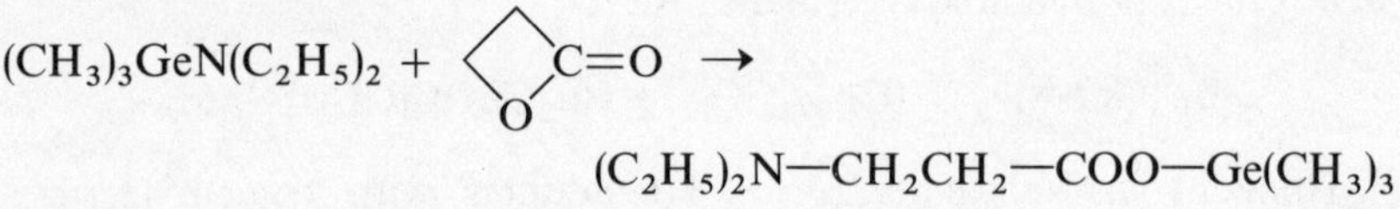

$$(CH_3)_3GeN(C_2H_5)_2 + \underset{O}{\square}C=O \rightarrow$$

$$(C_2H_5)_2N-CH_2CH_2-COO-Ge(CH_3)_3$$

The same reaction is noted with Bu_3GeNEt_2 (63).

Nucleophilic attack of the nitrogen atom of trimethylgermylamine on the β-carbon of β-propiolactone may play an important role in this reaction through a dipolar transition state giving a betaine-type intermediate:

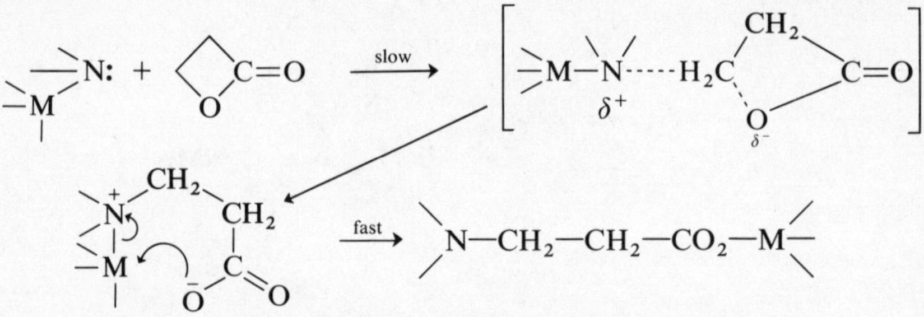

Kinetic study shows the higher reactivity of Ge—N compounds compared with that of Si—N compounds. The nucleophilicity of Si—N compounds is lower than that of Ge—N compounds on account of the larger contribution of $d\pi$–$p\pi$ conjugation in Si—N bonds (66).

The basicities of bis-(diethylamino)dialkyl derivatives of elements in Group IV B (C, Si, Ge, Sn), and of imidazolidine derivatives (C, Si, Ge)

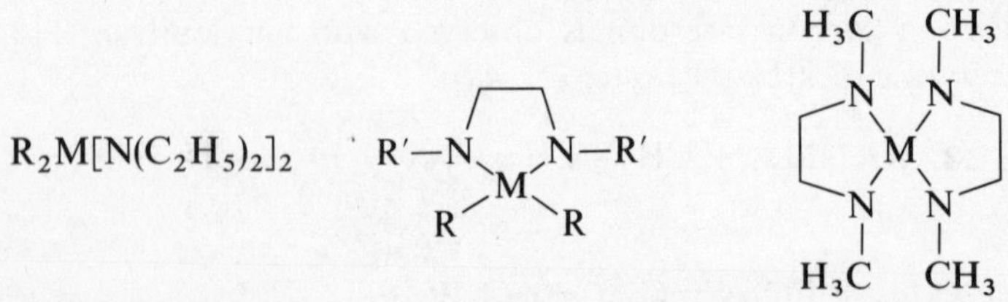

have been investigated by a variety of current methods. NMR chemical shifts and spin–spin coupling constants both for the base and for chloroform in the presence of the base have been used. Changes in infrared stretching frequencies for chloroform have also been measured (101). General agreement has been reached, establishing that the silicon compound is very different from the analogous germanium compound and is the least basic member for each series of compounds studied. It appears that the effective electronegativity of the nitrogen in the silicon compounds is greater than that in the analogous germanium compounds. Since the electronegativities of silicon and germanium are very similar, this difference can be ascribed to a greater amount of $(p-d)\pi$ bonding in the Si—N bond as compared with the Ge—N bond (101, 186).

The absorption maximum λ_{max} of the $n \rightarrow \pi^*$ transition in N-trimethylsilyl and trimethylgermyl-benzophenone-imine occurs at 364 nm and 347 nm respectively.

The variation of λ_{max} for the compounds $(C_6H_5)_2C\!=\!NM(CH_3)_3$ (M = Si, Ge and Sn) and $(C_6H_5)_2C\!=\!NM(C_6H_5)_3$ (M = C, Si, Ge, Sn and Pb) has been found to be a combined result of the inductive effect and the relative amount of π-interaction of the Group IV element (87).

[(Trialkylgermyl)imino]trialkylphosphoranes form stable 1/1 addition compounds with trimethylaluminium, -gallium and -indium (185):

$$(CH_3)_3Ge\!-\!N\!=\!PR_3 + (CH_3)_3M\cdot O(C_2H_5)_2 \rightarrow$$

$$
(C_2H_5)_2O +
\begin{array}{c}
Ge(CH_3)_3 \\
| \\
N \\
| \quad \backslash \\
(CH_3)_3M \quad PR_3
\end{array}
$$

$$(R = CH_3, C_2H_5) \qquad (M = Al, Ga, In)$$

The NMR and IR spectra provide information regarding relative acceptor and donor strengths of the components.

A comparative NMR study of a series of N-germanium, N-silicon and N-alkylated derivatives of pyrazole allowed the observation of a progressive deshielding of all signals of the pyrazole cycle in the order N-C, N-Ge, N-Si, as well as a more and more marked difference between $J_{3,4}$ and $J_{4,5}$ corresponding to a clearer character of the simple and double bonds of the pyrazole ring (38):

$$
\begin{array}{c}
{}^3\!\!\overline{}\,{}^4 \\
N^2 \,{}_1\, {}^5 \\
N \\
| \\
M \\
\diagup | \diagdown
\end{array}
\qquad (M = C, Ge, Si)
$$

On the basis of the usually accepted electronegativities C > Ge > Si, the inverse result could be expected. It can be concluded that the inductive effects are largely counterbalanced by the $p\pi$–$d\pi$ interactions between the free nitrogen lone pair in 1 and the 3d orbitals of silicon and 4d orbitals of germanium. These interactions, which deplete the pyrazolic cycle's π electron density, reduce its aromatic character (38).

Randall and Zuckerman (102) have carried out experiments designed to test the hypothesis of $(p \rightarrow d)$-π interactions in the bonds formed between nitrogen and the fourth group elements silicon, germanium and tin. Infrared and proton magnetic resonance spectra of the aniline-^{14}N and ^{15}N isotopomers and their N-trimethylsilyl, -germyl and -stannyl derivatives $(CH_3)_3MNHC_6H_5$ (M = Si, Ge, Sn) have been measured and compared. NMR assignments have been checked by comparisons of proton spectra at 60 and 100 Mcps; by proton -^{15}N heteronuclear experiments at 9400 gauss; and by observation of ^{15}N satellites. The one bond ^{15}N—H couplings can be interpreted (on the assumption of a dominant Fermi spin–spin interaction, small radial variations for nitrogen wave functions, and equal distributions of s character in the σ bonds) in terms of pyramidal arrangements of bonds at nitrogen. The relation of the stereochemical situation at the nitrogen to the question of (p → d)-π or (p → p)-π bonding involving nitrogen is discussed.

These authors deem it possible to conclude from their results that there is no stereochemical evidence for (p → d)-π bonding in these compounds.

The IR (4000–33 cm^{-1}) and Raman spectra of the dimethylamides $M[N(CH_3)_2]_4$ with M = Si, Ge, Sn, have been reported and assigned. The M–N force constants were calculated to be 3.618, 3.391, 3.110 and 3.111 mdyn/Å for the Si, Ge, Sn and analogous Ti compounds (20).

8-1-3 Organogermanium–nitrogen compounds

Table 8-1(a–l) showing some physical properties of organogermanium–nitrogen compounds appears on the following pages.

Table 8-1
Organogermanium–Nitrogen Compounds

(a) $\diagup$Ge—NH$_2$

COMPOUND	EMPIRICAL FORMULA	M.P. (°C)	B.P. (°C/mm)	n_D^{20}	d_4^{20}	REFERENCES
H$_3$GeNH$_2$	H$_5$NGe	—	—	—	—	109, 166, 167
H$_2$Ge(NH$_2$)$_2$	H$_6$N$_2$Ge	—	—	—	—	109
(C$_2$H$_5$)H$_2$GeNH$_2$	C$_2$H$_9$NGe	—	—	—	—	54
(C$_2$H$_5$)Ge(NH$_2$)$_3$	C$_2$H$_{11}$N$_3$Ge	—	—	—	—	54
(C$_6$H$_5$)$_3$GeNH$_2$	C$_{18}$H$_{17}$NGe	—	—	—	—	83, 84, 158, 159, 168
(C$_6$H$_5$)$_3$GeNHK	C$_{18}$H$_{16}$NKGe	—	—	—	—	84

(b) (R$_3$Ge)$_2$NH

COMPOUND	EMPIRICAL FORMULA	M.P. (°C)	B.P. (°C/mm)	n_D^{20}	d_4^{20}	REFERENCES
[(CH$_2$Cl)(CH$_3$)$_2$Ge]$_2$NH	C$_6$H$_{17}$Cl$_2$NGe$_2$	—	72–74/0.1	—	—	183
[(CH$_3$)$_3$Ge]$_2$NH	C$_6$H$_{19}$NGe$_2$	—	47/17	—	—	111
[(C$_2$**H$_5$**)$_2$(H)Ge]$_2$NH	C$_8$H$_{23}$NGe$_2$	—	59/2	1.4833	1.2491	90
[(C$_2$H$_5$)$_3$Ge]$_2$NH	C$_{12}$H$_{31}$NGe$_2$	—	90–92/0.2	1.4755	—	75, 81
[(C$_3$H$_7$)$_3$Ge]$_2$NH	C$_{18}$H$_{43}$NGe$_2$	—	128–30/1.5	1.4720	—	81
[(C$_4$H$_9$)$_3$Ge]$_2$NH	C$_{24}$H$_{55}$NGe$_2$	—	144–46/0.2	1.4727	—	88, 105, 106
[(C$_6$H$_5$)$_3$Ge]$_2$NH	C$_{36}$H$_{31}$NGe$_2$	—	—	—	—	84

(c) $(R_3Ge)_2NR'$

COMPOUND	EMPIRICAL FORMULA	M.P. (°C)	B.P. (°C/mm)	n_D^{20}	d_4^{20}	REFERENCES
$(CH_3)_3Ge-N-Ge(CH_3)_3$ $\quad\;\; CH_3$	$C_7H_{21}NGe_2$	—	172/735	—	—	137, 143
$(CH_3)_3Ge-N-Si(CH_3)_3$ $\quad\;\; CH_3$	$C_7H_{21}NSiGe$	—	42/12	—	—	137, 143
$(CH_3)_3Ge-N-Sn(CH_3)_3$ $\quad\;\; CH_3$	$C_7H_{21}NSnGe$	—	28/2	—	—	137, 143
$(CH_3)_3Ge-N-Pb(CH_3)_3$ $\quad\;\; CH_3$	$C_7H_{21}NPbGe$	—	49/2	—	—	137, 143
$[(CH_3)_3Ge]_2N-n-C_8H_{17}$	$C_{14}H_{35}NGe_2$	—	147/13	1.4604	—	113
$[(C_2H_5)_3Ge]_2NAl(C_2H_5)_2$	$C_{16}H_{40}NAlGe_2$	—	241/3	—	—	97
$[(C_2H_5)_3Ge]_2NC_6H_5$	$C_{18}H_{35}NGe_2$	—	120/0.2	1.5290	—	113
$[(C_2H_5)_3Ge]_2N-n-C_8H_{17}$	$C_{20}H_{47}NGe_2$	—	152/0.3	1.4769	1.0545	114

(d) $(R_3Ge)_3N$

COMPOUND	EMPIRICAL FORMULA	M.P. (°C)	B.P. (°C/mm)	n_D^{20}	d_4^{20}	REFERENCES
$[Cl_3Ge]_3N$	NGe_3Cl_9	—	—	—	—	164, 177
$[(CH_3)_2(Cl)Ge]_3N$	$C_6H_{18}N(Cl)_3Ge$	62	75/2	—	—	111
$[(C_2H_5)(H)_2Ge]_3N$	$C_6H_{21}NGe_3$	—	99.5/12	1.4891	1.3959	90
$[(CH_3)_3Ge]_3N$	$C_9H_{27}NGe_3$	—	60/2	—	—	111, 113
			103/13	1.4798	—	90
$[(C_2H_5)_2(H)Ge]_3N$	$C_{12}H_{33}NGe_3$	—	159/15	1.4918	1.2550	—
$[(C_2H_5)_3Ge]_3N$	$C_{18}H_{45}NGe_3$	—	148–51/0.04	1.5108	—	75
$[n\text{-}C_3H_7)_3Ge]_3N$	$C_{27}H_{63}NGe_3$	—	198–200/0.50	1.4937	—	75
$[n\text{-}C_4H_9)_3Ge]_3N$	$C_{36}H_{81}NGe_3$	—	222–23/0.05	1.4880	—	75
$[(C_6H_5)_3Ge]_3N$	$C_{54}H_{45}NGe_3$	163–64	—	—	—	82, 84

(e) R_3GeNHR'

COMPOUND	EMPIRICAL FORMULA	M.P. (°C)	B.P. (°C/mm)	n_D^{20}	d_4^{20}	REFERENCES
$(CH_3)_3GeNHSi(CH_3)_3$	$C_6H_{19}NSiGe$	—	133–34/752	1.4183	0.9587	96
$(CH_3)_3GeNHC_6H_5$	$C_9H_{15}NGe$	—	85–86/4	—	—	102
$(CH_3)_3GeNHSi(C_2H_5)_3$	$C_9H_{25}NSiGe$	—	81–83/10	—	—	125
$(CH_3)_3GeNHSi(CH_3)_2\text{—}N\text{—}Si(CH_3)_3$ (with CH_3 on N)	$C_9H_{28}N_2Si_2Ge$	—	45–47/0.2	—	—	124
$(CH_3)_3GeNH\text{—}P[C(CH_3)_3]_2$	$C_{11}H_{28}PNGe$	—	49–50/1	—	—	129
$(C_2H_5)_3GeNHC_6H_5$	$C_{12}H_{21}NGe$	—	141/11	1.5373	1.1282	121

Table 8-1(e)—continued

COMPOUND	EMPIRICAL FORMULA	M.P. (°C)	B.P. (°C/mm)	n_D^{20}	d_4^{20}	REFERENCES
$(C_2H_5)_3GeNHC_6H_{13}$-n	$C_{12}H_{29}NGe$	—	135/20	1.4526	0.9863	121
$(C_2H_5)_3GeNHC_8H_{17}$-n	$C_{14}H_{33}NGe$	—	160-62/20	1.4550	0.9756	114
$(C_4H_9)_3GeNHC_8H_{17}$-n	$C_{20}H_{45}NGe$	—	130-34/0.06	1.4594	—	88, 105, 106

(f) $R_3GeNR'_2$

COMPOUND	EMPIRICAL FORMULA	M.P. (°C)	B.P. (°C/mm)	n_D^{20}	d_4^{20}	REFERENCES
$Cl_3GeN(CH_3)_2$	$C_2H_6NGeCl_3$	—	—	—	—	19
$(CH_3)_3GeNCH_3Li$	$C_4H_{12}NGeLi$	—	—	—	—	137, 143
$(CH_3)_3GeN(CH_3)_2$	$C_5H_{15}NGe$	—	103/760	1.4246	—	21, 113
$(CH_3O)_3GeN(CH_3)_2$	$C_5H_{15}NO_3Ge$	—	—	—	—	19
$(CH_3)_2(Cl)GeN(C_2H_5)_2$	$C_6H_{16}NGeCl$	—	177/743	—	—	189
$(CH_3)_3GeN(C_2H_5)_2$	$C_7H_{19}NGe$	—	138-39/760	1.4304	1.01	2, 191
$(CH_3)_2Ge-N-(CH_2)_2N-Ge(CH_3)_2$ (with Cl, CH₃, CH₃, Cl substituents)	$C_8H_{12}N_2Ge_2Cl_2$	—	90/9	—	—	189
$(C_2H_5)_3GeN(CH_3)_2$ $C(CH_3)_3$	$C_8H_{21}NGe$	—	176/760	1.4498	1.0235	121
$Cl(CH_3)_2Ge-N$ $Si(CH_3)_3$	$C_9H_{24}ClNSiGe$	—	72/0.5	—	—	144
$(CH_3)_3Ge-N[Si(CH_3)_3]_3$	$C_9H_{27}NSi_2Ge$	29-32	54-56/1	—	—	127, 128, 133

Formula	Structure	m.p.	b.p./mm	n_D	d	Ref.
$C_{10}H_{25}NGe$	$(C_2H_5)_3GeN(C_2H_5)_2$	—	86/10	1.455	1.001	121
$C_{10}H_{27}NSiGe$	$(CH_3)_3Ge-N\!\!<\!\!{C(CH_3)_3 \atop Si(CH_3)_3}$	—	47/1	—	—	144
$C_{12}H_{33}NO_3Si_2Ge$	$(C_2H_5O)_3GeN[Si(CH_3)_3]_2$	—	87/1	1.4394	—	76
$C_{12}H_{33}NSiSnGe$	$(CH_3)_3GeN\!\!<\!\!{Si(C_2H_5)_3 \atop Sn(CH_3)_3}$	108-10	138-41/10	—	—	125
$C_{12}H_{36}N_2Si_2Ge$	$(CH_3)_3GeN\!\!<\!\!{Si(CH_3)_2-N-Si(CH_3)_3 \atop \quad\quad\quad\; |\;\; \atop \quad\quad\quad\;CH_3}$	—	84-86/0.2	—	—	124
$C_{14}H_{29}NOSi_2Ge$	$(CH_3)_2(C_6H_5O)GeN[Si(CH_3)_3]_2$	—	135/1	1.5260	—	76
$C_{14}H_{33}NGe$	$(C_4H_9)_3GeN(CH_3)_2$	—	144-45/19	1.4583	0.9607	121
$C_{16}H_{37}NGe$	$(C_4H_9)_3GeN(C_2H_5)_2$	—	84-88/ 0.15-0.20	1.4606	—	88, 105, 106, 121
$C_{30}H_{25}NGe$	$(C_6H_5)_3GeN(C_6H_5)_2$	153.5-55	—	—	—	14

(g) R_3GeN⟩

COMPOUND	EMPIRICAL FORMULA	M.P. (°C)	B.P. (°C/mm)	n_D^{20}	d_4^{20}	REFERENCES
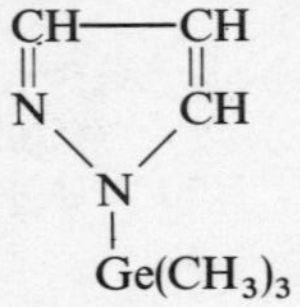 [structure: pyrazolyl–Ge(CH₃)₃]	$C_6H_{12}N_2Ge$	—	106/65	1.4859	—	38
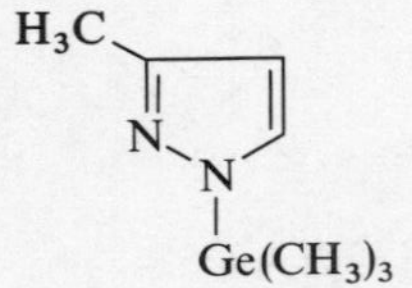 [structure: H₃C-pyrazolyl–Ge(CH₃)₃]	$C_7H_{14}N_2Ge$	—	95/25	1.4860	—	38
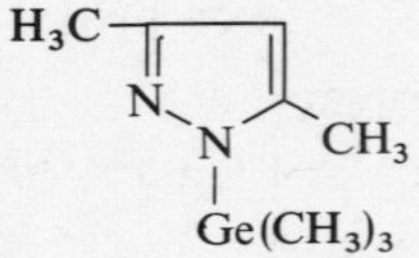 [structure: H₃C-, CH₃-pyrazolyl–Ge(CH₃)₃]	$C_8H_{16}N_2Ge$	—	109/28	1.4900	—	38
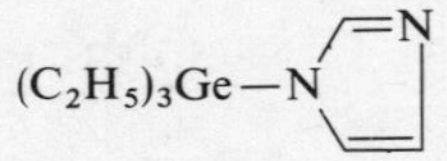 [structure: $(C_2H_5)_3Ge-N$ imidazolyl]	$C_9H_{18}N_2Ge$	—	155–56/13	1.4943	—	88, 105, 106

Structure	Formula		bp °C/mm	n_D		Ref.
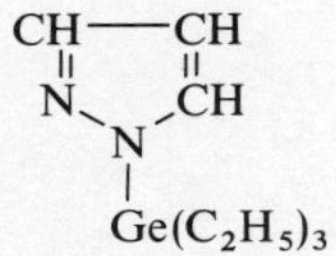 pyrazolyl-N–N–Ge(C₂H₅)₃	$C_9H_{18}N_2Ge$	—	121/23	1.4878	—	38
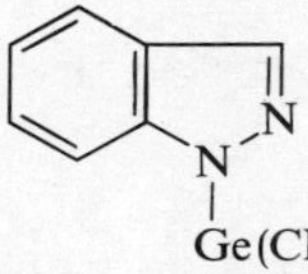 benzotriazolyl–N–Ge(CH₃)₃	$C_{10}H_{14}N_2Ge$	—	146/15	1.5668	—	38
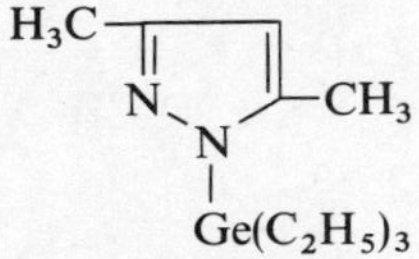 3,5-dimethylpyrazolyl–N–Ge(C₂H₅)₃	$C_{11}H_{22}N_2Ge$	—	121/10	1.4939	—	38
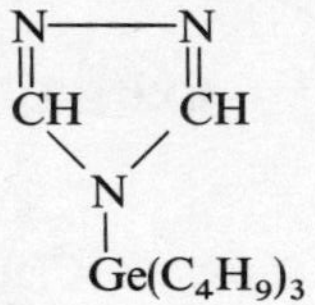 triazolyl–N–Ge(C₄H₉)₃	$C_{14}H_{29}N_3Ge$	—	123/0.25	1.4825	—	88, 105, 106
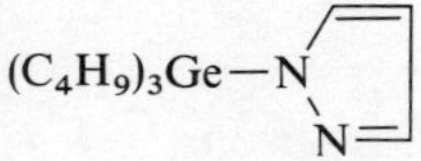 $(C_4H_9)_3Ge$–N pyrazolyl	$C_{15}H_{30}N_2Ge$	—	105–9/0.53	1.4828	—	88, 105, 106
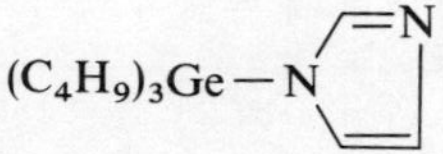 $(C_4H_9)_3Ge$–N imidazolyl	$C_{15}H_{30}N_2Ge$	—	117/0.07	1.4892	—	88, 105, 106

Table 8-1(g)—continued

COMPOUND	EMPIRICAL FORMULA	M.P. (°C)	B.P. (°C/mm)	n_D^{20}	d_4^{20}	REFERENCES
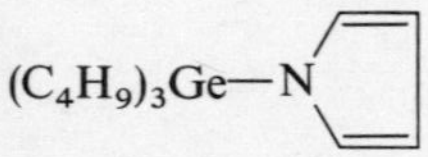 (C₄H₉)₃Ge—N (pyrrole ring)	$C_{16}H_{31}NGe$	—	110–14/0.53	1.4882	—	88, 105, 106
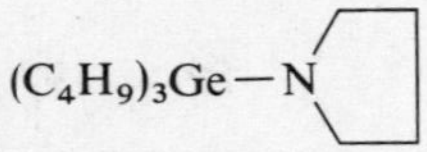 (C₄H₉)₃Ge—N (pyrrolidine ring)	$C_{16}H_{35}NGe$	—	158–60/12	1.4729	—	88, 105, 106
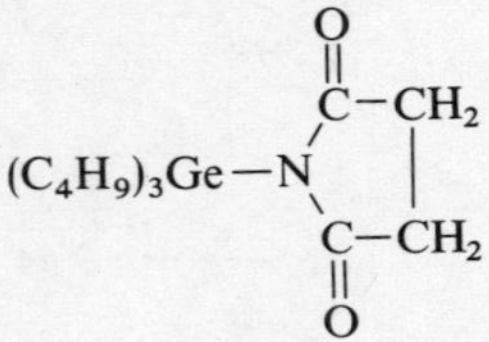 (C₄H₉)₃Ge—N (succinimide ring)	$C_{16}H_{31}O_2NGe$	—	177/2	1.4870	—	88, 105, 106
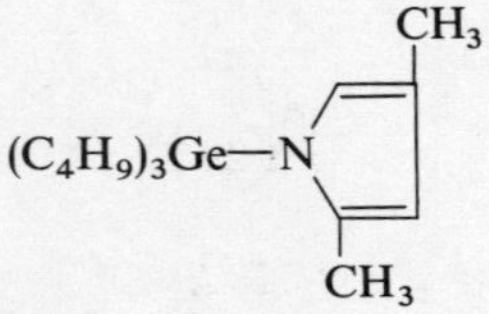 (C₄H₉)₃Ge—N (2,5-dimethylpyrrole ring)	$C_{18}H_{35}NGe$	—	172–74/10	1.4932	—	88, 105, 106

COMPOUND	EMPIRICAL FORMULA	M.P. (°C)	B.P. (°C/mm)	n_D^{20}	d_4^{20}	REFERENCES
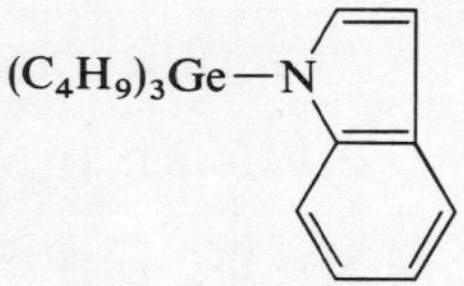 $(C_4H_9)_3Ge{-}N$ [indol-1-yl]	$C_{20}H_{33}NGe$	—	125–30/0.0015	1.5364	—	88, 105, 106
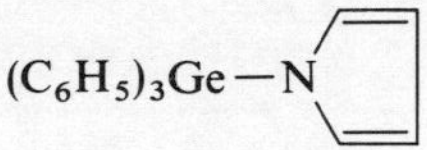 $(C_6H_5)_3Ge{-}N$ [pyrrol-1-yl]	$C_{22}H_{19}NGe$	198–99	—	—	—	88, 105, 106
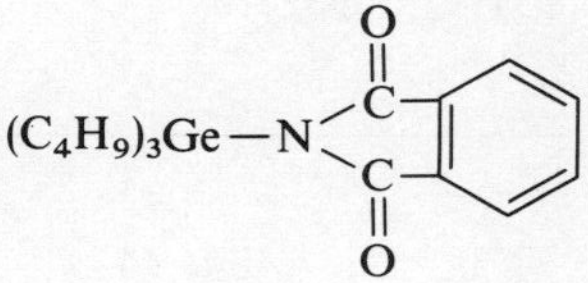 $(C_4H_9)_3Ge{-}N$ [phthalimido]	$C_{20}H_{31}NO_2Ge$	—	159/0.3	1.5208	—	88, 105, 106

(h) $R_2Ge(NR'_2)_2$

COMPOUND	EMPIRICAL FORMULA	M.P. (°C)	B.P. (°C/mm)	n_D^{20}	d_4^{20}	REFERENCES
$Cl_2Ge[N(CH_3)_2]_2$	$C_4H_{12}N_2GeCl_2$	—	—	—	—	19
$(CH_3O)_2Ge[N(CH_3)_2]_2$	$C_6H_{18}N_2O_2Ge$	—	—	—	—	19
$(CH_3)_2Ge[NSi(CH_3)_3]_2$	$C_8H_{24}N_2Si_2Ge$	—	87/15	—	—	127
$(CH_3)_2Ge[N(C_2H_5)_2]_2$	$C_{10}H_{26}N_2Ge$	—	106–8/2	1.5025	—	76, 101, 188
$Cl_2Ge(NHC_6H_5)_2$	$C_{12}H_{12}N_2GeCl_2$	—	—	—	—	28
$Cl_2Ge(NHC_6H_{11})_2$	$C_{12}H_{24}N_2GeCl_2$	—	—	—	—	28

Table 8-1(h)—continued

COMPOUND	EMPIRICAL FORMULA	M.P. (°C)	B.P. (°C/mm)	n_D^{20}	d_4^{20}	REFERENCES
$(C_4H_9)_2Ge[N(CH_3)_2]_2$	$C_{12}H_{30}N_2Ge$	—	249/760	1.4605	1.001	13, 189
$(CH_3)_2Ge[NHSi(C_2H_5)_3]_2$	$C_{14}H_{38}N_2Si_2Ge$	—	115–17/7	—	—	125
$(C_6H_5)_2Ge[N(CH_3)_2]_2$	$C_{16}H_{22}N_2Ge$	—	95–98/0.2	—	—	189
			181–82/13			
$(C_4H_9)_2Ge\left[\text{N}\langle\!\!\!\square\rangle\right]_2$ 	$C_{16}H_{26}N_2Ge$	—	156–60/1.1	1.5222	—	88, 105, 106

(i) RGe(NR'$_2$)$_3$

COMPOUND	EMPIRICAL FORMULA	M.P. (°C)	B.P. (°C/mm)	n_D^{20}	d_4^{20}	REFERENCES
$ClGe[N(CH_3)_2]_3$	$C_6H_{18}N_3GeCl$	—	—	—	—	19
$(CH_3O)Ge[N(CH_3)_2]_3$	$C_7H_{21}N_3OGe$	—	—	—	—	19
$C_2H_5Ge[N(CH_3)_2]_3$	$C_8H_{23}N_3Ge$	–46	191/760	—	1.049 (22°)	9
			105–7/34			
$C_2H_5Ge[N(C_2H_5)_2]_3$	$C_{14}H_{35}N_3Ge$	—	249/760	—	1.108 (22°)	9
$ClGe(NHC_6H_5)_3$	$C_{18}H_{18}N_3GeCl$	—	117–18/12	—	—	28

(j) Ge(NR₂)₄

COMPOUND	EMPIRICAL FORMULA	M.P. (°C)	B.P. (°C/mm)	n_D^{20}	d_4^{20}	REFERENCES
Ge[N(CH₃)₂]₄	C₈H₂₄N₄Ge	14	203/760 87–89/15	—	1.069 (22°)	9, 19
Ge[N(C₂H₅)₂]₄	C₁₆H₄₀N₄Ge	—	72/0.7 266/760 108–10/2	1.4726	1.215 (22°)	9
$\left[\begin{array}{c} Ge \diagdown_{N}^{CH=CH}\diagup CH=CH \end{array}\right]_4$	C₁₆H₁₆N₄Ge	202	—	—	—	153
$\left[\begin{array}{c} Ge \diagdown_{N}^{CH_2-CH_2}\diagup CH_2-CH_2 \diagdown_{CH_2}\diagup^{CH_2} \end{array}\right]_4$	C₂₀H₄₀N₄Ge	—	—	—	—	176
Ge(NHC₆H₅)₄	C₂₄H₂₄N₄Ge	—	—	—	—	176
Ge(NHC₆H₁₁)₄	C₂₄H₄₈N₄Ge	—	—	—	—	28

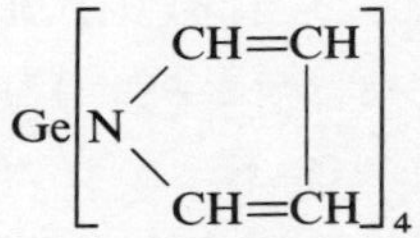

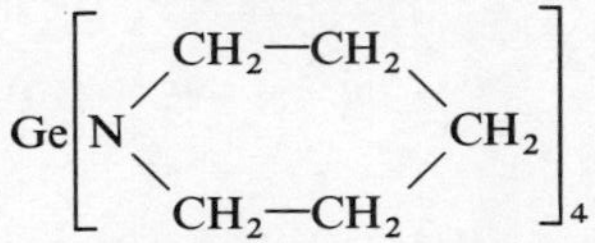

(k) R₃GeN—N⟨

COMPOUND	EMPIRICAL FORMULA	M.P. (°C)	B.P. (°C/mm)	n_D^{20}	d_4^{20}	REFERENCES
(C₂H₅)₃GeNH—NH—C₆H₅	C₁₂H₂₂N₂Ge	—	124–26/0.6	1.5400	—	115
(C₆H₅)₃GeN(C₆H₅)—NH(C₆H₅)	C₃₀H₂₆N₂Ge	142–43	—	—	—	50

(I) *Other Ge—N derivatives*

COMPOUND	EMPIRICAL FORMULA	M.P. (°C)	B.P. (°C/mm)	n_D^{20}	d_4^{20}	REFERENCES
H₃Ge—N=C=N—GeH₃	CH₆N₂Ge₂	—	10 ± 0.5	—	—	24
C₂H₅GeN	C₂H₅NGe	—	—	—	—	44
(C₂H₅N)₂Ge	C₄H₁₀N₂Ge	—	—	—	—	176
(CH₃)₂Ge—N=P(CH₃)₃ (N₃)	C₅H₁₅N₄PGe	87–89	80–85/0.1	—	—	134
(CH₃)₃Ge—N=P(CH₃)₃	C₆H₁₈NPGe	—	69–70/12	—	—	134
(C₂H₅)₃GeN—C—H (H O)	C₇H₁₇NOGe	—	129–31/12	1.4701	1.1619	114
(CH₃)Ge—N=C·Fe(CO)₄	C₈H₉O₄NFeGe	69–70	—	—	—	157
(C₂H₅)₃GeN—C—CH₃ (H O)	C₈H₁₉NOGe	—	82/0.2	1.4713	1.1381	121
[(C₂H₅)₂N]₂Ge	C₈H₂₀N₂Ge	—	—	—	—	176
(CH₃)₂Ge[N=P(CH₃)₃]₂	C₈H₂₄N₂P₂Ge	9–10	72–75/0.01	—	—	134
(CH₃)₂[(CH₃)₃SiO]GeN=P(CH₃)₃	C₈H₂₄ONPSiGe	—	41–42/0.2	—	—	134
(CH₃)₃Ge—N=P(C₂H₅)₃	C₉H₂₄NPGe	—	55–57/0.5	—	—	134
(CH₃)₃GeN⟨P(CH₃)₃ / Al(CH₃)₃⟩	C₉H₂₇NPAlGe	75–77	112–13/0.2	—	—	185

Structure	Molecular formula	mp	bp/pressure			Ref.
$(CH_3)_3Ge{-}N$ with $P(CH_3)_3$ / $Ga(CH_3)_3$	$C_9H_{27}NPGaGe$	41–43	87–89/0.2	—	—	185
$(CH_3)_3Ge{-}N$ with $P(CH_3)_3$ / $In(CH_3)_3$	$C_9H_{27}NPInGe$	46–48	93–95/0.2	—	—	185
$(CH_3)_3GeN{=}P(Cl)[C(CH_3)_3]_2$	$C_{11}H_{28}NPGe$	$-4-{-}6$	46/0.5	—	—	130
$(CH_3)_3GeNH{-}P[C(CH_3)_3]_2$	$C_{11}H_{27}ClNPGe$	4	46/0.05	—	—	130
$(CH_3)_3GeNHP(S)[C(CH_3)_3]_2$	$C_{11}H_{28}NPSGe$	96–98	—	—	—	130
$(C_6H_5N)_2Ge$	$C_{12}H_{10}N_2Ge$	—	—	—	—	176
$(C_2H_5)_3GeN{=}N{-}C_6H_5$	$C_{12}H_{20}N_2Ge$	—	107–14/0.8	—	—	115
$(CH_3)_3Ge{-}N{=}C{-}N(CH_3)_2$ / C_6H_5	$C_{12}H_{20}N_2Ge$	30–32	76–77/0.1	—	—	126
$(CH_3)_3GeN{=}P{-}[C(CH_3)_3]_2$ / CH_3	$C_{12}H_{30}NPGe$	—	50/0.1	—	—	130
$[(CH_3)_3C]_2P$ $N{-}Ge(CH_3)_3$ / $NH{-}CH_3$	$C_{12}H_{31}N_2PGe$	-6	74/0.05	—	—	131
$(CH_3)_3GeN$ with $P(C_2H_5)_3$ / $Al(CH_3)_3$	$C_{12}H_{33}NPAlGe$	182–83	150/0.2 (subl.)	—	—	185

Table 8-1(l)—continued

COMPOUND	EMPIRICAL FORMULA	M.P. (°C)	B.P. (°C/mm)	n_D^{20}	d_4^{20}	REFERENCES
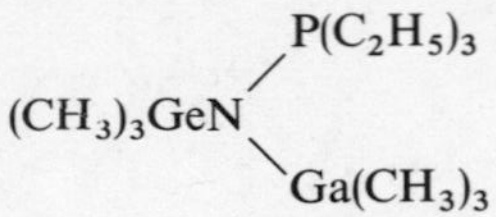 $(CH_3)_3GeN$ with $P(C_2H_5)_3$ and $Ga(CH_3)_3$	$C_{12}H_{33}NPGaGe$	112–13	110/0.2 (subl.)	—	—	185
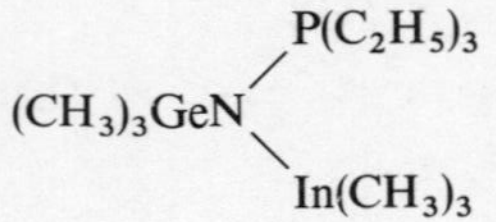 $(CH_3)_3GeN$ with $P(C_2H_5)_3$ and $In(CH_3)_3$	$C_{12}H_{33}NPInGe$	149–50	140/0.2 (subl.)	—	—	185
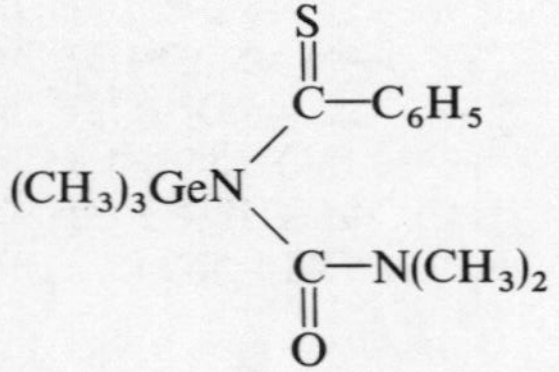 $(CH_3)_3GeN$ with $S=C-C_6H_5$ and $C-N(CH_3)_2$ ($=O$)	$C_{13}H_{20}N_2OSGe$	47–49	—	—	—	64
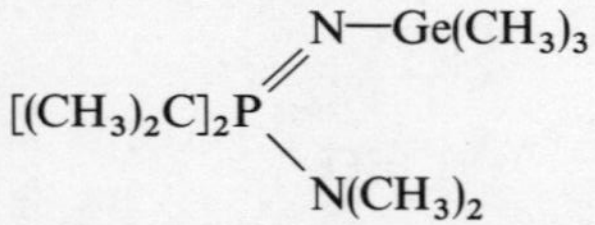 $[(CH_3)_2Cl]_2P$ with $N-Ge(CH_3)_3$ and $N(CH_3)_2$	$C_{13}H_{33}N_2PGe$	15	61/0.05	—	—	131
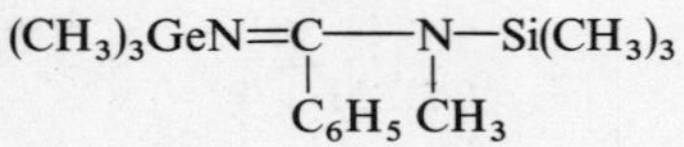 $(CH_3)_3GeN=C$ with $N-Si(CH_3)_3$ and C_6H_5 CH_3	$C_{14}H_{26}N_2SiGe$	—	79–81/0.1	—	—	126

Structure	Molecular Formula	mp	bp			Ref.
$[(C_2H_5)_3Ge]_2N-C(=O)-CH_3$	$C_{14}H_{33}NOGe_2$	51.5–52	191–92/18	—	—	114
$(C_4H_9)_2Ge[N=P(OCH_3)_3]_2$	$C_{14}H_{36}N_2O_6P_2Ge$	—	—	—	—	181
$(CH_3)_3Ge-N=P[C(CH_3)_3]_2Si(CH_3)_3$	$C_{14}H_{36}NPSiGe$	—	80/0.1 (subl.)	—	—	129
$(CH_3)_3Ge-N=P[C(CH_3)_3]_2Ge(CH_3)_3$	$C_{14}H_{36}NPGe_2$	—	90/0.1 (subl.)	—	—	129
$[(CH_3)_2Cl]_2P\langle \begin{smallmatrix} N-Si(CH_3)_3 \\ NH-Ge(CH_3)_3 \end{smallmatrix}$	$C_{14}H_{37}N_2PSiGe$	81	93/0.05	—	—	131
$(C_2H_5)_3GeN-\underset{\underset{C_6H_5}{\overset{\parallel}{S}}}{C}-N(CH_3)_2$	$C_{15}H_{26}N_2SGe$	—	135–36/0.5	1.577	—	107, 121
$[(CH_3)_3C]_2P\langle \begin{smallmatrix} N=Ge(CH_3)_3 \\ N-Ge(CH_3)_3 \\ CH_3 \end{smallmatrix}$	$C_{15}H_{39}N_2PGe_2$	93	95/0.05	—	—	131
$[(CH_3)_3C]_2P\langle \begin{smallmatrix} N=Si(CH_3)_3 \\ N-Ge(CH_3)_3 \\ CH_3 \end{smallmatrix}$	$C_{15}H_{39}N_2PSiGe$	107	110/0.05	—	—	131

Table 8-1(l)—continued

COMPOUND	EMPIRICAL FORMULA	M.P. (°C)	B.P. (°C/mm)	n_D^{20}	d_4^{20}	REFERENCES
$[(CH_3)_3C]_2P$ with $=N—Ge(CH_3)_3$ and $—N—Sn(CH_3)_3$ / CH_3	$C_{15}H_{39}N_2PSnGe$	103	105/0.05 (subl.)	—	—	131
$(CH_3)_3GeN=C(C_6H_5)_2$	$C_{16}H_{19}NGe$	—	109–10/0.17	—	—	87
$(C_2H_5)_3GeN—C=N(C_2H_5)_2$, C_6H_5, O	$C_{17}H_{30}N_2OGe$	—	119/0.2	1.520	1.1120	121
$(C_2H_5)_3GeN—C=N(C_2H_5)_2$, C_6H_5, S	$C_{17}H_{30}N_2SGe$	—	143/0.4	1.566	1.1288	121
$(C_2H_5)_3GeN—C=P(C_2H_5)_2$, C_6H_5, O	$C_{17}H_{30}NOPGe$	—	118/0.15	1.5787	1.1311	114
$(C_2H_5)_3GeN—C=P(C_2H_5)_2$, C_6H_5, S	$C_{17}H_{30}NSPGe$	—	92/0.2	1.5745	1.1270	114
$(C_2H_5)_3GeN—C—N—Ge(C_2H_5)_3$, C_6H_5 O H	$C_{19}H_{36}N_2OGe_2$	—	134/0.1	1.5147	—	114
$(C_6H_5)_3GeN(CN)_2$	$C_{20}H_{15}N_3Ge$	145	—	—	—	78
$(C_4H_9)_3Ge—N=C=N—Ge(C_4H_9)_3$	$C_{25}H_{54}N_2Ge_2$	—	146–54/0.003	1.4823	—	88, 105, 106
$(C_6H_5)_3GeN=C(C_6H_5)_2$	$C_{31}H_{25}NGe$	127–32	—	—	—	87
$(C_6H_5)_3GeN=P(C_6H_5)_3$	$C_{36}H_{30}NPGe$	192–93	—	—	—	103, 172

8-1-4   Organogermanium azides

The most general and successful synthetic method for organometallic azides employs an ionic azide as starting material. NaN_3 has been most often reported (169):

$$R_nGeX_{4-n} + {}_{4-n}NaN_3 \rightarrow R_nGe(N_3)_{4-n} + {}_{4-n}NaX$$

Methylgermanium azides, $(CH_3)_3GeN_3$ and $(CH_3)_2Ge(N_3)_2$, are prepared by the direct action of $(CH_3)_3GeCl$ and $(CH_3)_2GeCl_2$ on an ether suspension of NaN_3 (112). Triphenylgermanium azide $(C_6H_5)_3GeN_3$ is obtainable by the action of triphenylgermanium bromide on NaN_3 in water/ether (172), or by a rather similar method which involves refluxing triphenylgermanium bromide and NaN_3 in benzene with $LiAlH_4$ as catalyst (103). Recently germyl azide, H_3GeN_3, has been prepared from germyl fluoride and trimethylsilyl azide, and some physical properties were reported (23).

But the usual preparation involves a water/ether system in which HN_3 is generated in the aqueous layer from an ionic azide (NaN_3) and acid, and passes into the ethereal layer, where it reacts with the organometallic hydroxide or oxide (172, 173):

$$R_3GeOH \text{ or } (R_3Ge)_2O + HN_3 \rightarrow R_3GeN_3 + H_2O$$

$$(R = CH_3, C_6H_5)$$

The thermal stability of these azides is higher than is generally believed. Thayer and West (172, 173) and Reichle (103) have made comparative studies on the thermal stability of Group IV azides. The thermal stability of the organometallic azides decreases as the atomic weight of the metal increases. For example, there is a steady drop in the decomposition temperature of the $(C_6H_5)_3MN_3$ series (M = Si, Ge, Sn, Pb). The approximate decomposition temperatures of azides are: $(C_6H_5)_3SiN_3$, 380°C; $(C_6H_5)_3GeN_3$, 375°C; $(C_6H_5)_3SnN_3$, 300°C; $(C_6H_5)_3PbN_3$, 200°C (103). Triphenylgermanium azide decomposes on heating to yield products having phenyl groups bound to nitrogen. Some authors have attributed the stability of Si and Ge azides to dative π-bonding between the nitrogen and the central atom (103, 172, 173).

The fundamental IR vibrations of organometallic azides are: $v_1(v_{AS}NNN)$, $v_2(v_{sy}NNN)$, $v_3(vM–N)$, $v_4(\pi\text{-}NNN)$, $v_5(\delta NNN)$, $v_6(\delta MNN)$. v_1, v_2, v_3, the three stretching modes, are usually easy to distinguish. The three bending modes v_4, v_5, and v_6 are much weaker and only v_4 is usually reported (169):

$$IR\ frequencies,\ cm^{-1}$$

	v_1	v_2	v_3	v_4	REFERENCES
$(CH_3)_3GeN_3$	2103	1286	456	675	171,173
$(CH_3)_2Ge(N_3)_2$	2110	1282		680	112
$(C_6H_5)_3GeN_3$	2100	1280		660	172

Thayer and West investigated the UV spectra of $(CH_3)_3MN_3$ (173) (M = Si, Ge, Sn, Pb) and found bands arising from $\pi-\pi^*$ transitions. One of these bands was shifted to higher frequency in the Si and Ge compounds relative to aliphatic azides. This shift is interpreted in terms of dative π-bonding from the azide group onto the central atom Si or Ge.

Hydrolysis of $(CH_3)_3GeN_3$ and $(CH_3)_2Ge(N_3)_2$ is immediate (112):

$$2\,(CH_3)_3GeN_3 + 2\,HOH \rightarrow [(CH_3)_3Ge]_2O + 2\,HN_3$$

$$4\,(CH_3)_2Ge(N_3)_2 + 4\,HOH \rightarrow [(CH_3)_2GeO]_4 + 8\,HN_3$$

$(C_6H_5)_3GeN_3$, on the other hand, is stable to water and atmospheric moisture, undergoing hydrolysis only slowly. Reaction is much more rapid in mixed organic-aqueous media such as acetone–water (172).

Most organic azides react with triphenylphosphine to give nitrogen and phosphineimines:

$$RN_3 + (C_6H_5)_3P \rightarrow RN{=}P(C_6H_5)_3 + N_2$$

$(C_6H_5)_3GeN_3$ also forms a stable phosphineimine,

$$(C_6H_5)_3GeN{=}P(C_6H_5)_3$$

(103, 172). $Bu_2Ge[N{=}P(OMe)_3]_2$ is also reported (181). The N-trimethyl-germyl-iminotrimethylphosphorane, which had already been isolated by the action of trimethylchlorogermane on $LiN{=}P(CH_3)_3$ (135), can be obtained in very high yields in the reaction between the trimethylgermanium azide and the trimethylphosphine (134):

$$(CH_3)_3GeCl + LiN{=}P(CH_3)_3 \rightarrow (CH_3)_3GeN{=}P(CH_3)_3 + LiCl$$

$$(CH_3)_3GeN_3 + P(CH_3)_3 \rightarrow (CH_3)_3GeN{=}P(CH_3)_3 + N_2$$

From $(CH_3)_3GeN_3$ and triethylphosphine it is possible in an analogous way to obtain the N-trimethylgermyl-iminotriethylphosphorane. The reaction of $(CH_3)_2Ge(N_3)_2$ with $P(CH_3)_3$ yields in a first step the

N-dimethylazidogermyl-iminotrimethylphosphorane:

$$(CH_3)_2Ge(N_3)_2 + P(CH_3)_3 \rightarrow (CH_3)_2\!\!\underset{\underset{N_3}{|}}{Ge}\!\!-N{=}P(CH_3)_3 + N_2$$

Under more severe conditions and with more phosphine, formation of the bis-(phosphineimine) $(CH_3)_2Ge-[N{=}P(CH_3)_3]_2$ is observed (134). The action of lithium trimethylsilanolate on the monophosphineimine leads to N-(trimethylsiloxy-dimethylgermyl)iminomethylphosphorane with formation of lithium nitride:

$$(CH_3)_2\!\!\underset{\underset{N_3}{|}}{Ge}\!\!-N{=}P(CH_3)_3 + (CH_3)_3SiOLi \rightarrow$$

$$LiN_3 + (CH_3SiO)(CH_3)_2GeN{=}P(CH_3)_3$$

$(C_6H_5)_3GeN_3$ forms stable 1:1 adducts with Lewis acids, $SnCl_4$ and BBr_3. These N-complexes $(C_6H_5)_3GeN_3 \cdot SnCl_4$ and $(C_6H_5)_3GeN_3 \cdot BBr_3$ are white solids which decompose without melting (174).

8-1-5 Organogermanium azides

Table 8-2 showing some physical properties of organogermanium azides appears on the following page.

Table 8-2
Organogermanium Azides

COMPOUND	EMPIRICAL FORMULA	M.P. (°C)	B.P. (°C/mm)	n_D^{20}	d_4^{20}	REFERENCES
H_3GeN_3	—	—	—	—	—	23
$(CH_3)_2Ge(N_3)_2$	$C_2H_6N_6Ge$	−14	43.5/2	—	—	112
$(CH_3)_3GeN_3$	$C_3H_9N_3Ge$	−65	138	—	—	112, 173
$(C_6H_5)_3GeN_3$	$C_{18}H_{15}N_3Ge$	107–7.5	—	—	—	103, 172
		102–3				
$(C_6H_5)_3GeN_3 \cdot BBr_3$	$C_{18}H_{15}N_3Br_3BGe$	—	—	—	—	174
$(C_6H_5)_3GeN_3 \cdot SnCl_4$	$C_{18}H_{15}N_3Cl_4SnGe$	—	—	—	—	174

8-1-6   Organogermanium pseudohalides

8-1-6-1   Organogermanium Isocyanates, Isothiocyanates, (iso)Cyanides and Phthalocyanines. Et_3GeNCO, $Et_2Ge(NCO)_2$ and $EtGe(NCO)_3$ were the first organogermanium isocyanates to be described in the literature (4). They were obtained by the reaction of the three ethylchlorogermanes with silver isocyanate in boiling benzene. Germanium tetraisocyanate was prepared by the same method from $GeCl_4$ (45, 86). From that time onwards, isocyanates were generally prepared by reacting organogermanium halides with silver isocyanate (7, 8, 11–13, 99, 160–162, 171) in benzene in the presence of nitromethane, in ether or in the absence of a solvent. Hydrolysis is rapid and exothermic for $Ge(NCO)_4$, and progressively becomes slower for the following derivatives: $EtGe(NCO)_3$, $Et_2Ge(NCO)_2$, Et_3GeNCO (4). $[GeO(CN)_2]_n$ or $Ge_2O(CN)_6$ would be the products resulting from partial hydrolysis of $Ge(NCO)_4$ in moist air (4).

The following stages may be expected from the hydrolysis of diarylgermanium dipseudohalides (162):

$$R_2GeX_2 \rightarrow R_4Ge_2X_2(OH)_2 \rightarrow R_8Ge_4X_4O_2 \rightarrow$$

$$R_8Ge_4X_2(OH)_2O_2 \rightarrow R_2GeO$$

$$(R = C_6H_5 \text{ or } p\text{-}C_6H_4CH_3 ; X = NCS, NCO \text{ or } CN).$$

All four isocyanates react with alcohols, yielding the alkoxy compounds (4):

$$(C_2H_5)_{4\text{-}n}Ge(NCO)_n + 2nROH \rightarrow (C_2H_5)_{4\text{-}n}Ge(OR)_n + nNH_2CO_2R$$

At 500°C $GeCl_4$ and $Ge(NCO)_4$ undergo a redistribution with the formation, probably, of germanium trichloroisocyanate Cl_3GeNCO which is unstable (46).

Germanium tetraisocyanate and SbF_3, when heated without solvent, give germanium tetrafluoride and $Sb(NCO)_3$ (5).

The action of silver isothiocyanate on organogermanium halides seems to be a less general method for organogermanium isothiocyanates (4). Anderson reported the formation of $(i\text{-}C_3H_7)_3Ge(NCS)$ (12) and $EtGe(NCS)_3$ (6) but these products have not been isolated in a pure state.

However, Srivastava and Tandon recently have prepared the triphenyl- and tri-p-tolylgermanium(iso)thiocyanates and (iso)cyanates by the interaction of triarylgermanium bromide with the corresponding silver or lead pseudohalide in an inert solvent (161). Diarylgermanium dipseudohalides (aryl = phenyl or p-tolyl; pseudohalide = CNS^-, CNO^- or CN^-) are synthesized by interactions between diaryldibromogermanes and silver or lead pseudohalides (162).

Reacting a concentrated aqueous solution of potassium isothiocyanate with Me_3GeBr and Et_3GeBr leads to Me_3GeNCS and to Et_3GeNCS in 81 and 84% yields respectively (95); and adding an acetonitrile solution of germanium tetrachloride to an ammonium thiocyanate solution gives germanium tetraisothiocyanate $Ge(NCS)_4$ (57).

The cleavage of bis-(trialkylgermanium) oxide and dialkylgermanium oxide by isothiocyanic acid leads to tri- and dialkylgermanium isothiocyanates in high yields (6, 10, 12, 13):

$$(R_3Ge)_2O + 2\,HNCS \rightarrow 2\,R_3GeNCS + H_2O$$

Germylisocyanate H_3GeNCO and germylisothiocyanate H_3GeNCS have been prepared in almost quantitative yield by the interaction of germyl bromide H_3GeBr with silver cyanate and thiocyanate (99, 160).

Thermal stability study of these compounds shows the following decompositions:

$$H_3GeNCO \rightarrow Ge + H_2 + HNCO \text{ at } 200\text{--}220°C$$

$$H_3GeNCS \rightarrow GeH_2 + HNCS \text{ at } 55°C$$

The structure of isocyanates and isothiocyanates has been investigated, particularly by IR spectroscopy (57, 59, 93, 162, 171). The presence of absorption bands at $2271\,cm^{-1}$ (antisym. $N{=}C{=}O$ stretch), $1420\,cm^{-1}$ (sym-$N{=}C{=}O$ stretch), $495\,cm^{-1}$ (Ge—N stretch) in the spectrum of H_3GeNCO is a conclusive evidence that the isocyanate is the major chemical species formed in the reaction, but this does not preclude the presence of the normal cyanate (58, 59). The microwave spectrum of H_3GeNCO is also reported (100).

The same basic bands are found in $Ge(NCO)_4$ (2304 and $2247\,cm^{-1}$, $1432\,cm^{-1}$, $492\,cm^{-1}$) (93) and in $(CH_3)_3GeNCO$ (2240, 1415 and 454 cm^{-1}) (171) and are comparable with the absorption bands of organic isocyanates, cf. CH_3NCO (2232 and $1412\,cm^{-1}$). These spectroscopic results clearly show that the —Ge—$N{=}C{=}O$ structure is widely prevailing in organogermanium isocyanates. The isostructure in $Ge(NCO)_4$ has also been confirmed by Foster and Goodgame (47). Similarly the presence of two strong IR absorption bands in the 1940–2080 cm^{-1} and 920–1070 cm^{-1} ranges seems to show that the thiocyanate group is bonded to the nitrogen atom rather than the sulphur atom in organogermanium isothiocyanates (57, 161, 162). Rochow and Allred (108) however have reported the formation of $(CH_3)_2Ge(SCN)_2$ on adding $(CH_3)_2GeCl_2$ to a solution containing CNS^- thiocyanate ions. The presence of the isomeric isothiocyanate form, however, is not precluded.

Triphenylfulminatogermane has been prepared by the action of triphenylbromogermane on silver fulminate (15):

$$(C_6H_5)_3GeBr + AgCNO \rightarrow AgBr + (C_6H_5)_3GeCNO$$

This compound is hydrolysed in water, in contrast to the corresponding silicone derivative $(C_6H_5)_3SiCNO$.

Organogermanium (iso)cyanides are prepared by reacting organogermanium halides with silver cyanide (6, 12, 132, 156, 161). Anderson has obtained by this method $EtGe(CN)_3$ (6) and $(i\text{-}C_3H_7)_3GeCN$ but impure. Seyferth and Kahlen have prepared trimethyl(iso)cyanogermane in 70–75 % yield by reacting $(CH_3)_3GeI$ with AgCN in boiling benzene (156). Trimethyl(iso)cyanogermane, on the basis of its infrared spectrum and chemical properties, consists of an equilibrium mixture of the normal and the isocyanoisomers:

$$(CH_3)_3Ge{-}C{\equiv}N \rightleftarrows (CH_3)_3Ge{-}N{\equiv}C$$

A strong band at 2197 cm^{-1} has been assigned to the normal stretching frequency while a much weaker band at 2100 cm^{-1} was assigned to the isocyanide bond. $(CH_3)_3GeC{\equiv}N$ is the predominant isomer in this mixture (156).

Trimethyl(iso)cyanogermane reacts with sulphur, forming trimethylisothiocyanatogermane $(CH_3)_3GeNCS$. While sulphur appears to react only with the isocyanogermane in the mixture, continuously displacing the equilibrium $(CH_3)_3GeC{\equiv}N \rightleftarrows (CH_3)_3GeNC$ to the right, boron trifluoride apparently reacts with both isomers. Trimethyl(iso)cyanogermane and boron trifluoride diethyl etherate react with formation of the adduct $(CH_3)_3GeCN{\cdot}BF_3$. In the IR spectrum of that adduct, the presence of bands in the $C{\equiv}N$ and $-N{\equiv}C$ region suggests that both isomers act as donor molecules (156).

Trimethyl(iso)cyanogermane reacts with iron pentacarbonyl at 65–75°C liberating carbon monoxide and forming $(CH_3)_3GeN{\equiv}C{\cdot}Fe(CO)_4$ in good yield:

$$(CH_3)_3GeN{\equiv}C + Fe(CO)_5 \rightarrow (CH_3)_3GeN{\equiv}C{\cdot}Fe(CO)_4 + CO$$

This reaction with iron carbonyl is a reaction characteristic of the isonitrile structure and shows the presence of the isocyano isomer (157). However, crystallographic data are in favour of the cyanide form (132). The cyanide form is also predominant in the triarylgermanium (iso)cyanide series (161) and in diarylgermanium (iso)cyanides (162).

Trialkylcyanogermanes are converted to the isochalcocyanates R_3GeNCX (X = O, S, Se). Introduction of oxygen and sulphur requires heating but selenium reacts at room temperature. There is no reaction with tellurium. The mechanism of chalcogenation is discussed in connection with the long-standing "normal-iso" structural uncertainty of cyanogermanes (170).

Hydrocyanic acid reacts with bis-(trialkylgermanium) oxides only to give trialkylgermanium cyanides; isocyanic acid does not react (6):

$$(R_3Ge)_2O + {}_2HC{\equiv}N \rightarrow 2\,R_3GeCN + H_2O$$

Gradual addition of a deficiency of $Hg(CN)_2$ or $Hg(SCN)_2$ to dibutylgermane $n\text{-}Bu_2GeH_2$ gives 74% yields of $n\text{-}Bu_2Ge(H)CN$ or $n\text{-}Bu_2Ge(H)NCS$ respectively; these compounds are the first examples of the R_2GeHX type in which R is an alkyl group and X is the pseudohalide CN or NCS (13).

The formation of germyl(iso)cyanide, $H_3Ge(CN)$, has been achieved by reactions of H_3GeBr (99, 160, 165) or H_3GeI (56) with silver cyanide. The reactions of germyl(iso)cyanide studied may be summarized by the equations (165):

$$H_3GeNC + NH_3 \rightarrow GeH_2 + NH_4CN$$

$$H_3GeNC + H_2O \rightarrow H_3GeOH + HCN$$

$$H_3GeNC + HBr \rightarrow H_3GeBr + HCN$$

Germyl (iso)cyanide did not react with trimethylborane but gives with BF_3 a stable complex (165). Germyl (iso)cyanide reacts with trimethylgallium to give methylgermane CH_3GeH_3 and $(CH_3)_2GaCN$ (99). The infrared spectra of H_3GeCN and D_3GeCN in the region of the CN stretching fundamental indicates that these molecules are the normal cyanide forms. No evidence was obtained for the presence of the isocyanide (56). Assuming that the NC intensity would be roughly twice that of the CN band a maximum of 2% isocyanide could be in equilibrium with H_3GeCN and D_3GeCN and escape detection. In the alkylated germyl compounds a shift in the equilibrium in favour of the isocyanide (156) may be attributed to the enhancement of the resonance contribution to the isocyanide structure due to alkylation of the germyl cation (56):

$$\overset{\displaystyle +}{\underset{\diagdown}{\diagup}}Ge\ \overset{\displaystyle -}{\underset{\cdot\cdot}{N}} = C:$$

Tetracyanogermane $Ge(CN)_4$ has been prepared by germanium tetraiodide action on silver cyanide in boiling benzene (92) or by an exchange reaction between trimethyl(iso)cyanosilane and $GeCl_4$ in dry xylene (16).

Recently, Joyner and Kenney (72) described several new stable germanium phthalocyanines. These compounds include a chloride, a hydroxide, two phenoxides and a siloxide. Dichlorogermanium phthalocyanine, $PcGeCl_2$, is produced when phthalonitrile and $GeCl_4$ are reacted together in refluxing quinoline. Two additional methods for the synthesis of dichlorogermanium phthalocyanine were described by the same authors in 1962 (73). One method depends on the reaction of o-cyanobenzamide with $GeCl_4$ in refluxing 1-chloronaphthalene and the other on the reaction occurring between phthalocyanine and $GeCl_4$ in refluxing quinoline. The second of these methods appears to be especially advantageous in terms of yield, convenience and purity of product. In dichlorogermanium phthalocyanine, the germanium is surrounded by two chlorine and four nitrogen atoms in what may be assumed to be an octahedral fashion. This somewhat unusual hexacoordination and the size of the organic residue correlate well with the relative difficulty observed in dichloride hydrolysis. This hydrolysis is slow with water, refluxing concentrated ammonia, live steam and refluxing pyridine. The sublimed dichloride, however, was completely hydrolysed with a refluxing 1:1 pyridine–concentrated ammonia solution in six hours. Treatment of the dichloride with concentrated sulphuric acid gave dihydroxygermanium phthalocyanine, $PcGe(OH)_2$. The reaction of dihydroxygermanium phthalocyanine with phenol, p-phenylphenol and triphenylsilanol in refluxing benzene gives diphenoxygermanium phthalocyanine, $PcGe(OC_6H_5)_2$, bis-(phenylphenoxy)germanium phthalocyanine,

$$PcGe(O-C_6H_4-C_6H_5)_2$$

and bis-(triphenylsiloxy)germanium phthalocyanine,

$$PcGe[O-Si(C_6H_5)_3]_2$$

(72). Dihydrogermanium phthalocyanine reacts with diphenylsilanediol, and bis-(diphenylhydroxysiloxy)germanium phthalocyanine,

$$PcGe[OSi(OH)(C_6H_5)_2]_2$$

is obtained. This compound heated to reflux with benzyl alcohol gives bis-(diphenylbenzyloxysiloxy)germanium phthalocyanine,

$$PcGe[OSi(OCH_2C_6H_5)(C_6H_5)_2]_2$$

A polygermanosiloxane, $[PcGe(OSi(C_6H_5)_2O)_2]_n$, is formed by thermal decomposition of diphenylhydroxysiloxide (73).

$PcGeCl_2$ reacts with silver cyanate and silver thiocyanate in 1,2-dichlorobenzene to give $PcGe(NCO)_2$ and $PcGe(NCS)_2$ respectively in high yields. Only $PcGeBr_2$ reacts with silver selenocyanate in 1-chloronaphthalene and leads to $PcGe(NCSe)_2$ (163). A spectral investigation of these compounds shows that the pseudohalo groups are bonded to germanium through nitrogen.

Infrared and NMR studies have been carried out on a series of germanium hemiporphyrazines and phthalocyanines. The NMR studies have shown that the hemiporphyrazine ring, in contrast to the phthalocyanine ring, is not aromatic (40, 41).

8-1-6-2 Organogermanium Isocyanates

Tables 8-3, 8-4, 8-5 and 8-6 showing some physical properties of organogermanium pseudohalides appears on the following pages.

Table 8-3

Organogermanium Isocyanates

COMPOUND	EMPIRICAL FORMULA	M.P. (°C)	B.P. (°C/mm)	n_D^{20}	d_4^{20}	REFERENCES
Cl$_3$GeNCO	CCl$_3$NOGe	—	≈112	—	—	46
H$_3$GeNCO	CH$_3$NOGe	−44	71.5	—	—	59, 160
Ge(NCO)$_4$	C$_4$N$_4$O$_4$Ge	−8	204	1.4824	1.7714	4, 45, 86
(CH$_3$)$_3$GeNCO	C$_4$H$_9$NOGe		122			171
(C$_2$H$_5$)Ge(NCO)$_3$	C$_5$H$_5$N$_3$O$_3$Ge	−31	225.4	1.4739	1.5344	4
(C$_2$H$_5$)$_2$Ge(NCO)$_2$	C$_6$H$_{10}$N$_2$O$_2$Ge	−32	226	1.4619	1.330	4
(C$_2$H$_5$)$_3$GeNCO	C$_7$H$_{15}$NOGe	−26.4	200.4	1.4519	1.1514	4
(i-C$_3$H$_7$)$_2$Ge(NCO)$_2$	C$_8$H$_{14}$N$_2$O$_2$Ge	—	239	1.464	1.225	12
			72–73/1			
(n-C$_4$H$_9$)$_2$Ge(NCO)$_2$	C$_{10}$H$_{18}$N$_2$O$_2$Ge	—	273	1.4634	1.179	13
			93–95/1			
(n-C$_3$H$_7$)$_3$GeNCO	C$_{10}$H$_{21}$NOGe	−19	247	1.4575	1.055	7
			114/10			
(i-C$_3$H$_7$)$_3$GeNCO	C$_{10}$H$_{21}$NOGe	—	238	1.4602	1.097	11
			65–66/1			
(n-C$_4$H$_9$)$_3$GeNCO	C$_{13}$H$_{27}$NOGe	−47	283	1.4595	1.044	8
			109–10/2			
(C$_6$H$_5$)$_3$GeNCO	C$_{19}$H$_{15}$NOGe	117–18	—	—	—	161
(p-CH$_3$C$_6$H$_4$)$_3$GeNCO	C$_{22}$H$_{21}$NOGe	222–23	—	—	—	161

8-1-6-3   Table: Organogermanium Isothiocyanates

Table 8-4

Organogermanium Isothiocyanates

COMPOUND	EMPIRICAL FORMULA	M.P. (°C)	B.P. (°C/mm)	n_D^{20}	d_4^{20}	REFERENCES
H₃GeNCS	CH₃NSGe	18.6	150	—	—	160
Ge(NCS)₄	C₄N₄S₄Ge	—	decomp.	—	—	57
(CH₃)₃GeNCS	C₄H₉NSGe	—	191.5–93 63/8	1.4960 1.5145	1.2676 —	95, 156, 171
(C₂H₅)₂Ge(NCS)₂	C₆H₁₀N₂S₂Ge	16	298 113–16/1	— —	— 1.356	6
(C₂H₅)₃GeNCS	C₇H₁₅NSGe	−46	252 113–14/8	1.517 —	1.184 —	6, 95
(i-C₃H₇)₂Ge(NCS)₂	C₈H₁₄N₂S₂Ge	—	296 105–7/1	1.558 —	1.234 —	12
(n-C₄H₉)₂Ge(H)NCS	C₉H₁₉NSGe	—	96–98/1	1.5097	1.123	13
(n-C₄H₉)₂Ge(NCS)₂	C₁₀H₁₈N₂S₂Ge	—	337 140–41/1	— 1.5501	— 1.210	13
(n-C₃H₇)₃GeNCS	C₁₀H₂₁NSGe	−56	287 143–44/9	1.5063 —	1.105 —	6
(i-C₃H₇)₃GeNCS	C₁₀H₂₁NSGe	18	277	1.512	1.112	10, 12
(n-C₄H₉)₃GeNCS	C₁₃H₂₇NSGe	—	319 135–36/2	1.5039 —	1.071 —	6
(C₆H₅)₃GeNCS	C₁₉H₁₅NSGe	106–7	—	—	—	161
(p-CH₃C₆H₄)₃GeNCS	C₂₂H₂₁NSGe	213–14	—	—	—	161

8-1-6-4   Table: Organogermanium Isoselenocyanates

Table 8-5

Organogermanium Isoselenocyanates

COMPOUND	EMPIRICAL FORMULA	M.P. (°C)	B.P. (°C/mm)	n_D^{20}	d_4^{20}	REFERENCES
$(CH_3)_3GeNCSe$	C_4H_9NSeGe	13–14	118–9/232	—	—	170
$(C_2H_5)_3GeNCSe$	$C_7H_{15}NSeGe$	—	175–76/240	—	—	170
$(C_4H_9)_3GeNCSe$	$C_{13}H_{27}NSeGe$	—	221–22/100	—	—	170

8-1-6-5   *Organogermanium (iso)Cyanides*

Table 8-6

Organogermanium (iso)Cyanides

COMPOUND	EMPIRICAL FORMULA	M.P. (°C)	B.P. (°C/mm)	n_D^{20}	d_4^{20}	REFERENCES
H_3GeCN	CH_3NGe	—	—	—	—	56, 160
H_3GeNC	CH_3NGe	—	—	—	—	165
D_3GeCN	CD_3NGe	—	—	—	—	56
$H_3GeNC \cdot BF_3$	CH_3NBF_3Ge	—	—	—	—	165
$Ge(CN)_4$	C_4N_4Ge	—	—	—	—	16, 92
$(CH_3)_3GeCN$	C_4H_9NGe	38–38.5	150	—	—	132, 156
$(CH_3)_3GeCN \cdot BF_3$	$C_4H_9BF_3NGe$	85–87	—	—	—	156
$(C_2H_5)_3GeCN$	$C_7H_{15}NGe$	18	213	1.4509	1.111	6
$(n\text{-}C_4H_9)_2Ge(H)CN$	$C_9H_{19}NGe$	—	108–10/8	1.4527	1.050	13
$(n\text{-}C_3H_7)_3GeCN$	$C_{10}H_{21}NGe$	–13	115–17/10	1.4544	1.041	6
$(n\text{-}C_4H_9)_3GeCN$	$C_{13}H_{27}NGe$	—	292–94	—	—	170
$(C_6H_5)_3GeCN$	$C_{19}H_{15}NGe$	140–41	—	—	—	161
$(p\text{-}CH_3C_6H_4)_3GeCN$	$C_{22}H_{21}NGe$	151–53	—	—	—	161

8-2 ORGANOGERMANIUM–PHOSPHORUS COMPOUNDS

8-2-1 Organogermanium–phosphorus compounds with Ge—P bond

In 1963 Glockling and Hooton reported the preparation and some chemical reactions of diphenyl(triethylgermyl)phosphine, Et_3GePPh_2. This compound was the first example of an organogermanium compound in which the germanium is bonded to phosphorus (55). It is obtained by interaction of triethylbromogermane and lithium diphenylphosphide in tetrahydrofuran. Lithium diphenylphosphide reacts also with triphenyl-bromo- and diphenyldibromogermane, giving diphenyl(triphenylgermyl)-phosphine, Ph_3GePPh_2, and bis-(diphenylphosphino)diphenylgermane, $Ph_2Ge(PPh_2)_2$ (18).

$PhGeBr_3$ and $GeCl_4$ react differently with lithium diphenylphosphide (or $Ph_2PH + Et_3N$), giving tetraphenyldiphosphine and water-stable, polymeric diphenylphosphinogermane (18, 55).

Hydrolysis of the triethyl- and triphenylgermyl-compounds, R_3GePPh_2, gives $(R_3Ge)_2O$ and Ph_2PH. Oxidation by dry oxygen cleaves the Ge—P bond, forming the monomeric phosphorus (V) ester $R_3Ge—O—P(O)Ph_2$. The identity of this ester (R = Et) is confirmed by its preparation from diphenylphosphinic acid and bis-(triethylgermanium) oxide (18):

$$2\,Ph_2P(O)OH + (Et_3Ge)_2O \rightarrow Et_3GeOP(O)Ph_2 \xleftarrow{O_2} Et_2GePPh_2$$

Bromination of the Ge—P bond in Et_3GePPh_2 proceeds rapidly, giving triethylbromogermane and diphenylbromophosphine.

Cleavage by n-butyllithium is also rapid at 0°C :

$$Et_3GePPh_2 + BuLi \rightarrow Et_3GeBu + Ph_2PLi$$

In a similar reaction, phenyllithium shows only 50 % cleavage of the Ge—P bond after 2 hours in refluxing ether.

Methyl iodide also cleaves the Ge—P bond, giving triethyliodo-germane and dimethyldiphenylphosphonium iodide (18, 55):

$$Et_3GePPh_2 + MeI \rightarrow Et_3GeI + Ph_2PMe \xrightarrow{MeI} [Ph_2Me_2P]^+I^-$$

This contrasts with the behaviour of similar silicon and tin compounds which have been converted into the phosphorus (IV) complexes $[Me_3SiPEt_3]I$ and $[Ph_3SnP(Me)Ph_2]I$. The phosphorus atom in the triethylgermyl derivative retains its donor character to an appreciable degree since with silver iodide it readily forms a colourless, crystalline 1:1 complex $[Et_3GePPh_2 \cdot AgI]_4$. The triphenylgermyl analogue fails to yield the same silver iodide complex (18).

Cleavage of Ge—N bond in trialkylgermylamines by diphenyl-phosphine represents a new synthesis of trialkylgermyldiphenylphosphines (113, 121):

$$Et_3GeNMe_2 + Ph_2PH \rightarrow Et_3GePPh_2 + Me_2NH$$

$$(Me_3Ge)_2NH \text{ or } (Me_3Ge)_2N + Ph_2PH \rightarrow Me_3GePPh_2 + NH_3$$

Me_3GeNMe_2 reacts with $MePH_2$ in a 2:1 molar ratio in Et_2O at $-30°C$ to afford $(Me_3Ge)_2PMe$ (146):

$$2(CH_3)_3GeN(CH_3)_2 + CH_3PH_2 \rightarrow [(CH_3)_3Ge]_2PCH_3 + 2(CH_3)_2NH$$

$Me_3GePHPh$ and $(Me_3Ge)_2PPh$ were also prepared:

$$(CH_3)_3GeN(CH_3)_2 + C_6H_5PH_2 \rightarrow$$
$$(CH_3)_3GeP(H)C_6H_5 + (CH_3)_2NH$$

$$2(CH_3)_3GeN(CH_3)_2 + C_6H_5PH_2 \rightarrow$$
$$[(CH_3)_3Ge]_2PC_6H_5 + 2(CH_3)_2NH$$

$$(CH_3)_3GeCl + C_6H_5P(H)Li \rightarrow$$
$$(CH_3)_3GeP(H)C_6H_5 + LiCl$$

These compounds were characterized by NMR and IR spectroscopy (146). Diethyl(triethylgermyl)phosphine, Et_3GePEt_2, is obtained by the action of triethyl chlorogermane on lithium diethylphosphide in THF medium (116). This compound reacts exothermally with water, alcohols, thiols, and organic acids (116):

$$Et_3GePEt_2 + AH \rightarrow Et_3GeA + Et_2PH$$

$$(A = OH, RO, RS, RCOO)$$

and with aniline at about 100°C:

$$Et_3GePEt_2 + C_6H_5NH_2 \rightarrow Et_3Ge-NH-C_6H_5 + Et_2PH$$

The action of ammonia at 150°C in an autoclave causes ammonolysis of the Ge—P bond:

$$2Et_3GePEt_2 + NH_3 \rightarrow (Et_3Ge)_2NH + 2Et_2PH$$

The Ge—P bond is quantitatively cleaved by HCl (116) and HBr (25). Et_3GePEt_2 and Et_3GePPh_2 with CS_2, PhNCO and PhNCS produce insertion reactions (116) similar to those described by Schumann and coworkers for the Sn—P bond (141):

$$Et_3GePEt_2 + CS_2 \rightarrow Et_3Ge-S-\underset{\underset{S}{\|}}{C}-PEt_2$$

$$Et_3GePEt_2 + PhNCX \rightarrow Et_3Ge-\underset{\underset{Ph}{|}}{N}-\underset{\underset{X}{\|}}{C}-PEt_2 \qquad (X = O, S)$$

$$Et_3GePPh_2 + PhNCX \underset{t^o}{\rightleftarrows} Et_3Ge-\underset{\underset{Ph}{|}}{N}-\underset{\underset{X}{\|}}{C}-PEt_2 \qquad (X = O, S)$$

Acrylonitrile also gives an insertion reaction:

$$Et_3GePEt_2 + CH_2=CH-CN \rightarrow Et_3Ge-\underset{\underset{CN}{|}}{CH}-CH_2PEt_2$$

Phenylacetylene in the presence of azobis-isobutyronitrile produces simultaneously the addition reaction with formation of four ethylenic isomers and the cleavage reaction in the relative proportions of 60 and 40% (116):

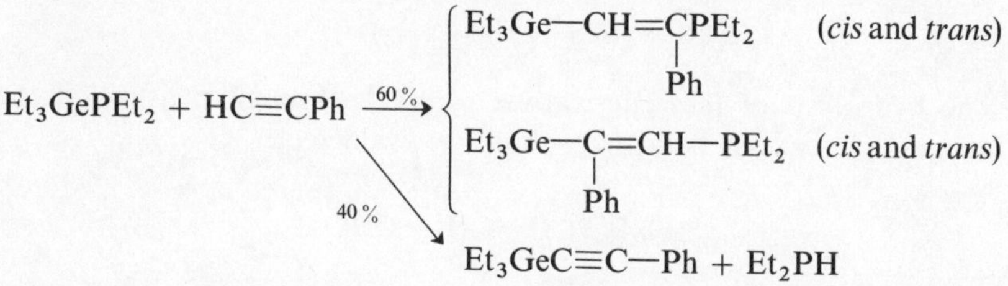

$$Et_3GePEt_2 + HC\equiv CPh \xrightarrow{60\%} \begin{cases} Et_3Ge-CH=\underset{\underset{Ph}{|}}{C}PEt_2 \quad (cis \text{ and } trans) \\[2ex] Et_3Ge-\underset{\underset{Ph}{|}}{C}=CH-PEt_2 \quad (cis \text{ and } trans) \end{cases}$$

$$\xrightarrow{40\%} Et_3GeC\equiv C-Ph + Et_2PH$$

Saturated aldehydes such as butanal or benzaldehyde add exothermally to the Ge—P bond of the germylphosphines with formation of phosphorus-containing alkoxygermanes (117, 119):

$$Et_3GePEt_2 + R'CHO \rightarrow Et_3Ge-O-\underset{\underset{R'}{|}}{CH}-PEt_2$$

$$(R' = n\text{-}C_3H_7, C_6H_5)$$

The insertion reactions of CS_2 PhNCO, PhNCS and the saturated aldehydes probably proceed *via* a polar, four-centred concerted mechanism involving the nucleophilic attack of phosphorus on the positive carbon of the unsaturated substrate with a simultaneous coordination between sulphur, nitrogen, oxygen and germanium through the free lone electron pairs of these atoms and the vacant germanium 4d orbitals:

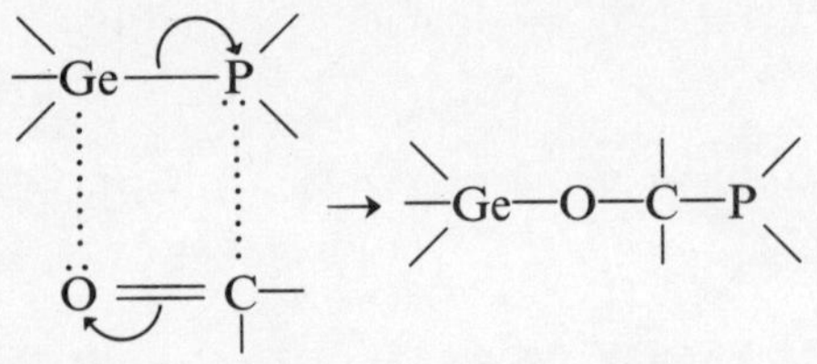

On the other hand, the phenylacetylene addition catalysed by AIBN seems to be of the radical type.

In the action of the α-ethylenic aldehydes, such as crotonaldehyde or cinnamaldehyde, on diethyl(triethylgermyl)phosphine, an addition of the 1,4-type is observed with formation of enoxygermanium products in their two *cis* and *trans* forms (yield ≈75%) (117, 119):

$$RCH{=}CH{-}C(H){=}O + Et_3GePEt_2 \rightarrow Et_3Ge{-}O{-}CH{=}CH{-}CH(PEt_2){-}R$$

$$(R = CH_3, C_6H_5)$$

The hydrolysis of these derivatives proceeds through cleavage of the germanium–oxygen bond to γ-phosphoryl aldehydes

$$Et_2P{-}CH(R){-}CH_2{-}CHO$$

The same type of addition is noted with cyclopent-2-ene-1-one (119). Ketenes also add to the Ge—P bond with opening of the C=O double bond (118, 119):

$$Et_3GePEt_2 + R_2C{=}C{=}O \rightarrow Et_3Ge{-}O{-}C(PEt_2){=}CR_2$$

$$(R = H \text{ or } C_6H_5)$$

The hydrolysis of these derivatives leads to the "α-phosphorylketones" $H_3C{-}CO{-}PEt_2$ and $Ph_2CH{-}COPEt_2$. However, in the hydrolysis of

$$Et_3Ge{-}O{-}C(PEt_2){=}CPh_2$$

noticeable percentages of Et_2PH and $Et_3Ge{-}O{-}COCHPh_2$ resulting from the cleavage of the carbon–phosphorus bond in this addition derivative also appear:

$$Et_3Ge-O-\underset{\underset{CPh_2}{\|}}{C}-PEt_2 \xrightarrow[H_2O]{ether} \begin{cases} \xrightarrow{80\%} (Et_3Ge)_2O + Ph_2CHCOPEt_2 \\ \\ \xrightarrow{20\%} Et_3GeO-COCHPh_2 + Et_2PH \end{cases}$$

Another interesting method of synthesis of the α-phosphorylketones is the action of the germylphosphines on the acid anhydrides (118, 119):

$$Et_3GePEt_2 + (R_2CHCO)_2O \rightarrow Et_3GeOCOCHR_2 + R_2CHCOPEt_2$$

$$(R = H \text{ or } Ph)$$

Trigermylphosphine $(H_3Ge)_3P$ has been obtained by treating trisilyl-phosphine with a small excess of germyl bromide. This compound is a colourless liquid which decomposes slowly at room temperature and rapidly at 55°C (22). On being set aside at room temperature, $(H_3Ge)_3P$ evolves GeH_4 and leaves a colourless liquid that presumably contains $GeH_3-P-GeH_2$ chains (25). Raman spectra of trigermylphosphine show, besides features near 2100 cm^{-1} (Ge—H stretching), 800 cm^{-1} (GeH$_3$ bending), and 550 cm^{-1} (GeH$_3$ rocking), lines at 366, 322, 112 and 88 cm^{-1}, assigned respectively to antisymmetric stretching, symmetric stretching, and the two bending modes of a pyramidal Ge$_3$P skeleton (22). In contrast, the infrared and Raman spectra of trisilylphosphine, $(H_3Si)_3P$, provide evidence that the PSi$_3$ skeleton is planar like the NSi$_3$ skeleton of trisilylamine, $(H_3Si)_3N$ (29). These results, recently confirmed by Cradock and coworkers (25), imply that $(p \rightarrow d)\pi$ bonding is less strong in trigermyl-phosphine than in trisilylphosphine. Moreover, substantial π-bonding is possible in a pyramidal molecule: thus there was no evidence of interaction between trigermylphosphine and BF$_3$, BEt$_3$, MeI or SiH$_3$Br.

Schumann and Blass recently described the synthesis of tris-[trimethyl-germyl]phosphine and arsine, $[(CH_3)_3Ge]_3P$ and $[(CH_3)_3Ge]_3As$ (140). These derivatives are colourless, distillable liquids and were obtained by the action of phosphine, PH$_3$, and arsine, AsH$_3$, on trimethylgermyl-dimethylamine, $(CH_3)_3GeN(CH_3)_2$, in ethereal solution, or by the action of trimethylchlorogermane on sodium phosphide or arsenide in liquid ammonia:

$$3(CH_3)_3GeCl + Na_3X \rightarrow [(CH_3)_3Ge]_3X + 3NaCl \qquad (X = P; As)$$

In the H—NMR spectrum of $[(CH_3)_3Ge]_3P$ a doublet is observed for the methyl protons bound to germanium at -24.75 Hz, $J_{HCGe^{31}P} = 3.65$ Hz, while in $[(CH_3)_3Ge]_3As$ a singlet is observed at -27.4 Hz. The IR spectra show evidence, against the oscillations of the trimethyl-Ge group, of the antisymmetric and symmetric stretching of a pyramidal Ge_3P skeleton:

$$\nu_{As}Ge_3P\ (397\ cm^{-1}) \qquad \nu_SGe_3P\ (320\ cm^{-1}) \quad and$$

$$\nu_{As}Ge_3As\ (275\ cm^{-1}) \quad (39, 140)$$

The conformity of IR and Raman spectra confirms a pyramidal structure for $[(CH_3)_3Ge]_3P$ (39).

The action of triphenylchlorogermane, -stannane, -plumbane on phosphine, PH_3, and phenylphosphine, $C_6H_5PH_2$, in benzene solution in the presence of triethylamine leads to the phosphines: $[R_3M]_3P$ and $[R_3M]_2PC_6H_5$ (M = Ge, Sn, Pb) (149):

$$2\,(C_6H_5)_3GeCl + C_6H_5PH_2 + 2\,(C_2H_5)_3N \rightarrow$$

$$[(C_6H_5)_3Ge]_2PC_6H_5 + 2\,(C_2H_5)_3N, HCl$$

$$3\,(C_6H_5)_3GeCl + PH_3 + 3\,(C_2H_5)_3N \rightarrow$$

$$[(C_6H_5)_3Ge]_3P + 3\,(C_2H_5)_3N, HCl$$

These derivatives are not attacked by water, but are split by alcoholic KOH with formation of triphenylgermanol and benzylphosphinic acid or phosphoric acid (149):

$$[(C_6H_5)_3Ge]_2PC_6H_5 \xrightarrow{KOH/O_2} 2\,(C_6H_5)_3GeOH + C_6H_5PO(OH)_2$$

$$[(C_6H_5)_3Ge]_3P \xrightarrow{KOH/O_2} 3\,(C_6H_5)_3GeOH + H_3PO_4$$

In the infrared spectrum of these derivatives there appear the following vibrational bands: $\nu_{As}Ge$—P—Ge (392 cm^{-1}), ν_sGe—P—Ge (368 cm^{-1}), $\nu_{As}PGe_3$ (375 cm^{-1}) and ν_sPGe_3 (308 cm^{-1}). The presence of the ν_sPGe_3 band shows that the molecular skeleton Ge_3P is pyramidal. The $(p \rightarrow d)\pi$ interactions do not result in a planar molecule (39, 149).

The cleavage by butyllithium of a Sn—P bond of bis-(triphenylstannyl)-phenyl-phosphine, $[(C_6H_5)_3Sn]_2PC_6H_5$, and tris-(triphenylstannyl)-phosphine, $[(C_6H_5)_3Sn]_3P$, leads respectively to lithium triphenyl-stannylphenylphosphine, $(C_6H_5)_3SnP(Li)C_6H_5$, and lithium bis-(tri-phenylstannyl)phosphine, $[(C_6H_5)_3Sn]_2PLi$. The action of triphenyl-chlorogermane on these lithium derivatives in benzene solution produces respectively triphenylgermyltriphenylstannylphenylphosphine and triphenylgermyl-bis-(triphenylstannyl)phosphine (150):

$$(C_6H_5)_3Sn\diagdown$$
$$P-Li + (C_6H_5)_3GeCl \rightarrow$$
$$\underset{\displaystyle C_6H_5}{|}$$

$$LiCl + \underset{\diagup}{\overset{(C_6H_5)_3Sn\diagdown}{}}PGe(C_6H_5)_3$$
$$C_6H_5$$

$$[(C_6H_5)_3Sn]_2P-Li + (C_6H_5)_3GeCl \rightarrow$$

$$LiCl + [(C_6H_5)_3Sn]_2PGe(C_6H_5)_3$$

The same reactions are observed with triphenylchloroplumbane. In the alkaline cleavage of these derivatives by alcoholic KOH in the presence of air, the formation of triphenylgermanol, $(C_6H_5)_3GeOH$, triphenylstannol, $(C_6H_5)_3SnOH$, and benzylphosphinic and phosphoric acid is observed. The infra-red spectra of these derivatives show the expected bands: $\nu_{GeP} = 369\ cm^{-1}$; $\nu_{As}PSn_2Ge$: $349\ cm^{-1}$; $\nu_s PSn_2Ge$: $297\ cm^{-1}$.

The stability of organometallophosphines of type $(R_3M)_nPR_{3-n}(M =$ Si, Ge, Sn or Pb) is ascribed to participation of the free electron-pair on the phosphorus in the covalent P—M bond (151). All previous attempts to prepare organometallophosphines with tetra- or pentavalent phosphorus have been unsuccessful. Fission of the metal–phosphorus bond occurs on utilization of the free electron-pair, for instance by quaternization of the phosphorus or by oxidation.

However, by treatment of tetracarbonylnickel with tris-(trimethylsilyl)-, tris-(trimethylgermyl)-, tris-(trimethylstannyl)- or tris-(trimethylplumbyl)-phosphine in tetrahydrofuran at room temperature, Schumann and Stelzer (151) synthesized carbonyl nickel (0) complexes which represent the first stable organometal-substituted phosphines with tetravalent phosphorus:

$$Ni(CO)_4 + [(CH_3)_3M]_3P \rightarrow \underset{OC\diagup}{\overset{OC\diagdown}{OC-}}Ni \leftarrow P\underset{\diagdown M(CH_3)_3}{\overset{\diagup M(CH_3)_3}{-}}M(CH_3)_3 + CO$$

(M = Si, Ge, Sn, Pb)   (I)

The considerable increase in the coupling constants $J^1HCM^{31}P$ of the complexes (Si: 5.2 Hz, Ge: 4.65 Hz, Sn: 3.35 Hz, Pb: 2.9 Hz) in comparison with those of the free organometallophosphines (Si: 4.62 Hz, Ge: 4.00 Hz,

Sn: 1.95 Hz, Pb: 0 Hz) confirms the increase in s-character of the P—M bonds on going from the p^3 structure of phosphorus atom of the free phosphines to the sp^3 hybridized phosphorus in the complexes.

The stability of the new compounds towards oxygen confirms the hypothesis that on oxidation of organometallophosphines, the oxygen attacks the free electron-pair of the phosphorus. In the complex (I) this electron-pair is no longer open to electrophilic attack owing to coordination to the nickel.

In reactions of the same type, tris-(organometallo)phosphine-dicarbonyl-nitrosylcobalt complexes are obtained almost quantitatively when tricarbonylnitrosylcobalt is allowed to react with tris-(trimethylsilyl)-, tris-(trimethylgermyl)- or tris-(trimethylstannyl)phosphine in tetrahydrofuran at room temperature (152):

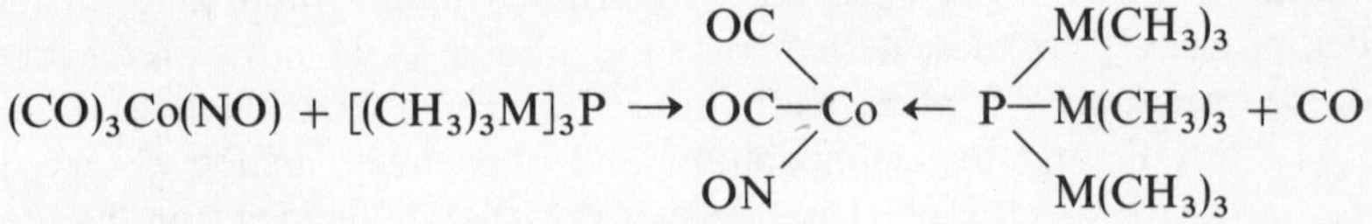

$$(CO)_3Co(NO) + [(CH_3)_3M]_3P \rightarrow \begin{array}{c} OC \\ \\ OC-Co \\ \\ ON \end{array} \longleftarrow \begin{array}{c} M(CH_3)_3 \\ \\ P-M(CH_3)_3 \\ \\ M(CH_3)_3 \end{array} + CO$$

(M = Si, Ge, Sn)

When a solution of hexacarbonylchromium and tris-(trimethylsilyl)-, tris-(trimethylgermyl)- or tris-(trimethylstannyl)phosphine in THF at room temperature is irradiated for 5 hours with UV light, carbon monoxide is evolved and the tris-(organometallo)phosphine penta-carbonyl-chromium (0) complex is formed (152):

$$Cr(CO)_6 + [(CH_3)_3M]_3P \rightarrow (CO)_5Cr \longleftarrow P[M(CH_3)_3]_3 + CO$$

(M = Si, Ge, Sn)

Acidic hydrolysis of annealed mixtures of germanides, phosphides and arsenides of calcium gives a mixture of gases identified by mass spectroscopy. From $CaGe/Ca_3P_2$, P_3H_5 and H_3GePH_2 are formed and from $CaGe/Ca_3As_2$, As_3H_5 and H_3GeAsH_2 are characterized (109, 110):

Germylphosphine, H_3GePH_2, can be prepared in high yield (88%) from the reaction of lithium tetrakis(dihydrogenphosphinido)aluminate, $[LiAl(PH_2)_4]$, and germyl bromide (184):

$$4 H_3GeBr + LiAl(PH_2)_4 \xrightarrow{\text{ether}} 4 H_3GePH_2 + LiBr + AlBr_3$$

The reactions between $LiAl(PH_2)_4$ and halogeno-silanes and germanes in bis-(2-methoxyethyl)ether (diglyme) is a new general method for the preparation of mono and bis-phosphino-silanes and germanes (98):

$$LiAl(PH_2)_4 + 4 Me_3GeX \rightarrow$$

$$LiX + AlX_3 + 4 Me_3GePH_2 \quad \text{(Yield 69\%)}$$

$$LiAl(PH_2)_4 + 2 Me_2GeX_2 \rightarrow$$

$$LiX + AlX_3 + 2 Me_2Ge(PH_2)_2 \quad \text{(Yield 46\%)}$$

Monosilylphosphine undergoes an "exchange reaction" with mono-chlorogermane to give monogermylphosphine. When the appropriately deuteriated species are used as starting materials the following reactions take place at $-78°C$ (32):

$$SiD_3PD_2 + GeH_3Cl \rightarrow SiD_3Cl + GeH_3PD_2$$

$$SiH_3PH_2 + GeD_3Cl \rightarrow SiH_3Cl + GeD_3PH_2$$

The ^1H NMR spectra of the deuteriated monogermylphosphines confirm these results.

Monogermylphosphine reacts quantitatively, with diborane at $-20°C$ in a sealed tube forming an adduct according to the reaction:

$$H_3GePH_2 + \tfrac{1}{2} B_2H_6 \rightarrow H_3GePH_2 \cdot BH_3$$

The comparison of the ^1H NMR spectrum of $H_3GePH_2 \cdot BH_3$ with those of the deuteriated species $H_3GePD_2 \cdot BH_3$, $D_3GePH_2 \cdot BH_3$ shows some interesting features.

The ^1H NMR spectrum of the adduct formed by monogermylphosphine confirms that it too is a triple mixed hydride, by exchange between the germanium and boron protons, and the spectra of the deuteriated adducts confirm the proton exchange (32).

Drake and Jolly have described the formation of "mixed hydrides" in a silent electric discharge. An equimolar mixture of germane, GeH_4, and phosphine, PH_3, led to a mixture of hydrides detected by mass spectroscopy: H_3GePH_2, Ge_2PH_7, Ge_3PH_9, GeP_2H_6, Ge_2H_6, Ge_3H_8, Ge_4H_{10} and P_2H_4.

H_3GePH_2 decomposed slowly at room temperature to give hydrogen, phosphine and a solid yellow polymeric hydride. Ge_2PH_7 may have either of the following structures: $H_3GePHGeH_3$ or $H_3GeGeH_2PH_2$ (30). With the GeH_4—AsH_3 system the only mixed hydride formed was H_3GeAsH_2. Germylarsine is thermally more stable than germylphosphine and is only very slowly decomposed at room temperature, both arsine and germane being formed (30). The proton magnetic resonance spectra of silyl- and germyl-phosphine and -arsine have been recorded and the values of the coupling constants and the chemical shifts are given. The relative chemical shifts of the resonance peaks may be explained by differences in electronegativity; when an SiH_3 group in Si_2H_6, or a GeH_3 group in Ge_2H_6, is replaced by a more electronegative group such as PH_2 or AsH_2 the shielding of the hydrogen atoms in the SiH_3 or $—GeH_3$ group decreases and the resonance is shifted to a weaker field. This shift is less important for the germyl compounds, which is related to the fact that germanium is more electronegative than silicon (31).

Germanium (II) iodide reacts with various alkyl and aryl phosphine derivatives in hydrocarbon solvents at 130°C to form adducts of the general formula R_3P, GeI_2; these compounds being formally analogous to the phosphinedihalomethylenes, form yellow air-sensitive solids soluble in certain organic solvents (74).

8-2-2   Organogermanium–phosphorus compounds with Ge—P bond

Table 8-6 showing some physical properties of organogermanium–phosphorus compounds with Ge—P bond appears on the following pages.

Table 8-6
Organogermanium–Phosphorus Compounds with Ge—P Bond

COMPOUND	EMPIRICAL FORMULA	M.P. (°C)	B.P. (°C/mm)	n_D^{20}	d_4^{20}	REFERENCES
D_3GePH_2	H_2D_3PGe	—	—	—	—	32
H_3GePD_2	H_3D_2PGe	—	—	—	—	32
H_3GePH_2	H_5PGe	—	—	—	—	30, 31, 109, 110, 184
$H_3GePH_2 \cdot BH_3$	H_8BPGe	—	—	—	—	32
$(H_3Ge)_3P$	H_9PGe_3	−83.8	—	—	—	22, 25
$(CH_3)_2Ge(PH_2)_2$	$C_2H_{10}P_2Ge$	—	—	—	—	98
$(CH_3)_3GePH_2$	$C_3H_{11}PGe$	—	—	—	—	98
$[(CH_3)_3GeI]_2PCH_3$	$C_7H_{12}PGe_2$	—	53/1	—	—	146
$(CH_3)_3GePHC_6H_5$	$C_9H_{15}PGe$	60/2	—	—	—	146
$[(CH_3)_3Ge]_3P$	$C_9H_{27}PGe_3$	—	62–63/0.1	—	—	39, 140
$(C_2H_5)_3GeP(C_2H_5)_2$	$C_{10}H_{25}PGe$	—	120/15	1.4845	—	116
$[(CH_3)_3Ge]_3P \rightarrow Co—(CO)_2$ NO	$C_{11}H_{27}O_3PNCoGe_3$ 132	—	—	—	—	152
$(C_6H_5)_2PH \cdot GeI_2$	$C_{12}H_{11}PGeI_2$	—	—	—	—	74
$[(CH_3)_3GeI]_2PC_6H_5$	$C_{12}H_{14}PGeI_2$	—	—	—	—	146
$[(CH_3)_3Ge]_3P \rightarrow Ni(CO)_3$	$C_{12}H_{27}O_3PNiGe_3$	100	107/2	—	—	151
$(n\text{-}C_4H_9)_3P \cdot GeI_2$	$C_{12}H_{27}PGeI_2$	—	—	—	—	74
$(C_6H_5)_2PC_2H_5 \cdot GeI_2$	$C_{14}H_{15}PGeI_2$	—	—	—	—	74
$[(CH_3)_3Ge]_3P \rightarrow Cr(CO)_5$	$C_{14}H_{27}O_5PCrGe_3$	150	—	—	—	152
$(C_6H_5)_2PCH_3 \cdot GeI_2$	$C_{15}H_{13}PGeI_2$	—	—	—	—	74
$(C_6H_5)_2PCH(CH_3)_2 \cdot GeI_2$	$C_{15}H_{17}PGeI_2$	—	—	—	—	74

Table 8-6—continued

COMPOUND	EMPIRICAL FORMULA	M.P. (°C)	B.P. (°C/mm)	n_D^{20}	d_4^{20}	REFERENCES
$(CH_3)_3GeP(C_6H_5)_2$	$C_{15}H_{19}PGe$	—	185–87/12	1.6089	—	113
$(C_6H_5)_2P(n\text{-}C_4H_9)\cdot GeI_2$	$C_{16}H_{19}PGeI_2$	—	—	—	—	74
$(C_6H_5)_3P\cdot GeI_2$	$C_{18}H_{15}PGeI_2$	—	—	—	—	74
$(C_2H_5)_3GeP(C_6H_5)_2$	$C_{18}H_{25}PGe$	—	$146/10^{-3}$	—	—	18, 55, 121
$(C_6H_5)_2P\cdot CH_2CH_2P(C_6H_5)_2\cdot 2\ GeI_2$	$C_{26}H_{24}P_2Ge_2I_2$	—	—	—	—	74
$(C_6H_5)_3GeP(C_6H_5)_2$	$C_{30}H_{25}PGe$	—	154–56 108°	—	—	18, 148
$(C_6H_5)_2Ge[P(C_6H_5)_2]_2$	$C_{36}H_{30}P_2Ge$	—	182/185	—	—	18, 148
$[(C_6H_5)_3Ge]_2PC_6H_5$	$C_{42}H_{35}PGe_2$	—	110	—	—	149
$(C_6H_5)_3GePSn(C_6H_5)_3$ C_6H_5	$C_{42}H_{35}PSnGe$	—	115–19	—	—	150
$[(C_6H_5)_3Ge]_3P$	$C_{54}H_{45}PGe_3$	—	128	—	—	39, 148, 149
$(C_6H_5)_3GeP[Sn(C_6H_5)_3]_2$	$C_{54}H_{45}PSn_2$	—	160	—	—	150
$[(C_2H_5)_3GeP(C_6H_5)_2,\ AgI]_4$	$C_{72}H_{100}P_4Ag_4Ge_4I_4$	—	183	—	—	18, 55

8-2-3   Organogermanium–phosphorus compounds without Ge—P bond

Germyl phosphonites, R_3Ge—O—PR'_2, can be prepared by the action of trialkylchlorogermanes on the potassium salts of R_2POH (67). Tris-(trialkylgermanyl)phosphates, $(R_3GeO)_3PO$, are obtained by reaction of hexaalkyldigermoxanes, R_3Ge—O—GeR_3, with phosphorus pentoxide (136). A yield of 48% of diethyl(trimethylgermyl)methylphosphonate, $(CH_3)_3GeCH_2P(O)(OC_2H_5)$, is obtained after nine hours of reflux when (chloromethyl)trimethylgermane, $(CH_3)_3GeCH_2Cl$, and $P(OC_2H_5)_3$ are allowed to react. (Trimethylgermyl)methylphosphonic acid,

$$(CH_3)_3GeCH_2P(O)(OH)_2,$$

is obtained by the action of hydrochloric acid on

$$(CH_3)_3GeCH_2P(O)(OC_2H_5)_2 \quad (94).$$

$[(CH_3)_3Ge]_2O$ and $POCl_3$ gave Me_3GeCl and $Me_3GeOPOCl_2$. $[(CH_3)_3Ge]_2O$ and $P_2O_3Cl_4$ gave a high yield of $(CH_3)_3GeOPOCl_2$. $(CH_3)_3SiOGe(CH_3)_3$ and $POCl_3$ yielded 80% $(CH_3)_3GeCl$, $(CH_3)_3SiO-POCl_2$ and 20% $(CH_3)_3SiCl$, $(CH_3)_3GeOPOCl_2$, whereas $(CH_3)_3SiOGe-(CH_3)_3$ and $P_2O_3Cl_4$ gave equimolecular amounts of $(CH_3)_3GeOPOCl_2$ and $(CH_3)_3SiOPOCl_2$ (138). $(Me_3Ge)_2O$, $(Et_3Ge)_2O$ and methylphosphonic difluoride, CH_3POF_2, after heating at 90–100°C gave $(CH_3)_3$-$GeOP(O)FCH_3$ or $(C_2H_5)_3GeOP(O)FCH_3$, with $(CH_3)_3GeF$ or $(C_2H_5)_3GeF$ (53).

Triphenylphosphinemethylene reacts with triphenylbromogermane in ether solution to form a phosphonium salt in which the germanium atom is linked to phosphorus by a methylene group (155):

$$(C_2H_5)_3\overset{\oplus}{P}-\overset{\ominus}{C}H_2 + -\overset{|}{\underset{|}{Ge}}-Br \rightarrow (C_6H_5)_3\overset{\oplus}{P}-CH_2\overset{|}{Ge}-Br$$

$$\downarrow$$

$$[(C_6H_5)_3\overset{\oplus}{P}-CH_2-\overset{|}{\underset{|}{Ge}}]Br^{\ominus}$$

Organogermanium compounds containing one to four halogens were treated with an aldehyde and a tervalent phosphorus ester to yield tetravalent germanium compounds containing one to four phosphinyl-hydrocarbyloxy groups —OCHP(O)$\diagdown^{\diagup}$ (104). Alternatively the compounds were prepared by treating the germanium compounds with α-hydroxyhydrocarbylphosphorus ester in the presence of a basic material. Thus triethylbromogermane with triethylphosphine and PrCHO give [1-

diethoxyphosphinyl)propoxy]triethylgermane:

$$Et_3GeOCH(Et)P(O)(OEt)_2$$

$GeCl_4$ treated with a mixture of triethylphosphite and PhCHO and treated afterwards with propylene oxide yield

$$CH_3CHClCH_2OGe[OCH(C_6H_5)P(O)(OC_2H_5)_2]_3.$$

A mixture of diethyl phenylphosphinite, furfuraldehyde and $PhGeCl_3$ treated with triethylphosphine and PrCHO, and afterwards with propylene oxide, yields [α(ethoxyphenylphosphinyl)furfuryloxy][1-(diethoxyphosphinyl)propoxy] (2-chloropropoxy)phenylgermane:

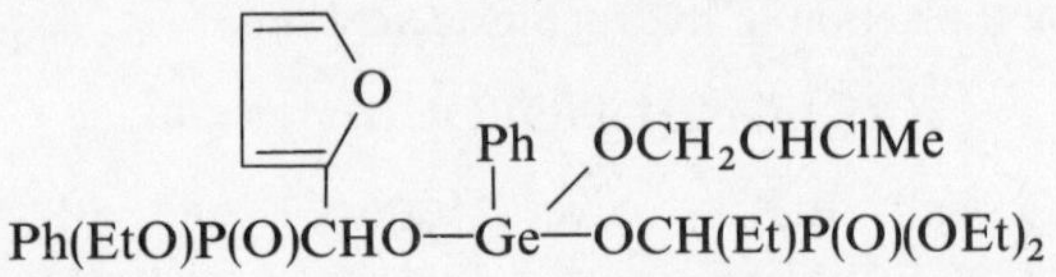

Eight other compounds were similarly prepared by the same procedure (104).

Tetraalkylgermanes and alkyl- and cycloalkyltrihalogermanes react with PCl_3 and oxygen with formation of the corresponding germanium-substituted alkyl- and cycloalkyl-phosphonic dichlorides (33):

$$Cl_3GeC_2H_5 + 2\ PCl_3 \xrightarrow{\ O_2\ } Cl_3GeC_2H_4POCl_2 + HCl + POCl_3$$

Like benzene, phenyltrichlorogermane does not undergo this reaction.

8-2-4   Organogermanium–phosphorus compounds without Ge–P bond

Table 8-7 showing organogermanium-phosphorus compounds without Ge—P bond appears on the following pages.

Table 8-7

Organogermanium–Phosphorus Compounds without Ge—P Bond

COMPOUND	EMPIRICAL FORMULA	M.P. (°C)	B.P. (°C/mm)	n_D^{20}	d_4^{20}	REFERENCES
$Cl_3GeC_2H_4POCl_2$	$C_2H_4Cl_5OPGe$	—	96/1.5	1.5305	1.8067	33
$Cl_3GeC_3H_6POCl_2$	$C_3H_6Cl_5OPGe$	—	123–25/1	1.5290	1.7700	33
$(CH_3)_3GeOP(O)Cl_2$	$C_3H_9O_2PCl_2Ge$	—	61	—	—	138
$(CH_3)_3GeOP(O)FCH_3$	$C_4H_{12}O_2PFGe$	—	49–49.5/5	1.4123	1.3656	53
$(CH_3)_3GeCH_2P(O)(OH)_2$	$C_4H_{13}O_3PGe$	—	93–94	—	—	94
$Cl_3GeC_6H_{10}POCl_2$	$C_6H_{10}Cl_5OPGe$	—	196–98/5	1.5488	1.6891	33
$(C_2H_5)_3GeOP(O)FCH_3$	$C_7H_{18}O_2PFGe$	—	111–12/9	1.4378	1.2610	53
$(C_2H_5)_3GeC_2H_4POCl_2$	$C_8H_{19}Cl_2OPGe$	—	118–19/1.5	1.5012	1.3564	33
$(CH_3)_3GeCH_2P(O)(OC_2H_5)_2$	$C_8H_{21}O_3PGe$	—	86–87/4	1.4469	1.1669	94
$Cl_3GeC_5H_{18}POCl_2$	$C_9H_{18}Cl_5OPGe$	—	170/1	1.5133	1.4536	33
$(C_2H_5)_3Ge(CH_2)_3POCl_2$	$C_9H_{21}Cl_2OPGe$	—	77/20	1.4610	1.1952	34
$[(CH_3)_3GeO]_3PO$	$C_9H_{27}PO_4Ge_3$	—	50/1			136
$(C_2H_5)_3GeSC(S)P(C_2H_5)_2$	$C_{11}H_{25}S_2PGe$	—	120/15	1.4845	—	116
$(C_2H_5)_3GeO—C—P(C_2H_5)_2$ $\|\|$ CH_2	$C_{12}H_{27}OPGe$	—	89/0.35		—	118
$(C_2H_5)_3GeC_2H_4PO(OC_2H_5)_2$	$C_{12}H_{29}O_3PGe$	—	106–7/3.5	1.4550	1.1074	33
$(C_2H_5)_3GeCH(CN)CH_2P(C_2H_5)_2$	$C_{13}H_{28}NPGe$	—	100/0.1	1.5006	1.0730	116
$(C_2H_5)_3Ge(CH_2)_3PO(OC_2H_5)_2$	$C_{13}H_{31}O_3PGe$	—	150/2–3	1.4560	1.0976	34
$(C_2H_5)_3GeOCH(C_2H_5)P(O)(OC_2H_5)_2$	$C_{13}H_{31}PO_4Ge$	—	98–101/0.1	1.4475	—	104
$(C_2H_5)_3GeO—CH=CH—CH—P(C_2H_5)_2$ CH_3	$C_{14}H_{31}OPGe$	—	93–95/0.15	—	—	117

Table 8-7—continued

COMPOUND	EMPIRICAL FORMULA	M.P. (°C)	B.P. (°C/mm)	n_D^{20}	d_4^{20}	REFERENCES
$(C_2H_5)_3GeO—CH—P(C_2H_5)_2$, with $|(CH_2)_2CH_3$	$C_{14}H_{33}OPGe$	—	85/1.5	—	—	117
$(C_2H_5)_3GeO—CH—P(C_2H_5)_2$, with C_6H_5	$C_{17}H_{31}OPGe$	—	117/0.9	—	—	117
$(C_2H_5)_3Ge(CH_2)_3PO(OC_4H_9)_2$	$C_{17}H_{39}O_3PGe$	—	157/2	1.4577	1.0510	34
$(C_2H_5)_3GeOP(O)(C_6H_5)_2$	$C_{18}H_{25}O_2PGe$	—	$160/10^{-3}$	—	—	18, 55
$(CH_3CHClCH_2O)_3GeOCH(o\text{-}ClC_6H_4)P(O)(OCH_3)_2$	$C_{18}H_{29}PO_7Cl_4Ge$	—	—	—	—	94
$(C_2H_5)_3GeOCH=CH—CH—P(C_2H_5)_2$, with C_6H_5	$C_{19}H_{33}OPGe$	—	113/0.15	—	—	117
$(CH_3CHClCH_2O)_3GeOCH(C_6H_5)P(O)(OC_2H_5)_2$	$C_{20}H_{34}PO_4Cl_3Ge$	—	—	—	—	104
$(CH_3CHClCH_2O)_3GeOCH(p\text{-}CH_3OC_6H_4)P(O)(OC_2H_5)_2$	$C_{20}H_{35}PO_8Cl_3Ge$	—	—	—	—	104
$(CH_3CHClCH_2O)_3GeOCH(C_6H_{13})P(O)(OCH_2CH_2Cl)_2$	$C_{20}H_{40}PO_7Cl_5Ge$	—	—	—	—	104
$(CH_3CHClCH_2O)_3GeOCH(CH_2CH_2CO_2C_2H_5)P(O)(OC_2H_5)_2$	$C_{20}H_{40}O_9PGe$	—	—	—	—	104
$(C_4H_9)_3Ge—O—P(C_4H_9)_2$	$C_{20}H_{45}POGe$	—	136–38/1	—	—	67
$(CH_3CHClCH_2O)_3GeOCH(p\text{-}CH_3C_6H_4)P(O)(OC_2H_5)_2$	$C_{21}H_{36}PO_7ClGe$	—	—	—	—	104
$(C_2H_5)_3GeO—C—P(C_2H_5)_2$, with $\|C(C_6H_5)_2$	$C_{24}H_{35}OPGe$	—	154/0.04	1.5793	—	118

Compound		m.p.	Ref.
(CH₃CHClCH₂O)₃GeOCH(CH=CH)P(O)(OC₆H₁₁)₂	C₂₄H₄₃PO₇Cl₃Ge	—	104
C₆H₅Ge[OCH(CH₃)P(O)(OC₂H₅)₂]₃	C₂₄H₄₇PO₁₂Ge	—	104
Ge[OCH(C₂H₅)P(O)(OC₂H₅)₂]₄	C₂₈H₆₄P₄O₁₆Ge	—	104

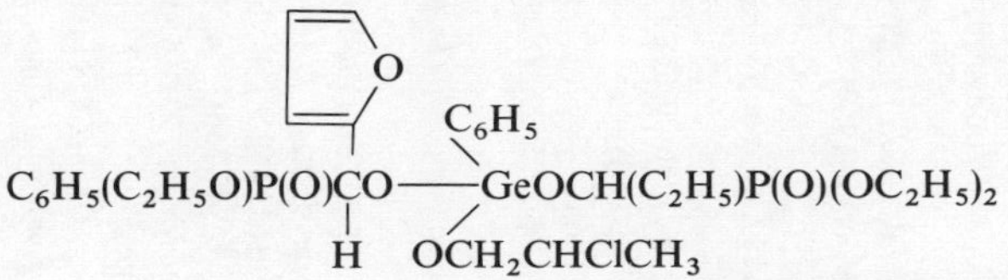

$$C_6H_5(C_2H_5O)P(O)CO\!-\!GeOCH(C_2H_5)P(O)(OC_2H_5)_2$$

with furan ring bearing C₆H₅, and H / OCH₂CHClCH₃

Compound		m.p.	Ref.
	C₂₉H₃₃P₂O₉ClGe	—	104
(C₆H₅)₃GeOP(O)(C₆H₅)₂	C₃₀H₂₅O₂PGe	—	18
(CH₃CHClCH₂O)Ge[OCH(C₆H₅)P(O)(OC₂H₅)₂]₃	C₃₆H₅₄P₃O₁₃Ge	—	104
[(C₆H₅)₃GeCH₂P(C₆H₅)₃]Br	C₃₇H₃₂PGeBr	121-22	155
[(C₆H₅)₃GeCH₂P(C₆H₅)₃]B(C₆H₅)₄	C₆₁H₅₂PBGe	84-85	155

REFERENCES

1. Abel, E. W., *J. Chem. Soc.*, 3746 (1958).
2. Abel, E. W., D. A. Armitage, and D. B. Brady, *J. Organometal. Chem.*, **5**, 130 (1966).
3. Abel, E. W., and J. P. Crow, *J. Chem. Soc. A*, 1361 (1968).
4. Anderson, H. H., *J. Am. Chem. Soc.*, **71**, 1799 (1949).
5. Anderson, H. H., *J. Am. Chem. Soc.*, **72**, 193 (1950).
6. Anderson, H. H., *J. Am. Chem. Soc.*, **73**, 5439 (1951).
7. Anderson, H. H., *J. Am. Chem. Soc.*, **73**, 5440 (1951).
8. Anderson, H. H., *J. Am. Chem. Soc.*, **73**, 5800 (1951).
9. Anderson, H. H., *J. Am. Chem. Soc.*, **74**, 1421 (1952).
10. Anderson, H. H., *J. Am. Chem. Soc.*, **75**, 814 (1953).
11. Anderson, H. H., *J. Org. Chem.*, **20**, 536 (1955).
12. Anderson, H. H., *J. Am. Chem. Soc.*, **78**, 1692 (1956).
13. Anderson, H. H., *J. Am. Chem. Soc.*, **83**, 547 (1961).
14. Baum, G., W. L. Lehn, and C. Tamborski, *J. Org. Chem.*, **29**, 1264 (1964).
15. Beck, W., and E. Shuierer, *Chem. Ber.*, **97**, 3517 (1964).
16. Bither, T. A., W. H. Knoth, R. V. Lindsey, and W. H. Sharkey, *J. Am. Chem. Soc.*, **80**, 4151 (1958).
17. Breederveld, H., *Rec. Trav. Chim.*, **79**, 1126 (1960); **81**, 276 (1962).
18. Brooks, E. H., F. Glockling, and K. A. Hooton, *J. Chem. Soc.*, 4283 (1965).
19. Burch, G. M., and J. R. van Wazer, *J. Chem. Soc. A*, 586 (1966).
20. Bürger, H., and W. Sawodny, *Spectrochimica Acta*, **23A**, 2841 (1967).
21. Chandra, G., T. A. George, and M. F. Lappert, *Chem. Commun.*, 116 (1967).
22. Cradock, S., G. Davidson, E. A. V. Ebsworth, and L. A. Woodward, *Chem. Commun.*, 515 (1965).
23. Cradock, S., and E. A. V. Ebsworth, *J. Chem. Soc. A*, 1420 (1968).
24. Cradock, S., and E. A. V. Ebsworth, *J. Chem. Soc. A*, 1423 (1968).
25. Cradock, S., E. A. V. Ebsworth, G. Davidson, and L. A. Woodward, *J. Chem. Soc. A*, 1229 (1967).
26. Creemers, H. M. J. C., and J. G. Noltes, *J. Organometal. Chem.*, **7**, 237 (1967).
27. Cuvigny, Th., and H. Normant, *Compt. Rend. C*, **268**, 834 (1969).
28. Davidson, W. E., Thesis, Braunschweig, 1961.
29. Davidson, G., E. A. V. Ebsworth, G. M. Sheldrick, and L. A. Woodward, *Spectrochim. Acta*, **22**, 67 (1966).
30. Drake, J. E., and W. L. Jolly, *Chem. Ind.* (*London*), **21b**, 1470 (1962).
31. Drake, J. E., and W. L. Jolly, *J. Chem. Phys.*, **38**, 1033 (1963).
32. Drake, J. E., and C. Riddle, *J. Chem. Soc. A*, 1675 (1968).
33. Dzhurinskaya, N. G., S. A. Mikhailyants, and V. P. Evdakov, *Zh. Obshch. Khim.*, **37**, 2278 (1967).
34. Dzhurinskaya, N. G., S. A. Mikhailyants, and V. P. Evdakov, *Zh. Obshch. Khim.*, **38**, 1267 (1968).
35. Eaborn, C., *Organosilicon Compounds*. Butterworths, London (1960), 345.
36. Eisenbuth, W., and J. R. van Wazer, *Inorg. Nucl. Chem. Letters*, **3**, 359 (1967).
37. Eisenhuth, W., and J. R. van Wazer, *Inorg. Chem.*, **7**, 1642 (1968).
38. Elguero, J., M. Riviere-Baudet, and J. Satgé, *Compt. Rend.*, **266C**, 44 (1968).
39. Engelhardt, G., D. Reich, and H. Schumann, *Z. Naturforsch.*, **22B**, 352 (1967).
40. Esposito, J. N., *Dissertation Abstr. B*, **27**, 2284 (1967).
41. Esposito, J. N., L. E. Sutton, and M. E. Kenney, *Inorg. Chem.*, **6**, 1116 (1967).
42. Findeiss, W., W. Davidsohn, and M. C. Henry, *J. Organometal. Chem.*, **9**, 435 (1967).

43. Flood, E. A., *J. Am. Chem. Soc.*, **54**, 1663 (1932).
44. Flood, E. A., *J. Am. Chem. Soc.*, **55**, 4935 (1933).
45. Forbes, G. S., and H. H. Anderson, *J. Am. Chem. Soc.*, **65**, 2271 (1943).
46. Forbes, G. S., and H. H. Anderson, *J. Am. Chem. Soc.*, **67**, 1911 (1945).
47. Forster, D., and D. M. L. Goodgame, *J. Chem. Soc.*, 262 (1965).
48. George, T. A., K. Jones, and M. F. Lappert, *J. Chem. Soc.*, 2157 (1965).
49. George, T. A., and M. F. Lappert, *J. Organometal. Chem.*, **14**, 327 (1968).
50. George, M. V., P. B. Talukdar, C. Q. Gerow, and H. Gilman, *J. Am. Chem. Soc.*, **82**, 4562 (1960).
51. George, M. V., P. B. Talukdar, and H. Gilman, *J. Organometal. Chem.*, **5**, 397 (1966).
52. Gielen, M., and N. Sprecher, *Organometal. Chem. Rev.*, **1**, 455 (1966).
53. Gladshtein, B. M., I. P. Kulgulin, and L. Z. Soborovskii, *Zh. Obshch. Khim.*, **36**, 488 (1966).
54. Glarum, S. N., and C. A. Kraus, *J. Am. Chem. Soc.*, **72**, 5398 (1950).
55. Glockling, F., and K. A. Hooton, *Proc. Chem. Soc.*, 146, May (1963).
56. Goldfarb, T. D., *J. Chem. Phys.*, **37**, 642 (1962).
57. Green, B. S., D. B. Sowerby, and K. J. Wihksne, *Chem. Ind. (London)*, 1306 (1960).
58. Griffiths, J. E., *J. Chem. Phys.*, **48**, 278 (1968).
59. Griffiths, J. E., and A. L. Beach, *Chem. Commun.*, 437 (1965).
60. Haiduc, I., and H. Gilman, *J. Organometal. Chem.*, **18**, P 5 (1969).
61. Highsmith, R. E., and H. H. Sisler, *Inorg. Chem.*, **8**, 996 (1969).
62. Highsmith, R. E., and H. H. Sisler, *Inorg. Chem.*, **8**, 1029 (1969).
63. Ishii, Y., *Asahi Garasu Kogyo Gijutsu Shorei-kai, Kenkyu Hokoku*, **13**, 479 (1967); *CA*, **69**, 9075 (1969).
64. Itoh, K., I. Matsuda, T. Katsuura, and Y. Ishii, *J. Organometal. Chem.*, **19**, 347 (1969).
65. Itoh, K., S. Sakai, and Y. Ishii, *Tetrahedron Letters*, 4941 (1966).
66. Itoh, K., S. Sakai, and Y. Ishii, *Chem. Commun.*, 36 (1967).
67. Issleib, K., and B. Walther, *Angew. Chem.*, **79**, 59 (1967).
68. Jappy, J., and P. N. Preston, *J. Organometal. Chem.*, **19**, 196 (1969).
69. Johnson, O. H., *Chem. Rev.*, **48**, 259 (1951).
70. Johnson, W. C., and A. E. Sidwell, *J. Am. Chem. Soc.*, **55**, 1884 (1933).
71. Jones, K., and M. F. Lappert, *Proc. Chem. Soc.*, 358 (1962).
72. Joyner, R. D., and M. E. Kenney, *J. Am. Chem. Soc.*, **82**, 5790 (1960).
73. Joyner, R. D., R. G. Linck, J. N. Esposito, and M. E. Kenney, *J. Inorg. Nucl. Chem.*, **24**, 299 (1962).
74. King, R. B., *Inorg. Chem.*, **2**, 199 (1963).
75. Köster-Pflugmacher, A., and A. Hirsch, *J. Organometal. Chem.*, **12**, 349 (1968).
76. Köster-Pflugmacher, A., and E. Termin, *Naturwiss.*, **51**, 554 (1964).
77. Kostyuk, A. S., L. N. Kalinina, Yu. I. Baukov, and I. F. Lutsenko, *Zh. Obshch. Khim.*, **38**, 413 (1968).
78. Köhler, V. H., and W. Beck, *Z. Anorg. Allg. Chem.*, **359**, 241 (1968).
79. Kraus, C. A., and C. L. Brown, *J. Am. Chem. Soc.*, **52**, 3690 (1930).
80. Kraus, C. A., and C. L. Brown, *J. Am. Chem. Soc.*, **52**, 4031 (1930).
81. Kraus, C. A., and E. A. Flood, *J. Am. Chem. Soc.*, **54**, 1635 (1932).
82. Kraus, C. A., and L. S. Foster, *J. Am. Chem. Soc.*, **49**, 457 (1927).
83. Kraus, C. A., and H. S. Nutting, *J. Am. Chem. Soc.*, **54**, 1622 (1932).
84. Kraus, C. A., and C. B. Wooster, *J. Am. Chem. Soc.*, **52**, 372 (1930).
85. Kruger, C., and U. Wannagat, *Z. Anorg. Allgem. Chem.*, **326**, 296 (1964).
86. Laubengayer, A. W., and L. Reggel, *J. Am. Chem. Soc.*, **65**, 1783 (1943).
87. Lui-Heung Shum, Thesis Harvard University, 1966. *Dissertation Abst.*, **B27**, 1800 (1966).

88. Luitjen, J. G. A., F. Rijkens, and G. J. M. van der Kerk, *Advances in Organometallic Chemistry*, Academic Press, New York, London, **3**, (1965), 397–446.
89. Lutsenko, I. F., V. L. Foss, and N. M. Semenenko, *Zh. Obshch. Khim.*, **39**, 1174 (1969).
90. Massol, M., and J. Satgé, *Bull. Soc. Chim. France*, 2737 (1966).
91. Mehrotra, R. C., and G. Chandra, *Indian J. Chem.*, **3**, 497 (1965).
92. Menzer, W., *Angew. Chem.*, **70**, 656 (1958).
93. Miller, F. A., and G. L. Carlson, *Spectrochim. Acta*, **17**, 977 (1961).
94. Mironov, V. F., and A. L. Kravchenko, *Izv. Akad. Nauk SSSR Ser. Khim.*, 1563 (1963).
95. Mironov, V. F., and A. L. Kravchenko, *Izv. Akad. Nauk SSSR Ser. Khim.*, 1026 (1965).
96. Mironov, V. F., E. S. Sobolev, and L. M. Antipin, *Zh. Obshch. Khim.*, **37**, 1707 (1967).
97. Mironov, V. F., E. S. Sobolev, and L. M. Antipin, *Zh. Obshch. Khim.*, **37**, 2573 (1967).
98. Norman, A. D., *Chem. Commun.*, 812 (1968).
99. Onyszchuk, M., *Angew. Chem.*, **75**, 577 (1963).
100. Ramaprasad, K. R., R. Varma, and R. Nelson, *J. Am. Chem. Soc.*, **90**, 6247 (1968).
101. Randall, E. W., C. H. Yoder, and S. S. Zuckerman, *Inorg. Chem.*, **6**, 744 (1967).
102. Randall, E. W., and J. J. Zuckerman, *J. Am. Chem. Soc.*, **90**, 3167 (1968).
103. Reichle, W. T., *Inorg. Chem.*, **3**, 402 (1964).
104. Richardson, G. A., and G. H. Birum, U.S. Pat. 3,190,892; *CA*, **63**, 11613 (1965).
105. Rijkens, F., M. J. Janssen, and G. J. M. van der Kerk, *Rec. Trav. Chim.*, **84**, 1597 (1965).
106. Rijkens, F., and G. J. M. van der Kerk, *Investigations in the Field of Organogermanium Chemistry*, TNO, Utrecht, 1964.
107. Rivière-Baudet, M., and J. Satgé, *Bull. Soc. Chim. France*, 1356 (1969).
108. Rochow, E. G., and A. L. Allred, *J. Am. Chem. Soc.*, **77**, 4489 (1955).
109. Royen, P., C. Rocktäschel, and W. Mosch, *Angew. Chem.*, **76**, 860 (1964).
110. Royen, P., and C. Rocktäschel, *Z. Anorg. Allgem. Chem.*, **346**, 290 (1966).
111. Ruidisch, I., and M. Schmidt, *Angew. Chem.*, **76**, 229 (1964).
112. Ruidisch, I., and M. Schmidt, *J. Organometal. Chem.*, **1**, 493 (1964).
113. Satgé, J., and M. Baudet, *Compt. Rend.*, **263C**, 435 (1966).
114. Satgé, J., and M. Baudet, Unpublished work. Baudet, M., Thesis, Toulouse, France, 1966.
115. Satgé, J., M. Baudet, and M. Lesbre, *Bull. Soc. Chim. France*, 2133 (1966).
116. Satgé, J., and C. Couret, *Compt. Rend.*, **264**, 2169 (1967).
117. Satgé, J., and C. Couret, *Compt. Rend.*, **267**, 173 (1968).
118. Satgé, J., and C. Couret, *Bull. Soc. Chim. France*, 333 (1969).
119. Satgé, J., and C. Couret, *Fourth Int. Conf. on Organometallic Chemistry, Bristol*, 1969. *Progress in Organometallic Chemistry*, Ed. Bruce, M. I., and F. G. A. Stone. M6.
120. Satgé, J., and M. Lesbre, *Bull. Soc. Chim. France*, 2578 (1965).
121. Satgé, J., M. Lesbre, and M. Baudet, *Compt. Rend.*, **259**, 4733 (1964).
122. Satgé, J., and M. Rivière-Baudet, *Bull. Soc. Chim. France*, 4093 (1968).
123. Scherer, O. J., *Organometal. Chem. Rev. Sect. A*, **3**, 281 (1968).
124. Scherer, O. J., and D. Biller, *Angew. Chem.*, **79**, 410 (1967).
125. Scherer, O. J., and D. Biller, *Z. Naturforsch.*, **B22**, 1079 (1967).
126. Scherer, O. J., and P. Hornig, *Chem. Ber.*, **101**, 2533 (1968).
127. Scherer, O. J., and M. Schmidt, *Angew. Chem.*, **75**, 642 (1963).
128. Scherer, O. J., and M. Schmidt, *J. Organometal. Chem.*, **1**, 490 (1964).
129. Scherer, O. J., and G. Schieder, *Angew. Chem.*, **80**, 83 (1968).
130. Scherer, O. J., and G. Scheider, *Chem. Ber.*, **101**, 4184 (1968).
131. Scherer, O. J., and G. Scheider, *J. Organometal. Chem.*, **19**, 315 (1969).
132. Schlemper, E. O., and D. Britton, *Inorg. Chem.*, **5**, 511 (1966).

133. Schmidbaur, H., *J. Am. Chem. Soc.*, **85**, 2336 (1963).
134. Schmidbaur, H., and W. Wolfsberger, *Chem. Ber.*, **101**, 1664 (1968).
135. Schmidbaur, H., and G. Jonas, *Chem. Ber.*, **100**, 1120 (1967).
136. Schmidt, M., and I. Ruidisch, Ger. Pat. 1,194,858; *CA*, **63**, 13315 (1965).
137. Schmidt, M., and I. Ruidisch, *Angew. Chem.*, **76**, 686 (1964); internat. edn **3**, 637 (1964).
138. Schmidt, M., H. Schmidbaur, and I. Ruidisch, *Angew. Chem.*, **73**, 408 (1961).
139. Schmitz-Dumont, V. O., and W. Jansen, *Z. Anorg. Allg. Chem.*, **363**, 140 (1968).
140. Schumann, H., and H. Blass, *Z. Naturforsch.*, **B21**, 1105 (1966).
141. Schumann, H., P. Jutzi, and M. Schmidt, *Angew. Chem.*, **77**, 812 (1965).
142. Schumann, H., P. Jutzi, and M. Schmidt, *Angew. Chem.*, **77**, 912 (1965).
143. Schumann-Ruidisch, I., and B. Jutzi-Mebert, *J. Organometal. Chem.*, **11**, 77 (1968).
144. Schumann-Ruidisch, I., W. Kalk, and R. Z. Brüning, *Z. Naturforsch.*, **23b**, 307 (1968).
145. Schumann-Ruidisch, I., and H. Blass, *Z. Naturforsch*, **B22**, 1081 (1967).
146. Schumann-Ruidisch, I., and J. Kuhlmey, *J. Organometal. Chem.*, **16P**, 26 (1969).
147. Schumann, H., and M. Schmidt, *Inorg. Nucl. Chem. Letters*, **1**, 1, (1965).
148. Schumann, H., and M. Schmidt, *Angew. Chem.*, **77**, 1049 (1965).
149. Schumann, H., P. Schwabe, and M. Schmidt, *Inorg. Nucl. Chem. Letters*, **2**, 309 (1966).
150. Schumann, H., P. Schwabe, and M. Schmidt, *Inorg. Nucl. Chem. Letters*, **2**, 313 (1966).
151. Schumann, H., and O. Stelzer, *Angew. Chem.*, **79**, 692 (1967).
152. Schumann, H., and O. Stelzer, *Angew. Chem.*, **80**, 318 (1968).
153. Schwarz, R., and W. Reinhard, *Chem. Ber.*, **65**, 1743 (1932).
154. Schwarz, R., and P. W. Schenk, *Chem. Ber.*, **63**, 296 (1930).
155. Seyferth, D., and S. O. Grim, *J. Am. Chem. Soc.*, **83**, 1610 (1961).
156. Seyferth, D., and N. Kahlen, *J. Org. Chem.*, **25**, 809 (1960).
157. Seyferth, D., and N. Kahlen, *J. Am. Chem. Soc.*, **82**, 1080 (1960).
158. Smith, F. B., Thesis, Brown University, May, 1934.
159. Smith, F. B., and C. A. Kraus, *J. Am. Chem. Soc.*, **74**, 1418 (1952).
160. Srivastava T. N., S. E. Griffiths, and M. Onyszchuk, *Can. J. Chem.*, **40**, 739 (1962).
161. Srivastava, T. N., and S. K. Tandon, *J. Inorg. Nucl. Chem.*, **30**, 1399 (1968).
162. Srivastava, T. N., and S. K. Tandon, *J. Prakt. Chem.*, **311**, 190 (1969).
163. Starshak, A. J., R. D. Joyner, and M. E. Kenney, *Inorg. Chem.*, **5**, 330 (1966).
164. Storr, R., A. N. Wright, and C. A. Winkler, *Can. J. Chem.*, **40**, 1296 (1962).
165. Sujishi, S., and J. N. Keith, Abstracts, 34th meeting, Am. Chem. Soc., Chicago, 44N (1958).
166. Teal, G. K., Thesis, Brown University, May, 1931.
167. Teal, G. K., and C. A. Kraus, *J. Am. Chem. Soc.*, **72**, 4706 (1950).
168. Teal, G. K., and C. A. Kraus, *J. Am. Chem. Soc.*, **72**, 4706 (1950).
169. Thayer, J. S., *Organometal. Chem. Rev.*, **1**, 157 (1966).
170. Thayer, J. S., *Inorg. Chem.*, **7**, 2599 (1968).
171. Thayer, J. S., and D. P. Strommen, *J. Organometal. Chem.*, **5**, 383 (1966).
172. Thayer, J. S., and R. West, *Inorg. Chem.*, **3**, 406 (1964).
173. Thayer, J. S., and R. West, *Inorg. Chem.*, **3**, 889 (1964).
174. Thayer, J. S., and R. West, *Inorg. Chem.*, **4**, 114 (1965).
175. Thomas, J. S., and Th. Pugh, *J. Chem. Soc.*, 60 (1931).
176. Thomas, J. S., and W. W. Southwood, *J. Chem. Soc.*, 2083 (1931).
177. Tittle, B., Thesis, Cambridge, 1961.
178. Vyazankin, N. S., V. T. Bychkov, O. V. Linzina, and G. A. Razuvaev, *Dokl. Akad. Nauk SSSR*, **186**, 101 (1969).
179. Wannagat, U., E. Bogusch, and R. Braun, *J. Organometal. Chem.*, **19**, 367 (1969).

180. Wannagat, U., and C. Krüger, *Z. Anorg. Allgem. Chem.*, **326**, 288 (1964).
181. Washburn, R. M., and R. A. Baldwin, U.S. Pat. 3,112,331, Nov. 26, 1963; *CA*, **60**, 5554h (1964).
182. Wieber, M., and M. Schmidt, *Z.Naturforsch.*, **18b**, 849 (1963); *Angew. Chem.*, **75**, 1116 (1963).
183. Wieber, M., and G. Schwarzmann, *Monatsch. Chem.*, **99**, 255 (1968).
184. Wingleth, D. C., and A. D. Norman, *Chem. Commun.*, 1218 (1967).
185. Wolfsberger, W., and H. Schmidbaur, *J. Organometal. Chem.*, **17**, 41 (1969).
186. Yoder, C. H., Thesis, Cornell University, 1966. *Dissertation Abstr.*, **B27**, 1801 (1966).
187. Yoder, C. H., and J. J. Zuckermann, *Inorg. Chem.*, **3**, 1329 (1964).
188. Yoder, C. H., and J. J. Zuckermann, *J. Am. Chem. Soc.*, **88**, 2170 (1966).
189. Yoder, C. H., and J. J. Zuckermann, *J. Am. Chem. Soc.*, **88**, 4831 (1966).
190. Yoder, C. H., and J. J. Zuckermann, *Inorg. Chem.*, **5**, 2055 (1966).
191. Yoder, C. H., and J. J. Zuckermann, *Chem. Commun.*, 694 (1966).
192. Zhinkin, D. Ya., M. M. Morgunova, K. K. Popkov, and K. A. Andrianov, *Dokl. Akad. Nauk SSSR*, **158**, 641 (1964).

9 Polygermanes

9-1 DIGERMANES

A digermane was first prepared in 1925 by Morgan and Drew (56). Digermanes fall into two categories:

(a) Genuine digermanes characterized by the Ge—Ge bond (α didigermanes).

(b) Bridged digermanes ($\beta, \gamma, \delta, \ldots$ digermanes) in which the two germanium atoms are interspaced by one or several carbon atoms forming either a saturated or unsaturated aliphatic bridge, or a cyclic arrangement.

9-1-1 Digermanes with Ge—Ge bond (α digermanes)

9-1-1-1 Physical Properties and Characteristics. The Ge—Ge bond in α-digermanes is remarkably unreactive and of high thermal stability. Hexaphenyldigermane melts without decomposition at 336°C (arylated digermanes are solids) (20). Hexaethyldigermane may be distilled in air at 265°C (37). Selwood has measured the magnetic susceptibility of hexaphenyldigermane in benzene solution and in powder form (68). α Digermanes have absorption in the ultraviolet attributed to the Ge—Ge bond acting as a chromophore, probably through the use of vacant d-orbitals of the germanium atom. Numerous spectroscopic investigations have been carried out, among which are the recent studies of Glockling and coworkers resulting in the location of the ultraviolet Ge—Ge absorption band within the range 210–240 nm (11, 20), of Hague and Prince (24), Ryan and Tamborski (67) concerning hexaphenyl digermane (λ_{max} 239 nm), of Cawley and Danyluc (NMR spectra of hexavinyldigermane) (10), of Brown and Fowles (Raman spectra of hexamethyldigermane) (7), etc.

In the series of elements Si—Ge—Sn, the average energies of the M—M bond in alkyl compounds decrease with increasing atomic weight of the element:

Si—Si   70 ± 10 kcal/mol
Ge—Ge   62 ± 5 calculated from the heat of combustion of hexaethyl-
         digermane (Rabinovitch, Razuvaev and coworkers (66))
Sn—Sn   50 ± 10

In fact, the literature data on the energy of the Ge—Ge bond differ greatly: through measurement of the heat of explosive decomposition of digermane in the presence of stibine, Gunn and Green (23) calculated a value for the Ge—Ge bond of 37.9 kcal/mol. Known compounds are summarized in Table 9-1.

9-1-1-2   Physical Properties and Characteristics of α Digermanes (*with* Ge—Ge *Bond*)
Table 9-1 showing some physical properties and characteristics of α digermanes (with Ge—Ge bond) appears on the following pages.

Table 9-1

Physical Properties and Characteristics of α Digermanes (with Ge—Ge Bond)

COMPOUND	EMPIRICAL FORMULA	M.P. (°C)	B.P. (°C mm)	n_D^{20}	REFERENCES
Hexamethyldigermane	$C_6H_{18}Ge_2$	−40	137/772	1.4564	8, 9
Dichlorotetraethyldigermane	$C_8H_{20}Cl_2Ge_2$	—	130-32/16	1.5197	6
1,1,1-Trimethyltriethyldigermane	$C_9H_{24}Ge_2$	—	80-82/14	1.4804	5
Dichloropentaethyldigermane	$C_{10}H_{25}ClGe_2$	—	127/16	1.5092	6, 77
1,1,2,2-Tetrabromo-1,2 diphenyldigermane	$C_{12}H_{10}Br_4Ge_2$	115-19	—	—	49
Hexavinyldigermane	$C_{12}H_{18}Ge_2$	—	55/0.35	1.5217	70
Hexaethyldigermane	$C_{12}H_{30}Ge_2$	—	260/760 132/16	1.4960 1.4975	5, 6, 18, 38
Chloropentanepropyldigermane	$C_{15}H_{35}ClGe_2$	—	110-12/0.4	1.5007	6
1,1,1-Trimethyltributyldigermane	$C_{15}H_{36}Ge_2$	—	78-80/0.07	1.4800	5
Dichlorotetrabutyldigermane	$C_{16}H_{36}Cl_2Ge_2$	—	133-38/0.16	1.5027	6
Benzoylpentaethyldigermane	$C_{17}H_{30}Ge_2O_2$	140	146/2	1.5287	74
Hexaisopropyldigermane	$C_{18}H_{42}Ge_2$	235-40	—	—	21, 31
Chloropentanebutyldigermane	$C_{20}H_{45}ClGe_2$	—	130-32/0.06	1.4932	6
Bis(cyclotetramethylene)diphenylgermane	$C_{20}H_{26}Ge_2$	—	176/0.6	1.6085	47
Penta-n-butyldigermane	$C_{20}H_{46}Ge_2$	—	115-16/0.09	1.4851	6
Pentaethyldigermyl-oxide	$C_{20}H_{50}Ge_2O$	—	162-65/0.3	1.5185	6
Methylpentabutyldigermane	$C_{21}H_{48}Ge_2$	—	125-26/0.1	1.4841	6
1,2-Dibromo-1,1,2,2-tetraphenyldigermane	$C_{24}H_{20}Br_2Ge_2$	167-69	—	—	49
1,1,1-Triethyltriphenyldigermane	$C_{24}H_{30}Ge_2$	89.5/90.5	—	—	39
Hexabutyldigermane	$C_{24}H_{54}Ge_2$	—	131/0.2	1.4858	5
1,1,1-Tribenzyltriethyldigermane	$C_{27}H_{36}Ge_2$	63.5-64.5	—	—	11

Table 9-1—continued

COMPOUND	EMPIRICAL FORMULA	M.P. (°C)	B.P. (°c mm)	n_D^{20}	REFERENCES
1,2-Diethyltetraphenyldigermane	$C_{28}H_{30}Ge_2$	125–26.5	—	—	11, 20
Hexaphenyldigermane	$C_{36}H_{30}Ge_2$	331–46	—	—	2, 25, 30, 36, 56
Hexacyclohexyldigermane	$C_{36}H_{66}Ge_2$	316 (decomp.)	—	—	29
Hexabenzyldigermane	$C_{42}H_{42}Ge_2$	183–84	—	—	2
Hexa-o-tolyldigermane	$C_{42}H_{42}Ge_2$	284–86	—	—	20
Hexa-m-tolyldigermane	$C_{42}H_{42}Ge_2$	177–79	—	—	19
Hexa-p-tolyldigermane	$C_{42}H_{42}Ge_2$	345	—	—	20
Hexa-β-styryldigermane	$C_{48}H_{42}Ge_2$	230–32	—	—	3

9-1-1-3   Chemical Properties. The Ge—Ge bond is far more stable and less reactive than the Sn—Sn bond. The conventional test designed for the latter, consisting of alcoholic silver nitrate, often proves inconclusive owing to lack of reactivity when applied to digermanes (53). Digermanes usually show high stability on exposure to air: oxygen will not react with a solution of hexaphenyldigermane in benzene (16). A highly unsaturated compound such as hexavinyldigermane will not turn yellow until after several weeks exposure to air (70). The most suitable reagent for the Ge—Ge bond is unquestionably bromine, in chloroform or carbon tetrachloride solution. It allows hexaphenyldigermane to be cleaved quantitatively (34):

$$Ph_3Ge\text{—}GePh_3 + Br_2 \rightarrow 2\,Ph_3GeBr$$

In 1,2 dibromoethane, bromination is carried a step further and yields diphenyldibromogermane, Ph_2GeBr_2, resulting from the cleavage of a Ge–phenyl bond. Unlike bromine, iodine does not cleave the hexaphenyl derivative; even in refluxing xylene the Ge—Ge bond is not broken. Slight oxidation, however, takes place which results in the formation of hexaphenyldigermoxane (Gilman and Gerow (16)).

Alkyldigermanes react more readily with halogens. Using iodine in refluxing chloroform as a reagent, Seyferth has achieved the cleavage of hexavinyldigermane to yield iodotrivinylgermane (70). Alkali metals also behave as very satisfactory cleaving agents for the Ge—Ge bond:

$$R_6Ge_2 + 2\,Na \rightarrow 2\,R_3GeNa$$

Kraus and coworkers employed sodium in liquid ammonia to cleave hexaphenyldigermane (36). An Na–K alloy is also suitable for hexaphenyl-digermane cleavage in THF (Brown and Fowles (8)). Potassium derivatives of the R_3GeK type are obtained when potassium is used in ethylamine or xylene medium in a sealed tube (38, 39):

$$Et_6Ge_2 + 2\,K \xrightarrow[\text{EtNH}_2]{} 2\,Et_3GeK$$

It has recently been shown by Bulten and Noltes that the yield becomes quantitative when hexamethylphosphotriamide is used as a solvent at room temperature (6). Normant and his coworkers (64) have established that this reagent is the most suitable solvent in metalation processes and especially so in the case of hexaphenyldigermane. Lithium derivatives are easily obtained, $(C_6H_5CH_2)_3GeLi$ resulting from the reaction of lithium with hexabenzyldigermane at 0°C in DME: as a side reaction, there is also evidence of cleavage of the Ge–benzyl bond, since toluene is characterized among the products of hydrolysis (Cross and Glockling) (11).

In contrast, alkylgermanes undergo practically no cleavage when treated with metallic lithium (Hughes (27)). Cesium in ether medium has also been used, as well as sodium–potassium alloy in various solvents: THF, benzene, DME, and a mixed ether–THF solvent (15, 69).

Cold concentrated sulphuric acid, and aqueous sodium hydroxide, seem to be only weak reagents whose action is restricted to special cases (hexa-β-styryldigermane (3, 8)). Aluminium bromide and Friedel–Crafts catalysts cleave hexaethyldigermane at 200°C, giving by disproportionation reactions tetraethylgermane and a high-molecular-weight polygermane (Vyazankin, Razuvaev and coworkers (77)):

$$(C_2H_5)_3Ge-Ge(C_2H_5)_3 \rightarrow (C_2H_5)_4Ge + [(C_2H_5)_2Ge]_n$$

Cleavage of the Ge—Ge bond by nucleophilic reagents occurs with some facility: hexaalkyldigermanes react instantaneously with potassium ethoxide or phenyllithium in HMPT at room temperature (Bulten and Noltes (6)):

$$R_3Ge-GeR_3 + EtOK \xrightarrow{\text{HMPT}} R_3GeK + R_3GeOEt$$

$$R_3Ge-GeR_3 + PhLi \xrightarrow{\text{HMPT}} R_3GeLi + R_3GePh$$

Industrial interest appears with hexaaryldigermanes $Ar_3Ge-GeAr_3$ (phenyl, naphthyl or halogen-substituted derivatives thereof): added in a very small proportion to filaments of polyamide, polyacrylonitrile or polyethylene terephthalate, they increase their heat stability and reduce the collection of electrostatic charges by the fibres (65).

9-1-1-4 Preparations. Würtz reaction is commonly used in organogermanium chemistry: alkali metal coupling trialkylgermanium halides leads to α-hexaorganodigermanes (8, 13, 69):

$$2 R_3GeX + 2 M \rightarrow R_3Ge-GeR_3 + 2 MX$$

Examples are:

Hexamethyldigermane: K and trimethylbromogermane refluxing (without solvent).
Hexaethyldigermane: Li and triethylbromogermane in THF medium.
Hexabutyldigermane: K and tributylchlorogermane in HMPT medium (80% yield).
Hexaisopropyldigermane: sodium–potassium alloy in toluene with triisopropylchlorogermane (refluxing for seven days). After aqueous ethanol treatment, hexaisopropyldigermane separates as colourless needles. This reaction gives also some triisopropylgermane and 1,1,2,2-tetraisopropyldigermane by partial cleavage of the isopropyl groups (13).

Sym-tetrabromodiphenyldigermane: action of lithium amalgam on phenyltribromogermane in ether solution at room temperature. This process was exploited further in preparing 1,2-dibromo-1,1,2,2-tetraphenyldigermane from diphenyldibromodigermane (Metlesics and Zeiss (49)).

Hexaorganodigermanes are by-products in the synthesis of tetraorganogermanes from germanium tetrahalides. Reaction of germanium tetrachloride with sodium and *p*-bromotoluene gives *hexa-p-tolyldigermane* along with some tetra-*p*-tolylgermane (2). In the same type of reaction with *β*-bromostyrene, Birr and Kraft (4) obtained *hexa-β-styryldigermane* in only 0.5 % yield. Another method is the interaction of germanium tetrahalides with a Grignard reagent in slight excess, or in the presence of excess magnesium. In such reactions, the tetraalkylgermane is also formed. Phenyl, *p*-tolyl, vinyl and ethylmagnesium bromides (slight excess) react with germanium tetrachloride to give respectively 69 %, 60 %, 26 % and 8 % yields of the corresponding digermanes, in addition to the tetraalkylgermanes (30, 70):

$$GeCl_4 + RMgBr \xrightarrow[Mg]{} GeR_4 + R_3Ge\!-\!GeR_3$$

With *o*-tolylmagnesium bromide, only the digermane was isolated (steric hindrance). Glockling and Hooton (20) interpreted this reaction as proceeding primarily through the formation of a Ge—Mg bond:

(1) $\qquad GeX_4 + RMgX \longrightarrow R_3GeX$

(2) $\qquad R_3GeX + Mg \longrightarrow R_3GeMgX$

(3) $\quad R_3GeMgX + R_3GeX \longrightarrow MgX_2 + R_3Ge\!-\!GeR_3$

(4) $\qquad R_3GeX + RMgX \longrightarrow R_4Ge + MgX_2$

(5) $\qquad R_3GeX + RMgX \xrightarrow[slow]{} R_3GeMgX + RX$

Effectively, in the reaction of $GeCl_4$ with isopropylmagnesium bromide, Mazerolles (45) obtained triisopropylgermanium hydride as the main product after hydrolysis but no hexaisopropyldigermane was detected.

In a few cases, formation of the digermane in the absence of free Mg may be observed. The reaction between $GeCl_4$ and vinylmagnesium bromide in THF seems to require the reduction to germanous dichloride, followed by conversion into the Grignard reagent R_3GeMgX (reductive alkylation)—see Seyferth (70):

$$GeCl_4 + 2\,RMgX \rightarrow R_2 + GeCl_2 + 2\,MgXCl$$

$$GeCl_2 + 3\,RMgX \rightarrow R_3GeMgX + 2\,MgXCl$$

$$R_3GeMgX + R_3GeCl \rightarrow R_6Ge_2 + MgXCl$$

For hexaphenyldigermane, the process is almost identical with the method described for the preparation of tetraphenylgermane (action of $GeCl_4$ in refluxing toluene on phenylmagnesium bromide in ether) (*Inorganic Syntheses*) (28). The sole difference is that ether is not removed immediately after the addition of the toluene solution of $GeCl_4$ (yield 60%). For hexabenzyldigermane, Glockling and Hooton suggest a similar action of germanium tetraiodide in ether–toluene on benzylmagnesium chloride (19).

Alkyllithium reactions also give rise to α-digermanes: Gilman and coworkers obtained an 8% yield of hexaethyldigermane from the reaction of ethyllithium with germanium tetrabromide (18). Hexaphenyldigermane is obtained by interaction of phenyllithium with triphenylgermane, and diethyltetraphenylgermane by the action of diphenylgermane upon butyllithium, followed by ethylation with ethyl bromide (yield 28%) (30).

Mixed hexaalkyldigermanes were recently obtained in 7–40% yield by a modified reaction of alkyllithium compounds. In 1965, Semlyen, Walker and Phillips (69) prepared some mixed hexaalkyl digermanes $Ge_2Me_xR_{6-x}$ (R = Et, Pr, i-Pr), by the reaction of $GeCl_4$ with mixed alkyllithium reagents (RLi + R'Li). For example, two moles of methyllithium and four moles of ethyllithium in diethyl ether with one mole of $GeCl_4$ gave 18% $Me_nEt_{4-n}Ge$, 30% $Me_nEt_{6-n}Ge_2$. There was also evidence for the presence of mixed octaalkyltrigermanes in the reaction products (8% $Me_nEt_{8-n}Ge_3$). Tetramethyldi-n-propyldigermane, tetramethyldiethyldigermane, tetrapropyldimethyldigermane were thus isolated.

Reaction of $GeCl_4$ with a mixture of Grignard reagents is more complex: for example, in the case of the $MeMgBr + i-PrMgBr + GeCl_4$ reaction, the formation of hydrides such as $i-Pr_2MeGeH$ and $i-Pr_3GeH$ was also detected.

The triphenylgermylmetallic reagents (e.g., Ph_3GeNa) are also useful for preparing unsymmetrically substituted digermanes (11, 39):

$$Ph_3GeNa + Et_3GeBr \rightarrow Ph_3Ge—GeEt_3 + NaBr$$

A solution of R_3GeK in HMPT at low temperature (even at $-60°C$) reacts upon trialkylchlorogermanes, in 60–70% yields (Bulten and Noltes (5)):

$$R_3GeK + R'_3GeCl \rightarrow KCl + R_3Ge—GeR'_3$$

Mazerolles has recently prepared an α cyclic organodigermane, with intracyclic germanium atoms, viz., a germacyclopentane containing two joining diethylgermyl groups.

9-1-1-5 Alkyldigermane Halides. Vyazankin first isolated bromopentaethyldigermane by reacting hexaethyldigermane with i-PrBr in the presence of AlCl$_3$ (78). Partial dehalogenation of Ph$_{4-n}$GeBr$_n$ with lithium amalgam in diethyl ether leads to the derivatives Ph$_2$Ge—GePh$_2$,

$$\underset{Br}{|} \quad \underset{Br}{|}$$

M.P. 169°C, and Br$_2$Ge—GeBr$_2$, M.P. 115–119°C. (Metlesics and Zeiss

$$\underset{Ph}{|} \quad \underset{Ph}{|}$$

(49)). The compound Bu$_2$Ge(I)—Ge(I)Bu$_2$ was isolated by Jacobs from the alkylation reaction of germanous iodide with Bu$_2$Hg (29).

Compounds of the type R$_3$Ge—GeR$_2$Cl and R$_2$ClGe—GeClR$_2$ are readily accessible according to the following reactions:

(1)   GeCl$_4$ + R$_3$Ge—GeR$_3$ $\xrightarrow{\text{200°C}}$ RGeCl$_3$ + R$_3$Ge—GeR$_2$Cl

(2)   2 GeCl$_4$ + R$_3$Ge—GeR$_3$ $\longrightarrow$ 2 RGeCl$_3$ + R$_2$ClGe—GeR$_2$Cl

The first reaction is catalyzed by germanium diiodide (yield c. 85%) (Bulten and Noltes (5)). Tin tetrachloride reacts essentially in the same way as GeCl$_4$, but is much more reactive and leads directly to tetraalkyldigermane dichlorides (reaction 2).

Alkyldigermane halides differ markedly from their tin analogues in thermal and oxidative stability. Whereas pentaethylditinchloride Et$_3$Sn—SnEt$_2$Cl decomposes at room temperature with formation of triethyltin chloride, pentaethyldigermane chloride is recovered unchanged after being heated for six hours at 200°C, and under the same conditions tetraethyldigermane dichloride was only slightly decomposed (about 8%) according to (5):

$$n\, \text{Et}_2\text{ClGe—GeClEt}_2 \rightarrow n\, \text{Et}_2\text{GeCl}_2 + (\text{Et}_2\text{Ge})_n$$

The chlorinated digermanes are also relatively inert towards oxidation.

The Ge—Ge bond in alkyldigermane halides is remarkably stable; some reactions can be performed without rupture of the Ge—Ge bond, e.g.:

(a) Reduction of Bu$_3$Ge—GeBu$_2$Cl with LiAlH$_4$ in ether affords Bu$_3$Ge—GeHBu$_2$ (methylation with Grignard reagent MeMgBr gives Bu$_3$Ge—GeMeBu$_2$).

(b) Hydrolysis of alkyldigermane halides affords the oxides:

$$\text{R}_3\text{Ge—GeR}_2\text{Cl} \xrightarrow{\text{H}_2\text{O}} (\text{R}_3\text{Ge—GeR}_2)_2\text{O}$$

Physical constants of these compounds have been included in Table 9-1.

9-1-2 Digermanes without Ge—Ge bond (bridged or β, γ ... digermanes)

These compounds exhibit altogether different properties from α digermanes owing to the lack of a Ge—Ge bond; physical properties of the most typical are indicated in Table 9-2.

9-1-2-1   Physical Properties and Characteristics of Bridged Digermanes (without Ge—Ge Bond)

Table 9-2 showing some physical properties and characteristics of bridged digermanes (without Ge—Ge bond) appears on the following page.

Table 9-2

Physical Properties and Characteristics of Bridged Digermanes (without Ge—Ge Bond)

COMPOUND	EMPIRICAL FORMULA	M.P. (°C)	B.P. (°C/mm)	n_D^{20}	REFERENCES
bis-(trichlorogermyl)methane	$CH_2Cl_6Ge_2$	—	110/18	—	70
bis-(trimethylgermyl)methane	$C_7H_{20}Ge_2$	—	63/15	1.4502	51
1,2-bis-(trimethylgermyl)acetylene	$C_8H_{18}Ge_2$	—	162/760	—	52
1,1-bis-(trimethylgermyl)ethylene	$C_8H_{20}Ge_2$	—	72/28	1.4655	54, 55
1,2-bis-(trimethylgermyl)ethylene	$C_8H_{20}Ge_2$	—	79/35	1.4628	54, 55
1,3-bis-(trimethylgermyl)propane	$C_9H_{24}Ge_2$	—	90/28	1.4500	50
para-bis-(chlorodimethylgermyl)benzene	$C_{10}H_{16}Cl_2Ge_2$	97–102	—	—	43a
1,2-bis-(triethylgermyl)acetylene	$C_{14}H_{30}Ge_2$	—	50/14	—	26
1,2-bis-(triethylgermyl)ethane	$C_{14}H_{30}Ge_2$	—	126/1.5	1.4473	41
1,3-bis-(triethylgermyl)propane	$C_{15}H_{36}Ge_2$	—	129/1.5	1.4759	41
1,3-bis-(trimethylgermyl)1,2,3,4-tetrahydronaphthalene	$C_{16}H_{28}Ge_2$	—	102/0.2	1.4765	33
1,4-bis-(triethylgermyl)butane	$C_{16}H_{38}Ge_2$	—	136/0.8	—	44
1,5-bis-(triethylgermyl)pentyne	$C_{17}H_{36}Ge_2$	—	140/0.8	1.4815	45
triethylgermyl-3-propyl-dibutylgermane	$C_{17}H_{40}Ge_2$	—	143/0.9	1.4760	47
para-bis-(triethylgermyl)benzene	$C_{18}H_{34}Ge_2$	—	153.3	1.5218	14
meta-bis-(triethylgermyl)benzene	$C_{18}H_{34}Ge_2$	—	149/1.8	1.5177	14
1,6-bis-(triethylgermyl)hexene	$C_{18}H_{40}Ge_2$	—	148/0.8	1.4818	45
1-(triethylgermyl)-2-(terbutylgermyl)acetylene	$C_{20}H_{42}Ge_2$	—	119/0.1	1.4749	44
di(triethylgermylpentyl)amine	$C_{22}H_{51}NGe_2$	—	232/0.2	1.4860	43
1,2-bis-(tributylgermyl)ethylene	$C_{26}H_{56}Ge_2$	—	150/0.2	1.4798	42
bis-(triphenylgermyl)methane	$C_{37}H_{32}Ge_2$	132–33	—	—	36
1,2-bis-(triphenylgermyl)acetylene	$C_{38}H_{30}Ge_2$	127	—	—	12, 26
1,2-bis-(tricyclohexylgermyl)acetylene	$C_{38}H_{66}Ge_2$	134–36	—	—	17, 40
1,3-bis-(triphenylgermyl)propane	$C_{39}H_{36}Ge_2$	158	—	—	26
para-bis-(triphenylgermyl)benzene	$C_{42}H_{34}Ge_2$	349	—	—	43a

9-1-2-2   Chemical Properties. Digermanes of type $(CH_3)_3Ge(CH_2)_nGe$-$(CH_3)_3$ (especially with $n = 1$ and 2) readily undergo nearly quantitative disproportionation when heated in the presence of aluminium bromide, releasing tetramethylgermane and forming a germanium polymer. For analogous organosilicon reactions see (22) and (74):

$$(CH_3)_3Ge(CH_2)_nGe(CH_3)_3 \xrightarrow[AlBr_3]{} (CH_3)_4Ge + \left[\begin{array}{c} CH_3 \\ | \\ (CH_2)_n{-}Ge{-} \\ | \\ CH_3 \end{array} \right]_n$$

Ethylenic-bridged digermanes are observed to behave in the same way and so are the mixed organosilicogermanium compounds:

$$(CH_3)_3GeCH{=}CHGe(CH_3)_3 \xrightarrow[AlBr_3]{}$$

$$(CH_3)_4Ge + \left[\begin{array}{c} CH_3 \\ | \\ CH{=}CH{-}Ge{-} \\ | \\ CH_3 \end{array} \right]_n$$

$$(CH_3)_3Ge{-}(CH_2)_m{-}Si(CH_3)_3 \rightarrow (CH_3)_4Ge + \left[\begin{array}{c} CH_3 \\ | \\ (CH_2)_mSi{-} \\ | \\ CH_3 \end{array} \right]_n$$

$$(m = 1 \text{ to } 6)$$

The latter reaction shows that heterolytic cleavage is more readily achieved with the germanium–carbon than with the silicon–carbon bond (Mironov and Kravchenko (52, 53). In contrast, acetylenic-bridged digermanes such as $(CH_3)_3GeC{\equiv}C{-}Ge(CH_3)_3$ are not liable to disproportionation in the presence of aluminium halides; only polymerization occurs without the liberation of tetramethylgermane.

9-1-2-3   Preparations. With regard to saturated aliphatic-bridged digermanes, investigations by Kraus and Brown (35) are of interest. These have been concerned with the action of triphenylgermylsodium in liquid ammonia on halides such as methyl chloride or 1,3-dibromopropane:

$$Br(CH_2)_3Br + 2\,NaGe(C_6H_5)_3 \rightarrow$$

$$2\,NaBr + (C_6H_5)_3Ge{-}(CH_2)_3{-}Ge(C_6H_5)_3$$

The resulting yields are widely variable: according to Kraus and Nutting (36) this is due to side-reactions occurring with the solvent and giving rise to triphenylgermylamine, which appears after hydrolysis as bis-(triphenyl-germanium) oxide:

$$RBr + NaGe(C_6H_5)_3 + NH_3 \rightarrow NaBr + RH + (C_6H_5)_3GeNH_2$$

It seems more convenient to use the addition reaction of hydrogeno-germanes on the double bond of allyl and vinylgermanes (Lesbre and Mazerolles) (41):

$$(C_2H_5)_3Ge—CH_2CH=CH_2 + (C_2H_5)_3GeH \rightarrow$$
$$(C_2H_5)_3Ge(CH_2)_3Ge(C_2H_5)_3$$

Germanocyclobutanes are readily cleaved by hydrogenogermanes (48):
Mironov and Kravchenko (52, 53) have recently prepared bis-(trimethyl-germyl)methane, $Me_3GeCH_2GeMe_3$, in good yield by the reaction of trimethylchlorogermane with the Grignard reagent Me_3GeCH_2MgCl. Another methylene-bridged organodigermane $CH_2(GeCl_3)_2$ has been obtained by the direct reaction of methylene chloride with Ge/Cu at elevated temperature (51). Concerning ethylene-bridged digermanes, Lesbre and Satgé (43) have shown that hydrogenogermanes undergo thermal addition to the triple bond of tributylethynylgermane, yielding 1-2 bis-(tributyl-germyl)ethylene;

$$(C_4H_9)_3GeH + (C_4H_9)_3GeC\equiv CH \rightarrow (C_4H_9)_3GeCH=CH—Ge(C_4H_9)_3$$

The synthesis of 1,1- and 1,2-bis-(trimethylgermyl)ethylenes (vinylidene and vinylene-bis-(trimethyl germanes) was also effected (54):

$$Me_3GeCH=CHCl + ClGeMe_3 \underset{Na}{\rightarrow} Me_3GeCH=CHGeMe_3$$

$$\underset{\underset{Cl}{|}}{Me_3Ge—C}=CH_2 + ClGeMe_3 \underset{Na}{\rightarrow} \underset{\underset{CH_2}{||}}{Me_3Ge—C}—GeMe_3$$

It was shown that 1,2-digermylethylenes have the *trans* configuration and show considerable exaltation of molecular refraction owing to the interaction of the π electrons of the multiple bond with the vacant d-orbitals of germanium atoms (d_π-p_π conjugation).

As for acetylene bridged digermanes, compounds of the type $R_3GeC\equiv CGeR'_3$ have been prepared by the reaction of organogermanium bromides with the acetylenic Grignard reagents (53):

$$R_3GeBr + R'_3Ge—C\equiv C—MgBr \rightarrow MgBr_2 + R_3Ge—C\equiv C—GeR'_3$$

or with sodium acetylide in THF medium (12):

$$2 (C_6H_5)_3GeBr + NaC{\equiv}CNa \rightarrow$$

$$2 NaBr + (C_6H_5)_3Ge{-}C{\equiv}C{-}Ge(C_6H_5)_3$$

but the most suitable process appears to be the one employed by Hartmann and Ahrens (26): ethynyl(trialkyl or triaryl)germanes disproportionate in THF medium, releasing acetylene and forming 1,2-bis-(trialkyl or triarylgermyl)-acetylenes:

$$2 R_3GeC{\equiv}CH \rightarrow C_2H_2 + R_3GeC{\equiv}CGeR_3$$

9-1-2-4 Cyclic Digermanes. Mazerolles (46) first isolated six-membered cyclic digermanes of type:

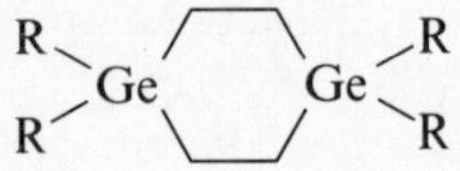

Vol'pin and coworkers have prepared the tetramethyl derivative (R = R' = CH_3), B.P. 70–72°C/12 mm, via hydrogenation of the corresponding cyclenic compounds (75). Eaborn has obtained *para*- and *meta*-bis-(triethylgermyl)benzene by reacting triethylbromogermane with the organolithium reagent of bromophenyltriethylgermane in ether (14). Some *para*-phenylene digermanes were synthesized successfully by Wurtz-type condensation reactions. Thus, organogermanium chlorides were reacted with *para*-chlorophenylgermanes, using sodium sand in refluxing toluene as the condensing agent:

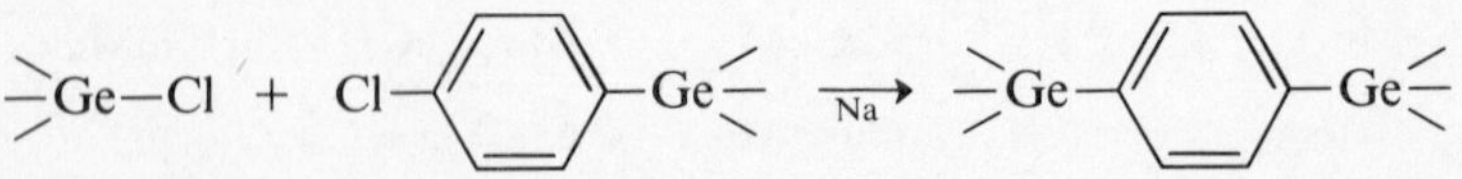

Van der Kerk and coworkers obtained *p*-bis-(chlorodimethylgermyl)-benzene by addition of the di-Grignard reagent of *p*-dibromobenzene to a slight excess of dimethylgermanium dichloride in THF (43a):

$$2 Me_2GeCl_2 + p\text{-}BrMgC_6H_4MgBr \xrightarrow[THF]{}$$

$$p\text{-}ClMe_2GeC_6H_4GeMe_2Cl$$

The compounds reported by Vol'pin and coworkers as unsaturated three-membered heterocyclic rings containing germanium ("germirenes") have been recently shown to be, in fact, derivatives of the ring system

1,4-digermin with the structure:

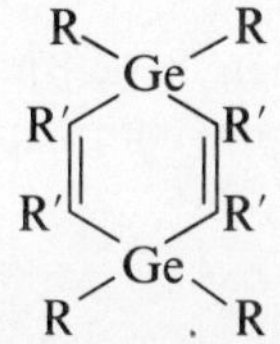

R—Ge(R)—R / R'—(ring)—R' / R'—(ring)—R' / R—Ge—R

In a more recent publication, Vol'pin (76) described an X-ray crystallographic study of "1,1-dichlorogermirene" and concluded that this compound has in reality the cyclohexadienic structure:

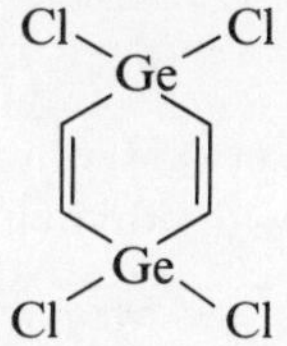

Similarly, "1,1-diiododiphenyl-2,3-germirene", the derivative obtained by addition of GeI_2 to diphenylacetylene, is in reality 1,1-4,4-tetraiodo-2,3,5,6-tetraphenyl-1,4-digermin (M.P. 302–3°C). 1,1-4,4-Tetramethyl-2,3,5,6-tetraphenyl-1,4-digermin has been also reported (M.P. 305°C). Osmometric determination of molecular weights and mass spectral measurements of several compounds of this type made in 1964 by Johnson, Gohlke and Nasutavicus also proved a six-membered ring structure with two atoms of germanium (32).

Hydrogermylation of naphthalene with trichlorogermane and subsequent methylation of the mixture with methylmagnesium bromide leads to a 52% yield of 1,3-bis-(trimethylgermyl)1,2,3,4-tetrahydronaphthalene. Under identical conditions, 1-methylnaphthalene leads to a bis-(trichlorogermyl) derivative, which on methylation affords 1-methyl-bis-(trimethylgermyl)-1,2,3,4-tetrahydronaphthalene (B.P. 104°C/0.2 mm) in an overall yield of 62% (Kolesnikov and Nefedov) (33):

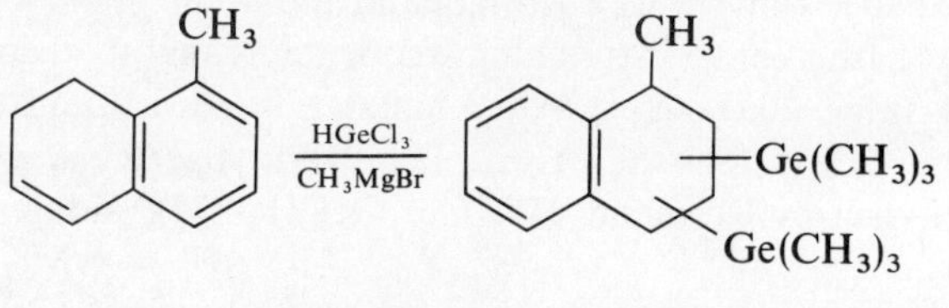

9-2 ORGANOPOLYGERMANES

Polygermanes referred to in the pertinent literature are far less numerous than polysilanes. The former fall into two categories:

(a) Acylic polygermanes.

(b) Cyclopolygermanes (with intracyclic germanium atoms).

9-2-1   Acyclic polygermanes

9-2-1-1   Trigermanes. Octaphenyltrigermane has been prepared by reacting diphenyldichlorogermane with triphenylgermylsodium (35):

$$2\,Ph_3GeNa + Ph_2GeCl_2 \;\rightarrow\; Ph_3Ge\!-\!\underset{\underset{\displaystyle Ph}{|}}{\overset{\overset{\displaystyle Ph}{|}}{Ge}}\!-\!GePh_3$$

The two Ge—Ge bonds of this compound are readily cleaved by bromine:

$$Ph_3Ge\!-\!GePh_2\!-\!GePh_3 + 2\,Br_2 \;\rightarrow\; 2\,Ph_3GeBr + Ph_2GeBr_2$$

Recently Mironov and coworkers (54) obtained two bridged trigermanes by reacting trichlorogermane with ethynyltrimethylgermane:

$$Me_3Ge\!-\!C\!\equiv\!CH + 2\,GeHCl_3 \;\rightarrow\; Me_3GeCH_2CH(GeCl_3)_2 \xrightarrow{\;MeMgBr\;}$$

$$Me_3GeCH_2CH(GeMe_3)_2$$

Leusink obtained dimethyl-bis-(*p*-trimethylgermylphenyl)germane, $Me_3Ge(C_6H_4GeMe_2)_2Me$, from the Würtz-type condensation reaction between trimethylchlorogermane and bis-(*p*-chlorophenyl)dimethylgermane. (43a).

9-2-1-2   Tetragermanes. Germanous iodide has been used by Glockling and Hooton to synthesize tris-triphenylgermylgermane, by allowing it to react with a triphenylgermyllithium solution in dimethylether/ethylene glycol. Hydrolysis was then effected and the tetragermane extracted with methylcyclohexane (20):

$$3\,Ph_3GeLi + GeI_2 \;\rightarrow\; (Ph_3Ge)_3GeLi + 2\,LiI \xrightarrow{\;hydrolysis\;}$$

$$(Ph_3Ge)_3GeH$$

The infrared spectrum shows a strong band at 228 cm^{-1} and a sharp band at 1953 cm^{-1} (the Ge—H stretching frequency is slightly displaced from its normal value 2037 cm^{-1}). After metalation (BuLi) and subsequent methylation (CH$_3$I) steps the tris-(triphenylgermyl)germane gave methyl-tris-(triphenylgermyl)germane, $(Ph_3Ge)_3GeCH_3$ (77), along with some

butylmethyl-bis-(triphenylgermyl)germane, $(Ph_3Ge)_2Ge\overset{\displaystyle\diagup Me}{\diagdown Bu}$ (78). Other

tetragermanes have recently been obtained by Neumann and coworkers (60, 61) using iodine as a cleaving agent for such cyclogermanes as octaphenylcyclotetragermane. These are as follows: 1,4-diiodooctaphenyl-

tetragermane, $I(Ph_2Ge)_4I$; decaphenyltetragermane, $Ph_3Ge-(GePh_2)_2-$
$GePh_3$; 1,4-dimethyloctaphenyltetragermane, $Me(GePh_2)_4Me$.

Catalytic cracking of bridged digermanes involving a saturated or an unsaturated bridge leads to dimethylated polygermanes in very good yield (Mironov and Kravchenko) (53):

$$(Me_3Ge)(CH_2)_2GeMe_3 \xrightarrow[AlBr_3]{} Me_4Ge + \left(-CH_2-CH_2-\overset{\displaystyle Me}{\underset{\displaystyle Me}{\overset{|}{\underset{|}{Ge}}}} \right)_n$$

The thermal decomposition of polydimethylgermanes has been studied by Russian chemists (57).

Alkylation of etherates of trichlorogermane $HGeCl_3 \cdot R_2O$ (I) (Et_2O or THF) with an excess of CH_3MgBr in ether leads to telomers (mixture of liquid and solid homologues) of the type $CH_3-[Ge(CH_3)_2]_nCH_3$ (II) where $n > 2$. Similar telomers ($n > 7$) are formed by the action of magnesium on (I), followed by methylation of the reaction products. In addition to (II), small amounts of cyclopolymers $[(CH_3)_2Ge]_n$ where $n = 6$ and probably $n = 4$ are formed (Nefedov and Kolesnikov (58)). Acyclic polygermanes are summarized in Table 9-3.

9-2-1-3 Physical Properties and Characteristics of Acyclic Polygermanes
Table 9-3 showing some physical properties and characteristics of acyclic polygermanes appears on the following page.

Table 9-3

Physical Properties and Characteristics of Acyclic Polygermanes

COMPOUND	EMPIRICAL FORMULA	M.P. (°C)	B.P. (°C/mm)	n_D^{20}	REFERENCES
Octamethyl-1,2,3-trigermane	$C_8H_{24}Ge_3$	—	43–45/0.05	1.4940	58
Decamethyl-1,2,3,4-tetragermane	$C_{10}H_{30}Ge_4$	—	81–84/0.8	1.5161	58
Tris-(trimethylgermyl)ethane	$C_{11}H_{30}Ge_3$	—	110/12	—	55
Octadecamethyl heptagermane	$C_{16}H_{48}Ge_7$	—	95–100/0.2	1.5356	58
Dimethylbis-(p-trimethylgermylphenyl)germane	$C_{20}H_{32}Ge_3$	94–96	—	—	43a
Dimethyl tris-(p-trimethylgermylphenyl)germane	$C_{28}H_{42}Ge_4$	137–39	—	—	43a
Butylmethylbis-(triphenyl)germane	$C_{41}H_{42}Ge_3$	187–90	—	—	20
Octaphenyltrigermane	$C_{48}H_{40}Ge_3$	247–48	—	—	35
1,4-dimethyloctaphenyltetragermane	$C_{50}H_{46}Ge_4$	189–91	—	—	62
Tris-(triphenylgermyl)germane	$C_{54}H_{46}Ge_4$	192–94	—	—	20, 62
Methyltris-(triphenylgermyl)germane	$C_{55}H_{48}Ge_4$	194–96	—	—	20
Decaphenyltetragermane	$C_{60}H_{50}Ge_4$	274–76	—	—	61, 62

9-2-2 Cyclopolygermanes

A cyclic trigermane, Ge-hexachloro-1,3,5-trigermacyclohexane, has been shown by Mironov and Gar to be among the products of the reaction of dichloromethane with Ge–Cu mixtures at temperatures of 350° (51):

$$Ge + CH_2Cl_2 \rightarrow CH_3GeCl_3 + Cl_3GeCH_2GeCl_3 +$$

27%   23%

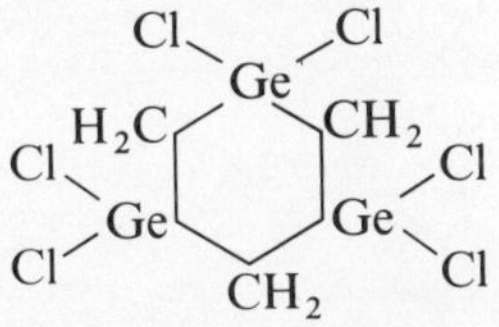

19%

This compound was alkylated with CH_3MgI reagent, giving the corresponding hexamethylated derivative.

9-2-2-1 Tetragermanes. In 1930, Kraus and Brown (35) first investigated the reaction of diphenyldichlorogermane with sodium in xylene. The main product obtained was identified in 1963 by Neumann and coworkers (61) as a crystalline cyclic tetramer of diphenylgermanium, octaphenylcyclotetragermane. On the other hand, the same cyclogermane was recently obtained in 34% yield from diphenylgermane and diethylmercury as follows:

$$Ph_2GeH_2 + Et_2Hg \rightarrow 2 C_2H_6 + (Ph_2GeHg)_n$$

This germyl-mercury polymer decomposes under the influence of UV light or heat to give mercury and octaphenylcyclotetragermane along with polymeric germanes:

$$(Ph_2GeHg)_n \rightarrow nHg + (Ph_2Ge)_4 + (Ph_2Ge)_{n-4}$$

Cleavage of the cyclic tetramer with sodium in liquid ammonia gives diphenylgermylsodium (Kraus and Brown) (35). Cleavage with iodine gives 1,4-diiodooctaphenyltetragermane from which can be prepared other polygermanes (Neumann and Kühlein) (62):

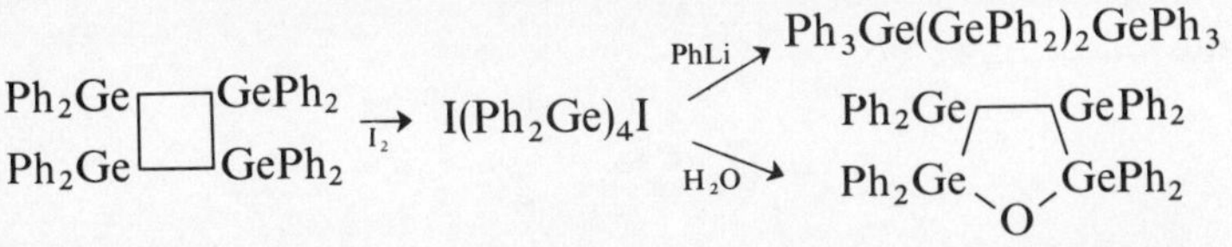

Decaphenyltetragermane is also obtained directly from octaphenylcyclotetragermane by cleavage with lithium followed by treatment of the cleavage product with bromobenzene.

9-2-2-2   Penta- and Hexagermanes. When diphenyldichlorogermane is treated with lithium in THF, a resin is formed together with high-melting pentameric and hexameric cyclopolydiphenylgermanes (yields 33 % and 3 %). The former is readily soluble in benzene, the latter only sparingly soluble (61):

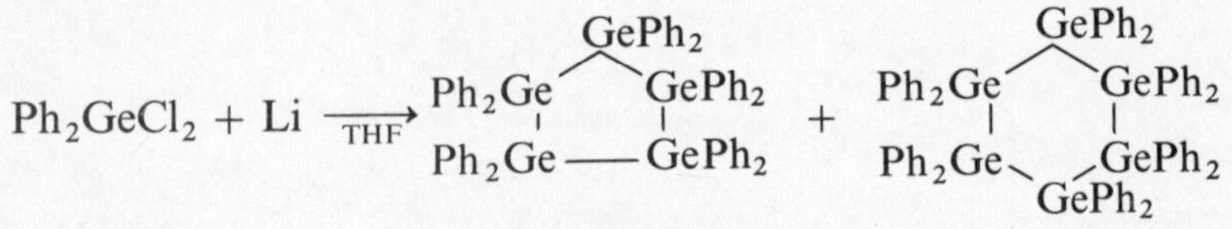

$$
Ph_2GeCl_2 + Li \xrightarrow{THF}
\begin{array}{c}
GePh_2 \\
Ph_2Ge \diagup \diagdown GePh_2 \\
| \quad\quad | \\
Ph_2Ge \text{——} GePh_2
\end{array}
+
\begin{array}{c}
GePh_2 \\
Ph_2Ge \diagup \diagdown GePh_2 \\
| \quad\quad | \\
Ph_2Ge \diagdown \diagup GePh_2 \\
GePh_2
\end{array}
$$

$$33\% \qquad\qquad 3\%$$

The same compounds were recently prepared in good yield (37 % and 17 %) by treatment of diphenyldichlorogermane with sodium naphthalenide in DME (63):

$$n\,Ph_2GeCl_2 + 2n\,C_{10}H_8Na \rightarrow$$

$$2n\,NaCl + 2n\,C_{10}H_8 + (Ph_2Ge)_5 + (Ph_2Ge)_6$$

$$37\% \qquad 17\%$$

In contrast to the tetramer, deca- and dodecaphenylcyclopentagermanes are inert toward iodine (in benzene). Their UV spectra are very similar to those of the analogous polysilanes. Sodium naphthalenide in diglyme solution cleaves Ge—Ge bonds but not Ph—Ge bonds in these perphenyl-cyclogermanes. In each instance there is found a germyl-sodium compound, e.g. Ph_2GeNa_2 (Neumann and Kühlein (62)).

Similar processes can be used for the preparation of aliphatic cyclic germanium dialkyls. For instance, Russian chemists prepared dodeca-methylcyclohexagermane by reaction of dimethyldichlorogermane with lithium wire in THF (54, 55).

Known cyclopolygermanes are summarized in Table 9-4.

9-2-2-3   Physical Properties of Cyclopolygermanes
Table 9-4 showing some physical properties of cyclopolygermanes appears on the following page.

Table 9-4

Physical Properties of Cyclopolygermanes

COMPOUND	FORMULA	M.P. (°C)	B.P. (°C/mm)	REFERENCES
Hexachloro-1,3,5-trigermacyclohexane	$C_3H_6Ge_3Cl_6$	91–92	152/5	51
Octamethylcyclotetragermane	$C_8H_{24}Ge_4$	89–90	—	35
Hexamethyl-1,3,5-trigermacyclohexane	$C_9H_{24}Ge_3$	—	73/4	51
Dodecamethylcyclohexagermane	$C_{12}H_{36}Ge_6$	211–13	—	57
Octaisopropylcyclotetragermane	$C_{24}H_{56}Ge_4$	—	—	21
Octaphenylcyclotetragermane	$C_{48}H_{40}Ge_4$	238 (decomp.)	—	61, 62, 63
Octaphenyloxacyclopentagermane	$C_{48}H_{40}GeO$	206–8	—	62
Decaphenylcyclopentagermane	$C_{60}H_{50}Ge_5$	360	—	61, 62
Dodecaphenylcyclohexagermane	$C_{72}H_{60}Ge_6$	360	—	61, 62

REFERENCES

1. *"Annual Surveys of Organometallic Chemistry"*, **2**, 153 (1966).
2. Bauer, H., and K. Burcshkies, *Chem., Ber.*, 1041, 67 (1934).
3. Birr, K. H., and D. Kraft, *Z. Anorg. Allgem. Chem.*, **311**, 235 (1961).
4. Birr, K. H., and D. Kraft, *Z. Anorg. Allgem. Chem.*, **311**, 238 (1961).
5. Bulten, E. J., and J. G. Noltes, *Tetrahedron Letters*, **36**, 4389 (1966).
6. Bulten, E. J., and J. G. Noltes, *J. Organometal. Chem.*, **11**, 22 (1968).
7. Brown, M. P., and G. W. A. Fowles, *J. Chem. Soc.*, 506 (1960).
8. Brown, M. P., and G. W. A. Fowles, *J. Chem. Soc.*, 2811 (1958).
9. Brown, M. P., E. Cartwell, and G. W. A. Fowles, 508 (1960).
10. Cawley, S., and S. S. Danyluc, *Can. J. Chem.*, **41**, 1850 (1963).
11. Cross, R. J., and F. J. Glockling, *J. Chem. Soc.*, 4125 (1964).
12. Davidsohn, W., and M. C. Henry, *J. Organometal. Chem.*, **5**, 29 (1966).
13. Eaborn, C., and K. C. Pande, *J. Chem. Soc.*, 3200 (1960).
14. Eaborn, C., A. Leyshon, and K. C. Pande, *J. Chem. Soc.*, 3423 (1960).
15. Gilman, H., and A. G. Brook, *J. Am. Chem. Soc.*, **76**, 77 (1954).
16. Gilman, H., and C. W. Gerow, *J. Am. Chem. Soc.*, **77**, 5509 (1955).
17. Gilman, H., and C. W. Gerow, *J. Am. Chem. Soc.*, **79**, 342 (1957).
18. Gilman, H., M. B. Hughes, and C. W. Gerow, *J. Org. Chem.*, **24**, 352 (1959).
19. Glockling, F. J., and K. A. Hooton, *J. Chem. Soc.*, 3509 (1962).
20. Glockling, F. J., and K. A. Hooton, *J. Chem. Soc.*, 1849 (1963).
21. Glockling, F. J., and A. Carrick, *J. Chem. Soc.*, 623 (1966).
22. Gorov, E., S. Puschevaya, D. Lubuzh, M. Udovin, and A. D. Petrov, *Izv. Akad. Nauk SSSR*, 822 (1963).
23. Gunn, S. R., and L. G. Green, *J. Phys. Chem.*, **65**, 779 (1961).
24. Hague, D. N., and R. H. Prince, *Proc. Chem. Soc.*, 300 (1962).
25. Harris, D. M., W. H. Nebergall, and O. H. Johnson, *Inorg. Synth.*, **5**, 72 (1957).
26. Hartmann, H., and J. U. Ahrens, *Angew. Chem.*, **70**, 75 (1958).
27. Hughes, M. B., *Dissertation Abstr.*, **19**, 1921 (1958).
28. *Inorganic Syntheses*, McGraw-Hill edit., **5**, 73 (1957).
29. Johnson, O. H., and W. H. Nebergall, *J. Am. Chem. Soc.*, **70**, 1706 (1948).
30. Johnson, O. H., and D. M. Harris, *J. Am. Chem. Soc.*, **72**, 5563 (1950).
31. Johnson, O. H., *Chem. Rev.*, **48**, 259 (1956).
32. Johnson, F., R. S. Gohlke, and W. A. Nasutavicus, *J. Organometal. Chem.*, **3**, 233 (1964).
33. Kolesnikov, G. S., and O. M. Nefedov, *Angew. Chem.*, **77**, 345 (1965).
34. Kraus, C. A., and L. S. Foster, *J. Am. Chem. Soc.*, **49**, 457 (1927).
35. Kraus, C. A., and C. L. Brown, *J. Am. Chem. Soc.*, **52**, 4031 (1930).
36. Kraus, C. A., and L. S. Nutting, *J. Am. Chem. Soc.*, **54**, 1622 (1932).
37. Kraus, C. A., and E. A. Flood, *J. Am. Chem. Soc.*, **54**, 1635 (1932).
38. Kraus, C. A., and E. A. Flood, *J. Am. Chem. Soc.*, **54**, 1688 (1932).
39. Kraus, C. A., and S. Sherman, *J. Am. Chem. Soc.*, **55**, 4694 (1933).
40. Kraus, C. A., and B. Smith, *J. Am. Chem. Soc.*, **74**, 1418 (1952).
41. Lesbre, M., and P. Mazerolles, *Compt. Rend.*, **248**, 2058 (1959).
42. Lesbre, M., and J. Satgé, *Compt. Rend.*, **250**, 2220 (1960).
43. Lesbre, M., and J. Satgé, *Compt. Rend.*, **248**, 2018 (1959).
43a. Leusink, A. J., J. G. Noltes, H. A. Budding, and G. J. M. Vanderkerk, *Rec. Trav. Chim.*, **83**, 845 (1964).
44. Mathis, F., and P. Mazerolles, *Bull. Soc. Chim. France*, 1955 (1961).

45. Mazerolles, P., M. Lesbre, and S. Marre, *Compt. Rend.*, **262**, 4134 (1965).
46. Mazerolles, P., *Bull. Soc. Chim. France*, 1907 (1962).
47. Mazerolles, P., and J. Dubac, *Compt. Rend.*, **257**, 1103 (1963).
48. Mazerolles, P., J. Dubac, and M. Lesbre, *Tetrahedron Letters*, **3**, 255 (1967).
49. Metlesics, W., and H. Zeiss, *J. Am. Chem. Soc.*, **82**, 3321 (1960).
50. Mironov, V. F., N. G. Dzurinskaya, T. K. Gar, and A. D. Petrov, *Izv. Akad. Nauk SSSR*, 460 (1962).
51. Mironov, V. F., and T. K. Gar, *Izv. Akad. Nauk SSSR*, **11**, 2067 (1963).
52. Mironov, V. F., and A. L. Kravchenko, *Izv. Akad. Nauk SSSR*, 768 (1964).
53. Mironov, V. F., and A. L. Kravchenko, *Izv. Akad. Nauk SSSR*, **6**, 1026 (1965).
54. Mironov, V. F., A. L. Kravchenko, and L. A. Leites, *Izv. Akad. Nauk SSSR*, **57**, 1177 (1966).
55. Mironov, V. F., A. L. Kravchenko, and A. D. Petrov, *Dokl. Akad. Nauk SSSR*, **155**, 843 (1964).
56. Morgan, G. T., and H. D. Drew, *J. Chem. Soc.*, **127**, 1760 (1925).
57. Nefedov, O. M., G. Garzo, and I. Shiriaev, *Dokl. Akad. Nauk SSSR*, **164**, 822 (1965).
58. Nefedov, O. M., and S. P. Kolesnikov, *Izv. Akad. Nauk SSSR*, **4**, 773 (1964).
59. Neumann, W. P., *Angew. Chem.*, **75**, 679 (1963).
60. Neumann, W. P., *Angew. Chem.*, **75**, 679 (1963).
61. Neumann, W. P., and K. Kuhlein, *Tetrahedron Letters*, 1541 (1963).
62. Neumann, W. P., and K. Kuhlein, *Ann. Chem.*, **683**, 1 (1965).
63. Neumann, W. P., and K. Kuhlein, *Ann. Chem.*, **702**, 13 (1967).
64. Normant, H., T. Cuvigny, J. Normant, and G. Angelo, *Bull. Soc. Chim. France*, 3346 (1965).
65. U.S. Pat. 2,924,586 (Feb. 9, 1960).
66. Rabinovitch, J. B., V. I. Tell'noi, N. V. Karyakin, and G. A. Razuvaev, *Dokl. Akad. Nauk SSSR*, **149**, 324 (1963).
67. Ryan, T., and C. Tamborski, *Spectrochim. Acta*, **18**, 21 (1962).
68. Selwood, P. W. J., *J. Am. Chem. Soc.*, **61**, 3168 (1939).
69. Semlyen, J. A., G. R. Walker, and C. S. G. Phillips, *J. Chem. Soc.*, 1197 (1965).
70. Seyferth, D., *J. Am. Chem. Soc.*, **79**, 2738 (1957).
71. Schackelford, J. M., H. de Schmertzing, C. H. Heuther, and H. Podall, *J. Org. Chem.*, **28**, 1700 (1963).
72. Tamborski, C., F. E. Ford, G. J. Moore, and E. J. Soleski, *J. Org. Chem.*, **27**, 619 (1962).
73. Tcharikian, A., *J. Am. Chem. Soc.*, **12**, 440 (1939).
74. Vdovin, V. M., S. Puschevaya, and A. D. Petrov, *Izv. Akad. Nauk SSSR*, **281**, 1275 (1961).
75. Vol'pin, M. E., Yu. T. Struchlov, V. Vilkov, S. Mastyukov, V. G. Dulova, and D. N. Kursanov, *Izv. Akad. Nauk SSSR*, **11**, 2067 (1963).
76. Vol'pin, M. E., V. G. Dulova, Yu. T. Struchlov, N. K. Bokii, and D. N. Kursanov, *J. Organometal. Chem.*, **8**, 87 (1967).
77. Vyazankin, N. S., G. A. Razuvaev, S. P. Korneva, and F. Guiliulina, *Dokl. Akad. Nauk SSSR*, **155**, 839 (1964).
78. Vyazankin, N. S., G. A. Razuvaev, S. P. Korneva, and F. Guliulina, *Dokl. Akad. Nauk SSSR*, **158**, 884 (1964).
79. Vyazankin, N. S., *Zh. Obshch. Khim.*, **34**, 1645 (1965).

10 Germanium–metal bond

10-1 GROUP IA ORGANOGERMANIUM–(ALKALI) METAL COMPOUNDS

10-1-1 Trialkylgermyl–alkali metal compounds

The preparation of R_3GeLi compounds by hydrogen–metal exchange with organolithium reagents gives varying results:

$$R_3GeH + R'Li \rightarrow R_3GeLi + R'H$$

Metalation of triethylgermane with phenyllithium or butyllithium in ethereal solvents gives triethylgermyllithium in less than 10% yield (23, 27). In recent years two routes to trialkylgermyl–alkali metal compounds have been disclosed.

(a) Vyazankin and coworkers have developed a convenient method for the synthesis of triethylgermyllithium in THF or benzene medium (40). Bis-(triethylgermyl)mercury (or cadmium) reacts quantitatively with metallic lithium under mild conditions, according to the equation:

$$(Et_3Ge)_2Hg + 2 Li \xrightarrow{\text{THF}} Hg + 2 Et_3GeLi$$

Similarly, tris-(triethylgermyl)thallium reacts by a metal exchange reaction with lithium wire in the absence of air (40). These reactions in THF are complete at room temperature in about 24 h. Like other compounds of this type sensitive to air, triethylgermyllithium was not isolated from the reaction mixtures; its existence was proved directly by coupling with ethyl bromide to give a nearly quantitative yield of tetraethylgermane (28).

(b) Bulten and Noltes obtained such compounds by reacting an alkali metal either with an alkylchlorogermane or with a hexaalkyldigermane

646

in HMPT medium (2):

$$R_3GeCl + 2M \xrightarrow[HMPT]{} R_3GeM + MCl$$

(M = Li, Na, K); for example:

$$Bu_2GeCl_2 + 4K \xrightarrow[HMPT]{} Bu_2GeK_2 + 2KCl$$

Hydrolysis of the brown reaction mixture leads to Bu_2GeH_2 (22 % yield). An alternative procedure involves the reaction:

$$R_6Ge_2 + 2M \xrightarrow[HMPT]{} 2R_3GeM$$

HMPT is a good solvent both for alkali metals and for hexaalkyldigermanes (33). Bulten and Noltes observed a 100 % conversion in 3–5 hours at 20°C with Li or K, 6–8 hours with Na (lower solubility in HMPT). Solutions of R_3GeK in HMPT are quite stable at low temperatures. In 1932, Kraus and Flood obtained Et_3GeK by shaking hexaethyldigermane with potassium in ethylamine in a sealed tube for several weeks (28) (ethereal solvents do not appear convenient for this use (6)):

$$Et_6Ge_2 + 2K \xrightarrow[EtNH_2]{} 2Et_3GeK$$

The trialkylgermyl–alkali metal derivatives are very reactive nucleophilic reagents which can be used for the preparation of a range of organogermanium derivatives.

The asymmetric α-digermanes, R_3Ge—GeR'_3, can be readily prepared in 70 % yield at low temperature ($-60°C$):

$$R_3GeK + R'_3GeCl \rightarrow KCl + R_3Ge\!-\!GeR'_3$$

With Me_3SiCl or Me_3SnCl, compounds of the type R_3Ge—MMe_3 may be obtained in about 60 % yield (1):

$$Et_3GeK + Me_3SnCl \rightarrow Et_3Ge\!-\!SnMe_3$$

With CH_2Cl_2, the bridged digermane was isolated in 65 % yield:

$$2Et_3GeK + CH_2Cl_2 \rightarrow Et_3GeCH_2GeEt_3$$

Trigermanes may be prepared conveniently by reacting triethylgermylpotassium with a dialkyldichlorogermane (3):

$$2Et_3GeK + Me_2GeCl_2 \rightarrow Et_3Ge\!-\!GeMe_2\!-\!GeEt_3$$

Carboxylation and subsequent careful acidification gives $Et_3GeCOOH$ which slowly decomposes on treatment with mineral acids (3):

$$Et_3GeK + CO_2 \rightarrow Et_3GeCOOK \xrightarrow[H^+]{} Et_3GeCOOH$$

$$Et_3GeCOOH + HCl \rightarrow Et_3GeCl + CO + H_2O$$

Et_3GeLi solutions also are highly reactive, in particular towards unactivated double bonds in a benzene medium (ethylene and its homologues (39)):

$$Et_3GeLi + RCH{=}CH_2 \xrightarrow{\;50°C\;} Et_3GeCH_2CHLiR \xrightarrow{\;H_2O\;}$$

$$Et_3GeCH_2CH_2R$$

$$(R = H, CH_3, C_4H_9, C_6H_5)$$

With styrene the addition is exothermic and affords triethylphenylethyl-germane in 43% yield (41).

In THF, triethylstannane and triethylsilane react as follows (40):

$$Et_3GeLi + Et_3MH \xrightarrow[THF]{} LiH + Et_3Ge{-}MEt_3$$

$$(M = Si, Sn, Ge)$$

Other derivatives with Ge—Si bond may be readily prepared at room temperature in THF medium, e.g.:

$$3\,Et_3GeLi + SiHCl_3 \rightarrow 3\,LiCl + (Et_3Ge)_3SiH \qquad (32\% \text{ yield})$$

$$2\,Et_3GeLi + Ph_2SiCl_2 \rightarrow 2\,LiCl + (Et_3Ge)_2SiPh_2 \qquad (66\% \text{ yield})$$

$$Et_3GeLi + EtMe_2SiCl \rightarrow LiCl + Et_3GeSiMe_2Et$$

10-1-2   Triarylgermyl–alkali metal compounds

Triphenylgermyl–alkali metal solutions have been prepared by the cleavage of:

Ge—C bonds   (10, 11)
Ge—Ge bonds   (13)
Ge—X bonds   (8)
Ge—H bonds   (26)

with the alkali metals in different solvents: ammonia, amines and more recently strongly donating ethereal solvents such as THF and ethylene glycol dimethylether.

10-1-2-1 Ph_3GeNa. Triphenylgermylsodium was originally obtained by the method of Kraus and Foster (29) who isolated a benzene-soluble crystalline complex, $Ph_3GeNa(NH_3)_3$, by treating hexaphenyldigermane with sodium in liquid ammonia. Reaction proceeds slowly on account of the low solubility of the digermane. The ammonia solution is highly conducting and reacts with oxygen to give sodium triphenylgermanolate. In diethyl ether, Ph_3GeNa reacts with trichlorosilane in substantially quantitative yield:

$$SiHCl_3 + Ph_3GeNa \rightarrow 3\,NaCl + (Ph_3Ge)_3SiH \quad (30)$$

Bromobenzene was found to undergo halogen–metal interconversion when allowed to react with triphenylgermylsodium (29).

10-1-2-2 Ph_3GeK. Triphenylgermylpotassium is obtained in 83% yield when hexaphenyldigermane is cleaved by Na–K alloy in THF. Other solvents such as diethyl and dibutyl ether have been used (12). Cleavage of Ge—C bond in Ph_3Ge—CPh_3 by Na–K alloy is much slower and gives only a 28% yield (10). Suspensions of Ph_3GeK in organic solvents are yellow-brown or grey-green (10). They react with phenyl-substituted ethylenic bonds, and with activated multiple bonds (16).

10-1-2-3 Ph_3GeLi. Gilman and coworkers have reported the preparation of triphenylgermyllithium from triphenylchloro- or -bromogermane and Li wire in THF (8):

$$Ph_3GeCl + 2\,Li \rightarrow Ph_3GeLi + LiCl$$

This direct method of synthesis was found to be a quite general reaction for Group IV and V elements (37).

Cleavage of a C—Ge bond in tetraphenylgermane, or metalation of triphenylgermane by use of Li wire in ethereal solvents such as THF or DME, lead to triphenylgermyllithium in variable yields (7, 11, 15–17, 22, 24, 35):

$$Ph_4Ge + 2\,Li \rightarrow Ph_3GeLi + PhLi$$

$$Ph_3GeH + 2\,Li \rightarrow Ph_3GeLi + LiH$$

In the first of the above reactions, phenyllithium decomposes via reaction with the solvent.

Lithium wire in THF also cleaves hexaphenyldisilane and hexaphenyldigermane to yield respectively triphenylsilyl- and triphenylgermyllithium (4):

$$Ph_6Ge_2 + 2\,Li \rightarrow 2\,Ph_3GeLi \quad (75\% \text{ yield})$$

This method offers some advantage in that the desired compound can be obtained in a high degree of purity and the yields are somewhat higher.

Metalation of triphenylgermane with an equivalent amount of an organolithium reagent is an even more satisfactory method. It can be carried out either in refluxing diethyl ether solution or in THF (14):

$$Ph_3GeH + RLi \rightarrow Ph_3GeLi + RH$$

In these reactions, triphenylgermane acts like triphenylmethane which gives triphenylmethyllithium with MeLi (9). However, some alkylation has also been observed in the reaction of triphenylgermane with methyllithium. Butyllithium is the most convenient reagent: for a typical preparation, 0.016 mol of Ph_3GeH was dissolved in 150 ml of THF, the solution was cooled to $-23°C$ and 0.016 mol of n-butyllithium was added slowly with stirring (32). The resulting pale green solution is not quite stable, and reacts very slowly with the solvent (20):

$$Ph_3GeLi + \underbrace{(CH_2)_4O} \rightarrow Ph_3Ge(CH_2)_4OLi \xrightarrow{H_2O} Ph_3Ge(CH_2)_4OH$$

Similar reactions were observed with saturated small ring oxygen heterocycles:

$$Ph_3GeLi + \underbrace{(CH_2)_nO} \xrightarrow{reflux} Ph_3Ge(CH_2)_nOLi \xrightarrow{H_2O}$$
$$Ph_3Ge(CH_2)_nOH$$

Germylmetallic compounds react with epoxides in the same way as silylmetallic compounds (18).

Triphenylgermyllithium has been used extensively as a synthetic reagent in organogermanium chemistry. Oxidation leads to triphenylgermanol, carbonation gives the acid $Ph_3Ge—COOH$, which decarbonylates at its melting point to triphenylgermanol partially esterified by undecomposed acid (10):

$$Ph_3GeCOOH \rightarrow CO + Ph_3GeOH$$

$$Ph_3GeOH + Ph_3GeCOOH \rightarrow H_2O + Ph_3GeOCOGePh_3$$

The reaction of bromine in DME at $-20°C$ caused the formation of small amounts of bromotriphenylgermane in addition to larger amounts of hexaphenyldigermane, presumably formed from coupling with bromotriphenylgermane:

$$Ph_3GeLi + Br_2 \rightarrow LiBr + Ph_3GeBr$$

$$Ph_3GeBr + Ph_3GeLi \rightarrow Ph_3Ge—GePh_3$$

Triphenylgermyllithium reacts slowly with triphenylgermane in diethyl ether, giving hexaphenyldigermane (14):

$$Ph_3GeLi + Ph_3GeH \rightarrow LiH + Ph_3Ge—GePh_2$$

The following reaction is quite general (8):

$$Ph_3GeLi + R_3MX \rightarrow LiX + Ph_3GeMR$$

$$(M = C, Si, Ge, Sn, Pb)$$

Ph$_3$GeLi and germanous iodide react exothermally in DME medium, forming a deep red solution, from which, after hydrolysis, the crystalline tris-(triphenylgermyl)germane may be isolated in 36% yield (25):

$$2\,Ph_3GeLi + GeI_2 \rightarrow (Ph_3Ge)_2Ge \xrightarrow[Ph_3GeLi]{} (Ph_3Ge)_3GeLi \xrightarrow{H_2O}$$

$$(Ph_3Ge)_3GeH$$

The reaction of Ph$_3$GeLi with ketones yields either an α-hydroxygermane or triphenylgermane, depending on the solvent. The sole product isolated from the reaction with acetone or acetophenone in THF was triphenyl-germane:

(1) $Ph_3GeLi + PhCOCH_3 \xrightarrow{THF} Ph_3GeH + PhCOCH_2Li$

With diethyl ether as the solvent, the corresponding carbinols are obtained in good yields after hydrolysis (32):

(2) $Ph_3GeLi + CH_3COCH_3 \xrightarrow[(2)\ H_2O]{(1)\ ether} Ph_3Ge\!-\!\underset{\underset{OH}{|}}{C}(CH_3)_2$

Benzophenone which has no α-hydrogen available, for reaction (1) gives only the carbinol when allowed to react in THF:

$$Ph_3GeLi + PhCOPh \xrightarrow[(2)\ H_2O]{(1)\ THF} Ph_3Ge\!\underset{\underset{OH}{|}}{C}\!-\!Ph_2$$

Sterically hindered ketones such as t-butyl-methyl-ketone do not react (32). Addition of Ph$_3$GeLi to the carbonyl group of formaldehyde gives triphenyl(hydroxymethyl)germane:

$$Ph_3GeLi \xrightarrow[(2)\ H_2O]{(1)\ HCHO} Ph_3GeCH_2OH$$

With benzaldehyde, triphenylgermylphenylcarbinol is obtained (4, 5). With benzalacetophenone, reaction occurs by means of a 1,4 addition to give 2-phenyl-2-triphenylgermylethyl-phenyl-ketone (13):

$$Ph_3GeLi + PhCH\!=\!CHCOPh \rightarrow Ph_3Ge\!-\!\underset{\underset{Ph}{|}}{C}H\!-\!CH_2COPh$$

Like triphenylgermylpotassium, Ph$_3$GeLi adds to some activated olefinic bonds: with 1,1-diphenylethylene, 1,1-diphenyl-2-triphenylgermylethane was obtained along with small amounts of hexaphenyldigermane and hexaphenyldigermoxane (14):

$$Ph_3GeLi + Ph_2C{=}CH_2 \rightarrow Ph_2\underset{\underset{Li}{|}}{C}{-}CH_2GePh_3 \xrightarrow[H_2O]{}$$

$$LiOH + Ph_2CHCH_2GePh_3$$

With octadecene-1 dissolved in DME (16), Gilman and Gerow isolated triphenyloctadecylgermane together with some triphenylgermanol. However, Ph_3GeLi does not react with octene-1, styrene, cyclohexene and stilbene. In contrast, triphenylsilyllithium adds to the olefinic linkage of trans-stilbene (31).

Ph_3GeLi was found to be a metalating agent of intermediate strength. For example it undergoes a hydrogen–lithium exchange reaction with acidic hydrogen in the 9-position of fluorene, to give on carbonation 69 % yield of 9-fluorenylcarboxylic acid, with small amounts of hexaphenyldigermane and hexaphenyldigermanoxane, whereas phenyllithium gives a 78 % yield under similar conditions (10, 21):

$$Ph_3GeLi + \text{[fluorene]} \xrightarrow{DME} Ph_3GeH + \text{[9-lithiofluorene]}$$

Insertion additions with elementary sulphur, selenium or tellurium in THF medium lead to the highly reactive compounds Ph_3GeMLi (M = S, Se, Te) (34).

Reaction of the Ph_3GeLi reagent and a series of aroyl halides has been shown to yield α-germyl ketones and 1,1-bis-(triphenylgermyl)carbinols as major products. In general, ketones are formed from reaction of Ph_3GeLi with aromatic carboxylic acid chlorides, and alcohols from its reaction with aliphatic carboxylic acid chlorides (32). For instance, addition of Ph_3GeLi to benzoyl chloride at low temperature leads to benzoyltriphenylgermane:

$$Ph_3GeLi + PhCOCl \xrightarrow{-78^\circ C} LiCl + Ph_3GeCOPh$$

At room temperature, with an excess of Ph_3GeLi, 1,1-bis-(triphenylgermyl)phenylcarbinol was obtained in 49 % yield:

$$2\,Ph_3GeLi \xrightarrow[\text{(2) } H_2O]{\text{(1) PhCOCl}} (Ph_3Ge)_2\underset{\underset{OH}{|}}{C}{-}Ph$$

This compound results from reaction of Ph_3GeLi with benzoyltriphenylgermane in equimolecular ratio.

10-1-2-4 $(PhCH_2)_3GeLi$. The best preparation of tribenzylgermyllithium seems to be via reaction of tetrabenzylgermane with lithium shot in ethylene glycol dimethyl ether. The resulting deep-brown solution contains largely tribenzylgermyllithium and also a small amount of the dilithium compound $(PhCH_2)_2GeLi_2$ according to (7):

$$(PhCH_2)_4Ge + 2\,Li \xrightarrow[DME]{0°C} (PhCH_2)_3GeLi + PhCH_2Li$$

$$PhCH_2Li + (PhCH_2)_3GeLi \rightarrow (PhCH_2)_2GeLi_2 + PhCH_2CH_2Ph$$

Hydrolysis gives di- and tribenzylgermanes and also tribenzylmethylgermane. Ge—benzyl bonds actually are much more easily cleaved than Ge—phenyl bonds; given time, the benzyllithium produced in reaction (1) may react with the solvent, and thus with DME simultaneous formation of toluene, ethylbenzene and methyl vinyl ether was reported:

$$PhCH_2Li + MeOCH_2CH_2OMe \nearrow PhCH_2CH_3 + LiOCH_2CH_2OMe$$

$$\searrow PhCH_3 + MeOCH—CH_2OMe$$

$$(\beta\text{-elimination}) \downarrow \overset{|}{Li}$$

$$LiOMe + H_2C{=}CHOMe$$

On carbonation, tribenzylgermyllithium gives the acid $(PhCH_2)_3GeCOOH$ which is unstable at room temperature; partial decarbonylation of this leads to tribenzylgermanol which is esterified by the remaining acid:

$$(PhCH_2)_3Ge—CO_2H \rightarrow CO + (PhCH_2)_3GeOH$$

$$(PhCH_2)_3GeOH + (PhCH_2)_3Ge—CO_2H \rightarrow$$

$$H_2O + (PhCH_2)_3GeOCOGe(CH_2Ph)_3$$

In this reaction a small amount of the ester $(PhCH_2)_3GeOCOCH_2Ph$ is also isolated. It originates in the phenylacetic acid produced from benzyllithium which always accompanies $(PhCH_2)_3GeLi$ when tetrabenzylgermane is cleaved.

Other reactions of tribenzylgermyllithium are:

REACTANT	PRODUCTS	M.P. or B.P. (°C) (°C/mm)
H_2O	$(PhCH_2)_3GeH$	80–82
D_2O	$(PhCH_2)_3GeD$	81
MeI	$(PhCH_2)_3GeMe$	82–85
EtBr	$(PhCH_2)_3GeEt$	34–35
Me_3SiCl	$(PhCH_2)_3GeSiMe_3$	63.5–64.5
Et_3GeBr	$(PhCH_2)_6Ge_2 + Et_6Ge_2$ or (reverse addition)	
	$(PhCH_2)_3Ge—GeEt_3$	$220–30/10^{-3}$

10-1-2-5   PhMe(1-$C_{10}H_7$)GeLi. Methyl-α-naphthylphenylgermyllithium has been prepared by metalation of (−)-methyl-α-naphthylphenylgermane with one equivalent of *n*-butyllithium in ether. The dark brown solution is optically stable; on treatment with benzophenone it gives in 68 % yield the carbinol (5):

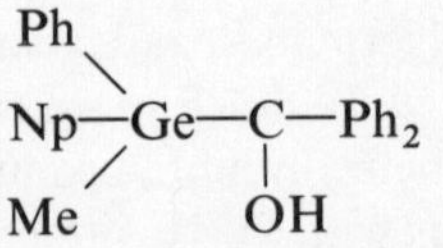

$$\begin{array}{c} \text{Ph} \\ \diagdown \\ \text{Np}\!-\!\text{Ge}\!-\!\text{C}\!-\!\text{Ph}_2 \qquad [\alpha]_D^{20} = 6.3° \\ \diagup \qquad | \\ \text{Me} \qquad \text{OH} \end{array}$$

By carbonation, the carboxylic acid $\rangle$ Ge—COOH was obtained: $[\alpha]_D^{22} = 5.1°$.

10-1-2-6 $Ph_2(PhCH_2CH_2)$GeLi. By cleavage of triphenyl-2-phenylethyl-germane with lithium wire in DME, Gilman and coworkers obtained a red-brown solution of diphenyl-2-phenylethylgermyllithium (18).

10-1-3   Diarylgermyl–alkali metal compounds

10-1-3-1 Ph_2GeNa_2. Tetraphenylgermane reacts with sodium dissolved in liquid ammonia to give benzene, sodium amide and triphenylgermyl-sodium. In concentrated solution, a second phenyl group is cleaved, giving the compound Ph_2GeNa_2 which imparts a deep red colour to the solution (36).

10-1-3-2 Ph_2GeLi_2. Cross and Glockling believe Ph_2GeLi_2 to be produced in small quantity in the reaction of butyllithium with diphenyl-germane, according to:

$Ph_2GeH_2 + 2\,BuLi \longrightarrow$

$\quad Ph_2GeLi_2 + Ph_2(Li)Ge\!-\!Ge(Li)Ph_2 + Ph_2Ge(Bu)Li + Ph_2GeBu_2$

Addition of ethyl bromide to the mixture enabled these workers to separate four compounds:

Ph_2GeEt_2	$Ph_2(Et)Ge\!-\!Ge(Et)_2$	$Ph_2Ge(Bu)Et$	Ph_2GeBu_2
2%	28%	20%	12%

10-1-3-3 $(PhCH_2)_2GeLi_2$. This compound is produced along with tribenzylgermyllithium when cleaving tetrabenzylgermane with lithium in DME medium (7).

References (10-1)

1. Bulten, E. J., and J. G. Noltes, *Tetrahedron Letters*, **36**, 4389 (1966).
2. Bullen, E. J., and J. G. Noltes, *Tetrahedron Letters*, **29**, 3471 (1966).
3. Bulten, E. J., and J. G. Noltes, *Tetrahedron Letters*, **16**, 1443 (1967).
4. Brook, A. G., M. A. Quigley, G. J. D. Peddle, N. V. Schwartz, and C. M. Wagner, *J. Am. Chem. Soc.*, **82**, 5102 (1960).
5. Brook, A. G., and G. J. D. Peddle, *J. Am. Chem. Soc.*, **85**, 2338 (1963).
6. Carrick, A., and F. Glockling, *J. Chem. Soc.*, 623 (1966).
7. Cross, J., and F. Glockling, *J. Chem. Soc.*, 4125 (1964).
8. George, J., J. Peterson, and H. Gilman, *J. Am. Chem. Soc.*, **82**, 403 (1960).
9. Gilman, H., and V. Young, *J. Org. Chem.*, **1**, 315 (1936).
10. Gilman, H., and G. Brook, *J. Am. Chem. Soc.*, **76**, 77 (1954).
11. Gilman, H., and C. W. Gerow, *J. Am. Chem. Soc.*, **77**, 4675 (1955).
12. Gilman, H., and C. W. Gerow, *J. Am. Chem. Soc.*, **77**, 5509 (1955).
13. Gilman, H., and C. W. Gerow, *J. Am. Chem. Soc.*, **77**, 5740 (1955).
14. Gilman, H., and C. W. Gerow, *J. Am. Chem. Soc.*, **78**, 5435 (1956).
15. Gilman, H., and C. W. Gerow, *J. Am. Chem. Soc.*, **78**, 5823 (1956).
16. Gilman, H., and C. W. Gerow, *J. Am. Chem. Soc.*, **79**, 342 (1957).
17. Gilman, H., B. Hughes, and W. Gerow, *J. Org. Chem.*, **24**, 342 (1959).
18. Gilman, H., D. Adke, and D. Wittenberg, *J. Am. Chem. Soc.*, **81**, 1107 (1959).
19. Gilman, H., J. George, and J. Peterson, *J. Am. Chem. Soc.*, **82**, 403 (1960).
20. Gilman, H., and E. A. Zuech, *J. Org. Chem.*, **26**, 3035 (1961).
21. Gilman, H., O. L. Mans, W. J. Trepka, and J. W. Diehl, *J. Org. Chem.*, **27**, 1260 (1962).
22. Gilman, H., F. K. Cartledge, and S. Y. Sim, *J. Organometal. Chem.*, **1**, 8 (1963).
23. Glanum, S. N., and C. R. Kraus, *J. Am. Chem. Soc.*, **72**, 5398 (1956).
24. Glockling, F., and K. A. Hooton, *J. Chem. Soc.*, 2658 (1962).
25. Glockling, F., and K. A. Hooton, *J. Chem. Soc.*, 1849 (1963).
26. Gorsich, R. D., *Organometallic Chemistry*, Reinhold, N.Y., 331.
27. Hughes, B., *Dissertation Abstr.*, **19**, 1921 (1958).
28. Kraus, C. A., and E. A. Flood, *J. Am. Chem. Soc.*, **54**, 1635 (1932).
29. Kraus, C. A., and L. S. Foster, *J. Am. Chem. Soc.*, **49**, 457 (1927).
30. Milligan, J. G., and C. A. Kraus, *J. Am. Chem. Soc.*, **72**, 5297 (1950).
31. Mook, G., H. Tai, and H. Gilman, *J. Am. Chem. Soc.*, **77**, 649 (1955).
32. Nicholson, D. A., and A. L. Allred, *Inorg. Chem.*, **4**, 1747 (1965).
33. Normant, H., T. Cuvigny, J. Normant, and B. Angelo, *Bull. Soc. Chim. France*, 3441 (1965).
34. Schumann, H., K. F. Thom, and M. Schmidt, *J. Organometal. Chem.*, **4**, 22 (1965).
35. Seyferth, D., G. Raab, and S. O. Grim, *J. Org. Chem.*, **26**, 3034 (1961).
36. Smith, F. B., and C. A. Kraus, *J. Am. Chem. Soc.*, **74**, 1418 (1952).
37. Tamborski, C., F. E. Ford, W. L. Lehn, G. J. Moore, and E. J. Soloski, *J. Org. Chem.*, **27**, 619 (1962).
38. Vyazankin, N. S., E. V. Mitrofanova, G. A. Razuvaev, and O. A. Kruglaya, *Zh. Obshch. Khim.*, **36**, 160 (1966).
39. Vyazankin, N. S., E. N. Gladyshev, S. P. Korneva, and G. A. Razuvaev, *Zh. Obshch. Khim.*, **36**, 2025 (1966).
40. Vyazankin, N. S., E. N. Gladyshev, S. P. Korneva, and G. A. Razuvaev, *Zh. Obshch. Khim.*, **36**, 952 (1966).
41. Vyazankin, N. S., E. N. Gladyshev, E. A. Arkhangel'skaya, G. A. Razuvaev, and S. P. Korneva, *Izv. Akad. Nauk SSSR*, **9**, 2081 (1968).

10-2   OTHER GERMYL–METALLIC COMPOUNDS

10-2-1   Group II organogermanium–metal compounds

10-2-1-1   Ge—Mg *Bond.* Many reactions can be rationalized in terms of transient germyl Grignard reagents, but none has been isolated to date. In 1957, Seyferth explained the formation of hexavinyldigermane in the alkylation of germanium tetrachloride with vinylic Grignard reagents by a reaction mechanism involving the formation of an unstable intermediate, trivinylgermylmagnesium bromide, which had its origin in germanous chloride (reductive alkylation process (118)):

$$GeCl_4 + 3\,RMgX \xrightarrow[\text{alkylation}]{} R_3GeCl + 3\,MgXCl$$

$$GeCl_4 + 2\,RMgX \xrightarrow[\text{reduction}]{} GeCl_2 + R\!-\!R + 2\,MgXCl$$

$$GeCl_2 + 2\,RMgX \longrightarrow GeR_2 + 2\,MgX_2$$

$$GeR_2 + RMgX \longrightarrow R_3GeMgX$$

$$R_3GeMgX + R_3GeCl \longrightarrow R_6Ge_2 + MgXCl$$

Recently Mendelsohn and coworkers showed that in the alkylation of $GeCl_4$ by certain Grignard reagents, the formation of large quantities of trialkylgermanes after hydrolysis can be explained by a similar mechanism, previous reduction in germanous dichloride followed by successive alkylation and formation of an intermediate germyl Grignard reagent (84, 85):

$$GeCl_4 + 2\,RMgX \rightarrow GeCl_2 + R(+H) + R(-H) + 2\,MgXCl$$

$$GeCl_2 + 3\,RMgX \rightarrow R_3GeMgX + 2\,MgXCl$$

$$R_3GeMgX + H_2O \rightarrow R_3GeH + MgXOH$$

Thus, in the reaction of cyclohexylmagnesium chloride on $GeCl_4$ these authors showed the formation of cyclohexene and tricyclohexylgermane. The latter compound can be observed only in the case of a stoichiometric ratio $RMgX/GeCl_4$ which is clearly above 2, the yield increasing with the molar ratio. An additional confirmation of the existence of the compound $(C_6H_{11})_3GeMgCl$ has been obtained by the study of its reactivity *in situ* with halogen derivatives such as benzyl chloride (17):

$$(C_6H_{11})_3GeMgCl + PhCH_2Cl \rightarrow MgCl_2 + (C_6H_{11})_3GeCH_2PH$$

These results correspond to those of Carrick and Glockling who have examined the reaction between $GeCl_4$ and an excess of isopropylmagnesium chloride, in the presence or absence of free magnesium. In both cases an appreciable amount (up to 36%) of triisopropylgermane was formed after hydrolysis:

$$GeCl_4 + 3\,i\text{-}PrMgCl \rightarrow i\text{-}Pr_3GeCl + 3\,MgCl_2$$

$$i\text{-}Pr_3GeCl + i\text{-}PrMgCl \rightarrow i\text{-}Pr_3GeMgCl + i\text{-}PrCl$$

$$i\text{-}Pr_3GeCl + Mg \rightarrow i\text{-}Pr_3GeMgCl$$

$$i\text{-}Pr_3GeMgCl + H_2O \rightarrow i\text{-}Pr_3GeH + MgClOH$$

Further examination of the complex mixtures provides evidence for a halogen–Grignard exchange reaction:

$$i\text{-}Pr_2GeCl_2 + i\text{-}PrMgCl \rightarrow i\text{-}Pr_2Ge(Cl)MgCl + i\text{-}PrCl$$

$$i\text{-}Pr_2Ge(Cl)MgCl + i\text{-}Pr_2GeCl_2 \rightarrow Cl(i\text{-}Pr_2)Ge\text{—}Ge(i\text{-}Pr_2)Cl + MgCl_2$$

The formation of germyl Grignard reagents seems generally to be favoured by the presence of bulky organic groups. Sterically hindered aryl Grignard reagents such as o-tolyl magnesium bromide were studied by Glockling and Hooton; in the absence of free magnesium the germyl Grignard reagent is formed by the equilibrium reaction (55, 56):

$$ArMgBr + Ar_3GeBr \rightleftarrows ArBr + Ar_3GeMgBr$$

In 1961, Gilman and Zuech suggested the formation of $Ph_3GeMgCl$ in the metalation reaction of triphenylgermane with allyl magnesium chloride in THF medium: hydrolysis leads to a cleavage product of the solvent, 4-hydroxybutyltriphenylgermane, while carbonation affords triphenyl-germane-carboxylic acid (52):

$$Ph_3GeH + H_2C\text{=}CHCH_2MgCl \rightarrow Ph_3GeMgCl + H_2C\text{=}CH\text{—}CH_3$$

$$\underset{\text{CO}_2\text{(H}_2\text{O)}}{\swarrow} \qquad \underset{\text{THF(H}_2\text{O)}}{\searrow}$$

$$Ph_3GeCOOH \qquad\qquad Ph_3Ge(CH_2)_4OH$$

It is possible to obtain, directly from magnesium and triphenylchloro-germane in hexamethylphosphotriamide solution, a reagent exhibiting all the properties of a Grignard reagent:

$$2\,Ph_3GeCl + Mg \xrightarrow[\text{HMPT}]{} Ph_6Ge_2 + MgCl_2$$

$$Ph_6Ge_2 + Mg \xrightarrow[\text{HMPT}]{} (Ph_3Ge)_2Mg$$

Hydrolysis by heavy water gives triphenyldeuteriogermane (55% yield) while allyl chloride leads to triphenylallylgermane (83).

10-2-1-2 Ge—Ba, Ge—Sr Bonds. Hexaphenyldigermane reacts with strontium or barium in the molecular ratio (1:1) in liquid ammonia at

$-40°C$, giving bis-(triphenylgermyl) metal derivatives (4):

$$Ph_3Ge\!\!-\!\!GePh_3 + M \longrightarrow (Ph_3Ge)_2M$$

$$(M = Sr, Ba)$$

After removal of the ammonia, the products are extracted with THF and recovered as THF adducts (1/1):

$(Ph_3Ge)_2Sr.THF$ 40–45°C decomp.

$(Ph_3Ge)_2Ba.THF$ 60–70°C decomp.

These adducts are very sensitive to air and light and decompose with blackening without melting at 40–60°C. On hydrolysis with moist methanol, THF is liberated, while the triphenylgermane formed slowly decomposes with evolution of hydrogen:

$$(Ph_3Ge)_2M + H_2O \xrightarrow{\text{fast}} Ph_3GeH + M(OH)_2$$

$$Ph_3GeH + H_2O \xrightarrow{\text{slow}} Ph_3GeOH + H_2$$

No reports of germanium–calcium compounds are available.

10-2-1-3 Ge—Zn, Ge—Cd *Bonds.* Preparation of bis-(triphenylgermyl)-zinc was achieved by reaction of zinc chloride with triphenylgermylsodium or -potassium, in liquid ammonia (4):

$$2\,Ph_3GeNa + ZnCl_2 \longrightarrow 2\,NaCl + (Ph_3Ge)_2Zn$$

With triphenylgermyllithium and THF as solvent, a pale yellow adduct with THF was obtained:

$$(Ph_3Ge)_2Zn\cdot0.5THF\cdot0.5\,NH_3$$

Bis-(triethylgermyl)zinc, when refluxed with EtBr, gives Et_4Ge and $ZnBr_2$. An exothermic reaction with 1,2-dibromoethane leads to C_2H_4, Et_3GeBr and $ZnBr_2$; benzoyl peroxide in C_6H_6 gives zinc benzoate and Et_3GeOBz (40). Mixed polygermanes such as $Et_3Ge\!\!-\!\!Zn(GeEt_2)_nGeEt_3$ were obtained by Vyazankin and coworkers in the reaction of diethylzinc with triethylgermane (127, 142).

Triethylgermane reacts with diethylcadmium at 80–85°C in sealed tubes, forming ethane and bis-(triethylgermyl)cadmium:

$$2\,Et_3GeH + Et_2Cd \longrightarrow 2\,C_2H_6 + (Et_3Ge)_2Cd$$

Under the action of UV light, this derivative decomposes readily with quantitative liberation of cadmium and hexaethyldigermane:

$$(Et_3Ge)_2Cd \underset{h\nu}{\rightarrow} Et_6Ge_2 + Cd$$

While bis-(triethylgermyl)mercury is stable to the action of water and alcohol, the cadmium analogue reacts with water, propanol and acetic acid, with cleavage of the Ge—Cd—Ge bond (106):

$$(Et_3Ge)_2Cd + RH \rightarrow Et_3GeH + Et_3GeR + Cd$$

$$(R = OH, C_3H_7O, CH_3COO)$$

With triethyltin hydride or triethyltin chloride, the reaction proceeds with intermediate formation of bis-(triethylstannyl)cadmium (137):

$$(Et_3Ge)_2Cd + 2\,Et_3SnH \rightarrow 2\,Et_3GeH + (Et_3Sn)_2Cd$$
$$\downarrow$$
$$Cd + Et_6Sn_2$$

With mercuric chloride, 1,2-dibromoethane and other bromides, it may be noted that the cleavage of Ge—Cd—Ge bonds gives cadmium chloride or bromide and not metallic cadmium:

$$(Et_3Ge)_2Cd + HgCl_2 \rightarrow Et_3GeCl + CdCl_2 + Hg$$

$$(Et_3Ge)_2Cd + 2\,RBr \rightarrow 2\,Et_3GeR + CdBr_2$$

$$(R = C_2H_5, C_6H_5CH_2)$$

$$(Et_3Ge)_2Cd + 2\,C_2H_4Br_2 \rightarrow 2\,Et_3GeBr + CdBr_2 + 2\,C_2H_4$$

Lithium, sodium or mercury in THF medium displace metallic cadmium from bis-(triethylgermyl)cadmium:

$$(Et_3Ge)_2Cd + Hg \rightarrow Cd + (Et_3Ge)_2Hg$$

With bromobenzene under the influence of UV light, formation of diphenyl-cadmium and triethylbromogermane is also observed (137, 138). With CCl_4 in diethyl ether at temperatures below $-75°C$, cadmium chloride is formed in quantitative yield:

$$2\,(Et_3Ge)_2Cd + 3\,CCl_4 \rightarrow$$
$$2\,Et_3GeCl + (Et_3Ge)_2CCl_2 + C_2Cl_4 + 2\,CdCl_2$$

Oxidation of bis-(triethylgermyl)cadmium gives a mixture of free cadmium and bis-(triethylgermanium) oxide:

$$2\,(Et_3Ge)_2Cd + O_2 \rightarrow 2\,Cd + 2\,(Et_3Ge)_2O$$

10-2-1-4   Ge—Hg *Bond.* Eaborn and coworkers described the preparation of bis-(trimethylgermyl)mercury, $(Me_3Ge)_2Hg$, by two methods:

(a) The reaction of trimethylbromogermane with sodium amalgam in cyclohexane (analogous to the preparation of bis-(trimethylsilyl)mercury (148).

Table 10-1
Ge—Cd, Ge—Zn *Compounds*

COMPOUND	M.P. (°C)	d_4^{20}	REFERENCES
$(Et_3Ge)_2Cd$	yellow oil	1.446	40, 137–139
$(Ph_3Ge)_2Cd$	(105 decomp.)		15
$(Ph_3Ge)_2Cd \cdot 2\,THF$	98		15
$(Ph_3Ge)_2Cd \cdot bipy$	(205 decomp.)		25
$(Ph_3Ge)_2Zn \cdot THF$	(110–120 decomp.)		4
$(Ph_3Ge)_2Zn \cdot 0.5\,THF\,0.5\,NH_3$	(110–120 decomp.)		4
$(Ph_3Ge)_2Zn \cdot bipy$	(205 decomp.)		25

(b) The reaction of trimethylgermane with diethylmercury, in the absence of solvent and with exclusion of air (39).

Vyazankin and coworkers have prepared bis-(triethylgermyl)mercury by the same process (140, 142):

$$2\,Et_3GeH + Et_2Hg \xrightarrow{120°C} 2\,C_2H_6 + (Et_3Ge)_2Hg$$

This compound is quite stable thermally, stable to water and alcohols, but sensitive to light and readily oxidized to hexaethyldigermoxane on exposure to air, liberating free mercury (135, 141):

$$(Et_3Ge)_2Hg \begin{array}{c} \xrightarrow{h\nu} Hg + Et_6Ge_2 \\ \xrightarrow{air} Hg + (Et_3Ge)_2O \end{array}$$

The photolysis in ethyl bromide, benzyl bromide or α-bromonaphthalene proceeds differently, according to the scheme:

$$(Et_3Ge)_2Hg + 2\,RBr \xrightarrow{h\nu} R_2Hg + 2\,Et_3GeBr$$

$$(R = C_2H_5,\ C_6H_5CH_2,\ \alpha C_{10}H_7)$$

some metallic mercury being formed as a side product. Bis-(triethylgermyl)-mercury reacts exothermally even at room temperature with iodine and 1,2-dibromoethane, respectively:

$$(Et_3Ge)_3Hg + I_2 \rightarrow Hg + 2\,Et_3GeI$$

$$(Et_3Ge)_2Hg + C_2H_4Br_2 \rightarrow Hg + 2\,Et_3GeBr + C_2H_4$$

With acetic acid, the reaction is as follows (80, 96, 140):

$$(Et_3Ge)_2Hg + CH_3COOH \rightarrow Hg + Et_3GeH + Et_3GeOCOCH_3$$

It was recently found that bis-(triethylgermyl)mercury reacts readily with unsaturated compounds containing C=C, C≡C, or N=N bonds: the triethylgermyl groups add to the multiple bonds with deposition of metallic mercury:

$$HC\equiv C-COOCH_3 + (Et_3Ge)_2Hg \xrightarrow[90°C]{}$$

$$Hg + Et_3GeCH=C \begin{array}{c} \diagup COOCH \\ \diagdown GeEt_3 \end{array}$$

$$H_5C_2OOC_5C\equiv C-COOC_2H_5 + (Et_3Ge)_2Hg \xrightarrow[120°C]{}$$

$$Hg + H_5C_2OOCC\!\!=\!\!=\!\!CCOOC_2H_5$$
$$\underset{GeEt_3}{|} \quad \underset{GeEt_3}{|}$$

$$Ph-N=N-Ph + (Et_3Ge)_2Hg \xrightarrow[160°C]{}$$

$$Hg + \begin{array}{c} Et_3Ge \diagdown \qquad \diagup Ph \\ N-N \\ Ph \diagup \qquad \diagdown GeEt_3 \end{array} \qquad (95\%)$$

These additions occur by way of homolytic fission giving triethylgermyl radicals as intermediates; ESR spectroscopy signals were found during the course of the reactions (75):

$$(Et_3Ge)_2Hg \rightarrow 2\,Et_3Ge\cdot \xrightarrow[RC\equiv CR']{} Et_3Ge-\underset{R}{\underset{|}{C}}=\underset{R'}{\underset{|}{C}}-GeEt_3$$

The Ge—Hg bond is easily cleaved by lithium in THF medium, and also by diacyl peroxides (35, 104):

$$(R_3Ge)_2Hg + 2\,Li \rightarrow Hg + 2\,R_3GeLi$$

$$(R_3Ge)_2Hg + (R'COO)_2 \rightarrow Hg + 2\,R_3GeOCOR'$$

Seyferth has recently reported on dichlorocarbene insertion into the Ge—Hg bond of bis-(trimethylgermyl)mercury, and bis-(tributylgermyl)-mercury:

$$(Me_3Ge)_2Hg + 2\,PhHgCCl_2Br \xrightarrow[C_6H_6]{}$$

$$2\,PhHgBr + Me_3GeCl + Me_3GeCCl=CCl_2 + Hg$$

A small amount of tetrachloroethylene and several other minor products were also identified. This process offers a convenient route to the preparation of trihalovinylgermanes (119).

A mixed germanium–silicon compound $Et_3Ge—Hg—SiEt_3$ was obtained according to the scheme (141):

$$Et_3GeH + EtHg—SiEt_3 \rightarrow C_2H_6 + Et_3Ge—Hg—SiEt_3$$
$$\downarrow$$
$$Hg + Et_3Ge—SiEt_3$$

Auto-oxidation of bis-(triethylgermyl)mercury in n-octane results in the formation of an intermediate compound, triethylgermyloxytriethylgermyl-mercury. Thermal decomposition of this compound gives free mercury and bis-triethylgermanium oxide:

$$Et_3GeHgOGeEt_3 \rightarrow Hg + Et_3GeOGeEt_3 \quad (92\%)$$

Oxidation is autocatalysed by bis-triethylgermanium oxide, this catalytic action is obviously due to the formation of a complex:

$$Et_3GeHgGeEt_3 + Et_3GeOGeEt_3 \rightleftarrows Et_3GeHgGeEt_3 \cdot Et_3GeOGeEt_3$$

Razuvaev and coworkers suppose that the primary product of oxidation of bis-(triethylgermyl)mercury is a peroxide compound (105):

$$Et_3GeHgGeEt_3 + O_2 \rightarrow Et_3GeHgOOGeEt_3$$

$$Et_3GeHgOOGeEt_3 + Et_3GeHgGeEt_3 \rightarrow 2\,Et_3GeHgOGeEt_3$$

Bis-(triethylgermyl)mercury and related compounds possess a high reactivity useful for synthetic purposes. For example, to obtain organobi-metallic compounds with Ge—Li bonds, or with covalent Ge-transition metal bonds, e.g. Ge—Fe, Ge—Mo, Ge—W, Ge—Ni, Ge—Pt, etc. (53, 59, 142). Thus, when equimolecular amounts of bis-(triethylgermyl)-mercury and $\pi[—C_5H_5Fe(Co)_2]_2$ are heated in benzene at 100°C, tri-ethyl(dicarbonyl π-cyclopentadienyl iron) germane is formed in 48 % yield:

$$(Et_3Ge)_2Hg + [\pi\text{-}C_5H_5Fe(CO)_2]_2 \rightarrow Hg + 2\,Et_3GeFe(CO)_2C_5H_5\text{-}\pi$$

Table 10-2
Ge—Hg *Compounds*

COMPOUND	M.P. (°C)	B.P. (°C/mm)	REFERENCES
$(Me_3Ge)_2Hg$	(83/0.01 sublime) 60/0.005	—	19, 39, 57,
$(Et_3Ge)_2Hg$	—	118–20/1.5	40, 41, 74, 128, 140–144
$Et_3GeHgEt$	—	80–83/1	40, 129
$(Ph_3Ge)_2Hg$	(175 decomp.)	—	74
$Et_3SiHgGeEt_3$	—	130–31/1.5	129, 141
$Et_5Si_2HgGeEt_3$	—	159–63/1	141

10-2-2   Group III organogermanium–metal compounds

10-2-2-1   Ge—B Bond. From the addition of triethylgermyllithium to a solution of triphenylboron in HMPT, a solid germanium–boron compound was obtained, i.e. lithium (triethylgermyl) triphenylborate complexed with four molecules of the solvent (9):

$$Et_3GeLi + Ph_3B \xrightarrow{HMPT} Et_3GeBPh_3Li\cdot4\,HMPT$$

Seyferth and coworkers had reported previously a similar reaction with triphenylgermyllithium:

$$Ph_3GeLi + Ph_3B \xrightarrow{DME} Li[Ph_3Ge—BPh_3]$$

This complex forms insoluble compounds with large organic cations, such as $[Me_4N][Ph_3Ge—BPh_3]$ (in methanol) and $[Ph_3PMe][Ph_3Ge—BPh_3]$. Bromine readily cleaves the Ge—B bond.

10-2-2-2   Ge—Al Bond. The alkylation reactions of germanium tetrachloride with aluminium alkyls involve the formation of Ge—Al reaction intermediates analogous to the germyl Grignard reagents (61):

$$R_3GeAlR_2 + R_3GeCl \rightarrow R_3GeGeR_3 + R_2AlCl$$

10-2-2-3   Ge—Tl Bond. By heating triethylthallium with triethylgermane Razuvaev and coworkers obtained ethane and tris-(triethylgermyl)-thallium (91 % yield) (81):

$$3\,Et_3GeH + Et_3Tl \rightarrow (Et_3Ge)_3Tl + 3\,C_2H_6$$

They isolated a dark-red involatile liquid ($d_4^{20} = 1.535$) which is readily oxidized and decomposes quantitatively at 170°C into thallium and hexaethyldigermane. Hydrolysis gives thallium, triethylgermane and hexaethyldigermoxane. Tris-(triethylgermyl)thallium reacts with mercury and with lithium in THF medium, liberating free metallic thallium:

$$2\,(Et_3Ge)_3Tl + 3\,Hg \rightarrow 2\,Tl + 3\,(Et_3Ge_2)_2Hg$$

$$(Et_3Ge)_3Tl + 3\,Li \rightarrow Tl + 3\,Et_3GeLi$$

Its reaction with 1,2-dibromoethane is exothermic and complete at 20°C:

$$(Et_3Ge)_3Tl + 2\,C_2H_4Br_2 \rightarrow 2\,C_2H_4 + TlBr + 3\,Et_3GeBr$$

10-2-3   Group IV B organogermanium–metal compounds

10-2-3-1   Ge—Si Bond. Reactions of triethylgermyllithium in THF solution with trichlorosilane, diphenyldichlorosilane, dimethylethylchlorosilane have been reported (cf. "trialkylgermylalkali metal compounds"). These reactions take place at room temperature; the yields in general are about 40%, and a small amount of hexaethyldigermane is produced as a by-product (128–130). The reaction with triethylsilane requires heating at 90°C for 13 h (48% yield) (142):

$$Et_3GeLi + Et_3SiH \rightarrow LiH + Et_3GeSiEt_3$$

With $(EtO)_3SiH$ in benzene solution, triethylgermylsilane is prepared in 77% yield:

$$3\,Et_3GeLi + (EtO)_3SiH \xrightarrow{C_6H_6} 3\,EtOLi + (Et_3Ge)_3SiH$$

Trimethylsilylgermane, Me_3SiGeH_3, is formed by a coupling reaction between GeH_3K and trimethylchlorosilane. Bulten and Noltes used coupling reactions of trialkylchlorosilanes with R_3GeK derivatives in HMPT solution, e.g. (10, 11):

$$Et_3GeK + Me_3SiCl \xrightarrow{HMPT} Et_3Ge{-}SiMe_3 \quad (60\% \text{ yield})$$

Kraus and Milligan obtained tris-(triphenylgermyl)silane from trichlorosilane and triphenylgermylsodium in diethyl ether (86):

$$SiHCl_3 + 3\,Ph_3GeNa \rightarrow 3\,NaCl + (Ph_3Ge)_3SiH$$

The hydrogen atom is replaceable by bromine in ethyl bromide medium; subsequent ammonolysis by liquid ammonia leads to tris-(triphenylgermyl)silylamine:

$$(Ph_3Ge)_3SiBr + 2\,NH_3 \rightarrow NH_4Br + (Ph_3Ge)_3SiNH_2$$

Gilman and Gerow prepared triphenylsilyltriphenylgermane from triphenylsilylpotassium and triphenylchlorogermane (46–48):

$$Ph_3GeCl + Ph_3SiK \rightarrow KCl + Ph_3Ge{-}SiPh_3$$

Like hexaphenyldigermane, this compound has high stability towards water, alcohols, acids, oxygen and iodine, and melts at 357°C without decomposition. Nevertheless, it is cleaved by sodium–potassium alloy in diethyl ether, while hexaphenyldigermane does not react under the same conditions:

$$Ph_3Si{-}GePh_3 \xrightarrow[Et_2O]{Na/K} Ph_3SiK + Ph_3GeK \xrightarrow[H_2O]{CO_2}$$

$$Ph_3GeCOOH + CO + Ph_3SiOH$$

The Ge—Si bond appears to be stronger than the Ge—Ge bond in related compounds.

Tetrakis-(p-trimethylsilylphenyl)germane has been synthesized from $GeCl_4$ and p-trimethylsilylphenyllithium (49% yield):

$$4\,Me_3SiC_6H_4Li + GeCl_4 \xrightarrow[\text{toluene}]{} 4\,LiCl + (Me_3SiC_6H_4)_4Ge$$

In 1968, Bürger and Goetze prepared tetrakis-(trimethylsilyl)germane by a Würtz-type reaction (12):

$$4\,Me_3SiCl + 8\,Li + GeCl_4 \rightarrow 8\,LiCl + (Me_3Si)_4Ge$$

In fact, 50% of the germanium used is recovered as metal by reduction of the tetrachloride.

Table 10-3
Ge—Si *Compounds*

COMPOUND	M.P. (°C)	B.P. (°C/mm)	n_D^{20}	REFERENCES
$Et_3GeSiMe_3$	—	89–91/30	1.4670	10, 11
$Et_3GeSiMe_2Et$	—	65/1	1.4729	129, 130, 142
$Et_3GeSiEt_3$	—	254.5/760	1.4860	112
		92/1.5	1.4858	142
		87/3	1.4889	129, 130, 142
$(Et_3Ge)_3SiH$	—	160–61/1	1.5425	128, 130, 142
		177/1	1.5380	129
$Ph_3GeSiEt_3$	95–96	—	—	43
	93.5	—	—	79
	97–98	—	—	46
$(PhCH_2)_3GeSiMe_3$	63.5–64.5	—	—	30
$(Et_3Ge)_2SiPh_2$	—	199–200/1.5	1.5853	129, 130, 142
$Ph_3GeSiPh_3$	357–59	—	—	48
	357–59	—	—	45
$(Ph_3Ge)_3SiH$	α 187.5–88.5	—	—	86
	β 170–71	—	—	
$(Ph_3Ge)_3SiCl$	230–31	—	—	86
$(Ph_3Ge)_3SiBr$	241.5–42.5	—	—	86
$(Ph_3Ge)_3SiOH, C_6H_6$	196.5–97.5	—	—	86
$(Ph_3Ge)_3SiNH_2$	206–6.5	—	—	86
$(Ph_3Ge)_3SiEt$	(283–84.5)	—	—	86
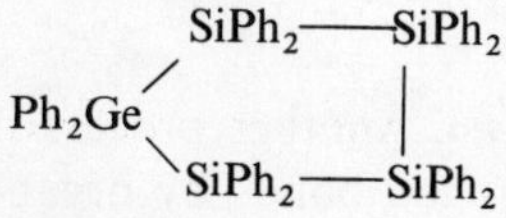 $Ph_2Ge\begin{smallmatrix} SiPh_2\!-\!SiPh_2 \\ \mid \quad\;\; \mid \\ SiPh_2\!-\!SiPh_2 \end{smallmatrix}$	316	—	—	65
$(Me_3Si)_4Ge$	295	—	—	12
$(Me_3SiC_6H_4)_4Ge$	351–54	—	—	64

10-2-3-2 Ge—Sn Bond. Some derivatives containing a Ge—Sn bond have been obtained by the interaction of triphenylgermyl–alkali metal compounds with a triorganotin halide (50, 78):

$$Ph_3GeM + R_3SnX \rightarrow MX + Ph_3Ge—SnR_3 \qquad (M = Na, Li, K)$$

Slow addition of Ph_3GeK to triphenyltin chloride gives $Ph_3Ge—SnPh_3$, but when the reverse addition is used, hexaphenyldigermane is predominantly formed by halogen–metal exchange and subsequent coupling. Triphenylgermyltriphenyltin is easily cleaved by oxygen in boiling xylene (102, 147). Unlike what occurs with the compound $Ph_3Si—GePh_3$, the reaction of $Ph_3GeSnPh_3$ with anhydrous acetic acid leads quantitatively to the hexacetate, $(AcO)_3Ge—Sn(OAc)_3$

Tetrakis-(triphenylgermyl)tin and tetrakis-(triphenylstannyl)germanium were respectively prepared from tin or germanium tetrahalide in poor yields, because of the predominant occurrence of lithium/chlorine exchange reaction (150):

$$4 Ph_3GeLi + SnCl_4 \rightarrow 4 LiCl + (Ph_3Ge)_4Sn$$

$$4 Ph_3SnLi + GeCl_4 \rightarrow 4 LiCl + (Ph_3Sn)_4Ge$$

In the aliphatic series, Vyazankin and coworkers investigated the reaction of trialkyltin hydrides with triethylgermyllithium in THF solution:

$$Et_3GeLi + R_3SnH \xrightarrow[\text{THF}]{20°C} LiH + Et_3Ge—SnR_3$$

Similarly, with methoxytriethyltin in benzene this gave $Et_3Sn—GeEt_3$ in 87% yield (128). The same compounds were obtained in about 60% yield by reacting triorganotin halides with R_3GeK derivatives in hexamethylphosphotriamide as solvent (10):

$$Et_3GeK + Me_3SnCl \xrightarrow{\text{HMPT}} KCl + Et_3Ge—SnMe_3$$

The hydrogenolytic fission of germanium–nitrogen bonds by organotin hydrides is the most important synthetic method for the preparation of various types of compounds containing the Sn—Ge bond (27):

$$R_3GeNEt_2 + Ph_3SnH \xrightarrow{\Delta} R_3Ge—SnPh_3 + Et_2NH$$

The diethylamine formed is stripped off *in vacuo*. Another process involves similar hydrogenolytic fission of tin–nitrogen bonds by organogermanium hydrides, e.g. (27, 98):

$$R_3SnNEt_2 + Ph_3GeH \xrightarrow{100°C} R_3Sn—GePh_3 + Et_2NH$$

Polymetal-derivatives may be obtained according to the reactions (95):

$$R_3Ge-NEt_2 + Bu_2SnH_2 \longrightarrow R_3GeSnBu_2H + Et_2NH$$

$$2\,Bu_3Ge-SnBu_2H \xrightarrow[Et_2NH]{}$$

$$Bu_3Ge-SnBu_2-SnBu_2-GeBu_3 + H_2$$

and also

$$2\,R_3Sn-NEt_2 + Ph_2GeH_2 \longrightarrow R_3Sn-GePh_2-SnR_3 + 2\,Et_2NH$$

$$(R = Et, Ph)$$

The reaction of triphenylgermane with a bis-(dialkylamino)tin derivative in a 1:1 ratio affords the versatile intermediate:

$$Et_2Sn(NEt_2)_2 + Ph_3GeH \rightarrow Ph_3Ge-SnEt_2-NEt_2 + Et_2NH$$

With an excess of triphenylgermane, a compound with three catenated metal atoms is obtained (27):

$$Ph_3GeSnEt_2-NEt_2 + Ph_3GeH \rightarrow Ph_3GeSnEt_2GePh_3 + Et_2NH$$

and similarly with triphenyltin hydride:

$$Ph_3Ge-SnEt_2-NEt_2 + Ph_3SnH \rightarrow Ph_3GeSnEt_2SnPh_3 + Et_2NH$$

while diphenylgermane leads to a penta-metal derivative:

$$2\,Ph_3GeSnEt_2NEt_2 + Ph_2GeH_2 \rightarrow$$

$$Ph_3Ge-SnEt_2-GePh_2-SnEt_2-GePh_3 + 2\,Et_2NH$$

Transamination reaction with N-phenyl formamide forms an essential step in the preparation of longer-chain derivatives with up to eight catenated metal atoms, according to the following scheme:

$$Ph_3GeSnEt_2NEt_2 + HCONHPh \rightarrow$$

$$Et_2NH + Ph_3GeSnEt_2N-PhCOH \xrightarrow{Ph_2SnH_2}$$

$$Ph_3GeSnEt_2SnPh_2H + HCONHPh$$

$$2\,Ph_3GeSnEt_2SnPh_2H \xrightarrow[Et_2NH]{}$$

$$H_2 + Ph_3GeSnEt_2SnPh_2SnPh_2SnEt_2GePh_3$$

The UV spectra of these polymetal-derivatives extend into the visible region, giving these compounds a yellow-red colour. Polymeric products with alternatively tin and germanium atoms have been synthesized starting from bifunctional reactants:

$$n\,Ph_2Sn(NEt_2)_2 + n\,Ph_2GeH_2 \rightarrow 2n\,Et_2NH + [-SnPh_2-GePh_2-]_n$$

Table 10-4

Ge—Sn *Compounds*

COMPOUND	M.P. (°C)	B.P. (°C/mm)	n_D^{20}	REFERENCES
$Me_3Sn-GeMe_3$	—	154–56/760	—	115
$Et_3Ge-SnMe_3$	—	90/93/14	1.5040	10
$Et_3Ge-SnEt_3$	—	105–9/1.5	1.5130	128–131, 143
		120–22/6	1.5151	
$Bu_3Ge-SnPh_3$	24–25.5	—	—	25, 27
$Ph_3Ge-SnMe_3$	89	—	—	78, 98
	88	—	—	16
$Ph_3Ge-SnEt_3$	45–52	—	—	16, 25, 27
$Ph_3Ge-SnBu_3$	23	—	—	25, 27
$Ph_3Ge-SnEt_2-GePh_3$	133–37	—	—	25, 27
$(Ph_3Ge-SnEt_2)_2O$	85–95	—	—	25, 27
$Ph_3Ge-SnEt(C{=}CPh)_2$	40	—	—	25, 27
$Ph_3Ge-SnEt_2-OPh$	125–29	—	—	25, 27
$Ph_3Ge-SnEt_2-SnPh_2-SnEt_2-GePh_3$				
	40	—	—	25, 27
$(Ph_3Sn)_4Ge$	324 decomp.	—	—	38, 150
$(Ph_3Sn)_3Sn-GePh_3$	315 decomp.	—	—	44
$EtSn(GePh_3)_3$	330	—	—	25, 27
$[-Sn(Et)(GePh_3)-O-]_n$	200 decomp.	—	—	25, 27
$EtSn(GePh_3)(SnPh_3)_2$	250	—	—	25, 27
$EtSn(GePh_3)_2OH$	260	—	—	25, 27
$EtSn(GePh_3)_2-SnPh_2-Sn(GePh_3)_2Et$				
	175	—	—	27
$Ph_3Sn-GeMe_3$	110–16	—	—	25, 27
	282–94	—	—	50, 147
$Ph_3Sn-GeBu_3$	284–86	—	—	
$Ph_3Sn-GePh_3$	288	—	—	
$Ph_3Sn-GePh_2-SnPh_3$	169–78	—	—	25, 27
$(-SnPh_2-GePh_2-)_n$	260	—	—	50, 147
$(-SnEt_2-GePh_2-)_n$	261	—		
$(Et_3Sn)_2GeBuEt$			1.5798	

10-2-3-3 Ge—Pb *Bond.* The first compound with a lead–germanium bond to be reported was the neopentane-like molecule $(Ph_3Pb)_4Ge$; it was obtained in 1964 by Willemsens and van der Kerk by reacting at low temperatures $GeCl_4$ with triphenylplumbyllithium prepared from the reaction of $PbCl_2$ with phenyllithium in diethyl ether (44, 50, 149):

$$PbCl_2 + 3\,PhLi \xrightarrow[\text{ether}]{10°C} Ph_3PbLi + 2\,LiCl$$

$$GeCl_4 + 4\,Ph_3PbLi \xrightarrow[\text{ether}]{} (Ph_3Pb)_4Ge + 4\,LiCl$$

producing yellow crystals melting about 210°C with decomposition. A small amount of hexaphenyldilead was formed as a by-product. The near ultraviolet absorption spectrum of the product was discussed.

In 1966, Neumann and Kuhlein used trialkylplumbyl diethylamines for synthesis of compounds of R_3Pb—$GePh_3$ type (hydrogenolytic fission of lead–nitrogen bonds) (97):

$$R_3Pb\text{—}NEt_2 + Ph_3GeH \rightarrow R_3Pb\text{—}GePh_3 + Et_2NH$$

$\quad$ (R = isobutyl, M.P. 86°C

$\qquad$ cyclohexyl, M.P. 174°C (decomp.)

$\qquad$ phenyl, M.P. 227°C (decomp.)

With diphenylgermane and $Ph_2Pb(NEt_2)_2$, the three-catenated-metal-atom derivative Ph_3Pb—$GePh_2$—$PbPh_3$ (decomp 154°C) was obtained.

Halogenation of compounds containing the Ge—Pb bond can be carried out under very mild conditions, so that titration with iodine is possible:

$$R_3Pb\text{—}GePh_3 + I_2 \rightarrow R_3PbI + Ph_3GeI$$

10-2-4 Group V organogermanium–metal compounds

10-2-4-1 Ge—As, Ge—Sb, Ge—Bi Bonds. Very few compounds with a Ge—As bond have been recorded in the literature. Acid hydrolysis of the mixture of the arsenide, As_2Ca_3, and germanide, CaGe, of calcium gives a mixture of gases (identified by mass spectroscopy); the most important is H_3GeAsH_2 (36, 37, 108, 109).

Tris-(trimethylgermyl) arsine, $(Me_3Ge)_2As$, may be prepared according to the two schemes (116):

$$Me_3GeCl + Me_2NLi \longrightarrow LiCl + Me_3GeNMe_2$$

$$3\,Me_3GeNMe_2 + AsH_3 \xrightarrow[\text{ether}]{15°C} 3\,Me_2NH + (Me_3Ge)_3As$$

or directly

$$3\,Me_3GeCl + AsNa_3 \xrightarrow[\text{NH}_3\text{ liq.}]{} 3\,NaCl + (Me_3Ge)_3As$$

The trichlorogermanate $Ph_4As[GeCl_3]$ is formed by adding Ph_4AsCl to HCl solutions of divalent germanium. The anion forms adducts with

Lewis acids like BF_3 or BCl_3, involving $Ge \rightarrow B$ donor–acceptor bonds (68).

The photolytic reaction of $Ph_4As[GeCl_3]$ with the Group VI hexacarbonyls $M(CO)_6$, investigated by Ruff, was found to give high yields (56–75%) of the corresponding mono-substituted products, with displacement of carbon monoxide by the $GeCl_3^-$ anion (this method relies on the high degree of Lewis basicity possessed by the $GeCl_3^-$ anion):

$$Ph_4As[GeCl_3] + M(CO)_6 \xrightarrow[\text{CH}_2\text{Cl}_2]{h\nu} CO + Ph_4As[M(CO)_5GeCl_3]$$

$$(M = Cr, Mo \text{ or } W)$$

$Fe(CO)_5$ reacts similarly, but gives lower yields (Ruff (110)):

$$Ph_4As[GeCl_3] + Fe(CO)_5 \rightarrow CO + Ph_4As[Fe(CO)_4GeCl_3]$$

The compound $Ph_3GeAsPh_2$ may be prepared from triphenylchlorogermane, sodium and triphenylarsine. It oxidizes readily to $Ph_3Ge-O-AsPh_2$ (M.P. 178°C). The latter can also be formed from

$$Ph_3Ge-O-\underset{\underset{O}{\downarrow}}{As}Ph_2$$

diphenylarsenic acid and triphenylchlorogermane in the presence of triethylamine (113):

$$Ph_3GeCl + NaAsPh_2 \xrightarrow{NH_3 \text{ liq.}} NaCl + Ph_3GeAsPh_2$$

$$\downarrow \text{air}$$

$$Ph_3GeCl + HOAsOPh_2 + Et_3N \rightarrow Et_3N, HCl + Ph_3GeOAs(O)Ph_2$$

Amberger and Salazar have synthesized tris-(trimethylgermyl)antimony by a coupling reaction between the highly reactive trilithium antimonide, Li_3Sb, and trimethylchlorogermane in liquid ammonia (3):

$$3\, Me_3GeCl + Li_3Sb \xrightarrow{NH_3 \text{ liq.}} (Me_3Ge)_3Sb + 3\, LiCl \qquad (85\% \text{ yield})$$

Tris-(trimethylgermyl)antimony is a low-melting solid (M.P. 11–13°C). On contact with air it rapidly turns to a white solid substance, probably the stibine oxide.

Tris-(trimethylgermyl)bismuth, prepared in a similar way, slowly decomposes when exposed to moist air, liberating free metal (117):

$$2\,(Me_3Ge)_3Bi + \tfrac{3}{2}O_2 \rightarrow 2\, Bi + 3\,(Me_3Ge)_2O$$

$$2\,(Me_3Ge)_3Bi + 3\, H_2O \rightarrow 2\, Bi + 3\, H_2 + 3\,(Me_3Ge)_2O$$

In 1966, Vyazankin and coworkers reported a novel synthesis of tris–(tri-ethylgermyl)antimony and bismuth (144):

$$Et_3M + 3 Et_3GeH \xrightarrow{120-130°C} 3 C_2H_6 + (Et_3Ge)_3M$$

$$(M = Sb, Bi)$$

In this reaction, the reactivity of triethylbismuth is greater than that of triethylantimony; the former gives also $Et_3GeBiEt_2$ and $(Et_3Ge)_2BiEt$ as by-products. These compounds are yellow liquids relatively stable thermally, but extremely sensitive to air and oxygen. Tris-(triethylgermyl)-bismuth decomposes at 270°C to metallic bismuth and hexaethyl digermane:

$$2 (Et_3Ge)_3Bi \rightarrow 2 Bi + 3 Et_3Ge\text{—}GeEt_3$$

The reactions with benzoyl peroxide in benzene solution are very exothermic and proceed by cleavage of the Ge—M bond:

$$2 (Et_3Ge)_3M + 3 (PhCOO)_2 \rightarrow 2 M + 6 Et_3GeOCOPh$$

Interactions with alkyl and aryl bromides, in the absence of a solvent, are nearly quantitative (113, 133, 145):

$$(Et_3Ge)_3M + 3 RBr \rightarrow MR_3 + 3 Et_3GeBr$$

$$2 (Et_3Ge)_3Bi + 3 CH_2Br\text{—}CH_2Br \rightarrow 2 Bi + 6 Et_3GeBr + 3 C_2H_6$$

Tris-(triethylgermyl)bismuth gives an exchange reaction with triethyltin hydride:

$$(Et_3Ge)_3Bi + 3 Et_3SnH \rightarrow 3 Et_3GeH + (Et_3Sn)_3Bi \rightarrow Bi + Et_6Sn_2$$

Acetic acid cleaves the Ge—Bi bond (138):

$$2(Et_3Ge)_3Bi + 3 CH_3COOH \rightarrow 2 Bi + 3 Et_3GeH + 3 Et_3GeOCOCH_3$$

In the aromatic series, Schumann and Schmidt obtained the compound $Ph_3Ge\text{—}SbPh_2$ by treating triphenyl stibine with sodium in liquid ammonia, adding ammonium bromide and finally triphenylchlorogermane (113):

$$Ph_3GeCl + Ph_2SbNa \xrightarrow{NH_3 \text{ liq.}} Ph_3Ge\text{—}SbPh_2 + NaCl$$

By exposure to air or to alkaline aqueous H_2O_2 solution this compound is oxidized to $Ph_3Ge\text{—}O\text{—}SbPh_2$ (M.P. 280°C).

10-2-4-2   Table: Group V Organogermanium–Metal Compounds

Table 10-5
Group V Organogermanium–Metal Compounds

COMPOUND	M.P. (°C)	B.P. (°C/mm)	REFER-ENCES
$(Me_3Ge)_3As$	—	67–68/0.1	111, 116
$Ph_3GeAsPh_2$	114	—	114
$(Me_3Ge)_3Sb$	11–13	—	2, 3, 111
$(Et_3Ge)_3Sb$	—	157–61/1	145, 146
$(Et_3Ge)_2SbCH_2Ph$	—	155–59/1.5	145, 146
$Ph_3GeSbPh_2$	120	—	114
$(Ph_3Ge)_3Sb$	208–9	—	131
$(Me_3Ge)_3Bi$	—	114–16/1	117
$Et_3GeBiEt_2$	—	132–36/2	80
$(Et_3Ge)_3Bi$	—	167–68/2.5	40, 80, 145, 146
$(Ph_3Ge)_3Bi$	225–27	—	131

10-3   ORGANOGERMANIUM–TRANSITION-METAL COMPLEXES

Many transition metals form σ-bonds with germanium atoms: among the most stable complexes are those involving Ge—Mo, Ge—W, Ge—Pt bonds. The heavier transition metals generally form the more stable compounds, as appears in the sequences:

$$Pt > Pd > Ni \qquad W \geqslant Mo > Cr \qquad Au > Ag > Cu$$

The effect of coordination numbers is evident in the latter series: copper and silver complexes are only stable when the metal is tetracoordinate. Reactivity and thermal stability are greatly dependent upon the nature of π-bonding ligands attached to the transition metal; the majority of the organogermanium–transition metal complexes comprise carbonyl, phosphine and cyclopentadienyl complexes. Thus, the triphenylphosphine–gold–germanium complex Ph_3Ge—$AuPPh_3$ is quite stable to heat, air and hydrolysis but the trimethylphosphine analogue is thermally unstable (60). The stabilizing effect of phosphine ligands has been largely utilized in the synthesis of a wide range of Ge–transition-metal complexes, mainly those formed from the metals to the right of the transition series.

Many ligand exchange reactions normally occur without scission of the Ge—M bond (67):

$$Ph_3GeRe(CO)_5 + LR_3 \rightarrow Ph_3GeRe(CO)_4LR_3 + CO(L = P, As, Sb)$$

The transition metal is often the centre of chemical reactivity of the complexes.

A great variety of methods is available for the synthesis of organogermanium compounds containing the Ge–transition-metal bond; two groups of classical processes are most frequently used:

(a) interaction of organogermanium halides with alkali metal salts of transition metal carbonyls, in an inert atmosphere using polar solvents (120):

$$Ph_3GeX + NaMn(CO)_5 \rightarrow NaX + Ph_3GeMn(CO)_5$$

(b) insertion reactions of GeI_2 or $GeCl_2$ into carbonyl compounds (100):

$$GeI_2 + Co_2(CO)_8 \rightarrow I_2Ge(Co(CO)_4)_2$$

Trichlorogermane acting as a source of $GeCl_2$ reacts similarly with carbonyl compounds containing a metal–halogen bond (62):

$$GeHCl_3 + ClMn(CO)_5 \rightarrow HCl + Cl_3GeMn(CO)_5$$

10-3-1 Group IV A organogermanium–metal compounds

Coutts and Wailes recently reported on the preparation of several complexes of titanium (III) and (IV) and also zirconium (IV) with metal–germanium bonds. Triphenylgermyllithium reacts at low temperature in THF medium with the cyclopentadienyl–metal derivatives $\pi(C_5H_5)_2TiCl$, $\pi(C_5H_5)_2TiCl_2$, $\pi(C_5H_5)_2ZrCl_2$ (24):

$$\pi(C_5H_5)_2TiCl_2 + Ph_3GeLi \xrightarrow[THF]{} LiCl + \pi(C_5H_5)_2ClTiGePh_3$$

The Ti (III) derivative was isolated as the solvate $(C_5H_5)_2Ti\text{—}GePh_3$. THF (M.P. 110°C, decomp.), the fourth coordination position on the titanium atom being occupied by THF. The Ti (IV) compound was isolated as dark green, diamagnetic crystals $\pi(C_5H_5)_2Ti(Cl)GePh_3$ (M.P. 194–196°C).

Further, in 1968, Kingston and Lappert obtained in the pure state the Zr (IV) compound $\pi(C_5H_5)_2ClZrGePh_3$ (orange crystals, subl. 190°C/5.10^{-4} mm, and also the Hf (IV) isologue (yellow crystals, subl. 200°C/5.10^{-4} mm) (73). The mass-spectral breakdown patterns indicate that stability decreases in the series Hf > Zr > Ti, and Sn > Ge > Si. In all cases Ge-metal bonds are easily cleaved by hydrohalic acids giving triphenylgermane:

$$\pi(C_5H_5)_2M(Cl)GePh_3 + HX \rightarrow \pi(C_5H_5)_2MClX + Ph_3GeH$$

10-3-2 Group VI A organogermanium–metal compounds

Graham and coworkers described the preparation of compounds in which germanium is covalently linked to Group VI transition metals in the form of their tricarbonyl π-cyclopentadienyl complexes. By reaction of the anion $(\pi C_5H_5(CO)_3M)^-$ with trialkyl or triarylhalogermanes in diglyme solution, several complexes have been obtained (99):

$$\pi\text{-}C_5H_5(CO)_3MNa + R_3GeCl \rightarrow NaCl + \pi\text{-}C_5H_5(CO)_3MGeR_3$$

$$(M = Cr, Mo, W)$$
$$(R = Me, Et, Pr, Ph)$$

Starting from bis-(triethylgermyl)mercury, Razuvaev has developed a new method for synthesizing the same compounds (Mo, W) (53):

$$(Et_3Ge)_2Hg + [\pi\text{-}C_5H_5(CO)_3W] \xrightarrow[\text{benzene}]{} Hg + 2\,Et_3GeW(CO)_3C_5H_5$$

Examination of the infrared spectra revealed three strong terminal CO bonds. Ultraviolet absorption data and PMR spectra were also reported (17). Their general stability increases markedly from Cr to W; the chromium complexes decompose rapidly in air, tungsten derivatives are more stable towards oxygen and heat, but phenyl derivatives seem to be more stable than alkyl derivatives. All the complexes are unreactive to hydrogen, even at high temperatures and pressures. Molybdenum complexes are completely oxidized in benzene solution to give finally bis-(trialkyl-germanium)oxides and the blue form of molybdenum trioxide:

$$C_5H_5(CO)_3MoGeR_3 \xrightarrow[\text{benzene}]{O_2} CO + CO_2 + (R_3Ge)_2O + MoO_3$$

Chemical reactions of both Ge—Mo and Ge—W compounds proceed similarly: cleavage reactions of the Ge–metal bond were observed with iodine, hydrogen chloride, ethyl bromide, 1,2-dibromoethane and mercuric chloride (18):

$$\pi\text{-}C_5H_5(CO)_3MGeR_3$$

$$\xrightarrow{I_2} C_5H_5(CO)_2I_2MI + R_3GeI + CO$$

$$\xrightarrow{HCl} C_5H_5(CO)_3MH + R_3GeCl$$

$$\xrightarrow{C_2H_4Br_2} C_5H_5(CO)_3MBr + R_3GeBr + C_2H_4$$

$$\xrightarrow{HgCl_2} C_5H_5(CO)_3MHgCl + R_3GeCl$$

In reactions with polar reagents, the polarity of the Ge–metal bond seems to be $\overset{-\delta}{M}\text{—}\overset{+\delta}{Ge}$. Magnesium bromide also cleaves the W—Ge bond in THF medium, forming the unstable germyl Grignard reagent

π-$C_5H_5(CO)_3WMgBr(THF)$ which is readily hydrolysed to the corresponding hydride π-$C_5H_5(CO)_3WH$. Ligand exchange reactions with diethylphosphine have been reported, with displacement of one carbonyl group:

$$\pi\text{-}C_5H_5(CO)_3MGeR_3 + Et_2PH \rightarrow$$

$$CO + \pi\text{-}C_5H_5(CO)_2(Et_2PH)M\text{—}GeR_3$$

Germanium tetrahalides and phenyltrichlorogermane have been found to react readily with several substituted carbonyls of molybdenum and tungsten (not hexacarbonyls), in hot diglyme solutions (76):

$$GeCl_4 + bipyMo(CO)_4 \rightarrow CO + bipy(CO)_3ClMoGeCl_3$$

$$(bipy = 2{,}2'\text{-bipyridyl } C_{10}H_8N_2)$$

These complexes are brown to red crystalline solids which exhibit unusually great stability to air and light for carbonyl derivatives. Conductivity measurements in acetone solutions establish that these compounds are non-ionic and hence may be formulated as hepta-coordinate complexes of bivalent molybdenum or tungsten. No reaction is observed with bipy-$Cr(CO)_4$, so that chromium seems unlikely to form a heptacoordinate derivative of this type.

10-3-2-1   Table: Ge—(Cr, Mo, W) *Compounds*

Table 10-6
Ge—(Cr, Mo, W) *Compounds*

COMPOUND	M.P. (°C)	B.P. (°C/mm)	REFERENCES
$Ph_3GeCr(CO)_3C_6H_5$	229–230 (decomp.)	—	53
$Me_3GeMo(CO)_3C_5H_5$	87–88	—	18
$Et_3GeMo(CO)_3C_5H_5$	26.5	148–50/1	99
$Ph_3GeMo(CO)_3C_5H_5$	250	—	18
$Ph_3GeMo(CO)_3C_5H_5$	219–22 (decomp.)	—	53
$bipy(CO)_3ClMoGeCl_3$	135 (decomp.)	—	76
$bipy(CO)_3IMoGeI_3$	164 (decomp.)	—	76
$Me_3GeW(CO)_3C_5H_5$	106–7	—	18
$Et_3GeW(CO)_3C_5H_5$	36–39	156–58/0.02	99
$Ph_3GeW(CO)_3C_5H_5$	240 (decomp.)	—	53
$bipy(CO)_3ClWGeCl_3$	210 (decomp.)	—	76
$bipy(CO)_3BrWGeBr_3$	180 (decomp.)	—	76
$bipy(CO)_3ClWGePhCl_2$	190 (decomp.)	—	76

10-3-3   Group VII A organogermanium–metal compounds

10-3-3-1   Ge—Mn Bonds. The first compound containing a germanium–transition-metal bond to be reported was the triphenylgermylmanganese pentacarbonyl prepared in 1962 by Seyferth and coworkers, by interaction of triphenylbromogermane with sodium pentacarbonylmanganate in THF medium (120):

$$Ph_3GeBr + NaMn(CO)_5 \xrightarrow{\text{THF}} NaBr + Ph_3GeMn(CO)_5$$

The infrared spectrum in the carbonyl stretching region shows a strong band at 2100 cm^{-1} and broad, strong absorption at 1995 cm^{-1} (67, 120). Chlorination leads to the highly stable trichlorogermyl complex with cleavage of the phenyl groups only:

$$Ph_3GeMn(CO)_5 + 3 Cl_2 \rightarrow Cl_3GeMn(CO)_5 + 3 PhCl$$

The derivative $Me_3GeMn(CO)_5$ shows a remarkable degree of stability to heat and to light (under UV irradiation, only a very small amount of carbon monoxide was detected (62). The reaction of trimethylgermyl-manganese pentacarbonyl with tetrafluoroethylene gives the stable insertion product $Me_3GeCF_2CF_2Mn(CO)_5$ (23).

In 1963, Massey and coworkers reported on bis-(pentacarbonyl-manganese)germane, obtained by action of monogermane upon manganese pentacarbonyl hydride in THF solution under a stream of argon (nearly quantitative) (82):

$$GeH_4 + 2 HMn(CO)_5 \xrightarrow{\text{THF}} 2 H_2 + H_2Ge[Mn(CO)_5]_2$$

The first step of the reaction seems to be reduction of monogermane to GeH_2 and subsequent insertion of GeH_2 into $Mn_2(CO)_{10}$. In 1967, Nesmeyanov and coworkers obtained the derivative $Ph_2BrGeMn(CO)_5$ in 85% yield by the reaction (89):

$$Ph_2GeBr_2 + NaMn(CO)_5 \rightarrow NaBr + Ph_2BrGeMn(CO)_5$$

Bromine cleaves at $-72°C$ one or two germanium–phenyl bonds:

$$Ph_2BrGeMn(CO)_5 \xrightarrow[-PhBr]{} PhBr_2GeMn(CO)_5 \xrightarrow[-PhBr]{} PB_3GeMn(CO)_5$$

Nevertheless, cleavage of the Ge—Mn bond by excess bromine was observed in the triphenylgermyl compound:

$$Ph_3GeMn(CO)_5 + Br_2 \rightarrow Ph_3GeBr + BrMn(CO)_5$$

Consequently, the Ge—Mn bond seems to be stabilized by the introduction of a bromine atom.

The action of phenyltrichlorogermane on excess $NaMn(CO)_5$ results in substitution of only two halogen atoms by pentacarbonyl groups:

$$PhGeCl_3 + 2\,NaMn(CO)_5 \rightarrow 2\,NaCl + PhClGe[Mn(CO)_5]_2$$

A compound containing three pentacarbonyl groups was also synthesized from trichlorogermane (87, 92):

$$HGeCl_3 + ClMn(CO)_5 \rightarrow HCl + Cl_3GeMn(CO)_5$$

$$Cl_3GeMn(CO)_5 + 2\,NaMn(CO)_5 \rightarrow 2\,NaCl + ClGe[Mn(CO)_5]_3$$

Anisimov and coworkers recently showed that in the compound $Cl_2Ge[Mn(CO)_5]_2$ the chlorine atoms are readily replaced by acetoxy or thiocyanate groups without cleavage of the Ge—Mn bond, e.g.

$$2\,CH_3COOK + Cl_2Ge[Mn(CO)_5]_2 \rightarrow$$

$$2\,KCl + (CH_3COO)_2Ge[Mn(CO)_5]_2$$

This diacetate reacts with hydrogen halides HF, HBr, HI, the Ge—Mn bond being retained:

$$(CH_3COO)_2Ge[Mn(CO)_5]_2 + 2\,HX \rightarrow$$

$$2\,CH_3COOH + X_2Ge[Mn(CO)_5]$$

The IR spectra of these crystalline complexes contain intense absorptions in the 2000–$2150\,cm^{-1}$ region corresponding to the valency vibration of CO in metal carbonyl compounds (5).

10-3-3-2 Ge—Re Bonds. In 1966, Nesmeyanov and coworkers synthesized compounds with covalent germanium–rhenium bonds and investigated their properties. Derivatives of the $R_{4-n}Ge(Re(CO)_5)_n$ type were produced by reacting organogermanium halides with the sodium salt of rhenium pentacarbonyl (88):

$$n\,NaRe(CO)_5 + R_{4-n}GeX_4 \xrightarrow{\text{THF}} n\,NaX + R_{4-n}Ge[Re(CO)_5]_n$$

$$(R = C_6H_5, X = Br, Cl, n = 1, 2)$$

The action of anhydrous HCl on the compound $Ph_2Ge[Re(CO)_5]_2$ leads to the splitting out of one $Re(CO)_5$ group with the formation of $Ph_2Ge(Cl)Re(CO)_5$. The carbon–germanium bond is more stable to halogens and hydrogen halides than the corresponding tin–carbon bond in isologous compounds. Bromine cleaves progressively the phenyl groups of the complex $Ph_3GeRe(CO)_5$:

$$Ph_3GeRe(CO)_5 \xrightarrow{Br_2} Ph_2BrGeRe(CO)_5 \xrightarrow{Br_2}$$

$$PhBr_2GeRe(CO)_5 \xrightarrow{Br_2} Br_3GeRe(CO)_5$$

Table 10-7
Ge—Mn *Compounds*

COMPOUND	M.P. (°C)	REFERENCES
$Cl_3GeMn(CO)_5$	168.5	62
$Me_3GeMn(CO)_5$	34.5	23
$Bu_3GeMn(CO)_5$	165	23, 87
$H_2Ge[Mn(CO)_5]_2$	87–88	82
$PhBr_2GeMn(CO)_5$	68.5	88, 89, 93
$Ph_2BrGeMn(CO)_5$	88.5–89	88, 89, 93
$Ph_2Ge[Mn(CO)_5]_2$	145–47	66, 88, 93
$Ph_3GeMn(CO)_5$	162–64	67, 88, 93, 120, 126
$Ph_3GeMn(CO)_4AsPh_3$	229.5	88
$Ph_3GeMn(CO)_4SbPh_3$	214.5	88
$Ph_3GeMn(CO)_4PPh_3$	249–51	88, 93
$F_2Ge[Mn(CO)_5]_2$	153–54	5
$Cl_2Ge[Mn(CO)_5]_2$	119–20	5
$Br_2Ge[Mn(CO)_5]_2$	128–29	5
$I_2Ge[Mn(CO)_5]_2$	123.5	5
$(NCS)_2Ge[Mn(CO)_5]_2$	149–50	5
$(CH_3COO)_2Ge[Mn(CO)_5]_2$	112–13	5
$Ph(Cl)Ge[Mn(CO)_5]_2$	101–1.5	89, 93
$[PhGe[Mn(CO)_5]_2]_2$	155–57	89
$[Ph_2GeMn(CO)_5]_2$	205–6	88, 93

Table 10-8
Ge—Re *Compounds*

COMPOUND	M.P. (°C)	REFERENCES
$Ph_3GeRe(CO)_5$	155	67, 126
$Ph_2Ge[Re(CO)_5]_2$	167–68	88
$Br_3GeRe(CO)_5$	198–99	88
$Ph_2ClGeRe(CO)_5$	83	88
$Ph_3GeRe(CO)_4Ph_3$	231–33	88
$Ph_3GeRe(CO)_4AsPh_3$	225–27	88
$Ph_3GeRe(CO)_4SbPh_3$	216–18	88

Under the action of triphenylphosphine, triphenylarsine or triphenyl-stibine upon the compounds $R_3GeRe(CO)_5$, one mol of carbon monoxide is exchanged for the corresponding ligand, the rate of exchange decreasing in the sequence $Ph_3P > Ph_3As > Ph_3Sb$:

$$R_3Ge-Re(CO)_5 + LPh_3 \rightarrow R_3GeRe(CO)_4LPh_3 + CO$$

$$(L = P, As, Sb)$$

Thus, the compound $Ph_3GeRe(CO)_4PPh_3$ was obtained in 92% yield (white crystals). The IR spectrum contains absorption bands in the region of 1435–1440 cm^{-1} ascribed to the bonds of P with the phenyl ring.

10-3-4   Group VIII organogermanium–metal compounds

10-3-4-1   Ge—Fe Bond. Reaction of 1,1'-ferrocenylenedisodium with triphenylbromogermane gives a significant amount of the monosubstituted product, triphenylgermylferrocene (orange crystals, M.P. 155–156°C) in addition to the expected 1,1'-bis-(triphenylgermyl)ferrocene (M.P. 244°C), hexaphenyldigermane and minor amounts of hexaphenyldigermoxane and tetraphenylgermane (70). The reaction of triphenylbromogermane with a THF solution of the derivative $NaFe(CO)_2C_5H_5$-π gives triphenylgermyl-π-cyclopentadienyliron dicarbonyl (70):

$$Ph_3GeBr + NaFe(CO)_2C_5H_5 \rightarrow NaBr + Ph_3GeFe(CO)_2C_5H_5$$

This complex is stable in the solid state (M.P. 160–161°C), the IR spectrum shows two strong absorption bands corresponding to carbonyl groups at 1940 and 1995 cm^{-1} (120). Starting from triethylbromogermane, a similar process leads to the triethyl derivative (85% yield). The same compound may be synthesized by an independent route from bis-(triethylgermyl)mercury and the dimer of dicarbonyl-π-cyclopentadienyliron, or π-cyclopentadienyl dicarbonyliron chloride, in benzene solution (53):

$$(Et_3Ge)_2Hg + [\pi\text{-}C_5H_5Fe(CO)_2]_2 \xrightarrow{100°C} Hg + 2\,Et_3GeFe(CO)_2C_5H_5$$

$$(Et_3Ge)_2Hg + ClFe(CO)_2C_5H_5 \longrightarrow$$

$$Hg + Et_3GeCl + Et_3GeFe(CO)_2C_5H_5$$

The Ge—Fe bond is readily cleaved by bromine with evolution of carbon monoxide.

Insertion reactions of GeI_2 and $GeCl_2$ into the dicarbonyl-π-cyclopentadienyliron dimer have been reported:

$$GeI_2 + [\pi\text{-}C_5H_5Fe(CO)_2]_2 \xrightarrow{benzene} I_2Ge[Fe(CO)_2C_5H_5]_2$$

Nesmeyanov and coworkers described the preparation of bis-(π-cyclopentadienyliron dicarbonyl)germanium dichloride:

$$[\pi\text{-}C_5H_5Fe(CO)_2]_2 + C_4H_8O_2,GeCl_2 \rightarrow Cl_2Ge[Fe(CO)_2C_5H_5]_2$$

dioxan complex

In the latter compound, the Ge—Fe bond length is 2.36 Å, and the Ge—Cl bond length 2.26 Å (13). The chlorine atoms bonded to germanium atom can be replaced by acetoxy, alkoxy, alkyl, phenyl, allyl and isothiocyanato groups. Trichlorogermane may be used instead of germanium dichloride, e.g.:

$$GeHCl_3 + \pi\text{-}C_5H_5Fe(CO)_2Cl \rightarrow HCl + \pi\text{-}C_5H_5Fe(CO)_2GeCl_3$$

The IR spectra of the two series of compounds $(R_3Ge)_2Fe(CO)_4$ and $[R_2GeFe(CO)_4]_2$ (R = Cl, OCH$_3$, Ph, Me, Et) have been reported by Kahn and Bigorgne (69). Recently, Zimmer and Huber investigated the structure of the derivative $[Et_2GeFe(CO)_4]_2$ by X-ray crystallography (153).

In 1968, Brooks and coworkers showed the general ability of R_2Ge groups (R = Me, Ph) to bridge transition metals: insertion of diphenyl-germane into triirondodecacarbonyl affords the cyclic compound (6):

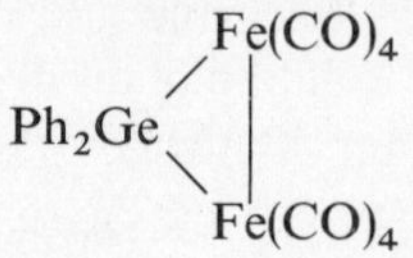

$$Ph_2Ge \overset{\displaystyle Fe(CO)_4}{\underset{\displaystyle Fe(CO)_4}{\diagup \; | \; \diagdown}}$$

Table 10-9
Ge—Fe Compounds

COMPOUND (X = $\pi C_5H_5(CO)_2Fe$)	M.P. (°C)	COLOUR	REFERENCES
X_2GeCl_2	192–97 (decomp.)	Orange	13, 14, 42, 90 91
$X_2Ge(OCOCH_3)_2$	160–62	Yellow	91
X_2GeBr_2	198–201 (decomp.)	Orange red	91
X_2GeI_2	231–33 (decomp.)	Red	91
X_2GeF_2	196–98 (decomp.)	Yellow	91
$X_2Ge(OCH_3)_2$	122–23	Yellow	91
$X_2Ge(SC_2H_5)_2$	139–41	Orange	91
$X_2Ge(C_5H_5)_2$	139–41 (decomp.)	Reddish brown	91
$X_2Ge(NCS)_2$	204–7 (decomp.)	Yellow	91
X_2GeH_2	110 (decomp.)	Yellow	91
$X_2Ge(CH_3)_2$	129–30	Yellow	91
$X_2Ge(C_2H_5)_2$	122–24	Yellow	91
$X_2Ge(C_4H_9)_2$	117–18	Yellow	91
$X_2Ge(C_3H_5)_2$	98–99	Orange red	91
$XGe(C_6H_5)_3$	160–61	Yellow	120
$XGeCH_3Cl_2$	74–75	Yellow	42
$XGeCl_3$	139–40	Yellow	42

The IR and NMR spectra of the latter derivative were consistent with a molecule derived from diironenneacarbonyl by replacing the bridging carbonyls with dimethylgermanium groups.

10-3-4-2   Ge—Co Bond. The first reported compound containing a germanium–cobalt bond was obtained in 1966 by an insertion reaction of germanium diiodide into dicobalt octacarbonyl in THF medium (100):

$$GeI_2 + Co_2(CO)_8 \rightarrow I_2Ge[Co(CO)_4]_2$$

The complex is rapidly oxidized by air. Other derivatives were obtained by a direct reaction of germanium tetrahalides with dicobalt octacarbonyl, in the correct ratio:

$$4\,GeX_4 + 3\,Co_2(CO)_8 \xrightarrow{THF} 2\,CoX_4 + 8\,CO + 4\,X_3GeCo(CO)_4$$

$$2\,GeX_4 + 3\,Co_2(CO)_8 \longrightarrow 2\,CoX_2 + 8\,CO + 2\,X_2Ge[Co(CO)_4]_2$$

The conventional halide displacement method was also used (101):

$$PhGeCl_3 + NaCo(CO)_4 \xrightarrow{THF} NaCl + Cl_2PhGe—Co(CO)_4$$

$$Ph_3GeBr + NaCo(CO)_4 \longrightarrow NaBr + Ph_3Ge—Co(CO)_4$$

Starting from methyltriiodogermane, a bis-cobalt derivative was obtained:

$$MeGeI_3 + 2\,NaCo(CO)_4 \rightarrow 2\,NaI + IMeGe[Co(CO)_4]_2$$

Kahn and Bigorgne have studied the v (CO) frequencies in the compounds (69) $R_3GeCo(CO)_4$ (R = alkyl, phenyl, Cl, OCH_3). Recently, cyclopentadienylcobaltdicarbonyl has been found to react in hot benzene with germanium tetrahalides, giving two types of compounds:

I π-C_5H_5(CO)XCoGeX$_3$ and   II π-C_5H_5(CO)Co(GeX$_3$)$_2$

These oxidative elimination reactions can be written:

$$GeX_4 + \pi\text{-}C_5H_5(CO)_2Co \rightarrow CO + C_5H_5(CO)XCoGeX_3$$

$$2\,GeX_4 + 2\,[\pi\text{-}C_5H_5(CO)_2Co] \rightarrow$$
$$2\,CO + CoX_2 + (C_5H_5) + \pi\text{-}C_5H_5(CO)Co(GeX_3)_2$$

With the tetraiodide, only the first type is found, while germanium tetrachloride forms only the second type.

The course of these reactions is quite similar to that established for iron pentacarbonyl. King and Stone have shown a strong chemical similarity between $Fe(CO)_5$ and π-$C_5H_5Co(CO)_2$ (71, 72). Compounds of type I form black air-stable crystals, soluble in the usual solvents, while compounds having two Ge—Co bonds are yellow to orange brown;

no melting points are listed because most of them decompose gradually over the range 10–40°C. Like germanium tetrahalides, organogermanium trihalides react in boiling benzene with cyclopentadienyldicarbonyl cobalt to form stable, crystalline compounds which exhibit two terminal carbonyl stretching bands in cyclohexane solution: this fact is attributed to the presence of more than one conformer in the molecule (77).

Table 10-10
Ge—Co *Compounds* (*100*)

COMPOUND	M.P. (°C)	COLOUR
$Cl_3GeCo(CO)_4$	71–73	Yellow
$Cl_2(Ph)GeCo(CO)_4$	59–61	Pale yellow
$I_2(Me)GeCo(CO)_4$	70–71.5	Yellow
$Cl(Me)_2GeCo(CO)_4$	51–54	Pale yellow
$Cl(Ph_2GeCo(CO)_4$	59–61.5	Yellow
$Ph_3GeCo(CO)_4$	130	White
$Cl_2Ge[Co(CO)_4]_2$	102–4	Red
$I_2Ge[Co(CO)_4]_2$	95 (decomp.)	Red
$IMeGe[Co(CO)_4]_2$	66–68	Orange
$Me_2Ge[Co(CO)_4]_2$	39–41	Yellow

10-3-4-3   Ge—Ni, Ge—Pd, Ge—Pt *Bonds*. The stability of σ-bonded germanium derivatives of Group VIII metals largely increases in the series Ni < Pd < Pt. Very few compounds with a Ge—Ni bond have been reported. There is some evidence for the formulation of a Ni—Ge complex in the reaction of trichlorogermane with nickel (II) chloride in nitromethane or acetone medium, a deep purple colour being observed (152).

By heating a benzene solution of bis-(triethylgermyl)mercury and π-C_5H_5—$Ni(CO)_2$ in equimolecular amounts, Razuvaev and coworkers recently obtained triethyl(dicarbonyl-π-cyclopentadienylnickel)germanium with a yield of 74% (orange liquid oxidizing in the air, B.P. 79°C/0.2, n_D^{20} 1.5932) (53):

$$(Et_3Ge)_2Hg + \pi\text{-}C_5H_5Ni(CO)_2 \rightarrow Hg + Et_3GeNi(CO)_2C_5H_5$$

Brooks and Glockling prepared triphenylgermylpalladium complexes by reacting triphenylgermyllithium with bis-(triethylphosphine) palladium dichloride in 1,2-dimethoxyethane as solvent, at low temperature (the same reaction is negative with the nickel isologue) (8):

$$(Et_3P)_2PdCl_2 + 2\ Ph_3GeLi \xrightarrow{-20°C} (Et_3P)_2Pd(GePh_3)_2 + 2\ LiCl$$

In the solid state this complex is relatively stable to 97°C, but in toluene solution extensive decomposition occurs at room temperature, leaving a Pd residue:

$$(Et_3P)_2Pd(GePh_3)_2 \longrightarrow$$

$$Ph_6Ge_2 + Ph_4Ge + Et_3P + C_6H_6 + C_2H_4 + H_2 + Pd$$

The formation of ethylene must be attributed to cleavage of C—P bonds. The Ge—Pd bond is easily cleaved by acids in ether:

$$(Et_3P)_2Pd(GePh_3)_2 + 2\,HX \longrightarrow 2\,Ph_3GeH + (Et_3P)_2PdX_2$$

With 1,2-dibromoethane the reaction follows the expected course already established for Ge—Au or Ge—Pt compounds (55):

$$(Et_3P)_2Pd(GePh_3)_2 + 2\,C_2H_4Br_2 \longrightarrow$$

$$2\,C_2H_4 + 2\,Ph_3GeBr + (Et_3P)_2PdBr_2$$

The complex reacts slowly with hydrogen, hydrogenolysis taking place at 100 atm over 1 week at room temperature. In this reaction only one Pd—Ge bond is cleaved:

$$(Et_3P)_2Pd(GePh_3)_2 + H_2 \longrightarrow Ph_3GeH + (Et_3P)_2Pd(H)GePh_3$$

With KCN, replacement of triethylphosphine by cyanide proceeds smoothly (7):

$$(Et_3P)_2Pd(GePh_3)_2 + 2\,KCN \longrightarrow 2\,Et_3P + K_2[(CN)_2Pd(GePh_3)_2]$$

The high stability of platinum–carbon compounds is well known, and comparison of Pt—Ge complexes with their Pt—C isologues shows many similarities of thermal stability and general reactivity. Complexes of the type $(R_3M)_2Pt(GeR_3)_nX_{2-n}$ where M = As, Sb, P, R = alkyl or aryl, $n = 1$ or 2, are generally prepared from platinum halides and alkali salts of (GeR_3) anions. Cross and Glockling have reported the preparation in high yields of bis-(triphenylgermyl)platinum complexes by reaction of triphenylgermyllithium with the trialkylphosphine platinum dichlorides:

$$(R_3P)_2PtCl_2 + 2\,Ph_3GeLi \longrightarrow 2\,LiCl + (R_3P)_2Pt(GePh_3)_2$$

cis or *trans*   *cis* or *trans*

(R = Et, Pr)

The diiodides react partly by halogen–metal exchange (29, 33):

$$(R_3P)_2PtI_2 + Ph_3GeLi \longrightarrow Ph_3GeI + (R_3P)_2Pr(I)Li$$
$$(R_3P)_2Pt(I)GePh_3$$
$$(R_3P)_2Pt(GePh_3)_2$$

For the preparation of trimethylgermylplatinum complexes, Glockling and Hooton found that bis-(trimethylgermyl)mercury prepared from sodium amalgam and trimethylbromogermane is a convenient source of Me_3Ge groups:

$$(Me_3Ge)_2Hg + cis\text{-}(Et_3P)_2PtCl_2 \longrightarrow$$
$$trans\text{-}(Et_3P)_2Pt(Cl)GeMe_3 + Me_3GeCl + Hg$$

This reaction is nearly quantitative in boiling benzene (59). The complexes typified by $(R_3P)_2Pt(GePh_2)_2$ are yellow (*trans*) or white (*cis*) crystalline solids, stable to hydrolysis, decomposing partially above 150°C by a radical process with cleavage of Ge—Ph, Ge—Pt and P—R bonds. They undergo a great variety of reactions under very mild conditions, which result in cleavage of the Ge—Pt bonds. One of the most striking reactions is the hydrogenolysis of only one Ge—Pt bond by molecular hydrogen at room temperature and atmospheric pressure:

$$(R_3P)_2Pt(GePh_3)_2 + H_2 \longrightarrow Ph_3GeH + (R_3P)_2Pt(H)GePh_3$$

The reaction is a reversible process requiring a low activation energy and probably proceeds by the formation of an octahedral hexacoordinate complex:

$$(R_3P)_2Pt(GePh_3)_2 + H_2 \longrightarrow (R_3P)_2Pt(H_2)(GePh_3)_2$$
$$\downarrow$$
$$Ph_3GeH + (R_3P)_2Pt(H)GePh_3$$

Cis-platinum complexes are less readily hydrogenated than the *trans* derivatives. Trimethylgermylplatinum compounds are even more sensitive to hydrogenolysis. Lithium aluminium hydride reduction results in cleavage of both Ge—Pt bonds. In diethyl ether or benzene, anhydrous hydrogen chloride readily cleaves the Ge—Pt bonds (57):

$$(Et_3P)_2Pt(Cl)GeMe_3 + HCl \longrightarrow Me_3GeCl + (Et_3P)_2Pt(Cl)H$$

$$(Et_3P)_2Pt(GePh_3)_2 + 2\,HCl \longrightarrow 2\,Ph_3GeH + (Et_3P)_2PtCl_2$$

Like the hydrogenolysis reaction, these hydrogen chloride cleavage reactions appear to involve an octahedral addition complex as intermediate, according to the following scheme:

$$(R_3P)_2Pt(GePh_3)_2 + HCl \xrightarrow[\text{benzene}]{} (R_3P)_2Pt(HCl)(GePh_3)_2$$

$$\downarrow$$

$$Ph_3GeCl + (R_3P)_2Pt(H)GePh_3$$

$$\downarrow HCl$$

$$Ph_3GeH + trans\text{-}(R_3P)_2PtHCl$$

Iodine reacts at room temperature:

$$(R_3P)_2Pt(GePh_3)_2 + 2\,I_2 \rightarrow (R_3P)_2PtI_2 + 2\,Ph_3GeI$$

Carbon tetrachloride reacts similarly:

$$(R_3P)_2Pt(GePh_3)_2 + CCl_4 \rightarrow 2\,Ph_3GeCl + (R_3P)_2PtCl_2$$

Methyl iodide in a sealed tube at 110°C leads to Ph_3GeI and also Ph_3GeMe. The reaction with 1,2-dibromoethane is essentially quantitative at room temperature:

$$(R_3P)_2Pt(GePh_3)_2 + 2\,C_2H_4Br_2 \rightarrow$$

$$trans\text{-}(R_3P)_2PtBr_2 + 2\,Ph_3GeBr + 2\,C_2H_4$$

Magnesium iodide gives a complex reaction involving the formation of platinum–Grignard reagents (31, 33):

$$(R_3P)_2Pt(GePh_3)_2 + MgI_2 \rightarrow (R_3P)_2Pt(I)GePh_3 + Ph_3GeIMg$$

$$\downarrow MgI_2 \qquad\qquad \downarrow H_2O$$

$$(R_3P)_2PtI_2 \qquad\qquad Ph_3GeH$$

Some nucleophilic agents such as phenyllithium slowly cleave the Pt—Ge bonds:

$$(R_3P)_2Pt(GePh_3)_2 + 2\,PhLi \rightarrow (R_3P)_2PtPh_2 + 2\,Ph_3GeLi$$

In contrast, ethanolic potassium hydroxide does not react. Several ligand exchange reactions may be carried out without cleavage of the germanium–platinum bond, e.g.:

$$(Et_3P)_2Pt(GePh_3)_2 + 2\,KCN \rightarrow K_2[(CN)_2Pt(GePh_3)_2] + 2\,Et_3P$$

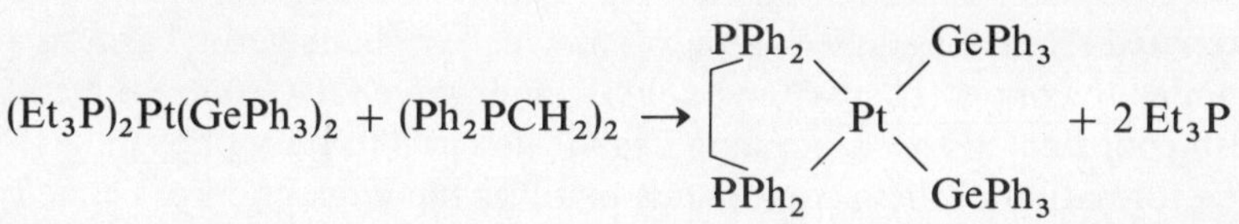

$$(Et_3P)_2Pt(GePh_3)_2 + (Ph_2PCH_2)_2 \rightarrow \left[\begin{array}{c} PPh_2 \\ \\ PPh_2 \end{array} \underset{\diagup \diagdown}{\overset{\diagdown \diagup}{Pt}} \begin{array}{c} GePh_3 \\ \\ GePh_3 \end{array}\right] + 2\,Et_3P$$

Hexacoordinate trichlorogermylplatinum complexes were obtained from trichlorogermane and potassium tetrachloroplatinate in the presence of triphenylphosphine:

$$PtCl_4H_2 + HGeCl_3 \xrightarrow[Ph_3P]{} (Ph_3P)_2Pt(Cl)_2(GeCl_3)_2$$

The complex anion $Cl_2Pt(GeCl_3)_2^{--}$ has been prepared in a red form and a yellow form, like the tin-isologues reported by Wilkinson (152). All of the platinum complexes containing a $GeCl_3$ group have a broad infrared bond at about $360–380 \text{ cm}^{-1}$ (151).

Table 10-11
Ge—Pt *Compounds*

COMPOUND	M.P. (°C)	REFERENCES
$(Et_3P)_2Pt(GePh_3)_2$	155 (decomp.)	30, 31
$(Et_3P)_2Pt(OH)GePh_3$	153–56 (decomp.)	31
$(Et_3P)_2Pt(OMeGePh_3$	172–80 (decomp.)	31
$(Et_3P)_2Pt(OEt)GePh_3$	160–70 (decomp.)	31
$(Et_3P)_2Pt(O\text{-}i\text{-}Pr)GePh_3$	162–72 (decomp.)	31
$(Et_3P)_2Pt(H)GePh_3$	150 (decomp.)	31
trans-$(Et_3P)_2Pt(Br)GeMe_3$	53–54	34, 59
trans-$(Et_3P)_2Pt(I)GeMe_3$	70–71	59
cis-$(Et_3P)_2Pt(CN)GeMe_3$	102–3	59
trans-$(Et_3P)_2Pt(SCN)GeMe_3$	71–72	34, 59
cis-$(Et_3P)_2Pt(Ph)GeMe_3$	109–11	34, 59
$(Et_3P)_2Pt(Ph_3Ge)GeMe_3 \cdot xEt_2O$	98–100	59
$(Pr_3P)_2Pt(GePh_3)_2$	120 (decomp.)	34, 59
$(Pr_3P)_2Pt(I)GePh_3$	148–49	34, 59
$(Ph_2PC_2H_4PPh_2)Pt(GePh_3)_2$	260–80 (decomp.)	20
$(Me_2PPh)_2Pt(Cl)GeMePh_2$	156–58	21
$(Me_2PPh)_2Pt(Cl)GePh_3$	170–72	22
$(Ph_3P)_2Pt(Cl)GeCl_3$	253–54	151
$(Ph_3P)_2Pt(Cl_2)(GeCl_3)_2$	245–49	151
$(Ph_3P)_2Pt(Cl_3)(GeCl_3)$	240	151
$(Me_4N)_2Pt(Cl_2)(GeCl_3)_2$	207–10	151

10-3-5   Group I B organogermanium–metal compounds

10-3-5-1   Ge—Cu, Ge—Ag, Ge—Au *Bonds*. Glockling and Hooton have prepared relatively stable complexes in which triphenylgermyl groups are bonded to copper (I), silver and gold (I), and where the Group I B metal is also coordinated to a π-bonding ligand such as a tertiary phosphine (55). The formation of these compounds involves the interaction of triphenyl-germyllithium and a tertiary phosphine–metal halide complex in an inert solvent:

$$Ph_3GeLi + (R_3P)_nMX \rightarrow LiX + Ph_3Ge\text{—}M(PR_3)_n$$

$$(M = Cu, Ag, Au, n = 1 \text{ or } 3)$$

The stability of these complexes is greatest for gold, particularly for the gold–triphenylphosphine complex Ph_3Ge—$AuPPh_3$, which is quite stable to air and water. The copper and silver complexes are susceptible to hydrolysis and oxidation at room temperature, and only isolable with three-coordinated phosphines $(Ph_3P)_3M$—$GePh_3$. The Ge–metal bond is quantitatively cleaved by 1,2-dibromethane, with simultaneous formation of ethylene, at room temperature:

$$Ph_3Ge\text{—}MPR_3 + C_2H_4Br_2 \rightarrow C_2H_4 + Ph_3GeBr + R_3PMBr$$

Phenyllithium also cleaves the Ge—Au bond in triphenylgermyltriphenylphosphine–gold:

$$Ph_3Ge\text{—}AuPPh_3 + PhLi \rightarrow Ph_3GeLi + PhAuPPh_3$$

This reaction is followed by rapid addition of triphenylgermyllithium to the original compound with displacement of the triphenylphosphine:

$$Ph_3Ge\text{—}AuPPh_3 + Ph_3GeLi \xrightarrow[Et_2O]{} Ph_3P + Li(Ph_3Ge)_2Au, 4\,Et_2O$$

The last compound provides the only example so far of a stable germanium transition metal complex without a strongly π-bonding ligand. The effect of different ligands has been examined for the series: the triphenylphosphine–gold complex Ph_3Ge—$AuPPh_3$ is more thermally stable and less reactive towards air and water than the corresponding trimethylphosphine complex $Ph_3GeAuPMe_3$ (32). In the reactions with hydrogen chloride and methyl iodide, the germanium–gold bond shows the polarity $\overset{+\delta}{Au}$—$\overset{-\delta}{Ge}$:

$$Ph_3Ge\text{—}AuPPh_3 + HCl \rightarrow Ph_3PAuCl + Ph_3GeH$$

$$Ph_3Ge\text{—}AuPPh_3 + MeI \rightarrow$$

$$Ph_3GeMe + Ph_3PAuI + (CH_4 + C_2H_6 + Au)$$

Cleavage by mercury chloride proceeds through a germylmercury intermediate:

$$Ph_3Ge\text{—}AuPPh_3 + HgCl_2 \rightarrow Ph_3GeHgCl + Ph_3PAuCl \rightarrow$$

$$Hg + Ph_3GeCl$$

Stannic chloride forms a gold–tin complex:

$$Ph_3Ge\text{—}AuPPh_3 + SnCl_4 \rightarrow Ph_3GeCl + Ph_3PAuSnCl_3$$

(orange crystals).

The cleavage of the germanium–gold bond by magnesium bromide is slow and reversible involving the formation of a germyl Grignard reagent:

$$Ph_3Ge\!\!-\!\!AuPPh_3 + MgBr_2 \rightleftarrows Ph_3GeMgBr + Ph_3PAuBr$$
$$\downarrow H_2O$$
$$Ph_3GeH$$

The trimethylgermyl complex was isolated from the reaction between triphenylphosphine-gold (I) chloride and bis-(trimethylgermyl)mercury at room temperature:

$$(Me_3Ge)_2Hg + Ph_3PAuCl \xrightarrow[15°C]{} Me_3Ge\!\!-\!\!AuPPh_3 + Me_3GeCl + Hg$$

Stable to air in the solid state, this complex decomposes rapidly in benzene solution (60).

Table 10-12
Ge(Cu, Ag, Au) *Compounds*

COMPOUND	M.P. (°C)	REFERENCES
$Ph_3GeCuPPh_3$	168 (decomp.)	55
$Ph_3GeCu(PPh_3)$	130 (decomp.)	55
$Ph_3GeAg(PPh_3)_3 2\ MeOCH_2CH_2OMe$	167–70 (decomp.)	55
$Ph_3GeAuMe_3$	125–30 (decomp.)	32, 55
$Ph_3GeAuPEt_3$	159 (decomp.)	60
$Ph_3GeAuPPh_3C_6H_6$	185 (decomp.)	55
$[(Ph_3Ge)_2Au]NEt_4$	195–200 (decomp.)	32, 55
$Ph_3GeAuPPh_3$	126 (decomp.)	60

References (10-2 and 10-3)

1. Amberger, E., and E. Muehlhofer, *J. Organometal. Chem.*, **12**, 55 (1968).
2. Amberger, E., and R. W. Salazar, *Int. Symp. Organosilicon Compounds*, Prague, 31, 1965.
3. Amberger, E., and R. W. Salazar, *J. Organometal. Chem.*, **8**, III (1967).
4. Amberger, E., W. Stoeger, and H. R. Honigschmid-Grossich, *Angew. Chem.*, **78**, 549 (1966).
5. Anisimov, K. N., N. E. Kolobova, A. B. Antonova, and B. V. Bezborodov, *Izv. Akad. Nauk SSSR*, 202 (1968).
6. Brooks, E. H., M. Elder, A. G. Graham, and D. Hall, *J. Am. Chem. Soc.*, **90**, 3587 (1968).
7. Brooks, E. H., and F. Glockling, *Chem. Commun.*, 510 (1965).
8. Brooks, E. H., and F. Glockling, *J. Chem. Soc.*, 1241 (1966).
9. Bulten, E. J., *Chemistry of Alkylpolygermanes*, Thesis, Utrecht, 51, 1969.
10. Bulten, E. J., and J. G. Noltes, *Tetrahedron Letters*, 4389 (1966).
11. Bulten, E. J., and J. G. Noltes, *Tetrahedron Letters*, 1443 (1967).

12. Bünger, H., and V. Goetze, *Angew. Chem. Int. Ed.*, **7**, 212 (1968).
13. Bush, M. A., and P. Woodward, *Chem. Commun.*, 166 (1967).
14. Bush, M. A., and P. Woodward, *J. Chem. Soc.*, 1833 (1967).
15. Bychkov, V. T., O. V. Linzana, N. S. Vyazankin, and G. A. Razuvaev, *Izv. Akad. Nauk SSSR*, in press.
16. Carey, N. A. D., and H. C. Clark, *Chem. Commun.*, 292 (1967).
17. Carrick, A., and F. Glockling, *J. Chem. Soc. C*, 623 (1966).
18. Carrick, A., and F. Glockling, *J. Chem. Soc. A*, 913 (1968).
19. Chambers, D. B., and F. Glockling, *J. Chem. Soc. A*, 735 (1968).
20. Chatt, J., *Chem. Rev.*, **48**, 7 (1951).
21. Chatt, J., C. Eaborn, and S. Ibekwe, *Chem. Commun.*, 700 (1966).
22. Chatt, J., C. Eaborn, S. Ibekwe, and P. N. Kapoor, *Chem. Commun.*, 869 (1967).
23. Clark, H. C., J. D. Cotton, and J. H. Tsai, *Inorg. Chem.*, **5**, 1582 (1966).
24. Coutts, R. S. P., and P. C. Wailes, *Chem. Commun.*, 260 (1968).
25. Creemers, H. M. J. C., *Hydrostannolysis*, Schotanus and Jens, Utrecht, 1967.
26. Creemers, H. M. J. C., *Rec. Trav. Chim.*, **84**, 1589 (1965).
27. Creemers, H. M. J. C., and J. G. Noltes, *J. Organometal. Chem.*, **7**, 237 (1967).
28. Cross, R. J., *Organometal. Chem. Rev.*, 99 (1967).
29. Cross, R. J., and F. Glockling, *Proc. Chem. Soc.*, 143 (1964).
30. Cross, R. J., and F. Glockling, *J. Chem. Soc.*, 4125 (1964).
31. Cross, R. J., and F. Glockling, *J. Chem. Soc.*, 5422 (1965).
32. Cross, R. J., and F. Glockling, *J. Organometal. Chem.*, **3**, 146 (1965).
33. Cross, R. J., and F. Glockling, *J. Organometal. Chem.*, **3**, 253 (1965).
34. Cross, R. J., F. Glockling, and K. A. Hooton, *Abstr. 3rd Int. Symp. Organometal. Chem. München*, 272 (1967).
35. Davies, A. G., *Angew. Chem. Int. Ed.*, **80**, 125 (1968).
36. Drake, J. E., and W. L. Jolly, *Chem. Ind. (London)*, **21**, 1470 (1962).
37. Drake, J. E., and W. L. Jolly, *J. Chem. Phys.*, **38**, 1033 (1963).
38. Drenth, W., M. J. Janssen, G. J. M. van der Kerk, and J. A. Wiegenthart, *J. Organometal. Chem.*, **2**, 265 (1964).
39. Eaborn, C., W. A. Dutton, F. Glockling, and K. A. Hooton, *J. Organometal. Chem.*, **9**, 175 (1967).
40. Egorochkin, A. N., S. Y. Khorshev, N. S. Vyazankin, E. N. Gladyshev, V. T. Bychkov, and O. A. Kruglaya, *Zh. Obshch. Khim.*, **38**, 276 (1968).
41. Egorochkin, A. N., N. S. Vyazankin, G. A. Razuvaev, O. A. Kruglaya, and M. N. Bochkarev, *Dokl. Akad. Nauk SSSR*, **170**, 333 (1966).
42. Flitcroft, N., P. J. Harbourne, I. Paul, P. M. Tucker, and F. G. A. Stone, *J. Chem. Soc.*, 1130 (1966).
43. George, M. V., D. J. Peterson, and H. Gilman, *J. Am. Chem. Soc.*, **82**, 403 (1960).
44. Gilman, H., and F. K. Cartledge, *J. Organometal. Chem.*, **5**, 48 (1966).
45. Gilman, H., and C. W. Gerow, *J. Am. Chem. Soc.*, **77**, 4675 (1955).
46. Gilman, H., and C. W. Gerow, *J. Am. Chem. Soc.*, **77**, 5509 (1955).
47. Gilman, H., and C. W. Gerow, *J. Am. Chem. Soc.*, **77**, 5740 (1955).
48. Gilman, H., and C. W. Gerow, *J. Am. Chem. Soc.*, **78**, 5485 (1956).
49. Gilman, H., and C. W. Gerow, *J. Am. Chem. Soc.*, **78**, 5823 (1956).
50. Gilman, H., and C. W. Gerow, *J. Org. Chem.*, **22**, 334 (1957).
51. Gilman, H., L. Summers, and R. W. Leepers, *J. Org. Chem.*, **17**, 630 (1952).
52. Gilman, H., and E. A. Zuech, *J. Org. Chem.*, **26**, 3035 (1961).
53. Gladyshev, E. N., V. I. Ermolaev, Y. A. Sorokin, O. A. Kruglaya, N. S. Vyazankin, and G. A. Razuvaev, *Dokl. Akad. Nauk SSSR*, **179**, 1333 (1968).

54. Gladyshev, E. N., V. I. Ermolaev, N. S. Vyazankin, and Y. A. Zorokin, *Zh. Obshch. Khim.*, **38**, 662 (1968).
55. Glockling, F., and K. A. Hooton, *J. Chem. Soc.*, 2658 (1962).
56. Glockling, F., and K. A. Hooton, *J. Chem. Soc.*, 3509 (1962).
57. Glockling, F., and K. A. Hooton, *Chem. Commun.*, 218 (1966).
58. Glockling, F., and K. A. Hooton, *Inorg. Synth.*, **8**, 31 (1966).
59. Glockling, F., and K. A. Hooton, *J. Chem. Soc.*, 1066 (1967).
60. Glockling, F., and M. D. Wilbey, *J. Chem. Soc. A*, 2168 (1968).
61. Glockling, F., and J. R. C. Light, *J. Chem. Soc.*, 623 (1967).
62. Gorsich, R. D., *J. Am. Chem. Soc.*, **84**, 2486 (1962).
63. Harbourne, A., and F. G. A. Stone, *Pure Appl. Chem.*, **10**, 37 (1967).
64. Heen, R. H., and H. Gilman, *J. Org. Chem.*, **22**, 564 (1957).
65. Hengge, E., and U. Brychcy, *Monatsh. Chem.*, **97**, 1309 (1966).
66. Jetz, W., and A. G. Graham, *J. Am. Chem. Soc.*, **89**, 2773 (1967).
67. Jetz, W., P. B. Simons, J. A. J. Thompson, and W. A. G. Graham, *Inorg. Chem.*, **5**, 2217 (1966).
68. Johnson, M. P., D. F. Shriver, and S. A. Shriver, *J. Am. Chem. Soc.*, **88**, 1588 (1966).
69. Kahn, O., and M. Bigorgne, *Compt. Rend.*, **262**, 906 (1966).
70. King, R. B., *Organometal. Synth. N.Y.*, 175 1965.
71. King, R. B., *Inorg. Chem.*, **5**, 82 (1966).
72. King, R. B., P. M. Treichel, and F. G. A. Stone, *J. Am. Chem. Soc.*, **83**, 3593 (1961).
73. Kingston, B. M., and M. F. Lappert, *2nd Symp. Int. sur la Chimie des Composés Organosiliciques, Bordeaux*, 107, 1968.
74. Kuehlein, K., Diplomarbeit geissen Universität, 1963.
75. Kuehlein, K., W. P. Neumann, and H. P. Becker, *Angew. Chem.*, **79**, 870 (1967).
76. Kummer, R., and W. A. G. Graham, *Inorg. Chem.*, **7**, 310 (1968).
77. Kummer, R., and W. A. G. Graham, *Inorg. Chem.*, **7**, 523 (1968).
78. Kraus, C. A., and L. S. Foster, *J. Am. Chem. Soc.*, **49**, 457 (1927).
79. Kraus, C. A., and W. K. Nelson, *J. Am. Chem. Soc.*, **56**, 195 (1934).
80. Kruglaya, O. A., N. S. Vyazankin, and G. A. Razuvaev, *Zh. Obshch. Khim.*, **35**, 394 (1965).
81. Kruglaya, O. A., N. S. Vyazankin, G. A. Razuvaev, and E. V. Mitrophanova, *Dokl. Akad. Nauk SSSR*, **173**, 834 (1967).
82. Massey, A. G., A. J. Park, and F. G. A. Stone, *J. Am. Chem. Soc.*, **85**, 2021 (1963).
83. Mendelsohn, J. C., Thesis Bordeaux 67, 1966.
84. Mendelsohn, J. C., F. Metras, J. C. Lahournere, and J. Valade, *J. Organometal. Chem.*, **12**, 327 (1968).
85. Mendelsohn, J. C., F. Metras, and J. Valade, *Compt. Rend.*, **261**, 756 (1965).
86. Milligan, J. G., and C. A. Kraus, *J. Am. Chem. Soc.*, **72**, 5297 (1950).
87. Nesmeyanov, A. N., K. N. Anisinov, N. E. Kolobova, and A. B. Antonova, *Izv. Akad. Nauk SSSR*, 1309 (1965).
88. Nesmeyanov, A. N., K. N. Anisimov, N. E. Kolobova, and A. B. Antonova, *Izv. Akad. Nauk SSSR, Ser. Khim.*, **1**, 160 (1966).
89. Nesmeyanov, A. N., K. N. Anisimov, N. E. Kolobova, and A. B. Antonova, *Dokl. Akad. Nauk SSSR*, **176**, 844 (1967).
90. Nesmeyanov, A. N., K. N. Anisimov, N. E. Kolobova, and J. S. Denisova, *Izv. Akad. Nauk SSSR*, 2246 (1966).
91. Nesmeyanov, A. N., K. N. Anisimov, N. E. Kolobova, and J. S. Denisova, *Izv. Akad. Nauk SSSR*, 142 (1968).

92. Nesmeyanov, A. N., K. N. Anisimov, N. E. Kolobova, and M. Y. Zakharova, *Izv. Akad. Nauk SSSR*, 1880 (1967).

93. Nesmeyanov, A. N., G. Dvoryantseva, T. N. Ulyanova, N. E. Kolobova, K. N. Anisimov, and A. B. Antonova, *Izv. Akad. Nauk SSSR*, 2241 (1967).

94. Nesmeyanov, A. N., N. E. Kolobova, K. N. Anisimov, and V. M. Khandozhko, *Izv. Akad. Nauk SSSR*, 163 (1966).

95. Neumann, W. P., and K. König, *J. Am. Chem. Soc.*, **1**, 677 (1964).

96. Neumann, W. P., and K. Kuehlein, *Ann. Chem.*, **1**, 683 (1965).

97. Neumann, W. P., and K. Kuehlein, *Tetrahedron Letters*, **29**, 3419 (1966).

98. Neumann, W. P., B. Schneider, and R. Sommer, *Ann. Chem.*, **1**, 692 (1966).

99. Patil, R. H., and W. A. G. Graham, *Inorg. Chem.*, **5**, 1401 (1966).

100. Patmore, D. J., and W. A. G. Graham, *Inorg. Chem.*, **5**, 1465 (1966).

101. Patmore, D. J., and W. A. G. Graham, *Inorg. Chem.*, **6**, 981 (1967).

102. Quane, D., and R. S. Bottei, *Chem. Rev.*, **63**, 403 (1963).

103. Quane, D., and G. W. Hunt, *J. Organometal. Chem.*, **13**, 16 (1968).

104. Razuvaev, G. A., *Int. Symp. in Organic Peroxides, Berlin*, 1967.

105. Razuvaev, G. A., A. Alexandrov, V. N. Glushackova, and G. N. Figurova, *J. Organometal. Chem.*, **14**, 339 (1968).

106. Razuvaev, G. A., S. P. Korneva, and O. A. Kruglaya, *Dokl. Akad. Nauk SSSR*, **158**, 884 (1964).

107. Rijkens, F., and G. J. M. van der Kerk, *Organogermanium Compounds*, 9 (1966).

108. Royen, P., and C. Rocktaeschel, *Z. Anorg. Allgem. Chem.*, **34**, 290 (1966).

109. Royen, P., C. Rocktaeschel, and W. Mosh, *Angew. Chem.*, **76**, 860 (1964).

110. Ruff, J. K., *Inorg. Chem.*, **6**, 1502 (1967).

111. Salazar, R. W., Dissertation, München, 1967.

112. Shackelford, J. M., H. de Schmertzing, C. H. Henther, and H. Podall, *J. Org. Chem.*, **28**, 1700 (1963).

113. Schumann, H., and M. Schmidt, *Inorg. Nucl. Chem. Letters*, **1**, 1 (1965).

114. Schumann, H., and M. Schmidt, *J. Organometal. Chem.*, **3**, 485 (1965).

115. Schumann, H., and S. Ronecker, *Z. Naturforsch.*, **22**, 452 (1967).

116. Schumann-Ruidisch, I., and H. Blass, *Z. Naturforsch.*, **21**, 1105 (1966).

117. Schumann-Ruidisch, I., and H. Blass, *Z. Naturforsch.*, **22**, 1081 (1967).

118. Seyferth, D., *J. Am. Chem. Soc.*, **79**, 2738 (1957).

119. Seyferth, D., R. J. Cross, and B. Prokai, *J. Organometal. Chem.*, **7**, 20 (1968).

120. Seyferth, D., H. P. Hofmann, R. Burton, and J. F. Helling, *Inorg. Chem.*, **1**, 227 (1962).

121. Seyferth, D., G. Raab, and S. O. Grim, *J. Org. Chem.*, **26**, 3034 (1961).

122. Sommer, R., W. P. Neumann, and B. Schneider, *Tetrahedron Letters*, 3875 (1964).

123. Stone, F. G. A., *J. Chem. Soc.*, 1130 (1966).

124. Struchkov, Y. T., K. N. Anisimov, N. E. Kolobova, and A. N. Nesmeyanov, *Dokl. Akad. Nauk SSSR*, **172**, 167 (1967).

125. Struchkov, Y. T., K. N. Anisimov, O. P. Osifova, N. E. Kolovova, and A. N. Nesmeyanov, *Dokl. Akad. Nauk SSSR*, **172**, 107 (1967).

126. Vyazankin, N. S., V. T. Bychkov, and I. A. Vostokov, *Zh. Obshch. Khim.*, **38**, 1345 (1968).

127. Vyazankin, N. S., E. N. Gladyshev, and S. P. Korneva, *Zh. Obshch. Khim.*, **37**, 1736 (1967).

128. Vyazankin, N. S., E. N. Gladyshev, S. P. Korneva, G. A. Razuvaev, and E. A. Arkhangel'skaya, *Zh. Obshch. Khim.*, **38**, 1803 (1968).

129. Vyazankin, N. S., E. N. Gladyshev, G. A. Razuvaev, and S. P. Korneva, *Zh. Obshch. Khim.*, **36**, 952 (1966).

130. Vyazankin, N. S., G. S. Kalinina, O. A. Kruglaya, and G. A. Razuvaev, *Zh. Obshch. Khim.,* in press.

131. Vyazankin, N. S., and O. A. Kruglaya, *Usp. Khim.,* **35**, 1388 (1966).

132. Vyazankin, N. S., O. A. Kruglaya, and G. A. Razuvaev, *Zh. Obshch. Khim.,* **35**, 394 (1954).

133. Vyazankin, N. S., O. A. Kruglaya, G. A. Razuvaev, and G. S. Semchikova, *Dokl. Akad. Nauk SSSR,* **166**, 99 (1966).

134. Vyazankin, N. S., O. A. Kruglaya, G. A. Razuvaev, and G. S. Semchikova, *J. Organometal. Chem.,* **6**, 474 (1966).

135. Vyazankin, N. S., E. V. Mitrophanova, O. A. Kruglaya, and G. A. Razuvaev, *Zh. Obshch. Khim.,* **36**, 160 (1966).

136. Vyazankin, N. S., G. A. Razuvaev, and V. T. Bychkov, *Dokl. Akad. Nauk SSSR,* **158**, 382 (1964).

137. Vyazankin, N. S., G. A. Razuvaev, and V. T. Bychkov, *Izv. Akad. Nauk SSSR,* 1665 (1965).

138. Vyazankin, N. S., G. A. Razuvaev, V. T. Bychkov, and V. L. Zvezlein, *Izv. Akad. Nauk SSSR,* 562 (1966).

139. Vyazankin, N. S., G. A. Razuvaev, and E. N. Gladyshev, *Dokl. Akad. Nauk SSSR,* **151**, 1326 (1963).

140. Vyazankin, N. S., G. A. Razuvaev, and E. N. Gladyshev, *Dokl. Akad. Nauk SSSR,* **155**, 830 (1964).

141. Vyazankin, N. S., G. A. Razuvaev, E. N. Gladyshev, and T. G. Gurikova, *Dokl. Akad. Nauk SSSR,* **155**, 1108 (1964).

142. Vyazankin, N. S., G. A. Razuvaev, E. N. Gladyshev, and S. P. Korneva, *J. Organometal. Chem.,* **7**, 353 (1967).

143. Vyazankin, N. S., G. A. Razuvaev, S. P. Korneva, O. A. Kruglaya, and R. F. Galuilina, *Dokl. Akad. Nauk SSSR,* **158**, 884 (1964).

144. Vyazankin, N. S., G. A. Razuvaev, O. A. Kruglaya, and G. S. Senchikova, *J. Organometal. Chem.,* **6**, 474 (1966).

145. Vyazankin, N. S., G. A. Razuvaev, O. A. Kruglaya, and G. S. Senchikova, *Dokl. Akad. Nauk SSSR,* **166**, 99 (1966).

146. Vyazankin, N. S., L. P. Sanina, G. S. Kalinina, and M. N. Bochkarev, *Zh. Obshch. Khim.,* **38**, 1800 (1968).

147. Wiberg, E., E. Amberger, and H. Cambensi, *Z. Anorg. Chem.,* **351**, 164 (1967).

148. Wiberg, E., O. Stecher, H. J. Andrascherk, L. Kreuzbichler, and E. Staude, *Angew. Chem.,* **75**, 516 (1963).

149. Willemsens, L. C., and G. J. M. van der Kerk, *J. Organometal. Chem.,* **2**, 260 (1964).

150. Willemsens, L. C., and G. J. M. van der Kerk, *Investigations in the Field of Organolead Chemistry,* Schotanus and Jens, Utrecht, 1965.

151. Wittle, J. K., and G. Urrey, *Inorg. Chem.,* **7**, 560 (1968).

152. Young, J. F., R. D. Gillard, and G. Wilkinson, *J. Chem. Soc.,* 5176 (1964).

153. Zimmer, Z. C., and M. Huber, *Compt. Rend.,* **267**, 1685 (1968).

Subject Index

693